An Introduction to Measure-Theoretic Probability

Second edition

This page is intentionally left blank

An Introduction to Measure-Theoretic Probability

Second edition

by

GEORGE G. ROUSSAS
Department of Statistics
University of California, Davis

AMSTERDAM • BOSTON • HEIDELBERG • LONDON
NEW YORK • OXFORD • PARIS • SAN DIEGO
SAN FRANCISCO • SINGAPORE • SYDNEY • TOKYO

Academic Press is an imprint of Elsevier

Academic Press is an imprint of Elsevier
The Boulevard, Langford Lane, Kidlington, Oxford OX5 1GB, UK
Radarweg 29, PO Box 211, 1000 AE Amsterdam, The Netherlands
225 Wyman Street, Waltham, MA 02451, USA
525 B Street, Suite 1800, San Diego, CA 92101-4495, USA

Second edition **2014**

Library of Congress Cataloging-in-Publication Data
Application submitted

British Library Cataloguing in Publication Data
A catalogue record for this book is available from the British Library

For information on all **Academic Press** publications
visit our web site at store.elsevier.com

ISBN: 978-0-12-800042-7

This book is dedicated to the memory of Edward W. Barankin, the probabilist, mathematical statistician, classical scholar, and philosopher, for his role in stimulating my interest in probability with emphasis on detail and rigor.
Also, to my dearest sisters, who provided material means in my needy student years, and unrelenting moral support throughout my career.

This page is intentionally left blank

Contents

This page is intentionally left blank

Pictured on the Cover

Carathéodory, Constantine (1873–1950)

He was born in Berlin to Greek parents and grew up in Brussels, Belgium. In high school, he twice won a prize as the best Mathematics student in Belgium. He studied Military Engineering in Belgium, and pursued graduate studies in Göttingen under the supervision of Hermann Minkowski.

He is known for his contributions to the theory of functions, the calculus of variations, and measure theory. His name is identified with the theory of outer measure, an application of which in measure theory is the so-called Carathéodory Extension Theorem. Also, he did work on the foundations of thermodynamics, and in 1909, he published the "first axiomatic rigid foundation of thermodynamics," which was acclaimed by Max Planck and Max Born.

From correspondence between Albert Einstein and Constantine Carathéodory, it may be deduced that Carathéodory's work helped Einstein in shaping some of his theories.

In 1924, he was appointed professor of Mathematics at the University of Munich, where he stayed until his death in 1950.

This page is intentionally left blank

Preface to First Edition

This book in measure-theoretic probability has resulted from classroom lecture notes that this author has developed over a number of years, by teaching such a course at both the University of Wisconsin, Madison, and the University of California, Davis. The audience consisted of graduate students primarily in statistics and mathematics. There were always some students from engineering departments, and a handful of students from other disciplines such as economics.

The book is not a comprehensive treatment of probability, nor is it meant to be one. Rather, it is an excursion in measure-theoretic probability with the objective of introducing the student to the basic tools in measure theory and probability as they are commonly used in statistics, mathematics, and other areas employing this kind of moderately advanced mathematical machinery. Furthermore, it must be emphasized that the approach adopted here is entirely classical. Thus, characteristic functions are a tool employed extensively; no use of martingale or empirical process techniques is made anywhere.

The book does not commence with probabilistic concepts, and there is a good reason for it. As many of those engaged in teaching advanced probability and statistical theory know, very few students, if any, have been exposed to a measure theory course prior to attempting a course in advanced probability. This has been invariably the experience of this author throughout the years. This very fact necessitates the study of the basic measure-theoretic concepts and results—in particular, the study of those concepts and results that apply immediately to probability, and also in the form and shape they are used in probability.

On the basis of such considerations, the framework of the material to be dealt with is therefore determined. It consists of a brief introduction to measure theory, and then the discussion of those probability results that constitute the backbone of the subject matter. There is minimal flexibility allowed, and that is exploited in the form of the final chapter of the book. From many interesting and important candidate topics, this author has chosen to present a brief discussion of some basic concepts and results of ergodic theory.

From the very outset, there is one point that must be abundantly clarified, and that is the fact that everything is discussed in great detail with all proofs included; no room is allowed for summary unproven statements. This approach has at least two side benefits, as this author sees them. One is that students have at their disposal a comprehensive and detailed proof of what are often deep theorems. Second, the instructor may skip the reproduction of such proofs by assigning their study to students.

In the experience of this author, there are no topics in this book which can be omitted, except perhaps for the final chapter. With this in mind, the material can be taught in two quarters, and perhaps even in one semester with appropriate calibration of the rate of presentation, and the omission of proofs of judiciously selected theorems. With all details presented, one can also cover an entire year of instruction, perhaps with some supplementation.

Most chapters are supplied with examples, and all chapters are concluded with a varying number of exercises. An unusual feature here is that an *Answers Manual* of all exercises will be made available to those instructors who adopt the book as the textbook in their course. Furthermore, an overview of each one of the 15 chapters is included in an appendix to the main body of the book. It is believed that the reader will benefit significantly by reviewing the overview of a chapter before the material in the chapter itself is discussed.

The remainder of this preface is devoted to a brief presentation of the material discussed in the 15 chapters of the book, chapter-by-chapter.

Chapter 1 commences with the introduction of the important classes of sets in an abstract space, which are those of a field, a σ-field, including the Borel σ-field, and a monotone class. They are illustrated by concrete examples, and their relationships are studied. Product spaces are also introduced, and some basic results are established. The discussion proceeds with the introduction of the concept of measurable functions, and in particular of random vectors and random variables. Some related results are also presented. This chapter is concluded with a fundamental theorem, Theorem 17, which provides for pointwise approximation of any random variable by a sequence of so-called simple random variables.

Chapter 2 is devoted to the introduction of the concept of a measure, and the study of the most basic results associated with it. Although a field is the class over which a measure can be defined in an intuitively satisfying manner, it is a σ-field—the one generated by an underlying field—on which a measure must be defined. One way of carrying out the construction of a measure on a σ-field is to use as a tool the so-called outer measure. The concept of an outer measure is then introduced, and some of its properties are studied in the second section of the chapter. Thus, starting with a measure on a field, utilizing the associated outer measure and the powerful Carathéodory extension theorem, one ensures the definition of a measure over the σ-field generated by the underlying field. The chapter is concluded with a study of the relationship between a measure over the Borel σ-field in the real line and certain point functions. A measure always determines a class of point functions, which are nondecreasing and right-continuous. The important thing, however, is that each such point function uniquely determines a measure on the Borel σ-field.

In Chapter 3, sequences of random variables are considered, and two basic kinds of convergences are introduced. One of them is the almost everywhere convergence, and the other is convergence in measure. The former convergence is essentially the familiar pointwise convergence, whereas convergence in measure is a mode of convergence not occurring in a calculus course. A precise expression of the set of pointwise convergence is established, which is used for formulating necessary and sufficient conditions for almost everywhere convergence. Convergence in measure is weaker than almost everywhere convergence, and the latter implies the former for finite measures. *Almost everywhere convergence* and *mutual almost everywhere convergence* are equivalent, as is easily seen. Although the same is true when convergence in measure is involved, its justification is fairly complicated and requires the introduction of the concept of *almost uniform convergence*. Actually, a substantial part of the chapter is devoted in proving the equivalence just stated. In closing, it is to be mentioned that, in the presence of a probability measure, *almost everywhere convergence* and *convergence in measure* become, respectively, *almost sure convergence* and *convergence in probability*.

Chapter 4 is devoted to the introduction of the concept of the integral of a random variable with respect to a measure, and the proof of some fundamental properties of the integral. When the underlying measure is a probability measure, the integral of a random variable becomes its expectation. The procedure of defining the concept of the integral follows three steps. The integral is first defined for a simple random variable, then for a nonnegative random variable, and finally for any random variable, provided the last step produces a meaningful quantity. This chapter is concluded with a result, Theorem 13, which transforms integration of a function of a random variable on an abstract probability space into integration of a real-valued function defined on the real line with respect to a probability measure on the Borel σ-field, which is the probability distribution of the random variable involved.

Chapter 5 is the first chapter where much of what was derived in the previous chapters is put to work. This chapter provides results that in a real sense constitute the workhorse whenever convergence of integrals is concerned, or differentiability under an integral sign is called for, or interchange of the order of integration is required. Some of the relevant theorems here are known by names such as the Lebesgue Monotone Convergence Theorem, the Fatou–Lebesgue Theorem, the Dominated Convergence Theorem, and the Fubini Theorem. Suitable modifications of the basic theorems in the chapter cover many important cases of both theoretical and applied interest. This is also the appropriate point to mention that many properties involving integrals are established by following a standard methodology; namely, the property in question is first proved for indicator functions, then for nonnegative simple random variables, next for nonnegative random variables, and finally for any random variables. Each step in this process relies heavily on the previous step, and the Lebesgue Monotone Convergence Theorem plays a central role.

Chapter 6 is the next chapter in which results of great utilitarian value are established. These results include the standard inequalities (Hölder (Cauchy–Schwarz), Minkowski, c_r, Jensen), and a combination of a probability/moment inequality, which produces the Markov and Tchebichev inequalities. A third kind of convergence—convergence in the rth mean—is also introduced and studied to a considerable extent. It is shown that convergence in the rth mean is equivalent to mutual convergence in the rth mean. Also, necessary and sufficient conditions for convergence in the rth mean are given. These conditions typically involve the concepts of uniform continuity and uniform integrability, which are important in their own right. It is an easy consequence of the Markov inequality that convergence in the rth mean implies convergence in probability. No direct relation may be established between convergence in the rth mean and almost sure convergence.

In Chapter 7, the concept of absolute continuity of a measure relative to another measure is introduced, and the most important result from utilitarian viewpoint is derived; this is the Radon–Nikodym Theorem, Theorem 3. This theorem provides the representation of a dominated measure as the indefinite integral of a nonnegative random variable with respect to the dominating measure. Its corollary provides the justification for what is done routinely in statistics; namely, employing a probability density function in integration. The Radon–Nikodym Theorem follows easily from the Lebesgue Decomposition Theorem, which is a deep result, and this in turn is based on the Hahn–Jordan Decomposition Theorem. Although all these results are proved in great detail, this is an instance where an instructor may choose to give the outlines of the first two theorems, and assign to students the study of the details.

Chapter 8 revolves around the concept of distribution functions and their basic properties. These properties include the fact that a distribution function is uniquely determined by its values on a set that is dense in the real line, that the discontinuities, being jumps only, are countably many, and that every distribution function is uniquely decomposed into two distribution functions, one of which is a step function and the other a continuous function. Next, the concepts of weak and complete convergence of a sequence of distribution functions are introduced, and it is shown that a sequence of distribution functions is weakly compact. In the final section of the chapter, the so-called Helly–Bray type results are established. This means that sufficient conditions are given under which weak or complete convergence of a sequence of distribution functions implies convergence of the integrals of a function with respect to the underlying distribution functions.

The purpose of Chapter 9 is to introduce the concept of conditional expectation of a random variable in an abstract setting; the concept of conditional probability then follows as a special case. A first installment of basic properties of conditional expectations is presented, and then the discussion proceeds with the derivation of the conditional versions of the standard inequalities dealt with in Chapter 6. Conditional versions of some of the standard convergence theorems of Chapter 5 are also derived, and the chapter is concluded with the discussion of further properties of conditional expectations, and an application linking the abstract definition of conditional probability with its elementary definition.

In Chapter 10, the concept of independence is considered first for events and then for σ-fields and random variables. A number of interesting results are discussed, including the fact that real-valued (measurable) functions of independent random variables are independent random variables, and that the expectation of the product of independent random variables is the product of the individual expectations. However, the most substantial result in this chapter is the fact that factorization of the joint distribution function of a finite number of random variables implies independence of the random variables involved. This result is essentially based on the fact that σ-fields generated by independent fields are themselves independent.

Chapter 11 is devoted to characteristic functions, their basic properties, and their usage for probabilistic purposes. Once the concept of a characteristic function is defined, the fundamental result, referred to in the literature as the inversion formula, is established in a detailed manner, and several special cases are considered; also, the applicability of the formula is illustrated by means of two concrete examples. One of the main objectives in this chapter is that of establishing the Paul Lévy Continuity Theorem, thereby reducing the proof of weak convergence of a sequence of distribution functions to that of a sequence of characteristic functions, a problem much easier to deal with. This is done in Section 3, after a number of auxiliary results are first derived. The multidimensional version of the continuity theorem is essentially reduced to the one-dimensional case through the so-called Cramér–Wold device; this is done in Section 4. Convolution of two distribution functions and several related results are discussed in Section 5, whereas in the following section additional properties of characteristic functions are established. These properties include the expansion of a characteristic function in a Taylor-like formula around zero with a remainder given in three different forms. A direct application of this expansion produces the Weak Law of Large Numbers and the Central Limit Theorem. In Section 8, the significance of the moments of a random variable is dramatized by showing that, under certain conditions, these moments completely determine the distribution of the random variable through its characteristic function. The rigorous proof of the relevant theorem makes use of a number of results from complex analysis, which for convenient reference are cited in the final section of the chapter.

In the next two chapters—Chapters 12 and 13—what may be considered as the backbone of classical probability is taken up: namely, the study of the central limit problem is considered under two settings, one for centered random variables and one for noncentered random variables. In both cases, a triangular array of row-wise independent random variables is considered, and, under some general and weak conditions, the totality of limiting laws—in the sense of weak convergence—is obtained for the row sums. As a very special case, necessary and sufficient conditions are given for convergence to the normal law for both the centered and the noncentered case. In the former case, sets of simpler sufficient conditions are also given for convergence to the normal law, whereas in the latter case, necessary and sufficient conditions are given for convergence to the Poisson law. The Central Limit Theorem in its usual simple form and the convergence of binomial probabilities to Poisson probabilities are also derived as very special cases of general results.

The main objective of Chapter 14 is to present a complete discussion of the Kolmogorov Strong Law of Large Numbers. Before this can be attempted, a long series of other results must be established, the first of which is the Kolmogorov inequalities. The discussion proceeds with the presentation of sufficient conditions for a series of centered random variables to convergence almost surely, the Borel–Cantelli Lemma, the Borel Zero–One Criterion, and two analytical results known as the Toeplitz Lemma and the Kronecker Lemma. Still the discussion of another two results is needed—one being a weak partial version of the Kolmogorov Strong Law of Large Numbers, and the other providing estimates of the expectation of a random variable in terms of sums of probabilities—before the Kolmogorov Strong Law of Large Numbers, Theorem 7, is stated and proved. In Section 4, it is seen that, if the expectation of the underlying random variable is not finite, as is the case in Theorem 7, a version of Theorem 7 is still true. However, if said expectation does not exist, then the averages are unbounded with probability 1. The chapter is concluded with a brief discussion of the tail σ-field of a sequence of random variables and pertinent results, including the Kolmogorov Zero–One Law for independent random variables, and the so-called Three Series Criterion.

Chapter 15 is not an entirely integral part of the body of basic and fundamental results of measure-theoretic probability. Rather, it is one of the many possible choices of topics that this author could have covered. It serves as a very brief introduction to an important class of discrete parameter stochastic processes—stationary and ergodic or nonergodic processes—with a view toward proving the fundamental result, the Ergodic Theorem. In this framework, the concept of a stationary stochastic process is introduced, and some characterizations of stationarity are presented. The convenient and useful coordinate process is also introduced at this point. Next, the concepts of a transformation as well as a measure-preserving transformation are discussed, and it is shown that a measure-preserving transformation along with an arbitrary random variable define a stationary process. Associated with each transformation is a class of invariant sets and a class of almost sure invariant sets, both of which are σ-fields. A special class of transformations is the class of ergodic transformations, which are defined at this point. Invariance with respect to a transformation can also be defined for a stationary sequence of random variables, and it is so done. At this point, all the required machinery is available for the formulation of the Ergodic Theorem; also, its proof is presented, after some additional preliminary results are established. In the final section of the chapter, invariance of sets and of random variables is defined relative to a stationary process. Also, an alternative form of the Ergodic Theorem is given for nonergodic as well as ergodic processes. In closing, it is to be pointed out that one direction of the Kolmogorov Strong Law of Large Numbers is a special case of the Ergodic Theorem, as a sequence of independent identically distributed random variables forms a stationary and ergodic process.

Throughout the years, this author has drawn upon a number of sources in organizing his lectures. Some of those sources are among the books listed in the Selected References Section. However, the style and spirit of the discussions in this book lie closest to those of Loève's book. At this point, I would like to mention a recent worthy addition to the literature in measure theory—the book by Eric Vestrup, not least because Eric was one of our Ph.D. students at the University of California, Davis.

The lecture notes that eventually resulted in this book were revised, modified, and supplemented several times throughout the years; comments made by several of my students were very helpful in this respect. Unfortunately, they will have to remain anonymous, as I have not kept a complete record of them, and I do not want to provide an incomplete list. However, I do wish to thank my colleague and friend Costas Drossos for supplying a substantial number of exercises, mostly accompanied by answers. I would like to thank the following reviewers: Ibrahim Ahmad, University of Central Florida; Richard Johnson, University of Wisconsin; Madan Puri, Indiana University; Doraiswamy Ramachandran, California State University at Sacramento; and Zongwu Cai, University of North Carolina at Charlotte. Finally, thanks are due to my Project Assistant Newton Wai, who very skillfully turned my manuscript into an excellent typed text.

George G. Roussas
Davis, California
November 2003

This page is intentionally left blank

Preface to Second Edition

This is a revised version of the first edition of the book with copyright year 2005.

The basic character of the book remains the same, although its style is slightly different. Whatever changes were effected were made to correct misprints and oversights; add some clarifying points, as well as insert more references to previous parts of the book in support of arguments made; make minor modifications in the formulation, and in particular, the proof of some results; and supply additional exercises.

Specifically, the formulation of Theorem 8 in Chapter 3 has been rearranged. The proof of Theorem 3, case 3, in Chapter 4, has been simplified. The proof of Theorem 12 in Chapter 5 has been modified. Proposition 1, replaces Remark 6(ii) in Chapter 7. The proof of Theorem 3(iii) in Chapter 8 has been modified, and so has the concluding part of the proof of Theorem 5 in the same chapter. Likewise for the proofs of Theorems 7 and 8 in the same chapter. Remark 2 was inserted in Chapter 9 in order to further illustrate the abstract definition and the significance of the conditional expectation.

Section 3 of Chapter 11 has been restructured. Theorem 3 has been split into two parts, Theorem 3 and Theorem 3^*. Part (i) is the same in both of these theorems, as well as in the original theorem. There is a difference, however, in the formulation of part (ii) in the new versions of the theorems. Theorem 3 here is formulated along the familiar lines involving distribution functions and characteristic functions of random variables. Its formulation is followed by two lemmas, which facilitate its proof. The formulation of the second part of Theorem 3^* is more general, and along the same lines as the converse of the original Theorem 3. Theorem 3^* is also followed by two lemmas and one proposition, which lead to its justification. This section is concluded with two propositions, Propositions 2 and 3, where some restrictions imposed in the formulation of Lemmas 1–4 and Proposition 1 are lifted. In the same chapter, Chapter 11, the proof of Theorem 8 is essentially split into two parts, with the insertion of a "Statement" (just a few lines below relation (11.28)), in order to emphasize the asserted uniformity in the convergence.

In Chapter 12, Example 3 was added, first to illustrate the process of checking the Lindeberg-Feller condition, and second to provide some insight into this condition.

In Chapter 14, the second part of Lemma 7 has been modified, and so has its proof .

Finally, a new chapter, Chapter 16, has been added. This Chapter discusses some material on statistical inference, and it was added for the benefit of statistically oriented users of the book and upon the suggestion of a reviewer of the revised edition of the book. Its main purpose, however, is to demonstrate as to how some of the theorems, corollaries, etc. discussed in the book apply in establishing statistically inference results.

For a chapter-by-chapter brief description of the material discussed in the book, and also advice us to how the book can be used, the reader should go over the preface of its first edition.

The *Answers Manual* has been revised along the same lines as the text of the book. Thus, misprints and oversights have been corrected, and a handful of solutions have been modified. Of course, solutions to all new exercises are supplied. Again, the *Answers Manual*, in its revised version, will be made available to all those instructors who adopt the book as the textbook in their course.

Misprints and oversights were located in the usual way; that is, by teaching from the book. Many of the misprints and oversights were pointed out by attentive students. In this respect, special mention should be made of my students Qiuyan Xu and Gabriel Becker. Clarifications, modifications, and rearrangement of material, as described earlier, were also stimulated, to a large extent, by observations made and questions posed by students. Warm thanks are extended to all those who took my two-quarter course the last two offerings. Also, I am grateful to Stacy Hill and Paul Ressel for constructive comments. In particular, I am indebted to Michael McAssey, a formerly graduate student in the Department of Statistics, for the significant role he played toward the revision of the book and the *Answers Manual*. The accuracy and efficiency by which he handled the material was absolutely exemplary. Thanks are also due to Chu Shing (Randy) Lai for most efficiently implementing some corrections and inserting additional material into the book and the Answer Manual.

In closing, I consider it imperative to mention the following facts. Each chapter is introduced by a brief summary, describing the content of the chapter. In addition, there is an appendix in the book, Appendix A, where a much more extensive description is provided, chapter-by-chapter. It is my opinion that the reader would benefit greatly by reading this appendix before embarking on the study of the chapters. In this revision, a new appendix, Appendix B, has been added, providing a brief review of the Riemann–Stieltjes integral, and its relationship to the Riemann integral and the Lebesgue integral on the real line. The Riemann–Stieltjes integral is used explicitly in parts of Chapter 8, and implicitly in part of Chapters 11 through 13. Finally, it is mentioned here that some notation and abbreviations have been added to refresh readers' memory and ensure uniformity in notation.

George G. Roussas
Davis, California
September 2013

CHAPTER

1 Certain Classes of Sets, Measurability, and Pointwise Approximation

In this introductory chapter, the concepts of a field and of a σ-field are introduced, they are illustrated by means of examples, and some relevant basic results are derived. Also, the concept of a monotone class is defined and its relationship to certain fields and σ-fields is investigated. Given a collection of measurable spaces, their product space is defined, and some basic properties are established. The concept of a measurable mapping is introduced, and its relation to certain σ-fields is studied. Finally, it is shown that any random variable is the pointwise limit of a sequence of simple random variables.

1.1 Measurable Spaces

Let Ω be an abstract set (or space) and let $\mathcal{C}$ be a class of subsets of Ω; i.e., $\mathcal{C} \subseteq \mathcal{P}(\Omega)$, the class of all subsets of Ω.

Definition 1. $\mathcal{C}$ is said to be a *field*, usually denoted by $\mathcal{F}$, if

(i) $\mathcal{C}$ is nonempty.
(ii) If $A \in \mathcal{C}$, then $A^c \in \mathcal{C}$.
(iii) If $A_1, A_2 \in \mathcal{C}$, then $A_1 \cup A_2 \in \mathcal{C}$. ■

Remark 1. In view of (ii) and (iii), the union $A_1 \cup A_2$ may be replaced by the intersection $A_1 \cap A_2$.

Examples. (Recall that a set is *countable* if it is either finite or it has the same cardinality as the set of integers. In the latter case it is *countably infinite*. A set is *uncountable* if it has the same cardinality as the real numbers.)

(1) $\mathcal{C} = \{\emptyset, \Omega\}$ is a field called the *trivial* field. (It is the smallest possible field.)
(2) $\mathcal{C} = \{\text{all subsets of } \Omega\} = \mathcal{P}(\Omega)$ is a field called the *discrete* field. (It is the largest possible field.)
(3) $\mathcal{C} = \{\emptyset, A, A^c, \Omega\}$ for some A with $\emptyset \subset A \subset \Omega$.
(4) Let Ω be infinite (countably or not) and let $\mathcal{C} = \{A \subseteq \Omega; A \text{ is finite or } A^c \text{ is finite}\}$. Then $\mathcal{C}$ is a field.
(5) Let $\mathcal{C}$ be the class of all (finite) sums (unions of pairwise disjoint sets) of the partitioning sets of a finite partition of an arbitrary set Ω (see Definition 12 below). Then $\mathcal{C}$ is a field (*induced* or *generated* by the underlying partition). ■

An Introduction to Measure-Theoretic Probability, Second Edition. http://dx.doi.org/10.1016/B978-0-12-800042-7.00001-3

Remark 2. In Example 4, it is to be observed that if Ω is finite rather than infinite, then $\mathcal{C} = \mathcal{P}(\Omega)$.

Consequences of Definition 1.

(1) $\Omega, \oslash \in \mathcal{F}$ for every $\mathcal{F}$.
(2) If $A_j \in \mathcal{F}, j = 1, \ldots, n$, then $\bigcup_{j=1}^{n} A_j \in \mathcal{F}$.
(3) If $A_j \in \mathcal{F}, j = 1, \ldots, n$, then $\bigcap_{j=1}^{n} A_j \in \mathcal{F}$.

Remark 3. It is shown by examples that $A_j \in \mathcal{F}, j \geq 1$, need not imply $\bigcup_{j=1}^{\infty} A_j \in \mathcal{F}$, and similarly for $\bigcap_{j=1}^{\infty} A_j$ (see Remark 5 below).

Definition 2. $\mathcal{C}$ is said to be a *σ-field*, usually denoted by $\mathcal{A}$, if it is a field and (iii) in Definition 1 is strengthened to

(iii′) If $A_j \in \mathcal{C}, j = 1, 2, \ldots$, then $\bigcup_{j=1}^{\infty} A_j \in \mathcal{C}$. ■

Remark 4. In view of (ii) and (iii), the union in (iii′) may be replaced by the intersection $\bigcap_{j=1}^{\infty} A_j$.

Examples.

(6) $\mathcal{C} = \{\oslash, \Omega\}$ is a σ-field called the *trivial σ-field*.
(7) $\mathcal{C} = \mathcal{P}(\Omega)$ is a σ-field called the *discrete σ-field*.
(8) Let Ω be uncountable and let $\mathcal{C} = \{A \subseteq \Omega;\ A$ is countable or A^c is countable$\}$. Then $\mathcal{C}$ is a σ-field. (Of course, if Ω is countably infinite, then $\mathcal{C} = \mathcal{P}(\Omega)$).
(9) Let $\mathcal{C}$ be the class of all countable sums of the partitioning sets of a countable partition of an arbitrary set Ω. Then $\mathcal{C}$ is a σ-field (*induced* or *generated* by the underlying partition). ■

Remark 5. A σ-field is always a field, but a field need not be a σ-field. In fact, in Example 4 take $\Omega = \Re$ (real line), and let $A_j = \{k$ integer; $-j \leq k \leq j\}$, $j = 0, 1, \ldots$ Then $A_j \in \mathcal{C}, \bigcup_{j=0}^{n} A_j \in \mathcal{C}$ for any $n = 0, 1, \ldots$ but $\bigcup_{j=0}^{\infty} A_j$ (= set of all integers) $\notin \mathcal{C}$.

Let I be any index set. Then

Theorem 1.

(i) *If $\mathcal{F}_j, j \in I$ are fields, so is* $\bigcap_{j \in I} \mathcal{F}_j = \{A \subseteq \Omega;\ A \in \mathcal{F}_j, j \in I\}$.
(ii) *If $\mathcal{A}_j, j \in I$ are σ-fields, so is* $\bigcap_{j \in I} \mathcal{A}_j = \{A \subseteq \Omega;\ A \in \mathcal{A}_j, j \in I\}$. ■

Proof. Immediate. ■

Let $\mathcal{C}$ be any class of subsets of Ω. Then

Theorem 2.

(i) There is a unique minimal field containing $\mathcal{C}$. This is denoted by $\mathcal{F}(\mathcal{C})$ and is called the *field generated by* $\mathcal{C}$.

(ii) There is a unique minimal σ-field containing $\mathcal{C}$. This is denoted by $\sigma(\mathcal{C})$ and is called the *σ-field generated by $\mathcal{C}$*. ■

Proof.

(i) $\mathcal{F}(\mathcal{C}) = \bigcap_{j\in I} \mathcal{F}_j$, where $\{\mathcal{F}_j, j \in I\}$ is the nonempty class of all fields containing $\mathcal{C}$.

(ii) $\sigma(\mathcal{C}) = \bigcap_{j\in I} \mathcal{A}_j$, where $\{\mathcal{A}_j, j \in I\}$ is the nonempty class of all σ-fields containing $\mathcal{C}$. ■

Remark 6. Clearly, $\sigma(\mathcal{F}(\mathcal{C})) = \sigma(\mathcal{C})$. Indeed, $\mathcal{C} \subseteq \mathcal{F}(\mathcal{C})$, which implies $\sigma(\mathcal{C}) \subseteq \sigma(\mathcal{F}(\mathcal{C}))$. Also, for every σ-field $\mathcal{A}_i \supseteq \mathcal{C}$ it holds $\mathcal{A}_i \supseteq \mathcal{F}(\mathcal{C})$, since $\mathcal{A}_i$ is a field (being a σ-field), and $\mathcal{F}(\mathcal{C})$ is the minimal field (over $\mathcal{C}$). Hence $\sigma(\mathcal{C}) = \bigcap_i \mathcal{A}_i \supseteq \mathcal{F}(\mathcal{C})$. Since $\sigma(\mathcal{C})$ is a σ-field, it contains the minimal σ-field over $\mathcal{F}(\mathcal{C})$, $\sigma(\mathcal{F}(\mathcal{C}))$; i.e., $\sigma(\mathcal{C}) \supseteq \sigma(\mathcal{F}(\mathcal{C}))$. Hence $\sigma(\mathcal{C}) = \sigma(\mathcal{F}(\mathcal{C}))$.

Application 1. Let $\Omega = \Re$ and $\mathcal{C}_0 = \{\text{all intervals in } \Re\} = \{(x, y), (x, y], [x, y), [x, y], (-\infty, a), (-\infty, a], (b, \infty), [b, \infty);\ x, y \in \Re, x < y, a, b \in \Re\}$. Then $\sigma(\mathcal{C}_0)$ is denoted by $\mathcal{B}$ and is called the *Borel σ-field* over the real line. The sets in $\mathcal{B}$ are called *Borel sets*. Let $\bar{\Re} = \Re \cup \{-\infty, \infty\}$. $\bar{\Re}$ is called the *extended* real line and the σ-field $\bar{\mathcal{B}}$ generated by $\mathcal{B} \cup \{-\infty\} \cup \{\infty\}$ the *extended* Borel σ-field.

Remark 7. $\{x\} \in \mathcal{B}$ for every $x \in \Re$. Indeed, $\{x\} = \bigcap_{n=1}^{\infty} \left[x, x + \frac{1}{n}\right]$ with $\left[x, x + \frac{1}{n}\right] \in \mathcal{B}$. Hence $\bigcap_{n=1}^{\infty} \left[x, x + \frac{1}{n}\right] \in \mathcal{B}$, or $\{x\} \in \mathcal{B}$. Alternatively, with $a < x < b$, we have $\{x\} = (a, x] \cap [x, b) \in \mathcal{B}$.

Definition 3. The pair $(\Omega, \mathcal{A})$ is called a *measurable space* and the sets in $\mathcal{A}$ *measurable sets*. In particular, $(\Re, \mathcal{B})$ is called the *Borel real line*, and $(\bar{\Re}, \bar{\mathcal{B}})$ the *extended Borel real line*. ■

Let $\mathcal{C}$ again be a class of subsets of Ω. Then

Definition 4. $\mathcal{C}$ is called a *monotone* class if $A_j \in \mathcal{C}$, $j = 1, 2, \ldots$ and $A_j \uparrow$ (i.e., $A_1 \subseteq A_2 \subseteq \cdots$) or $A_j \downarrow$ (i.e., $A_1 \supseteq A_2 \supseteq \cdots$), then $\lim_{j\to\infty} A_j \stackrel{def}{=} \bigcup_{j=1}^{\infty} A_j \in \mathcal{C}$ and $\lim_{j\to\infty} A_j \stackrel{def}{=} \bigcap_{j=1}^{\infty} A_j \in \mathcal{C}$, respectively. ■

Theorem 3. A σ-field $\mathcal{A}$ is a monotone field (i.e., a field that is also a monotone class) and conversely. ■

Proof. One direction is immediate. As for the other, let $\mathcal{F}$ be a monotone field and let any $A_j \in \mathcal{F}$, $j = 1, 2, \ldots$. To show that $\bigcup_{j=1}^{\infty} A_j \in \mathcal{F}$. We have: $\bigcup_{j=1}^{\infty} A_j = A_1 \cup (A_1 \cup A_2) \cup \cdots \cup (A_1 \cup \cdots \cup A_n) \cup \cdots = \bigcup_{n=1}^{\infty} B_n$, where $B_n = \bigcup_{j=1}^{n} A_j$, and hence $B_n \in \mathcal{F}$, $n = 1, 2, \ldots$ and $B_n \uparrow$. Thus $\bigcup_{n=1}^{\infty} B_n \in \mathcal{F}$. ■

Theorem 4. If $\mathcal{M}_j$, $j \in I$, are monotone classes, so is $\bigcap_{j\in I} \mathcal{M}_j = \{A \subseteq \Omega;\ A \in \mathcal{M}_j, j \in I\}$. ■

Proof. Immediate. ■

Theorem 5. There is a unique minimal monotone class $\mathcal{M}$ containing $\mathcal{C}$. ■

Proof. $\mathcal{M} = \bigcap_{j \in I} \mathcal{M}_j$, where $\{\mathcal{M}_j, j \in I\}$ is the nonempty class of all monotone classes containing $\mathcal{C}$. ■

Remark 8. $\{\mathcal{M}_j, j \in I\}$ is nonempty since $\sigma(\mathcal{C})$ or $\mathcal{P}(\Omega)$ belong in it.

Remark 9. It may be seen by means of examples (see Exercise 12) that a monotone class need not be a field.

However, see the next lemma, as well as Theorem 6.

Lemma 1. Let $\mathcal{C}$ be a field and $\mathcal{M}$ be the minimal monotone class containing $\mathcal{C}$. Then $\mathcal{M}$ is a field.

Proof. In order to prove that $\mathcal{M}$ is a field, it suffices to prove that relations (*) hold, where

$$(*) \left\{ \text{for every } A, B \in \mathcal{M}, \text{ we have: } \begin{array}{l} \text{(i) } A \cap B \in \mathcal{M} \\ \text{(ii) } A^c \cap B \in \mathcal{M} \\ \text{(iii) } A \cap B^c \in \mathcal{M} \end{array} \right\}.$$

(That is, for every $A, B \in \mathcal{M}$, their intersection is in $\mathcal{M}$, and so is the intersection of any one of them by the complement of the other.)

In fact, $\mathcal{M} \supseteq \mathcal{C}$, implies $\Omega \in \mathcal{M}$. Taking $B = \Omega$, we get that for every $A \in \mathcal{M}$, $A^c \cap \Omega = A^c \in \mathcal{M}$ (by (ii)). Since also $A \cap B \in \mathcal{M}$ (by (i)) for all $A, B \in \mathcal{M}$, the proof would be completed.

In order to establish (*), we follow the following three steps:

Step 1. For any $A \in \mathcal{M}$, define $\mathcal{M}_A = \{B \in \mathcal{M}; (*) \text{ holds}\}$, so that $\mathcal{M}_A \subseteq \mathcal{M}$. Obviously $A \in \mathcal{M}_A$, since $\oslash \in \mathcal{M}$. It is asserted that $\mathcal{M}_A$ is a *monotone class*. Let $B_j \in \mathcal{M}_A$, $j = 1, 2, \ldots$, $B_j \uparrow$. To show that $\bigcup_{j=1}^{\infty} B_j \stackrel{def}{=} B \in \mathcal{M}_A$; i.e., to show that (*) holds. We have: $A \cap B = A \cap (\bigcup_j B_j) = \bigcup_j (A \cap B_j) \in \mathcal{M}$, since $\mathcal{M}$ is monotone and $A \cap B_j \uparrow$. Next, $A^c \cap B = A^c \cap (\bigcup_j B_j) = \bigcup_j (A^c \cap B_j) \in \mathcal{M}$ since $A^c \cap B_j \in \mathcal{M}$, by (*)(ii), and $A^c \cap B_j \uparrow$. Finally, $A \cap B^c = A \cap (\bigcup_j B_j)^c = A \cap (\bigcap_j B_j^c) = \bigcap_j (A \cap B_j^c)$ with $A \cap B_j^c \in \mathcal{M}$ by (*)(iii) and $A \cap B_j^c \downarrow$, so that $\bigcap_j (A \cap B_j^c) \in \mathcal{M}$ since $\mathcal{M}$ is monotone. The case that $B_j \downarrow$ is treated similarly, and the proof that $\mathcal{M}_A$ is a monotone class is complete.

Step 2. If $A \in \mathcal{C}$, then $\mathcal{M}_A = \mathcal{M}$. As already mentioned, $\mathcal{M}_A \subseteq \mathcal{M}$. So it suffices to prove that $\mathcal{M} \subseteq \mathcal{M}_A$. Let $B \in \mathcal{C}$. Then (*) holds and hence $B \in \mathcal{M}_A$. Therefore $\mathcal{C} \subseteq \mathcal{M}_A$. By step 1, $\mathcal{M}_A$ is a monotone class and $\mathcal{M}$ is the minimal monotone class containing $\mathcal{C}$. Thus $\mathcal{M} \subseteq \mathcal{M}_A$ and hence $\mathcal{M}_A = \mathcal{M}$.

Step 3. If A is any set in $\mathcal{M}$, then $\mathcal{M}_A = \mathcal{M}$. We show that $\mathcal{C} \subseteq \mathcal{M}_A$, which implies $\mathcal{M} \subseteq \mathcal{M}_A$ since $\mathcal{M}_A$ is a monotone class containing $\mathcal{C}$ and $\mathcal{M}$ is the minimal monotone class over $\mathcal{C}$. Since also $\mathcal{M}_A \subseteq \mathcal{M}$, the result $\mathcal{M}_A = \mathcal{M}$ would follow. To show $\mathcal{C} \subseteq \mathcal{M}_A$, take $B \in \mathcal{C}$ and consider $\mathcal{M}_B$. Then $\mathcal{M}_B = \mathcal{M}$ by step 2. Since $A \in \mathcal{M}$, we have $A \in \mathcal{M}_B$, which implies that $B \cap A$, $B^c \cap A$, and $B \cap A^c$ all belong in $\mathcal{M}$; or $A \cap B$, $A^c \cap B$, and $A \cap B^c$ belong in $\mathcal{M}$, which means that $B \in \mathcal{M}_A$. ■

Theorem 6. Let $\mathcal{C}$ be a field and $\mathcal{M}$ be the minimal monotone class containing $\mathcal{C}$. Then $\mathcal{M} = \sigma(\mathcal{C})$. ■

Proof. Evidently, $\mathcal{M} \subseteq \sigma(\mathcal{C})$ since every σ-field is a monotone class. By Lemma 1, $\mathcal{M}$ is a field, and hence a σ-field, by Theorem 3. Thus, $\mathcal{M} \supseteq \sigma(\mathcal{C})$ and hence $\mathcal{M} = \sigma(\mathcal{C})$. ■

Remark 10. Lemma 1 and Theorem 6 just discussed provide an illustration of the intricate relation of fields, monotone classes, and σ-fields in a certain setting. As will also be seen in several places in this book, monotone classes are often used as tools in arguments meant to establish results about σ-fields. In this kind of arguments, the roles of a field and of a monotone class may be substituted by the so-called π-*systems* and λ-*systems*, respectively. The definition of these concepts may be found, for example, in page 41 in Billingsley (1995). A result analogous to Theorem 6 is then Theorem 1.3 in page 5 of the reference just cited, which states that: If $\mathcal{P}$ is a π-system and $\mathcal{G}$ is a λ-system, then $\mathcal{P} \subset \mathcal{G}$ implies $\sigma(\mathcal{P}) \subset \mathcal{G}$.

1.2 Product Measurable Spaces

Consider the measurable spaces $(\Omega_1, \mathcal{A}_1)$, $(\Omega_2, \mathcal{A}_2)$. Then

Definition 5. The *product space* of Ω_1, Ω_2, denoted by $\Omega_1 \times \Omega_2$, is defined as follows: $\Omega_1 \times \Omega_2 = \{\omega = (\omega_1, \omega_2); \omega_1 \in \Omega_1, \omega_2 \in \Omega_2\}$. In particular, for $A \in \mathcal{A}_1$, $B \in \mathcal{A}_2$ the product of A, B, denoted by $A \times B$, is defined by: $A \times B = \{\omega = (\omega_1, \omega_2); \omega_1 \in A, \omega_2 \in B\}$, and the subsets $A \times B$ of $\Omega_1 \times \Omega_2$ for $A \in \mathcal{A}_1$, $B \in \mathcal{A}_2$ are called (measurable) *rectangles*. A, B are called the *sides* of the rectangle. ■

From Definition 5, one easily verifies the following lemma.

Lemma 2. Consider the rectangle $E = A \times B$. Then, with "+" denoting union of disjoint events,

(i) $E^c = (A \times B^c) + (A^c \times \Omega_2) = (A^c \times B) + (\Omega_1 \times B^c)$.
Consider the rectangles $E_1 = A_1 \times B_1$, $E_2 = A_2 \times B_2$. Then

(ii) $E_1 \cap E_2 = (A_1 \cap A_2) \times (B_1 \cap B_2)$. Hence $E_1 \cap E_2 = \oslash$ if and only if at least one of the sets $A_1 \cap A_2$, $B_1 \cap B_2$ is $\oslash$.
Consider the rectangles E_1, E_2 as above, and the rectangles $F_1 = A_1' \times B_1'$, $F_2 = A_2' \times B_2'$. Then

(iii)
$$
\begin{aligned}
(E_1 \cap F_1) \cap (E_2 \cap F_2) &= [(A_1 \cap A_1') \times (B_1 \cap B_1')] \cap [(A_2 \cap A_2') \\
&\quad \times (B_2 \cap B_2')] \qquad \text{(by (ii))} \\
&= [(A_1 \cap A_1') \cap (A_2 \cap A_2')] \\
&\quad \times [(B_1 \cap B_1') \cap (B_2 \cap B_2')] \qquad \text{(by (ii))} \\
&= [(A_1 \cap A_2) \cap (A_1' \cap A_2')] \\
&\quad \times [(B_1 \cap B_2) \cap (B_1' \cap B_2')].
\end{aligned}
$$

Hence, the left-hand side is $\oslash$ if and only if at least one of $(A_1 \cap A_2) \cap (A_1' \cap A_2')$, $(B_1 \cap B_2) \cap (B_1' \cap B_2')$ is $\oslash$.

Theorem 7. Let $\mathcal{C}$ be the class of all finite *sums* (i.e., unions of pairwise disjoint) of rectangles $A \times B$ with $A \in \mathcal{A}_1$, $B \in \mathcal{A}_2$. Then $\mathcal{C}$ is a field (of subsets of $\Omega_1 \times \Omega_2$). ■

Proof. Clearly, $\mathcal{C} \neq \oslash$. Next, let $E, F \in \mathcal{C}$. Then we show that $E \cap F \in \mathcal{C}$. In fact, $E, F \in \mathcal{C}$ implies that $E = \sum_{i=1}^{m} E_i$, $F = \sum_{j=1}^{n} F_j$ with $E_i = A_i \times B_i, i = 1, \ldots, m$, $F_j = A'_j \times B'_j, j = 1, \ldots, n$. Thus $E \cap F = \bigcup_{i=1}^{m} \bigcup_{j=1}^{n} (E_i \cap F_j)$ and $E_i \cap F_j, E_{i'} \cap F_{j'}$ are disjoint for $(i, j) \neq (i', j')$ by Lemma 2 (ii), (iii). Indeed, in Lemma 2 (iii), make the identification: $A_1 = A_i, B_1 = B_i, A_2 = A_{i'}, B_2 = B_{i'}$, $A'_1 = A'_j, B'_1 = B'_j, A'_2 = A'_{j'}, B'_2 = B'_{j'}$ to get $(E_i \cap F_j) \cap (E_{i'} \cap F_{j'}) = [(A_i \cap A'_i) \cap (A'_j \cap A'_{j'})] \times [(B_i \cap B_{i'}) \cap (B'_j \cap B'_{j'})]$ by the third line on the right-hand side in Lemma 2(iii), and at least one of $(A_i \cap A_{i'}) \cap (A'_j \cap A'_{j'})$, $(B_i \cap B_{i'}) \cap (B'_j \cap B'_{j'})$ is equal to $\oslash$. Then, by Lemma 2(iii) again, $(E_i \cap F_j) \cap (E_{i'} \cap F_{j'}) = \oslash$, and therefore $E \cap F = \sum_{i=1}^{m} \sum_{j=1}^{n} (E_i \cap F_j)$. However, $E_i \cap F_j = (A_i \cap A'_j) \times (B_i \cap B'_j)$ (by Lemma 2(ii)), and $A_i \cap A'_j \in \mathcal{A}_1, B_i \cap B'_j \in \mathcal{A}_2, i = 1, \ldots, m, j = 1, \ldots, n$. Thus $E \cap F$ is the sum of finitely many rectangles and hence $E \cap F \in \mathcal{C}$. (By induction it is also true that if $E_k \in \mathcal{C}, k = 1, \ldots, \ell$, then $\bigcap_{k=1}^{\ell} E_k \in \mathcal{C}$.) Finally, $E^c = (\sum_{i=1}^{m} E_i)^c = \bigcap_{i=1}^{m} E_i^c = \bigcap_{i=1}^{m} [(A_i \times B_i^c) + (A_i^c \times \Omega_2)]$ (by Lemma 2(i)), and $A_i \times B_i^c$, $A_i^c \times \Omega_2$ are disjoint rectangles so that their sum is in $\mathcal{C}$. But then so is their intersection over $i = 1, \ldots, m$ by the induction just mentioned. The proof is completed. ■

Remark 11. Clearly, the theorem also holds true if we start out with fields $\mathcal{F}_1$ and $\mathcal{F}_2$ rather than σ-fields $\mathcal{A}_1$ and $\mathcal{A}_2$.

Definition 6. The σ-field generated by the field $\mathcal{C}$ is called the *product σ-field* of $\mathcal{A}_1, \mathcal{A}_2$ and is denoted by $\mathcal{A}_1 \times \mathcal{A}_2$. The pair $(\Omega_1 \times \Omega_2, \mathcal{A}_1 \times \mathcal{A}_2)$ is called the *product measurable space* of the (measurable) spaces $(\Omega_1, \mathcal{A}_1)$, $(\Omega_2, \mathcal{A}_2)$. ■

If we have $n \geq 2$ measurable spaces $(\Omega_i, \mathcal{A}_i), i = 1, \ldots, n$, the product measurable space $(\Omega_1 \times \cdots \times \Omega_n, \mathcal{A}_1 \times \cdots \times \mathcal{A}_n)$ is defined in an analogous way. In particular, if $\Omega_1 = \cdots = \Omega_n = \Re$ and $\mathcal{A}_1 = \cdots = \mathcal{A}_n = \mathcal{B}$, then the product space $(\Re^n, \mathcal{B}^n)$ is the *n-dimensional Borel space*, where $\Re^n = \Re \times \cdots \times \Re$, $\mathcal{B}^n = \mathcal{B} \times \cdots \times \mathcal{B}$ (n factors), and $\mathcal{B}^n$ is called the *n-dimensional Borel σ-field*. The members of $\mathcal{B}^n$ are called the *n-dimensional Borel sets*.

Now we consider the case of infinitely (countably or not) many measurable spaces $(\Omega_t, \mathcal{A}_t), t \in T$, where the ($\neq \oslash$) index set T will usually be the real line or the positive half of it or the unit interval (0, 1) or [0,1].

Definition 7. The *product space* of $\Omega_t, t \in T$, denoted by $\prod_{t \in T} \Omega_t$ or Ω_T, is defined by $\Omega_T = \prod_{t \in T} \Omega_t = \{\omega = (\omega_t, t \in T); \omega_t \in \Omega_t, t \in T\}$. ■

By forming the point $\omega = (\omega_t, t \in T)$ with $\omega_t \in \Omega_t, t \in T$, we tacitly assume, by invoking the *axiom of choice*, that there exists a function on T into $\bigcup_{t \in T} \Omega_t$ with $\Omega_t \neq \oslash, t \in T$, whose value at t, ω_t, belongs in Ω_t.

Now for $T = \{1, 2\}$, $\Omega_1 \times \Omega_2 = \{\omega = (\omega_1, \omega_2); \omega_1 \in \Omega_1, \omega_2 \in \Omega_2\}$. Also, let $f : T \to \Omega_1 \cup \Omega_2$ such that $f(1) \in \Omega_1, f(2) \in \Omega_2$. Then $(f(1), f(2)) \in \Omega_1 \times \Omega_2$. Conversely, any $(\omega_1, \omega_2) \in \Omega_1 \times \Omega_2$ is the (ordered) pair of values of a function f on T into $\Omega_1 \cup \Omega_2$ with $f(1) \in \Omega_1, f(2) \in \Omega_2$; namely, the function for which $f(1) = \omega_1, f(2) = \omega_2$. Thus, $\Omega_1 \times \Omega_2$ may be looked upon as the collection of *all* functions f on T into $\Omega_1 \cup \Omega_2$ with $f(1) \in \Omega_1, f(2) \in \Omega_2$. Similar interpretation holds for any finite collection of ($\neq \emptyset$) $\Omega_i, i = 1, \ldots, n$, as well as any collection of ($\neq \emptyset$) $\Omega_t, t \in T(\neq \emptyset)$ (by the axiom of choice). Thus, $\Omega_T = \prod_{t\in T}\Omega_t = \{f : T \to \bigcup_{t\in T}\Omega_t;\ f(t) \in \Omega_t, t \in T\}$. In particular, if $T = \Re$ and $\Omega_t = \Re, t \in T$, then $\Omega_T = \prod_{t\in T}\Omega_t$ is the set of all real-valued functions defined on $\Re$.

Remark 12. In many applications, we take $T = [0, 1], \Omega_t = \Re, t \in T$, and we consider subsets of $\prod_{t\in T}\Omega_t$, such as the set of all continuous functions, denoted by $\mathcal{C}([0, 1])$, or the set of all bounded and right-continuous functions, denoted by $\mathcal{D}([0, 1])$.

Next, for any positive integer N, let $T_N = \{t_1, \ldots, t_N\}$ with $t_i \in T, i = 1, \ldots, N$, and let $A_{T_N} = \prod_{t\in T_N} A_t$. Then A_{T_N} is a rectangle in $\Omega_{t_1} \times \cdots \times \Omega_{t_N}$. Furthermore,

Definition 8. The subset $A_{T_N} \times \prod_{t\in T_N^c}\Omega_t = \prod_{t\in T_N} A_t \times \prod_{t\in T_N^c}\Omega_t$ of $\prod_{t\in T}\Omega_t$ is called a *product cylinder* in $\Omega_T = \prod_{t\in T}\Omega_t$ with *basis* A_{T_N} and *sides* $A_t \in \mathcal{A}_t$, $t \in T_N$. ■

Theorem 8. Let $\mathcal{C}$ be the class of all finite sums of all product cylinders. Then $\mathcal{C}$ is a field (of subsets of $\prod_{t\in T}\Omega_t$). ■

The proof of this theorem is based on the same ideas as those used in proving Theorem 7.

Definition 9. The σ-field generated by $\mathcal{C}$ is called the *product σ-field* of $\mathcal{A}_t, t \in T$, and is denoted by $\mathcal{A}_T = \prod_{t\in T}\mathcal{A}_t$. The pair $(\Omega_T = \prod_{t\in T}\Omega_t, \mathcal{A}_T = \prod_{t\in T}\mathcal{A}_t)$ is called the *product measurable space* of the (measurable) spaces $(\Omega_t, \mathcal{A}_t), t \in T$. The space $(\Re^\infty, \mathcal{B}^\infty)$, the (countably) *infinite dimensional Borel space*, where $\Re^\infty = \Re \times \Re \times \cdots$, and $\mathcal{B}^\infty = \mathcal{B} \times \mathcal{B} \times \cdots$, is often of special interest. $\mathcal{B}^\infty$ is the (countably) *infinite-dimensional Borel σ-field.* The members of $\mathcal{B}^\infty$ are called (countably) *infinite dimensional Borel sets.* ■

For more information, see also page 62 of Loève (1963).

1.3 Measurable Functions and Random Variables

Let Ω, Ω' be two spaces and let X be a mapping such that $X : \Omega \to \Omega'$. Then the set *operator* X^{-1} associated with the mapping X is defined as follows:

Definition 10. $X^{-1} : \mathcal{P}(\Omega') \to \mathcal{P}(\Omega)$ and $X^{-1}(A') = A$, where $A = \{\omega \in \Omega;\ X(\omega) \in A'\}$; $X^{-1}(A')$ is the *inverse image* of A' under X. ■

From Definition 10 it follows that

Theorem 9.

(i) If $A' \cap B' = \emptyset$, then $[X^{-1}(A')] \cap [X^{-1}(B')] = \emptyset$.
(ii) $X^{-1}(A'^c) = [X^{-1}(A')]^c$.
(iii) $X^{-1}(\bigcup_{j\in I} A'_j) = \bigcup_{j\in I} X^{-1}(A'_j)$ and $X^{-1}(\sum_{j\in I} A'_j) = \sum_{j\in I} X^{-1}(A'_j)$.
(iv) $X^{-1}(\bigcap_{j\in I} A'_j) = \bigcap_{j\in I} X^{-1}(A'_j)$.

(v) $X^{-1}(A' - B') = X^{-1}(A') - X^{-1}(B')$ (equivalently, $X^{-1}(A' \cap B'^c) = X^{-1}(A') \cap [X^{-1}(B')]^c$).

(vi) If $A' \subseteq B'$, then $X^{-1}(A') \subseteq X^{-1}(B')$.

(vii) $\mathcal{C}' \subseteq \mathcal{C}''$, then $X^{-1}(\mathcal{C}') \subseteq X^{-1}(\mathcal{C}'')$, where $X^{-1}(\mathcal{C}') = \{A \subseteq \Omega;\ A = X^{-1}(A')$ for some $A' \in \mathcal{C}'\}$; and similarly for $\mathcal{C}''$. ■

Now let us assume that Ω' is supplied with a σ-field $\mathcal{A}'$. Then we have

Theorem 10. Define the class $\mathcal{C}$ of subsets of Ω as follows: $\mathcal{C} = X^{-1}(\mathcal{A}')$. Then $\mathcal{C}$ is a σ-field (i.e., the inverse image of a σ-field is a σ-field). ■

Remark 13. This σ-field is called the σ-field *induced* (in Ω) by X.

Proof of Theorem 10. This is immediate from (ii) and (iii) of Theorem 9. ■

Next assume that Ω is supplied with a σ-field $\mathcal{A}$. Then

Theorem 11. Define the class $\mathcal{C}'$ of subsets of Ω' as follows: $\mathcal{C}' = \{A' \subseteq \Omega';\ X^{-1}(A') \in \mathcal{A}\}$. Then $\mathcal{C}'$ is a σ-field. ■

Proof. Immediate from (ii) and (iii) of Theorem 9. ■

Theorem 12. Let $\mathcal{C}'$ be a class of subsets of Ω' and let $\mathcal{A}' = \sigma(\mathcal{C}')$. Then $\mathcal{A} \stackrel{def}{=} \sigma[X^{-1}(\mathcal{C}')] = X^{-1}(\mathcal{A}')$. ■

Proof. We have: $\mathcal{C}' \subseteq \mathcal{A}'$ implies $X^{-1}(\mathcal{C}') \subseteq X^{-1}(\mathcal{A}')$, and this implies $\mathcal{A} \subseteq X^{-1}(\mathcal{A}')$ because $X^{-1}(\mathcal{A}')$ is a σ-field by Theorem 10. Thus, to show $X^{-1}(\mathcal{A}') \subseteq \mathcal{A}$. Define $\mathcal{C}^*$ as follows: $\mathcal{C}^* = \{A' \subseteq \Omega';\ X^{-1}(A') \in \mathcal{A}\}$. Then, clearly, $\mathcal{C}' \subseteq \mathcal{C}^*$, and $\mathcal{C}^*$ is a σ-field by Theorem 11, and the fact that $X^{-1}(\mathcal{C}^*) \subseteq \mathcal{A}$. Hence $\mathcal{A}' \subseteq \mathcal{C}^*$ and therefore $X^{-1}(\mathcal{A}') \subseteq X^{-1}(\mathcal{C}^*) \subseteq \mathcal{A}$. Thus $X^{-1}(\mathcal{A}') \subseteq \mathcal{A}$ and $X^{-1}(\mathcal{A}') = \mathcal{A}$. ■

Now assume that both Ω and Ω' are supplied with σ-fields $\mathcal{A}$ and $\mathcal{A}'$, respectively. Then

Definition 11. If $X^{-1}(\mathcal{A}') \subseteq \mathcal{A}$ we say that X is *measurable* with respect to $\mathcal{A}$ and $\mathcal{A}'$, or just measurable if no confusion is possible. In particular, if $(\Omega', \mathcal{A}') = (\Re^n, \mathcal{B}^n)$ and X is measurable, we say that X is an n-dimensional *random vector* and if $n = 1$, a *random variable* (r.v.). If $(\Omega, \mathcal{A}) = (\Re^n, \mathcal{B}^n)$, $(\Omega', \mathcal{A}') = (\Re^m, \mathcal{B}^m)$ and $f : \Omega \to \Omega'$ is measurable, then f is called a *Borel function*, and for $m = 1$ a *Baire function*. ■

The meaning and significance of Theorem 12 are this: if we want to check measurability of X, it suffices only to check that $X^{-1}(\mathcal{C}') \subseteq \mathcal{A}$, where $\mathcal{C}'$ is a class generating $\mathcal{A}'$. Indeed, if $X : (\Omega, \mathcal{A}) \to (\Omega', \mathcal{A}')$, then $(\mathcal{A}, \mathcal{A}')-$ measurability of X means $X^{-1}(\mathcal{A}') \subseteq \mathcal{A}$. Let $X^{-1}(\mathcal{C}') \subseteq \mathcal{A}$ and let $\mathcal{A}' = \sigma(\mathcal{C}')$. Then $\sigma[X^{-1}(\mathcal{C}')] \subseteq \mathcal{A}$. But $\sigma[X^{-1}(\mathcal{C}')] = X^{-1}(\mathcal{A}')$. Thus $X^{-1}(\mathcal{A}') \subseteq \mathcal{A}$. In particular, in the Borel real line, X is a r.v. if only $X^{-1}(\mathcal{C}_0)$ or $X^{-1}(\mathcal{C}_j)$ or $X^{-1}(\mathcal{C}'_j) \subseteq \mathcal{A}$, $j = 1, \ldots, 8$, where $\mathcal{C}_0$ is as in Application 1, the classes $\mathcal{C}_j$, $j = 1, \ldots, 8$, are the classes of intervals each consisting of intervals from $\mathcal{C}_0$ of one type, and $\mathcal{C}'_j$ is the class taken from $\mathcal{C}_j$ when the endpoints of the intervals are restricted to be rational numbers $j = 1, \ldots, 8$.

Theorem 13. Let $X : (\Omega, \mathcal{A}) \to (\Omega', \mathcal{A}')$ be measurable and let $f : (\Omega', \mathcal{A}') \to (\Omega'', \mathcal{A}'')$ be measurable. Define $f(X) : \Omega \to \Omega''$ as follows: $f(X)(\omega) = f[X(\omega)]$.

Then the mapping $f(X)$ is measurable. That is, a measurable mapping of a measurable mapping is a measurable mapping. ■

Proof. For $A'' \in \mathcal{A}''$, $[f(X)]^{-1}(A'') = X^{-1}[f^{-1}(A'')] = X^{-1}(A')$ with $A' \in \mathcal{A}'$. Thus $X^{-1}(A') = A \in \mathcal{A}$. ■

Corollary 1. Borel functions of random vectors are random vectors.

Proof. Take $(\Omega', \mathcal{A}') = (\Re^n, \mathcal{B}^n)$, $(\Omega'', \mathcal{A}'') = (\Re^m, \mathcal{B}^m)$. ■

We now consider the measurable spaces $(\Omega, \mathcal{A})$, $(\Omega', \mathcal{A}')$ and assume that Ω and Ω' are also provided with topologies $\mathcal{T}$ and $\mathcal{T}'$, respectively. (Recall that $\mathcal{T}$ is a *topology* for Ω if $\mathcal{T}$ is a class of subsets of Ω with the following properties: (i) $\emptyset$, $\Omega \in \mathcal{T}$, (ii) $\mathcal{T}$ is closed under finite intersections of members of $\mathcal{T}$, and (iii) $\mathcal{T}$ is closed under arbitrary unions of members of $\mathcal{T}$.) The pair $(\Omega, \mathcal{T})$ is called a *topological space*, and the members of $\mathcal{T}$ are called *open* sets. Also, $f : (\Omega, \mathcal{T}) \to (\Omega', \mathcal{T}')$ is said to be *continuous* (with respect to the topologies $\mathcal{T}$ and $\mathcal{T}'$), if $f^{-1}(T') \in \mathcal{T}$ for every $T' \in \mathcal{T}'$.

Theorem 14. Let $f : \Omega \to \Omega'$ be continuous and let that $\mathcal{T} \subseteq \mathcal{A}$, $\mathcal{A}' = \sigma(\mathcal{T}')$. Then f is measurable. ■

Proof. Continuity of f implies $f^{-1}(T') \in \mathcal{T}$, $T' \in \mathcal{T}'$. Hence $f^{-1}(\mathcal{T}') \subseteq \mathcal{A}$. Since $\mathcal{T}'$ generates $\mathcal{A}'$, we have $f^{-1}(\mathcal{A}') = \sigma[f^{-1}(\mathcal{T}')] \subseteq \mathcal{A}$ by Theorem 12. ■

Application 2. Recall that a class of sets in $\mathcal{T}$ is a *base* for $\mathcal{T}$ if every T in $\mathcal{T}$ is the union of members of this class. A topology $\mathcal{T}$ and the corresponding topological space are called *separable* if there exists a countable base for $\mathcal{T}$. In the spaces $(\Re^k, \mathcal{B}^k)$, $k \geq 1$, the "usual" topology $\mathcal{T}_k$ is the one with base the class of all finite open intervals (rectangles) or only the class of all open intervals (rectangles) with rational endpoints. This second base is countable and the topology $\mathcal{T}_k$ and the space $(\Re^k, \mathcal{T}_k)$ are separable. Then, clearly, $\mathcal{B}^k$ is generated by $\mathcal{T}_k$ (see Theorem 7, Definition 6, and the paragraph following it). Thus we have

Corollary 2. Let $X : (\Omega, \mathcal{A}) \to (\Re^n, \mathcal{B}^n)$ be measurable and let $f : (\Re^n, \mathcal{B}^n) \to (\Re^m, \mathcal{B}^m)$ be continuous. Then $f(X) : \Omega \to \Re^m$ is measurable (i.e., continuous functions of a random vector are random vectors).

Proof. Follows by the fact that $\mathcal{T}_m$ and $\mathcal{T}_n$ generate $\mathcal{B}^m$ and $\mathcal{B}^n$, respectively. ■

This corollary implies that the usual operations applied on r.v.s, such as forming sums, products, or quotients, will give r.v.s.

Now if $X : \Omega \to \Re^n$, then X can be written as $X = (X_1, \ldots, X_n)$. In connection with this we have

Theorem 15. Let $X = (X_1, \ldots, X_n) : (\Omega, \mathcal{A}) \to (\Re^n, \mathcal{B}^n)$. Then X is a random vector (measurable function) if and only if X_j, $j = 1, \ldots, n$ are r.v.s. ■

Proof. Let $B_i \in \mathcal{B}$, $i = 1, \ldots, n$. Then $X^{-1}(B_1 \times \cdots \times B_n) = (X_1, \ldots, X_n)^{-1}$ $(B_1 \times \cdots \times B_n) = (X_1 \in B_1) \cap \cdots \cap (X_n \in B_n) = [X_1^{-1}(B_1)] \cap \cdots \cap [X_n^{-1}(B_n)]$. Thus, if X_j, $j = 1, \ldots, n$ are r.v.s, then $X_j^{-1}(B_j) \in \mathcal{A}$ for every j and hence so is

$\bigcap_{j=1}^{n} X^{-1}(B_j)$. So, if X_j, $j = 1, \ldots, n$, are measurable, so is X (by the definition of the product σ-field $\mathcal{B}^n$). Next, consider the *projection* functions $f_j : \Re^n \rightarrow \Re$ such that $f_j(x_1, \ldots, x_n) = x_j$, $j = 1, \ldots, n$. It is known that f_j, $j = 1, \ldots, n$, are continuous, hence measurable. Then $X_j = f_j(X)$, $j = 1, \ldots, n$, and the measurability of X implies the measurability of X_j, $j = 1, \ldots, n$. ■

Let X be a r.v. Then the *positive part* of X, denoted by X^+, and the *negative part* of X, denoted by X^-, are defined as follows:

$$X^+ = \begin{cases} X \text{ if } X \geq 0 \\ 0 \text{ if } X < 0 \end{cases}; \quad X^- = \begin{cases} 0 \text{ if } X \geq 0, \\ -X \text{ if } X < 0 \end{cases}.$$

Then, clearly, $X = X^+ - X^-$ and $|X| = X^+ + X^-$.

Now as a simple application of the Corollary to Theorem 14, we show that both X^+ and X^- are r.v.s and then, of course, so is $|X|$. To this end, take $n = m = 1$ and define f by: $f(x) = x^+$, which is a continuous function of x, and similarly for $f(x) = x^-$. Directly, the measurability of X^+ is established as follows. In order to prove the measurability of X^+, it suffices to show that $(X^+)^{-1}((-\infty, x]) \in \mathcal{A}$ for $x \in \Re$. For $x < 0$, $(X^+ \leq x) = \emptyset$. For $x = 0$, $(X^+ \leq 0) = (X \leq 0) \in \mathcal{A}$. For $x > 0$, $(X^+ \leq x) = (X^+ = 0) \cup (0 < X^+ \leq x) = (X \leq 0) \cup (0 < X \leq x) = (X \leq x) \in \mathcal{A}$.

(Recall that for a sequence $\{x_n\}$ of real numbers, and as $n \rightarrow \infty$:

(1) $\limsup_n x_n$ or $\overline{\lim}_n x_n = \bar{x}$ if for every $\varepsilon > 0$ there exists $n(\varepsilon) > 0$ integer such that $n \geq n(\varepsilon)$ implies $x_n \leq \bar{x} + \varepsilon$, and $x_n > \bar{x} - \varepsilon$ for at least one $n \geq n(\varepsilon)$.

(2) $\liminf_n x_n$ or $\underline{\lim}_n x_n = \underline{x}$ if for every $\varepsilon > 0$ there exists $n(\varepsilon) > 0$ integer such that $n \geq n(\varepsilon)$ implies $x_n \geq \underline{x} - \varepsilon$ and $x_n < \underline{x} + \varepsilon$ for at least one $n \geq n(\varepsilon)$.

Also,

(3) $\overline{\lim}_n x_n = \inf_n \sup_{i \geq n} x_i = \inf_n y_n$, $y_n \overset{def}{=} \sup_{i \geq n} x_i$, so that $y_n \downarrow$ and set $\inf_n y_n = \lim_n y_n \overset{def}{=} \bar{x}$.

(4) $\underline{\lim}_n x_n = \sup_n \inf_{i \geq n} x_i = \sup_n z_n$, $z_n \overset{def}{=} \inf_{i \geq n} x_i$, so that $z_n \uparrow$ and set $\sup_n z_n = \lim_n z_n \overset{def}{=} \underline{x}$.

For every $n \geq 1$, $z_n \leq y_n$ so that $\sup_n z_n \leq \inf_n y_n$ or equivalently $\underline{x} \leq \bar{x}$. If $\underline{x} \geq \bar{x}$, then the common value $\underline{x} = \bar{x} = x$ is the $\lim_n$ of x_n.)

Next let $X_n, n \geq 1$, be r.v.s. Then define the following mappings (which are assumed to be finite). The sup and inf are taken over $n \geq 1$ and all limits are taken as $n \to \infty$.

$$\left.\begin{array}{ll} \sup_n X_n & : \left(\sup_n X_n\right)(\omega) = \sup_n X_n(\omega) \\ \inf_n X_n & : \left(\inf_n X_n\right)(\omega) = \inf_n X_n(\omega) \\ \limsup\limits_n X_n \text{ or } \overline{\lim\limits_n} X_n & : \left(\limsup\limits_n X_n\right)(\omega) = \limsup\limits_n X_n(\omega) \\ \liminf\limits_n X_n \text{ or } \underline{\lim\limits_n} X_n & : \left(\liminf\limits_n X_n\right)(\omega) = \liminf\limits_n X_n(\omega) \end{array}\right\} \omega \in \Omega.$$

Then $\liminf_n X_n \leq \limsup_n X_n$ and if $\liminf_n X_n = \limsup_n X_n$, this defines the mapping $\lim_n X_n$. Then we have the following:

Theorem 16. If $X_n, n \geq 1$, are r.v.s, then the mappings just defined are also r.v.s. ■

Proof. We have $(\sup_n X_n \leq x) = (X_n \leq x, n \geq 1) = \bigcap_{n=1}^{\infty}(X_n \leq x) \in \mathcal{A}$. Thus $\sup_n X_n$ is a r.v. Now $\inf_n X_n = -\sup(-X_n)$ and then the measurability of $\sup_n X_n$ implies the measurability of $\inf_n X_n$.

Next, $\limsup_n X_n = \inf_n(\sup_{j \geq n} X_j)$. Thus, if $Y_n = \sup_{j \geq n} X_j$, then $Y_n, n \geq 1$, are r.v.s and then so is the $\inf_n Y_n$. Finally, $\liminf_n X_n = -\limsup_n(-X_n)$, and then the previous result implies the measurability of $\liminf_n X_n$. The measurability of $\lim_n X_n$, if the limit exists, is an immediate consequence of the last two results. ■

A measurable mapping X on $(\Omega, \mathcal{A})$ into $(\bar{\Re}, \bar{\mathcal{B}})$, the extended Borel real line, is an *extended r.v.* Then Theorem 16 still holds true if the operations applied on $X_n, n \geq 1$, produce extended r.v.s.

Definition 12. Consider the measurable space $(\Omega, \mathcal{A})$ and let $\{A_j, j \in I\}$ be a collection of sets in $\mathcal{A}$ such that $A_i \cap A_j = \emptyset, i, j \in I, i \neq j$, and $\sum_j A_j = \Omega$. Then this collection is called a (measurable) *partition* of Ω. The partition is *finite* if I is a finite set and *infinite* otherwise. ■

Definition 13. Let $\{A_j, j = 1, \ldots, n\}$ be a (finite, measurable) partition of Ω, and define the mapping $X : \Omega \to \Re$ as follows: $X = \sum_{j=1}^{n} \alpha_j I_{A_j}$, such that $\alpha_j \in \Re, j = 1, \ldots, n$ (which may be assumed to be distinct). Then X is called a *simple r.v.* If $\{A_j, j = 1, 2, \ldots\}$ is a (countably infinite measurable) partition of Ω, then the mapping $X : \Omega \to \Re$ such that $X = \sum_{j=1}^{\infty} \alpha_j I_{A_j}, \alpha_j \in \Re, j = 1, 2, \ldots$ (which may be assumed to be distinct) is called an *elementary r.v.* ■

Remark 14. By I_A we denote the *indicator* of the set A; i.e.,

$$I_A(\omega) = \begin{cases} 1 \text{ if } \omega \in A, \\ 0 \text{ if } \omega \in A^c. \end{cases}$$

It is evident that simple and elementary r.v.s are indeed r.v.s. What is more important, however, is that some kind of an inverse of this statement also holds true. More precisely,

Theorem 17. Every r.v. is the pointwise limit of a sequence of simple r.v.s. ■

Proof. Consider the r.v. X, the interval $[-n, n)$, and define the sets:

$$A_{nj} = \left(\frac{j-1}{2^n} \leq X < \frac{j}{2^n}\right), \quad j = -n2^n + 1, \quad -n2^n + 2, \ldots, n2^n,$$
$$A'_n = (X < -n),\ A''_n = (X \geq n), \quad n = 1, 2, \ldots$$

Then, clearly, $\{A_{nj}, j = -n2^n + 1, \ldots, n2^n, A'_n, A''_n\}$ is a (measurable) partition of Ω. Thus, if we define X_n by $X_n = \sum_{j=-n2^n+1}^{n2^n} \frac{j-1}{2^n} I_{A_{nj}} + (-n) I_{A'_n} + n I_{A''_n}$, then X_n is a simple r.v.

We are going to show next that $X_n(\omega) \underset{n\to\infty}{\to} X(\omega)$ for every $\omega \in \Omega$.

Let $\omega \in \Omega$. Then there exists $n_o = n_o(\omega)$ such that $|X(\omega)| < n_o$. It is asserted that $\omega \in A_{nj}$ for $n \geq n_o$, some $j = -n2^n + 1, \ldots, n2^n$. This is so because for $n \geq n_o, [-n, n) \supseteq [-n_o, n_o)$ and the intervals $[\frac{j-1}{2^n}, \frac{j}{2^n}),\ j = -n2^n + 1, \ldots, n2^n$ form a partition of $[-n, n)$. Let that $\omega \in A_{nj(n)}$. Then $\frac{j(n)-1}{2^n} \leq X(\omega) < \frac{j(n)}{2^n}$. But then $X_n(\omega) = \frac{j(n)-1}{2^n}$ so that

$$|X_n(\omega) - X(\omega)| < \frac{1}{2^n}. \quad \text{Thus } X_n(\omega) \underset{n\to\infty}{\to} X(\omega).$$ ■

To this theorem we have the following:

Corollary 3. If the r.v. $X \geq 0$, then there exists a sequence of simple r.v.s X_n such that $0 \leq X_n \uparrow X$ as $n \to \infty$.

Proof. If $X \geq 0$, then X_n of the theorem becomes as follows:
$X_n = \sum_{j=1}^{n2^n} \frac{j-1}{2^n} I_{A_{nj}} + nA''_n$, so that $0 \leq X_n \underset{n\to\infty}{\to} X$. We will next show that $X_n \uparrow$. For each n, we have that $[0, n)$ is divided into the $n2^n$ subintervals $[\frac{j-1}{2^n}, \frac{j}{2^n})$, $j = 1, 2, \ldots, n2^n$, and for $n + 1$, $[0, n + 1)$ is divided into $(n + 1)2^{n+1}$ subintervals $[\frac{j-1}{2^{n+1}}, \frac{j}{2^{n+1}})$, $j = 1, 2, \ldots, (n + 1)2^{n+1}$, and each one of the intervals in the first class of intervals is split into two intervals in the second class of intervals. Thus $X_n(\omega) \leq X_{n+1}(\omega)$ for every $\omega \in \Omega$ (see following picture).

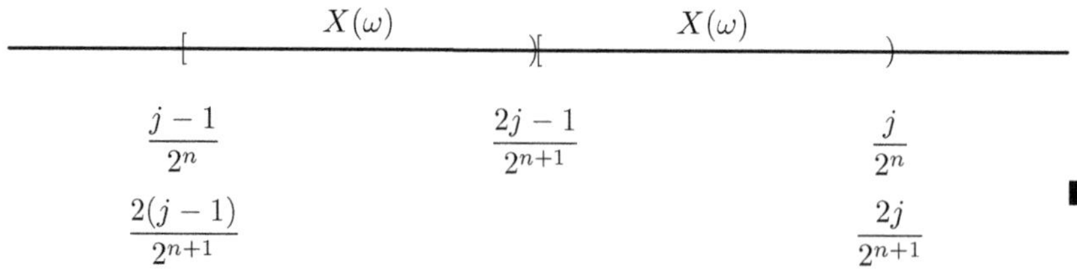

■

Remark 15. The significance of the corollary is that the nondecreasing simple r.v.s X_n are also ≥ 0. This point will be exploited later on in the so-called Lebesgue Monotone Convergence Theorem and elsewhere.

Remark 16. Theorem 17 and its corollary are, clearly, true even if X is an extended r.v.

Exercises.

1. Consider the measurable space $(\Omega, \mathcal{A})$ and let $A_n \in \mathcal{A}$, $n = 1, 2, \ldots$ Then recall that

$$\liminf_{n\to\infty} A_n = \underline{\lim}_{n\to\infty} A_n = \bigcup_{n=1}^{\infty}\bigcap_{j=n}^{\infty} A_j \text{ and } \limsup_{n\to\infty} A_n = \overline{\lim}_{n\to\infty} A_n = \bigcap_{n=1}^{\infty}\bigcup_{j=n}^{\infty} A_j.$$

(i) Show that $\underline{\lim}_{n\to\infty} A_n \subseteq \overline{\lim}_{n\to\infty} A_n$. (If also $\overline{\lim}_{n\to\infty} A_n \subseteq \underline{\lim}_{n\to\infty} A_n$, so that $\underline{\lim}_{n\to\infty} A_n = \overline{\lim}_{n\to\infty} A_n$, then this set is denoted by $\lim_{n\to\infty} A_n$ and is called the *limit* of the sequence $\{A_n\}, n \geq 1$.)

(ii) Show that $\left(\underline{\lim}_{n\to\infty} A_n\right)^c = \overline{\lim}_{n\to\infty} A_n^c$, $\left(\overline{\lim}_{n\to\infty} A_n\right)^c = \underline{\lim}_{n\to\infty} A_n^c$. Conclude that if $\lim_{n\to\infty} A_n = A$, then $\lim_{n\to\infty} A_n^c = A^c$.

(iii) Show that

$$\underline{\lim}_{n\to\infty} (A_n \cap B_n) = \left(\underline{\lim}_{n\to\infty} A_n\right) \cap \left(\underline{\lim}_{n\to\infty} B_n\right),$$

and

$$\overline{\lim}_{n\to\infty} (A_n \cup B_n) = \left(\overline{\lim}_{n\to\infty} A_n\right) \cup \left(\overline{\lim}_{n\to\infty} B_n\right).$$

(iv) Show that

$$\overline{\lim}_{n\to\infty} (A_n \cap B_n) \subseteq \left(\overline{\lim}_{n\to\infty} A_n\right) \cap \left(\overline{\lim}_{n\to\infty} B_n\right),$$

and

$$\underline{\lim}_{n\to\infty} (A_n \cup B_n) \supseteq \left(\underline{\lim}_{n\to\infty} A_n\right) \cup \left(\underline{\lim}_{n\to\infty} B_n\right).$$

(v) By a counterexample, show that the inverse inclusions in part (iv) do not hold, so that $\overline{\lim}_{n\to\infty}(A_n \cap B_n)$ need not be equal to $\left(\overline{\lim}_{n\to\infty} A_n\right) \cap \left(\overline{\lim}_{n\to\infty} B_n\right)$, and $\underline{\lim}_{n\to\infty}(A_n \cup B_n)$ need not be equal to $\left(\underline{\lim}_{n\to\infty} A_n\right) \cup \left(\underline{\lim}_{n\to\infty} B_n\right)$.

(vi) If $\lim_{n\to\infty} A_n = A$ and $\lim_{n\to\infty} B_n = B$, then show that $\lim_{n\to\infty}(A_n \cap B_n) = A \cap B$ and $\lim_{n\to\infty}(A_n \cup B_n) = A \cup B$.

(vii) If $\lim_{n\to\infty} A_n = A$, then show that for any set B, $\lim_{n\to\infty}(A_n \Delta B) = A \Delta B$, where $A_n \Delta B$ is the symmetric difference of A_n and B.

(viii) If $A_{2j-1} = B$ and $A_{2j} = C$, $j = 1, 2, \ldots$, determine $\underline{\lim}_{n\to\infty} A_n$ and $\overline{\lim}_{n\to\infty} A_n$. Under what condition on B and C does the limit exist, and what is it equal to?

Hint:

(i) Use the definition of $\underline{\lim}_{n\to\infty} A_n$ and $\overline{\lim}_{n\to\infty} A_n$, and show that each side is contained in the other.

(ii) Use the definition of $\underline{\lim}_{n\to\infty} A_n$, $\overline{\lim}_{n\to\infty} A_n$, and DeMorgan's laws.

(iii), (iv) Use the definition of $\underline{\lim}_{n\to\infty} A_n$ and $\overline{\lim}_{n\to\infty} A_n$, and then show that each side is included in the other.

(v) A choice of the A_ns and B_ns that does this is if one takes $A_{2j-1} = A$, $A_{2j} = A_0$, $B_{2j-1} = B$, $B_{2j} = B_0$, $j \geq 1$, then take $\Omega = \Re$, and finally select A, A_0, B and B_0 suitably.

(vi) Use parts (iii) and (iv).

(vii) It follows from parts (vi) and (ii).

(viii) It follows from part (v).

2. **(i)** Setting $\underline{A} = \liminf_{n\to\infty} A_n$, $\overline{A} = \limsup_{n\to\infty} A_n$, and $A = \lim_{n\to\infty} A_n$ if it exists, show that all $\underline{A}$, $\overline{A}$, and A are in $\mathcal{A}$.

(ii) If $A_n \uparrow$ as $n \to \infty$, show that $\lim_{n\to\infty} A_n$ exists and is equal to $\cup_{n=1}^{\infty} A_n$, and if $A_n \downarrow$ as $n \to \infty$, then $\lim_{n\to\infty} A_n$ exists and is equal to $\cap_{n=1}^{\infty} A_n$.

3. Carry out the details of the proof of Theorem 1.

4. By means of an example, show that $A_j \in \mathcal{F}, j \geq 1$, need not imply that $\bigcup_{j=1}^{\infty} A_j \in \mathcal{F}$, and similarly for $\bigcap_{j=1}^{\infty} A_j$.

5. Let $\mathcal{P} = \{A_n, n = 1, 2, \ldots\}$ be a partition of Ω where $A_n \neq \emptyset, n \geq 1$, and let $\mathcal{C}$ be the class of all sums of members in $\mathcal{P}$. Then show that $\mathcal{C}$ is the σ-field generated by the class $\mathcal{P}$.

6. Let $\mathcal{C}_0$ be the class of all intervals in $\Re$, and consider the eight classes $\mathcal{C}_j$, $j = 1, \ldots, 8$, each of which consists of all intervals in $\mathcal{C}_0$ of one type. Then $\mathcal{B} = \sigma(\mathcal{C}_j), j = 1, \ldots, 8$. Also, if $\mathcal{C}'_j$ denotes the class we get from $\mathcal{C}_j$ by considering intervals with rational endpoints, then $\sigma(\mathcal{C}'_j) = \mathcal{B}, j = 1, \ldots, 8$.

Hint: One may choose to carry out the detailed proof for just one of these classes, e.g., the class $\mathcal{C}_1 = \{(x, y); x, y \in \Re, x < y\}$ or the class $\mathcal{C}'_1 = \{(x, y); x, y$ rationals in $\Re$ with $x < y\}$.

7. **(i)** If $\mathcal{C}$ is the class of all finite sums of intervals in $\Re$ (unions of pairwise disjoint intervals) of the form: $(\alpha, \beta], \alpha, \beta \in \Re, \alpha < \beta$; $(-\infty, \alpha]$, $\alpha \in \Re$; (β, ∞), $\beta \in \Re$, $\emptyset$, then $\mathcal{C}$ is a field and $\sigma(\mathcal{C}) = \mathcal{B}$.

(ii) The same is true if $\mathcal{C}$ is the class of all finite sums of all kinds of intervals in $\Re$.

8. Consider the space $(\Omega, \mathcal{F})$ and for an arbitrary but fixed set A with $\emptyset \subset A \subset \Omega$, define $\mathcal{F}_A$ by: $\mathcal{F}_A = \{B \subseteq \Omega; B = A \cap C, C \in \mathcal{F}\}$. Then $\mathcal{F}_A$ is a field (of subsets of A).

Hint: Notice that the complement of a set in $\mathcal{F}_A$ is with respect to the set A rather than Ω.

9. Consider the space $(\Omega, \mathcal{A})$ and let A be as in Exercise 8. Define $\mathcal{A}_A$ by $\mathcal{A}_A = \{B \subseteq \Omega; B = A \cap C, C \in \mathcal{A}\}$. Then $\mathcal{A}_A$ is a σ-field (of subsets of $\mathcal{A}$). Furthermore, $\mathcal{A}_A = \sigma(\mathcal{F}_A)$, where $\mathcal{F}_A$ is as in Exercise 8 and $\mathcal{A} = \sigma(\mathcal{F})$.

Hint: First, show that $\mathcal{A}_A$ is a σ-field and $\sigma(\mathcal{F}_A) \subseteq \mathcal{A}_A$. Next, show that $\sigma(\mathcal{F}_A) \supseteq \mathcal{A}_A$ by showing that, for any σ-field $\mathcal{A}^*$ of subsets of A, with $\mathcal{A}^* \supseteq \mathcal{F}_A$, it holds that $\mathcal{A}^* \supseteq \mathcal{A}_A$. This is done by defining $\mathcal{M}$ by $\mathcal{M} = \{C \in \mathcal{A}; A \cap C \in \mathcal{A}^*\}$ and showing that $\mathcal{M}$ is a monotone class.

10. Show that, if $\{\mathcal{A}_n\}$, $n \geq 1$, is a nondecreasing sequence of σ-fields, then $\bigcup_{n=1}^{\infty} \mathcal{A}_n$ is always a field, but it may fail to be a σ-field.

11. Carry out the details of the proof of Theorem 4.

12. By means of an example, show that a monotone class need not be a field.

13. Carry out the details of the proof of Lemma 2.

14. Let Ω_1, Ω_2 be two spaces and let $A, A_i \subseteq \Omega_1$, $B, B_i \subseteq \Omega_2$, $i = 1, 2$. Then show that

(i) $(A_1 \times B_1) - (A_2 \times B_2) = [(A_1 \cap A_2) \times (B_1 - B_2)] + [(A_1 - A_2) \times B_1]$.

(ii) $A \times B = \emptyset$, if and only if at least one of A, B is $\emptyset$.

(iii) If $A_i \times B_i$, $i = 1, 2$ are $\neq \emptyset$, then $A_1 \times B_1 \subseteq A_2 \times B_2$, if and only if $A_1 \subseteq A_2$, $B_1 \subseteq B_2$.

(iv) If $A_1 \times B_1 = A_2 \times B_2 \neq \emptyset$, then $A_1 = A_2$ and $B_1 = B_2$.

(v) Let $A \times B$, $A_i \times B_i$, $i = 1, 2$ be $\neq \emptyset$. Then $A \times B = (A_1 \times B_1) + (A_2 \times B_2)$, if and only if $A = A_1 + A_2$ and $B = B_1 = B_2$, or $A = A_1 = A_2$ and $B = B_1 + B_2$.

15. **(i)** With $A \subseteq \Omega_1$, and $B \subseteq \Omega_2$, show that $A \times B = \emptyset$ if and only if at least one of A or B is equal to $\emptyset$.

(ii) With $A_1, A_2 \subseteq \Omega_1$ and $B_1, B_2 \subseteq \Omega_2$, set $E_1 = A_1 \times B_1$ and $E_2 = A_2 \times B_2$ and assume that E_1 and E_2 are $\neq \emptyset$. Then $E_1 \subseteq E_2$ if and only if $A_1 \subseteq A_2$ and $B_1 \subseteq B_2$. Explain why the assumption that E_1 and E_2 are $\neq \emptyset$ is essential.

16. **(i)** Let $A_i \subseteq \Omega_i$, $i = 1, 2, \ldots, n$, and set $E = A_1 \times \cdots \times A_n$. Then $E = \emptyset$ if and only if at least one of A_i, $i = 1, 2, \ldots, n$, is $= \emptyset$.

(ii) If also $B_i \subseteq \Omega_i$, $i = 1, 2, \ldots, n$, and $F = B_1 \times \cdots \times B_n$, then show that $E \cap F = (A_1 \times \cdots \times A_n) \cap (B_1 \times \cdots \times B_n) = (A_1 \cap B_1) \times \cdots \times (A_n \cap B_n)$.

17. For $i = 1, 2, \ldots, n$, let $A_i, B_i, C_i \subseteq \Omega_i$ and set $E = A_1 \times \cdots, \times A_n$, $F = B_1 \times \cdots \times B_n$, $G = C_1 \times \cdots \times C_n$. Suppose that E, F, and G are all $\neq \emptyset$ and that $E = F + G$. Then show that there exists a j with $1 \leq j \leq n$ such that $A_j = B_j + C_j$ while $A_i = B_i = C_i$ for all $i \neq j$.

18. In reference to Theorem 7, show that $\mathcal{C}$ is still a field, if $\mathcal{A}_i$ is replaced by a field $\mathcal{F}_i$, $i = 1, 2$.

19. Consider the measurable spaces $(\Omega_i, \mathcal{A}_i)$, $i = 1, 2$, and let $\mathcal{C}$ be the class of all countable sums of rectangles (unions of pairwise disjoint rectangles) in the product space $\Omega_1 \times \Omega_2$. Then by an example, show that $\mathcal{C}$ need not be a σ-field.

Remark: Compare it to Theorem 7 in this chapter.

Hint: Take $\Omega_1 = \Omega_2 = [0, 1]$ and show that the main diagonal D of the rectangle $[0, 1] \times [0, 1]$ belongs in the σ-field generated by the field of all finite rectangles, but it is not in $\mathcal{C}$.

20. Carry out the details of the proof of Theorem 10.

21. Carry out the details of the proof of Theorem 11.

22. Consider the mapping X defined on $(\Omega, \mathcal{A})$ onto $\Omega' = X(\Omega)$, the image of Ω under X, and let $\mathcal{C}' \subseteq \mathcal{P}(\Omega')$ be defined as follows:

$$\mathcal{C}' = \{B \subseteq \Omega'; \quad B = X(A), \quad A \in \mathcal{A}\}.$$

Then, by means of an example, show that $\mathcal{C}'$ need not be a σ-field.

Remark: Compare this result with Theorem 11 in this chapter.

23. Consider the measurable space $(\Omega, \mathcal{A})$ and let X be defined by $\sum_{i=1}^{n} \alpha_i I_{A_i}$ or $X = \sum_{i=1}^{\infty} \alpha_i I_{A_i}$, where $\alpha_i \in \Re$ are distinct for all i and $\{A_1, \ldots, A_n\}$ or $\{A_i, i \geq 1\}$ are partitions of Ω. Then show that X is a r.v. (a simple r.v. and an elementary r.v., respectively) if and only if the partitions are measurable (i.e., $A_i \in \mathcal{A}$ for all i).

24. If X and Y are mappings on Ω into $\Re$, show that

$$\{\omega \in \Omega; \quad X(\omega) + Y(\omega) < x\} = \bigcup_{r \in Q} \big[\{\omega \in \Omega; X(\omega) < r\}$$
$$\cap \{\omega \in \Omega; \quad Y(\omega) < x - r\}\big],$$

where Q is the set of rationals in $\Re$.

25. If X is a r.v. defined on the measurable space $(\Omega, \mathcal{A})$, then $|X|$ is also a r.v. By an example, show that the converse need not be true.

26. By a direct argument (that is, by using the definition of measurability), show that, if X and Y are r.v.s, then so are the mappings $X \pm Y$, XY, and X/Y ($Y \neq 0$ a.s.; i.e., $Y \neq 0$ with probability 1, or $Y \neq 0$ almost surely).

27. Carry out the details of the proof of the Corollary to Theorem 14.

28. If X and Y are r.v.s defined on $(\Omega, \mathcal{A})$, show that $(X + Y)^+ \leq X^+ + Y^+$ and $(X + Y)^- \leq X^- + Y^-$.

29. Let $A_1, A_2, \ldots$ be arbitrary events in $(\Omega, \mathcal{A})$, and define B_m by: B_m = "A_m is the first event which occurs among the events $A_1, A_2, \ldots$," $m \geq 1$. Then

(i) Express B_m in terms of A_ns, $m \geq 1$.
(ii) Show that $B_1, B_2, \ldots$ are pairwise disjoint.
(iii) Show that $\sum_{m=1}^{\infty} B_m = \cup_{n=1}^{\infty} A_n$.

30. For a sequence of events $\{A_n\}$, $n \geq 1$, show that

(i) $\underline{\lim}_{n\to\infty} A_n = \{\omega \in \Omega; \quad \omega \in A_n$ for all but finitely many ns$\}$,
(ii) $\overline{\lim}_{n\to\infty} A_n = \{\omega \in \Omega; \omega \in A_n$ for infinitely many ns$\}$ (to be denoted by $(A_n$ i.o.) and read (A_ns occur infinitely often)).

31. If A_n and B_n are events such that $A_n \subseteq B_n$, $n \geq 1$, then show that $(A_n$ i.o.$) \subseteq (B_n$ i.o.$)$.

32. In $\Re$, let Q be the set of rational numbers, and for $n = 1, 2, \ldots$, let A_n be defined by

$$A_n = \left\{r \in \left(1 - \frac{1}{n+1}, 1 + \frac{1}{n}\right); \quad r \in Q\right\}.$$

Examine whether or not the $\lim_{n\to\infty} A_n$ exists.

33. In $\Re$, define the sets $A_n, n = 1, 2, \ldots$ as follows:

$$A_{2n-1} = \left[-1, \frac{1}{2n-1}\right], \quad A_{2n} = \left[0, \frac{1}{2n}\right).$$

Examine whether or not the $\lim_{n\to\infty} A_n$ exists.

34. Take $\Omega = \Re$, and let $\mathcal{A}_n$ be the σ-field generated by the class $\{[0, 1), [1, 2), \ldots, [n-1, n)\}, n \geq 1$. Then show that

(i) $\mathcal{A}_n \subseteq \mathcal{A}_{n+1}, n \geq 1$, and indeed $\mathcal{A}_n \subset \mathcal{A}_{n+1}, n \geq 1$.
(ii) The class $\cup_{n=1}^{\infty} \mathcal{A}_n$ is not a σ-field.
(iii) Describe explicitly $\mathcal{A}_1$ and $\mathcal{A}_2$.

35. Let $A_1, \ldots, A_n$ be arbitrary subsets of an abstract set Ω, and let A_i' be either A_i or $A_i^c, i = 1, \ldots, n$. Define the class $\mathcal{C}$ of subsets of Ω as follows:

$$\mathcal{C} = \{\text{all unions of the intersections } A_1' \cap \cdots \cap A_n'\}.$$

Then show that

(i) The class $\mathcal{C}$ is a field (generated by the sets $A_1, \ldots, A_n$).
(ii) Compute the number of elements of $\mathcal{C}$.

36. If $f : \Omega \to \Omega'$, then show that

(i) $f^{-1}[f(A)] \supseteq A, A \subseteq \Omega$.
(ii) $f[f^{-1}(B)] \subseteq B, B \subseteq \Omega'$.
(iii) By concrete examples, show that the relations in (i) and (ii) may be strict.

37. **(i)** On the measurable space $(\Omega, \mathcal{A})$, define the function X as follows:

$$X(\omega) = \begin{cases} -1 & \text{on } A_1, \\ 1 & \text{on } A_1^c \cap A_2, \\ 0 & \text{on } A_1^c \cap A_2^c, \end{cases}$$

where $A_1, A_2 \in \mathcal{A}$. Examine whether or not X is a r.v.

(ii) On the measurable space $(\Omega, \mathcal{A})$ with $\Omega = \{a, b, c, d\}$ and $\mathcal{A} = \{\emptyset, \{a, b\}, \{c, d\}, \Omega\}$, define the function X as follows: $X(a) = X(b) = -1, X(c) = 1, X(d) = 2$. Examine whether or not X is a r.v.

(iii) If $\Omega = \{-2, -1, 0, 1, 2\}$ and X is defined on Ω by $X(\omega) = \omega$, determine the field induced by X and that induced by X^2. Verify that the latter is contained in the former.

38. For a sequence of r.v.s $\{X_n\}, n \geq 1$, set $\mathcal{B}_k = \sigma(X_k, X_{k+1}, \ldots), k \geq 1$. Then show that for every k and l with $k < l$, it holds that $\mathcal{B}_k \supseteq \mathcal{B}_l$.

39. For the r.v.s $X_1, X_2, \ldots, X_n$, set $S_k = \sum_{j=1}^{k} X_j, k = 1, \ldots, n$, and show that $\sigma(X_1, X_2, \ldots, X_n) = \sigma(S_1, S_2, \ldots, S_n)$.

40. For any set $B \subseteq \Re$, the set $B + c \overset{def}{=} B_c$ is defined by: $B_c = \{y \in \Re;\ y = x + c, x \in B\}$. Then show that if B is measurable, so is B_c.

41. Let Ω be an abstract set, and let $\mathcal{C}$ be an arbitrary nonempty class of subsets of Ω. Define the class $\mathcal{F}_1$ to consists of all members of $\mathcal{C}$ as well as all of their complements; i.e.,

$$\begin{aligned}\mathcal{F}_1 &= \{A \subseteq \Omega;\quad A \in \mathcal{C} \text{ or } A = C^c \text{ with } C \in \mathcal{C}\}\\ &= \{A \subseteq \Omega;\quad A \in \mathcal{C} \text{ or } A^c \in \mathcal{C}\} = \mathcal{C} \cup \{C^c;\quad C \in \mathcal{C}\},\end{aligned}$$

so that $\mathcal{F}_1$ is closed under complementation.
Next, define the class $\mathcal{F}_2$ as follows:

$$\begin{aligned}\mathcal{F}_2 &= \{\text{all finite intersections of members of } \mathcal{F}_1\}\\ &= \{A \subseteq \Omega;\quad A = A_1 \cap \cdots \cap A_m,\ A_i \in \mathcal{F}_1,\\ &\qquad i = 1, \ldots, m,\ m \geq 1\}.\end{aligned}$$

Also, define the class $\mathcal{F}_3$ by

$$\begin{aligned}\mathcal{F}_3 &= \{\text{all finite unions of members of } \mathcal{F}_2\}\\ &= \{A \subseteq \Omega;\quad A = \bigcup_{i=1}^{n} A_i \text{ with } A_i \in \mathcal{F}_2,\ i = 1, \ldots, n,\ n \geq 1\}\\ &= \{A \subseteq \Omega;\quad A = \bigcup_{i=1}^{n} A_i \text{ with } A_i = A_i^1 \cap \cdots \cap A_i^{m_i},\\ &\qquad A_i^1, \ldots, A_i^{m_i} \in \mathcal{F}_1,\ , m_i \geq 1 \text{ integers},\\ &\qquad i = 1, \ldots, n,\ n \geq 1\}.\end{aligned}$$

Set $\mathcal{F}_3 = \mathcal{F}$ and show that

(i) $\mathcal{F}$ is a field.
(ii) $\mathcal{F}$ is the field generated by $\mathcal{C}$; i.e., $\mathcal{F} = \mathcal{F}(\mathcal{C})$.

42. Refer to Exercise 41, and set $\mathcal{A}_1 = \mathcal{F}_1$. Then define the classes $\mathcal{A}_2$ and $\mathcal{A}_3$ instead of $\mathcal{F}_2$ and $\mathcal{F}_3$, respectively, by replacing finite intersections and finite unions by countable intersections and countable unions, respectively. Set $\mathcal{A}_3 = \mathcal{A}$ and examine whether or not $\mathcal{A}$ is a σ-field.
Hint: For $A \in \mathcal{A}$, check whether you can declare that $A^c \in \mathcal{A}$.

CHAPTER

Definition and Construction of a Measure and its Basic Properties

2

In this chapter, the concept of a measure is defined, and some of its basic properties are established. We then proceed with the introduction of an outer measure, study its relationship to the underlying measure, and determine the class of sets measurable with respect to the outer measure. These results are used as a basis toward obtaining an extension of a given measure from a field to the σ-field generated by this field. Next, by means of a measure, a class of point real-valued functions is defined, and their basic properties are studied. Finally, it is shown that any nondecreasing right-continuous function induces a unique measure in the Borel real line.

2.1 About Measures in General, and Probability Measures in Particular

Consider the measurable space $(\Omega, \mathcal{A})$. Then

Definition 1. A (set) function $\mu : \mathcal{A} \to \bar{\Re}$ is said to be a *measure* if

(i) $\mu(A) \geq 0$ for every $A \in \mathcal{A}$ (μ is *nonnegative*).

(ii) $\mu\left(\sum_{j=1}^{\infty} A_j\right) = \sum_{j=1}^{\infty} \mu(A_j)$, $A_j \in \mathcal{A}$, $j = 1, 2, \ldots$ (μ is *σ-additive*).

(iii) $\mu(\oslash) = 0$.

μ is said to be *infinite* if $\mu(\Omega) = \infty$; *σ-finite* if $\mu(\Omega) = \infty$ but there exists a partition $\{A_j, j = 1, 2, \ldots\}$ of Ω such that $\mu(A_j) < \infty$, $j = 1, 2, \ldots$; *finite* if $\mu(\Omega) < \infty$; and a *probability measure*, denoted by P, if $\mu(\Omega) = 1$. The triple $(\Omega, \mathcal{A}, \mu)$ is called a *measure space*, and in case $\mu = P$, $(\Omega, \mathcal{A}, P)$ is called a *probability space*. ■

Remark 1. It is possible that $\mu(A) = \infty$ for every $\oslash \neq A \in \mathcal{A}$, but this is a rather uninteresting case. So from now on, we will always assume that there exists at least one $\oslash \neq A \in \mathcal{A}$ such that $\mu(A) < \infty$. In such a case, $\mu(\oslash) = 0$ is a consequence of (i) and (ii). In fact, let $\oslash \neq A \in \mathcal{A}$ such that $\mu(A) < \infty$. Then $A = \sum_{j=1}^{\infty} A_j$, where $A_1 = A$, $A_j = \oslash$, $j = 2, 3, \ldots$ So $\mu(A) = \mu(\sum_{j=1}^{\infty} A_j) = \mu(A) + \sum_{j=2}^{\infty} \mu(\oslash)$ implies $\mu(\oslash) = 0$.

An Introduction to Measure-Theoretic Probability, Second Edition. http://dx.doi.org/10.1016/B978-0-12-800042-7.00002-5

Remark 2. Occasionally, we may be talking about a measure μ defined on a field $\mathcal{F}$ of subsets of Ω rather than a σ-field $\mathcal{A}$. This means that

(i) $\mu(A) \geq 0$ for every $A \in \mathcal{F}$.

(ii) $\mu\left(\sum_{j=1}^{\infty} A_j\right) = \sum_{j=1}^{\infty} \mu(A_j)$ for those $A_j \in \mathcal{F}$ for which $\sum_{j=1}^{\infty} A_j \in \mathcal{F}$.

(iii) $\mu(\oslash) = 0$.

Then Theorem 1(i), which follows, shows that

$$\mu\left(\sum_{j=1}^{n} A_j\right) = \sum_{j=1}^{n} \mu(A_j);$$

i.e., μ is *finitely additive* on $\mathcal{F}$.

Theorem 1. Consider the measure space $(\Omega, \mathcal{A}, \mu)$. Then

(i) μ is finitely additive; i.e., $\mu\left(\sum_{j=1}^{n} A_j\right) = \sum_{j=1}^{n} \mu(A_j)$, $A_j \in \mathcal{A}$, $j = 1, \ldots, n$.

(ii) μ is nondecreasing; i.e., $\mu(A_1) \leq \mu(A_2)$, $A_1, A_2 \in \mathcal{A}$, $A_1 \subseteq A_2$.

(iii) μ is sub-σ-additive; i.e., $\mu\left(\bigcup_{j=1}^{\infty} A_j\right) \leq \sum_{j=1}^{\infty} \mu(A_j)$, $A_j \in \mathcal{A}$, $j = 1, 2, \ldots$ ■

Proof.

(i) We have $\sum_{j=1}^{n} A_j = \sum_{j=1}^{\infty} B_j$, where $B_j = A_j, j = 1, \ldots, n$, $B_j = \oslash$, $j = n+1, \ldots$
Then $\mu(\sum_{j=1}^{n} A_j) = \mu(\sum_{j=1}^{\infty} B_j) = \sum_{j=1}^{\infty} \mu(B_j) = \sum_{j=1}^{n} \mu(B_j) = \sum_{j=1}^{n} \mu(A_j)$.

(ii) $A_1 \subseteq A_2$ implies $A_2 = A_1 + (A_2 - A_1)$, so that $\mu(A_2) = \mu[A_1 + (A_2 - A_1)] = \mu(A_1) + \mu(A_2 - A_1) \geq \mu(A_1)$.
From this, it also follows that: $A_1 \subseteq A_2$ implies $\mu(A_2 - A_1) = \mu(A_2) - \mu(A_1)$, provided $\mu(A_1)$ is finite.

(iii) $\bigcup_{j=1}^{\infty} A_j = A_1 + \left(A_1^c \cap A_2\right) + \cdots + \left(A_1^c \cap \cdots \cap A_n^c \cap A_{n+1}\right) + \cdots$, so that

$$\begin{aligned}\mu\left(\bigcup_{j=1}^{\infty} A_j\right) &= \mu(A_1) + \mu\left(A_1^c \cap A_2\right) + \cdots \\ &\quad + \mu\left(A_1^c \cap \cdots \cap A_n^c \cap A_{n+1}\right) + \cdots \\ &\leq \mu(A_1) + \mu(A_2) + \cdots + \mu(A_{n+1}) + \cdots \\ &= \sum_{j=1}^{\infty} \mu(A_j).\end{aligned}$$ ■

Definition 2. Consider the measurable space $(\Omega, \mathcal{A})$ and let μ be a measure on $\mathcal{A}$. We say that μ is *continuous from below*, if for every $A_j \in \mathcal{A}$, $j = 1, 2, \ldots$ with $A_j \uparrow$, we have $\mu(A_j) \uparrow \mu(\lim_{j\to\infty} A_j)(= \mu(\bigcup_{j=1}^{\infty} A_j))$. We say that μ is *continuous from above*, if for every $A_j \in \mathcal{A}$, $j = 1, 2, \ldots$ with $A_j \downarrow$ and for which there exists an A_n such that $\mu(A_n) < \infty$, we have $\mu(A_j) \downarrow \mu(\lim_{j\to\infty} A_j)(= \mu(\bigcap_{j=1}^{\infty} A_j))$. μ is said to be *continuous*, if it is both continuous from below and continuous from above. We say that μ is *continuous at* $\oslash$, if for every $A_j \in \mathcal{A}$, $j = 1, 2, \ldots$ with $A_j \downarrow \oslash$ and for which there exists an A_n such that $\mu(A_n) < \infty$, we have $\mu(A_j) \downarrow 0$. ■

Remark 3. In defining continuity from above and continuity at $\oslash$ for μ, one has got to assume that there exists an A_n such that $\mu(A_n) < \infty$ (then, of course, $\mu(A_j) < \infty$ for all $j \geq n$). In fact, consider the sets $A_j = [j, \infty)$, $j = 1, 2, \ldots$ in the real line with μ the Lebesgue measure (to be defined precisely later on and that assigns as measure to each interval its length). Then $\mu([j, \infty)) = \infty$, $j = 1, 2, \ldots$ but $\bigcap_{j=1}^{\infty}[j, \infty) = \oslash$. Thus $\mu([j, \infty)) \underset{j\to\infty}{\to} \infty$, $\mu(\oslash) = 0$.

The following theorem relates the concepts of additivity and continuity.

Theorem 2.

(i) A measure μ is finitely additive and continuous.
(ii) If the set function μ is nonnegative, $\mu(\oslash) = 0$, and finitely additive only, and either continuous from below, or finite and continuous at $\oslash$, then μ is σ-additive (hence a measure). ■

Proof.

(i) The finite additivity of μ was proved in Theorem 1(i). Now we will prove continuity. Let first $A_j \in \mathcal{A}$, $j = 1, 2, \ldots$, and A_j be $\uparrow$. If $\mu(A_n) = \infty$ for some n, then $\mu(A_j) = \infty$ for all $j \geq n$, so that $\mu(\bigcup_{j=1}^{\infty} A_j) = \infty$. Thus $\mu(A_j) \underset{j\to\infty}{\to} \mu(\bigcup_{j=1}^{\infty} A_j)$. So we may assume that $\mu(A_j) < \infty$ for all j. Then

$$\lim_{j\to\infty} A_j = \bigcup_{j=1}^{\infty} A_j = A_1 + (A_1^c \cap A_2) + \cdots$$
$$+ (A_1^c \cap \cdots \cap A_{n-1}^c \cap A_n) + \cdots$$
$$= A_1 + (A_2 - A_1) + \cdots + (A_n - A_{n-1}) + \cdots$$

Thus,

$$\mu(\lim_{j\to\infty} A_j) = \mu[A_1 + (A_2 - A_1) + \cdots + (A_n - A_{n-1}) + \cdots]$$
$$= \mu(A_1) + \mu(A_2 - A_1) + \cdots$$
$$+ \mu(A_n - A_{n-1}) + \cdots$$
$$= \lim_{n\to\infty} [\mu(A_1) + \mu(A_2 - A_1) + \cdots$$
$$+ \mu(A_n - A_{n-1})]$$

$$
\begin{aligned}
&= \lim_{n\to\infty} [\mu(A_1) + \mu(A_2) - \mu(A_1) + \cdots \\
&\quad + \mu(A_n) - \mu(A_{n-1})] \\
&= \lim_{n\to\infty} \mu(A_n).
\end{aligned}
$$

This establishes continuity from below. Let A_j now be $\downarrow$ as $j \to \infty$ and let $\mu(A_{n_0}) < \infty$. Then $A_{n_0} - A_j$ is $\uparrow$ for $j \geq n_0$, and $\bigcup_{j=n_0}^{\infty}(A_{n_0} - A_j) = A_{n_0} - \bigcap_{j=n_0}^{\infty} A_j$; i.e., as $j \to \infty$, $A_{n_0} - A_j \uparrow A_{n_0} - \bigcap_{j=n_0}^{\infty} A_j$ and thus $\mu(A_{n_0} - A_j) \uparrow \mu(A_{n_0} - \bigcap_{j=n_0}^{\infty} A_j)$ by the previous result. Since $\mu(A_{n_0} - A_j) = \mu(A_{n_0}) - \mu(A_j)$, $\mu(A_{n_0} - \bigcap_{j=n_0}^{\infty} A_j) = \mu(A_{n_0}) - \mu(\bigcap_{j=n_0}^{\infty} A_j)$, we have $\mu(A_j) \downarrow \mu(\bigcap_{j=n_0}^{\infty} A_j) = \mu(\bigcap_{j=1}^{\infty} A_j)$.

(ii) Assume first that μ is continuous from below and take limits as $n \to \infty$. Let $A_j \in \mathcal{A}$, $j = 1, 2, \ldots$ be pairwise disjoint. Then, clearly, $\sum_{j=1}^{n} A_j \uparrow \sum_{j=1}^{\infty} A_j$. Hence $\mu(\sum_{j=1}^{n} A_j) \uparrow \mu(\sum_{j=1}^{\infty} A_j)$ by continuity from below. But $\mu(\sum_{j=1}^{n} A_j) = \sum_{j=1}^{n} \mu(A_j)$, by finite additivity, and $\lim_n \sum_{j=1}^{n} \mu(A_j) = \sum_{j=1}^{\infty} \mu(A_j)$. Thus $\sum_{j=1}^{\infty} \mu(A_j) = \mu(\sum_{j=1}^{\infty} A_j)$; i.e., μ is σ-additive. Now assume that μ is finite and continuous at $\emptyset$, and let $A_j \in \mathcal{A}$, $j = 1, 2, \ldots$ be pairwise disjoint. Then $\sum_{j=1}^{\infty} A_j = \sum_{j=1}^{n} A_j + \sum_{j=n+1}^{\infty} A_j$ which implies that

$$
\mu\left(\sum_{j=1}^{\infty} A_j\right) = \mu\left(\sum_{j=1}^{n} A_j\right) + \mu\left(\sum_{j=n+1}^{\infty} A_j\right) = \sum_{j=1}^{n} \mu(A_j) + \mu\left(\sum_{j=n+1}^{\infty} A_j\right), \tag{2.1}
$$

by finite additivity. Next, $\sum_{j=n+1}^{\infty} A_j \downarrow \emptyset$, evidently, because of the disjointness of the A_js. Then $\mu(\sum_{j=n+1}^{\infty} A_j) \downarrow \mu(\emptyset) = 0$ by the finiteness of μ and its continuity at $\emptyset$. Thus, by taking the limits in (2.1), as $n \to \infty$, we get

$$
\mu\left(\sum_{j=1}^{\infty} A_j\right) = \lim_{n\to\infty} \sum_{j=1}^{n} \mu(A_j) = \sum_{j=1}^{\infty} \mu(A_j). \qquad \blacksquare
$$

Remark 4. In the case of μ being continuous from below, if $\mu(A_j) = \infty$ for at least one j, j_0, say, then $\mu(\sum_{j=1}^{\infty} A_j) = \infty$, so that $\mu(\sum_{j=1}^{\infty} A_j) = \sum_{j=1}^{\infty} \mu(A_j) (= \infty)$. So we may assume, if we wish, that $\mu(A_j) < \infty$ for all $j \geq 1$.

2.2 Outer Measures

Again, let $\mathcal{P}(\Omega)$ be the class of all subsets of Ω and let $\mathcal{C}, \mathcal{C}'$ be two subclasses of $\mathcal{P}(\Omega)$. Let φ, φ' also be two set functions defined on $\mathcal{C}, \mathcal{C}'$, respectively, and taking values in $\bar{\Re}$. Then

Definition 3. We say that φ' is an *extension* of φ, and φ is a *restriction* of φ', if $\mathcal{C} \subset \mathcal{C}'$ and $\varphi = \varphi'$ on $\mathcal{C}$. $\blacksquare$

Definition 4. A set function $\mu^\circ : \mathcal{P}(\Omega) \to \bar{\Re}$ is said to be an *outer measure*, if

(i) $\mu^\circ(\emptyset) = 0$.
(ii) μ° is nondecreasing; i.e., $A \subset B$ implies $\mu^\circ(A) \leq \mu^\circ(B)$.
(iii) μ° is sub-σ-additive; i.e., $\mu^\circ(\bigcup_{n=1}^{\infty} A_n) \leq \sum_{n=1}^{\infty} \mu^\circ(A_n)$. ■

Remark 5.

(i) $\mu^\circ(A) \geq 0$ for all A, since $\emptyset \subseteq A$ implies $0 = \mu^\circ(\emptyset) \leq \mu^\circ(A)$ by (i) and (ii).
(ii) It follows that μ° is finitely subadditive, since $\mu^\circ(\bigcup_{j=1}^{n} A_j) = \mu^\circ(\bigcup_{j=1}^{\infty} B_j)$, where $B_j = A_j, j = 1, \ldots, n, B_j = \emptyset, j \geq n+1$. Then

$$\mu^\circ\left(\bigcup_{j=1}^{n} A_j\right) = \mu^\circ\left(\bigcup_{j=1}^{\infty} B_j\right) \leq \sum_{j=1}^{\infty} \mu^\circ(B_j) = \sum_{j=1}^{n} \mu^\circ(B_j) = \sum_{j=1}^{n} \mu^\circ(A_j).$$

(iii) A measure is an outer measure restricted to $\mathcal{A} \subseteq \mathcal{P}(\Omega)$.

Now let $\mathcal{F}$ be a field of subsets of Ω, let μ be a measure on $\mathcal{F}$ and let $\mu^* : \mathcal{P}(\Omega) \to \bar{\Re}$ be defined as follows:

Definition 5. For $A \in \mathcal{P}(\Omega)$, $\mu^*(A) = \inf\{\sum_{j=1}^{\infty} \mu(A_j)\}$, where the inf is taken over all $A_j \in \mathcal{F}, j = 1, 2, \ldots$ such that $\bigcup_{j=1}^{\infty} A_j \supseteq A$; i.e., over all countable coverings of A by unions of members of $\mathcal{F}$. (Clearly, for every $A \in \mathcal{P}(\Omega)$ there exists such a covering, since $\Omega \in \mathcal{F}$.) ■

Then we have the following theorem.

Theorem 3. Let μ be a measure on $\mathcal{F}$, a field of subsets of Ω, and let μ^* be defined on $\mathcal{P}(\Omega)$ as before. Then

(i) μ^* is an extension of μ (from $\mathcal{A}$ to $\mathcal{P}(\Omega)$).
(ii) μ^* is an outer measure.
(iii) If μ is σ-finite on $\mathcal{F}$, then μ^* is σ-finite on $\mathcal{P}(\Omega)$.
(iv) If μ is finite on $\mathcal{F}$, then μ^* is finite on $\mathcal{P}(\Omega)$. ■

Proof.

(i) Let $A \in \mathcal{F}$. Then $A \subseteq A$ so that $\mu^*(A) \leq \mu(A)$ by the definition of μ^*. Thus, it suffices to show that $\mu^*(A) \geq \mu(A)$. Let $A_j \in \mathcal{F}, j = 1, 2, \ldots$, be a covering of A; i.e., $A \subseteq \bigcup_{j=1}^{\infty} A_j$. At this point we notice that $\bigcup_{j=1}^{\infty} A_j$ need not belong in $\mathcal{F}$ and hence $\mu(\bigcup_{j=1}^{\infty} A_j)$ need not be defined at all. So we work as follows: $A = A \cap (\bigcup_{j=1}^{\infty} A_j) = \bigcup_{j=1}^{\infty}(A \cap A_j)$, while $A \cap A_j \in \mathcal{F}$, since $A, A_j \in \mathcal{F}, j = 1, 2, \ldots$ Then $\mu(A) = \mu[\bigcup_{j=1}^{\infty}(A \cap A_j)] \leq \sum_{j=1}^{\infty} \mu(A \cap A_j)$ (see Remark 2(ii) and Theorem 1(iii)), and this is $\leq \sum_{j=1}^{\infty} \mu(A_j)$; i.e., $\mu(A) \leq \sum_{j=1}^{\infty} \mu(A_j)$ so that $\mu(A) \leq \mu^*(A)$. Thus $\mu^* = \mu$ on $\mathcal{F}$.

(ii) First, that $\mu^*(\emptyset) = 0$ follows from part (i). Next, let $A \subset B$. Since every covering of B is a covering of A, we get $\mu^*(A) \leq \mu^*(B)$. Thus it remains to prove sub-σ-additivity. Let $A_j \in \mathcal{P}(\Omega)$, $j = 1, 2, \ldots$, and let $\varepsilon > 0$. For each j, it follows from the definition of $\mu^*(A_j)$ that there exists a covering

$$A_{jk} \in \mathcal{F}, k = 1, 2, \ldots, \text{ such that } \mu^*(A_j) + \frac{\varepsilon}{2^j} > \sum_{k=1}^{\infty} \mu(A_{jk}). \tag{2.2}$$

Now, from $A_j \subseteq \bigcup_{k=1}^{\infty} A_{jk}$, $j = 1, 2 \ldots$, it follows that $\bigcup_{j=1}^{\infty} A_j \subseteq \bigcup_{j=1}^{\infty} \bigcup_{k=1}^{\infty} A_{jk}$; i.e., $\{A_{jk}, j, k = 1, 2, \ldots\}$ is a covering of $\bigcup_{j=1}^{\infty} A_j$. Hence

$$\mu^*\left(\bigcup_{j=1}^{\infty} A_j\right) \leq \sum_{j=1}^{\infty} \sum_{k=1}^{\infty} \mu(A_{jk}). \tag{2.3}$$

From (2.2), we have

$$\sum_{j=1}^{\infty} \sum_{k=1}^{\infty} \mu(A_{jk}) \leq \sum_{j=1}^{\infty} \mu^*(A_j) + \varepsilon \quad \left(\text{since } \sum_{j=1}^{\infty} \frac{1}{2^j} = 1\right). \tag{2.4}$$

From (2.3) and (2.4), we get

$$\mu^*\left(\bigcup_{j=1}^{\infty} A_j\right) \leq \sum_{j=1}^{\infty} \mu^*(A_j) + \varepsilon.$$

Letting $\varepsilon \to 0$, we get the desired result.

(iii) Since μ is σ-finite on $\mathcal{F}$, there exists a partition $\{A_j, j = 1, 2, \ldots\}$ of Ω with $A_j \in \mathcal{F}$, $j = 1, 2, \ldots$, such that $\mu(A_j) < \infty$, $j = 1, 2, \ldots$ But $A_j \in \mathcal{F}$ implies $\mu^*(A_j) = \mu(A_j)$, $j = 1, 2, \ldots$, by (i). Thus μ^* is also σ-finite.

(iv) Finally, $\mu^*(\Omega) = \mu(\Omega) < \infty$, by (i), since $\Omega \in \mathcal{F}$. ■

Theorem 3 exhibits the existence (and provides the construction) of an outer measure, namely μ^*. Then we may denote μ^* by μ^o. This outer measure μ^o is said to be *induced on* $\mathcal{P}(\Omega)$ by μ defined on $\mathcal{F}$.

Definition 6. Let μ^o be an outer measure. Then a set $A \subseteq \Omega$ is said to be μ^o-*measurable*, if for every $D \subseteq \Omega$, we have: $\mu^o(D) = \mu^o(A \cap D) + \mu^o(A^c \cap D)$ (i.e., μ^o is additive for $A \cap D$ and $A^c \cap D$). ■

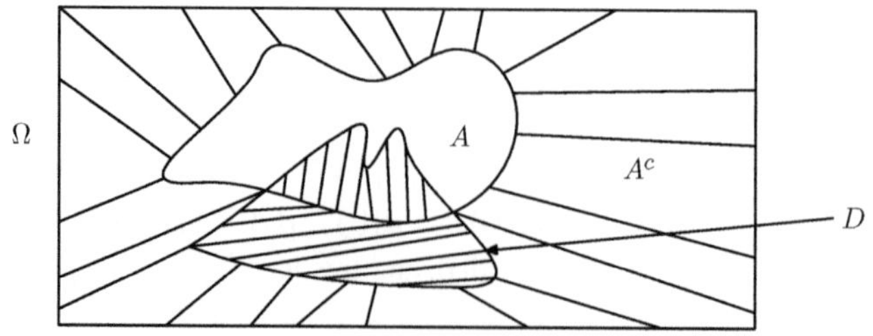

Remark 6.

(i) Since $D = (A \cap D) \cup (A^c \cap D)$ implies $\mu^o(D) \leq \mu^o(A \cap D) + \mu^o(A^c \cap D)$, in order to check μ^o-measurability for A, it suffices to check that

$$\mu^o(D) \geq \mu^o(A \cap D) + \mu^o(A^c \cap D) \quad \text{for every } D \subseteq \Omega.$$

(ii) There are μ^o-measurable sets. In fact Ω and $\oslash$ are such sets, because $D = (\Omega \cap D) \cup (\Omega^c \cap D)$, and $\mu^o(\Omega \cap D) = \mu^o(D)$, $\mu^o(\Omega^c \cap D) = \mu^o(\oslash) = 0$, so that $\mu^o(D) = \mu^o(\Omega \cap D) + \mu^o(\Omega^c \cap D)$, and similarly for the $\oslash$ set.

Theorem 4. Let μ^o be an outer measure. Then

(i) The class $\mathcal{A}^o$ of μ^o-measurable sets is a σ-field.
(ii) μ^o, restricted on $\mathcal{A}^o$, is a measure. ■

Proof.

(i) We first prove that $\mathcal{A}^o$ is a field. Let $A \in \mathcal{A}^o$. Then $\mu^o(D) = \mu^o(A \cap D) + \mu^o(A^c \cap D) = \mu^o(A^c \cap D) + \mu^o[(A^c)^c \cap D]$ for every $D \subseteq \Omega$, and this shows that $A^c \in \mathcal{A}^o$. Next, let $A, B \in \mathcal{A}^o$. To show that $(A \cap B) \in \mathcal{A}^o$. Since $B \in \mathcal{A}^o$, we get by writing $\mu^o(D) = \mu^o(B \cap D) + \mu^o(B^c \cap D)$ and taking D to be $A \cap D$ and $A^c \cap D$, successively,

$$\begin{cases} \mu^o(A \cap D) = \mu^o[B \cap (A \cap D)] + \mu^o[B^c \cap (A \cap D)], \\ \mu^o(A^c \cap D) = \mu^o[B \cap (A^c \cap D)] + \mu^o[B^c \cap (A^c \cap D)], \end{cases}$$

so that

$$\begin{aligned} &\mu^o(A \cap D) + \mu^o(A^c \cap D) \\ &= \mu^o[B \cap (A \cap D)] + \mu^o[B^c \cap (A \cap D)] \\ &+ \mu^o[B \cap (A^c \cap D)] + \mu^o[B^c \cap (A^c \cap D)]. \end{aligned} \tag{2.5}$$

But $A \in \mathcal{A}^o$ implies $\mu^o(D) = \mu^o(A \cap D) + \mu^o(A^c \cap D)$. Taking this into consideration and the fact that μ^o is finitely subadditive, (2.5) becomes

$$\begin{aligned} \mu^o(D) \geq\ &\mu^o(A \cap B \cap D) + \mu^o[(A \cap B^c \cap D) + (A^c \cap B \cap D) \\ &+ (A^c \cap B^c \cap D)]. \end{aligned} \tag{2.6}$$

Now $(A \cap B^c \cap D) + (A^c \cap B \cap D) + (A^c \cap B^c \cap D)$

$$\begin{aligned} &= D \cap [(A \cap B^c) + (A^c \cap B) + (A^c \cap B^c)], \\ &= D \cap [(A \triangle B) + (A \cup B)^c] = D \cap (A \cap B)^c. \end{aligned}$$

Therefore (2.6) becomes

$$\mu^o(D) \geq \mu^o[(A \cap B) \cap D] + \mu^o[(A \cap B)^c \cap D] \text{ so that } (A \cap B) \in \mathcal{A}^o.$$

So $\mathcal{A}^o$ is a field.

Finally, we prove that $\mathcal{A}^o$ is a σ-field. Let $A_j \in \mathcal{A}^o, j = 1, 2, \ldots$, and set $A = \bigcup_{j=1}^{\infty} A_j$. To show that $A \in \mathcal{A}^o$. Since $\bigcup_{j=1}^{\infty} A_j = A_1 + (A_1^c \cap A_2) + \cdots + (A_1^c \cap \cdots \cap A_{n-1}^c \cap A_n) + \cdots$ and $(A_1^c \cap \cdots \cap A_{n-1}^c \cap A_n) \in \mathcal{A}^o, n = 2, 3, \ldots$, by the fact that $\mathcal{A}^o$ is a field, it suffices to assume that the A_js are pairwise disjoint. Set $B_n = \sum_{j=1}^{n} A_j$, $B_0 = \oslash$. Then $B_n \in \mathcal{A}^o, n = 1, 2, \ldots$, and therefore

$$\mu^o(D) = \mu^o(B_n \cap D) + \mu^o\left(B_n^c \cap D\right) \quad \text{for every } D \subseteq \Omega. \tag{2.7}$$

Next, $A_n \in \mathcal{A}^o$. Thus, by writing $\mu^o(D) = \mu^o(A_n \cap D) + \mu^o(A_n^c \cap D)$ and taking D to be $B_n \cap D$, we have

$$\begin{aligned}\mu^o(B_n \cap D) &= \mu^o[A_n \cap (B_n \cap D)] + \mu^o\left[A_n^c \cap (B_n \cap D)\right], \\ &= \mu^o(A_n \cap D) + \mu^o(B_{n-1} \cap D),\end{aligned}$$

since $A_n \subseteq B_n$ and $A_n^c \cap B_n = A_n^c \cap (A_1 + \cdots + A_n) = A_1 + \cdots + A_{n-1} = B_{n-1}$.

That is, $\mu^o(B_n \cap D) = \mu^o(A_n \cap D) + \mu^o(B_{n-1} \cap D)$. Working in the same way with $\mu^o(B_{n-1} \cap D)$, etc. (or by using induction), we get $\mu^o(B_n \cap D) = \sum_{j=1}^{n} \mu^o(A_j \cap D)$. Then (2.7) becomes as follows:

$$\mu^o(D) = \sum_{j=1}^{n} \mu^o(A_j \cap D) + \mu^o\left(B_n^c \cap D\right).$$

But $\mu^o(B_n^c \cap D) \geq \mu^o(A^c \cap D)$, since $A \supseteq B_n$ or equivalently $A^c \subseteq B_n^c$, and μ^o is nondecreasing. Thus

$$\mu^o(D) \geq \sum_{j=1}^{n} \mu^o(A_j \cap D) + \mu^o(A^c \cap D).$$

Letting $n \to \infty$, we get

$$\begin{aligned}\mu^o(D) &\geq \sum_{j=1}^{\infty} \mu^o(A_j \cap D) + \mu^o(A^c \cap D), \\ &\geq \mu^o\left[\sum_{j=1}^{\infty} (A_j \cap D)\right] + \mu^o(A^c \cap D), \\ &= \mu^o(A \cap D) + \mu^o(A^c \cap D).\end{aligned} \tag{2.8}$$

Then $A \in \mathcal{A}^o$, and this completes the proof of part (i).

(ii) Consider the first line on the right-hand side in inequality (2.8): $\mu^o(D) \geq \sum_{j=1}^{\infty} \mu^o(A_j \cap D) + \mu^o(A^c \cap D)$, $D \subseteq \Omega$, and set A instead of D. Then

$$\mu^o(A) \geq \sum_{j=1}^{\infty} \mu^o(A_j \cap A) + \mu^o(\oslash) = \sum_{j=1}^{\infty} \mu^o(A_j),$$

since $A_j \cap A = A_j$ for all j.

Since the opposite inequality is always true, by sub-σ-additivity of μ^o, the proof is completed. (Observe that $\mathcal{A}_0$ may be contained strictly in $\mathscr{P}(\Omega)$; see Exercise 28(ii) in this chapter.) ■

In the following section, an outer measure will be instrumental in extending a given measure from a field to the σ-field generated by it.

2.3 The Carathéodory Extension Theorem

This section is devoted to the discussion of the Carathéodory extension theorem, which provides the basis for the construction and extension of measures.

Theorem 5 (Carathéodory Extension Theorem). Let μ be a measure on a field $\mathcal{F}$. Then

(i) μ can be extended to the σ-field $\mathcal{A}$ generated by $\mathcal{F}$.
(ii) If μ is finite on $\mathcal{F}$, then the extension is unique and finite.
(iii) If μ is σ-finite on $\mathcal{F}$, then the extension is unique and σ-finite. ■

Definition 7. The unique finite (σ-finite) extension of a finite (σ-finite) measure μ on $\mathcal{F}$ to $\mathcal{A} = \sigma(\mathcal{F})$ is called the *Carathéodory extension.* ■

Proof of Theorem 5. (i) Let μ^* be the set function defined just prior to Theorem 3. Then we saw that μ^* is an outer measure on $\mathcal{P}(\Omega)$, is an extension of μ on $\mathcal{F}$, and is σ-finite or finite, if μ is so, respectively. Also, μ^* is a measure on the σ-field $\mathcal{A}^*$ of μ^*-measurable sets, by Theorem 4. Then, all we have to prove is that $\mathcal{A} \subseteq \mathcal{A}^*$, or just that $\mathcal{F} \subseteq \mathcal{A}^*$. Let $A \in \mathcal{F}$ and $D \subseteq \Omega$. To show $\mu^*(D) \geq \mu^*(A \cap D) + \mu^*(A^c \cap D)$. From the definition of μ^*, for $\varepsilon > 0$, there exists a covering $\{A_j, j = 1, 2, \ldots\}$ of D in $\mathcal{F}$ such that

$$\mu^*(D) + \varepsilon > \sum_{j=1}^{\infty} \mu(A_j). \tag{2.9}$$

Now

$$\begin{aligned}\mu(A_j) &= \mu[(A \cap A_j) + (A^c \cap A_j)] = \mu(A \cap A_j) + \mu(A^c \cap A_j)\\ &= \mu^*(A \cap A_j) + \mu^*(A^c \cap A_j),\end{aligned}$$

since μ and μ^* agree on $\mathcal{F}$. Thus, (2.9) becomes

$$\mu^*(D) + \varepsilon > \sum_{j=1}^{\infty} \mu^*(A \cap A_j) + \sum_{j=1}^{\infty} \mu^*(A^c \cap A_j). \tag{2.10}$$

Next, $\bigcup_{j=1}^{\infty} A_j \supseteq D$ so that $(\bigcup_{j=1}^{\infty} A_j) \cap A = \bigcup_{j=1}^{\infty}(A \cap A_j) \supseteq A \cap D$ and $(\bigcup_{j=1}^{\infty} A_j) \cap A^c = \bigcup_{j=1}^{\infty}(A^c \cap A_j) \supseteq A^c \cap D$. Thus,

$$\sum_{j=1}^{\infty} \mu^*(A \cap A_j) \geq \mu^*\Big[\bigcup_{j=1}^{\infty}(A \cap A_j)\Big] \geq \mu^*(A \cap D),$$

$$\sum_{j=1}^{\infty} \mu^*(A^c \cap A_j) \geq \mu^*\Big[\bigcup_{j=1}^{\infty}(A^c \cap A_j)\Big] \geq \mu^*(A^c \cap D).$$

Adding up these relationships and using also (2.10), we get

$$\mu^*(D) + \varepsilon \geq \mu^*(A \cap D) + \mu^*(A^c \cap D).$$

Since this is true for every $\varepsilon > 0$, we get then $\mu^*(D) \geq \mu^*(A \cap D) + \mu^*(A^c \cap D)$. This proves that A is μ^*-measurable and hence $A \in \mathcal{A}^*$.

(ii) That the extension is finite has been seen in Theorem 3. So, all we have to do is to prove uniqueness. Let μ_1 be the above seen extension and let μ_2 be any other extension. Define $\mathcal{M} \subseteq \mathcal{A}$ as follows:

$$\mathcal{M} = \{A \in \mathcal{A};\ \mu_1(A) = \mu_2(A)\}.$$

We shall show that $\mathcal{M}$ is a monotone class and equals $\mathcal{A}$.

First $\mathcal{F} \subseteq \mathcal{M}$, since $\mu_1 = \mu_2 = \mu$ on $\mathcal{F}$. Let now $\{A_n\}$ be a monotone sequence of sets in $\mathcal{M}$. Then $\mu_1(\lim_{n\to\infty} A_n) = \lim_{n\to\infty}\mu_1(A_n) = \lim_{n\to\infty}\mu_2(A_n) = \mu_2(\lim_{n\to\infty} A_n)$ (by the finiteness of μ_1 and μ_2); i.e., $\mu_1(\lim_{n\to\infty} A_n) = \mu_2(\lim_{n\to\infty} A_n)$ and hence $\lim_{n\to\infty} A_n \in \mathcal{M}$. Thus, $\mathcal{M}$ is a monotone class. Then $\mathcal{M}$ contains the minimal monotone class over $\mathcal{F}$ that coincides with $\mathcal{A}$ (by Theorem 6 in Chapter 1). Thus, μ_1, μ_2 coincide on $\mathcal{A}$.

(iii) Again, the σ-finiteness of the extension follows from Theorem 3, and we only have to establish uniqueness. The σ-finiteness of μ implies the existence of a partition $\{A_j,\ j = 1, 2, \ldots\}$ of Ω in $\mathcal{F}$ such that $\mu(A_j) < \infty$, $j = 1, 2, \ldots$. For each A_j, consider the classes

$$\mathcal{F}_{A_j} = \{A_j \cap B;\ B \in \mathcal{F}\}, \quad \mathcal{A}_{A_j} = \{A_j \cap B;\ B \in \mathcal{A}\}.$$

Then $\mathcal{F}_{A_j}$ is a field and $\mathcal{A}_{A_j}$ is a σ-field. Furthermore, $\mathcal{A}_{A_j}$ is the σ-field generated by $\mathcal{F}_{A_j}$ (see Exercises 8 and 9 in Chapter 1). Let μ_1, μ_2 be as in (ii). Then $\mu_1 = \mu_2$, and finite on $\mathcal{A}_{A_j}$ by (ii). Next, let $A \in \mathcal{A}$. Then $A = \sum_{j=1}^{\infty}(A \cap A_j)$, while $A \cap A_j \in \mathcal{A}_{A_j}$, $j = 1, 2, \ldots$, so that $\mu_1(A \cap A_j) = \mu_2(A \cap A_j)$, $j = 1, 2, \ldots$ Thus $\mu_1(A) = \sum_{j=1}^{\infty}\mu_1(A \cap A_j) = \sum_{j=1}^{\infty}\mu_2(A \cap A_j) = \mu_2(A)$. Therefore $\mu_1 = \mu_2$ on $\mathcal{A}$. ■

Special cases. (1) Let $\Omega = \Re$ and let $\mathcal{C}$ be the class of all finite sums of intervals in $\Re$. Then $\mathcal{C}$ is a field (by Exercise 7(ii) in Chapter 1), and $\mathcal{B} = \sigma(\mathcal{C})$. Let $\mu(I) =$ length of I, where I is an interval, and let $\mu(A) = \sum_{j=1}^{n}\mu(I_j)$, if $A \in \mathcal{C}$ and hence $A = \sum_{j=1}^{n} I_j$, I_j, $j = 1, 2, \ldots, n$, intervals. The set function μ is σ-finite, since, for example, $\Re = \sum_{n=0}^{\infty}(-n-1, -n] + (0, 1) + \sum_{n=1}^{\infty}[n, n+1)$ and $\mu((-n-1, -n]) = \mu([n, n+1)) = 1$ (finite). Then, provided that μ is well defined and a measure on $\mathcal{C}$—which we will show later on (Theorem 7)—the unique extension of μ on $\mathcal{B}$ is called the *Lebesgue* measure. Let us denote it by λ.

(2) For $n \geq 2$, let $\mathcal{C}$ be the class of all finite sums of rectangles in $\Re^n$. Then $\mathcal{C}$ is a field and $\mathcal{B}^n = \sigma(\mathcal{C})$ (by Theorem 7 in Chapter 1 and its extension). If

$B = A_1 \times \cdots \times A_n$, A_j Borel sets in $\Re$, $j = 1, \ldots, n$, define $\mu(B)$ as follows: $\mu(B) = \prod_{j=1}^{n} \lambda(A_j)$, and if $E = \sum_{j=1}^{m} B_j$, B_j, $j = 1, \ldots, m$, rectangles in $\Re^n$, define $\mu(E)$ by $\mu(E) = \sum_{j=1}^{m} \mu(B_j)$. It is easily seen that μ is σ-finite on $\mathcal{C}$. Actually, there exists a denumerable partition of $\Re^n$ by rectangles in $\Re^n$ that are cartesian products of intervals in $\Re$ (recall that a set is *denumerable* if it has the same cardinality as the set of integers). Then, provided μ is well defined and a measure on $\mathcal{C}$ (that this is so is seen as a straightforward generalization of Theorem 7), the unique extension of μ on $\mathcal{B}^n$ is called the *n-dimension Lebesgue* measure. Let us denote it by λ_n.

Remark 7.

(i) Special case 1 indicates why the natural original class $\mathcal{C}$ on which a measure μ is defined is a field. There is also another reason, and that is that, if the class $\mathcal{C}$ is not a field, then the extension of μ to the minimal σ-field over $\mathcal{C}$ need not be unique although μ may be finite (σ-finite) on $\mathcal{C}$.

(ii) If μ on $\mathcal{C}$ is not finite (σ-finite), then the Carathéodory extension on $\mathcal{A} = \sigma(\mathcal{C})$ need not be unique.

(iii) An extension of μ to $\mathcal{A} = \sigma(\mathcal{C})$ may be σ-finite while μ is not so on $\mathcal{C}$. In connection with these remarks, see also Exercises 12 and 13.
Example 2 demonstrates the futility of starting out with an arbitrary class of sets in defining a measure on a field or a σ-field. Example 1 is a prelude to Example 2.

Example 1. Let $\mathcal{F}_i$, $i = 1, 2$, be fields of subsets of a set Ω and define the classes $\mathcal{C}_i$, $i = 1, 2, 3$, by:

$$\mathcal{C}_1 = \mathcal{F}_1 \cup \mathcal{F}_2 = \{A \subseteq \Omega; \quad A \in \mathcal{F}_1 \text{ or } A \in \mathcal{F}_2\},$$

so that $\mathcal{C}_1$ is closed under complementation;

$$\begin{aligned}
\mathcal{C}_2 &= \{\text{all finite intersections of members of } \mathcal{C}_1 \\
&= \{A \subseteq \Omega; \quad A = A_1 \cap \cdots \cap A_m, A_i \in \mathcal{C}_1, i = 1, \ldots, m, m \geq 1\}; \\
\mathcal{C}_3 &= \{\text{all finite unions of members of } \mathcal{C}_2\} \\
&= \{A \subseteq \Omega; \quad A = \bigcup_{i=1}^{n} A_i \text{ with } A_i \in \mathcal{C}_2, i = 1, \ldots, n, n \geq 1\} \\
&= \{A \subseteq \Omega; \quad A = \bigcup_{i=1}^{n} A_i \text{ with } A_i = A_i^1 \cap \cdots \cap A_i^{m_i}, A_i^1, \ldots, A_i^{m_i} \in \mathcal{C}_1, \\
&\qquad m_i \geq 1 \text{ integer}, i = 1, \ldots, n, n \geq 1\}.
\end{aligned}$$

Set $\mathcal{C}_3 = \mathcal{F}$. Then $\mathcal{F}$ is a field. This is so by Exercise 41(i) in Chapter 1, where the role of $\mathcal{C}$ and $\mathcal{F}_1$ is played by $\mathcal{C}_1$, the role of $\mathcal{F}_2$ is played by $\mathcal{C}_2$, and the role of $\mathcal{F}_3(=\mathcal{F})$ is played by $\mathcal{C}_3(=\mathcal{F})$. ■

Example 2. In reference to Example 1, take $\mathcal{F}_1 = \{\emptyset, A, A^c, \Omega\}$ and $\mathcal{F}_2 = \{\emptyset, B, B^c, \Omega\}(A \neq B, A \cap B \neq \emptyset)$, so that

$$\mathcal{C} = \mathcal{C}_1 = \mathcal{F}_1 \cup \mathcal{F}_2 = \{\emptyset, A, A^c, B, B^c, \Omega\}.$$

Then, as is easily seen,

$$\mathcal{C}_2 = \{\emptyset, A, A^c, B, B^c, A \cap B, A \cap B^c, A^c \cap B, A^c \cap B^c, \Omega\};$$

also,

$$\mathcal{C}_3 = \{\emptyset, A, A^c, B, B^c, A \cap B, A \cap B^c, A^c \cap B, A^c \cap B^c, A \cup B, A \cup B^c, A^c \cup B, A^c \cup B^c, (A \cap B) \cup (A^c \cap B^c), (A \cap B^c) \cup (A^c \cap B), \Omega\};$$

as it can be verified. Set $\mathcal{C}_3 = \mathcal{F}$. Then $\mathcal{F}$ is a field on account of Example 1. Alternatively, the assertion is checked directly, since $\mathcal{F}$ is closed under complementation and under the union of any two of its members (see Exercise 35). Clearly, $\mathcal{F}$ is the field generated by $\mathcal{C}(= \mathcal{C}_1 = \mathcal{F}_1 \cup \mathcal{F}_2)$. This is so by Exercise 41(ii) cited above or by direct considerations. On the class $\mathcal{C}$, define $\mu_i, i = 1, 2$, as follows: $\mu_i(\emptyset) = 0, \mu_i(A) = 0.40, \mu_i(B) = 0.35, \mu_i(A^c) = 0.60, \mu_i(B^c) = 0.65, \mu_i(\Omega) = 1$, so that $\mu_i, i = 1, 2$, are (probability) measures on $\mathcal{C}$.

Next, extend these measures from $\mathcal{C}$ to the field $\mathcal{F}$ generated by $\mathcal{C}$ in the following manner:

$$\begin{aligned}
\mu_1(A \cup B) &= 0.50 & \mu_2(A \cup B) &= 0.60,\\
\mu_1(A \cup B^c) &= 0.90 & \mu_2(A \cup B^c) &= 0.80,\\
\mu_1(A^c \cup B) &= 0.85 & \mu_2(A^c \cup B) &= 0.75,\\
\mu_1(A^c \cup B^c) &= 0.75 & \mu_2(A^c \cup B^c) &= 0.85,\\
\mu_1(A \cap B) &= 0.25 & \mu_2(A \cap B) &= 0.15,\\
\mu_1(A \cap B^c) &= 0.15 & \mu_2(A \cap B^c) &= 0.25,\\
\mu_1(A^c \cap B) &= 0.10 & \mu_2(A^c \cap B) &= 0.20,\\
\mu_1(A^c \cap B^c) &= 0.50 & \mu_2(A^c \cap B^c) &= 0.40
\end{aligned}$$

The assigned values are legitimate, because they satisfy the ordering property ($C \subset D$ implies $\mu_i(C) \leq \mu_i(D), i = 1, 2$) and the additivity property. Furthermore, whereas μ_1 and μ_2 coincide on $\mathcal{C}$, they do not coincide on $\mathcal{F}$. That is, a measure defined on a class $\mathcal{C}$ (which is not a field) may assume more than one extension to a measure on the field generated by $\mathcal{C}$. The same holds if $\mathcal{F}$ is replaced by a σ-field $\mathcal{A}$. ■

The interested reader may find a rather extensive treatment of measure theory in the reference Vestrup (2003).

2.4 Measures and (Point) Functions

Let μ be a measure on $\mathcal{B}$, the Borel σ-field in $\Re$ such that μ(finite interval) $< \infty$ (the Lebesgue measure, for example, does this). Then for a constant c, define a function

$F_c = F : \Re \to \Re$ as follows:

$$F(x) = \begin{cases} c + \mu((0, x]) \text{ if } x \geq 0, \\ c - \mu((x, 0]) \text{ if } x < 0. \end{cases}$$

(Then $F(0) = c$, since $(0, 0] = \oslash$ and $\mu(\oslash) = 0$.)
Then we have the following easy theorem.

Theorem 6. Let F be defined as above. Then F is

(i) Nondecreasing.
(ii) Continuous from the right. ■

Proof.

(i) Let $0 \leq x_1 < x_2$. Then $F(x_1) = c + \mu((0, x_1]) \leq c + \mu((0, x_2]) = F(x_2)$. Next, let $x_1 < 0 \leq x_2$. Then $F(x_1) = c - \mu((x_1, 0]) \leq c + \mu((0, x_2]) = F(x_2)$. Finally, let $x_1 < x_2 < 0$. Then $F(x_1) = c - \mu((x_1, 0]) \leq c - \mu((x_2, 0]) = F(x_2)$.
(ii) Let $x \geq 0$ and choose $x_n \downarrow x$ as $n \to \infty$ here and in the sequel. Then $(0, x_n] \downarrow (0, x]$ so that $\mu((0, x_n]) \downarrow \mu((0, x])$, or $c + \mu((0, x_n]) \downarrow c + \mu((0, x])$, or equivalently, $F(x_n) \downarrow F(x)$.

Next, let $x < 0$, and pick x_n such that $x_n \downarrow x$. Then $(x_n, 0] \uparrow (x, 0]$ so that $\mu((x_n, 0]) \uparrow \mu((x, 0])$, or equivalently, $-\mu((x_n, 0]) \downarrow -\mu((x, 0])$, or $c - \mu((x_n, 0]) \downarrow c - \mu((x, 0])$, or equivalently, $F(x_n) \downarrow F(x)$. ■

Thus, we proved that a measure μ on $\mathcal{B}$ with the property that μ(finite interval) $< \infty$ defines a class of (point) functions on $\Re \to \Re$ that are nondecreasing and continuous from the right. Each such function is called a *distribution function* (d.f.). If μ is finite, then each F is bounded. In particular, if μ is a probability measure and, if we take $c = \mu(-\infty, 0]$, then F is a *d.f. of* a *r.v.* X (i.e., in addition to (i) and (ii), $F(-\infty) = \lim_{x \to -\infty} F(x) = 0$, $F(\infty) = \lim_{x \to \infty} F(x) = 1$).

Now we will work the other way around. Namely, we will start with any function F that is nondecreasing and continuous from the right, and we will show that such a function induces a measure on $\mathcal{B}$. To this end, define the class $\mathcal{C} \subset \mathcal{B}$ as follows: $\mathcal{C} = \oslash \cup \{(\alpha, \beta]; \alpha, \beta \in \Re, \alpha < \beta\}$, and on this class, we define a function Δ as follows:

$$\Delta((\alpha, \beta]) \stackrel{def}{=} \Delta(\alpha, \beta) = F(\beta) - F(\alpha), \quad \Delta(\oslash) = 0.$$

Then we have the following easy lemma.

Lemma 1. Let $\mathcal{C}$ and Δ be defined as above. Then

(i) $\Delta \geq 0$.
(ii) $\Delta(\alpha, \beta) \downarrow 0$ as $\beta \downarrow \alpha$.
(iii) If $\alpha_1 \leq \alpha_2 \leq \cdots \leq \alpha_n$, then $\sum_{j=1}^{n-1} \Delta(\alpha_j, \alpha_{j+1}) = \Delta(\alpha_1, \alpha_n) = F(\alpha_n) - F(\alpha_1)$.
(iv) Δ is nondecreasing.

Proof.

(i) Obvious from the nondecreasing property of F.

(ii) Obvious by the continuity from the right of F.

(iii) $\sum_{j=1}^{n-1} \Delta(\alpha_j, \alpha_{j+1}) = \sum_{j=1}^{n-1}[F(\alpha_{j+1}) - F(\alpha_j)] = F(\alpha_n) - F(\alpha_1) = \Delta(\alpha_1, \alpha_n)$.

(iv) If $(\alpha_1, \alpha_2] \supset (\alpha_3, \alpha_4]$, then by (iii) and (i), $\Delta(\alpha_1, \alpha_2) = \Delta(\alpha_1, \alpha_3) + \Delta(\alpha_3, \alpha_4) + \Delta(\alpha_4, \alpha_2) \geq \Delta(\alpha_3, \alpha_4)$. ■

Next, we have the following less obvious lemma.

Lemma 2. The function Δ on $\mathcal{C}$ is a measure. That is, $\Delta(\oslash) = 0$, $\Delta((\alpha, \beta]) \geq 0$, and for $(\alpha, \beta] = \sum_{j=1}^{\infty}(\alpha_j, \beta_j]$, it holds $\Delta((\alpha, \beta]) = \sum_{j=1}^{\infty} \Delta((\alpha_j, \beta_j])$.

Proof. Since $\Delta(\oslash) = 0$, and $\Delta \geq 0$ (and Δ is nondecreasing), all that remains to prove is that Δ is σ-additive; i.e., if $(\alpha, \beta] = \sum_{j=1}^{\infty}(\alpha_j, \beta_j]$, then $\Delta(\alpha, \beta) = \sum_j \Delta(\alpha_j, \beta_j)$.

Consider the n intervals $(\alpha_j, \beta_j]$, $j = 1, \ldots, n$. These intervals are nonoverlapping and we may rearrange them $(\alpha_{j1}, \beta_{j1}], (\alpha_{j2}, \beta_{j2}], \ldots, (\alpha_{jn}, \beta_{jn}]$ so that

$$\alpha_{j1} < \beta_{j1} \leq \alpha_{j2} < \beta_{j2} \leq \cdots \leq \alpha_{jn} < \beta_{jn}.$$

Then $\sum_{i=1}^{n} \Delta(\alpha_{ji}, \beta_{ji}) \leq \Delta(\alpha_{j1}, \beta_{j1}) + \Delta(\beta_{j1}, \alpha_{j2}) + \cdots + \Delta(\alpha_{jn}, \beta_{jn}) = \Delta(\alpha_{j1}, \beta_{jn})$, by Lemma 1 (iii), and this is $\leq \Delta(\alpha, \beta)$, since $(\alpha_{j1}, \beta_{jn}] \subseteq (\alpha, \beta]$. So, for every finite n, $\sum_{i=1}^{n} \Delta(\alpha_{ji}, \beta_{ji}) = \sum_{j=1}^{n} \Delta(\alpha_j, \beta_j) \leq \Delta(\alpha, \beta)$, which implies

$$\sum_{j=1}^{\infty} \Delta(\alpha_j, \beta_j) \leq \Delta(\alpha, \beta). \tag{2.11}$$

We then have to establish the reverse inequality; i.e.,

$$\sum_{j=1}^{\infty} \Delta(\alpha_j, \beta_j) \geq \Delta(\alpha, \beta). \tag{2.12}$$

Consider $(\alpha, \beta]$, choose $0 < \varepsilon < \beta - \alpha$, and look at the interval $[\alpha + \varepsilon, \beta]$. By Lemma 1 (ii), $\Delta(\beta_j, \beta_j + \delta) \downarrow 0$ as $\delta \downarrow 0$. Thus, there exists $\delta_j > 0$ such that

$$\Delta\left(\beta_j, \beta_j + \delta_j\right) < \frac{\varepsilon}{2^j}, \; j = 1, 2, \ldots \tag{2.13}$$

Also, $\Delta(\alpha_j, \beta_j + \delta_j) = \Delta(\alpha_j, \beta_j) + \Delta(\beta_j, \beta_j + \delta_j)$, by Lemma 1 (iii). Thus, by using (2.13), one gets

$$\Delta\left(\alpha_j, \beta_j + \delta_j\right) < \Delta(\alpha_j, \beta_j) + \frac{\varepsilon}{2^j}, \; j = 1, 2, \ldots \tag{2.14}$$

The intervals $\{(\alpha_j, \beta_j + \delta_j), j = 1, 2, \ldots\}$ evidently cover $[\alpha + \varepsilon, \beta]$. Then there exists a finite number of them, n_0, (by the Borel–Heine Theorem) covering $[\alpha + \varepsilon, \beta]$.

From these n_0 intervals, select $m \leq n_0$ with the following properties:

$$\alpha_{j1} < \alpha + \varepsilon, \qquad \beta_{jm} + \delta_{jm} > \beta,$$
$$\alpha_{j,i-1} < \alpha_{ji} < \beta_{j,i-1} + \delta_{j,i-1} < \beta_{ji} + \delta_{ji}, \quad i = 2, \ldots, m.$$

Now look at the following intervals:

$$(\alpha_{j1}, \beta_{j1} + \delta_{j1}], (\beta_{j1} + \delta_{j1}, \beta_{j2} + \delta_{j2}], \ldots, (\beta_{j,m-1} + \delta_{j,m-1}, \beta_{jm} + \delta_{jm}]$$

(see the following picture).

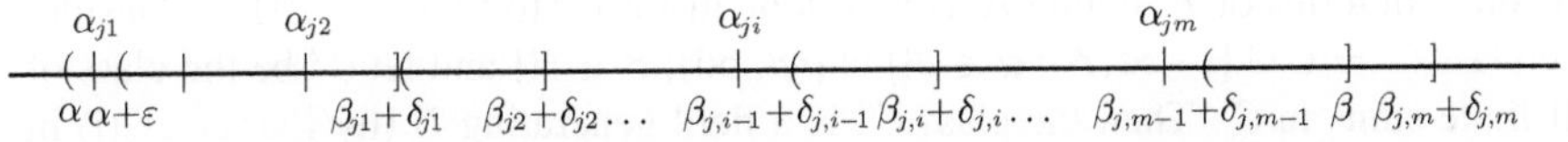

These intervals are nonoverlapping and their sum is $(\alpha_{j1}, \beta_{jm}+\delta_{jm}] \supset (\alpha+\varepsilon, \beta]$. Then

$$\begin{aligned}\Delta(\alpha + \varepsilon, \beta) &\leq \Delta(\alpha_{j1}, \beta_{jm} + \delta_{jm}) \\ &= \Delta\,(\alpha_{j1}, \beta_{j1} + \delta_{j1}) + \sum_{i=2}^{m} \Delta(\beta_{j,i-1} + \delta_{j,i-1}, \beta_{ji} + \delta_{ji}) \\ &\leq \sum_{i=1}^{m} \Delta(\alpha_{ji}, \beta_{ji} + \delta_{ji}),\end{aligned}$$

because $(\beta_{j,i-1}+\delta_{j,i-1}, \beta_{ji}+\delta_{ji}] \subset (\alpha_{ji}, \beta_{ji}+\delta_{ji}]$ and Δ is nondecreasing. That is,

$$\Delta(\alpha + \varepsilon, \beta) \leq \sum_{i=1}^{m} \Delta(\alpha_{ji}, \beta_{ji} + \delta_{ji}).$$

Next,

$$\sum_{j=1}^{m} \Delta(\alpha_{ji}, \beta_{ji} + \delta_{ji}) \leq \sum_{j=1}^{\infty} \Delta(\alpha_j, \beta_j + \delta_j),$$

(since $(\alpha_{ji}, \beta_{ji} + \delta_{ji}], i = 1, \ldots, m$ are only m of the intervals $(\alpha_j, \beta_j + \delta_j]$, $j = 1, 2, \ldots$)

$$\begin{aligned}&= \sum_{j=1}^{\infty} [\Delta(\alpha_j, \beta_j) + \Delta(\beta_j, \beta_j + \delta_j)] \\ &\leq \sum_{j=1}^{\infty} \left[\Delta(\alpha_j, \beta_j) + \frac{\varepsilon}{2^j}\right] \text{(by (2.13))}, \\ &= \sum_{j=1}^{\infty} \Delta(\alpha_j, \beta_j) + \varepsilon; \quad \text{i.e.,}\end{aligned}$$

$$\Delta(\alpha + \varepsilon, \beta) \leq \sum_{j=1}^{\infty} \Delta(\alpha_j, \beta_j) + \varepsilon.$$

But $\Delta(\alpha+\varepsilon, \beta) = F(\beta)-F(\alpha+\varepsilon)$ and hence $F(\beta)-F(\alpha+\varepsilon) \leq \sum_{j=1}^{\infty} \Delta(\alpha_j, \beta_j)+\varepsilon$. Letting $\varepsilon \downarrow 0$ and using Theorem 6(ii), we get then $F(\beta) - F(\alpha) = \Delta(\alpha, \beta) \leq \sum_{j=1}^{\infty} \Delta(\alpha_j, \beta_j)$ which is (2.12). Therefore (2.11) and (2.12) show that Δ is a measure on $\mathcal{C}$. ■

Theorem 7. Let $F : \Re \to \Re$ be a d.f. Then F uniquely determines a measure ν on $\mathcal{B}$ such that $\nu((\alpha, \beta]) = F(\beta) - F(\alpha)$, called the measure induced by F. If F is bounded, the measure is finite, and, in particular, if F is a d.f. of a r.v., then the measure is a probability measure. ■

Proof. In terms of F, define Δ as was done just prior to Lemma 1. Then consider the class $\mathcal{C}' = \mathcal{C} \cup \{(-\infty, \alpha]; \alpha \in \Re\} \cup \{(\beta, \infty); \beta \in \Re\}$ and let $\mathcal{C}''$ be the class of all finite sums in $\mathcal{C}'$. Then the class $\mathcal{C}''$ is a field generating $\mathcal{B}$ (by Exercise 7(i) in Chapter 1), and, clearly, the elements of $\mathcal{C}''$ are sums of countably many elements of $\mathcal{C}$; i.e., if $A \in \mathcal{C}''$, then $A = \sum_j I_j$ with $I_j \in \mathcal{C}, j = 1, 2, \ldots$

On $\mathcal{C}''$, we define a function ν as follows:

$$\nu(A) = \sum_j \Delta(I_j), \quad \text{if } A = \sum_j I_j, I_j \in \mathcal{C}, j = 1, 2, \ldots$$

We will first show that ν is well defined; i.e., if $A = \sum_i I_i$ and $A = \sum_j I'_j$, then $\sum_i \Delta(I_i) = \sum_j \Delta(I'_j)(= \nu(A))$. Clearly, $I_i \cap I'_j \in \mathcal{C}, i, j = 1, 2, \ldots$ Next, $I_i = \sum_j I_i \cap I'_j$, since $I_i \subseteq A = \sum_j I'_j$ so that $\Delta(I_i) = \Delta(\sum_j I_i \cap I'_j) = \sum_j \Delta(I_i \cap I'_j)$, since Δ is a measure on $\mathcal{C}$ (by Lemma 2). In a similar fashion, $\Delta(I'_j) = \sum_i \Delta(I_i \cap I'_j)$. Hence

$$\sum_j \Delta\left(I'_j\right) = \sum_j \sum_i \Delta\left(I'_j \cap I_i\right) = \sum_i \sum_j \Delta\left(I_i \cap I'_j\right) = \sum_i \Delta(I_i),$$

as was asserted.

Clearly, ν coincides with Δ on $\mathcal{C}$, and we will next show that ν is a measure on $\mathcal{C}''$. It suffices to prove σ-additivity. Let $A_j \in \mathcal{C}'', j = 1, 2, \ldots$, such that $A_i \cap A_j = \emptyset, i \neq j$, and let $A = \sum_j A_j \in \mathcal{C}''$. To show that $\nu(A) = \sum_j \nu(A_j)$. Since $A_j \in \mathcal{C}'', j = 1, 2, \ldots$, we have that $A_j = \sum_i I_{ji}$ with $I_{ji} \in \mathcal{C}, i = 1, 2, \ldots$ Clearly, $\{I_{ji}, j, i = 1, 2 \ldots\}$ form a partition of A. Thus $\nu(A) = \sum_{j,i} \Delta(I_{ji}) = \sum_j \sum_i \Delta(I_{ji}) = \sum_j \nu(A_j)$. The remaining assertions are obvious; e.g., uniqueness follows from Theorem 5 since ν is at least σ-finite. ■

Remark 8. The preceding derivations also justify the questions left open in the two special cases discussed earlier (special cases right after the proof of Theorem 5). The d.f. F which induces the Lebesgue measure λ is defined by $F : \Re \to \Re$, $F(x) = x$.

Remark 9. If X is a r.v., its probability distribution (or just distribution) is usually denoted by P_X and is defined by: $P_X(B) = P(X \in B)$, $B \in \mathcal{B}$, so that P_X is a probability measure on $\mathcal{B}$. Next, if F_X is the d.f. of X, then $F_X(x) = P_X((-\infty, x])$, $x \in \Re$, so that F_X is determined by P_X. Theorem 7 shows that the converse is also true; that is, F_X uniquely determines P_X.

Exercises.

1. If Ω is countable and μ is defined on $\mathcal{P}(\Omega)$ by: $\mu(A) =$ number of points of A, show that μ is a measure. Furthermore, μ is finite or σ-finite, depending on whether Ω is finite or denumerable, respectively. (This measure is called the *counting* measure.)
2. Refer to the field $\mathcal{C}$ of Example 4 in Chapter 1 and on $\mathcal{C}$, define the set function P as follows:
$P(A) = 0$ if A is finite, and $P(A) = 1$ if A^c is finite. Then show that

   (i) P is finitely additive.
   (ii) If Ω is denumerable, P is not σ-additive.
   (iii) If Ω is uncountable, P is σ-additive and a probability measure.

3. Refer to the σ-field $\mathcal{C}$ of Example 8 in Chapter 1 and on $\mathcal{C}$, define the function P as follows:
$P(A) = 0$ if A is countable, and $P(A) = 1$ if A^c is countable. Then show that P is a probability measure.
4. Let $A_n, n = 1, 2, \ldots$ be events in the probability space $(\Omega, \mathcal{A}, P)$ such that $P(A_n) = 1$ for all n. Then show that $P(\cap_{n=1}^{\infty} A_n) = 1$.
5. Let $\{A_i, i \in I\}$ be an uncountable collection of pairwise disjoint events in the probability space $(\Omega, \mathcal{A}, P)$. Then show that $P(A_i) > 0$ for countably many A_is only.
Hint: If $I_n = \{i \in I;\ P(A_i) > \frac{1}{n}\}$, then the cardinality of I_n is $\leq n - 1, n \geq 2$, and $I_0 = \{i \in I;\ P(A_i) > 0\} = \cup_{n \geq 2} I_n$.
6. Let Ω be an infinite set (countable or not) and let $\mathcal{A}$ be the discrete σ-field. Let $\{\omega_1, \omega_2, \ldots\} \subset \Omega$, and with each ω_n, associate a nonnegative number p_n (such that $\sum_{n=1}^{\infty} p_n \leq \infty$). On $\mathcal{A}$, define the set function μ by: $\mu(A) = \sum_{\omega_n \in A} p_n$. Then show that μ is a measure on $\mathcal{A}$.
7. In the measure space $(\Omega, \mathcal{A}, \mu)$ a set $A \in \mathcal{A}$ is called an *atom*, if $\mu(A) > 0$ and for any $B \subseteq A$ with $B \in \mathcal{A}$, it follows that $\mu(B) = 0$, or $\mu(B) = \mu(A)$. In reference to Exercise 6, identify the atoms of $\mathcal{A}$.
8. In any measure space $(\Omega, \mathcal{A}, \mu)$ and with any $A_n \in \mathcal{A}, n = 1, 2, \ldots$, show that
$$\mu\left(\liminf_{n \to \infty} A_n\right) \leq \liminf_{n \to \infty} \mu(A_n); \text{ also, } \mu\left(\limsup_{n \to \infty} A_n\right) \geq \limsup_{n \to \infty} \mu(A_n),$$
provided $\mu\left(\bigcup_{j=n}^{\infty} A_j\right) < \infty$ for some n.
In Exercises 9 and 10, show that the set function μ° defined on $\mathcal{P}(\Omega)$ is, indeed, an outer measure.
9. Ω is an arbitrary set, ω_0 is a fixed point of Ω, and μ° is defined by: $\mu^\circ(A) = I_A(\omega_0)$.
10. Ω is the set of 100 points arranged in a square array of 10 columns, each with 10 points, and μ° is defined by: $\mu^\circ(A) =$ number of columns that contain at least one point of A.

11. For an arbitrary set Ω containing at least two points, consider the trivial field $\mathcal{F} = \{\oslash, \Omega\}$, and let (the finite measure) μ be defined on $\mathcal{F}$ by: $\mu(\oslash) = 0$ and $\mu(\Omega) = 1$.

(i) Determine the set function μ^* on $\mathcal{P}(\Omega)$, as is given in the definition just prior to Theorem 3 (which μ^* is, actually, an outer measure, by Theorem 3(ii)).
(ii) Show that the σ-field of μ^*-measurable sets, $\mathcal{A}^*$, say, is the trivial σ-field, so that $\mathcal{A}^*$ is strictly contained in $\mathcal{P}(\Omega)$.

12. Let $\Omega = \{\omega_1, \omega_2, \omega_3, \omega_4\}$, let $\mathcal{C} = \{\oslash, \{\omega_1, \omega_2\}, \{\omega_1, \omega_3\}, \{\omega_2, \omega_4\}, \{\omega_3, \omega_4\}, \Omega\}$, and define μ on $\mathcal{C}$ as follows:

$$\mu(\{\omega_1, \omega_2\}) = \mu(\{\omega_1, \omega_3\}) = \mu(\{\omega_2, \omega_4\}) = \mu(\{\omega_3, \omega_4\}) = 3,\ \mu(\Omega) = 6,\ \mu(\oslash) = 0.$$

Next, on $\mathcal{P}(\Omega)$, define the measures μ_1 and μ_2 by taking

$$\mu_1(\{\omega_1\}) = \mu_1(\{\omega_4\}) = \mu_2(\{\omega_2\}) = \mu_2(\{\omega_3\}) = 1,$$
$$\mu_1(\{\omega_2\}) = \mu_1(\{\omega_3\}) = \mu_2(\{\omega_1\}) = \mu_2(\{\omega_4\}) = 2.$$

Then show that

(i) $\mathcal{C}$ is not a field.
(ii) μ is a measure on $\mathcal{C}$.
(iii) Both μ_1 and μ_2 are extensions of μ (from $\mathcal{C}$ to $\mathcal{P}(\Omega)$).
(iv) Construct the outer measure μ^* (as is defined in Definition 5) by means of μ defined on $\mathcal{C}$.
(v) Conclude that $\mu^* \neq \mu_1 \neq \mu_2$ (so that, if the class $\mathcal{C}$ is not a field, the extension of (even a finite measure μ on $\mathcal{C}$) need not be unique).

13. Let $\Omega = \{0, 1, 2, \ldots\}$, let $A = \{1, 3, 5, \ldots\}$, and let $\mathcal{C} = \{\oslash, A, A^c, \Omega\}$. Let μ be the counting measure on $\mathcal{C}$, and let μ_1, μ_2 be defined on $\mathcal{P}(\Omega)$ by

$$\mu_1(B) = \text{the number of points of } B, \quad \mu_2(B) = 2\mu_1(B).$$

Then show that

(i) $\mathcal{C}$ is a field.
(ii) μ is not σ-finite on $\mathcal{C}$.
(iii) Both μ_1 and μ_2 are extensions of μ and are also σ-finite.
(iv) Determine the outer measure μ^* (as is defined in Definition 5) by showing that $\mu^*(B) = \infty$ whenever $B \neq \oslash$.
(v) Show that the σ-field of μ^*-measurable sets, $\mathcal{A}^*$ say, is equal to $\mathcal{P}(\Omega)$.

(From this example, we conclude that if μ is not σ-finite on the field $\mathcal{C}$, then there need not be a unique extension. Also, there may be σ-finite extensions, such as μ_1 and μ_2 here, when the original measure on $\mathcal{C}$ is not σ-finite.)

14. Construct additional examples to illustrate the points made in Exercises 11(ii), 12, and 13.

15. Consider the measure space $(\Omega, \mathcal{A}, \mu)$, and define the classes $\mathcal{A}^*$ and $\bar{\mathcal{A}}$ as follows:

$$\mathcal{A}^* = \{A \Delta M;\ A \in \mathcal{A},\ M \subseteq N \text{ with } N \in \mathcal{A} \text{ and } \mu(N) = 0\},$$
$$\bar{\mathcal{A}} = \{A \cup M;\ A \in \mathcal{A},\ M \subseteq N \text{ with } N \in \mathcal{A} \text{ and } \mu(N) = 0\}.$$

Then show that

(i) $A \Delta M = (A - N) \cup [N \cap (A \Delta M)]$.
(ii) $A \cup M = (A - N) \Delta [N \cap (A \cup M)]$.
From parts (i) and (ii), conclude that
(iii) $\mathcal{A}^* = \bar{\mathcal{A}}$.

16. Refer to the classes $\mathcal{A}^*$ and $\bar{\mathcal{A}}$, defined in Exercise 15, and show that $\bar{\mathcal{A}}$ (and hence $\mathcal{A}^*$) is a σ-field.

17. Refer to Exercise 15 and on $\mathcal{A}^*$, define μ^* by: $\mu^*(A \Delta M) = \mu(A)$.

(i) Show that $\mu(A - N) = \mu(A)$.
By Exercise 15(ii), $\mu^*(A \cup M) = \mu(A - N)$. Therefore, by part (i), we may define μ^* on $\bar{\mathcal{A}}$ by: $\mu^*(A \cup M) = \mu(A)$.
(ii) Show that μ^* so defined is well defined; that is, if $A_1 \cup M_1 = A_2 \cup M_2$, where A_i and $M_i, i = 1, 2$ are as in Exercise 15, then $\mu(A_1) = \mu(A_2)$.
(iii) Show that μ^* is a measure on $\bar{\mathcal{A}}$ (and hence on $\mathcal{A}^*$, by Exercise 15(iii)).

Remark: The measure space $(\Omega, \bar{\mathcal{A}}, \mu^*)$ (and $(\Omega, \mathcal{A}^*, \mu^*)$) is called *completion* of $(\Omega, \mathcal{A}, \mu)$.

18. (i) Show that the class $\mathcal{A}^*(= \bar{\mathcal{A}})$ defined in Exercise 15 contains the class $\hat{\mathcal{C}}$ defined by
$\hat{\mathcal{C}} = \{B \subseteq \Omega;$ either $B = A$ for some $A \in \mathcal{A}$, or $B \subseteq N$ for some $N \in \mathcal{A}$ with $\mu(N) = 0\}$.
(That is, $\hat{\mathcal{C}}$ is taken from $\mathcal{A}$ by supplementing $\mathcal{A}$ with all subsets of all *null* sets in $\mathcal{A}$; that is, sets of measure 0.)
(ii) Also, show that $\sigma(\hat{\mathcal{C}}) \overset{\text{def}}{=} \hat{\mathcal{A}} = \mathcal{A}^*(= \bar{\mathcal{A}})$.

19. If $(\Omega, \mathcal{A}, \mu)$ is a measure space, by means of an example, show that there may be subsets $M \subset N$, where N are null sets in $\mathcal{A}$, with $M \notin \mathcal{A}$. (If it so happens, however, that for every $M \subseteq N$ for every null $N \in \mathcal{A}$, then $M \in \mathcal{A}$, we say that $\mathcal{A}$ is *complete* with respect to μ.)

20. If μ° is an outer measure and $\mathcal{A}^\circ$ is the σ-field of μ°-measurable sets, then show that $\mathcal{A}^\circ$ is complete (with respect to μ°).

21. Consider the measure space $(\Re, \mathcal{B}, \mu)$ where μ(finite interval) $< \infty$, and for $c \in \Re$, let $F_c : \Re \to \Re$ be defined by

$$F_c(x) = \begin{cases} c + \mu((0, x])\ x \geq 0, \\ c - \mu((x, 0])\ x < 0. \end{cases}$$

Then, by means of an example, show that F_c need not be left-continuous.

22. Consider the measurable space $(\Omega, \mathcal{A})$, and let μ be a set function defined on $\mathcal{A}$ and taking nonnegative values, and such that $\mu(\oslash) = c$ with $0 < c < \infty$. Then show that μ cannot be additive.

23. Let $\mu_1, \ldots, \mu_n$ be σ-finite measures on $(\Omega, \mathcal{A})$ and set $\mu = \mu_1 + \cdots + \mu_n$ (in the sense that $\mu(A) = \sum_{i=1}^n \mu_i(A)$, $A \in \mathcal{A}$). Then show that μ is also σ-finite.

24. In the probability space $(\Omega, \mathcal{A}, P)$, show that

(i) $P[(A \cap B^c) \cup (A^c \cap B)] = P(A) + P(B) - 2P(A \cap B)$.
(ii) $P(A_1 \cap \cdots \cap A_n) \geq \sum_{i=1}^n P(A_i) - (n-1)$.

25. Consider the probability space $(\Omega, \mathcal{A}, P)$, where $\Omega = \{\omega_1, \omega_2, \omega_3, \omega_4\}$, $\mathcal{A}$ is the discrete σ-field, and P is defined by

$$P(\{\omega_1\}) = \frac{1}{6},\ P(\{\omega_2\}) = \frac{1}{3},\ P(\{\omega_3\}) = \frac{1}{5},\ P(\{\omega_4\}) = \frac{3}{10}.$$

For $n = 1, 2, \ldots$, define A_n by: $A_{2n-1} = \{\omega_1, \omega_2\}$, $A_{2n} = \{\omega_2, \omega_3\}$. Determine the: $\underline{\lim}_{n\to\infty} A_n$, $\overline{\lim}_{n\to\infty} A_n$ and compute the probabilities:

$$P(\underline{\lim}_{n\to\infty} A_n),\ P(\overline{\lim}_{n\to\infty} A_n),\ \underline{\lim}_{n\to\infty} P(A_n),\ \overline{\lim}_{n\to\infty} P(A_n).$$

26. Let $\Omega = \{\omega_1, \omega_2, \ldots\}$ and let $\mathcal{A}$ be a σ-field of subsets of Ω. Then show that

(i) $\mathcal{A} = \mathcal{P}(\Omega)$ if and only if $\{\omega_i\} \in \mathcal{A}$ for all ω_i.
(ii) A measure μ on $\mathcal{A}(= \mathcal{P}(\Omega))$ is determined by defining $\mu(\{\omega_i\}), i \geq 1$.

27. Suppose that $\Omega = \{\omega_1, \omega_2, \ldots\}$, and let $\mathcal{A}$ be the discrete σ-field. On $\mathcal{A}$, define the set function μ by

$$\mu(A) = \begin{cases} 0 & \text{if } A \text{ is finite,} \\ \infty & \text{if } A \text{ is infinite.} \end{cases}$$

Then show that

(i) μ is nonnegative, $\mu(\oslash) = 0$ and is finitely additive, but it is not σ-additive.
(ii) $\Omega = \lim_{n\to\infty} A_n$ for a sequence $\{A_n\}$, with $A_n \subseteq A_{n+1}, n \geq 1$, and $\mu(A_n) = 0$ (and therefore $\mu(A_n^c) = \infty$), $n \geq 1$.

28. Let $\Omega = \{1, 2, \ldots\}$, and for any $A \subseteq \Omega$, let $a = \sup A$. On the discrete σ-field, define μ^0 by

$$\mu^0(A) = \begin{cases} \frac{a}{a+1} & \text{if } A \text{ is finite,} \\ 0 & \text{if } A = \oslash, \\ 1 & \text{if } A \text{ is infinite.} \end{cases}$$

Then

(i) Show that μ^0 is an outer measure.
(ii) Determine the class $\mathcal{A}^\circ$ of μ°-measurable sets.

29. Recall that m is a *median* of a r.v. X if $P(X \leq m) \geq \frac{1}{2}$ and $P(X \geq m) \geq \frac{1}{2}$. Then show that, if m is a median of the r.v. X, then $-m$ is a median of the r.v. $-X$.

30. Recall that a r.v. X is said to be *symmetric* about 0, if X and $-X$ have the same distribution. If X is symmetric (about 0), then show that 0 is a median of X; i.e., $P(X \leq 0) \geq \frac{1}{2}$ and $P(X \geq 0) \geq \frac{1}{2}$.

31. Let μ^0 be an outer measure, and suppose that $\mu^0(A) = 0$ for some $A \subset \Omega$. Then show that $\mu^0(A \cup B) = \mu^0(B)$ for every $B \subseteq \Omega$.

32. Consider the measure space $(\Omega, \mathcal{A}, \mu)$, and suppose that $\mathcal{A}$ is complete with respect to μ. Let $f, g : \Omega \to \Re$ be such that $\mu(f \neq g) = 0$. Then show that, if one of f or g is measurable, then so is the other.

33. Consider the measure space $(\Omega, \mathcal{A}, \mu)$, and let $\mathcal{A}$ be complete with respect to μ. Let $A \in \mathcal{A}$ with $\mu(A) = 0$, and let f be an arbitrary function $f : A \to \Re$. Then f is measurable.

34. **(i)** If μ_1 and μ_2 are two measures on $(\Omega, \mathcal{A})$, then show that $\mu = \mu_1 + \mu_2$ is also a measure, where $\mu(A) = \mu_1(A) + \mu_2(A)$, $A \in \mathcal{A}$.

(ii) If at least one of the measures μ_1 and μ_2 is complete, then so is μ.

35. Refer to Example 2 and:

(i) Show that the class $\mathcal{C}_3$ consists of the unions of any two members of the class $\mathcal{C}_2$.

(ii) Also, show that $\mathcal{C}_3$ is closed under complementation and the formation of the union of any two of its members.

(iii) Conclude that $\mathcal{C}_3$ is a field, and, indeed, the field generated by $\mathcal{C}$.

This page is intentionally left blank

CHAPTER 3

Some Modes of Convergence of Sequences of Random Variables and their Relationships

In this short chapter, we introduce two basic kinds of convergence, almost everywhere convergence and convergence in measure, and we then investigate their relationships. The mutual versions of these convergences are also introduced, and they are related to the respective convergences themselves. Furthermore, conditions for almost sure convergence are established. Finally, the concept of almost uniform convergence is defined and suitably exploited.

3.1 Almost Everywhere Convergence and Convergence in Measure

Consider the measure space $(\Omega, \mathcal{A}, \mu)$ and let $\{X_n\}$, $n = 1, 2, \ldots, X$ be r.v.s. Then

Definition 1. We say that $\{X_n\}$ converges *almost everywhere* (a.e.) to X and write $X_n \underset{n\to\infty}{\overset{\text{a.e.}}{\to}} X$, if $X_n(\omega) \underset{n\to\infty}{\to} X(\omega)$ except on a set $N(\epsilon\mathcal{A})$ such that $\mu(N) = 0$; i.e., a μ-*null* set. We also write $\mu(X_n \underset{n\to\infty}{\nrightarrow} X) = 0$. In particular, if μ is a probability measure, this convergence is called *almost sure* (a.s.) convergence. We say that $\{X_n\}$ converges *mutually* a.e., if $\mu(X_m - X_n \underset{m,n\to\infty}{\nrightarrow} 0) = 0$, or equivalently, $X_{n+\nu} - X_n \underset{n\to\infty}{\overset{\text{a.e.}}{\to}} 0$ uniformly in $\nu \geq 1$ (see also Exercise 20). Then, by the Cauchy criterion for sequences of numbers, $\{X_n\}$ converges a.e. to X, if and only if it converges mutually a.e. If we modify X_n to X'_n on a null set N_n and X to X' on null set N_0, then, since $\bigcup_{n=1}^{\infty} N_n \cup N_0 \overset{def}{=} N \in \mathcal{A}$ and $\mu(N) = 0$, we have that: $X'_n \underset{n\to\infty}{\overset{\text{a.e.}}{\to}} X'$, if $X_n \underset{n\to\infty}{\overset{\text{a.e.}}{\to}} X$, $\{X'_n\}$ converges mutually a.e., if $\{X_n\}$ converges mutually a.e., and the preceding statement about equivalence of a.e. convergence and mutual convergence of $\{X'_n\}$ is still valid. Of course, $\mu(X'_n \neq X_n) = \mu(X' \neq X) = 0, n = 1, 2, \ldots$. ■

Definition 2. We say that $\{X_n\}$ converges *in measure* to X and write $X_n \underset{n\to\infty}{\overset{\mu}{\to}} X$, if, for every $\varepsilon > 0$, $\mu(|X_n - X| \geq \varepsilon) \underset{n\to\infty}{\to} 0$. In particular, if μ is a probability measure, this convergence is called convergence *in probability*. We say that $\{X_n\}$ converges *mutually in measure*, if for every $\varepsilon > 0$, $\mu(|X_m - X_n| \geq \varepsilon) \underset{m,n\to\infty}{\to} 0$, or equivalently, $\mu(|X_{n+\nu} - X_n| \geq \varepsilon) \underset{n\to\infty}{\to} 0$ uniformly in $\nu \geq 1$ (see also Exercise 21). The

An Introduction to Measure-Theoretic Probability, Second Edition. http://dx.doi.org/10.1016/B978-0-12-800042-7.00003-7

convergence is *mutual convergence in probability*, if μ is a probability measure. ■

Obviously, these convergences remain intact when the X_ns and X are modified as before. We will show later on (see Theorems 2 and 6) that $X_n \underset{n\to\infty}{\overset{\mu}{\to}} X$, if and only if $\{X_n\}$ converges mutually in measure.

Clearly, if $X_n \underset{n\to\infty}{\overset{\text{a.e.}}{\to}} X$ and $X_n \underset{n\to\infty}{\overset{\text{a.e.}}{\to}} X'$, then $\mu(X' \neq X) = 0$. This is also true for convergence in measure but is less obvious. So

Theorem 1. Let $X_n \underset{n\to\infty}{\overset{\mu}{\to}} X$ and $X_n \underset{n\to\infty}{\overset{\mu}{\to}} X'$. Then $\mu(X' \neq X) = 0$. ■

Proof. We have $|X - X'| = |(X - X_n) + (X_n - X')| \leq |X_n - X| + |X_n - X'|$, so that

$$\left(|X - X'| \geq \varepsilon\right) \subseteq \left(|X_n - X| \geq \frac{\varepsilon}{2}\right) \cup \left(|X_n - X'| \geq \frac{\varepsilon}{2}\right), \text{ and hence}$$

$$\mu\left(|X - X'| \geq \varepsilon\right) \leq \mu\left(|X_n - X| \geq \frac{\varepsilon}{2}\right) + \mu\left(|X_n - X'| \geq \frac{\varepsilon}{2}\right) \underset{n\to\infty}{\to} 0.$$

Thus, $\mu(|X - X'| \geq \varepsilon) = 0$, for every $\varepsilon > 0$ and hence $\mu\left(\left(|X - X'| \geq \frac{1}{k}\right)\right) = 0, k = 1, 2, \ldots$. But $(X \neq X') = \bigcup_{k=1}^{\infty}\left(|X - X'| \geq \frac{1}{k}\right)$. Thus

$$\mu(X \neq X') \leq \sum_k \mu\left(|X - X'| \geq \frac{1}{k}\right) = 0; \text{ i.e., } \mu(X \neq X') = 0.$$ ■

Convergence in measure implies mutual convergence in measure. That is,

Theorem 2. Let $X_n \underset{n\to\infty}{\overset{\mu}{\to}} X$. Then $\{X_n\}$ converges mutually in measure. ■

Proof. We have:

$$|X_m - X_n| \leq |X_m - X| + |X_n - X|.$$

Thus,

$$\mu([|X_m - X_n| \geq \varepsilon]) \leq \mu\left(\left[|X_m - X| \geq \frac{\varepsilon}{2}\right]\right) + \mu\left(\left[|X_n - X| \geq \frac{\varepsilon}{2}\right]\right) \underset{m,n\to\infty}{\to} 0.$$ ■

Remark 1. It is obvious that any one of the modes of convergences introduced so far is true for any subsequence of a given sequence.

What we are going to do in the remaining part of this section is to find an expression for the set $X_n \underset{n\to\infty}{\to} X$, then use this result to formulate a criterion for a.e. convergence of X_n to X.

The set of convergence $\left(X_n \underset{n\to\infty}{\to} X\right)$ consists of all those $\omega \in \Omega$ for which: For every $\varepsilon > 0$, there is an $n = n(\varepsilon, \omega) \geq 1$ (integer) such that, for all $\nu \geq 1$, it holds

$$|X_{n+\nu}(\omega) - X(\omega)| < \varepsilon. \tag{3.1}$$

That (3.1) holds for all $\nu \geq 1$ means

$$\omega \in \bigcap_{\nu=1}^{\infty}(|X_{n+\nu} - X| < \varepsilon) \text{ for some } n \geq 1, \text{ and every } \varepsilon > 0. \tag{3.2}$$

That (3.2) holds for some $n \geq 1$ means

$$\omega \in \bigcup_{n=1}^{\infty}\bigcap_{\nu=1}^{\infty}(|X_{n+\nu} - X| < \varepsilon) \text{ for every } \varepsilon > 0. \tag{3.3}$$

That (3.3) holds for every $\varepsilon > 0$ means

$$\omega \in \bigcap_{\varepsilon>0}\bigcup_{n=1}^{\infty}\bigcap_{\nu=1}^{\infty}(|X_{n+\nu} - X| < \varepsilon).$$

Thus: $\left(X_n \underset{n\to\infty}{\to} X\right) = \bigcap_{\varepsilon>0} \bigcup_{n=1}^{\infty} \bigcap_{\nu=1}^{\infty}(|X_{n+\nu} - X| < \varepsilon)$. Clearly, $\varepsilon > 0$ may be replaced by $\varepsilon_k \downarrow 0$ or $\frac{1}{k}(\downarrow 0)$. Then

$$\left(X_n \underset{n\to\infty}{\to} X\right) = \bigcap_{k=1}^{\infty}\bigcup_{n=1}^{\infty}\bigcap_{\nu=1}^{\infty}\left(|X_{n+\nu} - X| < \frac{1}{k}\right).$$

In a similar way we find that the set for which $\{X_n\}$ converges mutually is

$$(X_{n+\nu} - X_n \underset{\substack{n\to\infty \\ \nu\geq 1}}{\to} 0) = \bigcap_{k=1}^{\infty}\bigcup_{n=1}^{\infty}\bigcap_{\nu=1}^{\infty}\left(|X_{n+\nu} - X_n| < \frac{1}{k}\right).$$

Both of these sets are measurable, since they are taken by countable operations on measurable sets. Thus we have

Theorem 3. We have

$$(X_n \underset{n\to\infty}{\to} X) = \bigcap_{k=1}^{\infty}\bigcup_{n=1}^{\infty}\bigcap_{\nu=1}^{\infty}\left(|X_{n+\nu} - X| < \frac{1}{k}\right) \in \mathcal{A},$$

$$(X_{n+\nu} - X_n \underset{\substack{n\to\infty \\ \nu\geq 1}}{\to} 0) = \bigcap_{k=1}^{\infty}\bigcup_{n=1}^{\infty}\bigcap_{\nu=1}^{\infty}\left(|X_{n+\nu} - X_n| < \frac{1}{k}\right) \in \mathcal{A},$$

$$\text{and hence, } (X_n \underset{n\to\infty}{\not\to} X) = \bigcup_{k=1}^{\infty}\bigcap_{n=1}^{\infty}\bigcup_{\nu=1}^{\infty}\left(|X_{n+\nu} - X| \geq \frac{1}{k}\right) \in \mathcal{A},$$

$$(X_{n+\nu} - X_n \underset{\substack{n\to\infty \\ \nu\geq 1}}{\not\to} 0) = \bigcup_{k=1}^{\infty}\bigcap_{n=1}^{\infty}\bigcup_{\nu=1}^{\infty}\left(|X_{n+\nu} - X_n| \geq \frac{1}{k}\right) \in \mathcal{A}. \quad \blacksquare$$

In the following, it is understood that $k, n = 1, 2, \ldots$ and $\nu = 1, 2, \ldots$ (or $\nu = 0, 1, \ldots$).

Now $X_n \underset{n\to\infty}{\overset{\text{a.e.}}{\to}} X$ means $\mu(X_n \underset{n\to\infty}{\not\to} X) = 0$, or equivalently,

$$\mu\left(\bigcup_k \bigcap_n \bigcup_\nu \left(|X_{n+\nu} - X| \geq \frac{1}{k}\right)\right) = 0.$$

But $\mu\left(\bigcup_k \bigcap_n \bigcup_\nu \left(|X_{n+\nu} - X| \geq \frac{1}{k}\right)\right) = 0$ implies $\mu\left(\bigcap_n \bigcup_\nu \left(|X_{n+\nu} - X| \geq \frac{1}{k}\right)\right) = 0, k = 1, 2, \ldots,$ since $\bigcap_n \bigcup_\nu \left[|X_{n+\nu} - X| \geq \frac{1}{k}\right] \uparrow \bigcup_k \bigcap_n \bigcup_\nu \left(|X_{n+\nu} - X| \geq \frac{1}{k}\right)$ as $k \to \infty$. In the other way around, if for all $k = 1, 2, \ldots,$ these sets have measure zero, then the measure of $(X_n \underset{n\to\infty}{\not\to} X)$ is zero. Thus $\mu(X_n \underset{n\to\infty}{\not\to} X) = 0$ if and only if $\mu\left(\bigcap_n \bigcup_\nu \left(|X_{n+\nu} - X| \geq \frac{1}{k}\right)\right) = 0, k = 1, 2, \ldots,$ and in a similar fashion $\mu(X_{n+\nu} - X_n \underset{\substack{n\to\infty\\ \nu\geq 1}}{\not\to} 0) = 0$ if and only if $\mu\left(\bigcap_n \bigcup_\nu \left(|X_{n+\nu} - X_n| \geq \frac{1}{k}\right)\right) = 0, k = 1, 2, \ldots.$ These statements are true for any μ.

Assume now that μ is finite. We have then, for a fixed k,

$$\bigcup_\nu \left(|X_{n+\nu} - X| \geq \frac{1}{k}\right) \downarrow \bigcap_n \bigcup_\nu \left(|X_{n+\nu} - X| \geq \frac{1}{k}\right), \text{ as } n \to \infty, \text{ so that}$$

$$\mu\left(\bigcup_\nu \left(|X_{n+\nu} - X| \geq \frac{1}{k}\right)\right) \downarrow \mu\left(\bigcap_n \bigcup_\nu \left(|X_{n+\nu} - X| \geq \frac{1}{k}\right)\right) \quad \text{as } n \to \infty.$$

Therefore, in this case,

$$\mu(X_n \underset{n\to\infty}{\not\to} X) = 0 \text{ if and only if } \mu\left(\bigcup_\nu \left(|X_{n+\nu} - X| \geq \frac{1}{k}\right)\right) \underset{n\to\infty}{\to} 0$$

for each $k = 1, 2, \ldots,$ and in a similar way,

$$\mu(X_{n+\nu} \underset{\substack{n\to\infty\\ \nu\geq 1}}{\not\to} X_n) = 0 \text{ if and only if } \mu\left(\bigcup_\nu \left(|X_{n+\nu} - X_n| \geq \frac{1}{k}\right)\right) \underset{n\to\infty}{\to} 0$$

for each $k = 1, 2, \ldots.$

Thus we have the following theorem.

Theorem 4.

$$\mu(X_n \underset{n\to\infty}{\not\to} X) = 0 \text{ if and only if } \mu\left(\bigcap_n \bigcup_\nu \left(|X_{n+\nu} - X| \geq \frac{1}{k}\right)\right) = 0, \quad k = 1, 2, \ldots,$$

and

$$\mu(X_{n+\nu} - X_n \underset{\substack{n\to\infty\\ \nu\geq 1}}{\not\to} 0) = 0 \text{ if and only if } \mu\left(\bigcap_n \bigcup_\nu \left(|X_{n+\nu} - X_n| \geq \frac{1}{k}\right)\right) = 0, \quad k = 1, 2, \ldots. \tag{3.4}$$

In particular, if μ is finite, then

$$\mu(X_n \underset{n\to\infty}{\nrightarrow} X) = 0 \text{ if and only if } \mu\left(\bigcup_\nu \left(|X_{n+\nu} - X| \geq \frac{1}{k}\right)\right) \underset{n\to\infty}{\to} 0,$$
$$k = 1, 2, \ldots,$$

and

$$\mu(X_{n+\nu} - X_n \underset{\substack{n\to\infty \\ \nu\geq 1}}{\nrightarrow} 0) = 0 \text{ if and only if}$$
$$\mu\left(\bigcup_\nu \left(|X_{n+\nu} - X_n| \geq \frac{1}{k}\right)\right) \underset{n\to\infty}{\to} 0, \quad k = 1, 2, \ldots$$ ■

Corollary. If μ is finite, then $X_n \underset{n\to\infty}{\overset{\text{a.e.}}{\to}} X$ implies $X_n \underset{n\to\infty}{\overset{\mu}{\to}} X$.

Proof. In fact, let $X_n \underset{n\to\infty}{\overset{\text{a.e.}}{\to}} X$, or $\mu(X_n \underset{n\to\infty}{\nrightarrow} X) = 0$. However,

$$\mu(X_n \underset{n\to\infty}{\nrightarrow} X) = 0 \text{ is equivalent to } \mu\left(\bigcup_\nu \left(|X_{n+\nu} - X| \geq \frac{1}{k}\right)\right) \underset{n\to\infty}{\to} 0,$$
$$k = 1, 2, \ldots.$$

But

$$\mu\left(|X_n - X| \geq \frac{1}{k}\right) \leq \mu\left(\bigcup_{\nu\geq 0} \left(|X_{n+\nu} - X| \geq \frac{1}{k}\right)\right).$$

Thus

$$\mu\left(|X_n - X| \geq \frac{1}{k}\right) \underset{n\to\infty}{\to} 0, \quad k = 1, 2, \ldots, \text{ or equivalently,}$$
$$X_n \underset{n\to\infty}{\overset{\mu}{\to}} X.$$ ■

Remark 2. This need not be true if μ is not finite. Also, the inverse need not be true even if μ is finite. These points can be illustrated by examples (see Exercises 2(i) and 2(ii)).

3.2 Convergence in Measure is Equivalent to Mutual Convergence in Measure

It is first shown that convergence in measure ensures the existence of a subsequence that converges almost everywhere. Then, by means of this result and almost uniform convergence, it is shown that convergence in measure and mutual convergence in measure are equivalent, as is the case in almost everywhere convergence.

Theorem 5.

(i) If $\{X_n\}$ converges mutually in measure, there is a subsequence $\{X_{n_k}\}$ of $\{X_n\}$ and a r.v. X' such that $X_{n_k} \overset{\text{a.e.}}{\to} X'$ as $n \to \infty$ (which implies $n_k \to \infty$).

(ii) If $X_n \underset{n\to\infty}{\overset{\mu}{\to}} X$, there is a subsequence $\{X_{n_k}\}$ of $\{X_n\}$ such that $X_{n_k} \underset{n\to\infty}{\overset{\text{a.e.}}{\to}} X'$, a r.v. X', and $\mu(X' \neq X) = 0$. ■

Proof. (i) $\{X_n\}$ converges mutually in measure implies that

$$\mu\left(|X_m - X_n| \geq \frac{1}{2^k}\right) < \frac{1}{2^k} \text{ for } m, n \geq n(k), \quad k = 1, 2, \ldots. \tag{3.5}$$

Define

$$\begin{aligned} n_1 &= n(1) \\ n_2 &= \max\{n_1 + 1, n(2)\} \\ n_3 &= \max\{n_2 + 1, n(3)\}. \\ &\vdots \end{aligned}$$

Then $n_1 < n_2 < n_3 < \cdots \to \infty$ since each term increases at least by 1. For $k = 1, 2, \ldots$, we set $X'_k = X_{n_k}$ and define

$$A_k = \left(\left|X'_{k+1} - X'_k\right| \geq \frac{1}{2^k}\right), \quad B_n = \bigcup_{k=n}^{\infty} A_k.$$

Then $\mu(A_k) = \mu\left(\left|X_{n_{k+1}} - X_{n_k}\right| \geq \frac{1}{2^k}\right) < \frac{1}{2^k}$, by (3.5), since $n_k, n_{k+1} \geq n(k)$ from their own definition, and hence $\mu(B_n) \leq \sum_{k=n}^{\infty} \mu(A_k) \leq \frac{1}{2^{n-1}}$; i.e.,

$$\mu(B_n) \leq \frac{1}{2^{n-1}}. \tag{3.6}$$

We are aiming at showing that $\{X'_k\}$ converges mutually a.e. To this end, for $\varepsilon > 0$, choose $n_0 = n(\varepsilon)$ such that $\frac{1}{2^{n_0-1}} < \varepsilon$. Then $\frac{1}{2^{n-1}} < \varepsilon, n \geq n_0$. Now for $k \geq n \geq n_0$ and $\nu \geq 1$, one has

$$\begin{aligned} \left|X'_{k+\nu} - X'_k\right| &= \left|(X'_{k+\nu} - X'_{k+\nu-1}) + (X'_{k+\nu-1} - X'_{k+\nu-2}) \right. \\ &\quad \left. + (X'_{k+\nu-2} - X'_{k+\nu-3}) + \cdots + (X'_{k+1} - X'_k)\right| \\ &\leq \left|X'_{k+\nu} - X'_{k+\nu-1}\right| + \left|X'_{k+\nu-1} - X'_{k+\nu-2}\right| \\ &\quad + \left|X'_{k+\nu-2} - X'_{k+\nu-3}\right| + \cdots + \left|X'_{k+1} - X'_k\right| \\ &= \sum_{j=k}^{k+\nu-1} \left|X'_{j+1} - X'_j\right| \leq \sum_{j=n}^{k+\nu-1} \left|X'_{j+1} - X'_j\right| \\ &\leq \sum_{j=n}^{\infty} \left|X'_{j+1} - X'_j\right|. \end{aligned}$$

Therefore, if $\omega \in B_n^c = \bigcap_{k=n}^{\infty} A_k^c$, or equivalently, $\omega \in A_j^c$, $j \geq n$, then

$$\left|X'_{j+1}(\omega) - X'_j(\omega)\right| < \frac{1}{2^j}$$

implies

$$\sum_{j=n}^{\infty} \left|X'_{j+1}(\omega) - X'_j(\omega)\right| \leq \frac{1}{2^{n-1}} < \varepsilon,$$

and this implies in turn that

$$\left|X'_{k+\nu}(\omega) - X'_k(\omega)\right| < \varepsilon, \quad k \geq n(\geq n_0), \quad \nu \geq 1. \tag{3.7}$$

This, in turn, gives that

$$\left(\left|X'_{k+\nu} - X'_k\right| \geq \varepsilon\right) \subseteq B_n, \quad k \geq n(\geq n_0), \quad \nu = 1, 2, \ldots.$$

Hence

$$\bigcup_{\nu=1}^{\infty} \left(\left|X'_{k+\nu} - X'_k\right| \geq \varepsilon\right) \subseteq B_n, \quad k \geq n(\geq n_0)$$

implies

$$\bigcap_{k=n}^{\infty} \bigcup_{\nu=1}^{\infty} \left(\left|X'_{k+\nu} - X'_k\right| \geq \varepsilon\right) \subseteq B_n \quad (n \geq n_0),$$

and therefore, by (3.6),

$$\mu\left(\bigcap_{k\geq n} \bigcup_{\nu\geq 1} \left(\left|X'_{k+\nu} - X'_k\right| \geq \varepsilon\right)\right) \leq \mu(B_n) \leq \frac{1}{2^{n-1}}, \quad (n \geq n_0),$$

so that $\mu\left(\bigcap_{k\geq 1} \bigcup_{\nu\geq 1} \left(\left|X'_{k+\nu} - X'_k\right| \geq \varepsilon\right)\right) \leq \frac{1}{2^{n-1}} \quad (n \geq n_0)$.

Letting $n \to \infty$, we get

$$\mu\left(\bigcap_{k\geq 1} \bigcup_{\nu\geq 1} \left(\left|X'_{k+\nu} - X'_k\right| \geq \varepsilon\right)\right) = 0 \quad \text{for every } \varepsilon > 0, \text{ or equivalently,}$$

$$\mu\left(\bigcap_{k\geq 1} \bigcup_{\nu\geq 1} \left(\left|X'_{k+\nu} - X'_k\right| \geq \frac{1}{m}\right)\right) = 0, \quad m = 1, 2, \ldots \text{ (see (3.4)).}$$

Then Theorem 4 applies and gives that $\{X'_k\}$ converges mutually a.e. and therefore there exists a r.v. X' such that $\{X'_k\} = \{X_{n_k}\}$ (subsequence of $\{X_n\}$) converges to X' a.e. as $k \to \infty$.

(ii) Let $X_n \underset{n\to\infty}{\overset{\mu}{\to}} X$. Then, by Theorem 2, $\{X_n\}$ converges mutually in measure, and hence, by part (i), there exists $\{X'_k\} \subseteq \{X_n\}$ such that $X'_k \underset{k\to\infty}{\overset{\text{a.e.}}{\to}} X'$. It remains for us to show that $\mu(X' \neq X) = 0$. This is done in the Proof of Theorem 5(ii) (continued later). ■

Remark 3. It is to be pointed out here that the subsequence, constructed in Theorem 5, that converges a.e. to a r.v. will be used extensively in many instances and for various purposes.

Theorem 6. If $\{X_n\}$ converges mutually in measure, then $\{X_n\}$ converges in measure to a r.v. X as $n \to \infty$. ■

Proof (For the case that μ is finite). By Theorem 5 (i), there exists a subsequence $\{X_{nk}\}$ of $\{X_n\}$ and a r.v. X such that $X_{n_k} \underset{k\to\infty}{\overset{\text{a.e.}}{\to}} X$. Also $X_{n_k} \underset{k\to\infty}{\overset{\mu}{\to}} X$ by the Corollary to Theorem 4 (since μ is finite). To show that $X_n \underset{n\to\infty}{\overset{\mu}{\to}} X$. Indeed,

$$|X_n - X| \leq |X_n - X_{n_k}| + |X_{n_k} - X|$$

implies

$$(|X_n - X| \geq \varepsilon) \subseteq \left(|X_n - X_{n_k}| \geq \frac{\varepsilon}{2}\right) \cup \left(|X_{n_k} - X| \geq \frac{\varepsilon}{2}\right),$$

so that

$$\mu(|X_n - X| \geq \varepsilon) \leq \mu\left(|X_n - X_{n_k}| \geq \frac{\varepsilon}{2}\right) + \mu\left(|X_{n_k} - X| \geq \frac{\varepsilon}{2}\right).$$

Letting $n \to \infty$ (which implies that $n_k \to \infty$), we get $\mu\left(|X_n - X_{n_k}| \geq \frac{\varepsilon}{2}\right) \to 0$ from mutual convergence in measure of $\{X_n\}$, and $\mu\left(|X_{n_k} - X| \geq \frac{\varepsilon}{2}\right) \to 0$. Hence, as $n \to \infty$, $\mu\left(|X_n - X| \geq \varepsilon\right) \to 0$, or equivalently, $X_n \underset{n\to\infty}{\overset{\mu}{\to}} X$. ■

Theorem 6 is valid even if μ is not finite, but for its proof we need some preliminary results. (See Proof of Theorem 6 (continued) later.)

Definition 3. We say that $\{X_n\}$ converges *almost uniformly* to X and write $X_n \underset{n\to\infty}{\overset{\text{a.u.}}{\to}} X$, if for every $\varepsilon > 0$, there exists $A_\varepsilon \in \mathcal{A}$ such that $\mu(A_\varepsilon) < \varepsilon$ and $X_n \underset{n\to\infty}{\to} X$ uniformly on A^c_ε. We say that $\{X_n\}$ *converges mutually almost uniformly* if for every $\varepsilon > 0$ there exists $A_\varepsilon \in \mathcal{A}$ such that $\mu(A_\varepsilon) < \varepsilon$ and $X_m - X_n \underset{m,n\to\infty}{\to} 0$ uniformly on A^c_ε. Of course, $X_n \underset{n\to\infty}{\overset{\text{a.u.}}{\to}} X$ if and only if $\{X_n\}$ converges mutually almost uniformly. (Observe that there exist such sequences; e.g., the subsequence $\{X_{n_k}\}$ constructed in Theorem 5 is such a sequence. See also Theorem 8 below.) ■

Theorem 7. If $X_n \underset{n\to\infty}{\overset{\text{a.u.}}{\to}} X$, then $X_n \underset{n\to\infty}{\overset{\text{a.e.}}{\to}} X$ and $X_n \underset{n\to\infty}{\overset{\mu}{\to}} X$. ■

Proof. We have that $X_n \underset{n\to\infty}{\overset{\text{a.u.}}{\to}} X$ implies that for $\frac{1}{k}$ there exists A_k such that $\mu(A_k) < \frac{1}{k}$ and $X_n \underset{n\to\infty}{\to} X$ uniformly on A^c_k, $k = 1, 2, \ldots$. Set $A = \bigcap_{k=1}^\infty A_k$. Then $A \in \mathcal{A}$ and $\mu(A) \leq \mu(A_k) < \frac{1}{k}$. Letting $k \to \infty$, we then get $\mu(A) = 0$. On the other hand, $A^c = \bigcup_{k=1}^\infty A^c_k$ and therefore for every $\omega \in A^c$, it follows that $\omega \in A^c_k$ for some $k = 1, 2, \ldots$, so that $X_n(\omega) \underset{n\to\infty}{\to} X(\omega)$. That is, $X_n \underset{n\to\infty}{\to} X$ on A^c or $X_n \underset{n\to\infty}{\overset{\text{a.e.}}{\to}} X$.

Again, $X_n \underset{n\to\infty}{\overset{\text{a.u.}}{\to}} X$ implies that for every $\varepsilon > 0$ there exists $A_\varepsilon \in \mathcal{A}$ such that $\mu(A_\varepsilon) < \varepsilon$ and $X_n \underset{n\to\infty}{\to} X$ uniformly on A_ε^c. This second part means that for every $\delta > 0$ there exists $N(\delta, \varepsilon) > 0$ (independent of $\omega \in A_\varepsilon^c$) such that for $n \geq N(\delta, \varepsilon)$, $|X_n(\omega) - X(\omega)| < \delta, \omega \in A_\varepsilon^c$. Thus, for $n \geq N(\delta, \varepsilon)$, $(|X_n - X| \geq \delta) \subseteq A_\varepsilon$, so that $\mu(|X_n - X| \geq \delta) \leq \mu(A_\varepsilon) < \varepsilon$; i.e., $X_n \underset{n\to\infty}{\overset{\mu}{\to}} X$. ■

Remark 4. Compare with Remark 2 where $X_n \underset{n\to\infty}{\overset{\text{a.e.}}{\to}} X$ does *not* necessarily imply that $X_n \underset{n\to\infty}{\overset{\mu}{\to}} X$.

Theorem 8.

(i) If $\{X_n\}$ converges mutually in measure, then there exists $\{X'_k\} \subseteq \{X_n\}$ and a r.v. X' such that $X'_k \underset{k\to\infty}{\overset{\text{a.u.}}{\to}} X'$.

(ii) If $X_n \underset{n\to\infty}{\overset{\mu}{\to}} X$, then there exists $\{X'_k\} \subseteq \{X_n\}$ and a r.v. X' such that $X'_k \underset{k\to\infty}{\overset{\text{a.u.}}{\to}} X'$ and $\mu(X' \neq X) = 0$. ■

Proof.

(i) Consider the subsequence $\{X'_k\}$ constructed in Theorem 5(i). By the arguments which led to relation (3.7), for every $\varepsilon > 0$, there is $n_0 = n(\varepsilon) > 0$ integer and a set B_n^c with $\mu(B_n) < \varepsilon$, $n \geq n_0$, such that, on B_n^c,

$$\left|X'_{k+\nu}(\omega) - X'_k(\omega)\right| < \varepsilon, \quad k \geq n_0, \quad \nu = 1, 2, \ldots.$$

Applying the above for $n = n_0$, we get: for every $\varepsilon > 0$, there exists $n_0 = n(\varepsilon)$ positive integer and a set B_{n_0} with $\mu(B_{n_0}) < \varepsilon$, such that, on $B_{n_0}^c$,

$$\left|X'_{k+\nu}(\omega) - X'_k(\omega)\right| < \varepsilon, \quad k \geq n_0, \quad \nu = 1, 2, \ldots. \tag{3.8}$$

But (3.8) is the definition of mutual a.u. convergence. Then there exists a r.v. X' such that $X'_k \underset{k\to\infty}{\overset{\text{a.u.}}{\to}} X'$.

(ii) If $X_n \underset{n\to\infty}{\overset{\mu}{\to}} X$, then $\{X_n\}$ converges mutually in measure (by Theorem 2). Then, by part (i), $X'_k \underset{k\to\infty}{\overset{\text{a.u.}}{\to}} X'$. However, $X'_k \underset{k\to\infty}{\overset{\mu}{\to}} X'$ by Theorem 7. So $X'_k \underset{k\to\infty}{\overset{\mu}{\to}} X'$ and $X'_k \underset{k\to\infty}{\overset{\mu}{\to}} X$ (since $\{X'_k\} \subseteq \{X_n\}$). Therefore $\mu(X' \neq X) = 0$ by Theorem 1. ■

Remark 5. See Exercise 11 for an implication of a.e. convergence with regards to a.u. convergence.

We are now in a position to complete the parts of various proofs left incomplete so far.

(ii)

Proof of Theorem 5(ii) (continued). Recall that in Theorem 5 (ii) we assumed $X_n \underset{n\to\infty}{\overset{\mu}{\to}} X$ and we proved the existence of a subsequence $\{X'_k\}$ such that $X'_k \underset{k\to\infty}{\overset{\text{a.e.}}{\to}} X'$, a r.v. X'. What remained to be verified was that $\mu(X' \neq X) = 0$. This is done in Theorem 8 (ii). ■

Proof of Theorem 6 (continued). We can now give the proof of Theorem 6 for not necessarily a finite μ. Consider the sequence $\{X_{n_k}\}$ of Theorem 5 or Theorem 8. Then $X_{n_k} \underset{k\to\infty}{\overset{\text{a.u.}}{\to}} X$ implies ($X_{n_k} \underset{k\to\infty}{\overset{\text{a.e.}}{\to}} X$ and) $X_{n_k} \underset{k\to\infty}{\overset{\mu}{\to}} X$ by Theorem 7. Hence $\mu(|X_n - X| \geq \varepsilon) \leq \mu\left(\left|X_n - X_{n_k}\right| \geq \frac{\varepsilon}{2}\right) + \mu\left(\left|X_{n_k} - X\right| \geq \frac{\varepsilon}{2}\right) \underset{k\to\infty}{\to} 0 + 0 = 0$, the first by assumption and the second by the fact that $X_{n_k} \underset{k\to\infty}{\overset{\mu}{\to}} X$. ■

Corollary to Theorem 2 and 6. $X_n \underset{n\to\infty}{\overset{\mu}{\to}} X$ if and only if $\{X_n\}$ converges mutually in measure. ■

A brief summary of the results obtained in this chapter is as follows. The sequence of r.v.s we refer to is $\{X_n\}$, and a limiting r.v. is X.

1. Convergence in measure implies ($\mu-$) uniqueness of the limit (Theorem 1).
2. Convergence in measure is equivalent to mutual convergence in measure (Theorems 2 and 6).
3. Expressions of the set of pointwise convergence (and hence of nonconvergence), as well as the set of pointwise mutual convergence (and hence of mutual nonconvergence) (Theorem 3).
4. Necessary and sufficient conditions for a.e. (and mutual a.e.) convergence for any μ, and in particular, for finite μ (Theorem 4).
5. Almost everywhere convergence and finiteness of μ imply convergence in measure. The converse need not be true even for finite μ (Corollary to Theorem 4, Exercise 2(i), (ii)).
6. Convergence in measure (or mutual convergence in measure) implies the existence of a subsequence that converges a.e. (or converges mutually a.e.) (Theorem 5).
7. Almost uniform convergence implies a.e. convergence and convergence in measure (Theorem 7).
8. Convergence in measure (or mutual convergence in measure) implies the existence of a subsequence that converges a.u. to a r.v. (or converges mutually a.u.) (Theorem 8).

Exercises.

1. If $X_n \underset{n\to\infty}{\overset{\mu}{\to}} X$, then show directly (that is, without reference to other results) that: $(X_n - X)^+ \underset{n\to\infty}{\overset{\mu}{\to}} 0$, $(X_n - X)^- \underset{n\to\infty}{\overset{\mu}{\to}} 0$, $X_n^+ \underset{n\to\infty}{\overset{\mu}{\to}} X^+$ and $X_n^- \underset{n\to\infty}{\overset{\mu}{\to}} X^-$.
   Hint: For the convergence $X_n^+ \underset{n\to\infty}{\overset{\mu}{\to}} X^+$, show that $|X_n^+ - X^+| \leq |X_n - X|$, and likewise for $X_n^- \underset{n\to\infty}{\overset{\mu}{\to}} X^-$. To this end, use Exercise 28 in Chapter 1.

2. By means of examples, show that:

(i) $X_n \underset{n\to\infty}{\overset{\text{a.e.}}{\to}} X$ need not imply $X_n \underset{n\to\infty}{\overset{\mu}{\to}} X$ if μ is not finite.

(ii) $X_n \underset{n\to\infty}{\overset{\mu}{\to}} X$ need not imply $X_n \underset{n\to\infty}{\overset{\text{a.e.}}{\to}} X$ even if μ is finite.

Hint: In (i), take $\Omega = \Re$, $\mathcal{A} = \mathcal{B}$, $\mu = \lambda$, the Lebesgue measure, and choose the r.v.s suitably. In part (ii), take $\Omega = (0, 1]$, $\mathcal{A} = \mathcal{B}_\Omega$, $\mu = \lambda$, the Lebesgue measure, and choose the r.v.s suitably.

3. For any sequence of events $\{A_n\}, n \geq 1$, show that $\sum_{n=1}^{\infty} P(A_n) < \infty$ implies $P(\limsup_{n\to\infty} A_n) = 0$.

4. The sequence $\{X_n\}, n \geq 1$, of r.v.s is said to converge *completely* to 0, if, for every $\varepsilon > 0$, $\sum_{n=1}^{\infty} P(|X_n| \geq \varepsilon) < \infty$.

(i) Show that, if $\{X_n\}, n \geq 1$, converges completely to 0, then $X_n \underset{n\to\infty}{\overset{\text{a.e.}}{\to}} 0$.

(ii) By means of an example, show that complete convergence is not necessary for a.s. convergence.

Hint: For part (i), use Exercise 3 here and Exercise 4 in Chapter 2. For part (ii), take $\Omega = (0, 1]$, $\mathcal{A} = \mathcal{B}_\Omega$, $P = \lambda$, the Lebesgue measure, and choose the r.v.s suitably.

Note: The most common way of establishing that $X_n \underset{n\to\infty}{\overset{\text{a.e.}}{\to}} X$ is to show that $\{X_n - X\}, n \geq 1$, converges completely to 0.

5. Show that $X_n \underset{n\to\infty}{\overset{P}{\to}} X$ if and only if, for every $\{n'\} \subseteq \{n\}$, there exists a further subsequence $\{n''\} \subseteq \{n'\}$ such that $X_{n''} \underset{n''\to\infty}{\overset{\text{a.s.}}{\to}} X$ and any two limiting r.v.s are a.s. equal.

6. By means of an example, show that $X_n \underset{n\to\infty}{\overset{\text{a.s.}}{\to}} c \neq 0$ need not imply that $P(X_n \neq 0) = 1$ for all n.

7. If for some $\varepsilon_n > 0$ with $\sum_{n=1}^{\infty} \varepsilon_n < \infty$, it holds that $\sum_{n=1}^{\infty} P(|X_{n+1} - X_n| \geq \varepsilon_n) < \infty$, then show that $X_n \underset{n\to\infty}{\overset{\text{a.s.}}{\to}}$ to a r.v.

Hint: It suffices to show that $\{X_n\}$ converges mutually a.s. To this end, set $A_n = (|X_{n+1} - X_n| \geq \varepsilon_n), n \geq 1$, and use Exercise 3 in this chapter in order to conclude that $P(\overline{\lim}_{n\to\infty} A_n) = 0$. Then, by setting $\overline{A} = \overline{\lim}_{n\to\infty} A_n$ and $N^c = \overline{A}^c$, it follows that the event N^c (with $P(N^c) = 1$) is the set over which $\{X_n\}$ converges mutually.

8. For $n = 1, 2, \ldots$, let X_n and X be (real-valued) r.v.s and let $g: \Re \to \Re$ be continuous. Then show that $X_n \underset{n\to\infty}{\overset{P}{\to}} X$ implies $g(X_n) \underset{n\to\infty}{\overset{P}{\to}} g(X)$ (so that continuity preserves convergence in probability).

Hint: One way of approaching this problem is to use Exercise 5.

9. For $n = 1, 2, \ldots$, let X_n, Y_n, X, and Y be (real-valued) r.v.s and let $g : \Re^2 \to \Re$ be continuous. Then show that $X_n \underset{n\to\infty}{\overset{P}{\to}} X$ and $Y_n \underset{n\to\infty}{\overset{P}{\to}} Y$ implies $g(X_n, Y_n) \underset{n\to\infty}{\overset{P}{\to}} g(X, Y)$ (so that continuity preserves convergence in probability).

10. Show that $X_n \underset{n\to\infty}{\overset{\text{a.s.}}{\to}} X$ if and only if there is a sequence $0 < \varepsilon_n \underset{n\to\infty}{\to} 0$ such that

$$P\left[\bigcup_{k\geq n}(|X_k - X| \geq \varepsilon_k)\right] \underset{n\to\infty}{\to} 0.$$

Hint: For the part $\varepsilon_n \underset{n\to\infty}{\to} 0$, show that, for every $\varepsilon > 0$, there exists $N = N(\varepsilon)$ such that $k \geq N$ and $n \geq N$ imply $\bigcup_{k\geq n}(|X_k - X| \geq \varepsilon) \subseteq \bigcup_{k\geq n}(|X_k - X| \geq \varepsilon_k)$ and then use Theorem 4 suitably. For the part $X_n \underset{n\to\infty}{\overset{\text{a.s.}}{\to}} X$, use Theorem 4 in order to conclude that $P\left[\bigcup_{k\geq n}(|X_k - X| \geq \varepsilon)\right] \underset{n\to\infty}{\to} 0$. Applying this conclusion for $m \geq 1$, show that there exists a sequence $n_m \uparrow \infty$ as $m \to \infty$ such that

$$P\left[\bigcup_{k\geq n_m}\left(|X_k - X| \geq \frac{1}{m}\right)\right] < \frac{1}{2^m}.$$

Finally, for $n_m \leq k < n_{m+1}$, set $\varepsilon_k = \frac{1}{m}$ and show that

$$P\left[\bigcup_{k\geq n}(|X_k - X| \geq \varepsilon_k)\right] \leq P\left[\bigcup_{k\geq n_m}(|X_k - X| \geq \varepsilon_k)\right] \leq \frac{1}{2^{n-1}}.$$

11. (*Egorov's Theorem*). Show that, if μ is finite, then $X_n \underset{n\to\infty}{\overset{\text{a.e.}}{\to}} X$ implies that $X_n \underset{n\to\infty}{\overset{\text{a.u.}}{\to}} X$.

Hint: For an arbitrary $\varepsilon > 0$ to be kept fixed throughout and $k \geq 1$ integer, use Theorem 4 in order to conclude that there exists $N_k = N(\varepsilon, k) > 0$ such that $\mu(A_{\varepsilon,k}) < \frac{\varepsilon}{2^k}$, $k \geq 1$, where $A_{\varepsilon,k} = \bigcup_{n\geq N_k}\left(|X_n - X| \geq \frac{1}{k}\right)$. Thus, if $A_\varepsilon = \bigcup_{k\geq 1} A_{\varepsilon,k}$, then $\mu(A_\varepsilon) \leq \varepsilon$. Finally, show that $X_n(\omega) \underset{n\to\infty}{\to} X(\omega)$ uniformly in $\omega \in A_\varepsilon^c$.

12. Show that the complement A^c of the set A of (pointwise) convergence of a sequence of r.v.s $\{X_n\}$, $n \geq 1$, is expressed as follows: $A^c = \bigcup_{r,s}\{\omega \in \Omega; \liminf_{n\to\infty} X_n(\omega) \leq r < s \leq \limsup_{n\to\infty} X_n(\omega)\}$, where the union is taken over all rationals r and s with $r < s$.

13. For $n = 1, 2, \ldots$, let X_n, X be r.v.s defined on the measurable space $(\Omega, \mathcal{A})$, and suppose that $X_n \underset{n\to\infty}{\longrightarrow} X$ (pointwise). Then show that

(i) X is $\sigma(X_1, X_2, \ldots)$-measurable.
(ii) $\sigma(X_1, X_2, \ldots) = \sigma(X_1, X_2, \ldots, X)$.

14. For a sequence of r.v.s $\{X_n\}$, $n \geq 1$, show that the set $A = \{\omega \in \Omega;\ \sum_{n=1}^{\infty} X_n(\omega)$ converges$\}$ is in the $\sigma(X_m, X_{m+1}, \ldots)$ for every $m \geq 1$.

15. Let X_n, $n \geq 1$, be r.v.s defined on the measure space $(\Omega, \mathcal{A}, P)$, and suppose that $\sum_{n=1}^{\infty} P(|X_n| > n) < \infty$. Then show that $\overline{\lim\limits_{n\to\infty}} \frac{|X_n|}{n} \leq 1$ a.s.
Hint: Refer to Exercise 3 in this chapter.

16. For $n = 1, 2, \ldots$, let $\mathbf{Z}_n = (Z_{n1}, \ldots, Z_{nk})$ and $\mathbf{Z} = (Z_1, \ldots, Z_k)$ be k-dimensional r.v.s. Then we say that $\mathbf{Z}_n \underset{n\to\infty}{\overset{P}{\to}} \mathbf{Z}$ if $||\mathbf{Z}_n - \mathbf{Z}|| \underset{n\to\infty}{\overset{P}{\to}} 0$, or $\left[\sum_{j=1}^{k} (Z_{nj} - Z_j)^2\right]^{1/2} \underset{n\to\infty}{\overset{P}{\to}} 0$. Then show that $\mathbf{Z}_n \underset{n\to\infty}{\overset{P}{\to}} \mathbf{Z}$ if and only if $Z_{nj} \underset{n\to\infty}{\overset{P}{\to}} Z_j$, $j = 1, \ldots, k$.

17. If $X_n \underset{n\to\infty}{\overset{P}{\to}} X$ and $g : \Re \to \Re$ is continuous, then $g(X_n) \underset{n\to\infty}{\overset{P}{\to}} g(X)$ (see Exercise 8). Show that this need not be true, if g is not continuous.
Hint: Consider the function $\delta_c(x) = 0$ for $x < c$, and $\delta_c(x) = 1$ for $x \geq c$, for some constant $c \in \Re$.

18. If $X_n \underset{n\to\infty}{\overset{\mu}{\longrightarrow}} X$ and $|X_n| \leq Y$ a.e. $n \geq 1$, then show that $|X| \leq Y$ a.e. with respect to the measure μ.

19. Refer to Exercise 13 and continue as follows: For $n = 1, 2, \ldots$, let X_n, X be r.v.s defined on the measure space $(\Omega, \mathcal{A}, \mu)$, and suppose that $X_n \underset{n\to\infty}{\overset{a.e.}{\to}} X$. Then, by means of concrete examples, show that:

(i) X is $\sigma(X_1, X_2, \ldots)$-measurable.
(ii) X is not $\sigma(X_1, X_2, \ldots)$-measurable.
(iii) If $X_n \underset{n\to\infty}{\overset{a.e.}{\to}} X$, show that the X_ns and X can be modified into X_ns and X', so that $X'_n \underset{n\to\infty}{\to} X'$ pointwise, X' is $\sigma\left(X'_1, X'_2, \ldots\right)$-measurable, and $\sum_{n=1}^{\infty} \mu\left(X'_n \neq X_n\right) = 0 = \mu(X' \neq X)$.
(As a consequence, instead of the X_ns and X one could use the X_ns and X', without loss of generality, and also ensure that X' is $\sigma(X'_1, X'_2, \ldots)$-measurable.)
(iv) Consider the measurable space $(\Omega, \mathcal{A}, \mu)$, and suppose that, for some $\omega_0 \in \Omega$, $\{\omega_0\} \in \mathcal{A}$ and $\mu(\{\omega_0\}) = 0$. Define $X_n(\omega) = 0$ on $\{\omega_0\}$, and $X_{2n-1}(\omega_0) = 2$, $X_{2n}(\omega_0) = 3$, $n \geq 1$; and $X(\omega) = 0$ on $\{\omega_0\}^c$, $X(\omega_0) = 1$.

Then verify that $X_n \underset{n\to\infty}{\overset{a.e.}{\to}} X$. Furthermore, modify the X_ns and X as indicated in part (iii), so that the conclusions of that part hold.

20. Show that the convergence $X_m - X_n \xrightarrow{a.e.} 0$ as $m, n \to \infty$ is equivalent to the convergence $X_{n+\nu} - X_n \underset{n\to\infty}{\overset{a.e.}{\longrightarrow}} 0$ uniformly in $\nu \geq 1$.

21. Show that, for every $\varepsilon > 0 : \mu(|X_m - X_n| \geq \varepsilon) \to 0$ as $m, n \to \infty$ if and only if $\mu(|X_{n+\nu} - X_n| \geq \varepsilon) \underset{n\to 0}{\longrightarrow} 0$ uniformly in $\nu \geq 1$.

CHAPTER

The Integral of a Random Variable and its Basic Properties

4

In this chapter, the concept of the integral of a r.v. X with respect to a measure μ is defined, and the basic properties of integrals are established. The definition of the integral is given first for simple r.v.s, then for nonnegative r.v.s, and finally for any r.v.s. These things are done in Section 4.1, whereas the basic properties of the integral are discussed in Section 4.2. A brief summary of the results in these sections is also presented at the end of the second section.

In the next short and final section of the chapter, the measure space $(\Omega, \mathcal{A}, \mu)$ is replaced by the probability space $(\Omega, \mathcal{A}, P)$, and the probability distribution of a r.v. X, P_X, is introduced. There is only one result here, Theorem 13 , whose significance lies in that integration over an abstract space Ω is transformed to integration over the real line $\Re$.

4.1 Definition of the Integral

Consider the measure space $(\Omega, \mathcal{A}, \mu)$ and let X be nonnegative simple r.v.; i.e., $X = \sum_{j=1}^{n} \alpha_j I_{A_j}$, where $\alpha_j \geq 0$, $j = 1, \ldots, n$, and $\{A_j,\ j = 1, \ldots, n\}$ is a partition of Ω. To such an X, we assign a possibly extended number, denoted by $I(X)$, as follows: $I(X) = \sum_{j=1}^{n} \alpha_j \mu(A_j)$. We also make throughout the convention that $\pm\infty \times 0 = 0$. Then,

Theorem 1. The function I: on the class of ≥ 0 simple r.v.s into $[0, \infty]$ as defined above is well defined. ■

Proof. If also $X = \sum_{i=1}^{m} \beta_i I_{B_i}$ is another representation of X, we have to prove that $\sum_{j=1}^{n} \alpha_j \mu(A_j) = \sum_{i=1}^{m} \beta_i \mu(B_i)$. Clearly, $\{A_j \cap B_i,\ j = 1, \ldots, n,\ i = 1, \ldots, m\}$ is a partition of Ω. For $j = 1, \ldots, n,\ i = 1, \ldots, m$, define c_{ij} as follows:

$$c_{ji} = \begin{cases} \alpha_j = \beta_i & \text{if } A_j \cap B_i \neq \oslash \\ \text{whatever (e.g., } = 0 \text{ for definiteness)} & \text{if } A_j \cap B_i = \oslash \end{cases}.$$

Then

$$\sum_{j,i} c_{ji}\mu(A_j \cap B_i) = \sum_{j}\sum_{i} c_{ji}\mu(A_j \cap B_i) = \sum_{j}\sum_{\substack{i \\ A_j \cap B_i \neq \oslash}} c_{ji}\mu(A_j \cap B_i)$$

An Introduction to Measure-Theoretic Probability, Second Edition. http://dx.doi.org/10.1016/B978-0-12-800042-7.00004-9

$$= \sum_{\substack{j \\ A_j \cap B_i \neq \oslash}} \sum_{i} \alpha_j \mu(A_j \cap B_i) = \sum_j \alpha_j \sum_{\substack{i \\ A_j \cap B_i \neq \oslash}} \mu(A_j \cap B_i) = \sum_j \alpha_j \mu(A_j);$$

i.e., $\sum_{j,i} c_{ji} \mu(A_j \cap B_i) = \sum_j \alpha_j \mu(A_j)$, and in a similar fashion

$\sum_{j,i} c_{ji} \mu(A_j \cap B_i) = \sum_i \beta_i \mu(B_i)$. Thus $\sum_j \alpha_j \mu(A_j)$

$= \sum_i \beta_i \mu(B_i)$ (see also Exercise 15). ∎

Theorem 2. Let X, Y be nonnegative simple r.v.s such that $X \geq Y$. Then $I(X) \geq I(Y)$; furthermore, if the measure μ is finite, then $I(X) > I(Y)$ if and only if $X > Y$ on a set of positive measure. In particular, $X \geq 0$ implies $I(X) \geq 0$; furthermore, if the measure μ is finite, then $I(X) > 0$ if and only if $X > 0$ on a set of positive measure. ∎

Proof. Let $X = \sum_{j=1}^{n} \alpha_j I_{A_j}$, $Y = \sum_{i=1}^{m} \beta_i I_{B_i}$. Then, clearly, $X = \sum_{j,i} \alpha_j \times I_{A_j \cap B_i}$, $Y = \sum_{i,j} \beta_i I_{A_j \cap B_i}$ since $I_{\oslash} = 0$, and hence $I(X) = \sum_{j,i} \alpha_j \mu(A_j \cap B_i)$, $I(Y) = \sum_{i,j} \beta_i \mu(A_j \cap B_i)$. Now $X \geq Y$ implies $\alpha_j \geq \beta_i$ on $A_j \cap B_i$ (which are $\neq \oslash$) and hence $I(X) \geq I(Y)$. Next, let μ be finite. Then if $X > Y$ on a set E with $\mu(E) > 0$, one has that $\alpha_j \geq \beta_i$ on all $A_j \cap B_i$ (which are $\neq \oslash$) and $\alpha_j > \beta_i$ on $E \cap (A_j \cap B_i)$ for some $A_j \cap B_i$ with $\mu[E \cap (A_j \cap B_i)] > 0$.

Since

$$\alpha_j \mu(A_j \cap B_i) = \alpha_j \mu[E \cap (A_j \cap B_i)] + \alpha_j \mu[E^c \cap (A_j \cap B_i)] >$$
$$\beta_i \mu[E \cap (A_j \cap B_i)] + \beta_i \mu[E^c \cap (A_j \cap B_i)] = \beta_i \mu(A_j \cap B_i),$$

we have that $I(X) > I(Y)$. Finally if $I(X) > I(Y)$, there must exist (by contradiction) a set E such that $\mu[E \cap (A_j \cap B_i)] > 0$ for some $A_j \cap B_i$ for which $\alpha_j > \beta_i$ (whereas always $\alpha_j \geq \beta_i$). Indeed, if $\alpha_j = \beta_i$ on all $A_j \cap B_i \neq \oslash$, then $I(X) = I(Y)$, a contradiction. If $\alpha_j > \beta_i$ on $A_j \cap B_i$ with $\mu(A_j \cap B_i) = 0$ and $\alpha_j = \beta_i$ on all $A_j \cap B_i$ with $\mu(A_j \cap B_i) > 0$, then again $I(X) = I(Y)$, a contradiction. Thus, there must exist $A_j \cap B_i \neq \oslash$ with $\mu(A_j \cap B_i) > 0$ on which $\alpha_j > \beta_i$; i.e., $X > Y$ on a set of positive measure; namely, $E = A_j \cap B_i$ for some j, i. The special case follows by taking $Y = 0$. ∎

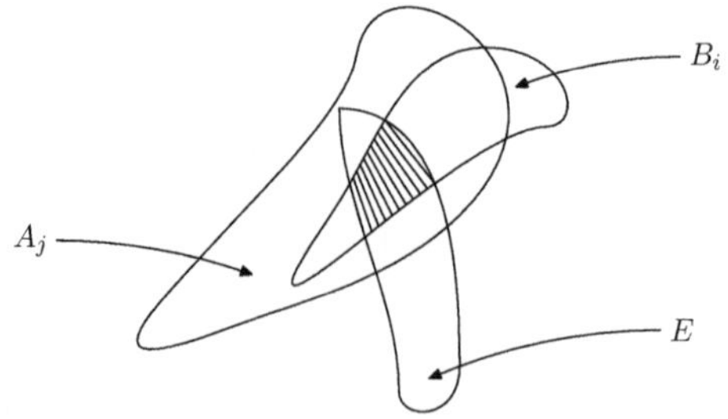

Next, let X be a nonnegative r.v. (not necessarily a simple r.v.). Then, by the Corollary to Theorem 17, Chapter 1, there exist X_n nonnegative, simple r.v.s such that $X_n \uparrow X$. To such an X we assign the quantity $I(X)$ as follows: $I(X) = \lim_{n\to\infty} I(X_n)$. This limit exists and, clearly, $I(X) \geq 0$ (but it may happen that $I(X) = \infty$). Then we have the following theorem.

Theorem 3. The (possibly extended) function I: on the class of ≥ 0 r.v.s into $[0, \infty]$ as defined earlier is well defined. ∎

Remark 1. Before we proceed with the proof of this theorem we observe that the I just defined coincides with the I defined before on the class of ≥ 0 simple r.v.s since in such a case we can take $X_n = X,\ n = 1, 2, \ldots$. Thus the I is an extension from the class of ≥ 0 simple r.v.s to the class of ≥ 0 r.v.s.

Proof of Theorem 3. In order to prove the theorem it suffices to prove that

$$\begin{aligned} &\text{If } Y \text{ is a } \geq 0 \text{ simple r.v. with } Y \leq X, \text{ then } I(Y) \leq \lim I(Z_n),\\ &\text{with } 0 \leq Z_n \text{ simple r.v.s } \uparrow X, \end{aligned} \tag{4.1}$$

where here and in the sequel, all limits are taken as $n \to \infty$, unless otherwise specified.

In fact, if (4.1) is true, then for $0 \leq X_n$ simple $\uparrow X$, $0 \leq Y_n$ simple $\uparrow X$, we have $Y_n \leq X$ implies $I(Y_n) \leq \lim I(X_n)$ and $\lim I(Y_n) \leq \lim I(X_n)$. Also $X_n \leq X$ implies $I(X_n) \leq \lim I(Y_n)$ and $\lim I(X_n) \leq \lim I(Y_n)$. Thus $\lim I(X_n) = \lim I(Y_n)$.

In order to establish (4.1) we distinguish three cases.

Case 1. $\mu(\Omega) < \infty$, $\min Y \overset{def}{=} m > 0$.

We have $X_n = \sum_{i=1}^{r_n} \alpha_{ni} I_{A_{ni}}$, $Y = \sum_{j=1}^{s} \beta_j I_{B_j}$. Choose $\varepsilon > 0$ such that $\varepsilon < m$ and define C_n as follows: $C_n = (X_n > Y - \varepsilon)$. Then $C_n \uparrow \Omega$, since $X_n \uparrow X$ and $X \geq Y$ imply $X > Y - \varepsilon$. Indeed, from the definition of C_n and the fact that $X_n \uparrow X$, we have $C_n \uparrow$, so that $\lim C_n \overset{def}{=} C = \bigcup_{n\geq 1} C_n$.

Then $C = \Omega$. If not, there exists $\omega \in \Omega$ but $\omega \notin C$ if and only if $\omega \in \bigcap_{n\geq 1} C_n^c$, which implies $X_n(\omega) \leq Y(\omega) - \varepsilon$ for every n, and hence $X(\omega) \leq Y(\omega) - \varepsilon$, a contradiction.

Next,

$$I(X_n) = \sum_{i=1}^{r_n} \alpha_{ni}\mu(A_{ni}) \geq \sum_{i=1}^{r_n} \alpha_{ni}\mu(A_{ni} \cap C_n),$$

since $A_{ni} \supseteq A_{ni} \cap C_n$.

Also $\mu(A_{ni} \cap C_n) = \sum_{j=1}^{s} \mu(A_{ni} \cap C_n \cap B_j)$. Hence $I(X_n) \geq \sum_{i=1}^{r_n}\sum_{j=1}^{s} \alpha_{ni} \times \mu(A_{ni} \cap C_n \cap B_j)$. Now, since $\omega \in (A_{ni} \cap C_n \cap B_j)$ implies $\alpha_{ni} > \beta_j - \varepsilon$, we get $I(X_n) \geq \sum_{i=1}^{r_n}\sum_{j=1}^{s}(\beta_j - \varepsilon)\mu(A_{ni} \cap C_n \cap B_j)$, and this is equal to: $\sum_{j=1}^{s}(\beta_j - \varepsilon)\mu(C_n \cap B_j)$; i.e., we have

$$I(X_n) \geq \sum_{j=1}^{s}(\beta_j - \varepsilon)\mu(C_n \cap B_j). \tag{4.2}$$

(Notice that up to now no use of the finiteness of μ has been made.)

Now, $C_n \cap B_j = B_j - B_j \cap C_n^c$. Hence $\mu(C_n \cap B_j) = \mu(B_j) - \mu(C_n^c \cap B_j)$, since $C_n^c \cap B_j \subseteq B_j$ and μ is finite. Thus

$$\begin{aligned} I(X_n) &\geq \sum_{j=1}^{s} (\beta_j - \varepsilon)[\mu(B_j) - \mu(C_n^c \cap B_j)] \\ &= \sum_{j=1}^{s} \beta_j \mu(B_j) - \varepsilon \sum_{j=1}^{s} \mu(B_j) - \sum_{j=1}^{s} (\beta_j - \varepsilon)\mu(C_n^c \cap B_j) \\ &= I(Y) - \varepsilon\mu(\Omega) - \sum_{j=1}^{s} (\beta_j - \varepsilon)\mu(C_n^c \cap B_j) \\ &\geq I(Y) - \varepsilon\mu(\Omega) - (\max Y) \sum_{j=1}^{s} \mu(C_n^c \cap B_j) \\ &= I(Y) - \varepsilon\mu(\Omega) - (\max Y)\mu(C_n^c); \text{ i.e.,} \\ I(X_n) &\geq I(Y) - \varepsilon\mu(\Omega) - (\max Y)\mu(C_n^c). \end{aligned} \tag{4.3}$$

Now, $C_n \uparrow \Omega$ if and only if $C_n^c \downarrow \oslash$, which implies (since μ is finite) $\mu(C_n^c) \downarrow \mu(\oslash) = 0$. Therefore, letting $n \to \infty$ in (4.3), we get

$$\lim I(X_n) \geq I(Y) - \varepsilon\mu(\Omega).$$

Now letting $\varepsilon \to 0$, we get $I(Y) \leq \lim I(X_n)$, which is (4.1).

Case 2. $\mu(\Omega) = \infty$, $\min Y \overset{def}{=} m > 0$.

Then $I(X_n) \geq \sum_{j=1}^{s} (\beta_j - \varepsilon)\mu(C_n \cap B_j)$ according to (4.2), where no use of the finiteness of μ was made. Now,

$$\begin{aligned} \sum_{j=1}^{s} (\beta_j - \varepsilon)\mu(C_n \cap B_j) &\geq \sum_{j=1}^{s} (m - \varepsilon)\mu(C_n \cap B_j) \\ &= (m - \varepsilon) \sum_{j=1}^{s} \mu(C_n \cap B_j) \\ &= (m - \varepsilon)\mu(C_n); \text{ i.e.,} \\ I(X_n) &\geq (m - \varepsilon)\mu(C_n). \end{aligned} \tag{4.4}$$

Since $\mu(C_n) \uparrow \mu(\Omega) = \infty$ and $m - \varepsilon > 0$, we get, by taking the limits in (4.4) as $n \to \infty$: $\lim I(X_n) = \infty$ and hence $I(Y) \leq \lim I(X_n)$, which is (4.1).

Case 3. $\mu(\Omega) \leq \infty$, $\min Y = m = 0$. Recall that $0 \leq Y = \sum_{j=1}^{s} \beta_j I_{B_j} \leq X$, and we wish to show that for any $0 \leq \sum_{i=1}^{r_n} \alpha_{ni} I_{A_{ni}} = X_n \uparrow X$, we have $I(Y) \leq \lim I(X_n)$.

If $A = (Y > 0)$, then $\min_A Y \overset{def}{=} m_A > 0$, whereas $A = \sum B_{j_k}$ where the summation is over those j_ks for which the corresponding β_{j_k}s are > 0, and of course,

$A^c = \sum B_{j_l}$, where the summation is over those j_ls for which the corresponding β_{j_l}s are $= 0$. Let $X^{(A)}$ be the restriction of X to A, and set

$$Y^{(A)} = \sum_{j=1}^{s} \beta_j I_{B_j \cap A}, \quad X_n^{(A)} = \sum_{i=1}^{r_n} \alpha_{ni} I_{A_{ni} \cap A}.$$

Then, clearly, $0 \leq X_n^{(A)}$ simple r.v.s $\uparrow X^{(A)}$, since $X_n^{(A)} = X_n$ and $X^{(A)} = X$ on A, and $X_n \uparrow X$. Also, $Y^{(A)} \leq X^{(A)}$ because $Y^{(A)} = Y$ and $X^{(A)} = X$ on A, and $Y \leq X$. Since $m_A = \min Y^{(A)} > 0$, case 1 or case 2 applies (depending on whether $\mu(\Omega) < \infty$ or $\mu(\Omega) = \infty$, respectively) and gives that $I(Y^{(A)}) \leq \lim I(X_n^{(A)})$. However,

$$Y = \sum_j \beta_j I_{B_j} = \sum \beta_{j_k} I_{B_{j_k}} + \sum \beta_{j_l} I_{B_{j_l}} = \sum \beta_{j_k} I_{B_{j_k}},$$

so that

$$\begin{aligned} I(Y) &= \sum \beta_{j_k} \mu(B_{j_k}) = \sum \beta_{j_k} \mu(B_{j_k} \cap A), \text{ since } B_{j_k} \cap A = B_{j_k} \\ &\leq \sum_j \beta_j \mu(B_j \cap A) = I(Y^{(A)}); \text{ i.e., } I(Y) \leq I(Y^{(A)}). \end{aligned}$$

Also,

$$\begin{aligned} I(X_n^{(A)}) &= \sum_i \alpha_{ni} \mu(A_{ni} \cap A) \\ &\leq \sum_i \alpha_{ni} \mu(A_{ni}) = I(X_n), \end{aligned}$$

so that $\lim I(X_n^{(A)}) \leq \lim I(X_n)$. Combining the results obtained, we have then

$$I(Y) \leq I(Y^{(A)}) \leq \lim I(X_n^{(A)}) \leq \lim I(X_n),$$

as was to be seen. ■

So if X is a ≥ 0 r.v. then $I(X)$ is well defined (but it may be $= \infty$). Now let X be any r.v. Then $X = X^+ - X^-$, where $X^+, X^- \geq 0$. Thus $I(X^+), I(X^-)$ are well defined. If not both $I(X^+), I(X^-)$ are $= \infty$, we define $I(X)$ by: $I(X) = I(X^+) - I(X^-)$. Clearly, if $X \geq 0$ then $X = X^+$ and $I(X) = I(X^+)$, since $I(0) = 0$; and if X is a nonnegative simple r.v., then $I(X)$ coincides with the $I(X)$ as defined at the beginning of this section, since we can take $X_n = X, n \geq 1$. Thus this I on the class of all r.v.s for which $I(X^+) - I(X^-)$ is defined is an extension of the I on the class of ≥ 0 r.v.s.

Definition 1. The possibly extended function I defined on the class of r.v.s X for which $I(X^+) - I(X^-)$ exists is called the *integral of X over the space Ω with respect to the measure μ* and is denoted by $\int_\Omega X d\mu$ or $\int X\, d\mu$, or $\int X$ if no confusion is possible. In particular, if μ is a probability measure P, then $\int X\, d\mu = \int X dP$ is denoted by $\mathcal{E}X$ and is called the (mathematical) *expectation* of X. ■

Thus if $X = \sum_{i=1}^{n} \alpha_i I_{A_i} \geq 0$ then $\int X\, d\mu = \sum_{i=1}^{n} \alpha_i \mu(A_i)$; if the r.v. X is ≥ 0 but not necessarily a simple r.v., then $\int X\, d\mu = \lim_{n\to\infty} \int X_n\, d\mu$ for any $0 \leq X_n$ simple r.v. $\uparrow X$ as $n \to \infty$; and for any r.v. X, $\int X\, d\mu = \int X^+\, d\mu - \int X^-\, d\mu$, provided that at least one of $\int X^+\, d\mu$, $\int X^-\, d\mu$ is $< \infty$.

If now $A \in \mathcal{A}$, then I_A is measurable, and hence so is XI_A for any r.v. Then,

Definition 2. The *integral of the r.v. X over* $A \in \mathcal{A}$, denoted by $\int_A X\, d\mu$, is defined by $\int_\Omega (XI_A) d\mu = \int (XI_A) d\mu$, provided this latter integral exists. ■

Definition 3. We say that X is *integrable* if both $\int X^+$, $\int X^-$ are $< \infty$; i.e., $\int X$ exists and is finite. ■

4.2 Basic Properties of the Integral

In this section, we discuss the basic properties of the integral stated in the form of theorems.

Theorem 4. If X and Y are ≥ 0 r.v.s, then $\int (X+Y) = \int X + \int Y$. ■

Proof. First, let X and Y be ≥ 0 simple r.v.s. That is, $X = \sum_{i=1}^{m} \alpha_i I_{A_i} = \sum_{i=1}^{m} \sum_{j=1}^{n} \alpha_i I_{A_i \cap B_j}$ and $Y = \sum_{j=1}^{n} \beta_j I_{B_j} = \sum_{j=1}^{n} \sum_{i=1}^{m} \beta_j I_{A_i \cap B_j}$, $\alpha_i \geq 0$, $i = 1, \ldots, m$, $\beta_j \geq 0$, $j = 1, \ldots, n$.

Then $X + Y = \sum_{i=1}^{m} \sum_{j=1}^{n} (\alpha_i + \beta_j) I_{A_i \cap B_j}$, and $\int X = \sum_{i=1}^{m} \alpha_i \mu(A_i)$, $\int Y = \sum_{j=1}^{n} \beta_j \mu(B_j)$, and

$$\int (X+Y) = \sum_{i=1}^{m} \sum_{j=1}^{n} (\alpha_i + \beta_j) \mu(A_i \cap B_j) = \sum_{i=1}^{m} \sum_{j=1}^{n} \alpha_i \mu(A_i \cap B_j)$$

$$+ \sum_{i=1}^{m} \sum_{j=1}^{n} \beta_j \mu(A_i \cap B_j) = \sum_{i=1}^{m} \alpha_i \sum_{j=1}^{n} \mu(A_i \cap B_j)$$

$$+ \sum_{j=1}^{n} \beta_j \sum_{i=1}^{m} \mu(A_i \cap B_j) = \sum_{i=1}^{m} \alpha_i \mu(A_i) + \sum_{j=1}^{n} \beta_j \mu(B_j) = \int X + \int Y.$$

Next, let X_n, Y_n be ≥ 0 simple r.v.s such that $X_n \uparrow X$, $Y_n \uparrow Y$. Then, clearly, $X_n + Y_n$ are ≥ 0 simple r.v.s such that $(X_n + Y_n) \uparrow (X + Y)$. Let $X_n = \sum_{j=1}^{r_n} \alpha_{ni} I_{A_{ni}}$, $Y_n = \sum_{j=1}^{s_n} \beta_{nj} I_{B_{nj}}$. Then $X_n + Y_n = \sum_{i=1}^{r_n} \sum_{j=1}^{s_n} (\alpha_{ni} + \beta_{nj}) I_{A_{ni} \cap B_{nj}}$, and $\int (X_n + Y_n) = \int X_n + \int Y_n$, as was just seen for the nonnegative simple r.v.s X and Y; i.e., $\int (X_n + Y_n) = \int X_n + \int Y_n$.

Taking the limits, as $n \to \infty$, we then get $\int (X+Y) = \int X + \int Y$. ■

Remark 2. The theorem generalizes in an obvious manner to any finite number of nonnegative r.v.s.

Theorem 5. Let $A, B \in \mathcal{A}$ with $A \cap B = \emptyset$. Then,

(i) If $\int_{A+B} X$ exists, then $\int_A X$, $\int_B X$ also exist and $\int_{A+B} X = \int_A X + \int_B X$.

(ii) If $\int_A X$, $\int_B X$ exist, and $\int_A X + \int_B X$ also exist (i.e., it is not of the form $\pm\infty \mp \infty$), then $\int_{A+B} X$ also exists and $\int_{A+B} X = \int_A X + \int_B X$. ■

Proof. (i) Clearly

$$(XI_{A+B})^+ = X^+ I_{A+B} = X^+ I_A + X^+ I_B = (XI_A)^+ + (XI_B)^+,$$

since $A \cap B = \oslash$. So

$$\begin{aligned}(XI_{A+B})^+ &= (XI_A)^+ + (XI_B)^+ \text{ and similarly,}\\ (XI_{A+B})^- &= (XI_A)^- + (XI_B)^-.\end{aligned} \tag{4.5}$$

Now that $\int_{A+B} X = \int XI_{A+B}$ exists implies that at least one of $\int (XI_{A+B})^+$, $\int (XI_{A+B})^-$ is $< \infty$. Let $\int (XI_{A+B})^+ < \infty$. By (4.5) and Theorem 4, we have $\int (XI_{A+B})^+ = \int (XI_A)^+ + \int (XI_B)^+$ and hence both $\int (XI_A)^+$, $\int (XI_B)^+$ are $< \infty$. Thus $\int_A X$, $\int_B X$ exist. Again by (4.5) and Theorem 4, $\int (XI_{A+B})^- = \int (XI_A)^- + \int (XI_B)^-$. Thus

$$\begin{aligned}\int_{A+B} X &= \int XI_{A+B} = \int (XI_{A+B})^+ - \int (XI_{A+B})^-\\ &= \left[\int (XI_A)^+ + \int (XI_B)^+\right] - \left[\int (XI_A)^- + \int (XI_B)^-\right]\\ &= \left[\int (XI_A)^+ - \int (XI_A)^-\right] + \left[\int (XI_B)^+ - \int (XI_B)^-\right]\\ &= \int XI_A + \int XI_B = \int_A X + \int_B X; \text{ i.e., } \int_{A+B} X = \int_A X + \int_B X.\end{aligned}$$

The same result follows if we assume that $\int (XI_{A+B})^- < \infty$.

(ii) Now, let $\int_A X$, $\int_B X$, $\int_A X + \int_B X$ exist. Then either $\int (XI_A)^+$ or $\int (XI_A)^-$ is $< \infty$, or both.

Also, either $\int (XI_B)^+$ or $\int (XI_B)^-$ is $< \infty$, or both. From the existence of $\int_A X + \int_B X$ we then have

$$\int_A X + \int_B X = \int (XI_A)^+ - \int (XI_A)^- + \int (XI_B)^+ - \int (XI_B)^-.$$

Set

$$\int (XI_A)^+ = a^+, \quad \int (XI_A)^- = a^-,$$
$$\int (XI_B)^+ = b^+, \quad \int (XI_B)^- = b^-.$$

Then

$$\int_A X + \int_B X = a^+ - a^- + b^+ - b^-,$$

and its existence means only the following cases may occur:

$$\begin{cases} a^{\pm}, b^{\pm} \text{ are all } < \infty \\ a^{+} \text{ and/or } b^{+} = \infty \text{ but } a^{-} \text{ and/or } b^{-} \text{ finite} \\ a^{-} \text{ and/or } b^{-} = \infty \text{ but } a^{+} \text{ and/or } b^{+} \text{ finite} \end{cases}.$$

For any of these cases, the grouping $(a^{+} + b^{+}) - (a^{-} + b^{-})$ is legitimate, and this is equal to $\int_{A+B} X^{+} - \int_{A+B} X^{-} = \int_{A+B} X$. Thus $\int_{A+B} X$ exists and $\int_{A+B} X = \int_A X + \int_B X$. ■

Remark 3. The theorem generalizes in an obvious manner to any finite number of pairwise disjoint events. (see also Exercise 5.)

Corollary. If $\int X$ exists, then for any $A \in \mathcal{A}$, $\int_A X$ exists and $\int X = \int_A X + \int_{A^c} X$. Furthermore, if X is integrable over Ω, then X is integrable over any $A \in \mathcal{A}$.

Proof. The first part of the corollary follows from Theorem 5(i) with $A = A$, $B = A^c$. As for the second part, we have: the fact that $\int X$ is finite implies that $\int X^{+}, \int X^{-} < \infty$. Since $\int X^{+} = \int_{A+A^c} X^{+} = \int_A X^{+} + \int_{A^c} X^{+}$, $\int X^{-} = \int_{A+A^c} X^{-} = \int_A X^{-} + \int_{A^c} X^{-}$, we then get $\int_A X^{+}, \int_A X^{-} < \infty$, and hence X is integrable over A. ■

Theorem 6. If $\int X$ exists and $c \in \Re$, then $\int cX$ exists and $\int cX = c \int X$. ■

Proof. We distinguish the following cases:
Case 1. Let $c = -1$. Then $cX = -X$. But: $(-X)^{+} = X^{-}$, $(-X)^{-} = X^{+}$, so that

$$\int X = \int X^{+} - \int X^{-} = \int (-X)^{-} - \int (-X)^{+}$$
$$= -\left[\int (-X)^{+} - \int (-X)^{-}\right] = -\int (-X) \text{ or } \int -X = -\int X.$$

Case 2. Let $c \geq 0$. If $c = 0$, then $c \times X = 0$ implies $\int cX = \int 0 = 0 = 0 \int X$, since $0 \times (\pm\infty)$ is also 0 (for the case that $\int X = \pm\infty$). We assume now that $c > 0$. Then $(cX)^{+} = cX^{+}$, $(cX)^{-} = cX^{-}$, so that $\int cX = \int (cX)^{+} - \int (cX)^{-} = \int cX^{+} - \int cX^{-}$; i.e.,

$$\int cX = \int cX^{+} - \int cX^{-}, \tag{4.6}$$

provided it exists.

Since $X^{+} \geq 0$, there exist $0 \leq X_n$ simple r.v.s $\uparrow X^{+}$ and hence $0 \leq cX_n$ simple r.v.s $\uparrow cX^{+}$ as $n \to \infty$, which implies $\int cX_n \underset{n\to\infty}{\to} \int cX^{+}$. But $\int cX_n = c \int X_n$ for simple r.v.s, as is readily seen. Since also $\int X_n \underset{n\to\infty}{\to} \int X^{+}$, we then get $\int cX^{+} = c \int X^{+}$. In a similar way, $\int cX^{-} = c \int X^{-}$, and then (4.6) exists and gives the desired result.
Case 3. Let $c < 0$. Then $\int cX = \int (-1)[(-c)X] = -1 \int (-c)X$, by case 1, and this is $= -(-c) \int X$, by case 2, since $-c \geq 0$; i.e., $\int cX = c \int X$. ■

Theorem 7. If $X \geq 0$, then $\int X \geq 0$. ■

Proof. Let $0 \leq X_n$ simple r.v.s $\uparrow X$ as $n \to \infty$ whereas $\int X_n \geq 0$ by Theorem 2. Since also $\int X_n \underset{n\to\infty}{\to} \int X$, we get then $\int X \geq 0$. ■

Theorem 8.

(i) If $X \leq Y$ and $\int X$, $\int Y$ exist, then $\int X \leq \int Y$.
(ii) If $X \leq Y$ and $\int X^- < \infty$ (so that $\int X$ exists), then $\int Y$ exists and $\int X \leq \int Y$ by part (i).
(iii) If $X \leq Y$ and $\int Y^+ < \infty$ (so that $\int Y$ exists), then $\int X$ exists and $\int X \leq \int Y$ by part (i).
(iv) If $\int X$ exists, then $\left|\int X\right| \leq \int |X|$. ■

Proof.

(i) Let $0 \leq X \leq Y$. Then $Y = X + (Y - X) = X + Z$, where $Z = Y - X \geq 0$. Hence $\int Y = \int X + \int Z$ by Theorem 4. Also, $\int Z \geq 0$ by Theorem 7. Thus $\int X \leq \int Y$. In general, $X \leq Y$ implies $X^+ \leq Y^+$, $X^- \geq Y^-$, clearly. Then $\int X^+ \leq \int Y^+$, $\int X^- \geq \int Y^-$ or $-\int X^- \leq -\int Y^-$, by what we have just established. Therefore $\int X^+ - \int X^- \leq \int Y^+ - \int Y^-$ or $\int X \leq \int Y$, since $\int X$, $\int Y$ exist.
(ii) From $\int X^- \geq \int Y^-$ above, we get that $\int X^- < \infty$, which implies $\int Y^- < \infty$, so that $\int Y$ exists and $\int X \leq \int Y$ by part (i).
(iii) From $\int X^+ \leq \int Y^+$ above, we get that $\int Y^+ < \infty$ so that $\int X^+ < \infty$, and hence $\int X$ exists and $\int X \leq \int Y$ by part (i).
(iv) Indeed, $-|X| \leq X \leq |X|$ implies $-\int |X| \leq \int X \leq \int |X|$ or $\left|\int X\right| \leq \int |X|$. ■

Theorem 9. If $X = Y$ a.e. and $\int X$ exists, then $\int Y$ exists and $\int X = \int Y$. ■

Remark 4. We first observe that if $\mu(B) = 0$, for some $B \in \mathcal{A}$, then $\int_B Z = 0$, for any r.v. Z. In fact, $Z^+ \geq 0$ implies that there exist $0 \leq Z_n I_B$ simple r.v.s $\uparrow Z^+ I_B$ as $n \to \infty$; hence $\int_B Z_n = \int Z_n I_B \underset{n\to\infty}{\to} \int Z^+ I_B = \int_B Z^+$. But $\int_B Z_n = 0$, because, if $Z_n = \sum_{j=1}^{r_n} \alpha_{nj} A_{nj}$, then $Z_n I_B = \sum_{j=1}^{r_n} \alpha_{nj} I_{B\cap A_{nj}}$ and $\int_B Z_n = \int Z_n I_B = \sum_{j=1}^{r_n} \alpha_{nj} \mu(B \cap A_{nj}) = 0$. This implies that $\int_B Z^+ = 0$. Similarly, $\int_B Z^- = 0$, so that $\int_B Z = 0$.

Proof of Theorem 9. Let $A = (X = Y)$. Then $A = (X - Y = 0)$ and A is measurable, since $A = (X - Y)^{-1}(\{0\})$ and $X - Y$ is a r.v.

Now, the existence of $\int X$ implies the existence of $\int_A X$, $\int_{A^c} X$ and $\int X = \int_A X + \int_{A^c} X$, by the Corollary to Theorem 5. But

$$\int_A X + \int_{A^c} X = \int X I_A + \int_{A^c} X = \int Y I_A + 0$$
$$= \int_A Y = \int_A Y + 0 = \int_A Y + \int_{A^c} Y$$

(by Remark 4), and this is equal to $\int Y$ by Theorem 5(ii). So $\int Y$ exists and $\int Y = \int X$. ■

Theorem 10. X is integrable if and only if $|X|$ is integrable. ■

Proof. Now X being integrable implies $\int X^+, \int X^- < \infty$. Since $|X| = X^+ + X^-$, we get $\int |X| = \int X^+ + \int X^-$ by Theorem 4, and then $\int |X| < \infty$; i.e., $|X|$ is integrable. Next, let $|X|$ be integrable. We have $|X| = X^+ + X^-$ and $\int |X| = \int X^+ + \int X^-$ by Theorem 4, and $\int |X| < \infty$ which implies $\int X^+ < \infty$ and $\int X^- < \infty$. Thus X is integrable. ■

Theorem 11. If $|X| \leq Y$ and $\int Y < \infty$, then X is integrable. ■

Proof. $|X| \leq Y$ implies $\int |X| \leq \int Y$ by Theorem 8. But $\int Y < \infty$; thus $\int |X| < \infty$ if and only if $\int X$ is finite, by Theorem 10. ■

Theorem 12. If $\int X, \int Y$, and $\int X + \int Y$ exist, then $\int (X+Y)$ exists and $\int (X+Y) = \int X + \int Y$. ■

Proof. The existence of $\int X + \int Y$ implies that if $\int X = \infty$, then $-\infty < \int Y \leq \infty$, and if $\int X = -\infty$, then $\infty > \int Y \geq -\infty$. So if $\int X = \int Y = \pm\infty$, then $\int X + \int Y = \pm\infty$, and, by means of simple r.v.s, it is seen (see also Exercise 6) that $\int (X+Y) = \pm\infty$. So $\int (X+Y) = \int X + \int Y$ in this case.

Next assume that at least one of $\int X, \int Y$ is finite and, in order to fix ideas, let $\int Y$ be finite. This implies that $\int_A Y$ is finite for any $A \in \mathcal{A}$ by the Corollary to Theorem 5.

Now, we consider the following partition of Ω by the sets A_j, $j = 1, \ldots, 6$, defined as follows:

$$\begin{aligned} A_1 &= \left(X \geq 0,\ Y \geq 0,\ X+Y \geq 0\right) \quad A_4 = \left(X < 0,\ Y \geq 0,\ X+Y \geq 0\right), \\ A_2 &= \left(X \geq 0,\ Y < 0,\ X+Y \geq 0\right); \quad A_5 = \left(X < 0,\ Y \geq 0,\ X+Y < 0\right), \\ A_3 &= \left(X \geq 0,\ Y < 0,\ X+Y < 0\right) \quad A_6 = \left(X < 0,\ Y < 0,\ X+Y < 0\right). \end{aligned}$$

The existence of $\int X, \int Y$ implies the existence of $\int_{A_j} X, \int_{A_j} Y$, $j = 1, \ldots, 6$, by the Corollary to Theorem 5. We will prove that $\int_{A_j} (X+Y) = \int_{A_j} X + \int_{A_j} Y$, $j = 1, \ldots, 6$, which will imply that

$$\begin{aligned} \int (X+Y) &= \int_{\sum_j A_j} (X+Y) \\ &= \sum_j \int_{A_j} (X+Y) \\ &= \sum_j \left(\int_{A_j} X + \int_{A_j} Y \right) \\ &= \sum_j \int_{A_j} X + \sum_j \int_{A_j} Y \end{aligned}$$

$$= \int_{\sum_j A_j} X + \int_{\sum_j A_j} Y$$
$$= \int X + \int Y$$

by Remark 3 and Exercise 5.

We have

A_1: $(X+Y)I_{A_1} = XI_{A_1} + YI_{A_1}$ with $XI_{A_1}, YI_{A_1} \geq 0$. Then $\int_{A_1}(X+Y) = \int(XI_{A_1}+YI_{A_1}) = \int XI_{A_1} + \int YI_{A_1}$, by Theorem 4, and this is $= \int_{A_1} X + \int_{A_1} Y$.

A_2: $XI_{A_2} = (X+Y)I_{A_2} + (-Y)I_{A_2}$ with $(X+Y)I_{A_2}, (-Y)I_{A_2} \geq 0$. Then, by Theorem 4, $\int_{A_2} X = \int XI_{A_2} = \int[(X+Y)I_{A_2} + (-Y)I_{A_2}] = \int(X+Y)I_{A_2} + \int -YI_{A_2} = \int_{A_2}(X+Y) - \int_{A_2} Y$. Since $\int_{A_2} Y$ is finite (by the Corollary to Theorem 5), we get $\int_{A_2}(X+Y) = \int_{A_2} X + \int_{A_2} Y$.

A_3: $-YI_{A_3} = -(X+Y)I_{A_3} + XI_{A_3}$ with $-(X+Y)I_{A_3}, XI_{A_3} \geq 0$. Then, by Theorem 6, $-\int_{A_3} Y = -\int_{A_3}(X+Y) + \int_{A_3} X$ or $\int_{A_3} Y = \int_{A_3}(X+Y) - \int_{A_3} X$, and since $\int_{A_3} Y$ is finite, so is $\int_{A_3} X$. Then we get $\int_{A_3}(X+Y) = \int_{A_3} X + \int_{A_3} Y$.

A_4: $YI_{A_4} = (X+Y)I_{A_4} + (-X)I_{A_4}$ and $\int_{A_4} Y = \int_{A_4}(X+Y) + \int_{A_4} -X = \int_{A_4}(X+Y) - \int_{A_4} X$. Since $\int_{A_4} Y$ is finite, so is $\int_{A_4} X$ and $\int_{A_4}(X+Y) = \int_{A_4} X + \int_{A_4} Y$.

A_5: $-XI_{A_5} = YI_{A_5} + (-X-Y)I_{A_5}$ and $-\int_{A_5} X = \int_{A_5} Y - \int_{A_5} X+Y$ or $-\int_{A_5} X - \int_{A_5} Y = -\int_{A_5} X+Y$, or $\int_{A_5} X + \int_{A_5} Y = \int_{A_5} X+Y$.

A_6: $-(X+Y)I_{A_6} = -XI_{A_6} - YI_{A_6}$ and $-\int_{A_6} X+Y = -\int_{A_6} X - \int_{A_6} Y$ or $\int_{A_6} X+Y = \int_{A_6} X + \int_{A_6} Y$. ■

The interested reader may find a rather extensive treatment of integration in the reference Vestrup (2003).

Here is a brief description of the results obtained in Sections 4.1 and 4.2.

For a simple r.v. X, the quantity $I(X)$ was defined, to be called eventually the integral of X.

For two nonnegative simple r.v.s X and Y with $X \geq Y$, it follows that $I(X) \geq I(Y)$; and $I(X) > I(Y)$ if and only if $X > Y$ on a set of positive μ-measure in the case that μ is finite.

For a nonnegative r.v. X, the quantity $I(X)$ is defined, and it is shown that it is well defined.

For any r.v. X, define $I(X)$, provided it exists, and set $I(X) = \int_\Omega X d\mu$ or just $\int X$. Also, for $A \in \mathcal{A}$, define $\int_A X d\mu$ or just $\int_A X$, and also define integrability of a r.v. X.

For two nonnegative r.v.s X, Y, it holds $\int(X+Y) = \int X + \int Y$.

For A, B in $\mathcal{A}$ with $A \cap B = \varnothing$, and any r.v. X for which $\int_{A+B} X$ exists, it follows that $\int_A X$, $\int_B X$ also exist and $\int_{A+B} X = \int_A X + \int_B X$. Furthermore, if the assumption is that all three $\int_A X$, $\int_B X$, and $\int_A X + \int_B X$ exist, then $\int_{A+B} X$ also exists and $\int_{A+B} X = \int_A X + \int_B X$.

If $\int X$ exists, then so does $\int_A X$ for every $A \in \mathcal{A}$. Furthermore, if $\int X$ is finite, so is $\int_A X$ for every $A \in \mathcal{A}$.

If $\int X$ exists, then $\int cX$ exists for every $c \in \Re$, and $\int cX = c \int X$.

$X \geq 0$ implies $\int X \geq 0$.

If $X \leq Y$ and $\int X$ and $\int Y$ exist, then $\int X \leq \int Y$; if $X \leq Y$ and $\int X^- < \infty$, then $\int X$ exists and $\int X \leq \int Y$; if $X \leq Y$ and $\int Y^+ < \infty$, then $\int X$ exists and $\int X \leq \int Y$.

If $X = Y$ a.s. and $\int X$ exists, then $\int Y$ also exists and $\int X = \int Y$.

$\left|\int X\right| < \infty$ if and only if $\int |X| < \infty$.

If $|X| \leq Y$ and $\int Y < \infty$, then X is integrable.

If all three $\int X$, $\int Y$, and $\int X + \int Y$ exist, then $\int (X + Y)$ exists and $\int (X + Y) = \int X + \int Y$.

4.3 Probability Distributions

In this short section, we consider a r.v. X defined on the probability space $(\Omega, \mathcal{A}, P)$, and on $\mathcal{B}$, define the set function P_X as follows: $P_X(B) = P(X^{-1}(B)) = P(X \in B)$. Then it is easy to see that P_X is a probability measure on $\mathcal{B}$.

Definition 4. The set function P_X on $\mathcal{B}$ associated with the r.v. X is called the (*probability*) *distribution* of X. The point function $F_X : \Re \to [0, 1]$ defined by $F_X(x) = P_X((-\infty, x]) = P(X \leq x)$ is called the *distribution function* (d.f.) of X. (See Theorem 6 and the comments following it in Chapter 2.) ■

It is easily seen that F_X satisfies the following properties: (1) is nondecreasing; (2) is continuous from the right; (3) $F_X(+\infty) \stackrel{def}{=} \lim_{x\to\infty} F_X(x) = 1$; (4) $F_X(-\infty) \stackrel{def}{=} \lim_{x\to-\infty} F_X(x) = 0$. (See Theorem 6 and the comments following it in Chapter 2.)

Conversely, a function on $\Re$ into $[0, 1]$ that satisfies properties (1)–(4) uniquely defines a probability measure Q on $\mathcal{B}$ (by Theorem 7 in Chapter 2), and it is also true that there exists a r.v. X whose P_X is this distribution Q. The simplest example would be that of taking $(\Omega, \mathcal{A}, P) = (\Re, \mathcal{B}, Q)$ and let $X(\omega) = \omega$, $\omega \in \Re$. Then, clearly, Q is the distribution of X, P_X. Also, if $Y \sim U(0, 1)$ and $X = F^{-1}(Y)$, then $X \sim F$, where $F^{-1}(y) = \inf\{x \in \Re; F(x) \geq y\}$. (See, e.g., Theorem 7, Chapter 2 in Roussas, 1997.)

The following theorem is an important one in statistics, since it allows integration over the real line rather than over the abstract space Ω.

Theorem 13. Let $g : \Re \to \Re$ be measurable. Then we have

$$\mathcal{E}g(X) = \int_\Omega g(X)dP = \int_\Re g(x)dP_X,$$

also denoted by $\int_\Re g(x)dF_X$, in the sense that if one side exists, so does the other, and they are equal. (See also the Appendix regrading the notation $\int_\Re g(x)dF_X$.) ■

Proof. We use the familiar method of proving the theorem in several steps starting with indicator functions. Let $g = I_B$ some $B \in \mathcal{B}$. Then $g(X) = I_A(X)$, where

$A = X^{-1}(B)$. Then,

$$\int_\Omega g(X)dP = \int_\Omega I_A(X)dP = P(A) = P(X^{-1}(B)) = P_X(B)$$
$$= \int_\Re I_B(x)dP_X = \int_\Re g(x)dP_X.$$

Next, let g be a nonnegative simple function; i.e., $g(x) = \sum_{i=1}^n \alpha_i I_{B_i}(x)$ with $\alpha_i > 0, i = 1, \ldots, n$, where $\{B_1, \ldots, B_n\}$ is a (measurable) partition of $\Re$, and let $A_i = X^{-1}(B_i), i = 1, \ldots, n$. Then $g(X) = \sum_{i=1}^n \alpha_i I_{A_i}(X)$, and by linearity of the integral (see also Exercise 8) and the previous step,

$$\int_\Omega g(X)dP = \sum_{i=1}^n \alpha_i \int_\Omega I_{A_i}(X)dP = \sum_{i=1}^n \alpha_i \int_\Re I_{B_i}(x)dP_X$$
$$= \int_\Re \left(\sum_{i=1}^n \alpha_i I_{B_i}(x) \right) dP_X = \int_\Re g(x)dP_X.$$

Now, let g be nonnegative. Then there exist $0 \leq g_n(x)$ simple $\uparrow g(x)$; i.e., $g_n(x) = \sum_{i=1}^{r_n} \alpha_{ni} I_{B_{ni}}(x)$ ($\alpha_{ni} \geq 0$, $i = 1, \ldots, r_n$), which implies that $0 \leq g_n(X) = \sum_{i=1}^{r_n} \alpha_{ni} I_{A_{ni}}(X)$ simple $\uparrow g(X)$ as $n \to \infty$ where $A_{ni} = X^{-1}(B_{ni})$. Then $\int_\Omega g_n(X)dP \underset{n\to\infty}{\to} \int_\Omega g(X)dP$, $\int_\Re g_n(x)dP_X \underset{n\to\infty}{\to} \int_\Re g(x)dP_X$, whereas $\int_\Omega g_n(X)\, dP = \int_\Re g_n(x)dP_X$ for all n, by the previous step. Hence $\int_\Omega g(X)dP = \int_\Re g(x)dP_X$.

Finally, for any g, write $g(x) = g^+(x) - g^-(x)$, which implies $g(X) = g^+(X) - g^-(X)$. Now, if $\int_\Omega g(X)dP$ exists, it then follows that either

$$\int_\Omega g^+(X)dP < \infty \quad \text{or} \quad \int_\Omega g^-(X)dP < \infty$$

or both. Since $\int_\Omega g^+(X)dP = \int_\Re g^+(x)dP_X$ and $\int_\Omega g^-(X)dP = \int_\Re g^-(x)dP_X$, by the previous step, it follows that either

$$\int_\Re g^+(x)dP_X < \infty \text{ or } \int_\Re g^-(x)dP_X < \infty$$

or both, respectively. Thus, $\int_\Re g(x)dP_X$ exists and

$$\int_\Omega g(X)dP = \int_\Omega g^+(X)dP - \int_\Omega g^-(X)dP$$
$$= \int_\Re g^+(x)dP_X - \int_\Re g^-(x)dP_X$$
$$= \int_\Re g(x)dP_X.$$

Likewise, the existence of $\int_\Re g(x)dP_X$ implies the existence of $\int_\Omega g(X)dP$ and their equality. ■

Remark 5. The proof is exactly the same for k-dimensional random vectors. However, we do not intend to present details.

Exercises.

1. Construct an example where r.v.s X_n and X, defined on a probability space $(\Omega, \mathcal{A}, P)$, are such that $X_n \underset{n\to\infty}{\to} X$ pointwise (hence $X_n \overset{\text{a.s.}}{\underset{n\to\infty}{\to}} X$ and $X_n \overset{P}{\underset{n\to\infty}{\to}} X$), but $\mathcal{E}X_n \underset{n\to\infty}{\nrightarrow} \mathcal{E}X$.
2. If the r.v. $X \geq 0$ and $\int X d\mu = 0$, then show that $\mu(X \neq 0) = 0$.
3. If the r.v. X takes on the values $n, n = 1, 2, \ldots$, with probability 1, so that $\sum_{n=1}^{\infty} P(X = n) = 1$, show that $\mathcal{E}X = \sum_{n=1}^{\infty} P(X \geq n)$.
4. **(i)** For a r.v. X and any two disjoint events A and B, show that

$$(XI_{A+B})^+ = (XI_A)^+ + (XI_B)^+,$$
$$(XI_{A+B})^- = (XI_A)^- + (XI_B)^-.$$

   (ii) More generally, for any finite collection of pairwise disjoint events A_i, $i = 1, \ldots, n$, show that

$$\left(XI_{\sum_{i=1}^n A_i}\right)^+ = \sum_{i=1}^{n} (XI_{A_i})^+, \left(XI_{\sum_{i=1}^n A_i}\right)^- = \sum_{i=1}^{n} (XI_{A_i})^-.$$

5. If for the pairwise disjoint events $A_i, i = 1, \ldots, n$, the integral $\int_{\sum_{i=1}^n A_i} X$ exists, then the integrals $\int_{A_i} X, i = 1, \ldots, n$, exist, and $\int_{\sum_{i=1}^n A_i} X = \sum_{i=1}^{n} \int_{A_i} X$.
6. Let X and Y be two simple r.v.s such that $\int X = \int Y = \infty$ or $\int X = \int Y = -\infty$. Then show that $\int(X + Y)$ exists and $\int(X + Y) = \int X + \int Y (= \infty$ or $-\infty)$.
7. Let μ_1 and μ_2 be two measures on $(\Omega, \mathcal{A})$ and suppose that $\int Xd(\mu_1 + \mu_2)$ exists. Then $\int Xd\mu_i$ exist, $i = 1, 2$, and $\int Xd(\mu_1 + \mu_2) = \int Xd\mu_1 + \int Xd\mu_2$.
8. If the r.v.s $X_i, i = 1, \ldots, n$, are integrable, then so is the r.v. $\sum_{i=1}^n X_i$ and $\int \sum_{i=1}^n X_i = \sum_{i=1}^n \int X_i$.
9. Let X and Y be r.v.s with finite second moments, and set $\mathcal{E}X = \mu_1, \mathcal{E}Y = \mu_2, 0 < Var(X) = \sigma_1^2, 0 < Var(Y) = \sigma_2^2$. Then the covariance and the correlation coefficient of X and Y, denoted respectively, by $Cov(X, Y)$ and $\rho(X, Y)$, are defined by: $Cov(X, Y) = \mathcal{E}[(X-\mu_1)(Y-\mu_2)] = \mathcal{E}(XY) - \mu_1\mu_2$ and $\rho(X, Y) = Cov(X, Y)/\sigma_1\sigma_2$.

   (i) Then show that $-\sigma_1\sigma_2 \leq Cov(X, Y) \leq \sigma_1\sigma_2$, and $Cov(X, Y) = \sigma_1\sigma_2$ if and only if $P\left[Y = \mu_2 + \frac{\sigma_2}{\sigma_1}(X - \mu_1)\right] = 1$, and $Cov(X, Y) = -\sigma_1\sigma_2$ if and only if $P\left[Y = \mu_2 - \frac{\sigma_2}{\sigma_1}(X - \mu_1)\right] = 1$.

   (ii) Also, $-1 \leq \rho(X, Y) \leq 1$, and $\rho(X, Y) = 1$ if and only if

$$P\left[Y = \mu_2 + \frac{\sigma_2}{\sigma_1}(X - \mu_1)\right] = 1,$$

and $\rho(X, Y) = -1$ if and only if $P\left[Y = \mu_2 - \frac{\sigma_2}{\sigma_1}(X - \mu_1)\right] = 1$.

Hint: Use Exercise 2.

10. Let $\Omega = \{-2, -1, 3, 7\}$, and let the measure μ be defined by: $\mu(\{-2\}) = 2$, $\mu(\{-1\}) = 1$, $\mu(\{3\}) = 3$, $\mu(\{7\}) = 7$. Let X be a r.v. defined by: $X(-2) = X(-1) = -1$, and $X(3) = X(7) = 1$. Then, for $A = \{-2, 3, 7\}$, compute the integral $\int_A X d\mu$.

11. Let $(\Omega, \mathcal{A}, \mu) = ((-5, 5), \mathcal{B}_{(-5,5)}, \lambda)$ where λ is the Lebesgue measure, and let the r.v. X be defined as follows:

$$X(\omega) = \begin{cases} \frac{1}{2}, & \omega \in (-5, 2) \\ \frac{1}{3}, & \omega = 2 \\ 1, & \omega \in (2, 3] \\ 0, & \omega \in (3, 5). \end{cases}$$

Then, for $A = [-1, 4]$, compute the integral $\int_A X d\mu$.

12. Let $\Omega = \{0, 1, 2, \ldots\}$, let $\mathcal{A}$ be the discrete σ-field of subsets of Ω, and on $\mathcal{A}$, define the function μ by: $\mu(A) =$ number of nonnegative integers in A. Then show that μ is a measure, and indeed, a σ-finite measure.

13. Let $g : \Re \to (0, \infty)$ be nondecreasing in $(0, \infty)$ and symmetric about 0 ($g(-x) = g(x)$, $x \in \Re$), and let X be a r.v. such that $\mathcal{E}g(X) < \infty$. Then show that: $P(|X| \geq c) \leq \mathcal{E}g(X)/g(c)$ for every $c > 0$.

14. For a r.v. X, show that

$$P(|X| \geq c) \leq \left(1 + \frac{1}{c}\right) \mathcal{E}\left(\frac{|X|}{1 + |X|}\right) \quad \text{for every } c > 0.$$

Hint: Use Exercise 13 above.

15. Let X be a simple r.v. defined on $(\Omega, \mathcal{A})$, so that (see Definition 13 in Chapter 1) $X = \sum_{j=1}^{n} \alpha_j I_{A_j}$, where $\{A_j, j = 1, \ldots, n\}$ is a (measurable) partition of Ω, and the α_j, $j = 1, \ldots, n$, are assumed to be distinct. Next, let $X = \sum_{i=1}^{m} \beta_i I_{B_i}$ be any other representation of X. Then show that each A_j, $j = 1, \ldots, n$, is the sum of some B_i's.

Remark 6. Under the assumption that the α_j, $j = 1, \ldots, n$, are distinct, the partition $\{A_j,\ j = 1, \ldots, n\}$ may be called *irreducible*. For such a partition, there is a unique representation of the respective simple r.v. and consequently, Theorem 1 is rendered superfluous.

16. If X and Y are two identically distributed integrable r.v.s, then

$$\mathcal{E}\left[X I_{(|X| \leq c)}\right] = \mathcal{E}\left[Y I_{(|Y| \leq c)}\right]$$

for any constant c.

This page is intentionally left blank

CHAPTER

Standard Convergence Theorems, The Fubini Theorem

5

This chapter consists of two sections. In the first section, we discuss the standard convergence theorems, such as the Lebesgue Monotone Convergence Theorem, the Fatou–Lebesgue Theorem, and the Dominated Convergence Theorem. The Lebesgue Monotone Convergence Theorem and the Dominated Convergence Theorem provide conditions under which, from the limit of an integral, one can pass to the integral of the limit. The Fatou–Lebesgue Theorem deals with inequalities, involving lim inf and lim sup, rather than equalities as was the case in the former two theorems. As an application, two theorems are discussed, on the basis of which the interchange of the operations of differentiation and integration is valid. This section is concluded with the introduction of the concept of convergence in distribution. Its relation to convergence in probability is mentioned, and a result, particularly useful in Statistics, is stated, involving convergence in distribution of two sequences of r.v.s.

The purpose of the second section is to establish the Fubini Theorem. This result gives conditions under which the interchange of the order of integration, involving a double integral, is valid. A number of auxiliary results have to be established first, and these include the so-called Product Measure Theorem.

5.1 Standard Convergence Theorems and Some of Their Ramifications

Theorem 1 (Lebesgue Monotone Convergence Theorem; interchange of the order of lim and $\int$ for ≥ 0 nondecreasing sequences of r.v.s). If $0 \leq X_n \uparrow X$ as $n \to \infty$, where X may be an extended r.v., then $\int X_n \uparrow \int X$ as $n \to \infty$ (i.e., $\lim\limits_{n\to\infty} \int X_n = \int \lim\limits_{n\to\infty} X_n$). ■

Proof. In the sequel all limits are taken as $n \to \infty$ unlesss otherwise specified. $0 \leq X_n \uparrow X$ implies that $\int X_n$, $\int X$ exist, and $X_n \uparrow$ implies that $\int X_n \uparrow$. So, to show $\int X_n \uparrow \int X$, let $0 \leq X_{nk}$ simple r.v.s $\uparrow X_n$, $k \to \infty$, and define the r.v.s Y_k by $Y_k = \max\limits_{1\leq n\leq k} X_{nk}$, $k = 1, 2, \ldots$:

$$\begin{array}{ccccccc} X_{11} & X_{12} & \cdots & X_{1k} & \cdots & \uparrow X_1 \\ X_{21} & X_{22} & \cdots & X_{2k} & \cdots & \uparrow X_2 \\ \cdot & \cdot & \cdot & \cdot & \cdot & \cdot \end{array}$$

An Introduction to Measure-Theoretic Probability, Second Edition. http://dx.doi.org/10.1016/B978-0-12-800042-7.00005-0

$$\begin{array}{cccccc} X_{k1} & X_{k2} & \cdots & X_{kk} & \cdots & \uparrow X_k \\ \cdot & \cdot & \cdot & \cdot & \cdot & \cdot \\ X_{n1} & X_{n2} & \cdots & X_{nk} & \cdots & \uparrow X_n \\ \cdot & \cdot & \cdot & \cdot & \cdot & \cdot \end{array}$$

Then, clearly, $0 \leq Y_k$ simple $\uparrow$, $X_{nk} \leq Y_k$, $n = 1, \ldots, k$. Next, $Y_k \leq X_k$. In fact, since $Y_k = \max_{1 \leq n \leq k} X_{nk}$, we have that for $\omega \in \Omega$ there exists $1 \leq n_o(\omega) \leq k$ such that $Y_k(\omega) = X_{n_o(\omega),k}$. But $X_{n_o(\omega),k}(\omega) \leq X_{n_o(\omega)}(\omega)$ since $X_{n_{o,k}} \uparrow X_{n_o}$ as $k \to \infty$, and $X_{n_o(\omega)}(\omega) \leq X_k(\omega)$, since $n_o(\omega) \leq k$ and $X_n \uparrow$. Thus $Y_k \leq X_k$ and therefore $X_{nk} \leq Y_k \leq X_k$, $n = 1, \ldots, k$. Keeping n fixed and letting $k \to \infty$, we get

$$X_n \leq \lim_{k \to \infty} Y_k \leq X.$$

Now letting $n \to \infty$, we get

$$X \leq \lim_{k \to \infty} Y_k \leq X, \text{ so that } \lim_{k \to \infty} Y_k = X. \text{ So } 0 \leq Y_k \text{ simpler r.v.s } \uparrow X$$

$$\text{implies that } \int Y_k \uparrow \int X \text{ as } k \to \infty.$$

From $Y_k \leq X_k \leq X$, we get

$$\int Y_k \leq \int X_k \leq \int X \text{ so that } \lim_{k \to \infty} \int Y_k = \int X \leq \limsup_{k \to \infty} \int X_k \leq \int X.$$

So $\int X_k \to \int X$ as $k \to \infty$, and of course, $\int X_k \uparrow \int X$. In other words, $\lim \int X_n = \int \lim X_n$. ■

Remark 1. Clearly, the convergence $X_n \uparrow X$ may be only a.e., and $X_n \geq 0$ may also be true only a.e.

Corollary 1 (Interchanging the $\int$ and $\sum$ in a series of ≥ 0 r.v.s).

(i) If $X_n \geq 0$, $n = 1, 2, \ldots$, then $\int \sum_n X_n = \sum_n \int X_n$.
(ii) If the r.v. $X \geq 0$ a.e. and $\{A_j, j \geq 1\}$ are pairwise disjoint, then $\int_{\sum_j A_j} X = \sum_j \int_{A_j} X$.

Proof.

(i) Let $Y_n = \sum_{j=1}^n X_j$. Then $0 \leq Y_n \uparrow \sum_{j=1}^\infty X_j$ and hence $\int Y_n \uparrow \int \sum_{j=1}^\infty X_j$. But $\int Y_n = \int \sum_{j=1}^n X_j = \sum_{j=1}^n \int X_j$. Thus $\lim \sum_{j=1}^n \int X_j = \sum_{j=1}^\infty \int X_j = \int \sum_{j=1}^\infty X_j$.
(ii) $\int_{\sum_j A_j} X = \int \left(X I_{\sum_j A_j}\right) = \int \left(X \sum_j I_{A_j}\right) = \int \sum_j (X I_{A_j}) = \sum_j \int (X I_{A_j})$ (by part (i)) and this equals to $\sum_j \int_{A_j} X$. ■

Corollary 2. If $Y \overset{\text{a.e.}}{\leq} X_n \underset{\text{a.e.}}{\uparrow} X$ (a possibly extended r.v.) and $\left|\int Y\right| < \infty$, then $\int X_n \uparrow \int X$ (i.e., here 0 is replaced by the integrable r.v. Y).

Proof. From $Y \leq X_n$ a.e., we have $Y^+ \leq X_n^+$ a.e. and $X_n^- \leq Y^-$ a.e., so that $\int X_n$ exists. Next, $Y \underset{\text{a.e.}}{\leq} X_n \underset{\text{a.e.}}{\uparrow} X$ implies $0 \underset{\text{a.e.}}{\leq} X_n - Y \underset{\text{a.e.}}{\uparrow} X - Y$ and this, by the theorem, implies $\int(X_n - Y) \uparrow \int(X - Y)$ or (by Theorem 12 in Chapter 4)

$$\int X_n - \int Y \uparrow \int X - \int Y \text{ or } \int X_n \uparrow \int X$$

since $\left|\int Y\right| < \infty$. ■

Theorem 2 (Fatou–Lebesgue Theorem; interchange of $\int$ and lim inf, lim sup, lim).

(i) If the r.v.s $Y, X_n, n = 1, 2, \ldots$ are such that $Y \overset{\text{a.e.}}{\leq} X_n$, $n = 1, 2, \ldots$, and Y is integrable, then

$$\int \liminf_{n\to\infty} X_n \leq \liminf_{n\to\infty} \int X_n.$$

(ii) If $X_n \underset{\text{a.e.}}{\leq} Z$, $n = 1, 2, \ldots$, and Z is integrable, then $\limsup_{n\to\infty} \int X_n \leq \int \limsup_{n\to\infty} X_n$.

(iii) If $Y \overset{\text{a.e.}}{\leq} X_n \overset{\text{a.e.}}{\leq} Z$, $n = 1, 2, \ldots$, where Y and Z are as above and $X_n \underset{n\to\infty}{\overset{\text{a.e.}}{\to}} X$, a possibly extended r.v., then $\int X_n \underset{n\to\infty}{\to} \int X$ and $\int |X_n| \underset{n\to\infty}{\to} \int |X|$. ■

Proof. In what follows, all limits are taken as $n \to \infty$.

(i) Assume first that $0 \underset{\text{a.e.}}{\leq} X_n$, $n = 1, 2, \ldots$, and define Y_n by $Y_n = \inf_{k\geq n} X_k$. Then $\lim Y_n = \lim \inf_{k\geq n} X_k = \lim \inf X_n$ and this convergence is, clearly, $\uparrow$; i.e., $0 \overset{\text{a.e.}}{\leq} Y_n \uparrow \liminf_n X_n$. Then the Lebesgue Monotone Convergence Theorem applies and gives: $\int Y_n \uparrow \int \liminf X_n$. On the other hand, $Y_n \leq X_n$. Thus

$$\int Y_n \leq \int X_n \text{ and } \lim \int Y_n \leq \liminf \int X_n, \text{ so that}$$
$$\int \liminf X_n \leq \liminf \int X_n.$$

In the general case, we consider $X_n - Y$ that are $\overset{\text{a.e.}}{\geq} 0$, $n = 1, 2, \ldots$, and apply the previous results. Specifically, $Y \leq \liminf X_n$ a.e. implies $(\liminf X_n)^- \leq Y^-$ a.e., so that $\int \liminf X_n$ exists. We then get

$$\int \liminf (X_n - Y) \leq \liminf \int (X_n - Y) \text{ or}$$
$$\int \liminf X_n - \int Y \leq \liminf \int X_n - \int Y$$

or

$$\int \liminf X_n \leq \liminf \int X_n, \text{ since } \int Y \text{ is finite.}$$

(ii) $X_n \leq Z$ a.e. or $-Z \leq -X_n$ a.e. and $-Z$ is integrable. Then, by part (i), $\int \liminf(-X_n) \leq \liminf \int -X_n = \liminf(-\int X_n)$. But $\liminf(-X_n) = -\limsup X_n$. Then

$$\int -\limsup X_n \leq \liminf \left(-\int X_n\right) = -\limsup \int X_n$$

or

$$-\int \limsup X_n \leq -\limsup \int X_n \quad \text{or}$$

$$\limsup \int X_n \leq \int \limsup X_n.$$

(iii) $Y \overset{\text{a.e.}}{\leq} X_n, n \geq 1, \left|\int Y\right| < \infty$, imply, by (i),

$$\int \liminf X_n \leq \liminf \int X_n.$$

Also, $X_n \overset{\text{a.e.}}{\leq} Z, n \geq 1, |\int Z| < \infty$; thus, by (ii), $\limsup \int X_n \leq \int \limsup X_n$. Hence: $\int \liminf X_n = \int X \leq \liminf \int X_n \leq \limsup \int X_n \leq \int \limsup X_n = \int X$. Thus $\int X_n \to \int X$. Next,

$$Y \leq X_n \leq Z \text{ or } -Z \leq -X_n \leq -Y \text{ so that } 0 \leq |X_n| \leq |Y| + |Z|$$

and $\int(|Y| + |Z|) < \infty$. Since $|X_n| \overset{\text{a.e.}}{\to} |X|$, we get $\int |X_n| \to \int |X|$. ■

We proceed with a definition and a lemma.

Definition 1. Let μ be a measure on $\mathcal{A}$ and let φ be a set function on $\mathcal{A}$. We say that φ is *absolutely continuous with respect to* μ, and write $\varphi \ll \mu$, if for every $A \in \mathcal{A}$ for which $\mu(A) = 0$, we have $\varphi(A) = 0$. ■

Lemma 1. Let φ be a nonnegative, σ-additive, and finite set function. Then $\varphi \ll \mu$ if and only if, for every $\varepsilon > 0$, there exists $\delta = \delta(\varepsilon) > 0$ such that $\mu(A) < \delta$ imply $\varphi(A) < \varepsilon$.

For its proof, refer to Exercise 6. Also, see Theorem 27.1 on page 191 in Munroe (1953).

Corollary. Let X be integrable and define the finite and ≥ 0 set function φ (actually, finite measure) as follows: $\varphi(A) = \int_A |X| \, d\mu$, $A \in \mathcal{A}$. Then $\varphi \ll \mu$.

Remark 2. φ as defined above is called the *indefinite integral of* $|X|$, and is σ-additive on account of Corollary 1(ii) to Theorem 1.

Proof of Corollary. See Remark 4 in Chapter 4.

Alternatively, let the r.v.s X_n be such that

$$X_n(\omega) = \begin{cases} X(\omega) \text{ if } |X(\omega)| < n \\ n \quad\quad \text{if } X(\omega) \geq n \\ -n \quad\ \text{if } X(\omega) \leq -n, \quad n = 1, 2, \ldots \text{ (so that } |X_n| \leq n). \end{cases}$$

Then, clearly, $0 \leq |X_n| \uparrow |X|$ as $n \to \infty$ and hence, as $n \to \infty$ $\int |X_n| \uparrow \int |X| < \infty$. Choose n_o such that $\int |X| < \int |X_n| + \frac{\varepsilon}{2}$, $n \geq n_o$. Set $\delta = \delta(\varepsilon) = \frac{\varepsilon}{2n_0}$ and let A be such that $\mu(A) < \delta$. Then

$$\varphi(A) = \int_A |X| = \int_A |X_{n_o}| + \int_A |X| - \int_A |X_{n_o}| \leq n_o \mu(A) + \frac{\varepsilon}{2},$$

by the fact that $|X_{n_o}| \leq n_o$ and the choice of n_o. This is $< n_o \times \frac{\varepsilon}{2n_o} + \frac{\varepsilon}{2}$; i.e., $\varphi(A) < \varepsilon$. ■

Theorem 3 (Dominated Convergence Theorem; interchange of $\int$ and lim). If $|X_n| \leq Y$ a.e., $n = 1, 2, \ldots$ with Y integrable and either (a) $X_n \underset{n\to\infty}{\overset{\text{a.e.}}{\to}} X$ or (b) $X_n \underset{n\to\infty}{\overset{\mu}{\to}} X$, then

(i) $\int_A X_n \underset{n\to\infty}{\to} \int_A X$ uniformly in $A \in \mathcal{A}$.
(ii) $\int X_n \underset{n\to\infty}{\to} \int X$.
(iii) $\int |X_n - X| \underset{n\to\infty}{\to} 0$.
(iv) (i) and (iii) are equivalent. ■

Remark 3.

(i) In the proof we will use the following property:
If $\int Z$ exists, then $\left|\int Z\right| \leq \int |Z|$, as was seen in Theorem 8 (iv) of Chapter 4.

(ii) Part (iv) is true under an integrability assumption only of the r.v.s involved. This integrability is ensured here by the assumptions made (see parts (a) and (b) in the proof of the theorem.)

Proof of Theorem 3. All limits that follow are taken as $n \to \infty$ unless otherwise specified. We first establish (iv) under either mode of convergence of X_n to X.
(iv) Let that $\int |X_n - X| \to 0$. Then

$$\left| \int_A X_n - \int_A X \right| = \left| \int_A (X_n - X) \right| \leq \int_A |X_n - X| \leq \int |X_n - X| \to 0$$

independently of A. So, $\int_A X_n \to \int_A X$ uniformly in $A \in \mathcal{A}$. Next,

$$\int |X_n - X| = \int (X_n - X)^+ + \int (X_n - X)^-$$
$$= \int_{(X_n - X \geq 0)} (X_n - X) - \int_{(X_n - X < 0)} (X_n - X)$$
$$= \int_{A_n} X_n - \int_{A_n} X - \int_{A_n^c} X_n + \int_{A_n^c} X,$$

where $A_n = (X_n - X \geq 0)$. So, if $\int_A X_n \to \int_A X$ uniformly in $A \in \mathcal{A}$, then

$$\int_{A_n} X_n - \int_{A_n} X \to 0, \int_{A_n^c} X_n - \int_{A_n^c} X \to 0 \text{ imply } \int |X_n - X| \to 0.$$

Now, since (i) and (iii) are equivalent and (i) implies (ii), it suffices to show (iii) only.

(a) Assume first that $X_n \overset{\text{a.e.}}{\to} X$. Then $|X_n| \leq Y$ a.e. implies $|X| \leq Y$ a.e., so that $0 \leq |X_n - X| \leq 2Y$ a.e. and $\int 2Y$ is finite. Then Theorem 2 (iii) gives $\int |X_n - X| \to 0$, since $|X_n - X| \overset{\text{a.e.}}{\to} 0$.

(b) Assume next that $X_n \overset{\mu}{\to} X$ or $X_n - X \overset{\mu}{\to} 0$ or $|X_n - X| \overset{\mu}{\to} 0$. By setting $Z_n = |X_n - X|$, $Z = 2Y$, we have then: $0 \leq Z_n \leq Z$ a.e. (by Exercise 18 in Chapter 3) integrable, $Z_n \overset{\mu}{\to} 0$, and we want to prove that $\int Z_n \to 0$.

For $r = 1, 2, \ldots$, define Y_r as follows:

$$Y_r(\omega) = \begin{cases} 1/r & \text{if } Z(\omega) > 1/r \\ Z(\omega) & \text{if } Z(\omega) \leq 1/r. \end{cases} \tag{5.1}$$

Then, clearly, $0 \leq Y_r \leq Z$ and $Y_r \underset{r\to\infty}{\to} 0$ (since $0 \leq Y_r \leq \frac{1}{r}$ everywhere), and this implies $\int Y_r \underset{r\to\infty}{\to} 0$ by Theorem 2 (iii). Now, let $Z_n(\omega) < \frac{1}{r}$. Then, if $Z(\omega) > \frac{1}{r}$, it follows from (5.1) that $Y_r(\omega) = 1/r$ and hence $Z_n(\omega) \leq Y_r(\omega)$. If $Z(\omega) \leq \frac{1}{r}$, then $Y_r(\omega) = Z(\omega)$ by (5.1), and hence $Z_n(\omega) \leq Y_r(\omega)$ since always $Z_n \leq Z$. To summarize, $Z_n < \frac{1}{r}$ implies $Z_n \leq Y_r$ and so

$$\int_{(Z_n < \frac{1}{r})} Z_n \leq \int_{(Z_n < \frac{1}{r})} Y_r \leq \int Y_r \underset{r\to\infty}{\to} 0. \tag{5.2}$$

Now, $Z_n \overset{\mu}{\to} 0$ implies $\mu(Z_n \geq 1/r) \underset{n\to\infty}{\to} 0$. So, if we define $\varphi(A) = \int_A Z d\mu$, $A \in \mathcal{A}$, then $\varphi \ll \mu$ and φ is finite because Z is integrable. Therefore

$$\int_{(Z_n \geq \frac{1}{r})} Z \underset{n\to\infty}{\to} 0 \quad \text{(by Lemma 1)}. \tag{5.3}$$

By $Z_n \leq Z$ and (5.2) and (5.3), we get then

$$\int Z_n = \int_{(Z_n < \frac{1}{r})} Z_n + \int_{(Z_n \geq \frac{1}{r})} Z_n \leq \int Y_r + \int_{(Z_n \geq \frac{1}{r})} Z \to 0$$

by letting $n \to \infty$ first and then by letting $r \to \infty$. ■

Remark 4. Theorems 1–3 remain true if the set $\{1, 2, \ldots\}$ is replaced by $T \subseteq \Re$ and $n \to \infty$ is replaced by $t \to t_o$, $t, t_o \in T$.

The following two theorems provide sufficient conditions that allow the interchange of the order of executing the operations of integration and differentiation.

Theorem 4 (Interchange of $\frac{d}{dt}$ and $\int$). Let $T \subseteq \Re$ and let t_o be an interior point of T. Let $X = X(\omega, t)$ be a real-valued function on $\Omega \times T$ that is an integrable r.v. for each $t \in T$ and such that $(\frac{\partial X}{\partial t})_{t_o}$ exists for a.e. $\omega \in \Omega$ and $\left|\frac{X(\cdot,t)-X(\cdot,t_o)}{t-t_o}\right| \overset{\text{a.e.}}{\leq} Y$ integrable for all t in some neighborhood of t_o. Then the $\frac{d}{dt}\int X(\cdot, t)$ and $\int \frac{\partial X(\cdot,t)}{\partial t}$ exist and are finite for $t = t_0$, and they are equal; i.e.,

$$\left[\frac{d}{dt}\int X(\cdot, t)\right]_{t_0} = \int \left(\frac{\partial X(\cdot, t)}{\partial t}\right)_{t_o}. \qquad \blacksquare$$

Proof. We have $\frac{1}{t-t_0}\left[\int X(\cdot, t) - \int X(\cdot, t_o)\right] = \int \frac{X(\cdot,t)-X(\cdot,t_o)}{t-t_o}$.

Now $\frac{X(\cdot,t)-X(\cdot,t_o)}{t-t_o} \underset{t\to t_o}{\overset{\text{a.e.}}{\to}} (\frac{\partial X(\cdot,t)}{\partial t})_{t_o}$ since $(\frac{\partial X}{\partial t})_{t_o}$ exists a.e., while for ts in a neighborhood of t_o, $\left|\frac{X(\cdot,t)-X(\cdot,t_o)}{t-t_o}\right| \overset{\text{a.e.}}{\leq} Y$ integrable, so that $\left|\left(\frac{\partial X(\cdot,t)}{\partial t}\right)_{t_o}\right| \leq Y$ a.e. and hence $\left(\frac{\partial X(\cdot,t)}{\partial t}\right)_{t_o}$ is integrable. Then Remark 4 applies and gives

$$\lim_{t\to t_o}\int \frac{X(\cdot, t) - X(\cdot, t_o)}{t - t_o} = \int \lim_{t\to t_o} \frac{X(\cdot, t) - X(\cdot, t_o)}{t - t_o} = \int \left(\frac{\partial X(\cdot, t)}{\partial t}\right)_{t_o}.$$

Since the left-hand side is equal to $\lim_{t\to t_o} \frac{\int X(\cdot,t)-\int X(\cdot,t_o)}{t-t_o}$, it follows that $\frac{d}{dt}\int X(\cdot, t)\big|_{t_o}$ is finite, and the asserted equality holds. $\blacksquare$

Theorem 5 (Interchange of $\frac{d}{dt}$ and $\int$). Let $T = [\alpha, \beta] \subset \Re$ and let $X = X(\omega, t)$ be defined on $\Omega \times T$ into $\Re$ and be such that: X is an integrable r.v. for each $t \in T$, $\frac{\partial X(\cdot,t)}{\partial t}$ exists for a.e. ω and all $t \in T$, $\left|\frac{\partial X(\cdot,t)}{\partial t}\right| \overset{\text{a.e.}}{\leq} Y$ integrable, for all $t \in T$. Then, for each $t_o \in [\alpha, \beta]$,

$$\left[\frac{d}{dt}\int X(\cdot, t)\right]_{t_o} = \int \left(\frac{\partial X(\cdot, t)}{\partial t}\right)_{t_o}. \qquad \blacksquare$$

Remark 5. For $t = \alpha$ or $t = \beta$, we mean derivative from the right or left, respectively.

Proof of Theorem 5. We have, by the Mean Value Theorem of Differential Calculus, a.e.:

$$X(\omega, t) - X(\omega, t_o) = (t - t_o)\left(\frac{\partial X(\omega, \cdot)}{\partial t}\right)_{t^*(\omega)},$$

where $t^*(\omega)$ lies between t, t_o.

Since $\left|\frac{\partial X(\cdot,t)}{\partial t}\right| \overset{\text{a.e.}}{\leq} Y$ integrable, for all $t \in T$, we get $\left|\frac{X(\cdot,t)-X(\cdot,t_o)}{t-t_o}\right| \leq Y$ a.e. Then Theorem 4 applies and gives the result. $\blacksquare$

Application 1. Results such as Theorems 4 and 5 have wide applicability in statistics. They are employed, e.g., when establishing the Cramér–Rao inequality, asymptotic normality of the Maximum Likelihood Estimate, and in many other instances.

This section is concluded with the concept of convergence in distribution, and two results involving convergence in distribution. To this end, let $X_n, n \geq 1$, and X be r.v.s defined on the probability space $(\Omega, \mathcal{A}, P)$, and let F_n and F be their d.f.s, respectively. Also, let $C(F)$ be the continuity set of F (the set on which F is continuous). Then

Definition 2. We say that $\{X_n\}$ *converges in distribution to* X and write $X_n \underset{n\to\infty}{\overset{d}{\to}} X$, if $F_n(x) \underset{n\to\infty}{\to} F(x)$ for all $x \in C(F)$. We also denote this convergence by writing $F_n \underset{n\to\infty}{\Rightarrow} F$ and call it *weak convergence* of $\{F_n\}$ to F. ■

The following theorem relates convergence in probability and convergence in distribution.

Theorem 6. If $X_n \underset{n\to\infty}{\overset{P}{\to}} X$, then $X_n \underset{n\to\infty}{\overset{d}{\to}} X$. The inverse is not true, in general. It is true, however, if $P(X = c) = 1$. ■

For its proof, see, e.g., page 168 in Loève (1963), or page 183 in Roussas (1997).

At this point, it should be mentioned that, although the book Loève (1963) is used as a standard reference here, there are of, course, other books which present a thorough treatment of probability; e.g., Shiryaev (1995) is such a book.

In the following example convergence in distribution does not imply convergence in probability.

Example 1. Let $\Omega = \{1, 2, 3, 4\}$, $\mathcal{A} = \mathcal{P}(\Omega)$, $P(\{1\}) = P(\{2\}) = P(\{3\}) = P(\{4\}) = \frac{1}{4}$. Define X_n, $n \geq 1$, and X as follows:

$$\begin{cases} X_n(1) = X_n(2) = 1 \\ X_n(3) = X_n(4) = 0, \quad n \geq 1 \end{cases}; \qquad \begin{cases} X(1) = X(2) = 0 \\ X(3) = X(4) = 1. \end{cases}$$

Then $|X_n - X| = 1$ for all $\omega \in \Omega$. Hence $X_n \underset{n\to\infty}{\overset{P}{\nrightarrow}} X$, clearly. Next

$$F_{X_n}(x) = \begin{cases} 0, & x < 0 \\ \frac{1}{2}, & 0 \leq x < 1 \\ 1, & x \geq 1 \end{cases}; \qquad F_X(x) = \begin{cases} 0, & x < 0 \\ \frac{1}{2}, & 0 \leq x < 1 \\ 1, & x \geq 1 \end{cases};$$

i.e., $F_{X_n}(x) = F_X(x), x \in \Re$ and hence $F_{X_n} \underset{n\to\infty}{\Rightarrow} F_X$, while $X_n \underset{n\to\infty}{\overset{P}{\nrightarrow}} X$. ■

The following theorem is very useful in statistics.

Theorem 7 (Slutsky). Let $X_n, Y_n, n \geq 1$, and X be r.v.s such that $F_{X_n} \underset{n\to\infty}{\Rightarrow} F_X$ (or $X_n \underset{n\to\infty}{\overset{d}{\to}} X$) and $Y_n \underset{n\to\infty}{\overset{P}{\to}} c$. Then

$$\begin{aligned} &\text{(i) } F_{X_n \pm Y_n} \underset{n\to\infty}{\Rightarrow} F_{X\pm c}, \\ &\text{(ii) } F_{X_n Y_n} \underset{n\to\infty}{\Rightarrow} F_{cX}, \\ &\text{(iii) } F_{\frac{X_n}{Y_n}} \underset{n\to\infty}{\Rightarrow} F_{\frac{X}{c}}, \quad c \neq 0. \end{aligned} \quad ; \text{ or } \quad \begin{aligned} &\text{(i}'\text{) } X_n \pm Y_n \underset{n\to\infty}{\overset{d}{\to}} X \pm c, \\ &\text{(ii}'\text{) } X_n Y_n \underset{n\to\infty}{\overset{d}{\to}} cX, \\ &\text{(iii}'\text{) } \frac{X_n}{Y_n} \underset{n\to\infty}{\overset{d}{\to}} \frac{X}{c}. \end{aligned}$$

■

Remark 6. $\frac{X_n}{Y_n}$ is well defined with probability $\underset{n\to\infty}{\to} 1$, since $Y_n \underset{n\to\infty}{\overset{P}{\to}} c \neq 0$.

For the proof of the theorem, see, e.g., pages 102–103 in Rao (1965), or page 201 in Roussas (1997). ■

Example 2. In this example, we use some of the results obtained so far to show that, under certain conditions and in a certain sense, convergence in distribution is preserved. To this end, let $X_n, n = 1, 2, \ldots$, be r.v.s, let $g : \Re \to \Re$ be differentiable, and let its derivative $g'(x)$ be continuous at a point d. Also, let c_n be constants such that $0 \neq c_n \underset{n\to\infty}{\to} \infty$, and let $c_n(X_n - d) \underset{n\to\infty}{\overset{d}{\to}} X$, a r.v. Then $c_n[g(X_n) - g(d)] \underset{n\to\infty}{\overset{d}{\to}} g'(d)X$.

All of the following limits are taken as $n \to \infty$. In the first place, by assumption, $c_n(X_n - d) \overset{d}{\to} X$ and $c_n^{-1} \to 0$, so that $X_n - d \overset{d}{\to} 0$, by Theorem 7 (ii), or $X_n - d \overset{P}{\to} 0$, by Theorem 6. Hence $|X_n - d| \overset{P}{\to} 0$, by Exercise 8 in Chapter 3. Next, expand $g(X_n)$ around d according to Taylor's formula in order to obtain

$$g(X_n) = g(d) + (X_n - d)g'(X_n^*),$$

where X_n^* is a r.v. lying between d and X_n. Hence

$$c_n[g(X_n) - g(d)] = c_n(X_n - d)g'(X_n^*).$$

However, $|X_n^* - d| \leq |X_n - d| \overset{P}{\to} 0$, so that $X_n^* \overset{P}{\to} d$, and hence $g'(X_n^*) \overset{P}{\to} g'(d)$, by the continuity of $g'(x)$ at d and the exercise cited above. Then, by Theorem 7 (ii), $c_n(X_n - d)g'(X_n^*) \overset{d}{\to} g'(d)X$, and therefore $c_n[g(X_n) - g(d)] \overset{d}{\to} g'(d)X$. ■

An application of the result discussed in the previous example is given below.

Example 3.

(i) Let $X_1, \ldots, X_n$ be i.i.d. r.v.s with mean $\mu \in \Re$ and variance $\sigma^2 \in (0, \infty)$, and let $g : \Re \to \Re$ be differentiable with derivative continuous at μ. Then

$$\sqrt{n}[g(\bar{X}_n) - g(\mu)] \underset{n\to\infty}{\overset{d}{\to}} N(0, [\sigma g'(\mu)]^2),$$

where $\bar{X}_n$ is the sample mean of the X_js.

(ii) In particular, if the X_js are distributed as $B(1, p)$, then

$$\sqrt{n}\,[\bar{X}_n(1 - \bar{X}_n) - pq] \underset{n\to\infty}{\overset{d}{\to}} N(0,\ pq(1 - 2p)^2),$$

where $q = 1 - p$.

(i) Indeed, the Central Limit Theorem (CLT) gives that $\sqrt{n}(\bar{X}_n - \mu) \underset{n\to\infty}{\overset{d}{\to}} X \sim N(0, \sigma^2)$. Then the assumptions of Example 2 are fulfilled, so that $\sqrt{n}[g(\bar{X}_n) - g(\mu)] \underset{n\to\infty}{\overset{d}{\to}} g'(\mu)X \sim N(0, [\sigma g'(\mu)]^2)$.

(ii) Here $\mu = p$ and $\sigma^2 = pq$, and take $g(x) = x(1 - x), 0 < x < 1$. Then $g(\bar{X}_n) = \bar{X}_n(1 - \bar{X}_n)$, and $g'(x) = 1 - 2x$, so that $g'(p) = 1 - 2p$. The result then follows from part (i). ■

The result obtained in Example 2 and its applications in Example 3 are often referred to as the *delta* method. See also Exercise 26 in Chapter 11.

Below, a simple application of an instance of Theorem 7 (iii) is given in a testing hypothesis problem.

Application 2. Let $X_1, \ldots, X_n$ be i.i.d. $N(\mu, \sigma^2)$ r.v.s. For testing $H : \mu = \mu_o$ against $A : \mu \neq \mu_o$, say, at level of significance α (where σ is unknown), one uses the t-test and determines the cutoff point c by the requirement that

$$P(|t_n| > c) = \alpha, \quad \text{where } t = \frac{\sqrt{n}(\bar{X}_n - \mu_o)}{\sqrt{\frac{1}{n-1}\sum_j (X_j - \bar{X}_n)^2}}$$

is t_{n-1} distributed under H. Now, whether the Xs are normal or not, we set

$$t_n = \frac{\sqrt{n}\frac{(\bar{X}_n - \mu_o)}{\sigma}}{\sqrt{\frac{1}{n-1}\sum_j (X_j - \bar{X}_n)^2/\sigma^2}} \text{ and have as } n \to \infty, \ \frac{\sqrt{n}(\bar{X}_n - \mu_o)}{\sigma} \underset{(H)}{\overset{d}{\to}} N(0, 1),$$

$$\frac{1}{\sigma^2}\frac{1}{n-1}\sum_j (X_j - \bar{X}_n)^2 \overset{P}{\to} 1,$$

provided, of course, μ, σ^2 are finite (in the nonnormal case). Then as $n \to \infty$, $t_n \underset{(H)}{\overset{d}{\to}} Z \sim N(0, 1)$, and hence the size α of the test will be intact for large n no matter whether the normality assumption is valid or not. This is known as the *robustness* property of the t-test.

In a probability framework, convergence almost everywhere (a.e.) becomes *almost sure* (a.s.) convergence, and convergence in measure becomes convergence *in probability*. These modes of convergences and convergence in distribution introduced earlier are related as follows, on the basis of the Corollary to Theorem 4 in Chapter 3, and Theorem 3 here:

$$X_n \overset{\text{a.s.}}{\to} X \text{ implies } X_n \overset{P}{\to} X \text{ implies } X_n \overset{d}{\to} X, \quad \text{as } n \to \infty.$$

Also,

$$X_n \overset{d}{\to} X \text{ implies } X_n \overset{P}{\to} X \text{ if } P(X = c) = 1, \text{ but not otherwise, and}$$

$$X_n \overset{P}{\to} X \text{ need not imply } X_n \overset{\text{a.s.}}{\to} X, \text{ as } n \to \infty.$$

(For the last statement, see Exercise 2 (ii) in Chapter 3.)

5.2 Sections, Product Measure Theorem, the Fubini Theorem

The content of this section is highly technical, and a brief outline of the basic concepts and results is as follows. Consider the σ-finite measure spaces $(\Omega_i, \mathcal{A}_i, \mu_i)$, $i = 1, 2$, and the product measurable space $(\Omega_1 \times \Omega_2, \mathcal{A}_1 \times \mathcal{A}_2)$.

First, for any $E \in \mathcal{A}_1 \times \mathcal{A}_2$, define Ω_2-sections of E at ω_1, to be denoted by E_{ω_1}, and Ω_1-sections of E at ω_2, to be denoted by E_{ω_2}, and show that they are measurable.

Second, for a measurable function $f : \Omega_1 \times \Omega_2 \to \Re$, define its Ω_2-section at ω_1, $f_{\omega_1}(\cdot) : E_{\omega_1} \to \Re$, and the Ω_1-section of f at ω_2, $f_{\omega_2}(\cdot) : E_{\omega_2} \to \Re$, and show that all these sections are measurable.

Third, define the functions $f : \Omega_1 \to \Re$ and $g : \Omega_2 \to \Re$ by: $f(\omega_1) = \mu_2(E_{\omega_1})$ and $g(\omega_2) = \mu_1(E_{\omega_2})$. Then show that f and g are (nonnegative and) measurable, and that $\int f d\mu_1 = \int g d\mu_2$.

Fourth, on $\mathcal{A}_1 \times \mathcal{A}_2$, define λ by $\lambda(E) = \int f d\mu_1 = \int g d\mu_2$. Then show that λ is a σ-finite measure, and that $\lambda(A_1 \times A_2) = \mu_1(A_1)\mu_2(A_2)$, $A_1 \in \mathcal{A}_1$, $A_2 \in \mathcal{A}_2$. Because of this, λ is referred to as the product measure (of μ_1 and μ_2) and is denoted by $\mu_1 \times \mu_2$.

Finally, consider the r.v. $X : (\Omega_1 \times \Omega_2, \mathcal{A}_1 \times \mathcal{A}_2, \mu_1 \times \mu_2) \to \Re$, and look at the following integrals (whose existence is assumed here):

$$\int X(\omega_1, \omega_2) d\lambda = \int X(\omega_1, \omega_2) d(\mu_1 \times \mu_2),$$

$$\iint X(\omega_1, \omega_2) d\mu_1 d\mu_2 = \int \left[\int X(\omega_1, \omega_2) d\mu_1 \right] d\mu_2,$$

$$\iint X(\omega_1, \omega_2) d\mu_2 d\mu_1 = \int \left[\int X(\omega_1, \omega_2) d\mu_2 \right] d\mu_1.$$

Then conditions are given under which the above three integrals exist, and they are all equal.

Definition 3. For $E \in \mathcal{A} = \mathcal{A}_1 \times \mathcal{A}_2$ and for $\omega_1 \in \Omega_1$, $\omega_2 \in \Omega_2$, we define

$$E_{\omega_1} = \{\omega_2 \in \Omega_2;\ (\omega_1, \omega_2) \in E\}, \quad E_{\omega_2} = \{\omega_1 \in \Omega_1;\ (\omega_1, \omega_2) \in E\}.$$

Clearly, $E_{\omega_1} \subseteq \Omega_2$, $E_{\omega_2} \subseteq \Omega_1$; E_{ω_1} is called an Ω_2-*section* of E at ω_1, and E_{ω_2} is called an Ω_1-*section* of E at ω_2. ■

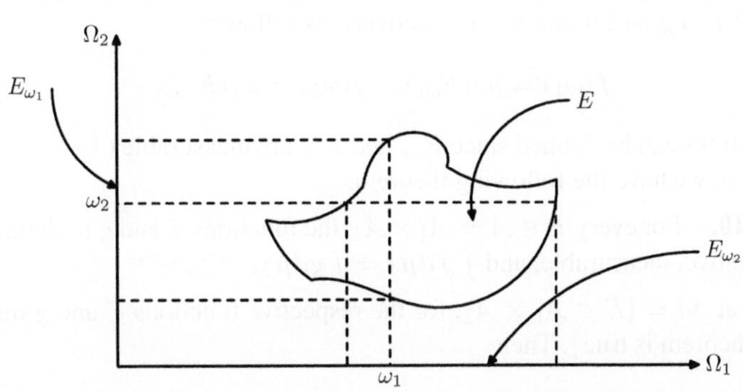

Remark 7. Clearly, if $E = A \times B$, then $E_{\omega_1} = B$ or $\oslash$ depending on whether $\omega_1 \in A$ or $\omega_1 \notin A$, and similarly $E_{\omega_2} = A$ or $\oslash$ depending on whether $\omega_2 \in B$ or $\omega_2 \notin B$.

Theorem 8. Every section of a measurable set is measurable. ∎

Proof. Let $\mathcal{C} = \{E \in \mathcal{A};$ every section of E is measurable$\}$. Then, by the previous remark, $\mathcal{C}$ contains all rectangles, and furthermore it is a σ-field. In fact, let $E_n \in \mathcal{C}, n = 1, 2, \ldots$, let $E = \bigcup_{n=1}^{\infty} E_n$, and let $\omega_1 \in \Omega_1$. Then $E_{\omega_1} = \bigcup_{n=1}^{\infty} E_{n,\omega_1}$, as is easily seen (see Exercise 7(iii)). Since $E_{n,\omega_1} \in \mathcal{A}_2, n = 1, 2, \ldots$, we have that $E_{\omega_1} \in \mathcal{A}_2$. Similarly, $E_{\omega_2} \in \mathcal{A}_1, \omega_2 \in \Omega_2$. Thus $E \in \mathcal{C}$. Now if $E \in \mathcal{C}$, then $E^c \in \mathcal{C}$. In fact, for $\omega_1 \in \Omega_1$ we have $(E^c)_{\omega_1} = (E_{\omega_1})^c$, as is easily seen (see Exercise 7(v)). Since $E_{\omega_1} \in \mathcal{A}_2$, we have $E^c_{\omega_1} \in \mathcal{A}_2$. Similarly, $E^c_{\omega_2} \in \mathcal{A}_1, \omega_2 \in \Omega_2$. Thus $E^c \in \mathcal{C}$. As mentioned already, $\mathcal{C}$ contains the class of all rectangles in $\Omega_1 \times \Omega_2$. Hence $\mathcal{A} \subseteq \mathcal{C}$. Since also $\mathcal{C} \subseteq \mathcal{A}$ by its definition, we have $\mathcal{C} = \mathcal{A}$. ∎

Let $f : E \subseteq \Omega_1 \times \Omega_2 \to \Re$.

Definition 4. For $\omega_1 \in \Omega_1$, the function f_{ω_1}, defined on E_{ω_1} into $\Re$ by $f_{\omega_1}(\omega_2) = f(\omega_1, \omega_2), (\omega_1, \omega_2) \in E$, is called an *$\Omega_2$-section of f at ω_1*. Similarly, for $\omega_2 \in \Omega_2$, the function f_{ω_2}, defined on E_{ω_2} into $\Re$ by $f_{\omega_2}(\omega_1) = f(\omega_1, \omega_2), (\omega_1, \omega_2) \in E$, is called an *$\Omega_1$-section of f at ω_2*. ∎

Theorem 9. If $f : (\Omega_1 \times \Omega_2, \mathcal{A}_1 \times \mathcal{A}_2) \to (\Re, \mathcal{B})$ is measurable, then every section of it is also measurable. ∎

Proof. Let $B \in \mathcal{B}$. Then

$$f_{\omega_1}^{-1}(B) = \{\omega_2 \in \Omega_2;\ f_{\omega_1}(\omega_2) \in B\} = \{\omega_2 \in \Omega_2;\ f(\omega_1, \omega_2) \in B\}$$
$$= \{\omega_2 \in \Omega_2;\ (\omega_1, \omega_2) \in f^{-1}(B)\} = (f^{-1}(B))_{\omega_1}.$$

Now $f^{-1}(B) \in \mathcal{A}_1 \times \mathcal{A}_2$ by the measurability of f, and $f_{\omega_1}^{-1}(B)$ is simply an Ω_2-section of $f^{-1}(B)$ at ω_1, thus an $\mathcal{A}_2$ measurable set. So, f_{ω_1} is $\mathcal{A}_2$-measurable and in a similar fashion f_{ω_2} is $\mathcal{A}_1$-measurable. ∎

Consider now the σ-finite measure spaces $(\Omega_i, \mathcal{A}_i, \mu_i), i = 1, 2$, and let $(\Omega_1 \times \Omega_2, \mathcal{A}_1 \times \mathcal{A}_2)$ be the product measurable space. For $E \in \mathcal{A}_1 \times \mathcal{A}_2$, define the functions f and g on Ω_1 and Ω_2, respectively, as follows:

$$f(\omega_1) = \mu_2(E_{\omega_1}), \quad g(\omega_2) = \mu_1(E_{\omega_2}). \tag{5.4}$$

These functions can be defined since E_{ω_1} and E_{ω_2} are measurable (Theorem 8). With this notation, we have the following theorem.

Theorem 10. For every $E \in \mathcal{A} = \mathcal{A}_1 \times \mathcal{A}_2$, the functions f and g as defined above are nonnegative, measurable, and $\int f d\mu_1 = \int g d\mu_2$. ∎

Proof. Let $\mathcal{M} = \{E \in \mathcal{A}_1 \times \mathcal{A}_2;$ for the respective functions f and g defined by (5.4), the theorem is true$\}$. Then

(i) $\mathcal{M} \neq \emptyset$ since clearly, $\Omega_1 \times \Omega_2 \in \mathcal{M}$.

(ii) $\mathcal{M}$ is closed under countable sums. In fact, let $E_n \in \mathcal{M}, n = 1, 2, \ldots$, with $E_i \cap E_j = \emptyset, i \neq j$, and set $E = \sum_n E_n$. To show that $E \in \mathcal{M}$. Since $E_{\omega_1} = (\sum_n E_n)_{\omega_1} = \sum_n E_{n,\omega_1}$ (see also Exercise 7(iii)), we have

$f(\omega_1) = \mu_2(E_{\omega_1}) = \mu_2(\sum_n E_{n,\omega_1}) = \sum_n \mu_2(E_{n,\omega_1}) = \sum_n f_n(\omega_1)$, where $f_n(\omega_1) = \mu_2(E_{n,\omega_1})$. That is, $f(\omega_1) = \sum_n f_n(\omega_1)$. Now $E_n \in \mathcal{M}$ implies that f_n is ≥ 0, measurable. Thus $f = \sum_n f_n \geq 0$ and since $\sum_{k=1}^n f_k \underset{n\to\infty}{\to} (\uparrow) f$ and $\sum_{k=1}^n f_k$ is measurable, it follows that f is measurable. Also, $\int f = \int \sum_n f_n = \sum_n \int f_n$ by Corollary 1 to Theorem 1. In a similar way, $g \geq 0$ and measure and $\int g = \int \sum_n g_n = \sum_n \int g_n$, where $g_n(\omega_2) = \mu_1(E_{n,\omega_2})$. But $\int f_n d\mu_1 = \int g_n d\mu_2, n = 1, 2, \ldots,$ because $E_n \in \mathcal{M}$. Hence $\int f d\mu_1 = \int g d\mu_2$.

(iii) If $E = A \times B$, then $E \in \mathcal{M}$. In fact, $f(\omega_1) = \mu_2(E_{\omega_1}) = \mu_2(B) I_A(\omega_1)$, $g(\omega_2) = \mu_1(E_{\omega_2}) = \mu_1(A) I_B(\omega_2)$. Thus f, g are ≥ 0, measurable. Next, $\int f d\mu_1 = \mu_2(B)\mu_1(A) = \int g \mu_2 = \mu_1(A)\mu_2(B)$. Hence $\int f d\mu_1 = \int g d\mu_2$.

(iv) If $\mathcal{C}$ is the field (by Theorem 7 in Chapter 1) of all finite sums of measurable rectangles in $\Omega_1 \times \Omega_2$, then $\mathcal{C} \subseteq \mathcal{M}$. This follows from (ii) and (iii).

(v) $\mathcal{M}$ is a monotone class. Let first $E_n \in \mathcal{M}, n = 1, 2, \ldots$. with $E_n \uparrow$. Then to show that $E \overset{def}{=} \lim_{n\to\infty} E_n = \bigcup_n E_n \in \mathcal{M}$. First, we notice that $E_n \uparrow$, implies $E_{n,\omega_1} \uparrow, \omega_1 \in \Omega_1$ (see also Exercise 7(i)), and $E_{\omega_1} = \left(\bigcup_n E_n\right)_{\omega_1} = \bigcup_n E_{n,\omega_1}$ (see also Exercise 7(iii)) or $E_{\omega_1} = (\lim_{n\to\infty} E_n)_{\omega_1} = \lim_{n\to\infty} E_{n,\omega_1}$. Next, with the limits taken everywhere as $n \to \infty$,

$$f(\omega_1) = \mu_2(E_{\omega_1}) = \mu_2\left(\lim E_{n,\omega_1}\right) = \lim \mu_2\left(E_{n,\omega_1}\right) = \lim f_n(\omega_1),$$

where $f_n(\omega_1) = \mu_2(E_{n,\omega_1})$; i.e., $f(\omega_1) = \lim f_n(\omega_1),\ \omega_1 \in \Omega_1$.

Since $f_n \geq 0$, measurable, so is f. But $f_n(\omega_1) = \mu_2(E_{n,\omega_1}) \leq \mu_2\left(E_{n+1,\omega_1}\right) = f_{n+1}(\omega_1)$; i.e., $0 \leq f_n \uparrow f$, and this implies $\int f_n d\mu_1 \uparrow \int f d\mu_1$.

In a similar fashion, $0 \leq g_n \uparrow g$, and this implies $\int g_n d\mu_2 \uparrow \int g d\mu_2$, where $g(\omega_2) = \mu_1(E_{\omega_2}),\ \ g_n(\omega_2) = \mu_1(E_{n,\omega_2})$.

But $\int f_n d\mu_1 = \int g_n d\mu_2$, since $E_n \in \mathcal{M}$. Hence $\int f d\mu_1 = \int g d\mu_2$ implies $E \in \mathcal{M}$.

Let now $E_n \in \mathcal{M}, n = 1, 2, \ldots, E_n \downarrow$. To show that $E \overset{def}{=} \lim E_n = \bigcap_n E_n \in \mathcal{M}$. First, assume that μ_1 and μ_2 are finite. Again, $E_n \downarrow E$ implies $E_{n,\omega_1} \downarrow$, and $E_{\omega_1} = \left(\bigcap_n E_n\right)_{\omega_1} = \bigcap_n E_{n,\omega_1}$ or $E_{\omega_1} = (\lim E_n)_{\omega_1} = \lim E_{n,\omega_1}, \omega_1 \in \Omega_1$ (see also Exercice 7(iv)). Next,

$$f(\omega_1) = \mu_2(E_{\omega_1}) = \mu_2\left((\lim E_n)_{\omega_1}\right) = \mu_2\left(\lim E_{n,\omega_1}\right) = \lim \mu_2\left(E_{n,\omega_1}\right)$$

(by finiteness of μ_2), $=\ \lim f_n(\omega_1)$; $f(\omega_1) = \ \lim f_n(\omega_1), \omega_1 \in \Omega_1$. Since $f_n \geq 0$, measurable, so is f. But

$$f_n(\omega_1) = \mu_2(E_{n,\omega_1}) \geq \mu_2\left(E_{n+1,\omega_1}\right) = f_{n+1}(\omega_1),\ \text{and}\ f_1(\omega_1)$$
$$= \mu_2(E_{1,\omega_1}) \leq \mu_2(\Omega_2) < \infty,\ \ \omega_1 \in \Omega_1.$$

Thus $0 \leq f_n \leq \mu_2(\Omega_2) < \infty$ with $f_n \to f$. Then Theorem 2 (iii) implies $\int f_n d\mu_1 \to \int f d\mu_1$. Similarly, $0 \leq g_n$ measurable $\leq \mu_1(\Omega_1) < \infty$ and $g_n \to g$, where $g(\omega_2) = \mu_1(E_{\omega_2}), g_n(\omega_2) = \mu_1(E_{n,\omega_2}), \omega_2 \in \Omega_2$, and hence $\int g_n d\mu_2 \to$

$\int g d\mu_2$. But $\int f_n d\mu_1 = \int g_n d\mu_2$, since $E_n \in \mathcal{M}$. Hence $\int f d\mu_1 = \int g d\mu_2$, which implies that $E \in \mathcal{M}$.

Now consider the case where μ_1 and μ_2 are σ-finite. Their σ-finiteness implies the existence of partitions $\{A_i, i = 1, 2, \ldots\}$ and $\{B_j, j = 1, 2, \ldots\}$ of Ω_1 and Ω_2, respectively, for which $\mu_1(A_i) < \infty$ and $\mu_2(B_j) < \infty$ for all i and j. For each i and j, define on $\mathcal{A}_1$ and $\mathcal{A}_2$, respectively, the finite measures $\mu_{1i}(A) = \mu_1(A \cap A_i)$ and $\mu_{2j}(B) = \mu_2(B \cap B_j)$ (see Exercise 8). For any $E \in \mathcal{A}_1 \times \mathcal{A}_2$, set

$$\begin{aligned}
f(\omega_1) &= \mu_2(E_{\omega_1}),\\
f_j(\omega_1) &= \mu_{2j}(E_{\omega_1}) \left(= \mu_2\left(\left[E \cap (A_i \times B_j)\right]_{\omega_1}\right) \text{ with } \omega_1 \in A_i\right)\\
&= \mu_2(E_{\omega_1} \cap B_j),\\
g(\omega_2) &= \mu_1(E_{\omega_2}),\\
g_i(\omega_2) &= \mu_{1i}(E_{\omega_2}) \left(= \mu_1\left(\left[E \cap (A_i \times B_j)\right]_{\omega_2}\right) \text{ with } \omega_2 \in B_j\right)\\
&= \mu_1(E_{\omega_2} \cap A_i).
\end{aligned}$$

Then observe that

$$f(\omega_1) = \sum_{j=1}^{\infty} f_j(\omega_1) \text{ and } g(\omega_2) = \sum_{i=1}^{\infty} g_i(\omega_2). \tag{5.5}$$

It follows that f_j, g_i are ≥ 0, f_j is $\mathcal{A}_1$-measurable, g_i is $\mathcal{A}_2$-measurable, and on account of (5.5), so are f and g, respectively.

Also, $\int f_j d\mu_{1i} = \int g_i d\mu_{2j}$ for all i and j.

It follows that

$$\sum_{i=1}^{\infty}\sum_{j=1}^{\infty} \int f_j d\mu_{1i} = \sum_{i=1}^{\infty}\sum_{j=1}^{\infty} \int g_i d\mu_{2j} = \sum_{j=1}^{\infty}\sum_{i=1}^{\infty} \int g_i d\mu_{2j}. \tag{5.6}$$

However,

$$\begin{aligned}
\sum_j \int f_j d\mu_{1i} &= \int \left(\sum_j f_j\right) d\mu_{1i} && \text{(by Corollary 1 to Theorem 1)}\\
&= \int f d\mu_{1i} && \text{(by the definition of } f\text{)}\\
&= \int_{A_i} f d\mu_1 && \text{(by Exercise 9)},
\end{aligned}$$

and

$$\sum_i\sum_j \int f_j d\mu_{1i} = \sum_i \int_{A_i} f d\mu_1 = \sum_i \int (f I_{A_i}) d\mu_1 = \int \sum_i (f I_{A_i}) d\mu_1$$

$$= \int \left(f \sum_i I_{A_i} \right) d\mu_1$$

$$= \int_{\Sigma_i A_i} f d\mu_1 \quad \text{(by Corollary 1 to Theorem 1)}$$

$$= \int f d\mu_1 \qquad \text{(since } \{A_i; i \geq 1\} \text{is a partition of } \Omega_1). \quad (5.7)$$

Likewise,

$$\sum_i \int g_i d\mu_{2j} = \int \left(\sum_i g_i \right) d\mu_{2j} = \int g d\mu_{2j} = \int_{B_j} g d\mu_2,$$

and

$$\sum_j \sum_i \int g_i d\mu_{2j} = \sum_j \int_{B_j} g d\mu_2 = \int_{\Sigma_j B_j} g d\mu_2 = \int g d\mu_2. \quad (5.8)$$

Relations (5.5)–(5.8) then yield $\int f d\mu_1 = \int g d\mu_2$. Since, as already mentioned, the f and g are nonnegative and $\mathcal{A}_1$-measurable and $\mathcal{A}_2$-measurable, respectively, the proof is completed. Thus, in all cases, $\int f d\mu_1 = \int g d\mu_2$ and this implies $E \in \mathcal{M}$. So we have that $\mathcal{M}$ is a nonempty monotone class containing $\mathcal{C}$; hence, $\mathcal{M}$ contains the minimal monotone class over $\mathcal{C}$, which is $\mathcal{A}_1 \times \mathcal{A}_2$ (by Theorem 6 in Chapter 1). Since also $\mathcal{M} \subseteq \mathcal{A}_1 \times \mathcal{A}_2$, it follows that $\mathcal{M} = \mathcal{A}_1 \times \mathcal{A}_2$. Thus the theorem is true for every $E \in \mathcal{A}_1 \times \mathcal{A}_2$. ■

Remark 8. The theorem need not be true if μ_1, μ_2 are not σ-finite, as the following example shows. This fact also has repercussions to Theorems 11 and 12 later.

Example 4. Let $\Omega_1 = \Omega_2 = [0, 1]$, let $\mathcal{A}_1 = \mathcal{A}_2 = \mathcal{B}_{[0,1]}$, let μ_1 be the Lebesgue measure on $\mathcal{A}_1$, and define the set function μ_2 on $\mathcal{A}_2$ by: $\mu_2(A) =$ number of points in A. Then μ_2 is an infinite measure since μ_2 is nondecreasing, $\mu_2(\oslash) = 0$, and $\infty = \mu_2(\sum_{i=1}^{\infty} A_i) = \sum_{i=1}^{\infty} \mu_2(A_i) = \infty$ for whatever A_is in $\mathcal{A}_2$.

However, μ_2 is not σ-finite. Indeed, if $\{A_1, A_2, \ldots\}$ is a partition of $[0, 1]$ with $\mu_2(A_i) < \infty$, then the A_is would have to be finite, and $\sum_{i=1}^{\infty} A_i = [0, 1]$, which is a contradiction as $\sum_{i=1}^{\infty} A_i$ is countable. Next, define the functions f^* and g^* as follows: $f^*, g^* : [0, 1] \times [0, 1] \rightarrow [0, 1]$, $f^*(x, y) = x$, and $g^*(x, y) = y$. Then f^* and g^* are measurable, since, e.g., for any $0 \leq x_1 < x_2 \leq 1$, $(f^*)^{-1}([x_1, x_2]) = [x_1, x_2] \times [0, 1]$, which is $(\mathcal{A}_1, \mathcal{A}_2)$-measurable, and likewise for g^*. It follows that $h = f^* - g^*$ is also measurable. Now, let D be the main diagonal of the square $[0, 1] \times [0, 1]$; i.e.,

$$D = \{(x, y) \in [0, 1] \times [0, 1]; x = y\}.$$

Then D is measurable, because $h^{-1}(\{0\}) = D$. Next, for each $x \in [0, 1]$, the $[0, 1]$-section of D at x, D_x, is $D_x = \{y \in [0, 1]; (x, y) \in D\} = \{y\}$ (with $y = x$), and likewise $D_y = \{x\}$ (with $x = y$). Therefore the functions f and g defined in relation (5.4) are here

$$f(x) = \mu_2(D_x) = \mu_2(\{y\}) = 1 \text{ and } g(y) = \mu_1(D_y) = \mu_1(\{x\}) = 0.$$

So, $f(x) = 1$ for all $x \in [0, 1]$ and $g(y) = 0$ for all $y \in [0, 1]$. It follows that

$$\int_{[0,1]} f d\mu_1 = 1 \neq 0 = \int_{[0,1]} g d\mu_2$$

(by following the convention that $0 \times \infty = 0$). ■

Theorem 11 (Product Measure Theorem). Let $(\Omega_i, \mathcal{A}_i, \mu_i), i = 1, 2$, be two σ-finite measure spaces. Define λ on $\mathcal{A}_1 \times \mathcal{A}_2$ as follows: for $E \in \mathcal{A}_1 \times \mathcal{A}_2, \lambda(E) = \int \mu_2(E_{\omega_1})d\mu_1 = \int \mu_1(E_{\omega_2})d\mu_2$. Then

(i) λ is a measure.
(ii) If $E = A \times B$, $A \in \mathcal{A}_1$, $B \in \mathcal{A}_2$, then $\lambda(A \times B) = \mu_1(A)\mu_2(B)$.
(iii) λ is σ-finite.
(iv) If μ is defined on the rectangles $A \times B$, $A \in \mathcal{A}_1$and $B \in \mathcal{A}_2$, by $\mu(A \times B) = \mu_1(A)\mu_2(B)$, and is extended to the field $\mathcal{C}$ of finite sums of (measurable) rectangles in $\Omega_1 \times \Omega_2$ by $\mu(E) = \sum_{i=1}^{r} \mu_1(A_i)\mu_2(B_i)$, where $E = \sum_{i=1}^{r} A_i \times B_i$, $A_i \in \mathcal{A}_1$, $B_i \in \mathcal{A}_2$, $i = 1, \ldots, r$, then μ is well defined on $\mathcal{C}$, is a σ-finite measure on $\mathcal{C}$, and λ is the unique extension of μ from $\mathcal{C}$ to $\mathcal{A}_1 \times \mathcal{A}_2$. ■

Remark 9. It should be mentioned at this point that the measure λ is instrumental in establishing the properties of the set function μ asserted in part (iv) of Theorem 11.

Proof of Theorem 11.

(i) From Theorem 10, if $f(\omega_1) = \mu_2(E_{\omega_1})$ and $g(\omega_2) = \mu_1(E_{\omega_2})$, then f, g are ≥ 0, measurable, and $\int f d\mu_1 = \int g d\mu_2$. Thus, for $E \in \mathcal{A}_1 \times \mathcal{A}_2, \lambda(E)$ is well defined. Next, λ is a measure. It suffices to prove that λ has the following properties:
(a) $\lambda(\emptyset) = 0$, (b) λ is σ-additive, (c) λ is nondecreasing or nonnegative. That $\lambda(\emptyset) = 0$ is obvious.
Let now $E_n \in \mathcal{A}_1 \times \mathcal{A}_2$, $n \geq 1$, $E_i \cap E_j = \emptyset, i \neq j$. To show that $\sum_{n=1}^{\infty} \lambda(E_n) = \lambda(E)$, where $E = \sum_{n=1}^{\infty} E_n$. Let f, f_n be defined by $f(\omega_1) = \mu_2(E_{\omega_1})$, $f_n(\omega_1) = \mu_2(E_{n,\omega_1})$. But, as in Theorem 10 (ii),

$$\mu_2(E_{\omega_1}) = \mu_2\left(\left(\sum_{n=1}^{\infty} E_n\right)_{\omega_1}\right) = \mu_2\left(\sum_{n=1}^{\infty} E_{n,\omega_1}\right) = \sum_{n=1}^{\infty} \mu_2(E_{n,\omega_1});$$

$$\text{i.e., } f(\omega_1) = \sum_{n=1}^{\infty} f_n(\omega_1), \quad \omega_1 \in \Omega_1.$$

Then,

$$\int f d\mu_1 = \int \sum_{n=1}^{\infty} f_n d\mu_1 = \sum_{n=1}^{\infty} \int f_n d\mu_1 \text{ (by Corollary 1 to Theorem 1)},$$

$$\text{or } \lambda(E) = \sum_{n=1}^{\infty} \lambda(E_n).$$

Finally, let $E, F \in \mathcal{A}_1 \times \mathcal{A}_2$ with $E \subset F$. To show that $\lambda(E) \leq \lambda(F)$. Indeed,

$E \subset F$ implies $E_{\omega_1} \subseteq F_{\omega_1}$ (see Exercise 7(i)), so that $\mu_2(E_{\omega_1}) \leq \mu_2(F_{\omega_1})$.

$$\text{Hence, } \int \mu_2(E_{\omega_1})d\mu_1 \leq \int \mu_2(E_{\omega_1})d\mu_1 \text{or } \lambda(E) \leq \lambda(F).$$

(Alternatively, $\lambda(E) \geq 0$, trivially.)

(ii) Let $E = A \times B,\ A \in \mathcal{A}_1,\ B \in \mathcal{A}_2$. Then $f(\omega_1) = \mu_2(B)I_A(\omega_1)$, and hence $\lambda(E) = \int f d\mu_1 = \mu_1(A)\mu_2(B)$.

(iii) Let $\{A_i,\ i = 1, 2, \ldots\}$, $\{B_j,\ j = 1, 2, \ldots\}$ be partitions of Ω_1, Ω_2, respectively, such that $\mu_1(A_i) < \infty, \mu_2(B_j) < \infty,\ i, j = 1, 2, \ldots$. Then $\{A_i \times B_j,\ i, j = 1, 2, \ldots\}$ is a partition of $\Omega_1 \times \Omega_2$ and $\lambda(A_i \times B_j) = \mu_1(A_i)\,\mu_2(B_j) < \infty, i, j = 1, 2, \ldots$.

(iv) By part (ii), $\mu(A \times B) = \lambda(A \times B)(= \mu_1(A)\mu_2(B))$. Next, if $E = \sum_{i=1}^{r} A_i \times B_i$ with $A_i \in \mathcal{A}_1$and $B_i \in \mathcal{A}_2, i = 1, \ldots, r$, then

$$\begin{aligned}\sum_{i=1}^{r} \mu(A_i \times B_i) &= \sum_{i=1}^{r} \mu_1(A_i)\mu_2(B_i) = \sum_{i=1}^{r} \lambda(A_i \times B_i)\\ &= \lambda\left(\sum_{i=1}^{r} A_i \times B_i\right)\\ &\quad \text{(since } \lambda \text{is a measure on } \mathcal{A}_1 \times \mathcal{A}_2 \supset \mathcal{C})\\ &= \lambda(E).\end{aligned}$$

Likewise, if $E = \sum_{j=1}^{s} A'_j \times B'_j$, then

$$\begin{aligned}\sum_{j=1}^{s} \mu(A'_j \times B'_j) &= \sum_{j=1}^{s} \mu_1(A'_j)\mu_2(B'_j)\\ &= \sum_{j=1}^{s} \lambda(A'_j \times B'_j)\\ &= \lambda(\sum_{j=1}^{s} A'_j \times B'_j)\\ &= \lambda(E)\end{aligned}$$

so that μ is well defined on $\mathcal{C}$. Furthermore, μ is a measure on $\mathcal{C}$ because $\mu = \lambda$ on $\mathcal{C}$ and λ is a measure on $\mathcal{A}_1 \times \mathcal{A}_2 \supset \mathcal{C}$. Finally, μ is σ-finite (by the proof of part (iii)). Hence by the Carathéodory Extension Theorem (Theorem 5 in Chapter 2), λ is the unique extension of μ from $\mathcal{C}$ to $\mathcal{A}_1 \times \mathcal{A}_2$. ■

Definition 5. The measure λ as defined previously is the *product measure* of μ_1 and μ_2 denoted by $\mu_1 \times \mu_2$. ■

Corollary. A set $E \in \mathcal{A}_1 \times \mathcal{A}_2$ is λ-null ($\lambda(E) = 0$) if and only if almost every section of it is null.

Proof. We assume $\mu_1(\Omega_1),\ \mu_2(\Omega_2) > 0$, since otherwise $\lambda = 0$. We have

$$\lambda(E) = \int \mu_2(E_{\omega_1})d\mu_1 = \int \mu_1(E_{\omega_2})d\mu_2.$$

Hence

$$\lambda(E) = 0 \text{ if and only if } \int \mu_2(E_{\omega_1})d\mu_1 = \int \mu_1(E_{\omega_2})d\mu_2 = 0,$$

and since the integrants are ≥ 0, these relations imply

$$\mu_2(E_{\omega_1}) = 0 \quad \text{a.e. } [\mu_1], \quad \mu_1(E_{\omega_2}) = 0 \quad \text{a.e. } [\mu_2].$$

(See Exercise 2 in Chapter 4.) ■

5.2.1 Preliminaries for the Fubini Theorem

Let $(\Omega_i, \mathcal{A}_i, \mu_i)$, $i = 1, 2$, be two σ-finite measure spaces and consider the product measure space $(\Omega_1 \times \Omega_2,\ \mathcal{A}_1 \times \mathcal{A}_2,\ \mu_1 \times \mu_2 = \lambda)$. Let X be a r.v. defined on $\Omega_1 \times \Omega_2$ and let that $\int X d\lambda = \int X d(\mu_1 \times \mu_2)$ exist. This integral is also called the *double integral* of X. Next, for fixed $\omega_1 \in \Omega_1$, set $X_{\omega_1}(\omega_2) = X(\omega_1, \omega_2)$, and for fixed $\omega_2 \in \Omega_2$, set $X_{\omega_2}(\omega_1) = X(\omega_1, \omega_2)$ (apply Definition 4 with $E = \Omega_1 \times \Omega_2$). By Theorem 9, $X_{\omega_1}(\cdot)$, $X_{\omega_2}(\cdot)$ are $\mathcal{A}_2, \mathcal{A}_1$-measurable, respectively. We assume now that $\int X_{\omega_1}(\cdot)d\mu_2$, $\int X_{\omega_2}(\cdot)d\mu_1$, exists and set $f(\omega_1) = \int X_{\omega_1}(\cdot)d\mu_2 = \int X(\omega_1, \cdot)d\mu_2$, $g(\omega_2) = \int X_{\omega_2}(\cdot)d\mu_1 = \int X(\cdot, \omega_2)d\mu_1$. Then f and g are measurable, as is seen in the proof of Theorem 12 later, and assume that $\int f d\mu_1 = \int [\int X(\omega_1, \omega_2)d\mu_2]d\mu_1$ and $\int g d\mu_2 = \int [\int X(\omega_1, \omega_2)d\mu_1]d\mu_2$, exist. These integrals are also called *iterated integrals*. The question then arises: under what conditions

$$\begin{aligned}
\int \left[\int X(\omega_1, \omega_2)d\mu_1\right] d\mu_2 &\overset{def}{=} \iint X(\omega_1, \omega_2)d\mu_1 d\mu_2 \\
&= \iint X(\omega_1, \omega_2)d\mu_2 d\mu_1 \\
&\overset{def}{=} \int \left[\int X(\omega_1, \omega_2)d\mu_2\right] d\mu_1?
\end{aligned}$$

The answer to this question is given by the Fubini Theorem.

Theorem 12 (The Fubini Theorem). Consider the product σ-finite measure space $(\Omega_1 \times \Omega_2, \mathcal{A}_1 \times \mathcal{A}_2,\ \mu_1 \times \mu_2 = \lambda)$ and let X be a r.v. defined on $\Omega_1 \times \Omega_2$ that is either nonnegative or λ-integrable. Then $\iint X d\mu_1 d\mu_2 = \iint X d\mu_2 d\mu_1$ and their common value is $\int X d\lambda$. ■

Proof. Assume first that X is nonnegative (Halmos). Then all $\int X d\lambda$, $\iint X d\mu_1 d\mu_2$, $\iint X d\mu_2 d\mu_1$ exist (since $X \geq 0$, $X_{\omega_1}(\cdot) \geq 0$, $X_{\omega_2}(\cdot) \geq 0$, $f(\omega_1) \geq 0$, $g(\omega_2) \geq 0$, and assuming that f and g are appropriately measurable). We will present the proof in three steps. Assume first that $X = I_E$ for some $E \in \mathcal{A}_1 \times \mathcal{A}_2$. Then $X(\omega_1, \omega_2) = I_E(\omega_1, \omega_2) = I_{E_{\omega_2}}(\omega_1)$, measurable for each fixed ω_2, and hence $\int X(\omega_1, \omega_2) d\mu_1 = \int I_{E_{\omega_2}}(\omega_1) d\mu_1 = \mu_1(E_{\omega_2})$ (measurable, by Theorem 10), and $\iint X(\omega_1, \omega_2) d\mu_1 d\mu_2 = \int \mu_1(E_{\omega_2}) d\mu_2$. Similarly, $\iint X(\omega_1, \omega_2) d\mu_2 d\mu_1 = \int \mu_2(E_{\omega_1}) d\mu_1$. But $\int \mu_2(E_{\omega_1}) d\mu_1 = \int \mu_1(E_{\omega_2}) d\mu_2 = \lambda(E) = \int X d\lambda$ by Theorems 10 and 11. Thus $\iint X d\mu_1 d\mu_2 = \iint X d\mu_2 d\mu_1 = \int X d\lambda$. By linearity of the integral, the theorem is then true for nonnegative simple r.v.s.

Next, let X be any nonnegative r.v., and in the sequel, take all limits as $n \to \infty$. Then there exist $0 \leq X_n$ simple r.v.s $\uparrow X$, and by the Monotone Convergence Theorem,

$$\int X_n d\lambda \to \int X d\lambda. \tag{5.9}$$

For each fixed $\omega_2 \in \Omega_2$, $X_n(\cdot, \omega_2)$ and $X(\cdot, \omega_2)$ are ($\mathcal{A}_1$-)measurable and $0 \leq X_n(\cdot, \omega_2) \uparrow X(\cdot, \omega_2)$, so that, by the Monotone Convergence Theorem,

$$(0 \leq) \int X_n(\cdot, \omega_2) d\mu_1 \uparrow \int X(\cdot, \omega_2) d\mu_1. \tag{5.10}$$

But $\int X_n(\cdot, \omega_2) d\mu_1$ is ($\mathcal{A}_2$-)measurable (as a finite linear combination of integrals of indicators) by the previous step; also, $\int X(\cdot, \omega_2) d\mu_1$ is ($\mathcal{A}_2$-)measurable as a limit of ($\mathcal{A}_2$-)measurable r.v.s. Then from (5.10) and the Monotone Convergence Theorem, it follows that

$$\int \left[\int X(\omega_1, \omega_2) d\mu_1 \right] d\mu_2 \uparrow \int \left[\int X(\omega_1, \omega_2) d\mu_1 \right] d\mu_2,$$

or

$$\iint X_n d\mu_1 d\mu_2 \to \iint X d\mu_1 d\mu_2. \tag{5.11}$$

Likewise,

$$\iint X_n d\mu_2 d\mu_1 \to \iint X d\mu_2 d\mu_1. \tag{5.12}$$

However, for each $n \geq 1$,

$$\iint X_n d\mu_1 d\mu_2 = \iint X_n d\mu_2 d\mu_1 = \int X_n d\lambda. \tag{5.13}$$

Then, from (5.9)–(5.13), we get

$$\iint X d\mu_1 d\mu_2 = \iint X d\mu_2 d\mu_1 = \int X_n d\lambda,$$

as was to be seen.

Now suppose that X is λ-integrable (Loève). Since $X = X^+ - X^-$ and the theorem is true for X^+ and X^-, it follows that it is true for X. In more detail, apply the result just obtained to X^+ and X^- to get

$$\iint X^+ d\mu_1 d\mu_2 = \iint X^+ d\mu_2 d\mu_1 = \int X^+ d\lambda,$$
$$\iint X^- d\mu_1 d\mu_2 = \iint X^- d\mu_2 d\mu_1 = \int X^- d\lambda,$$

so that

$$\iint X^+ d\mu_1 d\mu_2 - \iint X^- d\mu_1 d\mu_2 = \iint X^+ d\mu_2 d\mu_1$$
$$- \iint X^- d\mu_2 d\mu_1 = \int X^+ d\lambda - \int X^- d\lambda,$$

or

$$\iint (X^+ - X^-) d\mu_1 d\mu_2 = \iint (X^+ - X^-) d\mu_2 d\mu_1 = \int (X^+ - X^-) d\lambda,$$

or

$$\iint X d\mu_1 d\mu_2 = \iint X d\mu_2 d\mu_1 = \int X d\lambda. \quad \blacksquare$$

Remark 10. From the preceding elaboration, it also follows that the theorem is true even if both sides are either ∞ or $-\infty$, depending on whether $\int X^+ d\lambda = \infty$ or $\int X^- d\lambda = \infty$.

Remark 11. The Fubini Theorem also holds (and is established in a similar fashion) for any finite number of product σ-finite measure spaces.

Exercises.

1. If the r.v. X is integrable, then show that: $nP(|X| \geq n) \underset{n\to\infty}{\to} 0$. By a counterexample, show that the converse need not be true.
   Hint: For the converse part, let X take on the values $n = 3, 4, \ldots,$ with probabilities p_n proportional to $(\log n + 1)/(n \log n)^2$.
2. Let X be an integrable r.v. Then, for every $\varepsilon > 0$, there is a simple r.v. X_ε such that $\int |X - X_\varepsilon| < \varepsilon$.
   Hint: Write $X = X^+ - X^-$, consider nonnegative sequences of simple r.v.s $\{X_n\}$ and $\{Y_n\}$ converging to X^+ and X^-, appropriately employ the Dominated Convergence Theorem, and finally, define X_ε in terms of X_ns and Y_ns.
3. Establish the following generalized version of part (iii) of Theorem 2 (Fatou–Lebesgue Theorem). Namely, for $n = 1, 2, \ldots$, let X_n, U_n, V_n and X, U, V be r.v.s such that
   $U_n \leq X_n \leq V_n$ a.e., $n \geq 1$, and as $n \to \infty$, $X_n \overset{\text{a.e.}}{\to} X$, $U_n \overset{\text{a.e.}}{\to} U$, $V_n \overset{\text{a.e.}}{\to} V$, $\int U_n \to \int U$ finite and $\int V_n \to \int V$ finite. Then $\int X_n \underset{n\to\infty}{\to} \int X$ finite.

4. Establish the following generalized version of Theorem 3 (Dominated Convergence Theorem). Namely, for $n = 1, 2, \ldots$, let X_n, U_n, X, U be r.v.s such that $|X_n| \leq U_n$ a.e., $n \geq 1$, and as $n \to \infty$, $X_n \overset{\text{a.e.}}{\to} X$, $U_n \overset{\text{a.e.}}{\to} U$, and $\int U_n \to \int U$ finite. Then, as $n \to \infty$:
(i) $\int X_n \to \int X$ finite; (ii) $\int |X_n - X| \to 0$; (iii) $\int_A X_n \to \int_A X$ uniformly in $A \in \mathcal{A}$.

5. In reference to Exercise 4, leave the setup and the assumptions intact except that the assumptions $X_n \overset{\text{a.e.}}{\to} X$ and $U_n \overset{\text{a.e.}}{\to} U$ are replaced by $X_n \overset{\mu}{\to} X$ and $U_n \overset{\mu}{\to} U$ as $n \to \infty$. Then conclusions (i)–(iii) hold.
Hint: For part (i), use the following fact: A sequence of real numbers $\{x_n\}$ converges to a real number x as $n \to \infty$, if and only if for any subsequence $\{m\} \subseteq \{n\}$ there exists a further subsequence $\{r\} \subseteq \{m\}$ such that $x_r \underset{r\to\infty}{\to} x$.

6. Let μ be a σ-finite measure and let φ be a nonnegative σ-additive and finite function, both defined on the measurable space $(\Omega, \mathcal{A})$. Then show that $\varphi \ll \mu$ is equivalent to the following: for every $\varepsilon > 0$, there exists $\delta = \delta(\varepsilon) > 0$ such that $\mu(A) < \delta$ implies $\varphi(A) < \epsilon$.
Hint: If the assertion were not true when $\varphi \ll \mu$, there would exist an $\varepsilon > 0$ such that for every $\delta > 0$ and some A with $\mu(A) < \delta$, we would have $\varphi(A) \geq \varepsilon$. Apply this argument for $\delta_n = 1/2^n$, $\mu(A_n) < 1/2^n$, and $\varphi(A_n) \geq \varepsilon$ for some A_n, and set $A = \limsup_{n\to\infty} A_n$ in order to get $\mu(A) = 0$ and $\varphi(A) \geq \varepsilon$, a contradiction.

7. All sets figuring below are subsets of the product space $\Omega_1 \times \Omega_2$. Then show that

(i) $E \subseteq F$ implies $E_{\omega_1} \subseteq F_{\omega_1}$ and $E_{\omega_2} \subseteq F_{\omega_2}$, $\omega_1 \in \Omega_1$, $\omega_2 \in \Omega_2$.
(ii) $E \cap F = \varnothing$ implies $E_{\omega_1} \cap F_{\omega_1} = E_{\omega_2} \cap F_{\omega_2} = \varnothing$, $\omega_1 \in \Omega_1$, $\omega_2 \in \Omega_2$.
(iii) For $n = 1, 2, \ldots$, $\left(\bigcup_n E_n\right)_{\omega_1} = \bigcup_n E_{n,\omega_1}$, $\left(\bigcup_n E_n\right)_{\omega_2} = \bigcup_n E_{n,\omega_2}$, and in particular, $\left(\sum_n E_n\right)_{\omega_1} = \sum_n E_{n,\omega_1}$, $\left(\sum_n E_n\right)_{\omega_2} = \sum_n E_{n,\omega_2}$, $\omega_1 \in \Omega_1$, $\omega_2 \in \Omega_2$.
(iv) For $n = 1, 2, \ldots$, $\left(\bigcap_n E_n\right)_{\omega_1} = \bigcap_n E_{n,\omega_1}$, $\left(\bigcap_n E_n\right)_{\omega_2} = \bigcap_n E_{n,\omega_2}$, $\omega_1 \in \Omega_1$, $\omega_2 \in \Omega_2$.
(v) $(E^c)_{\omega_1} = (E_{\omega_1})^c$, $(E^c)_{\omega_2} = (E_{\omega_2})^c$, $\omega_1 \in \Omega_1$, $\omega_2 \in \Omega_2$.

8. **(i)** Consider the measure space $(\Omega, \mathcal{A}, \mu)$ and let C be a fixed set in $\mathcal{A}$. On $\mathcal{A}$, define the function $\mu_\circ$ by: $\mu_\circ(A) = \mu(A \cap C)$. Then show that $\mu_\circ$ is a measure.
(ii) Let X be a r.v. defined on $(\Omega, \mathcal{A}, \mu)$ and suppose that $\int X d\mu$ exists. Then show that $\int X d\mu_\circ$ also exists and that $\int X d\mu_\circ = \int_C X d\mu$.

9. Consider the (σ-finite) measure space $(\Omega, \mathcal{A}, \mu)$, and let $\{A_i, i = 1, 2, \ldots\}$ be a (measurable) partition of Ω. For each i, define the measure (see Exercise 8) μ_i by: $\mu_i(A) = \mu(A \cap A_i)$. Then, if X is a r.v. defined on $(\Omega, \mathcal{A}, \mu)$ for which the integral $\int X d\mu$ exists, show that the integrals $\int X d\mu_i$, $i \geq 1$, also exist and $\sum_{i=1}^{\infty} \int X d\mu_i = \int X d\mu$.

10. If X is a simple r.v. (not necessarily nonnegative), defined on the (product) space $(\Omega_1 \times \Omega_2, \mathcal{A}_1 \times \mathcal{A}_2, \lambda = \mu_1 \times \mu_2)$ with μ_1 and μ_2 σ-finite, for which $\int X d\lambda$ exists, then show directly (i.e., without appealing to Theorem 12 and Remark 10) that $\int\int X(\omega_1, \omega_2) d\mu_1 d\mu_2 = \int\int X(\omega_1, \omega_2) d\mu_2 d\mu_1 = \int\int X(\omega_1, \omega_2) d\lambda = \int X d\lambda$.

11. Let X be a r.v. defined on the product space $(\Omega_1 \times \Omega_2, \mathcal{A}_1 \times \mathcal{A}_2, \lambda = \mu_1 \times \mu_2)$ and suppose that $\int X d\lambda$ exists. Then show that the Fubini theorem holds true.

12. If the r.v.s $X_1, \ldots, X_n$ are i.i.d. with $\mathcal{E}X_1 = \mu \in \Re$ and $Var(X_1) = \sigma^2 \in (0, \infty)$, then by the CLT, $\frac{\sqrt{n}(\bar{X}_n - \mu)}{\sigma} \underset{n\to\infty}{\overset{d}{\to}} Z \sim N(0, 1)$, where $\bar{X}_n = \frac{1}{n}\sum_{j=1}^{n} X_j$ is the sample mean of the X_js. Show that the CLT implies the Weak Law of Large Numbers (WLLN); i.e., $\bar{X}_n \underset{n\to\infty}{\overset{P}{\to}} \mu$.

13. Let $(\Omega, \mathcal{A}, \mu) = ((0, 1], \mathcal{B}_{(0,1]}, \lambda)$ where λ is the Lebesgue measure, and consider the function $f : (0, 1] \to \Re$ defined by

$$f(x) = (-1)^n n, \quad \frac{1}{n+1} < x \leq \frac{1}{n}, \quad n = 1, 2, \ldots.$$

Then investigate whether or not the integral $\int_{(0,1]} f d\lambda$ exists. Also, compute it if it exists, as well as the $\int_{(0,1]} |f| d\lambda$.

14. For $n = 1, 2, \ldots$, consider the r.v.s $X_1, X_2, \ldots$ and X, and show that

$$X_n \underset{n\to\infty}{\overset{P}{\longrightarrow}} X \text{ if and only if } \mathcal{E}\left(\frac{|X_n - X|}{1 + |X_n - X|}\right) \underset{n\to\infty}{\to} 0.$$

Hint: Refer to Exercise 14 in Chapter 4.

15. Take $(\Omega, \mathcal{A}, \mathcal{P}) = ((0, 1), \mathcal{B}_{(0,1)}, \lambda)$, λ being the Lebesgue measure, and, for $n = 0, 1, \ldots$, define the r.v.s X_n by:

$$X_{2n+1} = I_{(0, \frac{1}{2})}, \quad X_{2n+2} = I_{[\frac{1}{2}, 1)}.$$

Then, as $n \to \infty$, show that:

$$\int \liminf X_n \leq \liminf \int X_n, \quad \limsup \int X_n \leq \int \limsup X_n$$

(as it should be, by Theorem 2, since $0 \leq X_n \leq 1$ for all n, and the bounds are integrable).

16. In reference to Exercise 12 in Chapter 4, define the r.v. X by $X(\omega) = \frac{1}{2^{\omega}}$ and then show that $\int_{\Omega} X \, d\mu = 2$.

17. For $n \geq 1$, let h_n, g_n, G_n, and h, g, G be real-valued measurable functions defined on $(\Re^k, \mathcal{B}^k, \lambda^k)$, where λ^k is the Lebesgue measure on $\mathcal{B}^k$, be such that:

(i) As $n \to \infty$,

$$h_n(x) \to h(x), \quad g_n(x) \to g(x), \quad G_n(x) \to G(x) \text{ a.e. } [\lambda^k].$$

(ii) For all $n \geq 1$,

$$g_n(x) \leq h_n(x) \leq G_n(x) \text{ a.e. } [\lambda^k].$$

(iii)

$$\int_{\Re^k} g_n d\lambda^k \to \int_{\Re^k} g\, d\lambda^k, \quad \int_{\Re^k} G_n d\lambda^k \to \int_{\Re^k} G\, d\lambda^k, \quad \text{as } n \to \infty,$$

and $\int_{\Re^k} g\, d\lambda^k, \int_{\Re^k} G\, d\lambda^k$ are finite.

Then show that $\int_{\Re^k} h_n d\lambda^k \to \int_{\Re^k} h\, d\lambda^k$ and $\int_{\Re^k} h\, d\lambda^k$ is finite.

18. For $n \geq 1$, let X_n, Y_n, and X be r.v.s defined on the probability space $(\Omega, \mathcal{A}, P)$, and suppose that $X_n \xrightarrow{d} X$ as $n \to \infty$. Then, as $n \to \infty$:

(i) $Y_n - X_n \xrightarrow{P} c$, a constant, implies $Y_n \xrightarrow{d} X + c$.

(ii) $Y_n \xrightarrow{P} c$ implies $X_n Y_n \xrightarrow{d} cX$.

19. As a variant of Exercise 6, consider the following situation. Let $(\Omega, \mathcal{A}, P)$ be a probability space and let X be an integrable r.v. On $\mathcal{A}$, define the (finite) measure ν by $\nu(A) = \int_A X\, dP$. Then show that $\nu << P$ if and only if for every $\varepsilon > 0$ there exists $\delta = \delta(\varepsilon)(> 0)$ such that $P(A) < \delta$ implies $\nu(A) < \varepsilon$.
Hint: Use appropriately the Dominated Convergence Theorem.

20. Let X be a nonnegative integrable r.v. Then:

(i) Use the Fubini Theorem to show that

$$\mathcal{E}X = \int_0^\infty P(X \geq t) dt.$$

(ii) Apply this result in case the d.f. of X is $F(x) = 1 - e^{-\lambda x}$ for $x \geq 0$ and 0 for $x < 0$ $(\lambda > 0)$; or $F(x) = 0$ for $x < 0$, $F(x) = x$ for $0 \leq x \leq 1$, and $F(x) = 1$ for $x \geq 1$.

This page is intentionally left blank

CHAPTER

Standard Moment and Probability Inequalities, Convergence in the rth Mean and its Implications

6

This chapter consists of two sections. The first section is devoted to the standard moment and probability inequalities. They include the Hölder (Cauchy–Schwarz) inequality, referring to the expectation of the product of r.v.s; the Minkowski and the c_r-inequality, referring to the expectation of the sum of r.v.s; and the Jensen inequality, concerning convex functions. The probability inequality in Theorem 6 provides both an upper and a lower bound, and the upper bound gives the Markov inequality and the Tchebichev inequality, as special cases.

In the second section, the concepts of convergence in the rth mean as well as mutual convergence in the rth mean are introduced, and it is shown that they are equivalent. Most of the remainder of this section is devoted to establishing various implications of convergence in the rth mean, and also to giving sufficient conditions that imply convergence in the rth mean. The concepts of uniform integrability and uniform continuity are instrumental in these derivations. These facts are summarized in the form of a table for easy reference.

6.1 Moment and Probability Inequalities

From now on the measure spaces to be considered will be probability spaces $(\Omega, \mathcal{A}, P)$.

Theorem 1. Let X be a r.v. whose rth absolute moment is finite; i.e., $\mathcal{E}\,|X|^r < \infty$. Then $\mathcal{E}\,|X|^{r'} < \infty$ for all $0 \le r' \le r$. ■

Proof. For each $\omega \in \Omega$, we have $|X(\omega)|^{r'} \le 1 + |X(\omega)|^r$ for $0 \le r' \le r$. In fact, this is clearly true if $|X(\omega)| \le 1$, while if $|X(\omega)| > 1$, this is also true because then $|X(\omega)|^{r'} < |X(\omega)|^r$; this inequality implies $\mathcal{E}\,|X|^{r'} \le 1 + \mathcal{E}\,|X|^r < \infty$. ■

Theorem 2 (The Hölder inequality). Let X, Y be two r.v.s and let $r > 1$. Then $\mathcal{E}\,|XY| \le \mathcal{E}^{1/r}\,|X|^r \times \mathcal{E}^{1/s}\,|Y|^s$ for $s > 0$ such that $\frac{1}{r} + \frac{1}{s} = 1$ (where it is assumed that $\mathcal{E}\,|X|^r$, $\mathcal{E}\,|Y|^r < \infty$, because otherwise the inequality is trivially true). In particular, for $r = s = 2$, we have

$$\mathcal{E}\,|XY| \le \mathcal{E}^{1/2}\,|X|^2 \times \mathcal{E}^{1/2}\,|Y|^2 \text{ or } \mathcal{E}^2\,|XY| \le \mathcal{E}\,|X|^2\,\mathcal{E}\,|Y|^2$$

(which is known as the *Cauchy–Schwarz inequality*). ■

In proving Theorem 2, we need the following lemma.

An Introduction to Measure-Theoretic Probability, Second Edition. http://dx.doi.org/10.1016/B978-0-12-800042-7.00006-2

Lemma 1. For $x, y \in \Re$ and $r, s > 0$ such that $\frac{1}{r}+\frac{1}{s} = 1$, we have $|xy| \leq \frac{|x|^r}{r}+\frac{|y|^s}{s}$.

Proof. Since $|xy| = |x|\,|y|$, it suffices to prove the lemma for $x, y > 0$; i.e., to show $xy \leq \frac{x^r}{r} + \frac{y^s}{s}$. For fixed x, consider the following function of y, $f(y) = \frac{x^r}{ry} + \frac{y^{s-1}}{s}$, $y > 0$. Hence $f'(y) = -\frac{x^r}{ry^2} + \frac{(s-1)y^{s-2}}{s}$. Thus $f'(y) = 0$ yields $x^r = \frac{r(s-1)}{s}y^s$; from $\frac{1}{r} + \frac{1}{s} = 1$, it follows that $\frac{r(s-1)}{s} = 1$. So $x^r = y^s$. Next, $f''(y) = \frac{2x^r}{ry^3} + \frac{(s-1)(s-2)y^{s-3}}{s}$, which for $x^r = y^s$ becomes $\frac{2y^r}{ry^3} + \frac{(s-1)(s-2)}{s}y^{s-3} > 0$. Thus $f(y)$ is minimized for $y^s = x^r$, and this minimum is

$$\min_{y>0} f(y) = x. \text{So, } f(y) \geq x \text{ or } \frac{x^r}{r}+\frac{y^s}{s} \geq xy. \qquad \blacksquare$$

Proof of Theorem 2. In the first place, if $\mathcal{E}\,|X|^r = 0$ then $|X|^r = 0$ a.s. so that $X = 0$ a.s. and $XY = 0$ a.s. or $|XY| = 0$ a.s. Then $\mathcal{E}\,|XY| = 0$; thus the inequality is true. Similarly, if $\mathcal{E}\,|Y|^s = 0$. So, we assume that $\mathcal{E}\,|X|^r$, $\mathcal{E}\,|Y|^r > 0$, and set $x = \frac{X(\omega)}{\mathcal{E}^{1/r}|X|^r}$, $y = \frac{Y(\omega)}{\mathcal{E}^{1/s}|Y|^s}$. Then $|xy| \leq \frac{|x|^r}{r} + \frac{|y|^s}{s}$ becomes

$$\frac{|X(\omega)Y(\omega)|}{\mathcal{E}^{1/r}\,|X|^r\,\mathcal{E}^{1/s}\,|Y|^s} \leq \frac{|X(\omega)|^r}{r\mathcal{E}\,|X|^r} + \frac{|Y(\omega)|^s}{sE\,|Y|^s} \text{ for all } \omega \in \Omega; \text{ hence}$$

$$\frac{\mathcal{E}\,|XY|}{\mathcal{E}^{1/r}\,|X|^r\,\mathcal{E}^{1/s}\,|Y|^s} \leq \frac{\mathcal{E}\,|X|^r}{r\mathcal{E}\,|X|^r} + \frac{\mathcal{E}\,|Y|^s}{s\mathcal{E}\,|Y|^s} = \frac{1}{r} + \frac{1}{s} = 1 \text{ or}$$

$$\mathcal{E}\,|XY| \leq \mathcal{E}^{1/r}\,|X|^r\,\mathcal{E}^{1/s}\,|Y|^s. \qquad \blacksquare$$

Corollary. $E^{1/r}\,|X|^r$ is nondecreasing in $r(> 0)$.

Proof. Consider the Cauchy–Schwarz inequality $\mathcal{E}^2\,|XY| \leq \mathcal{E}\,|X|^2 \times \mathcal{E}\,|Y|^2$ and for $0 < r' < r$, replace X by $|X|^{\frac{r-r'}{2}}$ and Y by $|X|^{\frac{r+r'}{2}}$. Then we get

$$\mathcal{E}^2\,|X|^r \leq \mathcal{E}\,|X|^{r-r'}\,\mathcal{E}\,|X|^{r+r'}, \quad \text{hence } 2\log\mathcal{E}\,|X|^r \leq \log\mathcal{E}\,|X|^{r-r'}$$
$$+\log\mathcal{E}\,|X|^{r+r'} \text{ or } \log\mathcal{E}\,|X|^r \leq \frac{1}{2}(\log\mathcal{E}\,|X|^{r-r'} + \log\mathcal{E}\,|X|^{r+r'}).$$

Thus, if we consider $\log\mathcal{E}\,|X|^r$ as a function of r, $g(r)$ say, then g is convex, since the last relationship above is equivalent to $g(r) \leq \frac{1}{2}[g(r-r') + g(r+r')]$, where $r = \frac{(r-r')+(r+r')}{2}$ and g is continuous. (To see that g is continuous, let $r \uparrow r_o$; then $|X|^r \leq 1+|X|^{r_o}$ (by Theorem 1), so that the Dominated Convergence Theorem applies and gives $\mathcal{E}\,|X|^r \underset{r\to r_0}{\to} \mathcal{E}\,|X|^{r_o}$ or $g(r) \underset{r\to r_0}{\to} g(r_o)$. Next, let $r \downarrow r_o$. Then $|X|^r \leq 1 + |X|^{r_1}$ for some $r_1 > r_o$ and all $r_o \leq r \leq r_1$, so that the previous argument applies.) Now $g(0) = 0$ and the slope of the line through the points $(0, 0)$ and $(r, g(r))$ is increasing in r. But this slope is $\frac{g(r)}{r}$ $(r > 0)$. Thus $\frac{1}{r}\log\mathcal{E}\,|X|^r \uparrow$ in r or $\log\mathcal{E}^{\frac{1}{r}}\,|X|^r \uparrow$ in r or $\mathcal{E}^{\frac{1}{r}}\,|X|^r \uparrow$ in r. $\blacksquare$

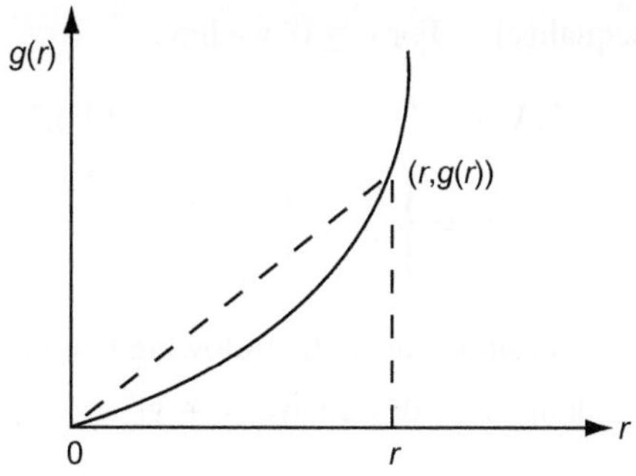

(For the facts on convex functions stated earlier, see, e.g., page 73 in Hardy et al. (1967))

Theorem 3 (The Minkowski inequality). For $r \geq 1$, we have

$$\mathcal{E}^{\frac{1}{r}} |X + Y|^r \leq \mathcal{E}^{\frac{1}{r}} |X|^r + \mathcal{E}^{\frac{1}{r}} |Y|^r$$

(where it is assumed that $\mathcal{E}|X|^r$, $\mathcal{E}|Y|^r < \infty$, because otherwise the inequality is trivially true). ■

Proof. For $r = 1$, we have

$$|X + Y| \leq |X| + |Y| \text{ so that } \mathcal{E}|X + Y| \leq \mathcal{E}|X| + \mathcal{E}|Y|.$$

So the inequality is true. Now let $r > 1$. Then

$$\begin{aligned} \mathcal{E}|X + Y|^r &= \mathcal{E}(|X + Y|\,|X + Y|^{r-1}) \\ &\leq \mathcal{E}(|X|\,|X + Y|^{r-1} + |Y|\,|X + Y|^{r-1}) \\ &= \mathcal{E}(|X|\,|X + Y|^{r-1}) + \mathcal{E}(|Y|\,|X + Y|^{r-1}). \end{aligned}$$

At this point, applying the Hölder inequality for the given $r(> 1)$ and s such that $\frac{1}{r} + \frac{1}{s} = 1$ (from which it follows that $s = r/(r-1)$), we get

$$\begin{aligned} &\mathcal{E}|X + Y|^r \leq \mathcal{E}^{\frac{1}{r}} |X|^r \, \mathcal{E}^{\frac{1}{s}} |X + Y|^{(r-1)s} + \mathcal{E}^{\frac{1}{r}} |Y|^r \, \mathcal{E}^{\frac{1}{s}} |X + Y|^{(r-1)s} \\ &= \mathcal{E}^{\frac{1}{s}} |X + Y|^{(r-1)s} \, (\mathcal{E}^{\frac{1}{r}} |X|^r + \mathcal{E}^{\frac{1}{r}} |Y|^r). \end{aligned}$$

Now, from $\frac{1}{r} + \frac{1}{s} = 1$ we get $(r-1)s = r$. Thus

$$\mathcal{E}|X + Y|^r \leq \mathcal{E}^{\frac{1}{s}} |X + Y|^r \, (\mathcal{E}^{\frac{1}{r}} |X|^r + \mathcal{E}^{\frac{1}{r}} |Y|^r).$$

Hence, if $\mathcal{E}|X + Y|^r > 0$, then divide both sides by $\mathcal{E}^{\frac{1}{s}}|X + Y|^r$ to get

$$\mathcal{E}^{\frac{1}{r}} |X + Y|^r \leq \mathcal{E}^{\frac{1}{r}} |X|^r + \mathcal{E}^{\frac{1}{r}} |Y|^r.$$

If $\mathcal{E}|X + Y|^r = 0$, the inequality is trivially satisfied. ■

Theorem 4 (The c_r-inequality). For $r \geq 0$, we have

$$\mathcal{E}|X+Y|^r \leq c_r(\mathcal{E}|X|^r + \mathcal{E}|Y|^r), \quad \text{where}$$

$$c_r = \begin{cases} 1 & \text{if } r \leq 1 \\ 2^{r-1} & \text{if } r > 1. \end{cases}$$ ■

For the proof of this theorem we need the following lemma.

Lemma 2. For $x, y \in \Re$ and $r \geq 0$, we have $|x+y|^r \leq c_r(|x|^r + |y|^r)$, where c_r is as in the theorem.

Proof. Since $|x+y| \leq |x| + |y|$ implies $|x+y|^r \leq (|x|+|y|)^r$, it suffices to show that $(|x|+|y|)^r \leq c_r(|x|^r + |y|^r)$. From this it also follows that it suffices to prove the lemma for $x, y > 0$.

Case 1: $0 \leq r \leq 1$. We have

$$\frac{x}{x+y}, \frac{y}{x+y} < 1 \text{ imply } \left(\frac{x}{x+y}\right)^r > \frac{x}{x+y}, \quad \left(\frac{y}{x+y}\right)^r > \frac{y}{x+y}$$

$$\text{and } \frac{x^r + y^r}{(x+y)^r} > 1$$

or $(x+y)^r \leq x^r + y^r$, as was to be seen.
Case 2: $r > 1$. Set $p = \frac{x}{x+y}$, $1-p = \frac{y}{x+y}$. Then p, $1-p < 1$. Set also $f(p) = p^r + (1-p)^r$ and minimize it. From

$$f'(p) = rp^{r-1} - r(1-p)^{r-1} = 0,$$

we get

$$p^{r-1} = (1-p)^{r-1}; \text{ hence } p = 1-p \text{ and } p = \frac{1}{2}.$$

Next, $f''(p) = r(r-1)[p^{r-2} + (1-p)^{r-2}]$ is > 0 for all $p(> 0)$ and hence for $p = \frac{1}{2}$. Since $\min_{0<p<1} f(p) = f(\frac{1}{2}) = \frac{1}{2^{r-1}}$, we have then:

$$f(p) \geq \frac{1}{2^{r-1}} \text{ or } \left(\frac{x}{x+y}\right)^r + \left(\frac{y}{x+y}\right)^r \geq \frac{1}{2^{r-1}} \text{ or } (x+y)^r \leq 2^{r-1}(x^r + y^r),$$

as was to be seen. ■

Proof of Theorem 4. For all $\omega \in \Omega$, we have

$$|X(\omega) + Y(\omega)|^r \leq c_r[|X(\omega)|^r + |Y(\omega)|^r].$$

Hence

$$\mathcal{E}|X+Y|^r \leq c_r(\mathcal{E}|X|^r + \mathcal{E}|Y|^r).$$ ■

Now, let I be an open interval $\subseteq \Re$ and assume $g : I \to \Re$, *convex*; i.e., for any $x, x' \in I$ and every $\alpha \in [0, 1]$, we have

$$g[\alpha x + (1-\alpha)x'] \leq \alpha g(x) + (1-\alpha)g(x').$$

For such a function g, we have the following facts:

(1) g is continuous (hence measurable; see Exercise 2).

(2) Through any point $(x_o, g(x_o))$ of the graph of g, there passes a straight line that stays beneath the graph of g or at most touches it. Such a line is called a line of *support* of the curve at the point $(x_o, g(x_o))$, and its equation is: $y - g(x_0) = \lambda(x_0)(x - x_0)$, for some number $\lambda(x_0)$, or $y = g(x_0) + \lambda(x_0)(x - x_0)$. But $g(x) \geq y,\ x \in I$. Thus $g(x) \geq g(x_0) + \lambda(x_0)(x - x_0)$ or $g(x) - g(x_0) \geq \lambda(x_0)(x - x_0),\ x \in I$.

Theorem 5 (The Jensen inequality). Let $g : I \to \Re$ be convex and for a r.v. X taking values in I, let $\mathcal{E}X \in I$ and $\mathcal{E}g(X)$ exist. Then

$$g(\mathcal{E}X) \leq \mathcal{E}g(X). \qquad \blacksquare$$

Proof. In the first place, g is continuous hence measurable and therefore $g(X)$ is a r.v. Next, in $g(x) - g(x_o) \geq \lambda(x_o)(x - x_o)$ replace x_o by $\mathcal{E}X$ and x by X. We then have $g(X) - g(\mathcal{E}X) \geq \lambda(\mathcal{E}X)(X - \mathcal{E}X)$. Taking the expectations of both sides, one gets $g(\mathcal{E}X) \leq \mathcal{E}g(X)$. $\blacksquare$

Applications.

1. Let $g(x) = x^2$. Then g is convex and therefore $\mathcal{E}^2 X \leq \mathcal{E}X^2$ (as also follows from the Cauchy–Schwartz inequality with $Y = 1$).
2. Let $g(x) = x^r,\ x \in (0, \infty),\ r \geq 1$. Then g is convex over the set $(0, \infty)$ and therefore $\mathcal{E}^r |X| \leq \mathcal{E}|X|^r$. In particular, $g(x) = x^{\frac{r_2}{r_1}},\ x \in (0, \infty)$ is convex in x for $r_2 \geq r_1$ (>0). Replacing X by $|X|^{r_1}$, we then get

$$\left(\mathcal{E}|X|^{r_1}\right)^{\frac{r_2}{r_1}} \leq \mathcal{E}\left(|X|^{r_1}\right)^{\frac{r_2}{r_1}} \text{ or } \mathcal{E}^{\frac{r_2}{r_1}}|X|^{r_1} \leq \mathcal{E}|X|^{r_2} \text{ or } \mathcal{E}^{\frac{1}{r_1}}|X|^{r_1} \leq \mathcal{E}^{\frac{1}{r_2}}|X|^{r_2};$$

i.e., $\mathcal{E}^{\frac{1}{r}}|X|^r \uparrow$ in r ($r > 0$), as was seen before.

Remark 1. This was also established in the Corollary to Theorem 2.

Definition 1. α is said to be the *almost sure sup* of the r.v. X ($\alpha =$ a.s. sup X) if $P(X > \alpha) = 0$ and for any $\beta < \alpha,\ P(X > \beta) > 0$. $\blacksquare$

Theorem 6. Let X be a r.v. and let $g : \Re \to [0, \infty)$, Borel function. Then

(i) If g is even, and nondecreasing on $[0, \infty)$, we have that

$$\frac{\mathcal{E}g(X) - g(c)}{\alpha} \leq P(|X| \geq c) \leq \frac{\mathcal{E}g(X)}{g(c)}, \quad c > 0 \text{ constant},$$

and $\alpha =$ a.s. sup $g(X)$.

(ii) If g is nondecreasing on $\Re$, then we have

$$\frac{\mathcal{E}g(X) - g(c)}{\alpha} \leq P(X \geq c) \leq \frac{\mathcal{E}g(X)}{g(c)}, \quad c \in \Re. \qquad \blacksquare$$

Remark 2. $\frac{\infty}{\infty}$ is interpreted as 0.

Proof.

(i) Let $A = (|X| \geq c)$. Then for $\omega \in A$ we have $|X(\omega)| \geq c$; equivalently, $X(\omega) \geq c$ or $-X(\omega) \geq c$. Hence $g(-X(\omega)) = g(X(\omega)) \geq g(c)$; i.e.,$g(X) \geq g(c)$ on A whether $X \geq c$ or $X \leq -c$. Similarly, $g(X) \leq g(c)$ on A^c. Next,

$$\mathcal{E}g(X) = \int g(X)dP = \int_A g(X)dP + \int_{A^c} g(X)dP$$

and

$$\begin{cases} g(c)P(A) \leq \int_A g(X)dP \leq \alpha P(A) \\ 0 \qquad\qquad \leq \int_{A^c} g(X)dP \leq g(c) \end{cases}.$$

Thus $g(c)P(A) \leq Eg(X) \leq \alpha P(A) + g(c)$ and hence

$$P(A) = P(|X| \geq c) \leq \frac{\mathcal{E}g(X)}{g(c)} \text{ and } P(|X| \geq c) \geq \frac{\mathcal{E}g(X) - g(c)}{\alpha}.$$

(ii) Let $B = (X \geq c)$. Then $g(X) \geq g(c)$ on B and $g(X) \leq g(c)$ on B^c. Since

$$\mathcal{E}g(X) = \int_B g(X)dP + \int_{B^c} g(X)dP,$$

we get

$$\begin{cases} g(c)P(B) \leq \int_B g(X)dP \leq \alpha P(B) \\ 0 \qquad\qquad \leq \int_{B^c} g(X)dP \leq g(c) \end{cases},$$

which leads to

$$g(c)P(B) \leq \mathcal{E}g(X) \leq \alpha P(B) + g(c),$$

and hence $P(B) = P(X \geq c) \leq \frac{\mathcal{E}g(X)}{g(c)}$, $P(X \geq c) \geq \frac{\mathcal{E}g(X) - g(c)}{\alpha}$. $\blacksquare$

Special Cases: By taking $g(x) = |x|^r$, $r > 0$, we get from the right-hand side of the inequality in (i): $P(|X| \geq c) \leq \frac{\mathcal{E}|X|^r}{c^r}$, which is the *Markov inequality*.

Also, $P(|X - \mathcal{E}X| \geq c) \leq \frac{\mathcal{E}|X - \mathcal{E}X|^r}{c^r}$, by replacing X by $X - \mathcal{E}X$ (by assuming that $\mathcal{E}X$ is finite); in particular, for $r = 2$ we have the *Tchebichev inequality*:

$$P(|X - \mathcal{E}X| \geq c) \leq \frac{\sigma^2(X)}{c^2}.$$

This section is concluded with a simple example regarding the Tchebichev inequality.

Example 1. When the distribution of X is not known, which is most often the case in statistics, the Tchebichev inequality lends itself handily to determining the smallest sample size n, so that the so-called sample mean will lie within a prescribed multiple of standard deviations σ from the population mean μ with probability no smaller than a preassigned value p. Thus, if $X_1, \ldots, X_n$ are i.i.d. r.v.s with expectation $\mu \in \Re$ and variance $\sigma^2 \in (0, \infty)$, then the sample mean of X_js is $\bar{X}_n = \frac{1}{n}\sum_{j=1}^{n} X_j$ with $\mathcal{E}\bar{X}_n = \mu$ and $\sigma^2(\bar{X}_n) = \sigma^2/n$. Then $P(|\bar{X}_n - \mu| < k\sigma) \geq 1 - \frac{1}{nk^2}$, and if we set $1 - \frac{1}{nk^2} \geq p$, then the required sample size is the smallest value of n that is greater than or equal to $\frac{1}{k^2(1-p)}$. The Markov inequality provides an upper bound for the probability $P(|X| \geq c)$ when $\mathcal{E}|X|^2 = \infty$ but $\mathcal{E}|X|^r < \infty$ for some $(0 <)r < 2$. ■

6.2 Convergence in the *r*th Mean, Uniform Continuity, Uniform Integrability, and Their Relationships

Definition 2. Let X, X_n, $n = 1, 2, \ldots$, be r.v.s such that $\mathcal{E}\,|X|^r, \mathcal{E}\,|X_n|^r < \infty$, $n = 1, 2, \ldots$, for some $r > 0$. We say that X_n converges *in the rth mean* to X, and write

$$X_n \underset{n\to\infty}{\overset{(r)}{\to}} X, \quad \text{if} \quad \mathcal{E}\,|X_n - X|^r \underset{n\to\infty}{\to} 0.$$

For $r = 2$, the convergence is referred to as convergence in *quadratic mean*, $X_n \underset{n\to\infty}{\overset{(2)}{\to}} X$, or $X_n \underset{n\to\infty}{\overset{q.m.}{\to}} X$. ■

Remark 3. Since $\mathcal{E}\,|X_n - X|^r \leq c_r(\mathcal{E}\,|X_n|^r + \mathcal{E}\,|X|^r)$, we have that $\mathcal{E}\,|X_n - X|^r < \infty$, $n = 1, 2, \ldots$.

At this point, it should be mentioned that the limit in the rth mean is a.s. uniquely defined. That is, we have the following

Proposition 1. Let $X_n \underset{n\to\infty}{\overset{(r)}{\to}} X$ and $X_n \underset{n\to\infty}{\overset{(r)}{\to}} Y$ or $\mathcal{E}\,|X_n - X|^r \underset{n\to\infty}{\to} 0$ and $\mathcal{E}\,|X_n - Y|^r \underset{n\to\infty}{\to} 0$. Then $X = Y$ a.s.

Proof. Indeed,

$$\begin{aligned}\mathcal{E}\,|X - Y|^r &= \mathcal{E}\,|(X_n - Y) - (X_n - X)|^r \leq \\ &c_r(\mathcal{E}|X_n - X|^r + \mathcal{E}|Y_n - Y|^r) \underset{n\to\infty}{\to} 0,\end{aligned}$$

so that $\mathcal{E}\,|X - Y|^r = 0$ and hence $|X - Y|^r = 0$ a.s. or $X = Y$ a.s. ■

The following theorem will prove useful in many cases.

Theorem 7. Let $\mathcal{E}\,|X_n|^r < \infty$ for all n. Then $X_n \underset{n\to\infty}{\overset{(r)}{\to}} X$ implies $X_n \underset{n\to\infty}{\overset{P}{\to}} X$ and $\mathcal{E}\,|X_n|^r \underset{n\to\infty}{\to} \mathcal{E}\,|X|^r$ finite. (However, see also Theorem 14.) ■

Proof. The first conclusion is immediate by the Markov inequality. As for the second, we have

Case 1: $0 < r \leq 1$. By the c_r-inequality we get

$$\begin{cases} \mathcal{E}\,|X_n|^r = \mathcal{E}\,|(X_n - X) + X|^r \leq \mathcal{E}\,|X_n - X|^r + \mathcal{E}\,|X|^r \\ \mathcal{E}\,|X|^r = \mathcal{E}\,|(X_n - X) + X_n|^r \leq \mathcal{E}\,|X_n - X|^r + \mathcal{E}\,|X_n|^r, \end{cases}$$

so that

$$\begin{cases} \mathcal{E}\,|X_n|^r - \mathcal{E}\,|X|^r \leq \mathcal{E}\,|X_n - X|^r \\ \mathcal{E}\,|X|^r - \mathcal{E}\,|X_n|^r \leq \mathcal{E}\,|X_n - X|^r, \text{ or } -\left(\mathcal{E}|X_n|^r - \mathcal{E}|X|^r\right) \leq \mathcal{E}|X_n - X|^r \end{cases}$$

and hence $\left|\mathcal{E}\,|X_n|^r - \mathcal{E}\,|X|^r\right| \leq \mathcal{E}\,|X_n - X|^r \underset{n\to\infty}{\to} 0$ so that $\mathcal{E}\,|X_n|^r \underset{n\to\infty}{\to} \mathcal{E}\,|X|^r$.

Case 2: $r > 1$. By the Minkowski inequality we get

$$\begin{cases} \mathcal{E}^{\frac{1}{r}}\,|X_n|^r = \mathcal{E}^{\frac{1}{r}}\,|(X_n - X) + X|^r \leq \mathcal{E}^{\frac{1}{r}}\,|X_n - X|^r + \mathcal{E}^{\frac{1}{r}}\,|X|^r \\ \mathcal{E}^{\frac{1}{r}}\,|X|^r = \mathcal{E}^{\frac{1}{r}}\,|(X_n - X) + X_n|^r \leq \mathcal{E}^{\frac{1}{r}}\,|X_n - X|^r + \mathcal{E}^{\frac{1}{r}}\,|X_n|^r, \end{cases}$$

so that

$$\begin{cases} \mathcal{E}^{\frac{1}{r}}\,|X_n|^r - \mathcal{E}^{\frac{1}{r}}\,|X|^r \leq \mathcal{E}^{\frac{1}{r}}\,|X_n - X|^r \\ \mathcal{E}^{\frac{1}{r}}\,|X|^r - \mathcal{E}^{\frac{1}{r}}\,|X_n|^r \leq \mathcal{E}^{\frac{1}{r}}\,|X_n - X|^r, \text{ or } -\left(\mathcal{E}^{\frac{1}{r}}|X_n|^r - \mathcal{E}^{\frac{1}{r}}|X|^r\right) \leq \mathcal{E}^{\frac{1}{r}}|X_n - X|^r \end{cases}$$

and hence $\left|\mathcal{E}^{\frac{1}{r}}\,|X_n|^r - \mathcal{E}^{\frac{1}{r}}\,|X|^r\right| \leq \mathcal{E}^{\frac{1}{r}}\,|X_n - X|^r \underset{n\to\infty}{\to} 0$; thus $\mathcal{E}^{\frac{1}{r}}\,|X_n|^r \underset{n\to\infty}{\to} \mathcal{E}^{\frac{1}{r}}\,|X|^r$ or $\mathcal{E}\,|X_n|^r \underset{n\to\infty}{\to} \mathcal{E}\,|X|^r$.

Finiteness of $\mathcal{E}\,|X|^r$ follows from $\left|\mathcal{E}\,|X_n|^r - \mathcal{E}\,|X|^r\right| \leq \mathcal{E}\,|X_n - X|^r$ (for $0 < r \leq 1$) or from $\left|\mathcal{E}^{\frac{1}{r}}\,|X_n|^r - \mathcal{E}^{\frac{1}{r}}\,|X|^r\right| \leq \mathcal{E}^{\frac{1}{r}}\,|X_n - X|^r$ (for $r > 1$). ■

Theorem 7 is supplemented now by the following result.

Theorem 8. If $X_n \overset{(r)}{\underset{n\to\infty}{\to}} X$, then $X_n \overset{P}{\underset{n\to\infty}{\to}} X$ (as was seen in Theorem 7). Conversely, if $X_n \overset{P}{\underset{n\to\infty}{\to}} X$ and $P(|X_n| \leq M < \infty) = 1$, then $X_n \overset{(r)}{\underset{n\to\infty}{\to}} X$, for every $r > 0$. ■

Proof. Assume that $X_n \overset{P}{\to} X$ and $P(|X_n| \leq M) = 1$. We have then

$X_n \overset{P}{\underset{n\to\infty}{\to}} X$ implies that there exists $\{m\} \subseteq \{n\}$ such that $X_m \overset{\text{a.s.}}{\underset{n\to\infty}{\to}} X$ (by Theorem 5 (ii) in Chapter 3).

Hence $P(|X_n| \leq M) = 1$ implies $P(|X| \leq M) = 1$ (by Exercise 18 in Chapter 3). Then, by Lemma 2,

$$|X_n - X|^r \leq c_r(|X_n|^r + |X|^r) \text{ and } P(|X_n - X|^r \leq 2c_r M^r) = 1,$$

and therefore

$$\mathcal{E}\,|X_n - X|^r = \int_{(|X_n - X| \geq \varepsilon)} |X_n - X|^r \, dP + \int_{(|X_n - X| < \varepsilon)} |X_n - X|^r \, dP$$
$$\leq 2c_r M^r P(|X_n - X| \geq \varepsilon) + \varepsilon^r.$$

Hence $\limsup_{n\to\infty} \mathcal{E}\,|X_n - X|^r \leq \varepsilon^r$. Letting now $\varepsilon \to 0$, we obtain $\mathcal{E}\,|X_n - X|^r \underset{n\to\infty}{\to} 0$. ■

Remark 4. In Definition 2 we defined $X_n \overset{(r)}{\underset{n\to\infty}{\to}} X$ by assuming that $\mathcal{E}\,|X_n|^r < \infty$, $n = 1, 2, \ldots$, and $\mathcal{E}\,|X|^r < \infty$. Now let us assume only that $\mathcal{E}\,|X_n|^r < \infty$, $n = 1, 2, \ldots$, and let $\mathcal{E}\,|X_n - X|^r \underset{n\to\infty}{\to} 0$ for some r.v. X. Then it follows that $\mathcal{E}\,|X|^r < \infty$. In fact, $\mathcal{E}\,|X|^r = \mathcal{E}\,|(X - X_n) + X_n|^r \leq c_r(\mathcal{E}\,|X_n - X|^r + \mathcal{E}\,|X_n|^r)$. Next, $\mathcal{E}\,|X_n - X|^r \underset{n\to\infty}{\to} 0$ implies that for $n \geq N$ (some N), $\mathcal{E}\,|X_n - X|^r < \infty$. Thus $\mathcal{E}\,|X|^r < \infty$.

Definition 3. Let $\mathcal{E}\,|X_n|^r < \infty$, $n = 1, 2, \ldots$. We say that $\{X_n\}$ *converges mutually in the rth mean* if $\mathcal{E}\,|X_m - X_n|^r \underset{m,n\to\infty}{\to} 0$. ■

Then we have the following theorem.

Theorem 9 (Completeness in the rth mean theorem). $X_n \overset{(r)}{\underset{n\to\infty}{\to}} X$, some r.v. X, if and only if $\{X_n\}$ converges mutually in the rth mean. ■

Proof. Let $X_n \overset{(r)}{\underset{n\to\infty}{\to}} X$. Then

$$\mathcal{E}\,|X_n - X_m|^r = \mathcal{E}\,|(X_n - X) + (X - X_m)|^r$$
$$\leq c_r(\mathcal{E}\,|X_n - X|^r + \mathcal{E}\,|X_m - X|^r) \underset{m,n\to\infty}{\to} 0.$$

Now let $X_m - X_n \overset{(r)}{\underset{m,n\to\infty}{\to}} 0$. Then $X_m - X_n \overset{P}{\underset{m,n\to\infty}{\to}} 0$ by the Markov inequality; i.e., $\{X_n\}$ converges mutually in probability. Then $X_n \overset{P}{\underset{n\to\infty}{\to}} X$, some r.v. X (by Theorem 6 in Chapter 3), which implies the existence of $\{k\} \subseteq \{n\}$ such that $X_k \overset{\text{a.s.}}{\underset{k\to\infty}{\to}} X'$ or $-X_k \overset{\text{a.s.}}{\underset{k\to\infty}{\to}} -X'$ with $P(X' \neq X) = 0$. Henceforth we treat X' as if it were X. Then, for every fixed m, we get $X_m - X_k \overset{\text{a.s.}}{\underset{k\to\infty}{\to}} X_m - X$. Thus we have

$$0 \leq |X_m - X_k|^r \text{ and } \liminf_{k\to\infty} |X_m - X_k|^r = \lim_{k\to\infty} |X_m - X_k|^r$$
$$= |X_m - X|^r \text{ a.s.}$$

Applying part (i) of the Fatou–Lebesgue Theorem, we get then

$$\int \liminf_{k\to\infty} |X_m - X_k|^r \le \liminf_{k\to\infty} \int |X_m - X_k|^r, \text{ or}$$

$$\int \lim_{k\to\infty} |X_m - X_k|^r \le \liminf_{k\to\infty} \int |X_m - X_k|^r, \text{ or}$$

$$\int |X_m - X|^r \le \liminf_{k\to\infty} \int |X_m - X_k|^r; \text{ i.e., } \mathcal{E}|X_m - X|^r \le \liminf_{k\to\infty} \mathcal{E}|X_m - X_k|^r.$$

Letting also $m \to \infty$, we get by our assumption:

$$\limsup_{m\to\infty} \mathcal{E}|X_m - X|^r \le \liminf_{m\to\infty} \liminf_{k\to\infty} \mathcal{E}|X_m - X_k|^r = \lim_{m,k\to\infty} \mathcal{E}|X_m - X_k|^r = 0.$$

So $\mathcal{E}|X_n - X|^r \underset{n\to\infty}{\to} 0$. ■

The following result is a characterization of integrability of a r.v. Namely,

Theorem 10. The r.v. X is integrable if and only if $\int_{(|X|\ge c)} |X|\, dP \underset{c\to\infty}{\to} 0$. ■

Proof. Let $\int |X|\, dP < \infty$. Then $P(|X| < \infty) = 1$, because otherwise $\int |X| dP = \infty$. Also, $|X|\, I_{(|X|\ge c)} \le |X|$ independent of c and integrable, and $|X|\, I_{(|X|\ge c)} \overset{\text{a.s.}}{\to} 0$ as $c \to \infty$. Therefore $\int [|X|\, I_{(|X|\ge c)}] dP \underset{c\to\infty}{\to} 0$ by the Dominated Convergence Theorem, or $\int_{(|X|\ge c)} |X|\, dP \underset{c\to\infty}{\to} 0$. Next, if $\int_{(|X|\ge c)} |X|\, dP \underset{c\to\infty}{\to} 0$, there exists c_o sufficiently large such that $\int_{(|X|\ge c_o)} |X|\, dP < 1$. Thus

$$\int |X|\, dP = \int_{(|X|\ge c_o)} |X|\, dP + \int_{(|X|<c_o)} |X|\, dP < 1 + c_o,$$

so that $\int |X|\, dP$ is finite. ■

Remark 5. The theorem need not be true if the measure is not finite. For example, let $(\Omega, \mathcal{A}, \mu) = (\Re, \mathcal{B}, \lambda)$, where λ is the Lebesgue measure, and let $X = 1$. Then for $c > 1$, $\int_{(|X|\ge c)} |X|\, d\lambda = \lambda(\oslash) = 0$, whereas $\int |X|\, d\lambda = \infty$.

Now, replace X by a sequence $\{X_n\}$, $n \ge 1$, and give the following definition.

Definition 4. The r.v.s X_n, $n \ge 1$, are said to be *uniformly integrable* if

$$\int_{(|X_n|\ge c)} |X_n| \underset{c\to\infty}{\to} 0 \textit{ uniformly} \text{ in } n \ge 1. \quad ■$$

The following concept will also be needed later.

Definition 5. If X_n, $n \geq 1$, are integrable, then $\int |X_n|$, $n \geq 1$, are said to be *uniformly* ($P-$) *absolutely continuous* if $P(A) \to 0$ implies $\int_A |X_n| \to 0$ *uniformly* in $n \geq 1$; i.e., for $\varepsilon > 0$, there exists $\delta(\varepsilon)$ independent of n such that

$$P(A) < \delta(\varepsilon) \text{ implies } \int_A |X_n| < \varepsilon \text{ for every } n \geq 1.$$

In this definition, the index n may be replaced by $t \in T \subseteq \Re$. (See also Exercise 6 in Chapter 5.) ■

Theorem 11. The r.v.s X_n, $n \geq 1$, are uniformly integrable if and only if the integrals of their absolute values are bounded and uniformly continuous. ■

Proof. Assume uniform integrability. Then for $\varepsilon > 0$, there exists $c = c(\varepsilon) > 0$ large enough such that $\int_{(|X_n| \geq c)} |X_n| < \varepsilon$ for all n. Now

$$\int |X_n| = \int_{(|X_n|<c)} |X_n| + \int_{(|X_n| \geq c)} |X_n| \leq cP(|X_n| < c) + \varepsilon \leq c + \varepsilon;$$

i.e., $\int |X_n|$ are bounded. Next, choose $c > 0$ such that $\int_{(|X_n| \geq c)} |X_n| < \frac{\varepsilon}{2}$ for all n. Take $\delta = \delta(\varepsilon) = \frac{\varepsilon}{2c}$ and let A be such that $P(A) < \delta$. Then

$$\int_A |X_n| = \int_{A \cap (|X_n|<c)} |X_n| + \int_{A \cap (|X_n| \geq c)} |X_n| \leq \int_{A \cap (|X_n|<c)} |X_n|$$
$$+ \int_{(|X_n| \geq c)} |X_n| \leq cP(A) + \frac{\varepsilon}{2} < c\frac{\varepsilon}{2c} + \frac{\varepsilon}{2} = \varepsilon;$$

i.e., $P(A) < \delta$ implies $\int_A |X_n| < \varepsilon$ and hence we have uniform continuity.

Now assume boundedness and uniform continuity. Boundedness implies $\int |X_n| \leq M$ for all n, so that $P(|X_n| \geq c) \leq \frac{M}{c}$ for all n. Then, by uniform continuity, for $\varepsilon > 0$, there exists $\delta(\varepsilon) > 0$, call it δ, such that if A has $P(A) < \delta$, we have $\int_A |X_n| < \varepsilon$ for all n. Taking $A = (|X_n| \geq c)$ and c large enough to make $\frac{M}{c} \leq \delta$, and since then $P(|X_n| \geq c) < \delta$, we will have $\int_{(|X_n| \geq c)} |X_n| < \varepsilon$ for all n, which proves uniform integrability. ■

The following theorem provides a criterion of uniform integrability.

Theorem 12. If $\mathcal{E}|X_n|^r < \infty$ for all n, and $X_n \underset{n\to\infty}{\overset{(r)}{\to}} X$, then $|X_n|^r$ are uniformly integrable. ■

Proof. By Theorem 11, it suffices to prove that $\int |X_n|^r$ are bounded and uniformly continuous. Now $X_n \underset{n\to\infty}{\overset{(r)}{\to}} X$ implies $\int |X_n|^r \underset{n\to\infty}{\to} \int |X|^r < \infty$, by Theorem 7, and hence $\int |X_n|^r$ are bounded. Next, for $A \in \mathcal{A}$, $X_n I_A = X I_A + (X_n - X) I_A$ and then, by the c_r-inequality, we get

$$\int |X_n I_A|^r \leq c_r \int |X I_A|^r + c_r \int |(X_n - X) I_A|^r \text{ or}$$
$$\int_A |X_n|^r \leq c_r \int_A |X|^r + c_r \int |X_n - X|^r.$$

Now $\int |X|^r$ is absolutely continuous, as is easily seen (see Exercise 10). Thus for $\varepsilon > 0$, there exists $\delta_o(\varepsilon) > 0$ such that if $P(A) < \delta_o(\varepsilon)$ then $\int_A |X|^r < \frac{\varepsilon}{2c_r}$. From $\int |X_n - X|^r \underset{n\to\infty}{\to} 0$ we have that there exists n_o such that $\int |X_n - X|^r < \frac{\varepsilon}{2c_r}$ for $n > n_o$. Therefore for $n > n_o$ and A such that $P(A) < \delta_o(\varepsilon)$, we have

$$\int_A |X_n|^r \leq c_r \frac{\varepsilon}{2c_r} + c_r \frac{\varepsilon}{2c_r} = \varepsilon.$$

Next consider $\int_A |X_n|^r$, $n = 1, \ldots, n_o$. Then for

$$\varepsilon > 0 \text{ there exists } \delta_n(\varepsilon),\ n = 1, \ldots, n_o, \text{ such that } P(A) < \delta_n(\varepsilon) \text{ implies } \int_A |X_n|^r < \varepsilon,$$

because $\int |X_n|^r$ is absolutely continuous. Set $\delta(\varepsilon) = \min\{\delta_o(\varepsilon), \delta_1(\varepsilon), \ldots, \delta_n(\varepsilon)\}$. Then for $\varepsilon > 0$ and A such that $P(A) < \delta(\varepsilon)$ we have $\int_A |X_n| < \varepsilon$ for all n. ■

Theorem 13 (Necessary and sufficient conditions for convergence in the rth mean). Let $\mathcal{E}|X_n|^r < \infty$ for all n. Then

(i) $X_n \underset{n\to\infty}{\overset{(r)}{\to}} X$ if and only if, either

(ii) $X_n \underset{n\to\infty}{\overset{P}{\to}} X$ and $\int |X_n|^r$ are uniformly continuous, or

(ii′) $X_n \underset{n\to\infty}{\overset{P}{\to}} X$ and $\int |X_n - X|^r$ are uniformly continuous. ■

Proof. The theorem is established by showing that: $(i) \Rightarrow (ii)$, $(ii') \Rightarrow (i)$ (which implies that $(ii') \Rightarrow (ii)$) and $(ii) \Rightarrow (ii')$ (which implies that $(ii) \Rightarrow (i)$). Then (i) and (ii) are equivalent, and (i) and (ii') are also equivalent. In the form of a diagram, we have

$$\begin{array}{ccc} (i) & \Rightarrow & (ii) \\ \Uparrow & \swarrow & \\ (ii') & & \end{array}$$

Indeed, $X_n \underset{n\to\infty}{\overset{(r)}{\to}} X$ implies $X_n \underset{n\to\infty}{\overset{P}{\to}} X$ (by Theorem 8) and that $|X_n|^r$, $n \geq 1$, are uniformly integrable (by Theorem 12), which, in turn, implies that $\int |X_n|^r$, $n \geq 1$, are uniformly continuous (and $\int |X_n| \leq M(< \infty), n \geq 1$) (by Theorem 11). So $(i) \Rightarrow (ii)$.

Next, $(ii') \Rightarrow (i)$ because, with $A_n = (|X_n - X| \geq \varepsilon)$, for any $\varepsilon > 0$,

$$\int |X_n - X|^r = \int_{A_n} |X_n - X|^r + \int_{A_n^c} |X_n - X|^r \leq \varepsilon + \varepsilon^r,$$

by (ii') for all $n > n_0 = n_0(\varepsilon)$, so that $\int |X_n - X|^r \to 0$; i.e., $(ii') \Rightarrow (i)$.

Finally, $(ii) \Rightarrow (ii')$ because, by the c_r-inequality,

$$\int |(X_n - X)I_A|^r \leq c_r \int |X_n I_A|^r + c_r \int |X I_A|^r, \quad A \in \mathcal{A},$$

or

$$\int_A |X_n - X|^r \leq c_r \int_A |X_n|^r + c_r \int_A |X|^r. \tag{6.1}$$

By (*ii*),

$$\int_A |X_n|^r < \frac{\varepsilon}{2c_r} \text{ for all } n, \text{ provided } P(A) < \delta(\varepsilon), \quad \text{some suitable } \delta(\varepsilon) > 0. \tag{6.2}$$

Next, we show that

$$\int_A |X|^r < \frac{\varepsilon}{2c_r}, \qquad \text{for the same } A \text{ as that in (6.2).} \tag{6.3}$$

(This would be true if $\mathcal{E}|X|^r < \infty$, which we do not know.) That (6.3) is true is seen as follows: By (*ii*), $X_n \underset{n\to\infty}{\overset{P}{\to}} X$. Hence there exists $\{m\} \subseteq \{n\}$ such that $X_m \underset{m\to\infty}{\overset{a.s.}{\to}} X'$ or $|X_m|^r I_A \underset{m\to\infty}{\overset{a.s.}{\to}} |X'|^r I_A$, $A \in \mathcal{A}$, with $P(X' \neq X) = 0$. Henceforth we treat X' as if it were X. So,

$0 \leq |X_m|^r I_A \underset{m\to\infty}{\overset{a.s.}{\to}} |X|^r I_A$. Then by part (iii) of the Fatou–Lebesgue Theorem (Theorem 2 in Chapter 5),

$$\int \liminf_{m\to\infty} |X_m|^r I_A = \int \lim_{m\to\infty} |X_m|^r I_A = \int |X|^r I_A \leq \liminf_{m\to\infty} \int |X_m|^r I_A$$

or

$$\int_A |X|^r \leq \liminf_{m\to\infty} \int_A |X_m|^r. \tag{6.4}$$

Again, by (ii) (see also (6.2)),

$$\int_A |X_m|^r \leq \frac{\varepsilon}{2c_r} \text{ for all } m, \text{ provided } P(A) < \delta(\varepsilon). \tag{6.5}$$

From (6.4) and (6.5), it follows that

$$\int_A |X|^r \leq \frac{\varepsilon}{2c_r} \quad \text{for the same } A \text{ as in (6.5).}$$

Then (6.2) and (6.3) hold simultaneously, and hence (6.1) yields

$$\int_A |X_n - X|^r \leq c_r \frac{\varepsilon}{2c_r} + c_r \frac{\varepsilon}{2c_r} = \varepsilon \text{ for all } n,$$

as was to be seen. ■

Corollary 1. If $X_n \underset{n\to\infty}{\overset{(r)}{\to}} X$ then $X_n \underset{n\to\infty}{\overset{(r')}{\to}} X$ for all $(0 <)r' \leq r$.

Proof. By the theorem, it suffices to prove that $X_n \underset{n\to\infty}{\overset{P}{\to}} X$ and $\int |X_n - X|^{r'}$ are uniformly continuous. Since $X_n \underset{n\to\infty}{\overset{(r)}{\to}} X$ implies $X_n \underset{n\to\infty}{\overset{P}{\to}} X$, we have to prove the latter part only. Let $A_n = (|X_n - X| \geq 1)$. Then, on A_n we have: $|X_n - X|^{r'} \leq |X_n - X|^r$. Thus, for any $A \in \mathcal{A}$,

$$\begin{aligned}\int_A |X_n - X|^{r'} &= \int_{A\cap A_n} |X_n - X|^{r'} + \int_{A\cap A_n^c} |X_n - X|^{r'} \\ &\leq \int_{A\cap A_n} |X_n - X|^{r} + \int_{A\cap A_n^c} |X_n - X|^{r'} \\ &\leq \int_A |X_n - X|^r + P(A).\end{aligned}$$

Hence, for $\varepsilon > 0$ and A with $P(A)$ as small as required but $< \frac{\varepsilon}{2}$, we have $\int_A |X_n - X|^r < \frac{\varepsilon}{2}$ for all n by uniform continuity of $\int |X_n - X|^r$ implied by $X_n \underset{n\to\infty}{\overset{(r)}{\to}} X$. The proof is completed. ■

Corollary 2. Let $X_n \underset{n\to\infty}{\overset{P}{\to}} X$ and $\mathcal{E}|X_n|^r \leq M(< \infty)$, $n \geq 1$. Then $X_n \underset{n\to\infty}{\overset{(r')}{\to}} X$ for all $(0 <)r' < r$.

Proof. By the theorem, it suffices to prove that $\int |X_n|^{r'}$ are uniformly continuous. For $c > 0$, let $A_n = (|X_n| \geq c)$. Then, for $|X_n| > 0$, we get

$$|X_n|^{r'} = |X_n|^{r'} \frac{|X_n|^r}{|X_n|^r} = \frac{1}{|X_n|^{r-r'}} |X_n|^r.$$

On A_n,

$$\frac{1}{|X_n|} \leq \frac{1}{c}, \quad \text{or} \quad \frac{1}{|X_n|^{r-r'}} \leq \frac{1}{c^{r-r'}}.$$

Hence

$$\int_A |X_n|^{r'} = \int_{A\cap A_n} |X_n|^{r'} + \int_{A\cap A_n^c} |X_n|^{r'}$$

$$= \int_{A\cap A_n} \frac{1}{|X_n|^{r-r'}} |X_n|^r + \int_{A\cap A_n^c} |X_n|^{r'}$$
$$\leq \frac{1}{c^{r-r'}} \int_{A\cap A_n} |X_n|^r + c^{r'} P(A) \leq \frac{1}{c^{r-r'}} M + c^{r'} P(A).$$

Then for $\varepsilon > 0$, take c sufficiently large so that $\frac{1}{c^{r-r'}} M < \frac{\varepsilon}{2}$. Also choose A such that $P(A) < \frac{\varepsilon}{2c^{r'}} = \delta(\varepsilon)$. Thus, for $\varepsilon > 0$, there exists $\delta(\varepsilon) > 0$ such that $P(A) < \delta(\varepsilon)$ implies $\int_A |X_n|^{r'} < \varepsilon$. ■

Corollary 3 (Compare with the converse part of Theorem 8). Let $X_n \underset{n\to\infty}{\overset{P}{\to}} X$ and $|X_n| \overset{\text{a.s.}}{\leq} Y$, for all n with $\mathcal{E}Y^r < \infty$. Then $X_n \underset{n\to\infty}{\overset{(r)}{\to}} X$.

Proof. $|X_n| \overset{\text{a.s.}}{\leq} Y$ implies $|X_n|^r \leq Y^r$ a.s. and hence $\int |X_n|^r \leq \int Y^r$. Then the absolute continuity of $\int Y^r$ (by Exercise 11) implies the uniform continuity of $\int |X_n|^r$, hence the result. ■

Lemma 3. Let $0 \leq X_n$, and $\mathcal{E}X_n < \infty$ for all n. Then $X_n \underset{n\to\infty}{\overset{(1)}{\to}} X$ if and only if $X_n \underset{n\to\infty}{\overset{P}{\to}} X$ and $\mathcal{E}X_n \underset{n\to\infty}{\to} \mathcal{E}X$ finite.

Proof. Throughout the proof all limits are taken as $n \to \infty$ unless otherwise specified. Let $X_n \overset{(1)}{\to} X$. Then $X_n \overset{P}{\to} X$ and $\mathcal{E}X_n \to \mathcal{E}X$ finite by Theorem 7 (applied with $r = 1$). Hence it suffices to prove the converse. Now $X_n \overset{P}{\to} X$ implies $X_n - X \overset{P}{\to} 0$ or $X - X_n \overset{P}{\to} 0$. Next, it is easily seen that $Y_n \overset{P}{\to} Y$ implies $g(Y_n) \overset{P}{\to} g(Y)$ for every continuous function g (see also Exercise 8 in Chapter 3). The function $g(x) = x^+$ is continuous. Hence $X - X_n \overset{P}{\to} 0$ implies $(X - X_n)^+ \overset{P}{\to} 0$. Now $0 \leq X_n$, $n \geq 1$, and $X_n \overset{P}{\to} X$ implies $X \geq 0$ a.s. (passing to a subsequence $\{m\} \subseteq \{n\}$ such that $X_m \underset{n\to\infty}{\overset{\text{a.s.}}{\to}} X$, so that $X \geq 0$ a.s. since $X_n \geq 0$ a.s. for all n), and $(X - X_n)^+ \leq X$, a.s. So we have

$$(0 \leq)(X - X_n)^+ \leq X \text{ with } \mathcal{E}X < \infty \text{ and } (X - X_n)^+ \overset{P}{\to} 0.$$

Then, the Dominated Convergence Theorem gives

$$\mathcal{E}(X - X_n)^+ \to 0. \tag{6.6}$$

It is also given that $\mathcal{E}X_n \to \mathcal{E}X$ finite, which implies

$$\mathcal{E}(X - X_n) \to 0. \tag{6.7}$$

From (6.6) and (6.7) and the relation $X - X_n = (X - X_n)^+ - (X - X_n)^-$ we get

$$\mathcal{E}(X - X_n)^- \to 0. \tag{6.8}$$

Adding (6.6) and (6.8), we get

$$\mathcal{E}|X - X_n| \to 0 \text{ or } X_n \overset{(1)}{\to} X. \qquad \blacksquare$$

Theorem 14 (Vitali's Theorem). Let $\mathcal{E}|X_n|^r < \infty$ for all n. Then $X_n \underset{n\to\infty}{\overset{(r)}{\to}} X$ if and only if $X_n \underset{n\to\infty}{\overset{P}{\to}} X$ and $\mathcal{E}|X_n|^r \underset{n\to\infty}{\to} \mathcal{E}|X|^r$ finite. $\blacksquare$

Proof. Throughout the proof all limits are taken as $n \to \infty$.

$$X_n \overset{(r)}{\to} X \text{ implies } X_n \overset{P}{\to} X \text{ and } \mathcal{E}|X_n|^r \to \mathcal{E}|X|^r$$
finite (by Theorem 7).

So it suffices to show the converse.

Now $X_n \overset{P}{\to} X$ implies $|X_n|^r \overset{P}{\to} |X|^r$ and we also have $\mathcal{E}|X_n|^r \to \mathcal{E}|X|^r$ finite. Set $Y_n = |X_n|^r$, $Y = |X|^r$. Then we have $0 \le Y_n$, and $\mathcal{E}Y_n < \infty$ for all sufficiently large n and $Y_n \overset{P}{\to} Y$, $\mathcal{E}Y_n \to \mathcal{E}Y$ finite. Then the lemma applies and gives $Y_n \overset{(1)}{\to} Y$ or $|X_n|^r \overset{(1)}{\to} |X|^r$. Hence $\int |X_n|^r$ are uniformly continuous by Theorem 13. So we have: $X_n \overset{P}{\to} X$ and $\int |X_n|^r$ are uniformly continuous. Then $X_n \overset{(r)}{\to} X$, by Theorem 13 again. $\blacksquare$

Remark 6. Lemma 3 is true if P is replaced by a σ-finite measure μ and X_n, X by probability densities functions f_n, f. Then the condition $\mathcal{E}X_n \to \mathcal{E}X$ is trivially true, since these quantities will be equal to 1. Then $f_n \overset{(1)}{\to} f$ (i.e., $\int |f_n - f|\, d\mu \to 0$) if and only if $f_n \overset{\mu}{\to} f$. In this form the lemma is known as Scheffé's Theorem. (See Scheffé (1947).) More precisely, we have

Theorem 15 (Scheffé's Theorem). If for $n = 1, 2, \ldots, f_n$ and $f : \Re^k \to \Re$ are p.d.f.s with respect to some σ-finite measure μ (e.g., the Lebesgue measure) in $\Re^k$, then as $n \to \infty$,

$$f_n \underset{(\mu)}{\overset{(1)}{\to}} f \quad \text{if and only if} \quad f_n \overset{\mu}{\to} f. \qquad \blacksquare$$

Proof. Markov's inequality is still true if P is replaced by μ, as is easily seen directly (see Exercise 12).

Thus, with all limits taken as $n \to \infty$, we have $f_n \underset{(\mu)}{\overset{(1)}{\to}} f$ implies $f_n \overset{\mu}{\to} f$ since $\mu(|f_n - f| \ge \varepsilon) \le \varepsilon^{-1} \int |f_n - f|\, d\mu$. Next, it is easily seen that $f_n \overset{\mu}{\to} f$ or $f - f_n \overset{\mu}{\to} 0$ implies $(f - f_n)^+ \overset{\mu}{\to} 0$ (see Exercise 1 in Chapter 3). As before, $0 \le (f - f_n)^+ \le f$ with $(f - f_n)^+ \overset{\mu}{\to} 0$ and $\int f d\mu = 1$ implies $\int (f - f_n)^+ d\mu \to 0$, by the Dominated Convergence Theorem, whereas $\int f_n d\mu = 1 \to 1 = \int f d\mu$ or $\int (f - f_n) d\mu = 0$. Thus $\int (f - f_n)^- d\mu \to 0$. So $\int (f - f_n)^+ d\mu \to 0$ and $\int (f - f_n)^- d\mu \to 0$ imply $\int |f_n - f|\, d\mu \to 0$ or $f_n \underset{(\mu)}{\overset{(1)}{\to}} f$. $\blacksquare$

Remark 7. Notice that $X_n \underset{n\to\infty}{\overset{(r)}{\to}} X$ is equivalent to $|X_n|^r \underset{n\to\infty}{\overset{(1)}{\to}} |X|^r$. Indeed, by Theorem 14, and with all limits taken as $n \to \infty$, we have $X_n \overset{(r)}{\to} X$ implies $X_n \overset{P}{\to} X$ and $\mathcal{E}|X_n|^r \to \mathcal{E}|X|^r$ finite. Hence $|X_n|^r \overset{P}{\to} |X|^r$ and $\mathcal{E}|X_n|^r \to \mathcal{E}|X|^r$ finite, and therefore, by Theorem 14 again, $|X_n|^r \overset{(1)}{\to} |X|^r$. Also, $|X_n|^r \overset{(1)}{\to} |X|^r$ implies $|X_n|^r \overset{P}{\to} |X|^r$ and $\mathcal{E}|X_n|^r \to \mathcal{E}|X|^r$ finite, and hence $X_n \overset{P}{\to} X$ and $\mathcal{E}|X_n|^r \to \mathcal{E}|X|^r$ finite. Therefore $X_n \overset{(r)}{\to} X$.

Another consequence of the convergence $X_n \overset{(r)}{\to} X$ is that $\int_A |X_n|^r \to \int_A |X|^r$ uniformly in $A \in \mathcal{A}$. Indeed,

$$\left|\int_A (|X_n|^r - |X|^r)\right| \le \int_A \left||X_n|^r - |X|^r\right| \le \mathcal{E}\left||X_n|^r - |X|^r\right|,$$

independent of A and converging to 0, since $|X_n|^r \overset{(1)}{\to} |X|^r$.

The following table presents in summary form the main implications of convergence in the rth mean, as well as conditions under which convergence in the rth mean holds. All limits are taken as $n \to \infty$.

$X_n \overset{(r)}{\to} X$ if and only if $\{X_n\}, n \ge 1$, converges mutually in the rth mean.

$$X_n \overset{(r)}{\to} X \text{ implies } \begin{cases} X_n \overset{P}{\to} X \\ \mathcal{E}|X_n|^r \to \mathcal{E}|X|^r \text{ finite} \\ |X_n|^r \text{ are uniformly integrable} \\ \int |X_n|^r \text{ are uniformly continuous} \\ \int |X_n - X|^r \text{ are uniformly continuous} \\ X_n \overset{(r')}{\to} X, \quad 0 < r' \le r. \end{cases}$$

$$X_n \overset{P}{\to} X \text{ and any one of } \begin{cases} P(|X_n| \le M < \infty) = 1, \quad n \ge 1 \\ \mathcal{E}|X_n|^r \to \mathcal{E}|X|^r \text{ finite} \\ \int |X_n|^r \text{ are uniformly continuous} \\ \int |X_n - X|^r \text{ are uniformly continuous} \\ |X_n| \le Y \text{ a.s., } n \ge 1, \text{ and } EY^r < \infty \end{cases}$$

$$\text{imply } X_n \overset{(r)}{\to} X.$$

$$X_n \overset{P}{\to} X \text{ and } \mathcal{E}|X_n|^r \le M < \infty,\ n \ge 1, \quad \text{imply } X_n \overset{(r')}{\to} X, \quad 0 < r' < r.$$

Finally, combining the converging implications at the end of Section 5.2 in Chapter 5 with Theorem 8 here, we have that the four modes of convergence: a.s., in probability, in distribution, and in the rth mean, are related as follows:

$$\begin{array}{c} X_n \overset{\text{a.s.}}{\to} X \Rightarrow X_n \overset{P}{\to} X \Rightarrow X_n \overset{d}{\to} X \\ \Uparrow \\ X_n \overset{r}{\to} X. \end{array}$$

Exercises.

1. Establish the following generalized version of Theorem 2 (the Hölder inequality). Namely, show that for any n (≥ 2) r.v.s $X_1, \ldots, X_n$ and any positive numbers $r_1, \ldots, r_n$ with $\frac{1}{r_1} + \cdots + \frac{1}{r_n} = 1$, it holds that

$$\mathcal{E}|X_1 \ldots X_n| \leq \mathcal{E}^{\frac{1}{r_1}}|X_1|^{r_1} \ldots \mathcal{E}^{\frac{1}{r_n}}|X_n|^{r_n}.$$

2. Let $g : I$ open interval in $\Re \rightarrow \Re$ be *convex*; i.e., for every $x, x' \in I$ and every $\alpha \in [0, 1]$, it holds $g(\alpha x + (1 - \alpha)x') \leq \alpha g(x) + (1 - \alpha)g(x')$. Then show that
   - **(i)** g is continuous.
   - **(ii)** For every $x_0 \in I$, there exists $\lambda(x_0) \in \Re$ such that $g(x) - g(x_0) \geq \lambda(x_0) \times (x - x_0)$, $x \in I$.

   Hint: For part (i), choose $x_1 < x_0 < x_2$, set $\alpha = (x_2 - x_0)/(x_2 - x_1)$, $\beta = (x_0 - x_1)/(x_2 - x_1)$, use convexity of g to get $g(x_0) \leq \alpha g(x_1) + \beta g(x_2)$, and take the lim inf by letting $x_2 \downarrow x_0$. Next, let $x_1 < x_2 < x_0$, take $\alpha = (x_0 - x_2)/(x_0 - x_1)$, $\beta = (x_2 - x_1)/(x_0 - x_1)$, use convexity of g to get $g(x_2) \leq \alpha g(x_1) + \beta g(x_0)$, and take the lim sup as $x_2 \uparrow x_0$.

   For part (ii), let $x_1 < x_0 < x$, and with $\alpha = (x - x_0)/(x - x_1)$ and $\beta = (x_0 - x_1)/(x - x_1)$, use convexity of g in order to get $g(x) - g(x_0) \geq [g(x_0) - g(x_1)](x - x_0)/(x_0 - x_1)$. The result then follows by taking $x_1 = cx_0$ for some $c > 0$ so that $cx_0 \in I$.
3. Establish the following generalized version of Theorem 3 (the Minkowski inequality). Namely, show that for any n (≥ 2) r.v.s $X_1, \ldots, X_n$ and any $r \geq 1$, it holds that

$$\mathcal{E}^{\frac{1}{r}}|X_1 + \cdots + X_n|^r \leq \mathcal{E}^{\frac{1}{r}}|X_1|^r + \cdots + \mathcal{E}^{\frac{1}{r}}|X_n|^r.$$

4. If $\sum_{n=1}^{\infty} \mathcal{E}|X_n - X|^r < \infty$ for some $r > 0$, then $X_n \underset{n\to\infty}{\overset{\text{a.s.}}{\rightarrow}} X$.

   Hint: Use Exercise 4 in Chapter 3.
5. If $X_n \underset{n\to\infty}{\overset{\text{a.s.}}{\rightarrow}} X$, and $X_n \underset{n\to\infty}{\overset{(r)}{\rightarrow}} Y$ for some $r > 0$, it follows that $P(X \neq Y) = 0$.
6. Construct r.v.s $X_n, n \geq 1$, and X on some probability space $(\Omega, \mathcal{A}, P)$ such that $X_n \underset{n\to\infty}{\overset{P}{\rightarrow}} X$, but $X_n \underset{n\to\infty}{\overset{(r)}{\nrightarrow}} X$ for any $r \geq 1$.
7. Construct r.v.s $X_n, n \geq 1$, and X on some probability space $(\Omega, \mathcal{A}, P)$ such that $X_n \underset{n\to\infty}{\overset{\text{a.s.}}{\nrightarrow}} X$, but $X_n \underset{n\to\infty}{\overset{(r)}{\rightarrow}} X$ for any $r > 0$.
8. Let $X_n, n = 1, 2, \ldots$, be r.v.s such that $\mathcal{E}|X_n|^\beta \leq M$ ($< \infty$), $n \geq 1$, for some $\beta > 0$. Then show that $|X_n|^\alpha, n \geq 1$, are uniformly integrable for all α with $0 < \alpha < \beta$.
9. For $n = 1, 2, \ldots$, let X_n be r.v.s such that $P(X_n = cn) = \frac{1}{n}$ and $P(X_n = 0) = 1 - \frac{1}{n}$, for some $c > 0$. Investigate whether or not these r.v.s are uniformly integrable.

10. If $\mathcal{E}|X| < \infty$, show that $\int |X|$ is absolutely continuous; that is, $\int_A |X| \to 0$ when $P(A) \to 0$.

11. For $n = 1, 2, \ldots$, let X_n, Y_n and X, Y be r.v.s such that $P(X_n \geq Y_n \geq 0) = 1$, $X_n \underset{n\to\infty}{\overset{P}{\to}} X$, $Y_n \underset{n\to\infty}{\overset{P}{\to}} Y$, and $\mathcal{E}X_n \underset{n\to\infty}{\to} \mathcal{E}X$ finite. Then show that $\mathcal{E}|Y_n - Y| \underset{n\to\infty}{\to} 0$.

12. For any r.v. X defined on the measure space $(\Omega, \mathcal{A}, \mu)$, show directly that the Markov inequality holds; namely, $\mu(|X| \geq c) \leq c^{-r} \int |X|^r d\mu$ for any $r > 0$ and any $c > 0$.

13. For a r.v. X and some $r > 0$, show that $\mathcal{E}|X|^r < \infty$ if and only if $\mathcal{E}|X - c|^r < \infty$ for every (finite) constant c.

14. For a r.v. X with $\mathcal{E}X = \mu \in \Re$ and $Var(X) = \sigma^2 = 0$, show that $P(X = \mu) = 1$.

15. For $n = 1, 2, \ldots$, consider the r.v.s X_n and X and show that

(i) $|X_n| \underset{n\to\infty}{\overset{P}{\to}} 0$ if and only if $\frac{|X_n|}{1+|X_n|} \underset{n\to\infty}{\overset{P}{\to}} 0$.

(ii) $|X_n| \underset{n\to\infty}{\overset{P}{\to}} 0$ if and only if $\mathcal{E}\left(\frac{|X_n|}{1+|X_n|}\right) \underset{n\to\infty}{\to} 0$.

Hint: For part (i), see Exercise 8 in Chapter 3.

16. **(i)** For $r > 1$ and $x_j \in \Re$, $j = 1, \ldots, n$, show that

$$\left| n^{-1} \sum_{j=1}^{n} x_j \right|^r \leq n^{-1} \sum_{j=1}^{n} |x_j|^r.$$

(ii) From part (i), deduce that

$$\mathcal{E}\left| \frac{1}{n} \sum_{j=1}^{n} X_j \right|^r \leq \frac{1}{n} \sum_{j=1}^{n} \mathcal{E}|X_j|^r.$$

(On the other hand, by the Minkowski inequality,

$$\mathcal{E}^{1/r}\left| \frac{1}{n} \sum_{j=1}^{n} X_j \right|^r \leq \frac{1}{n} \sum_{j=1}^{n} \mathcal{E}^{1/r}|X_j|^r.)$$

17. For $n = 1, 2, \ldots$, suppose that X_n and X are r.v.s defined on the probability space $(\Omega, \mathcal{A}, P)$, and suppose that $X_n \underset{n\to\infty}{\overset{P}{\to}} X$ and $\{X_n\}$ are uniformly integrable. Then show that $\int_A X_n dP \underset{n\to\infty}{\to} \int_A X dP$ uniformly in $A \in \mathcal{A}$.

18. If for $n = 1, 2, \ldots$, the r.v.s $\{X_n\}$ and $\{Y_n\}$ are uniformly integrable, show that $\{X_n + Y_n\}$ are also uniformly integrable.

19. For $n = 1, 2, \ldots$, let X_n be identically distributed with $\mathcal{E}X_n \in \Re$. Then show that $\{\bar{X}_n\}$ are uniformly integrable, where $\bar{X}_n = \frac{1}{n} \sum_{j=1}^{n} X_j$ is the sample mean of $X_1, \ldots, X_n$.

20. Let $Y = \sup_{n \geq 1} |X_n|$ with $\mathcal{E}Y^r < \infty$ (for some $r > 0$) and $X_n \underset{n\to\infty}{\overset{\text{a.s.}}{\to}} X$. Then show that $\mathcal{E}|X|^r < \infty$ and $X_n \underset{n\to\infty}{\overset{(r)}{\to}} X$.

21. For $n = 1, 2, \ldots$, let X_n be r.v.s such that $\mathcal{E}X_n = \mu_n \in \Re$ and $\text{Var}(X_n) = \sigma_n^2 \underset{n\to\infty}{\to} 0$. Then show that $X_n - \mu_n \underset{n\to\infty}{\overset{P}{\to}} 0$.

22. Let the r.v.s $X_n, n \geq 1$, be defined as follows: $X_n(\omega) = 2^n$ if $\omega \in (0, \frac{1}{n})$, and $X_n(\omega) = 0$ otherwise, where $(\Omega, \mathcal{A}, P) = ((0, 1], \mathcal{B}_{(0,1]}, \lambda)$ and λ is the Lebesgue measure. Then show that $X_n \underset{n\to\infty}{\to} 0$ pointwise, but $\mathcal{E}|X_n|^r \underset{n\to\infty}{\not\to}$ to any finite number for any $r > 0$; in fact, $\mathcal{E}|X_n|^r \underset{n\to\infty}{\to} \infty$ for any $r > 0$.

23. For $n = 1, 2, \ldots$, let the r.v. X_n be defined by

$$X_n = \begin{cases} n^c \text{ with probability } \dfrac{1}{n} \\ 0 \text{ with probability } 1 - \dfrac{2}{n} \\ n^{-c} \text{ with probability } \dfrac{1}{n}, \quad \text{where } c \text{ is a positive constant.} \end{cases}$$

Then show that

(i) $X_n \underset{n\to\infty}{\overset{P}{\to}} 0$.

(ii) $\mathcal{E}|X_n|^r \underset{n\to\infty}{\to} 0, 0 < cr < 1; \mathcal{E}|X_n|^r \underset{n\to\infty}{\to} \infty, cr > 1$.

(iii) $X_n \underset{n\to\infty}{\overset{(r)}{\to}} 0, cr < 1$.

24. Let $\mathcal{E}X_n^2 < \infty$ for all n. Then, if $\mathcal{E}|X_n - X|^2 \underset{n\to\infty}{\to} 0$, show that $\mathcal{E}|X_n^2 - X^2| \underset{n\to\infty}{\to} 0$.

25. For $n = 1, 2, \ldots$, let X_n, Y_n and X, Y be r.v.s defined on the probability space $(\Omega, \mathcal{A}, P)$, and suppose that $X_n \underset{n\to\infty}{\overset{(r)}{\to}} X$, $Y_n \underset{n\to\infty}{\overset{(s)}{\to}} Y$ where $r, s > 1$ with $\frac{1}{r} + \frac{1}{s} = 1$. Then show that $X_n Y_n \underset{n\to\infty}{\overset{(1)}{\to}} XY$.

26. Let $(\Omega, \mathcal{A}, P) = ((0, 1), \mathcal{B}_{(0,1)}, \lambda)$, where λ is the Lebesgue measure, and let X_n be a r.v. defined by

$$X_n = \begin{cases} 1, \ \omega \in \left(0, \dfrac{1}{n}\right) \\ 0, \ \omega \in \left[\dfrac{1}{n}, 1\right) \end{cases}, \quad n = 1, 2, \ldots.$$

Then show that $X_n \underset{n\to\infty}{\overset{(r)}{\to}} 0$ for any $r > 0$, and $X_n \underset{n\to\infty}{\overset{\text{a.s.}}{\to}} 0$; indeed, $X_n \underset{n\to\infty}{\longrightarrow} 0$ pointwise.

(Compare it with Exercise 7.)

27. If $X_n \underset{n\to\infty}{\overset{(r)}{\to}} X$, show that there is a subsequence $\{X_{n_k}\} \subseteq \{X_n\}$ such that $X_{n_k} \underset{n\to\infty}{\overset{\text{a.s.}}{\to}} X$.

28. For $n = 1, 2, \ldots$, let the r.v. X_n be defined by

$$X_n = \begin{cases} e^n \text{ with probability } n^{-2} \\ 0 \text{ with probability } 1 - n^{-2}. \end{cases}$$

Then show that

(i) $X_n \underset{n\to\infty}{\overset{P}{\to}} 0$ (indeed, $X_n \underset{n\to\infty}{\overset{\text{a.s.}}{\to}} 0$).

(ii) $X_n \underset{n\to\infty}{\overset{(r)}{\nrightarrow}} 0$ for any $r > 0$ (in fact, $\mathcal{E}|X_n|^r \underset{n\to\infty}{\longrightarrow} \infty$ for all $r > 0$). (Compare it with Exercises 6 and Theorem 8.)

29. For $n = 1, 2, \ldots$, let the r.v. X_n be defined by

$$X_n = \begin{cases} 2^{cn} & \text{with probability } 2^{-n} \\ 0 & \text{with probability } 1 - 2^{-n+1} \\ -2^{cn} & \text{with probability } 2^{-n}, \text{ for some positive constant } c. \end{cases}$$

(i) Show that $X_n \underset{n\to\infty}{\overset{\text{a.s.}}{\to}} 0$.

(ii) Determine the condition c and $r (> 0)$ must satisfy, so that $X_n \underset{n\to\infty}{\overset{(r)}{\to}} 0$. (Compare it with Exercises 6 and 8.)

30. **(i)** For any r.v. X, show that $P(|X| \geq c) \leq \mathcal{E}|X|/c$ $(c > 0)$, by using the obvious inequality $|X| \geq cI(|X| \geq c)$. In particular, if $X \geq 0$, then $P(X \geq c) \leq \mathcal{E}X/c$.

(ii) For any r.v. X and any $c > 0$, show that

$$P(X \geq c) \leq e^{-tc}\mathcal{E}e^{tX} \ (t > 0), \ P(X \leq c) \leq e^{-tc}\mathcal{E}e^{tX} \ (t < 0).$$

(iii) If $\mathcal{E}X^2 < \infty$, then show that $P(|X| > 0) \geq (\mathcal{E}|X|)^2/\mathcal{E}X^2$. In particular, if $X \geq 0$, then $P(X > 0) \geq (\mathcal{E}X)^2/\mathcal{E}X^2$.

31. For $n \geq 1$, let X_n be r.v.s with $\mathcal{E}X_n = \mu_n, \sigma^2(X_n) = \sigma_n^2$, and suppose that, as $n \to \infty$, $\mu_n \to \mu \in \Re$, and $\sigma_n^2 \to 0$. Then show that $\mathcal{E}(X_n - \mu)^2 \to 0$ and $X_n \overset{P}{\to} \mu$.

32. For $n \geq 1$, let X_n and Y_n be r.v.s defined on the probability space $(\Omega, \mathcal{A}, P)$, and suppose that $|X_n|$ and $|Y_n|$, $n \geq 1$, are uniformly integrable and $X_n - Y_n \overset{P}{\to} 0$ as $n \to \infty$. Then show that:

(i) $|X_n - Y_n|$, $n \geq 1$, are uniformly integrable.

(ii) $\mathcal{E}|X_n - Y_n| \underset{n\to\infty}{\longrightarrow} 0$.

33. Consider the probability space $(\Omega, \mathcal{A}, P)$, and let Θ be an open subset of $\Re$. Let $g(\cdot;\cdot) : \Omega \times \Theta \to \Re$ be $(\mathcal{A} \times \mathcal{B}_\Theta)$-measurable, where $\mathcal{B}_\Theta$ is the σ-field of Borel subsets of Θ. We say that $g(\cdot;\theta)$ is *differentiable in q.m.* at θ, if there exists a $(\mathcal{A} \times \mathcal{B}_\Theta)$-measurable function $\dot{g}(\cdot;\theta)$ (the *quadratic mean derivative* of $g(\cdot;\theta)$ at θ) such that

$$h^{-1}[g(\cdot;\theta+h) - g(\cdot;\theta)] \underset{n\to\infty}{\overset{q.m.}{\longrightarrow}} \dot{g}(\cdot;\theta).$$

Now let $p(x;\theta) = \frac{1}{2}e^{-|x-\theta|}$, $x \in \Re$ $(\theta \in \Re)$ be the double exponential probability density function, and for $\theta, \theta^* \in \Re$, set

$$g(x;\theta,\theta^*) = [p(x;\theta^*)/p(x;\theta)]^{1/2} = \exp\left(\frac{1}{2}|x-\theta| - \frac{1}{2}|x-\theta^*|\right).$$

(i) Show that $g(x;\theta,\theta^*)$ is *not* pointwise differentiable with respect to θ^* at (θ,θ) for any $\theta = x \in \Re$.

(ii) If the r.v. X is distributed according to $p(\cdot;\theta)$, show that $g(X;\theta,\theta^*)$ is differentiable in q.m. with respect to θ^* at $(\theta,\theta), \theta \in \Re$, with q.m. derivative $\dot{g}(X;\theta)$ given by:

$$\dot{g}(X;\theta) = \begin{cases} -1/2 & \text{if } X < \theta \\ 0 \text{ (for example)} & \text{if } X = \theta \\ 1/2 & \text{if } X > \theta \end{cases}.$$

34. All r.v.s appearing below are defined on the probability space $(\Omega, \mathcal{A}, P)$.

(i) If $X \geq 0$, then $\mathcal{E}X \geq 0$ and $\mathcal{E}X = 0$ only if $P(X = 0) = 1$.

(ii) Let $X \geq Y$ with finite expectations. Then $\mathcal{E}X \geq \mathcal{E}Y$ and $\mathcal{E}X = \mathcal{E}Y$ only if $P(X = Y) = 1$.

(iii) Let $X > Y$ with finite expectations. Then $\mathcal{E}X > \mathcal{E}Y$.

(iv) Let $g : I$ open subset of $\Re$ be strictly convex (i.e.,

$$g[\alpha x + (1-\alpha)x'] < \alpha g(x) + (1-\alpha)g(x'),$$

$x, x' \in I, 0 < \alpha < 1$), let Z be a r.v. taking values in I, let $\mathcal{E}Z \in I$, and let $\mathcal{E}g(Z)$ exist. Then $\mathcal{E}g(Z) > g(\mathcal{E}Z)$ unless $P(Z = \text{constant}) = 1$.

CHAPTER

7

The Hahn–Jordan Decomposition Theorem, The Lebesgue Decomposition Theorem, and the Radon–Nikodym Theorem

This chapter revolves around three classical theorems. In Section 7.1, it is shown that any σ-additive set function φ may be decomposed into the difference of two measures (i.e., φ is a signed measure). This is a brief description of the Hahn–Jordan Decomposition Theorem.

In the following section, this decomposition is used in order to obtain the Lebesgue Decomposition Theorem. This theorem states that, if μ and ν are any two σ-finite measures, then ν can be decomposed uniquely into two components ν_c and ν_s. The component ν_c is absolutely continuous with respect to μ, and ν_s is singular with respect to μ.

In the third and final section of the chapter, the Lebesgue Decomposition Theorem is specialized to the case where $\nu \ll \mu$ in order to establish the Radon–Nikodym Theorem. Namely, to show that ν is the indefinite integral, with respect to μ, of a r.v. that is nonnegative, a.e. $[\mu]$ finite, and a.e. $[\mu]$ unique. A corollary to this theorem justifies the replacement in integrals of a probability distribution by its probability density function (p.d.f.) with respect to a dominating measure, as is routinely done in Statistics.

7.1 The Hahn–Jordan Decomposition Theorem

In all that follows, φ is a σ-additive *set function* defined on $(\Omega, \mathcal{A})$ and taking values in $(-\infty, \infty]$. The value $-\infty$ is excluded in order to avoid expressions of the form $\infty - \infty$. It will also be assumed that φ is finite for at least one set A. Then from $A = A + \oslash + \oslash + \cdots$ and the σ-additivity of φ it follows that $\varphi(\oslash) = 0$. From this fact it also follows that φ is *finitely additive.*

Definition 1. The continuity of φ is defined the same way the continuity of a measure has been defined; i.e., φ is *continuous from below* if $A_n \uparrow A$ as $n \to \infty$ implies $\varphi(A_n) \to \varphi(A)$; it is *continuous from above* if $A_n \downarrow A$ as $n \to \infty$ with $|\varphi(A_{n_0})| < \infty$ for some n_0 implies $\varphi(A_n) \underset{n\to\infty}{\to} \varphi(A)$; and it is *continuous* if it is continuous from both below and above. ■

An Introduction to Measure-Theoretic Probability, Second Edition. http://dx.doi.org/10.1016/B978-0-12-800042-7.00007-4

Lemma 1. Every σ-additive set function φ is continuous.

Proof. Let $A_n \uparrow A$ as $n \to \infty$ and suppose first that $|\varphi(A_n)| < \infty$, for all n. Then

$$A = \bigcup_{j=1}^{\infty} A_j = A_1 + \sum_{j=2}^{\infty}(A_j - A_{j-1})$$

and hence

$$\varphi(A) = \varphi(A_1) + \sum_{j=2}^{\infty} \varphi(A_j - A_{j-1}).$$

But

$$A_{j-1} \subseteq A_j \text{ implies that } A_j = A_{j-1} + (A_j - A_{j-1}),$$

so that

$$\varphi(A_j) = \varphi(A_{j-1}) + \varphi(A_j - A_{j-1}),$$

and since $|\varphi(A_n)| < \infty$, for all n, we get

$$\varphi(A_j - A_{j-1}) = \varphi(A_j) - \varphi(A_{j-1}).$$

Therefore

$$\begin{aligned}\varphi(A) &= \varphi(A_1) + \lim_{n\to\infty} \sum_{j=2}^{n} \varphi(A_j - A_{j-1}) \\ &= \varphi(A_1) + \lim_{n\to\infty} \left[\varphi(A_n) - \varphi(A_1)\right] = \lim_{n\to\infty} \varphi(A_n).\end{aligned}$$

Now suppose that $\varphi(A_{n0}) = \infty$ for some n_0. Then for all $n \geq n_0$, $A_n = A_{n_0} + (A_n - A_{n_0})$, so that $\varphi(A_n) = \varphi(A_{n0}) + \varphi(A_n - A_{n_0}) = \infty$. Also $A = A_{n_0} + (A - A_{n_0})$ implies $\varphi(A) = \varphi(A_{n_0}) + \varphi(A - A_{n_0}) = \infty$, so that $\varphi(A_n) \underset{n\to\infty}{\to} \varphi(A)$.

Next, let $A_n \downarrow A$ as $n \to \infty$ and $|\varphi(A_{n_0})| < \infty$ for some n_o. Then $A_{n_0} = A + \sum_{j=n_0}^{\infty} (A_j - A_{j+1})$, so that

$$\begin{aligned}\varphi(A_{n_0}) &= \varphi(A) + \sum_{j=n_0}^{\infty} \varphi(A_j - A_{j+1}) = \varphi(A) \\ &\quad + \lim_{n\to\infty} \sum_{j=n_0}^{n} \varphi(A_j - A_{j+1}).\end{aligned}$$

For $j \geq n_0$, $A_j = A_{j+1} + (A_j - A_{j+1})$, so that $\varphi(A_j) = \varphi(A_{j+1}) + \varphi(A_j - A_{j+1})$, and since $\varphi(A_{n_0})$ is finite, so are $\varphi(A_{j+1})$ for all $j \geq n_0$ (as follows by induction). Hence

$$\varphi(A_{n_0}) = \varphi(A) + \lim_{n\to\infty} \left[\varphi(A_{n_0}) - \varphi(A_{n+1})\right] = \varphi(A) + \varphi(A_{n_0}) - \lim_{n\to\infty} \varphi(A_n).$$

Thus $\lim_{n\to\infty} \varphi(A_n) = \varphi(A)$ and φ is continuous. ■

Definition 2. For a set function φ on $(\Omega, \mathcal{A})$, define the set functions φ^+ and φ^- on $(\Omega, \mathcal{A})$, as follows:

$$\varphi^+(A) = \sup\{\varphi(B);\ B \subseteq A, B \in \mathcal{A}\}, \quad \varphi^-(A) = -\inf\{\varphi(B);\ B \subseteq A,\quad B \in \mathcal{A}\}.$$ ■

Remark 1. Both $\varphi^+(A)$ and $\varphi^-(A)$ are ≥ 0 for all A. This is so, because $\varphi(\emptyset) = 0$ implies $\varphi^+(A) \geq 0$ for all A. Also, $\inf\{\varphi(B);\ B \subseteq A\} = -\sup\{-\varphi(B);\ B \subseteq A\}$, or $-\inf\{\varphi(B);\ B \subseteq A\} = \sup\{-\varphi(B);\ B \subseteq A\}$, and since $-\varphi(\emptyset) = 0$, we have $\sup\{-\varphi(B);\ B \subseteq A\} \geq 0$, and hence $\varphi^-(A) \geq 0$ for all A.

Then we may formulate and prove the following result.

Theorem 1 (Hahn–Jordan Decomposition Theorem). Let φ be a σ-additive function defined on $(\Omega, \mathcal{A})$, let φ^+, φ^- be as in Definition 2, and let

$$-\infty < m = \inf\{\varphi(A);\ A \in \mathcal{A}\} \leq \sup\{\varphi(A);\ A \in \mathcal{A}\} = M \leq \infty.$$

Then

(i) There exists at least one set $D \in \mathcal{A}$ such that

$$\varphi(A) \geq 0 \text{ for every } A \subseteq D \quad \text{and} \quad \varphi(A) \leq 0 \text{ for every} A \subseteq D^c \qquad (A \in \mathcal{A}).$$

(ii) $\varphi^+(A) = \varphi(A \cap D)$, $\varphi^-(A) = -\varphi(A \cap D^c)$.

(iii) $\varphi^+, \varphi^-, |\varphi| = \varphi^+ + \varphi^-$ are measures and φ^- is finite.

(iv) $\varphi = \varphi^+ - \varphi^-$ (in the sense that $\varphi(A) = \varphi^+(A) - \varphi^-(A)$, $A \in \mathcal{A}$). ■

Proof.

(i) For $j = 1, 2, \ldots$, let $\varepsilon_j > 0$ and $\sum_j \varepsilon_j < \infty\left(\text{e.g., } \varepsilon_j = \frac{1}{j^2}\right)$. For each j, let $A_j \in \mathcal{A}$ be such that $m \leq \varphi(A_j) \leq m + \varepsilon_j$. Next, from

$$A_2 = (A_1 \cap A_2) + (A_2 - A_1), \quad A_1 \cup A_2 = A_1 + (A_2 - A_1),$$

it follows that

$$\varphi(A_1 \cap A_2) = \varphi(A_2) + \varphi(A_1) - \varphi(A_1 \cup A_2).$$

Therefore

$$\begin{aligned} m \leq \varphi(A_1 \cap A_2) &= \varphi(A_2) + \varphi(A_1) - \varphi(A_1 \cup A_2) \\ &\leq 2m + \varepsilon_1 + \varepsilon_2 - m = m + \varepsilon_1 + \varepsilon_2. \end{aligned}$$

Thus, $m \leq \varphi(A_1 \cap A_2) \leq m + (\varepsilon_1 + \varepsilon_2)$ and, clearly, this is true for any two sets A_n, A_{n+1}. Furthermore, it is easily seen by induction that

$$m \leq \varphi(\bigcap_{j=n}^{k} A_j) \leq m + \sum_{j=n}^{k} \varepsilon_j.$$

Letting $k \to \infty$ and utilizing the continuity of φ, we obtain

$$m \le \varphi(\bigcap_{j=n}^{\infty} A_j) \le m + \sum_{j=n}^{\infty} \varepsilon_j. \tag{7.1}$$

Now set $D^c = \liminf_{n\to\infty} A_n = \bigcup_{n=1}^{\infty}\bigcap_{j=n}^{\infty} A_j$, so that $D = \bigcap_{n=1}^{\infty}\bigcup_{j=n}^{\infty} A_j^c = \limsup_{n\to\infty} A_n^c$. Then, as $n \to \infty$, $\bigcap_{j=n}^{\infty} A_j \uparrow D^c$ and by continuity of φ,

$$\varphi(D^c) = \varphi(\lim \bigcap_{j=n}^{\infty} A_j) = \lim \varphi(\bigcap_{j=n}^{\infty} A_j) = m$$

on account of (7.1). That is, $\varphi(D^c) = m$. Next, let $A \subseteq D^c$. Then $D^c = A + (D^c - A)$ and hence

$$m = \varphi(D^c) = \varphi(A) + \varphi(D^c - A) \ge \varphi(A) + m,$$

so that $\varphi(A) \le 0$. Finally, if $A \subseteq D$, then

$$m \le \varphi(A + D^c) = \varphi(A) + \varphi(D^c) = \varphi(A) + m,$$

so that $\varphi(A) \ge 0$. This completes the proof of (i).

(ii) Let $B \subseteq A$. Then

$$\begin{aligned}\varphi(B) &= \varphi[(B\cap D) + (B\cap D^c)] = \varphi(B\cap D) + \varphi(B\cap D^c)\\ &\le \varphi(B\cap D) \quad (\text{since } \varphi(B\cap D^c) \le 0)\\ &\le \varphi(B\cap D) + \varphi[(A-B)\cap D] \quad (\text{since } \varphi[(A-B)\cap D] \ge 0)\\ &= \varphi(A\cap D) \qquad (\text{since } (B\cap D) + (A-B)\cap D = A\cap D).\end{aligned}$$

That is, for every $B \subseteq A$, $\varphi(B) \le \varphi(A\cap D)$ and this is, in particular, true for $B \subseteq (A\cap D)$. Since in forming $\sup_{B\subseteq A} \varphi(B)$ we may restrict ourselves to $B \subseteq A$ with $B \subseteq A \cap D$ (because $\varphi(C\cap D^c) \le 0$), we have that $\varphi(B) \le \varphi(A\cap D)$ for every $B \subseteq A\cap D$ and hence $\varphi(A\cap D) = \sup_{B\subseteq A} \varphi(B) = \varphi^+(A)$.

Next, for $B \subseteq A$,

$$\begin{aligned}\varphi(B) &= \varphi(B\cap D) + \varphi(B\cap D^c) \ge \varphi(B\cap D^c) \quad (\text{since } \varphi(B\cap D) \ge 0)\\ &\ge \varphi(B\cap D^c) + \varphi[(A-B)\cap D^c] \quad (\text{since } \varphi[(A-B)\cap D^c] \le 0)\\ &= \varphi(A\cap D^c) \quad (\text{since } (B\cap D^c) + (A-B)\cap D^c = A\cap D^c).\end{aligned}$$

That is, for every $B \subseteq A$, $\varphi(B) \ge \varphi(A\cap D^c)$ and this is, in particular, true for $B \subseteq A\cap D^c$. Since in forming $\sup_{B\subseteq A}[-\varphi(B)]$ we may restrict ourselves to $B \subseteq A$ with $B \subseteq A\cap D^c$ (because $\varphi(C\cap D) \ge 0$), we have that

$$-\varphi(B) \le -\varphi(A\cap D^c) \quad \text{for every } B \subseteq A\cap D^c,$$

and hence

$$-\varphi(A \cap D^c) = \sup_{B \subseteq A} [-\varphi(B)] = -\inf_{B \subseteq A} \varphi(B) = \varphi^-(A).$$

(iii) That φ^+ is a measure follows from the fact that

$$\varphi^+(A) = \varphi(A \cap D), \quad \varphi(\emptyset) = 0, \quad \varphi(B) \geq 0 \text{ for every } B \subseteq D$$

and the σ-additivity of φ; similarly for φ^-. Finally, from $(-\infty <)m \leq \varphi(A \cap D^c) \leq 0$ it follows that $0 \leq -\varphi(A \cap D^c) = \varphi^-(A) \leq -m$, so that φ^- is finite.

(iv) By (ii),

$$\begin{aligned}\varphi(A) &= \varphi[(A \cap D) + (A \cap D^c)] = \varphi(A \cap D) + \varphi(A \cap D^c) \\ &= \varphi^+(A) - \varphi^-(A).\end{aligned}$$

■

Remark 2.

(i) The theorem is true without the assumption that $m > -\infty$, but the proof is somewhat more complicated (see, e.g., pages 86–87 in Loève (1963), or pages 104–106 in Neveu (1965)).

(ii) If μ_1, μ_2 are two measures such that $\mu_1(A) - \mu_2(A)$ is defined for every $A \in \mathcal{A}$, then $\mu_1 - \mu_2$ is called a *signed measure*. Thus the theorem shows that every σ-additive function is a signed measure.

(iii) For the set D, one has $\varphi(D) = M$ and $\varphi(D^c) = m$. In fact, there exist $B_n \in \mathcal{A}$ such that $\varphi(B_n) \underset{n \to \infty}{\to} M$. Then

$$\begin{aligned}\varphi(D) &= \varphi(D \cap B_n) + \varphi(D \cap B_n^c) \geq \varphi(D \cap B_n) \\ &\quad (\text{since } \varphi(D \cap B_n^c) \geq 0), \\ &\geq \varphi(D \cap B_n) + \varphi(D^c \cap B_n) \quad (\text{since } \varphi(D^c \cap B_n) \leq 0) \\ &= \varphi(B_n) \quad (\text{by finite additivity of } \varphi).\end{aligned}$$

That is, $\varphi(D) \geq \varphi(B_n)$ and hence, as $n \to \infty$, $\varphi(D) \geq M$. Since also $\varphi(D) \leq M$, it follows that $\varphi(D) = M$. Next, recall that $m = \inf\{\varphi(A); A \in \mathcal{A}\}$. Then there exists $C_n \in \mathcal{A}$ such that $\varphi(C_n) \to m$ as $n \to \infty$. Furthermore,

$$\begin{aligned}\varphi(D^c) &= \varphi(D^c \cap C_n) + \varphi(D^c \cap C_n^c) \leq \varphi(D^c \cap C_n) \\ &\quad (\text{since } \varphi(D^c \cap C_n^c) \leq 0), \\ &\leq \varphi(D^c \cap C_n) + \varphi(D \cap C_n) \quad (\text{since } \varphi(D \cap C_n) \geq 0), \\ &= \varphi(C_n) \quad (\text{by finite additivity of } \varphi).\end{aligned}$$

That is, $\varphi(D^c) \leq \varphi(C_n)$ and hence, as $n \to \infty, \varphi(D^c) \leq m$. Since also $\varphi(D^c) \geq m$, it follows that $\varphi(D^c) = m$.

Corollary. Under the assumptions of the theorem, $|\varphi|$ is bounded if and only if $|\varphi|(\Omega) < \infty$, or if and only if $\varphi^+(\Omega) < \infty$.

Proof. The first assertion follows by the fact that $|\varphi|$ is a measure, so that $|\varphi|(A) \leq |\varphi|(\Omega)$ for all $A \in \mathcal{A}$. The second assertion follows by the fact that $\varphi^-(\Omega) \leq -m < \infty$, and the expression $|\varphi|(\Omega) = \varphi^+(\Omega) + \varphi^-(\Omega)$. ■

Definition 3. If μ, ν are two measures on $\mathcal{A}$, we recall that ν is said to be *μ-continuous* (or *absolutely continuous with respect to μ*), denoted by $\nu \ll \mu$, if $\mu(A) = 0$ implies that $\nu(A) = 0$. We also say that ν is *dominated* by μ or μ *dominates* ν. If $\nu \ll \mu$ and $\mu \ll \nu$, then μ and ν are said to be *mutually absolutely continuous* and write $\mu \approx \nu$. ν is said to be *μ-singular* (or *singular with respect to* μ) if there exists $N \in \mathcal{A}$ with $\mu(N) = 0$ such that $\nu(A) = \nu(A \cap N)$ for every $A \in \mathcal{A}$. We also say that ν and μ are *orthogonal* and write $\nu \perp \mu$. ■

7.2 The Lebesgue Decomposition Theorem

Theorem 2 (Lebesgue Decomposition Theorem). Let μ, ν be two σ-finite measures on $\mathcal{A}$. Then

(i) There exists a decomposition of ν into a μ-continuous measure ν_c and a μ-singular measure ν_s such that $\nu = \nu_c + \nu_s$ (in the sense that $\nu(A) = \nu_c(A) + \nu_s(A),\ A \in \mathcal{A}$).

(ii) The decomposition in (i) is unique.

(iii) ν_c is the indefinite integral of a nonnegative, a.e. $[\mu]$ finite r.v. X determined up to μ-equivalence (i.e., if for every $A \in \mathcal{A}$, $\nu_c(A) = \int_A X d\mu$ and if $\nu_c(A) = \int_A X' d\mu$ for another r.v. X', then $\mu(X \neq X') = 0$). ■

Proof.

Case 1: μ, ν finite.

(i) Let $\mathcal{X} = \{X \geq 0,\ \text{a.e. } [\mu] \text{ finite r.v.s}; \int_A X d\mu \leq \nu(A),\ A \in \mathcal{A}\}$. Then $X = 0 \in \mathcal{X}$, so that $\mathcal{X} \neq \oslash$, and for every $X \in \mathcal{X}$, $(0 \leq) \int X d\mu \leq \nu(\Omega) < \infty$. Thus

$$\sup_{X \in \mathcal{X}} \int X d\mu \overset{def}{=} \alpha < \infty \quad \text{and there exists } \{X_n\} \subseteq \mathcal{X} \text{ such that}$$

$$\int X_n d\mu \underset{n\to\infty}{\rightarrow} \alpha.$$

Define Y_n by: $Y_n = \max_{1 \leq k \leq n} X_k$. Then $Y_n \uparrow$ as $n \to \infty$ and let $X = \lim_{n\to\infty} Y_n$. Thus $0 \leq Y_n \uparrow X$ as $n \to \infty$ and hence $\int Y_n d\mu \uparrow \int X d\mu$ as $n \to \infty$. On the other hand, $X_n \leq Y_n$ and hence $\int X_n d\mu \leq \int Y_n d\mu$. Letting $n \to \infty$, we get then $\alpha \leq \int X d\mu$. Thus

$$\int X d\mu \geq \alpha. \tag{7.2}$$

Now define

$$A_1 = (Y_n = X_1)$$
$$A_2 = (Y_n = X_2) - A_1$$

$$A_3 = (Y_n = X_3) - (A_1 + A_2)$$
$$\cdots\cdots\cdots\cdots$$
$$A_n = (Y_n = X_n) - (A_1 + A_2 + \cdots + A_{n-1}).$$

Then, clearly, $\{A_1, \ldots, A_n\}$ is a partition of Ω and therefore for any $A \in \mathcal{A}$, one has

$$\int_A Y_n d\mu = \int_{\Sigma_{j=1}^n A \cap A_j} Y_n d\mu = \sum_{j=1}^n \int_{A \cap A_j} Y_n d\mu = \sum_{j=1}^n \int_{A \cap A_j} X_j d\mu$$
$$\leq \sum_{j=1}^n \nu(A \cap A_j) = \nu(A);$$

$$\text{i.e.,} \int_A Y_n d\mu \leq \nu(A) \text{ for every } A \in \mathcal{A}, \text{ and hence } Y_n \in \mathcal{X}.$$

Thus $\int Y_n d\mu \leq \alpha$ and since $\int Y_n d\mu \uparrow \int X d\mu$ as $n \to \infty$, one has $\int X d\mu \leq \alpha$. This result, together with (7.2), gives $\int X d\mu = \alpha$. Furthermore,

$$\int_A Y_n d\mu \leq \nu(A) \quad \text{for every } A \in \mathcal{A} \text{ and } \int_A Y_n d\mu \uparrow \int_A X d\mu \text{ as } n \to \infty$$
$$\text{(as follows from } 0 \leq Y_n \uparrow X \text{ as } n \to \infty)$$

imply that $\int_A X d\mu \leq \nu(A)$ for every $A \in \mathcal{A}$. Since $(0 \leq) \int X d\mu = \alpha < \infty$ implies that X is a.e. $[\mu]$ finite, we have that $X \in \mathcal{X}$ and $\int X d\mu = \alpha$.

Now define ν_c on $\mathcal{A}$ as follows: $\nu_c(A) = \int_A X d\mu$. Then ν_c is μ-continuous. Next, define ν_s by: $\nu_s(A) = \nu(A) - \nu_c(A)$, $A \in \mathcal{A}$. Since

$$\nu(A) - \nu_c(A) = \nu(A) - \int_A X d\mu \geq 0,$$

by the fact that $X \in \mathcal{X}$, we have that $\nu_s(A) \geq 0$, $A \in \mathcal{A}$. Since both ν and ν_c are measures, it follows that ν_s is itself a measure. The proof of (i) will be completed by showing that ν_s is μ-singular. To this end, for each n define the set function φ_n as follows:

$$\varphi_n(A) = \nu_s(A) - \frac{1}{n}\mu(A), \quad A \in \mathcal{A}.$$

Then, clearly, φ_n is a finite σ-additive set function. Hence, by Theorem 1, there exists $D_n \in \mathcal{A}$ such that

$$\varphi_n(A \cap D_n) \geq 0, \quad \varphi_n(A \cap D_n^c) \leq 0, \quad A \in \mathcal{A}. \tag{7.3}$$

Set $D = \bigcap_{j=1}^{\infty} D_j^c$. Then, as $r \to \infty$,

$$A \cap \bigcap_{j=1}^r D_j^c \downarrow A \cap D \quad \text{and hence} \quad \varphi_n(A \cap \bigcap_{\substack{j=1 \\ j \neq n}}^r D_j^c) \underset{r\to\infty}{\to} \varphi_n(A \cap D)$$

by Lemma 1. On the other hand, for $r \geq n$,

$$\varphi_n(A \cap \bigcap_{j=1}^{r} D_j^c) = \varphi_n[(A \cap \bigcap_{\substack{j=1 \\ j \neq n}}^{r} D_j^c) \cap D_n^c] \leq 0 \quad \text{by} \quad (7.3)$$

(where the role of A in (7.3) is played by $A \cap \bigcap_{\substack{j=1 \\ j \neq n}}^{r} D_j^c$ here) for all $n \leq r$; i.e., $\varphi_n(A \cap \bigcap_{j=1}^{r} D_j^c) \leq 0, \ n \leq r,$ and hence, as $r \to \infty$ $\varphi_n(A \cap D) \leq 0$, for all n. Equivalently,

$$\nu_s(A \cap D) - \frac{1}{n}\mu(A \cap D) \leq 0 \text{ for all } n, \text{ or } \nu_s(A \cap D) \leq \frac{1}{n}\mu(A \cap D)$$
for all n.

Letting $n \to \infty$, we get $\nu_s(A \cap D) = 0$, $A \in \mathcal{A}$. Since $\nu_s(A) = \nu_s(A \cap D) + \nu_s(A \cap D^c)$, we have that $\nu_s(A) = \nu_s(A \cap D^c)$. Thus ν_s will be μ-singular if we show that $\mu(D^c) = 0$. To this end, we have

$$\nu_c(A) = \nu(A) - \nu_s(A) = \nu(A) - \nu_s(A \cap D^c) \leq \nu(A) - \nu_s(A \cap D_n)$$
(since $D_n \subseteq D^c$).

Thus

$$\begin{aligned}
\int_A \left(X + \frac{1}{n}I_{D_n}\right) d\mu &= \nu_c(A) + \frac{1}{n}\mu(A \cap D_n) \\
&\quad \text{(from the definition of } \nu_c) \\
&\leq \nu(A) - \nu_s(A \cap D_n) + \frac{1}{n}\mu(A \cap D_n) \\
&\quad \text{(by the previous inequality)} \\
&= \nu(A) - [\nu_s(A \cap D_n) - \frac{1}{n}\mu(A \cap D_n)] \\
&= \nu(A) - \varphi_n(A \cap D_n) \\
&\quad \text{(from the definition of } \varphi_n), \\
&\leq \nu(A) \qquad (\text{since } \varphi_n(A \cap D_n) \geq 0).
\end{aligned}$$

So, for every $A \in \mathcal{A}$, one has $\int_A \left(X + \frac{1}{n}I_{D_n}\right) d\mu \leq \nu(A)$, so that $(X + \frac{1}{n}I_{D_n}) \in \mathcal{X}$. Hence $\int (X + \frac{1}{n}I_{D_n}) d\mu \leq \alpha$. But $\int_A (X + \frac{1}{n}I_{D_n}) d\mu \leq \alpha + \frac{1}{n}\mu(D_n)$, since $\int X d\mu = \alpha$; therefore $\alpha + \frac{1}{n}\mu(D_n) \leq \alpha$, so that $\mu(D_n) = 0$, for all n. Since $D = \bigcap_{j=1}^{\infty} D_j^c$ implies $\bigcup_{j=1}^{n} D_j \uparrow D^c$ as $n \to \infty$, we obtain that $\mu(D^c) = 0$. The proof of (i) is complete.

(ii) In (i) we proved that $\nu = \nu_c + \nu_s$, where

$$\begin{aligned}
\nu_c(A) &= \int_A X d\mu = \int_{A \cap D} X d\mu + \int_{A \cap D^c} X d\mu \\
&= \int_{A \cap D} X d\mu = \nu_c(A \cap D),
\end{aligned}$$

since $\mu(A \cap D^c) \leq \mu(D^c) = 0$. Also let $\nu = \nu'_c + \nu'_s$, where

$$\nu'_c(A) = \nu'_c(A \cap D_0), \quad \nu'_s(A) = \nu'_s(A \cap D_0^c) \text{ with } \mu(D_0^c) = 0.$$

We have

$$\begin{aligned} &\nu_c(A) + \nu_s(A) = \nu'_c(A) + \nu'_s(A) \quad (= \nu(A)) \quad \text{or} \\ &\nu_c(A) - \nu'_c(A) = \nu'_s(A) - \nu_s(A) \quad \text{for every } A \in \mathcal{A}. \end{aligned} \tag{7.4}$$

Set $N = D^c \cap D_0^c$. Then $\mu(N) = 0$ and hence $\nu_c(N) = \nu'_c(N) = 0$, since they are both μ-continuous. For $A \in \mathcal{A}$, write $A = (A \cap N) + (A \cap N^c)$. Then

$$\begin{aligned} \nu_c(A) - \nu'_c(A) &= [\nu_c(A \cap N) - \nu'_c(A \cap N)] \\ &\quad + [\nu_c(A \cap N^c) - \nu'_c(A \cap N^c)] \\ &= [\nu_c(A \cap N) - \nu'_c(A \cap N)] \\ &\quad + [\nu'_s(A \cap N^c) - \nu_s(A \cap N^c)] \quad (\text{by (7.4)}). \end{aligned}$$

But $\nu_c(A \cap N) = \nu'_c(A \cap N) = 0$ since $A \cap N \subseteq N$ and $\mu(N) = 0$. Next, $\nu'_s(A \cap N^c) = \nu_s(A \cap N^c) = 0$ because $A \cap N^c = A \cap D \cap D_0 \subseteq A \cap D$, $A \cap D_0$ and $\nu_s(A \cap D) = \nu'_s(A \cap D_0) = 0$. So $\nu_c = \nu'_c$ and therefore $\nu_s = \nu'_s$.

(iii) In (i) it was seen that $\nu_c(A) = \int_A X d\mu$ for every $A \in \mathcal{A}$. Let also $\nu_c(A) = \int_A X' d\mu$ for every $A \in \mathcal{A}$. To show that $X = X'$ a.e.$[\mu]$. In fact, if $\mu(X - X' > 0) > 0$, then there exists $\varepsilon > 0$ such that $\mu(X - X' > \varepsilon) > 0$. Hence $\int_{(X-X'>\varepsilon)}(X - X')d\mu \geq \varepsilon\mu(X - X' > \varepsilon) > 0$, which is a contradiction (to the assumption that $\int_A X d\mu = \int_A X' d\mu = \nu_c(A)$ for every $A \in \mathcal{A}$). Thus $\mu(X - X' > 0) = 0$ and similarly $\mu(X - X' < 0) = 0$, so that $\mu(X \neq X') = 0$.

Case 2: μ, ν σ-finite.

(i) From the σ-finiteness of μ, ν, there exist two countable partitions of Ω $\{A_j, j = 1, 2, \ldots\}$, $\{A'_j, j = 1, 2, \ldots\}$, say, such that $\mu(A_j)$, $\nu(A'_j) < \infty$, $j \geq 1$. Consider the intersection of these two partitions, which is another partition of Ω; call it $\{B_j, j \geq 1\}$. Then $\mu(B_j)$, $\nu(B_j) < \infty$, $j \geq 1$. Consider the restrictions μ_n, ν_n of μ, ν on B_n; i.e.,

$$\left.\begin{aligned} \mu_n(A) = \mu(A \cap B_n) \\ \nu_n(A) = \mu(A \cap B_n) \end{aligned}, \quad A \in \mathcal{A}. \right\} \quad \text{Then} \quad \left.\begin{aligned} \mu(A) = \textstyle\sum_n \mu_n(A) \\ \nu(A) = \textstyle\sum_n \nu_n(A). \end{aligned}\right\} \tag{7.5}$$

On each B_n, the theorem is true. Therefore $\nu_n = \nu_{nc} + \nu_{ns}$, where ν_{nc} is μ_n-continuous and ν_{ns} is μ_n-singular. We assert that ν_{nc} is, actually, μ-continuous. In fact, let $\mu(A) = 0$. Then $\mu_n(A) = 0$ for all n and hence $\nu_{nc}(A) = 0$ for all n. Next, from the μ_n-singularity of ν_{ns} we have that there exists $N_n \in \mathcal{A}$ such that $\mu_n(N_n) = 0$ and $\nu_{ns}(A) = \nu_{ns}(A \cap N_n)$, $A \in \mathcal{A}$, so that $\nu_{ns}(A \cap N_n^c) = 0$, $A \in \mathcal{A}$. Look at B_n^c. Then

$$\nu_n(A \cap B_n^c) = \mu(A \cap B_n^c \cap B_n) = \mu(\emptyset) = 0,$$

and similarly $\mu_n(A \cap B_n^c) = 0$. But ν_{nc} is μ_n-continuous. Hence $\nu_{nc}(A \cap B_n^c) = 0$ and this, together with $\nu_n(A \cap B_n^c) = 0$, implies that $\nu_{ns}(A \cap B_n^c) = 0$. To summarize,

$$\nu_{ns}(A \cap N_n^c) = 0, \quad \nu_{ns}(A \cap B_n^c) = 0, \quad A \in \mathcal{A}. \tag{7.6}$$

We assert that ν_{ns} is μ-singular. In fact,

$$\begin{aligned}\nu_{ns}[A\cap(B_n\cap N_n)^c] &= \nu_{ns}[A\cap(B_n^c\cup N_n^c)]\\ &= \nu_{ns}[(A\cap B_n^c)\cup(A\cap N_n^c)]\\ &= \nu_{ns}(A\cap B_n^c)+\nu_{ns}(A\cap N_n^c)\\ &\quad -\nu_{ns}(A\cap B_n^c\cap N_n^c)=0 \quad \text{by (7.6)}.\end{aligned}$$

So for the set $M_n = B_n \cap N_n$, we have that $\mu_n(M_n) = 0$ (since $\mu_n(N_n) = 0$) and $\nu_{ns}(A\cap M_n^c) = 0$. But $\mu_n(M_n) = \mu(M_n \cap B_n) = \mu(M_n)$ from (7.5) and the fact that $M_n \subseteq B_n$. Thus $\mu(M_n) = 0$ and $\nu_{ns}(A\cap M_n^c) = 0$, which is equivalent to saying that ν_{ns} is μ-singular. Up to this point we have shown that ν_{nc} is μ-continuous and ν_{ns} is μ-singular (on B_n).

The μ-singularity of ν_{ns} implies the existence of a set $N_n' \in \mathcal{A}$ such that $\mu(N_n') = 0$ and $\nu_{ns}(A\cap N_n^{'c}) = 0$. Set $N = \bigcup_{j=1}^{\infty} N_j'$. Then $\mu(N) = 0$ and

$$\begin{aligned}\nu_{ns}(A\cap N^c) &= \nu_{ns}[A\cap(\bigcup_{j=1}^{\infty} N_j')^c] = \nu_{ns}[A\cap(\bigcap_{j=1}^{\infty} N_j^{'c})]\\ &\le \nu_{ns}(A\cap N_n^{'c}) = 0;\end{aligned}$$

i.e., $\mu(N) = 0$, and hence $\nu_{nc}(N) = 0$, and $\nu_{ns}(A\cap N^c) = 0$. Next,

$$\begin{aligned}\nu_{nc}(A) &= \nu_{nc}(A\cap N^c)+\nu_{nc}(A\cap N) = \nu_{nc}(A\cap N^c)\\ &= \nu_{nc}(A\cap N^c)+\nu_{ns}(A\cap N^c)\\ &= \nu_n(A\cap N^c).\end{aligned}$$

That is, $\nu_{nc}(A) = \nu_n(A\cap N^c)$ and therefore

$$\nu_c(A) \overset{def}{=} \sum_n \nu_{nc}(A) = \sum_n \nu_n(A\cap N^c) = \nu(A\cap N^c) \text{ (by (7.5))}. \tag{7.7}$$

Next,

$$\begin{aligned}\nu_{ns}(A) &= \nu_{ns}(A\cap N)+\nu_{ns}(A\cap N^c) = \nu_{ns}(A\cap N)\\ &= \nu_{ns}(A\cap N)+\nu_{nc}(A\cap N) = \nu_n(A\cap N).\end{aligned}$$

That is, $\nu_{ns}(A) = \nu_n(A\cap N)$ and therefore

$$\nu_s(A) \overset{def}{=} \sum_n \nu_{ns}(A) = \sum_n \nu_n(A\cap N) = \nu(A\cap N) \text{ (by (7.5))}. \tag{7.8}$$

From (7.7) and (7.8), we have that $\nu(A) = \nu_c(A) + \nu_s(A)$, $A \in \mathcal{A}$, whereas ν_c is μ-continuous since every ν_{nc} is μ-continuous. Furthermore, ν_s is μ-singular since $\nu_s(A\cap N^c) = \nu(A\cap N^c\cap N) = \nu(\emptyset) = 0$. This completes the proof of (i).

(ii) By (i), $\nu = \nu_c + \nu_s$, where $\nu_c(A) = \nu(A \cap N^c)$, $\nu_s(A) = \nu(A \cap N)$, $A \in \mathcal{A}$, and $\mu(N) = 0$. Let $\nu = \nu_c' + \nu_s'$ be another decomposition of ν into a μ-continuous measure ν_c' and a μ-singular measure ν_s'. For each n, consider the restrictions ν_n, ν_{nc}, ν_{ns}, ν_{nc}' and ν_{ns}' of ν, ν_c, ν_s, ν_c', and ν_s', respectively, to B_n. Then we have $\nu_n = \nu_{nc} + \nu_{ns} = \nu_{nc}' + \nu_{ns}'$ and hence $\nu_{nc}' = \nu_{nc}$, $\nu_{ns}' = \nu_{ns}$ by part (ii) in Case 1. Since

$$\nu_c(A) = \sum_n \nu_{nc}(A \cap B_n), \quad \nu_s(A) = \sum_n \nu_{ns}(A \cap B_n)$$

and

$$\nu_c'(A) = \sum_n \nu_{nc}'(A \cap B_n), \quad \nu_s'(A) = \sum_n \nu_{ns}'(A \cap B_n)$$

we have $\nu_c = \nu_c'$, $\nu_s = \nu_s'$, as was to be seen.

(iii) From part (i) of Case 1, we have

$$\nu_{nc}(A) = \int_A X_n d\mu_n, \quad A \in \mathcal{A},$$

where X_n is ≥ 0 and an a.e. $[\mu_n]$ finite r.v. Actually, since μ_n assigns measure 0 outside B_n, we may assume that X_n is 0 on B_n^c. In the course of the proof of (i) in the present case, it was seen that $\nu_{nc}(A \cap B_n^c) = 0$. Therefore $\nu_{nc}(A) = \nu_{nc}(A \cap B_n)$ and hence

$$\int_A X_n d\mu_n = \int_{A \cap B_n} X_n d\mu_n.$$

On Ω, define X as follows: $X = \sum_n X_n$. Then, clearly, $X(\omega) = X_n(\omega)$ for $\omega \in B_n$, and

$$\nu_{nc}(A) = \int_A X_n d\mu_n = \int_{A \cap B_n} X_n d\mu_n = \int_{A \cap B_n} X_n d\mu = \int_{A \cap B_n} X d\mu.$$

Therefore

$$\nu_c(A) = \sum_n \int_{A \cap B_n} X d\mu = \int_A X d\mu.$$

That X is a.e. $[\mu]$ well defined follows as in part (iii) of Case 1. ■

Remark 3. The theorem is true if ν is only a σ-additive, σ-finite set function (see page 132 in Loève (1963)), but then the r.v. X need not be ≥ 0.

Remark 4. The theorem is still true if ν is not even a σ-finite set function; i.e., if ν is only a σ-additive set function the theorem is true, but then X need be neither ≥ 0 nor a.e. $[\mu]$ finite.

Remark 5. The theorem need not be true if μ is not σ-finite even if ν is a finite measure.

7.3 The Radon–Nikodym Theorem

Theorem 3 (Radon–Nikodym Theorem). Let μ, ν be two σ-finite measures on $\mathcal{A}$ such that ν is μ-continuous ($\nu \ll \mu$). Then ν is the indefinite integral of a nonnegative, a.e. $[\mu]$ finite r.v. X that is unique up to μ-equivalence. ■

Proof. By Theorem 2, $\nu = \nu_c + \nu_s$, where $\nu_c \ll \mu$ and ν_s is μ-singular. So there exists $N \in \mathcal{A}$ with $\mu(N) = 0$ such that $\nu_s(A) = \nu_s(A \cap N)$, $A \in \mathcal{A}$. Then, for $A \in \mathcal{A}$,

$$\nu(A) = \nu_c(A) + \nu_s(A). \tag{7.9}$$

In (7.9), take $A = N$ and use the assumption that $\nu \ll \mu$ and the fact that $\nu_c \ll \mu$ to obtain:

$$0 = \nu(N) = \nu_c(N) + \nu_s(N) = 0 + \nu_s(N) = \nu_s(N).$$

That is, $\nu_s(N) = 0$ and hence $\nu_s(A) = \nu_s(A \cap N) \leq \nu_s(N) = 0$. In other words, $\nu_s(A) = 0$ for all $A \in \mathcal{A}$, or to put it differently, $\nu_s = 0$. It follows that $\nu = \nu_c$, and therefore $\nu(A) = \nu_c(A) = \int_A X d\mu$, $A \in \mathcal{A}$, for a nonnegative r.v. X which is a.e. $[\mu]$ finite and a.e. $[\mu]$ uniquely determined. ■

Remark 6. It is to be emphasized that a.e. $[\mu]$ uniqueness of X means that, if Y is another nonnegative, a.e. $[\mu]$ finite r.v. such that $\int_A X d\mu = \int_A Y d\mu$, $A \in \mathcal{A}$, then $X = Y$ a.e. $[\mu]$.
Actually, this is true for any two r.v.s X and Y (that are finite everywhere, or if extended r.v.s., are a.e. $[\mu]$ finite) for which the integrals $\int X d\mu$ and $\int Y d\mu$ exist. That is, the following result holds.

Proposition 1. Let X and Y be (real-valued) r.v.s, or extended r.v.s but a.e. $[\mu]$ finite, and assume that the integrals $\int X \, d\mu$ and $\int Y \, d\mu$ exist. Then $\int_A X \, d\mu = \int_A Y \, d\mu$ for all $A \in \mathcal{A}$ implies that $X = Y$ a.e. $[\mu]$.

Proof. Indeed, the existence of $\int X d\mu$ and $\int Y d\mu$ implies the existence of $\int_A X d\mu$ and $\int_A Y d\mu$, $A \in \mathcal{A}$ (by the Corollary to Theorem 5 in Chapter 4). Assume first that $\int X d\mu$ is finite. Then $\int_A X d\mu$ is also finite, since $|\int_A X d\mu| = |\int X I_A d\mu| \leq \int |X| d\mu < \infty$, and so are $\int_A X^+ d\mu$ and $\int_A X^- d\mu$, because $\infty > \int |X| d\mu = \int X^+ d\mu + \int X^- d\mu$.

From $\int X d\mu = \int Y d\mu$, we have that $\int Y d\mu$ is finite and then so are $\int_A Y^+ d\mu$ and $\int_A Y^- d\mu$, $A \in \mathcal{A}$, as before. Next, from $\int_A X d\mu = \int_A Y d\mu$, we get $\int_A X^+ d\mu - \int_A X^- d\mu = \int_A Y^+ d\mu - \int_A Y^- d\mu$, or because of finiteness, $\int_A X^+ d\mu + \int_A Y^- d\mu = \int_A Y^+ d\mu + \int_A X^- d\mu$, or $\int_A (X^+ + Y^-) d\mu = \int_A (Y^+ + X^-) d\mu$. Since this is true for all $A \in \mathcal{A}$ and the integrands are ≥ 0 (and a.e. $[\mu]$ finite), we obtain $X^+ + Y^- = Y^+ + X^-$ a.e. $[\mu]$, or $X^+ - X^- = Y^+ - Y^-$ a.e. $[\mu]$, or $X = Y$ a.e. $[\mu]$. Similarly if $\int Y d\mu$ is finite.

Now, suppose that $\int X d\mu = \infty$. Then from $\infty = \int X d\mu = \int X^+ d\mu - \int X^- d\mu$, we get $\int X^+ d\mu = \infty$ and $\int X^- d\mu < \infty$. Then, for any $A \in \mathcal{A}$, $\int_A X^+ d\mu \leq \infty$ and $\int_A X^- d\mu < \infty$. From $\int X d\mu = \int Y d\mu$, we get $\int Y^+ d\mu = \infty$ and $\int Y^- d\mu < \infty$, so that $\int_A Y^+ d\mu \leq \infty$ and $\int_A Y^- d\mu < \infty$. Therefore the relation $\int_A X d\mu = \int_A Y d\mu$ becomes $\int_A X^+ d\mu - \int_A X^- d\mu = \int_A Y^+ d\mu - \int_A Y^- d\mu$, or, because of finiteness of

$\int_A X^- d\mu$ and $\int_A Y^- d\mu$, $\int_A (X^+ + Y^-) d\mu = \int_A (Y^+ + X^-) d\mu$. Hence, as before, $X^+ + Y^- = Y^+ + X^-$ a.e. $[\mu]$, or $X^+ - X^- = Y^+ - Y^-$ a.e. $[\mu]$, or $X = Y$ a.e. $[\mu]$. Likewise if $\int Y d\mu = \infty$.

If $\int X d\mu = -\infty$, then from $-\infty = \int X d\mu = \int X^+ d\mu - \int X^- d\mu$, it follows that $\int X^+ d\mu < \infty$ and $\int X^- d\mu = \infty$. Therefore $\int_A X^+ d\mu < \infty$ and $\int_A X^- d\mu \leq \infty$ for all $A \in \mathcal{A}$. Once again $-\infty = \int X d\mu = \int Y d\mu = \int Y^+ d\mu - \int Y^- d\mu$ implies $\int Y^+ d\mu < \infty$ and $\int Y^- d\mu = \infty$, so that $\int_A Y^+ d\mu < \infty$ and $\int_A Y^- d\mu \leq \infty$. Hence the relation $\int_A X d\mu = \int_A Y d\mu$ becomes $\int_A X^+ d\mu - \int_A X^- d\mu = \int_A Y^+ d\mu - \int_A Y^- d\mu$, or, because of finiteness of $\int_A X^+ d\mu$ and $\int_A Y^+ d\mu$, $-\int_A X^- d\mu - \int_A Y^+ d\mu = -\int_A Y^- d\mu - \int_A X^+ d\mu$, or $\int_A (X^- + Y^+) d\mu = \int_A (Y^- + X^+) d\mu$. It follows that $X^- + Y^+ = Y^- + X^+$ a.e. $[\mu]$, or $X^+ - X^- = Y^+ - Y^-$ a.e. $[\mu]$, or $X = Y$ a.e. $[\mu]$. Similarly if $\int Y d\mu = -\infty$.

So, in all cases as described above, $\int_A X d\mu = \int_A Y d\mu$ for all $A \in \mathcal{A}$ implies $X = Y$ a.e. $[\mu]$. ■

Remark 7. The r.v. X of the theorem is called a *Radon–Nikodym derivative* of ν with respect to μ and is denoted by $X = \frac{d\nu}{d\mu}$. Let now $f : \Re \rightarrow \Re$ and suppose that $\int_{-\infty}^{b} f(t)dt$ exists. Then for all $x \in (-\infty, b]$, $\int_{-\infty}^{x} f(t)dt = F(x)$ exists and $\frac{dF(x)}{dx} = f(x)$ at all continuity points $x \in (-\infty, b]$ of f. The point function F may also be thought of as a set function over the intervals $(-\infty, x]$, $x \in (-\infty, b]$. Thus, the notation $\frac{d\nu}{d\mu}$ generalizes the notation $\frac{dF(x)}{dx}$.

Remark 8. In most applications, $(\Omega, \mathcal{A}) = (\Re^n, \mathcal{B}^n)$, $\nu = P$ and $\mu = \lambda_n$, the n-dimension Lebesgue measure. This is the case, e.g., with all of the common distributions such as uniform, normal, gamma, beta, or Cauchy. In the discrete case, such as that of binomial, Poisson, negative binomial, or hypergeometric, the measure μ is the *counting measure*; i.e., the measure that assigns mass one to singletons.

Corollary. Let μ and λ be σ-finite measures on $\mathcal{A}$ such that $\mu \ll \lambda$, and let X be a r.v. for which $\int X d\mu$ exists. Then

$$\int_A X d\mu = \int_A X \frac{d\mu}{d\lambda} d\lambda \text{ for every } A \in \mathcal{A}.$$

Proof. First, let $X = I_B$ for some $B \in \mathcal{A}$. Then

$$\int_A X d\mu = \int_A I_B d\mu = \int_{A \cap B} d\mu = \mu(A \cap B). \tag{7.10}$$

Also,

$$\mu(A \cap B) = \int_{A \cap B} \frac{d\mu}{d\lambda} d\lambda = \int_A (I_B \frac{d\mu}{d\lambda}) d\lambda = \int_A \left(X \frac{d\mu}{d\lambda} \right) d\lambda. \tag{7.11}$$

From (7.10) and (7.11), we have then $\int_A X d\mu = \int_A \left(X \frac{d\mu}{d\lambda}\right) d\lambda,\ A \in \mathcal{A}$. Now, let $X = \sum_{i=1}^{r} \alpha_i I_{A_i}, \alpha_i \geq 0, i = 1, \ldots, r$. Then

$$\int_A X d\mu = \sum_{i=1}^{r} \alpha_i \int_A I_{A_i} d\mu = \sum_{i=1}^{r} \alpha_i \int_A \left(I_{A_i} \frac{d\mu}{d\lambda}\right) d\lambda$$

(by the previous step)

$$= \int_A \left[\left(\sum_{i=1}^{r} \alpha_i I_{A_i}\right) \frac{d\mu}{d\lambda}\right] d\lambda = \int_A \left(X \frac{d\mu}{d\lambda}\right) d\lambda,$$

so, again, the conclusion holds true.

Next, let $X \geq 0$. Then there exist $0 \leq X_n$ simple r.v.s $\uparrow X$ as $n \to \infty$, which implies $0 \leq X_n I_A \uparrow X I_A$ as $n \to \infty$. Therefore, by the Lebesgue Monotone Convergence Theorem,

$$\int X_n I_A d\mu \underset{n\to\infty}{\to} \int X I_A d\mu \text{ or } \int_A X_n d\mu \underset{n\to\infty}{\to} \int_A X d\mu. \tag{7.12}$$

However,

$$\begin{aligned}\int_A X_n d\mu &= \int_A \left(X_n \frac{d\mu}{d\lambda}\right) d\lambda \text{ (by the previous step)}\\ &\underset{n\to\infty}{\to} \int_A \left(X \frac{d\mu}{d\lambda}\right) d\lambda \left(\text{since } 0 \leq X_n \frac{d\mu}{d\lambda} \uparrow X \frac{d\mu}{d\lambda} \text{ as } n \to \infty\right).\end{aligned} \tag{7.13}$$

From (7.12) and (7.13), it follows that $\int_A X d\mu = \int_A \left(X \frac{d\mu}{d\lambda}\right) d\lambda,\ A \in \mathcal{A}$. Finally, for any r.v. X for which the $\int X d\mu$ exists, we have

$$\begin{aligned}\int_A X d\mu &= \int_A X^+ d\mu - \int_A X^- d\mu = \int_A \left(X^+ \frac{d\mu}{d\lambda}\right) d\lambda - \int_A \left(X^- \frac{d\mu}{d\lambda}\right) d\lambda\\ &= \int_A \left[(X^+ - X^-) \frac{d\mu}{d\lambda}\right] d\lambda = \int_A \left(X \frac{d\mu}{d\lambda}\right) d\lambda.\end{aligned}$$

Thus, in all cases, $\int_A X d\mu = \int_A (X \frac{d\mu}{d\lambda}) d\lambda,\ A \in \mathcal{A}$. ■

Remark 9. The result stated in the preceding corollary is what lies behind the routine replacement in integrals of probability distributions by their p.d.f.s. Thus, let X be a r.v. defined on the probability space $(\Omega, \mathcal{A}, P)$ with probability distribution function P_X, let $B \in \mathcal{B}$ and $A = X^{-1}(B)$, and let $g : \Re \to \Re$ be (measurable and) such that $\mathcal{E}g(X)$ exists. Then (by Theorem 13 in Chapter 4),

$$\int_A g(X)\, dP = \int_B g(x)\, dP_X = \int_B \left[g(x) \frac{dP_X}{d\lambda}(x)\right] d\lambda,$$

if $P_X << \lambda$ (which often is either the Lebesgue or the counting measure); i.e.,

$$\int_A g(X)\, dP = \int_B \left[g(x) \frac{dP_X}{d\lambda}(x) \right] d\lambda.$$

In the following remark a brief summary is presented of results leading to Theorem 3.

Remark 10. From among Theorems 1–3, the one that is most often used is Theorem 3, the Radon–Nikodym Theorem. Its proof, however, depends on Theorem 2, the Lebesgue Decomposition Theorem, whose proof is long and depends on Theorem 1, the Hahn–Jordan Decomposition Theorem. A shortcut to Theorem 3 may be as follows:

- Start out with a σ-additive set function φ defined on $(\Omega, \mathcal{A})$ into $(-\infty, \infty]$ for which there is at least one $A \in \mathcal{A}$ such that $|\varphi(A)| < \infty$.
- State Lemma 1, according to which φ is continuous (where continuity from above, continuity from below, and continuity of φ are as in Definition 1).
- Define φ^+ and φ^- as was done in Definition 2, and then state the following short version of the Hahn–Jordan Decomposition Theorem:

  1. φ^+, φ^- and $|\varphi| = \varphi^+ + \varphi^-$ are measures and φ^- is finite.
  2. $\varphi = \varphi^+ - \varphi^-$ (in the sense that $\varphi(A) = \varphi^+(A) - \varphi^-(A),\ A \in \mathcal{A}$).

- Recall the definition of absolute continuity, mutual absolute continuity, and singularity (see Definition 3).
- State the Lebesgue Decomposition Theorem, whose detailed proof is very long and is carried out first for the case that both μ and ν are finite, and then the case they are both σ-finite.
- Finally, state and prove the Radon–Nikodym Theorem.

This chapter is concluded with a useful inequality, both in information theory as well as in statistical inference.

Proposition 2 (Shannon–Kolmogorov Information Inequality). Let X be a k-dimensional random vector defined on the probability space $(\Omega, \mathcal{A}, P)$, and suppose that its probability distribution P_X has one of two possible probability density functions f_0 or f_1 with respect to a σ-finite measure ν on $\mathcal{B}^k$. Let $K(f_0, f_1)$ be the *mutual entropy* of f_0 and f_1; i.e.,

$$K(f_0, f_1) = -\mathcal{E}_0 \log \frac{f_1(X)}{f_0(X)} = \mathcal{E}_0 \log \frac{f_0(X)}{f_1(X)},$$

where $\mathcal{E}_0$ denotes expectation taken under f_0, and log is the natural logarithm.

Then $K(f_0, f_1)$ exists, is ≥ 0, and is $= 0$ if and only if $P_i[f_0(X) = f_1(X)] = 1$, where P_i is the probability measure induced by f_i, $i = 0, 1$.

Remark 11. By making the usual conventions $\frac{0}{0} = 0$, $\pm\infty \times 0 = 0$, and writing

$$K(f_0, f_1) = \mathcal{E}_0 \log \frac{f_0(X)}{f_1(X)} = \int_{\Re^k} \left[\log \frac{f_0(x)}{f_1(x)} \right] f_0(x) d\nu,$$

it is seen that $K(f_0, f_1) \geq 0$, although it may be ∞.

Proof. From the last relation above and the conventions made, it follows that:

(a) If $f_0 > 0$ and $f_1 = 0$, then $K(f_0, f_1) = \infty$.
(b) If $f_0 = 0$ and $f_1 > 0$, then $K(f_0, f_1) = 0$.
(c) If $f_0 = f_1 = 0$, then $K(f_0, f_1) = 0$.
(d) If $f_0 > 0$ and $f_1 > 0$, then the integrand is finite.

Thus, it suffices to focus on this last case alone. By setting $S_0 = (f_0 > 0)$ and $S_1 = (f_1 > 0)$, we may assume that $S_0 = S_1$, since the cases $S_0 - S_1 \neq \emptyset$ and $S_1 - S_0 \neq \emptyset$ are covered by (a) and (b) above. So, let $S = (f_0 > 0) = (f_1 > 0)$, $A = X^{-1}(S)$, so that $P_i(A) = 1, i = 0, 1$, where P_i is the probability measure induced by f_i, $i = 0, 1$. The function $g(z) = -\log z$ $(z > 0)$ is strictly convex (since $g''(z) = z^{-2} > 0$). Then with $Z = f_1(X)/f_0(X)$, $g(Z)$ is strictly convex on A (with $P_i(A) = 1, i = 0, 1$). Thus, $g(\mathcal{E}_0 Z) \leq \mathcal{E}_0 g(Z)$ and equality occurs if and only if $P_0(Z = c, \text{a constant}) = 1$. This is so by Exercise 35 in Chapter 6. Equivalently,

$$-\log \mathcal{E}_0 \frac{f_1(X)}{f_0(X)} \leq \mathcal{E}_0 \left[-\log \frac{f_1(X)}{f_0(X)}\right] = -\mathcal{E}_0 \log \frac{f_1(X)}{f_0(X)} = K(f_0, f_1),$$

or

$$-K(f_0, f_1) \leq \log \mathcal{E}_0 \frac{f_1(X)}{f_0(X)} = \log \int_S \frac{f_1(x)}{f_0(x)} f_0(x) d\nu = \log \int_S f_1(x) d\nu = \log 1 = 0,$$

and

$$-K(f_0, f_1) = 0 \text{ only if } P_0 \left[\frac{f_1(X)}{f_0(X)} = c\right] = 1.$$

From $\frac{f_1(X)}{f_0(X)} = c$ a.s.$[P_0]$, we get

$$c = \mathcal{E}_0 \frac{f_1(X)}{f_0(X)} = \int_S \frac{f_1(x)}{f_0(x)} f_0(x) d\nu = 1,$$

so that, first, $-K(f_0, f_1) \leq 0$, equivalently, $K(f_0, f_1) \geq 0$, and secondly, $K(f_0, f_1) = 0$ only if $P_0[f_1(X) = f_0(X)] = 1$; also, $P_1[f_1(X) = f_0(X)] = 1$. The proof is completed. ■

Exercises.

1. Verify the relation $A_{n_0} = A + \sum_{j=n_0}^{\infty}(A_j - A_{j+1})$ used in the proof of Lemma 1; namely, if $\{A_n\}$, $n \geq 1$, form a nonincreasing sequence and $A = \bigcap_{n=1}^{\infty} A_n$, then for any $n_0 \geq 1$, the given relation holds. In particular, $A_1 = \sum_{n=1}^{\infty}(A_n - A_{n+1})$ if $\bigcap_{n=1}^{\infty} A_n = \emptyset$.

2. Given the finite measures μ_n, $n = 1, 2, \ldots$, on the measurable space $(\Omega, \mathcal{A})$, define a probability measure μ such that $\mu_n \ll \mu$ for all n (i.e., $\mu(A) = 0$ for $A \in \mathcal{A}$ implies $\mu_n(A) = 0$, $n \geq 1$).

3. Let $\mathcal{C}$ be the class of all probability measures defined on the measurable space $(\Omega, \mathcal{A})$, and let

$$d(P, Q) = \|P - Q\| = 2\sup\{|P(A) - Q(A)|;\ A \in \mathcal{A}\}.$$

(i) Then show that d is a distance in $\mathcal{C}$; i.e., show that $d(P, Q) \geq 0$ and $d(P, Q) = 0$ only if $P = Q, d(P, Q) = d(Q, P)$, and $d(P, Q) \leq d(P, R) + d(R, Q)$ where $R \in \mathcal{C}$.

Next, let μ be a σ-finite measure in $\mathcal{A}$ such that $P << \mu$ for every $P \in \mathcal{C}$, and for P and Q in $\mathcal{C}$, let $f = dP/d\mu$ and $g = dQ/d\mu$.

(ii) Then show that

$$d(P, Q) = \int_\Omega |f - g| d\mu\ .$$

Hint: For part (ii) and for any $A \in \mathcal{A}$,

$$\int_A (f - g) d\mu = \int_{A\cap B} (f - g) d\mu - \int_{A\cap C} (g - f) d\mu,$$

where $B = (f - g > 0)$ and $C = (g - f > 0)$.

4. Let $U(\alpha, \beta)$ be the uniform distribution over the interval $[\alpha, \beta]$ $(\alpha < \beta)$, and for $n \geq 1$, let P_n and Q_n be the (probability) measures corresponding to $U(-\frac{1}{n}, 1)$ and $U(0, 1 + \frac{1}{n})$. Then show that $\|P_n - Q_n\| \to 0$ as $n \to \infty$, by using part (ii) of Exercise 3.

5. Let P and Q be two probability measures defined on the measurable space $(\Omega, \mathcal{A})$, and suppose that $P \approx Q$ (i.e., $P << Q$ and $Q << P$). Let $f = dP/d\mu$ and $g = dQ/d\mu$ for some σ-finite measure μ dominating both P and Q (e.g., $\mu = P + Q$), and let $Z = \log(g/f)$, where as always, log stands for the natural logarithm. Then (with reference to Exercise 3), show that

$$\|P - Q\| \leq 2(1 - e^{-\varepsilon}) + 2P(|Z| > \varepsilon) \text{ for every } \varepsilon > 0.$$

Hint: For $B = (f - g > 0)$, show that $\|P - Q\| = 2[P(B) - Q(B)]$, and then work with $P(B)$ and $Q(B)$ by introducing $C = (|Z| > \varepsilon)$.

6. Let P, Q and f, g be as in Exercise 3 and, without loss of generality, we suppose that the dominating measure μ is a probability measure (e.g., $\mu = (P + Q)/2$). Define ρ by $\rho = \int_\Omega (fg)^{1/2} d\mu$, and show that:

(i) $\rho \leq 1$.

(ii) $2(1 - \rho) \leq d(P, Q) \leq 2(1 - \rho^2)^{1/2}$, where (by Exercise 3), $d(P, Q) = \int_\Omega |f - g| d\mu = \|P - Q\|$.

Replacing P, Q and f, g, ρ by P_n, Q_n and $f_n, g_n, \rho_n, n \geq 1$, part (ii) becomes:

$$2(1 - \rho_n) \leq d(P_n, Q_n) \leq 2(1 - \rho_n^2)^{1/2}.$$

(iii) From this last relation, conclude that, as $n \to \infty$, $d(P_n, Q_n) \to 0$ if and only if $\rho_n \to 0$.

7. For $n \geq 1$, let f_n be real-valued measurable functions defined on an open set $E \subset \Re$ containing 0, and suppose that $|f_n(x)| \leq M(< \infty)$ for all n and all $x \in E$, and that $f_n(x) \underset{n\to\infty}{\longrightarrow} 0$ for $x \in E$. Then show that, for any sequence $x_n \underset{n\to\infty}{\longrightarrow} 0$, there exists a subsequence $\{x_m\} \subseteq \{x_n\}$ and a λ-null subset N of E (which may depend on $\{x_n\}$) such that $f_m(x + x_m) \underset{m\to\infty}{\longrightarrow} 0$, $x \in E - N$; here λ is the Lebesgue measure.

Hint: Consider the integral $\int_{\Re} |f_n(x + x_n)| d\Phi(x)$, where Φ is the d.f. of the $N(0, 1)$ distribution; refer to Exercise 17 in Chapter 5, and use the fact that the measure induced by Φ and λ are mutually absolutely continuous.

8. For $n \geq 1$, let f_n be real-valued measurable functions defined on $\Re$, and suppose that $|f_n(x)| \leq M(< \infty)$ for all n and $x \in \Re$, and that $\lim_{n\to\infty} f_n^-(x) = 0$ for $x \in \Re$. Then for any sequence $\{x_n\}$ with $x_n \underset{n\to\infty}{\longrightarrow} 0$ it holds $\limsup_{n\to\infty} f_n(x + x_n) \geq 0$ a.e. $[\lambda]$, where λ is the Lebesgue measure.

CHAPTER

8 Distribution Functions and Their Basic Properties, Helly–Bray Type Results

This chapter deals with some basic properties of d.f.s, the concept of weak and complete convergence of sequences of d.f.s, and some Helly–Bray type theorems. More precisely, in Section 8.1, it is shown that a d.f. is, actually, determined by its values on a dense set in $\Re$; that the set of its discontinuity points is countable; and that it is uniquely decomposed into three d.f.s, of which one is a step function, the second is a continuous d.f. whose induced measure is absolutely continuous with respect to the Lebesgue measure, and the third is a continuous d.f. whose induced measure is singular with respect to the Lebesgue measure.

In the following section, the concepts of weak and complete convergence of a sequence of d.f.s is introduced, and their relationships are discussed. It is also shown that for any sequence of d.f.s, there is always a subsequence that converges weakly to a d.f.

In the final section, the integral of a bounded and continuous real-valued function defined on $\Re$ with respect to a d.f. is considered; such integrals are to be interpreted in the Riemann–Stieltjes sense. Then conditions are given under which weak or complete convergence of a sequence of d.f.s implies convergence of associated integrals. Problems of this type arise often in research and applications.

8.1 Basic Properties of Distribution Functions

Definition 1. We say that F is a d.f. (not necessarily of a r.v.) if $F : \Re \to \Re$ and

(1) $0 \le F(x) \le 1, \quad x \in \Re.$
(2) $F \uparrow$.
(3) F is continuous from the right.

As usual, $F(+\infty) \overset{def}{=} \lim_{x\to\infty} F(x)$, $F(-\infty) \overset{def}{=} \lim_{x\to-\infty} F(x)$. Then $0 \le F(-\infty)$, $F(\infty) \le 1$, but $F(-\infty)$ need not be $= 0$ and $F(+\infty)$ need not be $= 1$.

Proposition 1. Defining property (3) is not essential in defining a d.f. in the following sense: if F^* satisfies (1) and (2), then one can construct a d.f. F (i.e., F satisfies properties (1)–(3)) which coincides with F^* whenever F^* is continuous or only right continuous.

An Introduction to Measure-Theoretic Probability, Second Edition. http://dx.doi.org/10.1016/B978-0-12-800042-7.00008-6

Proof. In the first place, F^* can have only jumps. Now if $x \in C(F^*)$, we set $F(x) = F^*(x)$, whereas if $x \notin C(F^*)$, we set $F(x) = F^*(x+0)$ which exists because of (2). Then, clearly, F satisfies properties (1)–(3). ■

Actually more is true, namely,

Proposition 2. Let D be a dense subset of $\Re$ (such as the set of rational numbers), and let $F_D : D \to \Re$ satisfy (1) and (2). Then F_D determines uniquely a d.f. F in $\Re$.

Proof. In the first place, we may assume that F_D is also continuous from the right in D, by Proposition 1. Next, for $x \in \Re$, let $x_n \in D$ such that $x_n \downarrow x$ as $n \to \infty$. Then define $F : \Re \to [0,1]$ as follows: $F(x) = F_D(x), x \in D$, and $F(x) = \lim_{n\to\infty} F_D(x_n), x \in D^c$. Clearly, F satisfies (1)–(3). ■

Corollary. If F_1, F_2 are two d.f.s such that $F_1(x) = F_2(x)$, $x \in D$, *then* $F_1 \equiv F_2$.

Proof. Let $x \in D^c$. Then there exists $x_n \in D$ such that $x_n \downarrow x$ as $n \to \infty$. It follows that, as $n \to \infty$, $F_1(x_n) \downarrow F_1(x)$, $F_2(x_n) \downarrow F_2(x)$. But $F_1(x_n) = F_2(x_n), n \geq 1$. Hence $F_1(x) = F_2(x)$. ■

We mentioned previously that a d.f. can have only discontinuities that are jumps. The following theorem refers to the number of these jumps.

Theorem 1. Any d.f. F has a countable number of discontinuous points (which, of course, can be equal to 0). ■

Proof. If F is continuous everywhere there is no problem. Let then F be discontinuous and let $(\alpha, \beta]$ be a finite interval in $\Re$. Let $x_1, \ldots, x_n$ be n points such that $\alpha < x_1 < \cdots < x_n \leq \beta$ at which F is discontinuous.

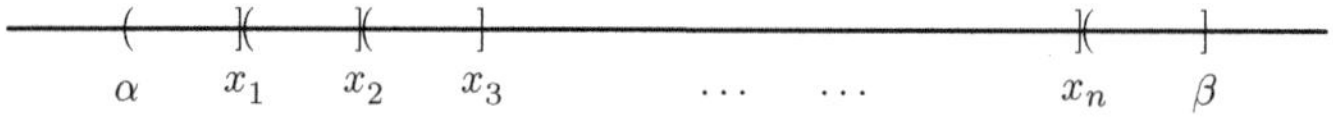

Then we have

$$\begin{aligned} F(\alpha) \leq F(x_1-) < F(x_1) \leq F(x_2-) < F(x_2) \leq \cdots \\ \leq F(x_n-) < F(x_n) \leq F(\beta). \end{aligned}$$

The lengths of the jumps are

$$F(x_j) - F(x_j-), \quad j = 1, \ldots, n.$$

Summing them up, we get

$$\begin{aligned} \sum_{j=1}^{n} [F(x_j) - F(x_j-)] &= [F(x_1) - F(x_1-)] + [F(x_2) - F(x_2-)] \\ &+ \cdots + [F(x_{n-1}) - F(x_{n-1}-)] + [F(x_n) - F(x_n-)] \\ &\leq [F(x_1) - F(x_1-)] + [F(x_2-) - F(x_1)] + [F(x_2) - F(x_2-)] \\ &+ \cdots + [F(x_n-) - F(x_{n-1})] + [F(x_n) - F(x_n-)] \end{aligned}$$

(by adding some nonnegative numbers),

$$= F(x_n) - F(x_1-) \leq F(\beta) - F(\alpha);\text{ i.e.,}$$

$$\sum_{j=1}^{n}[F(x_j) - F(x_j-)] \leq F(\beta) - F(\alpha). \tag{8.1}$$

Relation (8.1) implies the following: for $\varepsilon > 0$, the number of jumps in $(\alpha, \beta]$ with length of jump $> \varepsilon$ cannot be greater than $\frac{1}{\varepsilon}[F(\beta) - F(\alpha)]$. In fact, if this number is K, and $K > \frac{1}{\varepsilon}[F(\beta) - F(\alpha)]$, then the sum of lengths of the K jumps is $\geq K\varepsilon > F(\beta) - F(\alpha)$, which contradicts (8.1). Applying this for $m = 1, 2, \ldots$, we get the number of jumps in $(\alpha, \beta]$ with length of jump $> \frac{1}{m}$ is $\leq m[F(b) - F(a)]$, thus finite. Hence the total number of jumps in $(\alpha, \beta]$ is countable. Since $\Re$ can be written as the sum of denumerably many intervals, we have that the total number of jumps in $\Re$ is countable. ■

Definition 2. Consider the numbers x_n, $n = 1, 2, \ldots$, and let the positive numbers $p(x_n)$ be associated with x_n, $n \geq 1$, such that $\sum_n p(x_n) < \infty$. Define $F^* : \Re \rightarrow [0, \infty)$ as follows: $F^*(x) = \sum_{x_n \leq x} p(x_n)$. Then F^* is said to be a *step function*. ■

Remark 1. F^* as just defined is not necessarily a step function in the usual sense; i.e., constant over intervals, since the x_ns can form a set dense in $\Re$; e.g., the rational numbers. However, if $\{x_n, n \geq 1\}$ is not dense in $\Re$, these two concepts coincide.

Theorem 2 (Decomposition Theorem). Any d.f. F is uniquely decomposed into two d.f.s F_c, F_d such that F_c is continuous, F_d is a step function, and $F = F_c + F_d$. ■

Proof. Let x_n, $n \geq 1$, be the discontinuity points of F and set $p(x_n) = F(x_n) - F(x_n-)$ and $p(x) = 0$ for $x \neq x_n, n \geq 1$. Then, clearly, $\sum_n p(x_n) < \infty$(actually, ≤ 1, by (1) in Definition 1). Define:

$$F_d(x) = \sum_{x_n \leq x} p(x_n), \quad F_c(x) = F(x) - F_d(x).$$

We show first that F_d is a d.f. In the first place, $0 \leq F_d(x) \leq 1$, $x \in \Re$ (apply (8.1) for $\beta = x$ and let $\alpha \rightarrow -\infty$). Next, for $y > x$, we get

$$F_d(y) - F_d(x) = \sum_{x < x_n \leq y} p(x_n) \geq 0,$$

so that $F_d \uparrow$. So it remains to show continuity from the right. We have, by (8.1),

$$(0 \leq) F_d(y) - F_d(x) = \sum_{x < x_n \leq y} p(x_n) \leq F(y) - F(x). \tag{8.2}$$

Letting $y \downarrow x$ and utilizing the right continuity of F, we get

$$F_d(y) - F_d(x) \rightarrow 0 \text{ or } F_d(y) \rightarrow F_d(x).$$

We next prove that F_c is a continuous d.f. Since F, F_d are both d.f.s and $F_d(x) \leq F(x)$ (by (8.1)), $x \in \Re$, we have that $0 \leq F_c(x) \leq 1$ (by (8.1)), $x \in \Re$. Next, F_c is $\uparrow$. In fact, for $x < y$, we have, by (8.2),

$$F_d(y) - F_d(x) = \sum_{x<x_n\leq y} p(x_n) \leq F(y) - F(x), \text{ by (8.1)}.$$

Hence

$$[F(y) - F(x)] - [F_d(y) - F_d(x)] \geq 0.$$

Thus

$$\begin{aligned} F_c(y) - F_c(x) &= [F(y) - F_d(y)] - [F(x) - F_d(x)] \\ &= [F(y) - F(x)] - [F_d(y) - F_d(x)] \geq 0, \end{aligned}$$

so that $F_c \uparrow$. Now F_c is right continuous since both F, F_d are so. Thus it remains to prove that F_c is continuous from the left. Again for $x < y$, we have

$$\begin{aligned} F_c(y) - F_c(x) &= [F(y) - F_d(y)] - [F(x) - F_d(x)] \\ &= [F(y) - F(x)] - \sum_{x<x_n\leq y} p(x_n) \\ &= F(y) - F(x) - p(y) - \sum_{x<x_n<y} p(x_n) \\ &= F(y) - F(x) - F(y) + F(y-) - \sum_{x<x_n<y} p(x_n) \\ &= F(y-) - F(x) - \sum_{x<x_n<y} p(x_n). \end{aligned}$$

Thus, as $x \uparrow y$, we get $F(y-) - F(x) \to F(y-) - F(y-) = 0$ and

$$\begin{aligned} \sum_{x<x_n<y} p(x_n) = \sum_{x<x_n\leq y} p(x_n) - p(y) &\leq F(y) - F(x) \\ &\quad - [F(y) - F(y-)] = F(y-) - F(x) \to 0. \end{aligned}$$

Finally, we justify the uniqueness argument. We have $F = F_c + F_d$ and also let $F = F'_c + F'_d$, where F'_c is a continuous d.f. and F'_d is a step d.f.; i.e., $F'_d(x) = \sum_{x'_n\leq x} p'(x'_n)$, where $p'(x'_n)$, $n \geq 1$, are positive numbers (with $\sum_{x'_n} p'(x'_n) \leq 1$) assigned to the points $x'_1, x'_2, \ldots$, and $p'(x) = 0$ for $x \neq x'_n$, $n \geq 1$. It will first be shown that $F_d(x) - F_d(x-) = p(x)$ (the length of the jump of F_d at x), $x \in \Re$. To this end, let $\varepsilon > 0$. Then

$$\begin{aligned} F_d(x+\varepsilon) - F_d(x-\varepsilon) &= \sum_{x-\varepsilon<x_n\leq x+\varepsilon} p(x_n) \\ &= \sum_{x-\varepsilon<x_n\leq x} p(x_n) + \sum_{x<x_n\leq x+\varepsilon} p(x_n). \end{aligned}$$

But, by (8.1),

$$\sum_{x<x_n\leq x+\varepsilon} p(x_n) \leq F(x+\varepsilon) - F(x) \underset{\varepsilon\to 0}{\to} F(x) - F(x) = 0,$$

whereas

$$p(x) \leq \sum_{x-\varepsilon<x_n\leq x} p(x_n) \leq F(x) - F(x-\varepsilon) \underset{\varepsilon\to 0}{\to} F(x) - F(x-) = p(x).$$

Therefore

$$\sum_{x-\varepsilon<x_n\leq x} p(x_n) \underset{\varepsilon\to 0}{\to} p(x).$$

However,

$$\sum_{x-\varepsilon<x_n\leq x} p(x_n) = F_d(x) - F_d(x-\varepsilon) \underset{\varepsilon\to 0}{\to} F_d(x) - F_d(x-).$$

Hence $F_d(x) - F_d(x-) = p(x)$.

Next, we show that, for every $x \in \Re$:

$$\sum_{x'_n<x} p'(x'_n) = F'_d(x-),$$

$$F'_d(x) - F'_d(x-) = p'(x),$$
$$F'_d(x) - F'_d(x-) = p(x).$$

Indeed, let $\Omega = \{x'_1, x'_2, \ldots\}$, let $\mathcal{A}$ be the discrete σ-field, and on $\mathcal{A}$, define the (finite) measure μ by $\mu(A) = \sum_{x'_n\in A} p'(x'_n)$. For $k = 1, 2, \ldots$, set $A_k = \{x'_n; x'_n \leq x - \frac{1}{k}\}$, so that, as $k \uparrow \infty$, $A_k \uparrow \bigcup_{k=1}^{\infty} A_k \overset{def}{=} A = \{x'_n; x'_n < x\}$. Then $\mu(A_k) \underset{k\to\infty}{\to} \mu(A)$, or $\sum_{x'_n\in A_k} p'(x'_n) \underset{k\to\infty}{\to} \sum_{x'_n\in A} p'(x'_n)$, or $\sum_{x'_n\leq x-\frac{1}{k}} p'(x'_n) \underset{k\to\infty}{\to} \sum_{x'_n<x} p'(x'_n)$, or $F'_d(x-\frac{1}{k}) \underset{k\to\infty}{\to} \sum_{x'_n<x} p'(x'_n)$. However, $F'_d(x-\frac{1}{k}) \underset{k\to\infty}{\to} F'_d(x-)$. Hence $\sum_{x'_n<x} p'(x'_n) = F'_d(x-)$.

Next, for $\varepsilon > 0$,

$$\begin{aligned} F'_d(x) - F'_d(x-\varepsilon) &= \sum_{x'_n\leq x} p'(x'_n) - \sum_{x'_n\leq x-\varepsilon} p'(x'_n) \\ &= \sum_{x'_n<x} p'(x'_n) + p'(x) - \sum_{x'_n\leq x-\varepsilon} p'(x'_n) \\ &= F'_d(x-) + p'(x) - F'_d(x-\varepsilon) \\ &\underset{\varepsilon\to 0}{\to} F'_d(x-) + p'(x) - F'_d(x-) = p'(x). \end{aligned}$$

Also, $F'_d(x) - F'_d(x-\varepsilon) \underset{\varepsilon\to 0}{\to} F'_d(x) - F'_d(x-)$, so that $F'_d(x) - F'_d(x-) = p'(x)$.

Finally, from $F'_c + F'_d = F$, we get $F'_d = F - F'_c$, so that

$$\begin{aligned} F'_d(x) - F'_d(x-\varepsilon) &= [F(x) - F'_c(x)] - [F(x-\varepsilon) - F'_c(x-\varepsilon)] \\ &= [F(x) - F(x-\varepsilon)] - [F'_c(x) - F'_c(x-\varepsilon)] \\ &\underset{\varepsilon\to 0}{\to} F(x) - F(x-) = p(x), \end{aligned}$$

since F'_c is continuous, so that $F'_d(x) - F'_d(x-) = p(x)$.

Combining the last two results, and using also the result that $F_d(x) - F_d(x-) = p(x)$, we get $F'_d(x) - F'_d(x-) = p'(x) = p(x) = F_d(x) - F_d(x-)$.

From this it follows that F_d and F'_d have the same points of discontinuities and the same lengths of jumps. Thus, $x'_n = x_n$, and $p'(x'_n) = p(x_n), n \geq 1$, so that $F'_d(x) = \sum_{x'_n \leq x} p'(x'_n) = \sum_{x_n \leq x} p(x_n) = F_d(x), x \in \Re$, and then $F'_c(x) = F_c(x), x \in \Re$. (See also Exercise 10.) ■

Corollary. Any d.f. F is uniquely decomposed into three d.f.s F_d, F_{cc}, F_{cs} such that $F = F_d + F_{cc} + F_{cs}$, where F_d is a step function, singular with respect to Lebesgue measure λ; F_{cc} is λ-absolute continuous; and F_{cs} is continuous and λ-singular; by λ-absolute continuity and λ-singularity, we mean that the measures induced by the d.f.s have these properties.

Proof. Let μ_d, μ_c be the measures induced by F_d, F_c, respectively. We set $N = \{x_1, x_2, \ldots\}$ for the set of discontinuities of F, and we prove that μ_d is λ-singular by proving that $\mu_d(B) = \mu_d(B \cap N)$, $B \in \mathcal{B}$, while $\lambda(N) = 0$. Since $\mu_d(B) = \mu_d(B \cap N) + \mu_d(B \cap N^c)$, it suffices to prove that $\mu_d(B \cap N^c) = 0$, which would be the case if we would show that $\mu_d(N^c) = 0$. We have

$$\begin{aligned} \mu_d(N^c) = \mu_d(\Re - N) &= \mu_d\left[\lim_{n\to\infty}(\Re - \{x_1, \ldots, x_n\})\right] \\ &= \lim_{n\to\infty} \mu_d(\Re - \{x_1, \ldots, x_n\}) \\ &= \lim_{n\to\infty}\left[\mu_d(\Re) - \mu_d(\{x_1, \ldots, x_n\})\right] \\ &= \lim_{n\to\infty}\left[\alpha - \mu_d(\{x_1, \ldots, x_n\})\right], \end{aligned}$$

(where $\alpha = \mu_d(\Re)$); and this is

$$\begin{aligned} &= \alpha - \lim_{n\to\infty} \mu_d(\{x_1, \ldots, x_n\}) = \alpha - \lim_{n\to\infty}\sum_{j=1}^{n} p(x_j) \\ &= \alpha - \sum_{j=1}^{\infty} p(x_j) = \alpha - F_d(\infty) = \alpha - \alpha = 0. \end{aligned}$$

So, μ_d is λ-singular.

Next, μ_c is decomposed uniquely (by the Lebesgue Decomposition Theorem, Theorem 2 in Chapter 7) into a λ-continuous measure μ_{cc} and into a λ-singular measure μ_{cs} such that $\mu_c = \mu_{cc} + \mu_{cs}$. Then, for every $x \in \Re$, $\mu_c(-\infty, x] = \mu_{cc}(-\infty, x] + \mu_{cs}(-\infty, x]$ or $F_c(x) = F_{cc}(x) + F_{cs}(x)$, for the corresponding d.f.s, and $\mu_{cc}(B) = \int_B \varphi\, d\lambda$, where φ is a measurable, nonnegative, finite a.e. $[\lambda]$, and uniquely defined a.e. $[\lambda]$ function, and $F_{cc}(x) = \int_{(-\infty,x]} \varphi\, d\lambda$. Now F_{cc} is continuous since for $x < y$, we have $F_{cc}(y) - F_{cc}(x) = \int_{(x,y]} \varphi\, d\lambda \to 0$ as $x \uparrow y$, because $|\varphi I_{(x,y]}| \leq \varphi$ independent of x, λ-integrable, and as $x \uparrow y$, $\varphi I_{(x,y]} \to \varphi I_{\{y\}}$. Then, by the Dominated Convergence Theorem, $\int_{(x,y]} \varphi\, d\lambda = \int \varphi I_{(x,y]}\, d\lambda \to \int \varphi I_{\{y\}}\, d\lambda = \int_{\{y\}} \varphi\, d\lambda = 0$. From the expression $F_c = F_{cc} + F_{cs}$ and continuity of F_{cc}, it follows that F_{cs} is also continuous. Thus F is decomposed into a step d.f. F_d, a continuous d.f. F_{cc} with density, and a continuous d.f. F_{cs} whose corresponding measure vanishes outside a λ-null set. ■

8.2 Weak Convergence and Compactness of a Sequence of Distribution Functions

Definition 3. If F is a d.f. on $\Re$, then its *variation*, $Var F$, is defined to be $Var F = F(+\infty) - F(-\infty)$. Let now F and F_n, $n = 1, 2, \ldots$, be d.f.s. We have defined *weak* convergence, $F_n \underset{n\to\infty}{\Rightarrow} F$, to mean $F_n(x) \underset{n\to\infty}{\to} F(x)$, $x \in C(F)$. The convergence $F_n \underset{n\to\infty}{\Rightarrow} F$ need *not* imply that $Var F_n \underset{n\to\infty}{\to} Var F$. If this is so, we say that F_n *converges completely to* F and we denote it by writing $F_n \overset{c}{\underset{n\to\infty}{\to}} F$. ■

Clearly, if F, F_n, $n \geq 1$, are d.f.s of r.v.s, then $F_n \underset{n\to\infty}{\Rightarrow} F$ is equivalent to $F_n \overset{c}{\underset{n\to\infty}{\to}} F$.

Example 1. Let

$$F_n(x) = \begin{cases} 0 & \text{if } x < -n \\ \frac{1}{2} & \text{if } -n \leq x < n, \quad n \geq 1 \\ 1 & \text{if } x \geq n, \end{cases}$$

and $F(x) = \frac{1}{2}$. Then $F_n \underset{n\to\infty}{\Rightarrow} F$ but $F_n \overset{c}{\underset{n\to\infty}{\not\to}} F$. ■

Remark 2. It is to be noted that if $F_n \underset{n\to\infty}{\Rightarrow} F$, then F is uniquely determined. In fact, we have the following

Proposition 3. For two d.f.s F and F' and as $n \to \infty$, *suppose* $F_n \Rightarrow F$ and $F_n \Rightarrow F'$. *Then* $F(x) = F'(x)$, $x \in \Re$.

Proof. Indeed, for

$$x \in C(F) \cap C(F'), \quad F_n(x) \underset{n\to\infty}{\to} F(x), \quad F_n(x) \underset{n\to\infty}{\to} F'(x)$$

and hence $F(x) = F'(x)$ on $C(F) \cap C(F')$. Next, $[C(F) \cap C(F')]^c = C^c(F) \cup C^c(F')$ is countable, and for every x in this set there exists $x_n \in C(F) \cap C(F')$ such that $x_n \downarrow x$

as $n \to \infty$. Then $F(x_n) \underset{n\to\infty}{\to} F(x)$, $F'(x_n) \underset{n\to\infty}{\to} F'(x)$. But $F(x_n) = F'(x_n)$, $n \geq 1$. Hence $F(x) = F'(x)$, $x \in \Re$. ■

The following theorem relates the variations of d.f.s and their values at $\pm\infty$.

Theorem 3. Let $F_n \underset{n\to\infty}{\Rightarrow} F$. *Then*

(i) $\limsup_{n\to\infty} F_n(-\infty) \leq F(-\infty) \leq F(\infty) \leq \liminf_{n\to\infty} F_n(\infty)$.

(ii) $Var F \leq \liminf_{n\to\infty} Var F_n$.

(iii) $F_n(\pm\infty) \underset{n\to\infty}{\to} F(\pm\infty)$ if and only if $Var F_n \underset{n\to\infty}{\to} Var F$.

Thus $F_n \underset{n\to\infty}{\Rightarrow} F$ implies $F_n \overset{c}{\underset{n\to\infty}{\to}} F$ if and only if $F_n(\pm\infty) \underset{n\to\infty}{\to} F(\pm\infty)$.) ■

Proof.

(i) We have: $F_n(-\infty) \leq F_n(x) \leq F_n(\infty)$. Let now $x \in C(F)$ and take the limits as $n \to \infty$. Then

$$\limsup F_n(-\infty) \leq F(x) \leq \liminf F_n(\infty).$$

Now let $x \downarrow -\infty$ through continuity points. Then

$$\limsup F_n(-\infty) \leq F(-\infty) \leq \liminf F_n(\infty).$$

If $x \uparrow \infty$ through continuity points, then

$$\limsup F_n(-\infty) \leq F(\infty) \leq \liminf F_n(\infty).$$

Hence, we get (i).

(ii) From (i) and the fact that for any sequences of real numbers $\{x_n\}$ and $\{y_n\}$, $\limsup_{n\to\infty} x_n = -\liminf_{n\to\infty}(-x_n)$, and $\liminf_{n\to\infty}(x_n + y_n) \geq \liminf_{n\to\infty} x_n + \liminf_{n\to\infty} y_n$, we get

$$\begin{aligned} Var F = F(\infty) - F(-\infty) &\leq \liminf_{n\to\infty} F_n(\infty) - \limsup_{n\to\infty} F_n(-\infty) \\ &= \liminf_{n\to\infty} F_n(\infty) + \liminf_{n\to\infty}[-F_n(-\infty)] \\ &\leq \liminf_{n\to\infty}[F_n(\infty) - F_n(-\infty)] = \liminf_{n\to\infty} Var F_n, \end{aligned}$$

as was to be seen.

(iii) Assume first that

$$Var F_n \underset{n\to\infty}{\to} Var F \text{ or } F_n(\infty) - F_n(-\infty) \underset{n\to\infty}{\to} F(\infty) - F(-\infty).$$

Then for $\varepsilon > 0$ and all sufficiently large n, we have

$$\left[F_n(\infty) - F_n(-\infty)\right] - \left[F(\infty) - F(-\infty)\right] < \varepsilon,$$

or

$$F_n(\infty) - F(\infty) + F(-\infty) - \varepsilon < F_n(-\infty).$$

Hence as $n \to \infty$,

$$\liminf F_n(\infty) - F(\infty) + F(-\infty) - \varepsilon \le \liminf F_n(-\infty) \le \limsup F_n(-\infty),$$

or,

$$F(-\infty) - \varepsilon \le \liminf F_n(-\infty) \le \limsup F_n(-\infty) \le F(-\infty),$$

since by part (i), $\liminf F_n(\infty) - F(\infty) \ge 0$ and $\limsup F_n(-\infty) \le F(-\infty)$. Letting $\varepsilon \to 0$, we obtain $\liminf F_n(-\infty) = \limsup F_n(-\infty) = F(-\infty)$. Likewise, for $\varepsilon > 0$ and all sufficiently large n,

$$-\varepsilon < \left[F_n(-\infty) - F_n(\infty)\right] + \left[F(\infty) - F(-\infty)\right],$$

or

$$F_n(-\infty) - F(-\infty) + F(\infty) + \varepsilon > F_n(\infty).$$

Hence as $n \to \infty$,

$$\limsup F_n(-\infty) - F(-\infty) + F(\infty) + \varepsilon \ge \limsup F_n(\infty) \ge \liminf F_n(\infty),$$

or

$$F(\infty) + \varepsilon \ge \limsup F_n(\infty) \ge \liminf F_n(\infty) \ge F(\infty),$$

since by part (i), $\limsup F_n(-\infty) - F(-\infty) \le 0$ and $\liminf F_n(\infty) \ge F(\infty)$. Letting $\varepsilon \to 0$, we obtain $\liminf F_n(\infty) = \limsup F_n(\infty) = F(\infty)$. The proof of the other direction is immediate. ■

Theorem 4. $F_n \underset{n\to\infty}{\Rightarrow} F$ if and only if $F_n(x) \underset{n\to\infty}{\to} F(x)$ on a set D dense in $\Re$. ■

Proof. In all that follows, $n \to \infty$ unless otherwise noted. If $F_n \Rightarrow F$, then $F_n(x) \to F(x)$, $x \in C(F)$ which is a set dense in $\Re$. Now let $F_n(x) \to F(x)$, $x \in D$. We show that this convergence is true for $x \in C(F)$. For $x \in C(F)$, there exists x', $x'' \in D$ such that $x' < x < x''$. Hence $F_n(x') \le F_n(x) \le F_n(x'')$. Letting $n \to \infty$, we get

$$F(x') \le \liminf F_n(x) \le \limsup F_n(x) \le F(x'').$$

Now let $x' \uparrow x$, $x'' \downarrow x$ through values of D. Then we get

$$F(x-) = F(x) \le \liminf F_n(x) \le \limsup F_n(x) \le F(x+) = F(x).$$

Thus $F_n(x) \to F(x)$. ■

Remark 3. It is to be emphasized that in the proof of Theorem 4 only continuity points of F are relevant; discontinuity points and right-continuity points play no role in the proof.

Now let $\{F_n\}$ be any sequence of d.f.s. Of course, this sequence may or may not converge weakly. However, there is always a subsequence which converges weakly; i.e.,

Theorem 5 (Weak Compactness Theorem). If $\{F_n\}$ is a sequence of d.f.s, there exists a subsequence that converges weakly to a d.f. ■

Proof. Let $D = \{x_1, x_2, \ldots\}$ be a set dense in $\Re$, and look at the bounded sequence $\{F_n(x_1)\}, n \geq 1$. Then there exists a convergent subsequence; call it $\{F_{n1}(x_1)\}$. Here, and in the remainder of this proof, all limits are taken as $n \to \infty$. Next, look at the subsequence $\{F_{n1}\}$ (of $\{F_n\}$) and evaluate it at x_2 to get the bounded sequence $\{F_{n1}(x_2)\}$. Then there exists a convergent subsequence; call it $\{F_{n2}(x_2)\}$. Since $\{F_{n2}\} \subseteq \{F_{n1}\}$, it follows that $\{F_{n2}(x_1)\}$ is also convergent. Proceeding like this, at the nth step we consider the subsequence $\{F_{1,n-1}, F_{2,n-1}, F_{3,n-1}, \ldots\}$ and evaluate it at x_n to get the bounded sequence $\{F_{1,n-1}(x_n), F_{2,n-1}(x_n), F_{3,n-1}(x_n), \ldots\}$. Then there exists a convergent subsequence; call it $\{F_{1n}(x_n), F_{2n}(x_n), F_{3n}(x_n), \ldots\}$. Since $\{F_{1n}, F_{2n}, F_{3n}, \ldots\} \subseteq \{F_{1,n-1}, F_{2,n-1}, F_{3,n-1}, \ldots\}$, it follows that $\{F_{1n}(x_j), F_{2n}(x_j), F_{3n}(x_j), \ldots\}$ is also convergent for $j = 1, \ldots, n-1$. The process continues in this manner indefinitely.

In terms of an array, we have that

$$
\begin{array}{llllll}
F_{11} & F_{21} & F_{31} & \cdots & \text{converges at} & x_1 \\
F_{12} & F_{22} & F_{32} & \cdots & \text{converges at} & x_1, x_2 \\
\cdot & \cdot & \cdot & \cdot & \cdot & \cdot \\
F_{1n} & F_{2n} & F_{3n} & \cdots & \text{converges at} & x_1, x_2, \ldots, x_n \\
\cdot & \cdot & \cdot & \cdot & \cdot & \cdot
\end{array}
$$

At this point, consider the diagonal sequence $\{F_{nn}\}$ and argue that $\{F_{nn}\}$ converges at every point of D. To this end, let $x_n \in D$ and look at $\{F_{nn}(x_n)\}$. Clearly, $\{F_{nn}\} \subseteq \{F_{1n}, F_{2n}, F_{3n}, \ldots\}$ except for the first $n-1$ terms $F_{11}, F_{22}, \ldots, F_{n-1,n-1}$. Hence $\{F_{nn}(x_n)\}$ converges, since so does $\{F_{1n}(x_n), F_{2n}(x_n), F_{3n}(x_n), \ldots\}$. Thus, $\{F_{nn}\}$ converges on D and defines the function $F_D(x) = \lim F_{nn}(x), x \in D$. Clearly, F_D satisfies the properties of being between 0 and 1 and nondecreasing, since $0 \leq F_{nn}(x) \leq 1, x \in \Re$, and F_{nn} is nondecreasing. Next, extend F_D to F^* from D to $\Re$ by defining F^* thus: $F^*(x) = F_D(x), x \in D$, and $F^*(x) = \lim_{n\to\infty} F_D(x_n)$ with $x_n \in D, x_n \downarrow x$. Then, clearly, $0 \leq F^*(x) \leq 1, x \in \Re$, and F^* is nondecreasing. Furthermore, $F_{nn}(x) \underset{n\to\infty}{\longrightarrow} F^*(x), x \in C(F^*)$. Indeed, for $x \in C(F^*)$, there exist x' and x'' in D such that $x' < x < x''$, so that

$$F_{nn}(x') \leq F_{nn}(x) \leq F_{nn}(x''),$$

and this implies as $n \to \infty$,

$$
\begin{aligned}
\lim F_{nn}(x') = F_D(x') = F^*(x') &\leq \liminf F_{nn}(x) \\
&\leq \limsup F_{nn}(x) \leq \lim F_{nn}(x'') = F_D(x'') = F^*(x'')
\end{aligned}
$$

(since $F^* = F_D$ on D). Thus

$$F^*(x') \leq \liminf F_{nn}(x) \leq \limsup F_{nn}(x) \leq F^*(x'').$$

Next, let $x' \uparrow x$ and $x'' \downarrow x$ whereas always $x', x'' \in D$. Then

$$\begin{aligned}\lim F^*(x') &= F^*(x-) = F^*(x) \leq \liminf F_{nn}(x)\\ &\leq \limsup F_{nn}(x) \leq \lim F^*(x'') = F^*(x+) = F^*(x)\end{aligned}$$

(since $x \in C(F^*)$). Therefore, as $n \to \infty$,

$$\liminf F_{nn}(x) = \limsup F_{nn}(x) = \lim F_{nn}(x) = F^*(x);$$

i.e., $F_{nn}(x) \underset{n\to\infty}{\longrightarrow} F^*(x)$ for $x \in C(F^*)$, as asserted. Finally, modify F^* into F as follows: $F(x) = F^*(x)$, $x \in C(F^*)$, and for $x \notin C(F^*)$, $F(x) = \lim F^*(x_n)$ where $x_n \in C(F^*)$ with $x_n \downarrow x$ as $n \to \infty$, so that F inherits the set of continuity points of F^*; i.e., $C(F) = C(F^*)$, and is also right-continuous. Therefore F is a d.f. and $F_{nn} \Rightarrow F$, as was to be seen. ■

Remark 4. The method employed in the proof for constructing the subsequence $\{F_{nn}\}$ is known as the Cantor diagonal method.

8.3 Helly–Bray Type Theorems for Distribution Functions

Remark 5. The following integrals are to be understood in the Riemann–Stieltjes sense. For some comments, regrading such integrals and related results, the reader is referred to Appendix B.

As a prelude to this section, let us revisit briefly Chapter 5. In that chapter, it was seen that, if a sequence of r.v.s X_n, $n \geq 1$, converges to a r.v. X (as $n \to \infty$), either a.e. (as is the case in Theorem 1 and its corollaries, Theorem 2(iii) and Theorem 3 under (a)), or in measure (as is the case in Theorem 3 under (b)), then the integrals $\int_\Omega X_n \, d\mu$ converge to the integral $\int_\Omega X \, d\mu$, provided some additional requirements are met.

Now, let F_n and F be the d.f.s of X_n and X, respectively, and suppose that $F_n \Rightarrow F$ or $F_n \overset{c}{\longrightarrow} F$ as $n \to \infty$. Also, let $g : \Re \to \Re$ be a Borel function, so that $g(X_n)$ and $g(X)$ are r.v.s. Then, by Theorem 13 of Chapter 4, $\int_\Omega g(X_n)dP = \int_\Re g(x)dP_{X_n} = \int_\Re g(x)dF_n$, and $\int_\Omega g(X)dP = \int_\Re g(x)dP_X = \int_\Re g(x)dF$ (see also Remarks 6 of this chapter).

A question which may then arise is this: under what further conditions is it true that

$$\int_B g(x)dF_n \underset{n\to\infty}{\longrightarrow} \int_B g(x)dF, \text{ over a set } B \subseteq \Re?$$

It is this kind of questions for which the Helly–Bray type results provide an answer (see Theorems 6,7, and 8).

Theorem 6 (Helly–Bray Lemma). Let $F_n \underset{n\to\infty}{\Rightarrow} F$. Then for every α, β,

$$-\infty < \alpha < \beta < \infty, \text{ such that } F_n(\alpha) \underset{n\to\infty}{\to} F(\alpha), \quad F_n(\beta) \underset{n\to\infty}{\to} F(\beta)$$

and every $g : [\alpha, \beta] \to \Re$ continuous (and hence bounded), we have

$$\int_{(\alpha,\beta]} g dF_n \underset{n\to\infty}{\to} \int_{(\alpha,\beta]} g dF. \qquad \blacksquare$$

Proof. Consider the partition $\alpha = x_1 < x_2 < \cdots < x_{m+1} = \beta$ of $[\alpha, \beta]$ and on $(\alpha, \beta]$ define the function

$$g_m(x) = \sum_{j=1}^{m} g(\xi_j) I_{(x_j, x_{j+1}]}(x),$$

where $\xi_j \in (x_j, x_{j+1}]$, $j = 1, \ldots, m$. Then we assert that, as $m \to \infty$, so that $\max_{1\le j\le m} |x_{j+1} - x_j| \to 0$, we have $g_m(x) \to g(x)$, $x \in (\alpha, \beta]$. In fact, let $x \in (\alpha, \beta]$. Then $x \in (x_k, x_{k+1}]$ for some k, while also $\xi_k \in (x_k, x_{k+1}]$. Now, as $m \to \infty$,

$$\max_{1\le j\le m} |x_{j+1} - x_j| \to 0 \text{ implies } x_{k+1} - x_k \to 0 \text{ so that } \xi_k \to x$$

and hence $g(\xi_k) \to g(x)$ by continuity of g. But $g(\xi_k) = g_m(x)$. Thus $g_m(x) \underset{m\to\infty}{\to} g(x)$.

Next,

$$\left|\int_{(\alpha,\beta]} g\, dF_n - \int_{(\alpha,\beta]} g\, dF\right| \le \left|\int_{(\alpha,\beta]} (g - g_m) dF_n\right|$$
$$+ \left|\int_{(\alpha,\beta]} (g_m - g) dF\right| + \left|\int_{(\alpha,\beta]} g_m\, dF - \int_{(\alpha,\beta]} g_m\, dF_n\right|,$$

clearly, while all these integrals exist either because the integrand is continuous or because it is a step (in the usual sense) function. Then the right-hand side of the relation above is bounded by

$$2 \sup_{x\in(\alpha,\beta]} |g_m(x) - g(x)| + \left|\int_{(\alpha,\beta]} g_m\, dF - \int_{(\alpha,\beta]} g_m\, dF_n\right|,$$

since

$$F(\beta) - F(\alpha) \le 1, \quad F_n(\beta) - F_n(\alpha) \le 1.$$

Next,

$$\int_{(\alpha,\beta]} g_m\, dF = \sum_{j=1}^{m} \int_{(x_j,x_{j+1}]} g_m\, dF = \sum_{j=1}^{m} g(\xi_j)\left[F(x_{j+1}) - F(x_j)\right],$$

and similarly,

$$\int_{(\alpha,\beta]} g_m\, dF_n = \sum_{j=1}^{m} \int_{(x_j,x_{j+1}]} g_m\, dF_n = \sum_{j=1}^{m} g(\xi_j)\left[F_n(x_{j+1}) - F_n(x_j)\right].$$

Hence

$$\left|\int_{(\alpha,\beta]} g\,dF_n - \int_{(\alpha,\beta]} g\,dF\right| \leq 2 \sup_{x\in(\alpha,\beta]} |g_m(x) - g(x)| + \sum_{j=1}^{m} \{|g(\xi_j)| \times [|F_n(x_{j+1}) - F(x_{j+1})| + |F_n(x_j) - F(x_j)|]\}.$$

Taking the partitioning points $x_2 \ldots, x_m$ in $C(F)$, using weak convergence and the assumption that $F_n(\alpha) \to F(\alpha)$ and $F_n(\beta) \to F(\beta)$, we have $F_n(x_j) \underset{n\to\infty}{\to} F(x_j)$, $j = 1, \ldots, m$. Thus, the second summand at the right-hand side of the last inequality $\underset{n\to\infty}{\to} 0$. Next,

$$\begin{aligned}\sup_{x\in(\alpha,\beta]} |g_m(x) - g(x)| &= \max_{1\leq j\leq m} \{\sup |g_m(x) - g(x)|; \quad x \in (x_j, x_{j+1}]\} \\ &= \max_{1\leq j\leq m} \{\sup |g(\xi_j) - g(x)|; \quad x \in (x_j, x_{j+1}]\} \\ &\leq \max_{1\leq j\leq m} \{\sup |g(\xi_j) - g(x)|; \quad x \in [x_j, x_{j+1}]\} \\ &= \max_{1\leq j\leq m} \{|g(\xi_j) - g(y_j)|; \quad \text{some } y_j \in [x_j, x_{j+1}]\}\end{aligned}$$

by continuity of g, and this is equal to $|g(\xi_k) - g(y_k)|$ for some $k = 1, \ldots, m$. As $m \to \infty$, this last expression converges to zero by the uniform continuity of g in $[\alpha, \beta]$ and the fact that $\max_{1\leq j\leq m} |x_{j+1} - x_j| \to 0$ implies $|\xi_k - y_k| \to 0$. Thus, as $n \to \infty$,

$$\int_{(\alpha,\beta]} g\,dF_n \to \int_{(\alpha,\beta]} g\,dF. \qquad \blacksquare$$

Theorem 7 (Helly–Bray Extended Lemma). Let $F_n \underset{n\to\infty}{\Rightarrow} F$. Then, for every $g : \Re \to \Re$ continuous and such that $g(\pm\infty) = 0$ (in the sense that $\lim_{x\to\pm\infty} g(x) = 0$), we have $\int_{\Re} g\,dF_n \underset{n\to\infty}{\to} \int_{\Re} g\,dF$. ■

Proof. Since g is bounded and continuous on $\Re$, $\int_{\Re} g\,dF_n$, $\int_{\Re} g\,dF$ exist (finite) and are taken as follows:

$$\int_{\Re} g\,dF_n = \lim_{\substack{\alpha\to-\infty\\ \beta\to\infty}} \int_{(\alpha,\beta]} g\,dF_n, \quad \int_{\Re} g\,dF = \lim_{\substack{\alpha\to-\infty\\ \beta\to\infty}} \int_{(\alpha,\beta]} g\,dF.$$

Next, for any $-\infty < \alpha < \beta < \infty$, we have

$$\begin{aligned}\left|\int_{\Re} g\,dF_n - \int_{\Re} g\,dF\right| \leq &\left|\int_{(\alpha,\beta]} g\,dF_n - \int_{(\alpha,\beta]} g\,dF\right| \\ &+ \left|\int_{\Re} g\,dF - \int_{(\alpha,\beta]} g\,dF\right| \\ &+ \left|\int_{\Re} g\,dF_n - \int_{(\alpha,\beta]} g\,dF_n\right|. \qquad (8.3)\end{aligned}$$

In the sequel, α and β will be continuity points of F, and α and β will be taken as small as needed and as large as needed, respectively. Then $F_n(\alpha) \underset{n\to\infty}{\to} F(\alpha)$, $F_n(\beta) \underset{n\to\infty}{\to} F(\beta)$, and hence, by Theorem 6,

$$\left|\int_{(\alpha,\beta]} g\,dF_n - \int_{(\alpha,\beta]} g\,dF\right| < \frac{\varepsilon}{3} \quad \text{for all sufficiently large } n. \tag{8.4}$$

Also,

$$\left|\int_{\Re} g\,dF - \int_{(\alpha,\beta]} g\,dF\right| < \frac{\varepsilon}{3}. \tag{8.5}$$

Finally,

$$\left|\int_{\Re} g\,dF_n - \int_{(\alpha,\beta]} g\,dF_n\right| = \left|\int_{\Re-(\alpha,\beta]} g\,dF_n\right| \leq \sup_{\Re-(\alpha,\beta]} |g(x)| < \frac{\varepsilon}{3}, \tag{8.6}$$

by also taking into consideration that $g(x) \to 0$ as $x \to \pm\infty$. Then from (8.3)–(8.6), we get

$$\left|\int_{\Re} g\,dF_n - \int_{\Re} g\,dF\right| < \varepsilon \quad \text{for } n \text{ sufficiently large.}$$ ■

Theorem 8 (Helly–Bray Theorem). Let $F_n \underset{n\to\infty}{\overset{c}{\to}} F$. Then for every $g : \Re \to \Re$ bounded and continuous, we have $\int_{\Re} g\,dF_n \underset{n\to\infty}{\to} \int_{\Re} g\,dF$. ■

Proof. Since g is bounded and continuous on $\Re$, $\int_{\Re} g\,dF_n$, $\int_{\Re} g\,dF$ exist (finite). Next,

$$\begin{aligned}\left|\int_{\Re} g\,dF_n - \int_{\Re} g\,dF\right| &\leq \left|\int_{(\alpha,\beta]} g\,dF_n - \int_{(\alpha,\beta]} g\,dF\right| \\ &+ \left|\int_{\Re} g\,dF - \int_{(\alpha,\beta]} g\,dF\right| + \left|\int_{\Re} g\,dF_n - \int_{(\alpha,\beta]} g\,dF_n\right|.\end{aligned} \tag{8.7}$$

In the sequel, α and β will be continuity points of F; also, α and β will be as small as needed and as large as needed, respectively. With this in mind, we have

$$\left|\int_{(\alpha,\beta]} g\,dF_n - \int_{(\alpha,\beta]} g\,dF\right| < \frac{\varepsilon}{3} \quad \text{(by Theorem 6)}, \tag{8.8}$$

and, by the definition of $\int_{\Re} g\,dF$,

$$\left|\int_{\Re} g\,dF - \int_{(\alpha,\beta]} g\,dF\right| < \frac{\varepsilon}{3}. \tag{8.9}$$

Next,

$$
\begin{aligned}
&\left|\int_{\Re} g\,dF_n - \int_{(\alpha,\beta]} g\,dF_n\right| = \left|\int_{(-\infty,\alpha]} g\,dF_n + \int_{(\beta,\infty)} g\,dF_n\right| \\
&\le \int_{(-\infty,\alpha]} |g|\,dF_n + \int_{(\beta,\infty)} |g|\,dF_n \le M\left\{\left[F_n(\alpha) - F_n(-\infty)\right]\right. \\
&\quad \left. + \left[F_n(\infty) - F_n(\beta)\right]\right\} \\
&= M\left\{\left[F_n(\infty) - F_n(-\infty)\right] - \left[F_n(\beta) - F_n(\alpha)\right]\right\} \\
&= M\left\{Var F_n - \left[F_n(\beta) - F_n(\alpha)\right]\right\},
\end{aligned}
$$

where $|g(x)| \le M$. Let $n \to \infty$. From $F_n \overset{c}{\to} F$ it follows that $Var F_n \to Var F$. Also, $F_n(\alpha) \to F(\alpha)$, $F_n(\beta) \to F(\beta)$, so that $F_n(\beta) - F_n(\alpha) \to F(\beta) - F(\alpha)$. Thus, for all sufficiently large n,

$$
\begin{aligned}
&Var F_n - \left[F_n(\beta) - F_n(\alpha)\right] \le |Var F_n - Var F| \\
&\quad + \left|Var F - \left[F(\beta) - F(\alpha)\right]\right| + \left|\left[F_n(\beta) - F_n(\alpha)\right]\right. \\
&\quad \left. - \left[F(\beta) - F(\alpha)\right]\right| < \frac{\varepsilon}{9M} + \frac{\varepsilon}{9M} + \frac{\varepsilon}{9M} = \frac{\varepsilon}{3M},
\end{aligned}
$$

and therefore

$$\left|\int_{(\alpha,\beta]} g\,dF_n - \int_{(\alpha,\beta]} g\,dF\right| < \frac{\varepsilon}{3}. \tag{8.10}$$

By (8.8)–(8.10), (8.7) becomes

$$\left|\int_{\Re} g\,dF_n - \int_{\Re} g\,dF\right| < \varepsilon \quad \text{for all sufficiently large } n. \qquad \blacksquare$$

Remark 6. From the proof of Theorems 6–8, it is easily seen that they are also true if in the d.f.s involved property (1) in Definition 1 is replaced by the boundedness property $0 \le F(x) \le B$, $x \in \Re$, some $B > 0$.

Remark 7. In closing, it should be mentioned that the results discussed in this chapter, pertaining to weak convergence of d.f.s, are very special cases of the subject matter of weak convergence of probability measures. For an exposition of such material, see, e.g., Billingsley (1999).

Exercises.

1. Let $F : \Re \to [0, 1]$ be nondecreasing, right continuous, $F(-\infty) = 0$ and $F(\infty) = 1$ (i.e., a d.f. of a r.v.), and let F^{-1} be defined by

$$F^{-1}(y) = \inf\{x \in \Re;\ F(x) \ge y\},\ y \in [0, 1].$$

Next, consider the probability space $(\Omega, \mathcal{A}, P) = ([0, 1], \mathcal{B}_{[0,1]}, P)$, where $P = \lambda$ is the Lebesgue measure, and on Ω, define the function X by $X(\omega) = F^{-1}(\omega)$. Then show that X is a r.v. and that its d.f. is F.
Hint: Show that $F^{-1}(y) \le t$ if and only if $y \le F(t)$, $t \in \Re$.

2. **(i)** If $X_n \underset{n\to\infty}{\overset{P}{\to}} X$, then show that $F_{X_n} \underset{n\to\infty}{\Rightarrow} F_X$.

   (ii) By an example, show that the converse statement in part (i) need not be true.

   Hint: For part (i), show that, for every $\varepsilon > 0$ and any $x \in \Re$, $F_{X_n}(x) \leq P(|X_n - X| \geq \varepsilon) + F_X(x+\varepsilon)$, and $F_X(x-\varepsilon) \leq P(|X_n - X| \geq \varepsilon) + F_{X_n}(x)$. Next, let $x \in C(F_X)$ and take the limits first as $n \to \infty$ and then as $\varepsilon \to 0$ to obtain the desired result.
3. Let $F_n, n \geq 1$, be d.f.s of r.v.s and suppose that $F_n \underset{n\to\infty}{\Rightarrow} F$ where F is a d.f. with $VarF = 1$. Then show that F is the d.f. of a r.v.
4. Show that the Weak Compactness Theorem (Theorem 5) still holds, if condition 1 in Definition 1 is replaced by: $(0 \leq)F(x) \leq B, x \in \Re, n \geq 1$, for some $B < \infty$.
5. Show that Theorems 6–8 hold true if, for all $x \in \Re$, $0 \leq F(x) \leq B$ and $0 \leq F_n(x) \leq B, n = 1, 2, \ldots$, for some $B > 0$, as indicated in Remark 6.
6. For $n = 1, 2, \ldots$, consider the r.v.s X_n, X and Y such that $|X_n| \leq Y, \mathcal{E}Y^r < \infty$ (for some $r > 0$), and $X_n \underset{n\to\infty}{\overset{d}{\longrightarrow}} X$. Then show that $\mathcal{E}|X_n|^r \underset{n\to\infty}{\longrightarrow} \mathcal{E}|X|^r$.
7. Let X be a r.v. with d.f. F; i.e., $F(x) = P(X \leq x), x \in \Re$. Then show that

   (i) $P(X < x) = F(x-)$ (the left-hand side limit).

   (ii) $F(x)$ is continuous at x if and only if $P(X = x) = 0$.

8. In the following expression, determine the constants α and β, so that the function F defined is the d.f. of a r.v.

$$F(x) = 0, x \leq 0, \quad \text{and} \quad F(x) = \alpha + \beta e^{-x^2/2}, \quad x > 0.$$

9. For $n = 1, 2, \ldots$, let F_n and F be d.f.s such that $F_n \underset{n\to\infty}{\overset{c}{\longrightarrow}} F$, and let F be continuous. Then show that $F_n(x) \underset{n\to\infty}{\longrightarrow} F(x)$ uniformly in $x \in \Re$.

   Hint: For $\varepsilon > 0$, choose a and b sufficiently small and sufficiently large, respectively, so that $F(a) < \frac{\varepsilon}{3}$, $F(\infty) - F(b) < \frac{\varepsilon}{3}$. Next, partition $[a, b]$ by $a = x_0 < x_1 < \cdots < x_{k-1} < x_k = b$, so that $F(x_j) - F(x_{j-1}) < \frac{\varepsilon}{3}, j = 1, \ldots, k$. Finally, by taking $x \in [a, b]$, or $x < a$, or $x > b$, show that $|F_n(x) - F(x)| < \varepsilon$ for all x, provided $n \geq N$ some integer independent of $x \in \Re$.
10. Let X be a r.v. with continuous d.f. F_X, and define the r.v. Y as follows: $Y = X$ if $X \leq C$, and $Y = C$ if $X > C$, where C is a constant. Then

   (i) Determine the d.f. of the r.v. Y, F_Y.

   (ii) Show that $F_Y = F_1 + F_2$, where F_1 is a continuous d.f. and F_2 is a step function.

11. By a simple example demonstrate that it is possible that $F_n, n \geq 1$, are d.f.s of r.v.s with $F_n \underset{n\to\infty}{\Rightarrow} F$, a d.f., but not that of a r.v.
12. For $n \geq 1$, let X_n and X be r.v.s defined on the probability space $(\Omega, \mathcal{A}, P)$, and suppose that, as $n \to \infty$, $\mathcal{E}|X_n| \to \mathcal{E}|X| < \infty$ and $X_n \overset{d}{\to} X$. Then show that $|X_n|, n \geq 1$, are uniformly integrable.

13. For $n \geq 1$, let F_n and F be d.f.s of r.v.s, let f_n be real-valued measurable functions defined on $\Re$, let $f : \Re \to \Re$ be continuous, and let $g : \Re \to [0, \infty)$ be continuous. Assume that:

(i) $F_n \to F$ as $n \to \infty$.
(ii) $|f_n(x)| \leq g(x), x \in \Re, n \geq 1$.
(iii) $f_n(x) \underset{n\to\infty}{\longrightarrow} f(x)$ uniformly on finite intervals.
(iv) $\int_{\Re} g \, dF_n \underset{n\to\infty}{\longrightarrow} \int_{\Re} g \, dF$ and $\int_{\Re} g \, dF < \infty$.

Then show that $\int_{\Re} f_n \, dF_n \underset{n\to\infty}{\longrightarrow} \int_{\Re} f \, dF$ and $\int_{\Re} f \, dF$ is finite.
Hint: Use the Helly–Bray Lemma (Theorem 6 in this chapter).

This page is intentionally left blank

CHAPTER

Conditional Expectation and Conditional Probability, and Related Properties and Results

9

This chapter is primarily about conditional expectations and conditional probabilities defined in terms of σ-fields. The definitions are based on Theorems 2 and 3 of Chapter 7.

In Section 9.1, the concept of the conditional expectation, given a σ-field, is defined, and then the conditional probability is derived as a special case. In the following section, some basic properties of conditional expectations and of conditional probabilities are discussed.

In Section 9.3, versions for conditional expectations of standard convergence theorems proved in Chapter 5 are established. The same is done for the moment inequalities discussed in Chapter 6. Some additional properties of conditional expectations and of conditional probabilities are presented in the fourth section of the chapter. The section is concluded with an application, illustrating that the elementary definition of a conditional probability density function coincides with the more general definition used here.

In the course of derivations in Section 9.4, the concept of independence of r.v.s is needed, as well as a result regarding the expectations of the product of two independent r.v.s (see Lemma 1). An elaboration on independence and the proof of Lemma 1 are found in Chapter 10.

9.1 Definition of Conditional Expectation and Conditional Probability

Consider the probability space $(\Omega, \mathcal{A}, P)$ and let X be a r.v. defined on this space with $\mathcal{E}\,|X| < \infty$. Let $\mathcal{B}$ be a σ-field with $\mathcal{B} \subseteq \mathcal{A}$ and denote by $P_{\mathcal{B}}$ the restriction of P to $\mathcal{B}$. On $\mathcal{A}$, define the set functions φ^+ and φ^- as follows:

$$\varphi^+(A) = \int_A X^+\,dP, \quad \varphi^-(A) = \int_A X^-\,dP, \quad A \in \mathcal{A}.$$

Denoting by $\varphi^+_{\mathcal{B}}$ and $\varphi^-_{\mathcal{B}}$ the restrictions of φ^+, φ^- to $\mathcal{B}$, we clearly have

$$\varphi^+_{\mathcal{B}}(B) = \int_B X^+\,dP, \quad \varphi^-_{\mathcal{B}}(B) = \int_B X^-\,dP, \quad B \in \mathcal{B}.$$

An Introduction to Measure-Theoretic Probability, Second Edition. http://dx.doi.org/10.1016/B978-0-12-800042-7.00009-8

Then, clearly, $\varphi_{\mathcal{B}}^{+}, \varphi_{\mathcal{B}}^{-}$ are $P_{\mathcal{B}}$-continuous (finite) measures (see also Exercise 1) and therefore, by Theorem 3 in Chapter 7, there exists a.s. $[P_{\mathcal{B}}]$ well defined, ≥ 0, $\mathcal{B}$-measurable (a.s. $[P_{\mathcal{B}}]$ finite) r.v.s Y^{+}, Y^{-} such that

$$\varphi_{\mathcal{B}}^{+}(B) = \int_B Y^{+}\, dP_{\mathcal{B}}, \quad \varphi_{\mathcal{B}}^{-}(B) = \int_B Y^{-}\, dP_{\mathcal{B}}, \quad B \in \mathcal{B}.$$

Then, by setting $Y = Y^{+} - Y^{-}$, we have that Y is a.s. $[P_{\mathcal{B}}]$ well defined, $\mathcal{B}$-measurable, a.s. $[P_{\mathcal{B}}]$ finite, and

$$\int_B Y\, dP_{\mathcal{B}} = \int_B X\, dP, \quad B \in \mathcal{B}.$$

(However, see also Proposition 1 in Chapter 7.)

Definition 1. Given an integrable r.v. X defined on the probability space $(\Omega, \mathcal{A}, P)$, and a σ-field $\mathcal{B} \subseteq \mathcal{A}$, the *conditional expectation* of X, given $\mathcal{B}$, is any $\mathcal{B}$-measurable r.v. Y, denoted by $\mathcal{E}^{\mathcal{B}}X$ (or $\mathcal{E}(X \mid \mathcal{B})$) such that

$$\int_B \mathcal{E}^{\mathcal{B}}X\, dP_{\mathcal{B}} = \int_B X\, dP, \quad B \in \mathcal{B}. \tag{9.1}$$

By taking $X = I_A$, $A \in \mathcal{A}$, then $\mathcal{E}^{\mathcal{B}} I_A$ is called *conditional probability* of A, given $\mathcal{B}$, is denoted by $P^{\mathcal{B}}A$, and we have

$$\int_B \mathcal{E}^{\mathcal{B}} I_A dP_{\mathcal{B}} = \int_B P^{\mathcal{B}}A\, dP_{\mathcal{B}}$$
$$= \int_B I_A dP = P(A \cap B), \quad B \in \mathcal{B}. \quad \blacksquare \tag{9.2}$$

Special Cases: For $B \in \mathcal{A}$ with $P(B) > 0$ and a r.v. X with $\mathcal{E}\,|X| < \infty$, define the r.v. $\mathcal{E}_B X$ as follows:

$$(\mathcal{E}_B X)(\omega) = \frac{1}{P(B)} \int_B X\, dP, \quad \omega \in \Omega.$$

Now let $\{B_j,\, j = 1, 2, \ldots\}$ be a (measurable) partition of Ω with $P(B_j) > 0$, all j, and let $\mathcal{B}$ be the σ-field generated by this partition. Define the function Y on Ω as follows:

$$Y = \sum_{j=1}^{\infty} (\mathcal{E}_{B_j} X) I_{B_j}, \quad \mathcal{E}_{B_j} X = \frac{1}{P(B_j)} \int_{B_j} X\, dP, \quad j \geq 1.$$

Then, clearly, Y is $\mathcal{B}$-measurable and for any $B \in \mathcal{B}$, one has $B = \sum_{i \in I} B_i$ for some $I \subseteq \{1, 2, \ldots\}$. Therefore, by Exercise 1,

$$\int_B Y\, dP_{\mathcal{B}} = \int_{\Sigma_{i \in I} B_i} Y\, dP_{\mathcal{B}} = \sum_{i \in I} \int_{B_i} Y\, dP_{\mathcal{B}} = \sum_{i \in I} \int_{B_i} \left[\sum_j (\mathcal{E}_{B_j} X) I_{B_j} \right] dP_{\mathcal{B}}$$
$$= \sum_{i \in I} (\mathcal{E}_{B_i} X) P_{\mathcal{B}}(B_i) = \sum_{i \in I} (\mathcal{E}_{B_i} X) P(B_i) = \sum_{i \in I} \frac{P(B_i)}{P(B_i)} \int_{B_i} X\, dP$$

$$= \sum_{i\in I} \int_{B_i} X\,dP = \int_{\Sigma_{i\in I} B_i} X\,dP = \int_B X\,dP = \int_B \mathcal{E}^{\mathcal{B}} X\,dP_{\mathcal{B}};$$

$$\text{i.e.,} \quad \int_B Y\,dP_{\mathcal{B}} = \int_B \mathcal{E}^{\mathcal{B}} X\,dP_{\mathcal{B}}, \quad B \in \mathcal{B},$$

and therefore $Y = \mathcal{E}^{\mathcal{B}} X$ a.s. (by Proposition 1 in Chapter 7). That is, in the case that $\mathcal{B}$ is generated by a countable (measurable) partition $\{B_j, j \geq 1\}$ with $P(B_j) > 0$ for all j, then for any integrable r.v. X, one has that $(\mathcal{E}^{\mathcal{B}} X)(\omega) = \frac{1}{P(B_j)} \int_{B_j} X\,dP$, for a.a. (almost all) $\omega \in B_j$, $j \geq 1$. In particular, if $X = I_A$, $A \in \mathcal{A}$, then

$$(P^{\mathcal{B}} A)(\omega) = \frac{P(A \cap B_j)}{P(B_j)}, \quad \text{a.a. } \omega \in B_j, \quad j \geq 1. \tag{9.3}$$

If the partition of Ω consists of two sets B, $B^c \in \mathcal{A}$ with $0 < P(B) < 1$, then the σ-field generated by $\{B, B^c\}$ is $\mathcal{B} = \{\oslash, B, B^c, \Omega\}$ and

$$(P^{\mathcal{B}} A)(\omega) = \begin{cases} \frac{P(A\cap B)}{P(B)} & \text{a.a. } \omega \in B \\ \frac{P(A\cap B^c)}{P(B^c)} & \text{a.a. } \omega \in B^c. \end{cases}$$

Thus for a.a. $\omega \in B$, $(P^{\mathcal{B}} A)(\omega) = P(A \mid B)$, and for a.a. $\omega \in B^c$, $(P^{\mathcal{B}} A)(\omega) = P(A \mid B^c)$.

Now suppose that the σ-field $\mathcal{B} \subseteq \mathcal{A}$ is generated, as before, by a countable partition $\{B_j, j \geq 1\}$ with $P(B_j) > 0$ for all j, and that for $A \in \mathcal{A}$, the conditional probability of A, given $\mathcal{B}$, is defined by (9.3). In this context, it will be convenient to write $P_\omega^{\mathcal{B}} A$ rather than $(P^{\mathcal{B}} A)(\omega)$. Thus, we have

$$P_\omega^{\mathcal{B}} A = \frac{P(A \cap B_j)}{P(B_j)}, \quad \text{a.a. } \omega \in B_j, \quad j \geq 1. \tag{9.4}$$

From (9.4), it is obvious that for every fixed $\omega \in \Omega$ (lying outside a null set), $P_\omega^{\mathcal{B}}$ as a function on $\mathcal{A}$ is a probability measure. Then, if X is a r.v., we may talk about the integral of X with respect to $P_\omega^{\mathcal{B}}$, provided, of course, this integral exists. To this end, let X be a P-integrable r.v. Then for an arbitrary but fixed $\omega \in \Omega$ (lying outside a null set), we have that $\omega \in B_j$ for exactly one j. Therefore, by Exercise 2,

$$\int X(\omega')dP_\omega^{\mathcal{B}} = \frac{1}{P(B_j)} \int_{B_j} X\,dP = \mathcal{E}_{B_j} X, \quad \omega \in B_j, \quad j \geq 1,$$

and hence

$$\int X(\omega')dP_\omega^{\mathcal{B}} = \sum_j (\mathcal{E}_{B_j} X) I_{B_j} = \mathcal{E}_\omega^{\mathcal{B}} X, \quad \omega \in \Omega \text{ (lying outside a null set)}. \tag{9.5}$$

This result then shows that if $\mathcal{B}$ is generated as before, one may define conditional probability *first* and *then* get the conditional expectation as the integral of a P-integrable r.v. X with respect to the conditional probability.

Remark 1. The result just stated need not be true if $\mathcal{B}$ is not of the form just mentioned. (See, e.g., Loève (1963), pages 353–354.) This section is concluded with two examples illustrating some concepts discussed here.

Remark 2. Here is a possible explanation of the meaning/usefulness of the conditional probability and conditional expectation in their abstract setting.

From the definition of the conditional probability, we have, for $A \in \mathcal{A}$,

$$\int_B P^{\mathcal{B}} A \, dP_{\mathcal{B}} = \int_B I_A \, dP = P(A \cap B), \quad B \in \mathcal{B},$$

so that

$$\frac{1}{P(B)} \int_B P^{\mathcal{B}} A \, dP_{\mathcal{B}} = \frac{P(A \cap B)}{P(B)} = P(A|B)$$

Now, on the right-hand side in the last relation above, we have the conditional probability of A, given B, as is given by the familiar elementary definition. The left-hand side in the sane relation tells us that the actual calculation of $P(A|B)$ is obtained through the integration of the ($\mathcal{B}$-measurable) r.v. $P^{\mathcal{B}} A$ over B, and norming it by the $P(B)$.

Next, from the definition of the conditional expectation of a r.v. $X(\mathcal{E}X \in \Re)$, we have,

$$\int_B \mathcal{E}^{\mathcal{B}} X \, dP_{\mathcal{B}} = \int_B X \, dP, \quad B \in \mathcal{B},$$

so that

$$\frac{1}{P(B)} \int_B \mathcal{E}^{\mathcal{B}} X \, dP_{\mathcal{B}} = \frac{1}{P(B)} \int_B X \, dP = \int_\Omega X \, dP(\cdot|B),$$

where $P(\cdot|B)$ is the elementary definition of the conditional probability, given B (see also Exercise 2). So, $\mathcal{E}^{\mathcal{B}} X$ has the property that it is the ($\mathcal{B}$-measurable) r.v. whose weighted (by $P(B)$) integral over B gives the expectation of X with respect to the conditional probability measure $P(\cdot|B)$.

Example 1. Let $X_1, \ldots, X_n$ be i.i.d. r.v.s defined on the probability space $(\Omega, \mathcal{A}, P_{\boldsymbol{\theta}})$, where the parameter $\boldsymbol{\theta}$ takes values in the parameter space Θ, an open subset of $\Re^k$, $k \geq 1$, and set $\mathbf{X} = (X_1, \ldots, X_n)$. Also, let $\mathbf{T} = (T_1, \ldots, T_m)$ be an m-dimensional statistic of $\mathbf{X}$; i.e., a measurable function of $\mathbf{X}$ not involving any unknown quantities. We say that $\mathbf{T}$ is *sufficient* for $\boldsymbol{\theta}$, if the conditional distribution of $\mathbf{X}$, given the σ-field induced by $\mathbf{T}$, $\mathcal{B}_{\mathbf{T}}$, does not depend on $\boldsymbol{\theta}$. That is, if $\mathcal{A}_{\mathbf{X}}$ is the σ-field induced by $\mathbf{X}$ (so that $\mathcal{A}_{\mathbf{X}} \subseteq \mathcal{A}$ and $\mathcal{B}_{\mathbf{T}} \subseteq \mathcal{A}_{\mathbf{X}}$ since $\mathbf{T} = \mathbf{T}(\mathbf{X})$), then $\mathcal{E}_{\boldsymbol{\theta}}^{\mathcal{B}_{\mathbf{T}}} I_A = P_{\boldsymbol{\theta}}^{\mathcal{B}_{\mathbf{T}}} A$ is independent of $\boldsymbol{\theta}$ for all $A \in \mathcal{A}_{\mathbf{X}}$.

As an application of this concept, let $X_1, \ldots, X_n$ be i.i.d. r.v.s distributed as $B(1, \theta)$, $\theta \in \Theta = (0, 1)$, and set $T = \sum_{j=1}^n X_j$. Then T is sufficient for θ.

To this end, set $\mathbf{X} = (X_1, \ldots, X_n)$ and observe that $\mathbf{X}$ takes on 2^n values in $\Re^n$, call them $\mathbf{x}_i = (x_{i1}, \ldots, x_{in})$, $i = 1, \ldots, 2^n$. Thus, if $A_i = \mathbf{X}^{-1}(\{\mathbf{x}_i\})$, then $\mathcal{A}_{\mathbf{X}}$ is generated by $\{A_i, i = 1, \ldots, 2^n\}$. Next, $T \sim B(n, \theta)$ and $\mathcal{B}_T$ is generated by $\{B_t, t = 0, 1, \ldots, n\}$ where $B_t = T^{-1}(\{t\}) = \mathbf{X}^{-1}(C_t) = \sum_{j=1}^m A_{i_j}$, where $C_t = \{\mathbf{x}_{i_j} = (x_{i_j 1}, \ldots, x_{i_j n}) \in \Re^n$, each coordinate being either 0 or 1 and their sum being equal to t, $j = 1, \ldots, m = \binom{n}{t}\}$, since there are $\binom{n}{t}$ ways of choosing t positions out of n to place t 1s and 0s in the remaining $n - t$ positions; if $t = 0$, then $\binom{n}{0} = 1$, and this is the number of ways of placing 0s in all n places (and 1s nowhere). With $A \in \mathcal{A}_{\mathbf{X}}$, apply the special case following Definition 1 to get

$$P_\theta^{\mathcal{B}_T} A = \mathcal{E}_\theta^{\mathcal{B}_T} I_A = \sum_{t=0}^{n} (\mathcal{E}_{B_t,\theta} I_A) I_{B_t}, \quad \mathcal{E}_{B_t,\theta} I_A = \frac{P_\theta(A \cap B_t)}{P_\theta(B_t)}, \quad t = 0, 1, \ldots, n.$$

Clearly, for each $A \in \mathcal{A}_{\mathbf{X}}$, $A = \sum_{k=1}^{p} A_{j_k}$, for some p, so that

$$A \cap B_t = \left(\sum_{k=1}^{p} A_{j_k}\right) \cap \left(\sum_{j=1}^{m} A_{i_j}\right) = \sum_{l=1}^{q} A_{k_l}, \text{ for some } q,$$

and

$$\begin{aligned} P_\theta(A_{k_l}) &= P_\theta(X_1 = x_{k_l 1}, \ldots, X_n = x_{k_l n}) \\ &= P_\theta(X_1 = x_{k_l 1}) \ldots P_\theta(X_1 = x_{k_l n}) \\ &= \theta^{x_{k_l 1}}(1-\theta)^{1-x_{k_l 1}} \ldots \theta^{x_{k_l n}}(1-\theta)^{1-x_{k_l n}} \\ &= \theta^t (1-\theta)^{n-t}. \end{aligned}$$

It follows that $P_\theta(A \cap B_t) = q\theta^t(1-\theta)^{n-t}$. Since $P_\theta(T = t) = \binom{n}{t}\theta^t(1-\theta)^{n-t}$, it follows that

$$\mathcal{E}_{B_t,\theta} I_A = \frac{P_\theta(A \cap B_t)}{P_\theta(B_t)} = \frac{q\theta^t(1-\theta)^{n-t}}{\binom{n}{t}\theta^t(1-\theta)^{n-t}} = \frac{q}{\binom{n}{t}}$$

independent of θ; that is, $\mathcal{E}_{B_t,\theta} I_A$ is independent of θ, and then so is $P_\theta^{\mathcal{B}_T} I_A$. It follows that T is sufficient for θ. ■

Example 2. Refer to the previous example, and suppose that the independent r.v.s $X_1, \ldots, X_n$ are distributed as Poisson with parameter θ; i.e., $X_j \sim P(\theta)$, $\theta \in \Theta = (0, \infty)$, $j = 1, \ldots, n$. Then the statistic $T = \sum_{j=1}^{n} X_j$ is sufficient for θ.

As in Example 1, $\mathbf{X} = (X_1, \ldots, X_n)$ takes values $\mathbf{x}_i = (x_{i1}, \ldots, x_{in})$ in $\Re^n$ where each $x_{ij} = 0, 1, \ldots, j = 1, \ldots, n$. Thus, if $A_i = \mathbf{X}^{-1}(\{\mathbf{x}_i\})$, then $\mathcal{A}_{\mathbf{X}}$ is generated by $\{A_i, i = 1, 2, \ldots\}$. Next, $T \sim P(n\theta)$ and $\mathcal{B}_T$ is generated by $\{B_t, t = 0, 1, \ldots\}$ where $B_t = T^{-1}(\{t\}) = X^{-1}(C_t) = \sum_{j=1}^{m} A_{i_j}$, where $C_t = \{\mathbf{x}_{i_j} = (x_{i_j 1}, \ldots, x_{i_j n}) \in \Re^n$; each coordinate being ≥ 0 integer and their sum being equal to t, $j = 1, \ldots, m = \binom{n+t-1}{t}\}$. That $m = \binom{n+t-1}{t}$ follows from the fact that there are $\binom{n+t-1}{t}$ ways of selecting t positions out of n to place $x_{i_j 1}, \ldots, x_{i_j n} \geq 0$ integers with $x_{i_j 1} + \cdots + x_{i_j n} = t$ (see also Exercise 19). For $A \in \mathcal{A}_{\mathbf{X}}$, we have, as in Example 1,

$$P_\theta^{\mathcal{B}_T} A = \mathcal{E}_\theta^{\mathcal{B}_T} I_A = \sum_{t=0}^{\infty} (\mathcal{E}_{B_t,\theta} I_A) I_{B_t}, \quad \mathcal{E}_{B_t,\theta} I_A = \frac{P_\theta(A \cap B_t)}{P_\theta(B_t)}, \quad t = 0, 1, \ldots.$$

Furthermore, again as in Example 1, $A \cap B_t = \sum_{l=1}^{q} A_{k_l}$ for some q, and

$$P_\theta(A_{k_l}) = P_\theta(X_1 = x_{k_l 1}, \ldots, X_n = x_{k_l n})$$

$$= e^{-\theta}\frac{\theta^{x_{k_l1}}}{x_{k_l1}!}\dots e^{-\theta}\frac{\theta^{x_{k_ln}}}{x_{k_ln}!}$$
$$= e^{-n\theta}\frac{\theta^t}{x_{k_l1}!\dots x_{k_ln}!},$$

so that $P_\theta(A \cap B_t) = qe^{-n\theta}\theta^t/x_{k_l1}!\dots x_{k_ln}!$. Since $P_\theta(T = t) = e^{-n\theta}(n\theta)^t/t!$, it follows that

$$\mathcal{E}_{B_{t,\theta}} I_{A_i} = \frac{P_\theta(A \cap B_t)}{P_\theta(B_t)} = \frac{qe^{-n\theta}\theta^t/x_{k_l1}!\dots x_{k_ln}!}{e^{-n\theta}(n\theta)^t/t!}$$
$$= qt!/n^t x_{k_l1}!\dots x_{k_ln}!$$

independent of θ. Then so is $P_\theta^{\mathcal{B}_T} I_A$, and T is sufficient for θ. ■

9.2 Some Basic Theorems About Conditional Expectations and Conditional Probabilities

In all that follows the σ-field $\mathcal{B} \subseteq \mathcal{A}$. Also, Proposition 1 in Chapter 7 will be used throughout without explicit reference to it.

Theorem 1. Let $\mathcal{E}|X| < \infty$. Then

(i) $\mathcal{E}(\mathcal{E}^{\mathcal{B}}X) = \mathcal{E}X$.
(ii) $\mathcal{E}^{\mathcal{A}}X = X$ a.s.
(iii) If X is $\mathcal{B}$-measurable then $\mathcal{E}^{\mathcal{B}}X = X$ a.s. ■

Proof.

(i) From the definition of $\mathcal{E}^{\mathcal{B}}X$, one has

$$\int_B \mathcal{E}^{\mathcal{B}}X\, dP_{\mathcal{B}} = \int_B X\, dP, \quad B \in \mathcal{B}.$$

Then for $B = \Omega$, we get

$$\int \mathcal{E}^{\mathcal{B}}X\, dP_{\mathcal{B}} = \int \mathcal{E}^{\mathcal{B}}X\, dP = \int X\, dP \text{ or } \mathcal{E}(\mathcal{E}^{\mathcal{B}}X) = \mathcal{E}X.$$

(ii) Since both X and $\mathcal{E}^{\mathcal{A}}X$ are $\mathcal{A}$-measurable and

$$\int_A \mathcal{E}^{\mathcal{A}}X\, dP_{\mathcal{A}} = \int_A X\, dP_{\mathcal{A}}, \quad A \in \mathcal{A},$$

we have that $\mathcal{E}^{\mathcal{A}}X = X$ a.s.

(iii) As in (ii), both $\mathcal{E}^{\mathcal{B}}X$ and X are $\mathcal{B}$-measurable and

$$\int_B \mathcal{E}^{\mathcal{B}}X\,dP_{\mathcal{B}} = \int_B X\,dP_{\mathcal{B}}, \quad B \in \mathcal{B},$$

so that $\mathcal{E}^{\mathcal{B}}X = X$ a.s. ■

Theorem 2.

(i) If $X = c$ a.s., then $\mathcal{E}^{\mathcal{B}}X = c$ a.s.

(ii) If $X \geq Y$ a.s. and $\mathcal{E}\,|X|,\ \mathcal{E}\,|Y| < \infty$, then $\mathcal{E}^{\mathcal{B}}X \geq \mathcal{E}^{\mathcal{B}}Y$ a.s.

(iii) If for $j = 1, \ldots, n$, X_j are integrable and c_j are constants, one has

$$\mathcal{E}^{\mathcal{B}}\left(\sum_{j=1}^{n} c_j X_j\right) = \sum_{j=1}^{n} c_j \mathcal{E}^{\mathcal{B}} X_j \text{ a.s.}$$

(iv) $P^{\mathcal{B}}\Omega = 1$ a.s., $P^{\mathcal{B}}(\oslash) = 0$ a.s., $P^{\mathcal{B}}A \geq 0$ a.s., $\mathcal{E}^{\mathcal{B}}(\sum_{j=1}^{n} c_j I_{A_j}) = \sum_{j=1}^{n} c_j P^{\mathcal{B}} A_j$ a.s. ■

Proof.

(i) For every $B \in \mathcal{B}$, we have

$$\int_B \mathcal{E}^{\mathcal{B}}X\,dP_{\mathcal{B}} = \int_B X\,dP = \int_B c\,dP = \int_B c\,dP_{\mathcal{B}} \text{ (since } X = c \text{ a.s.).}$$

That is,

$$\int_B \mathcal{E}^{\mathcal{B}}X\,dP_{\mathcal{B}} = \int_B c\,dP_{\mathcal{B}} \text{ for every } B \in \mathcal{B}, \text{ and both } \mathcal{E}^{\mathcal{B}}X$$

and c are $\mathcal{B}$-measurable. Hence $\mathcal{E}^{\mathcal{B}}X = c$ a.s.

(ii) We have

$$\int_B \mathcal{E}^{\mathcal{B}}X\,dP_{\mathcal{B}} = \int_B X\,dP$$
$$\geq \int_B Y\,dP \text{ (since } X \geq Y\ a.s.)$$

and this equals $\int_B \mathcal{E}^{\mathcal{B}}Y\,dP_{\mathcal{B}},\ B \in \mathcal{B}$.
Since both $\mathcal{E}^{\mathcal{B}}X$ and $\mathcal{E}^{\mathcal{B}}Y$ are $\mathcal{B}$-measurable, we obtain that $\mathcal{E}^{\mathcal{B}}X \geq \mathcal{E}^{\mathcal{B}}Y$ a.s. (see also Exercise 4).

(iii) We have

$$\int_B \mathcal{E}^{\mathcal{B}}\Big(\sum_j c_j X_j\Big)dP_{\mathcal{B}} = \int_B \Big(\sum_j c_j X_j\Big)dP = \sum_j c_j \int_B X_j\,dP$$
$$= \sum_j c_j \int_B \mathcal{E}^{\mathcal{B}}X_j\,dP_{\mathcal{B}} = \int_B \Big(\sum_j c_j \mathcal{E}^{\mathcal{B}}X_j\Big)dP_{\mathcal{B}}, \quad B \in \mathcal{B}.$$

Since both $\mathcal{E}^{\mathcal{B}}(\sum_j c_j X_j)$ and $\sum_j c_j \mathcal{E}^{\mathcal{B}} X_j$ are $\mathcal{B}$-measurable, it follows that $\mathcal{E}^{\mathcal{B}}(\sum_j c_j X_j) = \sum_j c_j \mathcal{E}^{\mathcal{B}} X_j$ a.s.

(iv) We have

$$\int_B P^{\mathcal{B}}\Omega \, dP_{\mathcal{B}} = \int_B I_\Omega \, dP = P(B) = P_{\mathcal{B}}(B) = \int_B 1 \, dP_{\mathcal{B}}, \quad B \in \mathcal{B},$$

while both $P^{\mathcal{B}}\Omega$ and 1 are $\mathcal{B}$-measurable. Hence $P^{\mathcal{B}}\Omega = 1$ a.s.;

$$\int_B P^{\mathcal{B}} \oslash \, dP_{\mathcal{B}} = \int_B I_\oslash \, dP = 0 = \int_B 0 \, dP_{\mathcal{B}}, \quad B \in \mathcal{B},$$

and both $P^{\mathcal{B}}\oslash$ and 0 are $\mathcal{B}$-measurable. Thus $P^{\mathcal{B}}\oslash = 0$ a.s.;

$$\int_B P^{\mathcal{B}} A \, dP_{\mathcal{B}} = \int_B I_A \, dP = P(A \cap B) \geq 0 = \int_B 0 \, dP_{\mathcal{B}}, \quad B \in \mathcal{B},$$

and both $P^{\mathcal{B}}A$ and 0 are $\mathcal{B}$-measurable. Hence $P^{\mathcal{B}}A \geq 0$ a.s. (by Exercise 4). Finally,

$$\mathcal{E}^{\mathcal{B}}\Big(\sum_j c_j I_{A_j}\Big) = \sum_j c_j \mathcal{E}^{\mathcal{B}} I_{A_j} \text{ a.s. (by (iii)) and this equals}$$

$$\sum_j c_j P^{\mathcal{B}} A_j;$$

i.e., $\mathcal{E}^{\mathcal{B}}(\sum_j c_j I_{A_j}) = \sum_j c_j P^{\mathcal{B}} A_j$ a.s. ■

9.3 Convergence Theorems and Inequalities for Conditional Expectations

Theorem 3 (Lebesgue Monotone Convergence Theorem for Conditional Expectations). If, as $n \to \infty$, $0 \leq X_n \uparrow X$ a.s. and $\mathcal{E}X < \infty$, then $0 \overset{\text{a.s.}}{\leq} \mathcal{E}^{\mathcal{B}} X_n \uparrow \mathcal{E}^{\mathcal{B}} X$ a.s. In particular, $P^{\mathcal{B}}(\sum_{j=1}^{\infty} A_j) = \sum_{j=1}^{\infty} P^{\mathcal{B}} A_j$ a.s. ■

Proof. With all limits taken as $n \to \infty$, we have $0 \leq X_n \leq X_{n+1}$, so that

$$0 \leq \mathcal{E}^{\mathcal{B}} X_n \leq \mathcal{E}^{\mathcal{B}} X_{n+1} \text{ a.s.}$$

Therefore $\mathcal{E}^{\mathcal{B}} X_n \uparrow Y$(some Y) a.s. and Y is $\mathcal{B}$-measurable since it is the a.s. limit of $\mathcal{B}$-measurable r.v.s. (See also Exercise 19 in Chapter 3.) Then

$$\int_B \mathcal{E}^{\mathcal{B}} X_n dP_{\mathcal{B}} \uparrow \int_B Y \, dP_{\mathcal{B}}, \quad B \in \mathcal{B},$$

by the classical Monotone Convergence Theorem. But

$$\int_B \mathcal{E}^{\mathcal{B}} X_n\, dP_{\mathcal{B}} = \int_B X_n\, dP \text{ and } \int_B X_n\, dP \uparrow \int_B X\, dP = \int_B \mathcal{E}^{\mathcal{B}} X\, dP_{\mathcal{B}}.$$

Thus

$$\int_B Y\, dP_{\mathcal{B}} = \int_B \mathcal{E}^{\mathcal{B}} X\, dP_{\mathcal{B}}, \quad B \in \mathcal{B},$$

so that $Y = \mathcal{E}^{\mathcal{B}} X$ a.s. Therefore $0 \overset{\text{a.s.}}{\leq} \mathcal{E}^{\mathcal{B}} X_n \uparrow \mathcal{E}^{\mathcal{B}} X$ a.s.
For the second part, we have

$$P^{\mathcal{B}}(\sum_{j=1}^{n} A_j) = \mathcal{E}^{\mathcal{B}}\left(I_{\sum_{j=1}^{n} A_j}\right) = \mathcal{E}^{\mathcal{B}}\left(\sum_{j=1}^{n} I_{A_j}\right) \overset{\text{a.s.}}{=} \sum_{j=1}^{n} \mathcal{E}^{\mathcal{B}} I_{A_j}$$
$$\underset{\text{a.s.}}{\uparrow} \sum_{j=1}^{\infty} \mathcal{E}^{\mathcal{B}} I_{A_j} = \sum_{j=1}^{\infty} P^{\mathcal{B}} A_j.$$

On the other hand,

$$\sum_{j=1}^{n} A_j \uparrow \sum_{j=1}^{\infty} A_j \text{ implies } I_{\sum_{j=1}^{n} A_j} = \sum_{j=1}^{n} I_{A_j} \uparrow \sum_{j=1}^{\infty} I_{A_j} =$$

$$I_{\sum_{j=1}^{\infty} A_j}, \text{ and } \mathcal{E}^{\mathcal{B}}\left(I_{\sum_{j=1}^{n} A_j}\right) \uparrow \mathcal{E}^{\mathcal{B}}\left(I_{\sum_{j=1}^{\infty} A_j}\right) \text{ a.s., or } P^{\mathcal{B}}\left(\sum_{j=1}^{n} A_j\right) \uparrow$$

$$P^{\mathcal{B}}\left(\sum_{j=1}^{\infty} A_j\right) \text{ a.s.}$$

Comparing this with the last result obtained, we get

$$P^{\mathcal{B}}\left(\sum_{j=1}^{\infty} A_j\right) = \sum_{j=1}^{\infty} P^{\mathcal{B}}(A_j) \text{ a.s.} \qquad \blacksquare$$

Theorem 4. Suppose that $\mathcal{E}\,|Y|\,, \mathcal{E}\,|XY| < \infty$ and let X be $\mathcal{B}$-measurable. Then

$$\mathcal{E}^{\mathcal{B}}(XY) = X\mathcal{E}^{\mathcal{B}} Y \text{ a.s.} \qquad \blacksquare$$

Proof. First let $X = I_A$, $A \in \mathcal{B}$. Then $\mathcal{E}^{\mathcal{B}}(XY) = \mathcal{E}^{\mathcal{B}}(I_A Y)$ and hence

$$\int_B \mathcal{E}^{\mathcal{B}}(I_A Y) dP_{\mathcal{B}} = \int_B I_A Y\, dP = \int_{A\cap B} Y\, dP = \int_{A\cap B} \mathcal{E}^{\mathcal{B}} Y\, dP_{\mathcal{B}}$$
$$= \int_B (I_A \mathcal{E}^{\mathcal{B}} Y) dP_{\mathcal{B}}, \quad B \in \mathcal{B},$$

so that $\mathcal{E}^{\mathcal{B}}(I_A Y) = I_A \mathcal{E}^{\mathcal{B}} Y$ a.s., since both $\mathcal{E}^{\mathcal{B}}(I_A Y)$ and $I_A \mathcal{E}^{\mathcal{B}} Y$ are $\mathcal{B}$-measurable, or $\mathcal{E}^{\mathcal{B}}(XY) = X\mathcal{E}^{\mathcal{B}} Y$ a.s.

Next, let

$$X = \sum_{j=1}^{n} \alpha_j I_{A_j}, \quad A_j \in \mathcal{B}, \quad j = 1, \ldots, n.$$

Then

$$\mathcal{E}^{\mathcal{B}}(XY) = \mathcal{E}^{\mathcal{B}}[(\sum_j \alpha_j I_{A_j})Y] = \mathcal{E}^{\mathcal{B}}(\sum_j \alpha_j Y I_{A_j}) \overset{\text{a.s.}}{=} \sum_j \alpha_j \mathcal{E}^{\mathcal{B}}(Y I_{A_j})$$

(by Theorem 2(iii))

$$\overset{\text{a.s.}}{=} \sum_j \alpha_j I_{A_j} \mathcal{E}^{\mathcal{B}} Y \text{(by the previous step)}$$

$$= X\mathcal{E}^{\mathcal{B}} Y; \text{ i.e., } \mathcal{E}^{\mathcal{B}}(XY) = X\mathcal{E}^{\mathcal{B}} Y \text{ a.s.}$$

Next, suppose that $X, Y \geq 0$ and let $0 \leq X_n$ simple r.v.s $\uparrow X$ as $n \to \infty$. Then, as $n \to \infty$,

$$0 \leq X_n Y \uparrow XY \text{ implies } \mathcal{E}^{\mathcal{B}}(X_n Y) \uparrow \mathcal{E}^{\mathcal{B}}(XY) \text{ a.s.}$$

But

$$\mathcal{E}^{\mathcal{B}}(X_n Y) = X_n \mathcal{E}^{\mathcal{B}} Y \text{ a.s. and } X_n \mathcal{E}^{\mathcal{B}} Y \uparrow X\mathcal{E}^{\mathcal{B}} Y \text{ a.s.}$$

Thus $\mathcal{E}^{\mathcal{B}}(XY) = X\mathcal{E}^{\mathcal{B}} Y$ a.s.

Finally, for any r.v.s X and Y,

$$XY = (X^+ - X^-)(Y^+ - Y^-) = X^+Y^+ - X^+Y^- - X^-Y^+ + X^-Y^-,$$

and the previous results complete the proof of the theorem. ■

Theorem 5 (Conditional Inequalities). For $r, s > 0$, suppose that

$$\mathcal{E}|X|^r, \quad \mathcal{E}|Y|^r, \quad \mathcal{E}|Y|^s < \infty.$$

Then

(i) $\mathcal{E}^{\mathcal{B}}|XY| \leq (\mathcal{E}^{\mathcal{B}}|X|^r)^{1/r}(\mathcal{E}^{\mathcal{B}}|Y|^s)^{1/s}$ a.s., provided $\frac{1}{r} + \frac{1}{s} = 1$.

(ii) $(\mathcal{E}^{\mathcal{B}}|X + Y|^r)^{1/r} \leq (\mathcal{E}^{\mathcal{B}}|X|^r)^{1/r} + (\mathcal{E}^{\mathcal{B}}|Y|^r)^{1/r}$ a.s., provided $r \geq 1$.

(iii) $\mathcal{E}^{\mathcal{B}}|X + Y|^r \leq c_r(\mathcal{E}^{\mathcal{B}}|X|^r + \mathcal{E}^{\mathcal{B}}|Y|^r)$ a.s., where $c_r = 1$ if $(0 <)r \leq 1$, and $c_r = 2^{r-1}$ if $r > 1$.

(iv) $|\mathcal{E}^{\mathcal{B}}X| \leq \mathcal{E}^{\mathcal{B}}|X|$ a.s. and, more generally, $|\mathcal{E}^{\mathcal{B}}X|^r \leq \mathcal{E}^{\mathcal{B}}|X|^r$ a.s., $r \geq 1$. ■

Proof.

(i) Since $\mathcal{E}^{\mathcal{B}}|X|^r$ and $\mathcal{E}^{\mathcal{B}}|Y|^s$ are a.s. finite, the inequality is trivially true on a possibly exceptional null set; so we may assume that they are finite. More precisely, let $A = \{\omega \in \Omega; \mathcal{E}^{\mathcal{B}}|X|^r = 0\}$, $B = \{\omega \in \Omega; \mathcal{E}^{\mathcal{B}}|Y|^s = 0\}$. First, focus on the set A. If $P(A) = 0$, then there is no issue regarding $\omega \in A$. If $P(A) > 0$, then since $A \in \mathcal{B}$, we have

$$0 = \int_A \mathcal{E}^{\mathcal{B}}|X|^r dP_{\mathcal{B}} = \int_A |X|^r dP,$$

which implies that $|X|^r = 0$ a.s. on A (by Exercise 5(i)), or $X = 0$ a.s. on A, and this implies that $XY = 0$ a.s. on A, of $|XY| = 0$ a.s. on A. Hence $\mathcal{E}^{\mathcal{B}}|XY| = 0$

a.s. on A (by Exercise 5(ii)), and therefore the inequality is true on A. Likewise, the inequality is true on B, so that it is true on $A \cup B$.

Next, consider $(A \cup B)^c = A^c \cap B^c$ on which both $\mathcal{E}^{\mathcal{B}}|X|^r$ and $\mathcal{E}^{\mathcal{B}}|Y|^s$ are positive. Then in the inequality

$$|\alpha\beta| \le \frac{|\alpha|^r}{r} + \frac{|\beta|^s}{s},$$

and on $A^c \cap B^c$, set:

$$\alpha = X/\left(\mathcal{E}^{\mathcal{B}}|X|^r\right)^{1/r}, \quad \beta = Y/\left(\mathcal{E}^{\mathcal{B}}|Y|^s\right)^{1/s}.$$

Then, on $A^c \cap B^c$,

$$\frac{|XY|}{\left(\mathcal{E}^{\mathcal{B}}|X|^r\right)^{1/r}\left(\mathcal{E}^{\mathcal{B}}|Y|^s\right)^{1/s}} \le \frac{|X|^r}{r\mathcal{E}^{\mathcal{B}}|X|^r} + \frac{|Y|^s}{s\mathcal{E}^{\mathcal{B}}|Y|^s}$$

and hence, by taking the conditional expectations, given $\mathcal{B}$, and using $\mathcal{B}$-measurability of $\mathcal{E}^{\mathcal{B}}|X|^r$ and $\mathcal{E}^{\mathcal{B}}|Y|^s$, we get, a.s. on $A^c \cap B^c$,

$$\frac{\mathcal{E}^{\mathcal{B}}|XY|}{\left(\mathcal{E}^{\mathcal{B}}|X|^r\right)^{1/r}\left(\mathcal{E}^{\mathcal{B}}|Y|^s\right)^{1/s}} \le \frac{\mathcal{E}^{\mathcal{B}}|X|^r}{r\mathcal{E}^{\mathcal{B}}|X|^r} + \frac{\mathcal{E}^{\mathcal{B}}|Y|^s}{s\mathcal{E}^{\mathcal{B}}|Y|^s} = \frac{1}{r} + \frac{1}{s} = 1.$$

Therefore

$$\mathcal{E}^{\mathcal{B}}|XY| \le (\mathcal{E}^{\mathcal{B}}|X|^r)^{1/r}(\mathcal{E}^{\mathcal{B}}|Y|^s)^{1/s}$$

a.s. on $A^c \cap B^c$.

(ii) For $r = 1$, we have $|X+Y| \le |X| + |Y|$, so that

$$\mathcal{E}^{\mathcal{B}}|X+Y| \overset{\text{a.s.}}{\le} \mathcal{E}^{\mathcal{B}}(|X|+|Y|) \overset{\text{a.s.}}{=} \mathcal{E}^{\mathcal{B}}|X| + \mathcal{E}^{\mathcal{B}}|Y|.$$

For $r > 1$,

$$\begin{aligned}|X+Y|^r &= |X+Y|\,|X+Y|^{r-1} \le (|X|+|Y|)\,|X+Y|^{r-1}\\ &= |X|\,|X+Y|^{r-1} + |Y|\,|X+Y|^{r-1}.\end{aligned}$$

Hence, with $s > 1$ such that $\frac{1}{r} + \frac{1}{s} = 1$, we have

$$\begin{aligned}\mathcal{E}^{\mathcal{B}}|X+Y|^r &\overset{\text{a.s.}}{\le} \mathcal{E}^{\mathcal{B}}\left(|X|\,|X+Y|^{r-1}\right) + \mathcal{E}^{\mathcal{B}}\left(|Y|\,|X+Y|^{r-1}\right)\\ &\overset{\text{a.s.}}{\le} \left(\mathcal{E}^{\mathcal{B}}|X|^r\right)^{1/r} \times \left(\mathcal{E}^{\mathcal{B}}|X+Y|^{(r-1)s}\right)^{1/s}\\ &\quad + \left(\mathcal{E}^{\mathcal{B}}|Y|^r\right)^{1/r} \times \left(\mathcal{E}^{\mathcal{B}}|X+Y|^{(r-1)s}\right)^{1/s}\\ &= \left(\mathcal{E}^{\mathcal{B}}|X+Y|^{(r-1)s}\right)^{1/s}\left[\left(\mathcal{E}^{\mathcal{B}}|X|^r\right)^{1/r}\right.\\ &\quad \left.+ \left(\mathcal{E}^{\mathcal{B}}|Y|^r\right)^{1/r}\right]\\ &= \left(\mathcal{E}^{\mathcal{B}}|X+Y|^r\right)^{1/s}\left[\left(\mathcal{E}^{\mathcal{B}}|X|^r\right)^{1/r} + \left(\mathcal{E}^{\mathcal{B}}|Y|^r\right)^{1/r}\right],\end{aligned}$$

since $(r-1)s = r$ as follows from $\frac{1}{r}+\frac{1}{s}=1$. That is,

$$\mathcal{E}^{\mathcal{B}}|X+Y|^r \le (\mathcal{E}^{\mathcal{B}}|X+Y|^r)^{1/s}[(\mathcal{E}^{\mathcal{B}}|X|^r)^{1/r}+(\mathcal{E}^{\mathcal{B}}|Y|^r)^{1/r}] \text{ a.s.}$$

Now, if $(\mathcal{E}^{\mathcal{B}}|X+Y|^r)^{1/s}=0$ on a set A, then $\mathcal{E}^{\mathcal{B}}|X+Y|^r=0$ on A, whereas on A^c, $(\mathcal{E}^{\mathcal{B}}|X+Y|^r)^{1/s}>0$. Then on A, the desired inequality is true and so is on A^c by dividing through by $(\mathcal{E}^{\mathcal{B}}|X+Y|^r)^{1/s}$. Thus we get the result.

(iii) We have seen that $|X+Y|^r \le c_r(|X|^r+|Y|^r)$. Hence by Theorem 2 (ii), (iii), we get $\mathcal{E}^{\mathcal{B}}|X+Y|^r \le c_r(\mathcal{E}^{\mathcal{B}}|X|^r+\mathcal{E}^{\mathcal{B}}|Y|^r)$ a.s.

(iv) Let $X = X^+ - X^-$. Since $X^+, X^- \ge 0$, we get $\mathcal{E}^{\mathcal{B}}X^+, \mathcal{E}^{\mathcal{B}}X^- \ge 0$ a.s. and $\mathcal{E}^{\mathcal{B}}X = \mathcal{E}^{\mathcal{B}}(X^+-X^-) \overset{\text{a.s.}}{=} \mathcal{E}^{\mathcal{B}}X^+ - \mathcal{E}^{\mathcal{B}}X^-$, so that $\left|\mathcal{E}^{\mathcal{B}}X\right| \overset{\text{a.s.}}{\le} \mathcal{E}^{\mathcal{B}}X^+ + \mathcal{E}^{\mathcal{B}}X^- \overset{\text{a.s.}}{=} \mathcal{E}^{\mathcal{B}}(X^+ + X^-) \overset{\text{a.s.}}{=} \mathcal{E}^{\mathcal{B}}|X|$. (For the case that $r>1$, see Exercise 6.) ■

Theorem 6 (Convergence in the rth Mean for Conditional Expectations). For $n \ge 1$, let X_n, X be r.v.s such that $\mathcal{E}|X_n|, \mathcal{E}|X| < \infty$. Then if $X_n \underset{n\to\infty}{\overset{(r)}{\to}} X$, $r \ge 1$, it follows that $\mathcal{E}^{\mathcal{B}}X_n \underset{n\to\infty}{\overset{(r)}{\to}} \mathcal{E}^{\mathcal{B}}X$. ■

Proof. We have

$$\mathcal{E}\left|\mathcal{E}^{\mathcal{B}}X_n - \mathcal{E}^{\mathcal{B}}X\right|^r = \mathcal{E}\left|\mathcal{E}^{\mathcal{B}}(X_n - X)\right|^r \le \mathcal{E}(\mathcal{E}^{\mathcal{B}}|X_n - X|^r)$$

since for $r \ge 1$, $\left|\mathcal{E}^{\mathcal{B}}Z\right|^r \le \mathcal{E}^{\mathcal{B}}|Z|^r$ a.s., by Theorem 5 (iv), and the last expression equals $\mathcal{E}|X_n - X|^r \underset{n\to\infty}{\to} 0$. ■

Theorem 7 (Fatou–Lebesgue Theorem, Dominated Convergence Theorem for Conditional Expectations). For $n \ge 1$, let X_n, X, Y, Z be integrable r.v.s. Then, with $n \to \infty$ as appropriate:

(i) $Y \le X_n$ a.s., $n \ge 1$, implies $\mathcal{E}^{\mathcal{B}}(\liminf X_n) \le \liminf \mathcal{E}^{\mathcal{B}}X_n$ a.s., provided $\liminf X_n$ is integrable.

(ii) $X_n \le Z$ a.s., $n \ge 1$, implies $\limsup \mathcal{E}^{\mathcal{B}}X_n \le \mathcal{E}^{\mathcal{B}}\limsup X_n$ a.s., provided $\limsup X_n$ is integrable.
In particular,

(i′) $Y \le X_n \uparrow X$ a.s. implies $\mathcal{E}^{\mathcal{B}}X_n \uparrow \mathcal{E}^{\mathcal{B}}X$ a.s.

(ii′) $Y \le X_n \le Z$ a.s., $n \ge 1$, and $X_n \overset{\text{a.s.}}{\to} X$, imply $\mathcal{E}^{\mathcal{B}}X_n \overset{\text{a.s.}}{\to} \mathcal{E}^{\mathcal{B}}X$.

(iii′) $|X_n| \le U$ a.s., $n \ge 1$, U integrable and $X_n \overset{\text{a.s.}}{\to} X$ imply $\mathcal{E}^{\mathcal{B}}X_n \overset{\text{a.s.}}{\to} \mathcal{E}^{\mathcal{B}}X$. (See also Exercise 11.) ■

Proof. In all that follows, $n \to \infty$ as appropriate.

(i) We have $X_n - Y \ge 0$ a.s., $n \ge 1$, and

$$0 \overset{\text{a.s.}}{\le} \inf_{j\ge n}(X_j - Y) \uparrow \lim \inf_{j\ge n}(X_j - Y) = \liminf(X_n - Y) = \liminf X_n - Y.$$

Hence, by Theorem 3,

$$\mathcal{E}^{\mathcal{B}}\left[\inf_{j\geq n}(X_j - Y)\right] \uparrow \mathcal{E}^{\mathcal{B}}\left[\liminf(X_n - Y)\right] = \mathcal{E}^{\mathcal{B}}\left(\liminf X_n\right)$$
$$-\mathcal{E}^{\mathcal{B}}Y \text{ a.s.}$$

Next, $\inf_{j\geq n}(X_j - Y) \leq X_n - Y$, so that $\mathcal{E}^{\mathcal{B}}\left[\inf_{j\geq n}(X_j - Y)\right] \overset{\text{a.s.}}{\leq}$ $\mathcal{E}^{\mathcal{B}}(X_n - Y) \overset{\text{a.s.}}{=} \mathcal{E}^{\mathcal{B}}X_n - \mathcal{E}^{\mathcal{B}}Y$. Hence

$$\lim \mathcal{E}^{\mathcal{B}}\left[\inf_{j\geq n}(X_j - Y)\right] \leq \liminf \mathcal{E}^{\mathcal{B}}X_n - \mathcal{E}^{\mathcal{B}}Y \text{ a.s. or}$$
$$\mathcal{E}^{\mathcal{B}}\left(\liminf X_n\right) - \mathcal{E}^{\mathcal{B}}Y \leq \liminf \mathcal{E}^{\mathcal{B}}X_n - \mathcal{E}^{\mathcal{B}}Y \text{ a.s. or}$$
$$\mathcal{E}^{\mathcal{B}}\left(\liminf X_n\right) \leq \liminf \mathcal{E}^{\mathcal{B}}X_n, \text{ since } \mathcal{E}\,|Y| < \infty,$$

so that $\left|\mathcal{E}^{\mathcal{B}}Y\right| < \infty$ a.s. (See also Exercise 7.)

(ii) It is proved either similarly, or by utilizing the relations according to which $X_n \leq Z$ a.s. is equivalent to $-Z \leq -X_n$ a.s. and this implies that $\mathcal{E}^{\mathcal{B}}[\liminf(-X_n)] \leq \liminf \mathcal{E}^{\mathcal{B}}(-X_n)$ a.s. But $\mathcal{E}^{\mathcal{B}}[\liminf(-X_n)] = \mathcal{E}^{\mathcal{B}}(-\limsup X_n) \overset{\text{a.s.}}{=} -\mathcal{E}^{\mathcal{B}}(\limsup X_n)$ and $\liminf \mathcal{E}^{\mathcal{B}}(-X_n) \overset{\text{a.s.}}{=} \liminf(-\mathcal{E}^{\mathcal{B}} X_n) = -\limsup \mathcal{E}^{\mathcal{B}}X_n$. Thus, $-\mathcal{E}^{\mathcal{B}}(\limsup X_n) \leq -\limsup \mathcal{E}^{\mathcal{B}}X_n$a.s., or $\limsup \mathcal{E}^{\mathcal{B}}X_n \leq \mathcal{E}^{\mathcal{B}}(\limsup X_n)$ a.s.

(i′) We have

$$Y \leq X_n \uparrow X \text{ a.s. or } 0 \leq X_n - Y \uparrow X - Y \text{ a.s., so that}$$
$$\mathcal{E}^{\mathcal{B}}(X_n - Y) \uparrow \mathcal{E}^{\mathcal{B}}(X - Y) \text{ a.s. (by Theorem 3)}$$
$$\text{or } \mathcal{E}^{\mathcal{B}}X_n - \mathcal{E}^{\mathcal{B}}Y \uparrow \mathcal{E}^{\mathcal{B}}X - \mathcal{E}^{\mathcal{B}}Y \text{ a.s. and hence}$$
$$\mathcal{E}^{\mathcal{B}}X_n \uparrow \mathcal{E}^{\mathcal{B}}X, \text{ since } \left|\mathcal{E}^{\mathcal{B}}Y\right| < \infty \text{ a.s.}$$

(ii′) By (i) and (ii) we have

$$\mathcal{E}^{\mathcal{B}}\left(\liminf X_n\right) \overset{\text{a.s.}}{\leq} \liminf \mathcal{E}^{\mathcal{B}}X_n \leq \limsup \mathcal{E}^{\mathcal{B}}X_n$$
$$\overset{\text{a.s.}}{\leq} \mathcal{E}^{\mathcal{B}}\left(\limsup X_n\right).$$

But

$$\liminf X_n = \limsup X_n = X \text{ a.s.,}$$

so that $\mathcal{E}^{\mathcal{B}}X \overset{\text{a.s.}}{\leq} \liminf \mathcal{E}^{\mathcal{B}}X_n \leq \limsup \mathcal{E}^{\mathcal{B}}X_n \overset{\text{a.s.}}{\leq} \mathcal{E}^{\mathcal{B}}X$. Thus,

$$\mathcal{E}^{\mathcal{B}}X_n \overset{\text{a.s.}}{\to} \mathcal{E}^{\mathcal{B}}X.$$

(iii′) Apply *(ii′)* with $Y = -U$ and $Z = U$. ■

Theorem 8 (Jensen Inequality for Conditional Expectations). Let X be a r.v. taking values in I, an open interval in $\Re$, with $\mathcal{E}X \in I$, and let $g : I \to \Re$ be convex nd such that $\mathcal{E}\,|g(X)| < \infty$. Then

$$g(\mathcal{E}^{\mathcal{B}}X) \leq \mathcal{E}^{\mathcal{B}}g(X) \text{ a.s.}$$

In particular, if g is also nondecreasing, then

$$Y \leq \mathcal{E}^{\mathcal{B}}X \text{ a.s. implies } g(Y) \leq \mathcal{E}^{\mathcal{B}}g(X) \text{ a.s.}$$ ∎

Proof. In the proof, treat I as if it were the entire $\Re$ for convenient writing. It is well known that g is continuous and either monotone in $\Re$ or there exists $x_0 \in \Re$ such that g is nonincreasing in $(-\infty, x_0]$ and nondecreasing in $[x_0, \infty)$. (See Exercise 2 in Chapter 6, and the book by Hardy et al., (1967).) Consider this latter case first. Clearly, without loss of generality, we may assume that $x_0 = 0$ and that $g(x_0) = 0$. Thus $g(x) \geq 0$ for all $x \in \Re$. Next, let Z be a simple r.v.; i.e., $Z = \sum_{j=1}^{k} z_j I_{A_j}$, where $\{A_j, j = 1, \ldots, k\}$ is a (measurable) partition of Ω. Then, clearly,

$$g(Z) = g\left(\sum_j z_j I_{A_j}\right) = \sum_j g(z_j) I_{A_j},$$

so that

$$\begin{aligned}\mathcal{E}^{\mathcal{B}}g(Z) &= \mathcal{E}^{\mathcal{B}}\left[\sum_j g(z_j) I_{A_j}\right] \overset{\text{a.s.}}{=} \sum_j \mathcal{E}^{\mathcal{B}}\left[g(z_j) I_{A_j}\right] \\ &\overset{\text{a.s.}}{=} \sum_j g(z_j)\mathcal{E}^{\mathcal{B}} I_{A_j} \overset{\text{a.s.}}{\geq} g\left(\sum_j z_j \mathcal{E}^{\mathcal{B}} I_{A_j}\right) \quad \text{(see Exercise 8)}\end{aligned}$$

(by the convexity of g and the fact that $\sum_j \mathcal{E}^{\mathcal{B}} I_{A_j} = \sum_j P^{\mathcal{B}} A_j = 1$ a.s.)

$$\overset{\text{a.s.}}{=} g\left[\mathcal{E}^{\mathcal{B}}\left(\sum_j z_j I_{A_j}\right)\right] = g\left(\mathcal{E}^{\mathcal{B}}Z\right).$$

That is,

$$\mathcal{E}^{\mathcal{B}}g(Z) \geq g\left(\mathcal{E}^{\mathcal{B}}Z\right) \text{ a.s.} \tag{9.6}$$

Now for the given r.v. X, let $\{X_n\}, n \geq 1$, be a sequence of simple r.v.s such that, as $n \to \infty$ here and in the remainder of the proof,

$$X_n \to X, \quad |X_n| \leq |X|, \quad (|X_n| \leq n), \quad \text{and} \quad (0 \leq) g(X_n) \leq g(X).$$

(In Theorem 17, Chapter 1, take $X_n = \frac{j}{2^n}$ for $j = -n2^n + 1, \ldots, 0$, so that $X \leq X_n$, and take $X_n = \frac{j-1}{2^n}$ for $j = 1, \ldots, n2^n$, so that $X_n \leq X$.) We have then, by (9.6), with Z replaced by X_n,

$$\mathcal{E}^{\mathcal{B}} g(X_n) \geq g\left(\mathcal{E}^{\mathcal{B}} X_n\right) \text{ a.s.} \tag{9.7}$$

By the Dominated Convergence Theorem for conditional expectations (Theorem 7 (iii)),

$$\mathcal{E}^{\mathcal{B}} X_n \overset{\text{a.s.}}{\to} \mathcal{E}^{\mathcal{B}} X.$$

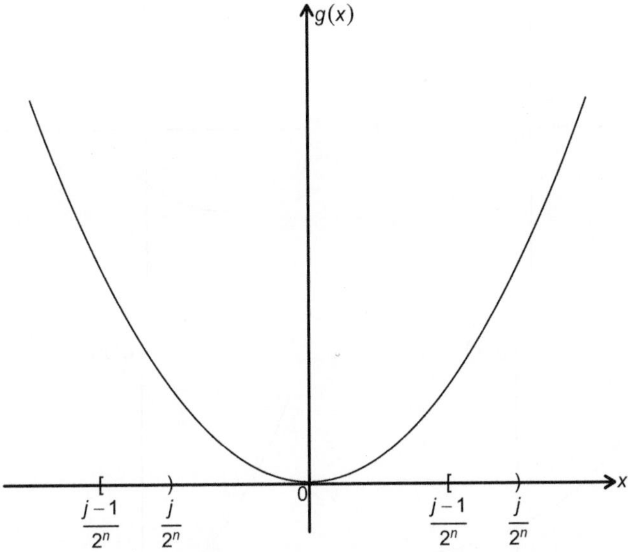

Hence

$$g(\mathcal{E}^{\mathcal{B}} X_n) \overset{\text{a.s.}}{\to} g(\mathcal{E}^{\mathcal{B}} X). \tag{9.8}$$

Also

$$0 \leq g(X_n) \leq g(X) \text{ and } g(X_n) \overset{\text{a.s.}}{\to} g(X).$$

Thus the Fatou–Lebesgue Theorem for conditional expectations again (Theorem 7(ii') or (iii')) gives

$$\mathcal{E}^{\mathcal{B}} g(X_n) \overset{\text{a.s.}}{\to} \mathcal{E}^{\mathcal{B}} g(X). \tag{9.9}$$

Relations (9.7)–(9.9) give that

$$\mathcal{E}^{\mathcal{B}} g(X) \geq g\left(\mathcal{E}^{\mathcal{B}} X\right) \text{ a.s.}$$

Next, let g be monotone. In this case, $\{X_n\}$ may be chosen so that

$$X_n \to X, \quad |X_n| \leq |X|, \quad (|X_n| \leq n), \quad \text{and } |g(X_n)| \leq |g(X)|.$$

Indeed, if g is either increasing or decreasing in $\Re$, then it will look like either one of the accompanying first four figures. In either case, we may assume (by switching the curve) that it looks like one of the last two pictures. Consider the figure at the left-hand side, as the other figure is treated similarly. For $j \leq 0$, take $X_n = j/2^n$.

Then $X \leq X_n$, $g(X) \leq g(X_n)$, $|X_n| \leq |X|$, and $|g(X_n)| \leq |g(X)|$. For $j \geq 1$, take $X_n = (j-1)/2^n$. Then $|X_n| = X_n \leq X = |X|$, and $|g(X_n)| = g(X_n) \leq g(X) = |g(X)|$. So, in all cases, $|X_n| \leq |X|$ and $|g(X_n)| \leq |g(X)|$.

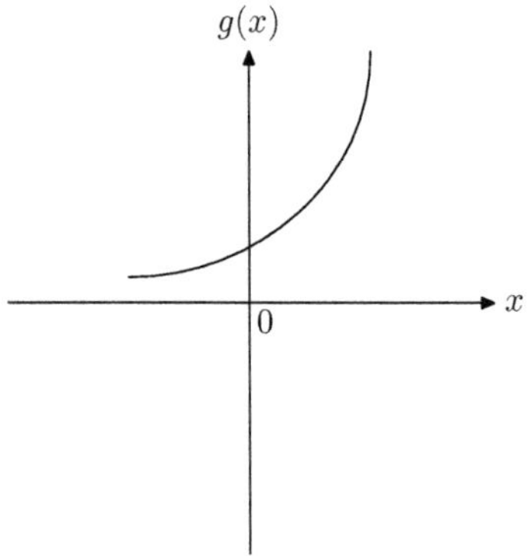

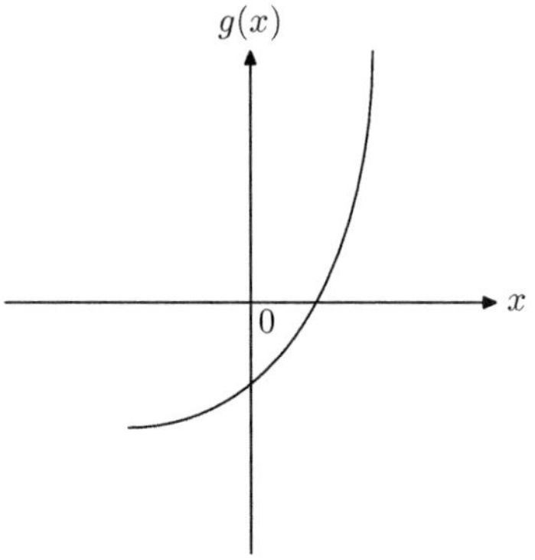

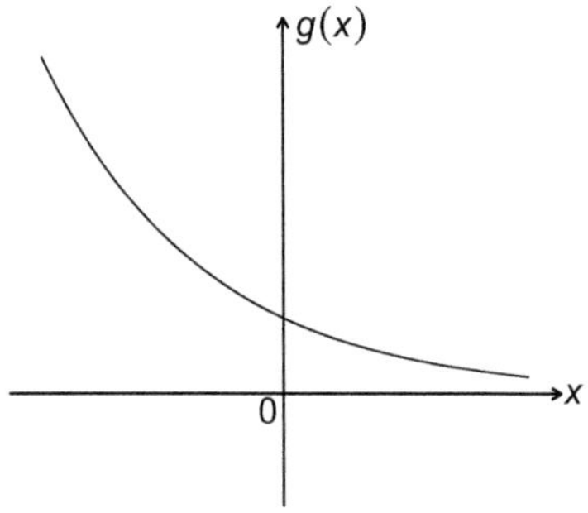

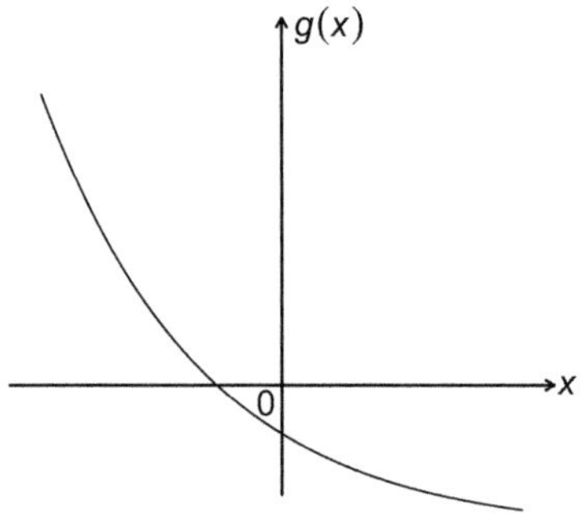

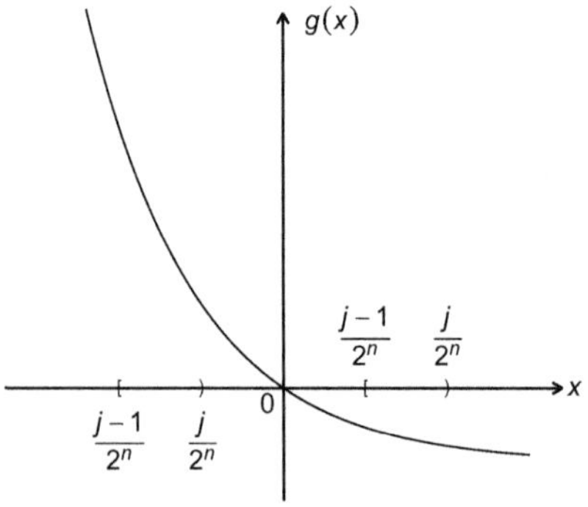

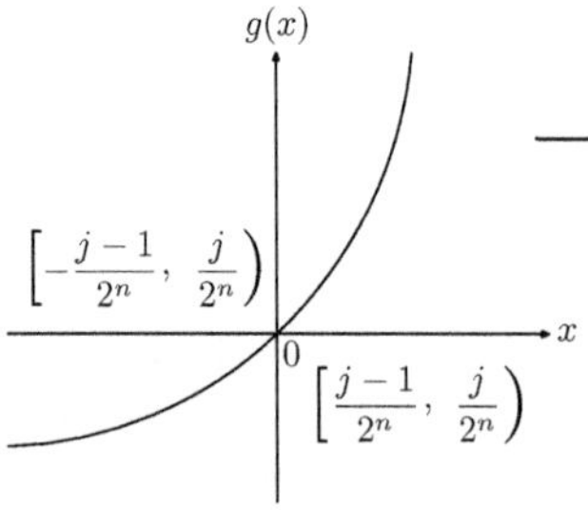

As before, the convexity of g implies (9.7), and thus we reach the same conclusion as before. Likewise if g is nonincreasing in $\Re$. For the second conclusion, one has

$$Y \leq \mathcal{E}^{\mathcal{B}} X \text{ a.s. implies } g(Y) \leq g\left(\mathcal{E}^{\mathcal{B}} X\right) \text{ a.s.}$$

Since $g\left(\mathcal{E}^{\mathcal{B}} X\right) \leq \mathcal{E}^{\mathcal{B}} g(X)$ a.s., we obtain the result. ■

9.4 Further Properties of Conditional Expectations and Conditional Probabilities

Theorem 9. If $\mathcal{B} \subseteq \mathcal{B}'(\subseteq \mathcal{A})$ and $\mathcal{E}|X| < \infty$, then

$$\mathcal{E}^{\mathcal{B}'}(\mathcal{E}^{\mathcal{B}} X) = \mathcal{E}^{\mathcal{B}} X = \mathcal{E}^{\mathcal{B}}(\mathcal{E}^{\mathcal{B}'} X) \text{ a.s.}$$ ■

Proof. The first equality follows by the fact that $\mathcal{E}^{\mathcal{B}} X$ is $\mathcal{B}'$-measurable, since it is $\mathcal{B}$-measurable and $\mathcal{B} \subseteq \mathcal{B}'$. As for the second equality, we have for all $B \in \mathcal{B}$,

$$\begin{aligned}\int_B \mathcal{E}^{\mathcal{B}}(\mathcal{E}^{\mathcal{B}'} X) dP_{\mathcal{B}} &= \int_B \mathcal{E}^{\mathcal{B}'} X \, dP_{\mathcal{B}'} \\ &= \int_B X \, dP = \int_B \mathcal{E}^{\mathcal{B}} X \, dP_{\mathcal{B}};\end{aligned}$$

i.e.,

$$\int_B \mathcal{E}^{\mathcal{B}}(\mathcal{E}^{\mathcal{B}'} X) dP_{\mathcal{B}} = \int_B \mathcal{E}^{\mathcal{B}} X \, dP_{\mathcal{B}}, \quad B \in \mathcal{B},$$

and hence $\mathcal{E}^{\mathcal{B}}(\mathcal{E}^{\mathcal{B}'} X) = \mathcal{E}^{\mathcal{B}} X$ a.s. since they are both $\mathcal{B}$-measurable. ■

Now, let X, Y be two r.v.s defined on the probability space $(\Omega, \mathcal{A}, P)$, and set

$$\mathcal{B}_X = X^{-1}(\text{Borel } \sigma\text{-field in } \Re), \quad \mathcal{B}_Y = Y^{-1}(\text{Borel } \sigma\text{-field in } \Re).$$

Then the elementary definition of independence of the r.v.s. X, Y is equivalent to independence of $\mathcal{B}_X$ and $\mathcal{B}_Y$; i.e.,

$$P(A \cap B) = P(A)P(B), \text{ for all } A \in \mathcal{B}_X \text{ and } B \in \mathcal{B}_Y.$$

A σ-field $\mathcal{B} \subseteq \mathcal{A}$ and Y are said to be *independent* if $\mathcal{B}$ and $\mathcal{B}_Y$ are independent. A formalization of these concepts and further elaboration on independence may be found in Section 1 of Chapter 10.

The following result will be needed in the sequel; its proof is deferred to Chapter 10 (see Section 10.3).

Lemma 1. If the r.v.s X and Y are independent and $\mathcal{E}|X|, \mathcal{E}|Y| < \infty$, then $\mathcal{E}|XY| < \infty$ and $\mathcal{E}(XY) = (\mathcal{E}X)(\mathcal{E}Y)$.

Remark 3. By induction, the lemma is true for any n independent integrable r.v.s.

Theorem 10. Let X be an integrable r.v. such that X and $\mathcal{B}$ are independent. Then $\mathcal{E}^{\mathcal{B}}X = \mathcal{E}X$ a.s. ■

Proof. For any $B \in \mathcal{B}$, we have that I_B and X are independent since $\{\varnothing, B, B^c, \Omega\} \subseteq \mathcal{B}$. Next,

$$\int_B \mathcal{E}^{\mathcal{B}}X\,dP_{\mathcal{B}} = \int_B X\,dP = \int (XI_B)dP = \mathcal{E}(XI_B) = (\mathcal{E}X)(\mathcal{E}I_B)$$
$$\text{(by independence of } X \text{ and } I_B)$$
$$= (\mathcal{E}X)P(B) = \int_B \mathcal{E}X\,dP = \int_B \mathcal{E}X\,dP_{\mathcal{B}}.$$

That is,

$$\int_B \mathcal{E}^{\mathcal{B}}X\,dP_{\mathcal{B}} = \int_B \mathcal{E}X\,dP_{\mathcal{B}}, \quad B \in \mathcal{B},$$

whereas both $\mathcal{E}^{\mathcal{B}}X$ and $\mathcal{E}X$ are $\mathcal{B}$-measurable. Thus $\mathcal{E}^{\mathcal{B}}X = \mathcal{E}X$ a.s. ■

Now let Y be a function defined on $(\Omega, \mathcal{A}, P)$ into Ω' and set

$$\mathcal{B}'_Y = \left\{A' \subseteq \Omega';\, Y^{-1}(A') = A \text{ for some } A \in \mathcal{A}\right\}.$$

Then, $\mathcal{B}'_Y$ is a σ-field of subsets of Ω' (by Theorem 11 in Chapter 1).

Let also $\mathcal{B}_Y = Y^{-1}(\mathcal{B}'_Y)$, so that $\mathcal{B}_Y$ is a σ-field $\subseteq \mathcal{A}$. On $\mathcal{B}_Y$ and $\mathcal{B}'_Y$ define the probability measures P_Y and P'_Y, respectively, as follows:

$$P_Y(B) = P(B),\ B \in \mathcal{B}_Y; \quad P'_Y(B') = P[Y^{-1}(B')]$$
$$= P_Y(B),\ B' \in \mathcal{B}'_Y,\ B = Y^{-1}(B').$$

Definition 2. If X is an integrable r.v. and Y is as above, the *conditional expectation of X, given Y*, is denoted by $\mathcal{E}^Y X$ or $\mathcal{E}(X \mid Y)$ and is defined to be $\mathcal{E}^{\mathcal{B}_Y}X$. ■

The reason for this notation will be explained hereafter. For this purpose, we need the following result, which is, actually, closely related to Theorem 13 in Chapter 4.

Lemma 2. Let $(\Omega, \mathcal{B}_Y, P_Y)$, $(\Omega', \mathcal{B}'_Y, P'_Y)$, be as before and let $g : (\Omega', \mathcal{B}'_Y) \to (\Re, \text{Borel } \sigma\text{-field})$ be measurable, so that $g(Y) : (\Omega, \mathcal{B}_Y) \to (\Re, \text{Borel } \sigma\text{-field})$ is measurable (a r.v.). Then

$$\int_{B'} g(y)dP'_Y = \int_B g(Y)\,dP_Y, \quad B' \in \mathcal{B}'_Y, \quad B = Y^{-1}(B'),$$

in the sense that, if one of these integrals exists, then so does the other, and both are equal.

Proof. The proof follows the familiar line. First, let $g(y) = I_{A'}(y)$, $A' \in \mathcal{B}'_Y$. Then, clearly, $g(Y) = I_A(Y)$, $A = Y^{-1}(A')$. Therefore

$$\int_{B'} g(y)dP'_Y = \int_{B'} I_{A'}(y)dP'_Y = \int_{A'\cap B'} dP'_Y = P'_Y(A' \cap B') = P[Y^{-1}(A' \cap B')]$$
$$= P(A \cap B) = P_Y(A \cap B) = \int_{A\cap B} dP_Y = \int_B I_A(Y)\,dP_Y = \int_B g(Y)\,dP_Y.$$

That is, the theorem is true for indicators. Then it is true for the case that g is a nonnegative simple r.v. Next, it is true for the case that g is a ≥ 0 r.v., by the Lebesgue Monotone Convergence Theorem, and finally for any g, as described earlier, by writing g as $g = g^+ - g^-$. (For the details, see Exercise 9; also, Theorem 13 in Chapter 4). ■

Theorem 11. Let $(\Omega, \mathcal{B}_Y, P_Y)$, $(\Omega', \mathcal{B}'_Y, P'_Y)$ be as before and let X be an integrable r.v. Then $\mathcal{E}^Y X (= \mathcal{E}^{\mathcal{B}_Y} X)$ is a.s. ($[P]$) a function of Y; i.e., it depends on $\omega \in \Omega$ only through Y for almost all (a.a.) $\omega \in \Omega$. More precisely, there exists a (measurable) function $g : (\Omega', \mathcal{B}'_Y) \to$ Borel real line, such that $(\mathcal{E}^Y X)(\omega) = g(Y(\omega))$ a.s. $[P]$. ■

Proof. On $\mathcal{B}_Y$, define the finite measures φ^+, φ^- and the finite set function φ as follows:

$$\varphi^+(A) = \int_A X^+ \, dP, \quad \varphi^-(A) = \int_A X^- \, dP,$$

$$\varphi(A) = \varphi^+(A) - \varphi^-(A) = \int_A X \, dP.$$

Then $\varphi^+, \varphi^-, \varphi$ are all P_Y-continuous finite measures and φ is a signed measure. On $\mathcal{B}'_Y$, define the measures φ'^+, φ'^- and the set function φ' as follows:

$$\varphi'^+(A') = \varphi^+(A), \quad \varphi'^-(A') = \varphi^-(A), \quad \varphi'(A') = \varphi(A), \quad A = Y^{-1}(A').$$

Then φ'^+, φ'^- are finite measures, φ' is a signed measure, and all three are P'_Y-continuous. Then, by Theorem 3 in Chapter 7, there exist $\mathcal{B}'_Y$-measurable, nonnegative (a.s. $[P'_Y]$ finite), and a.s. $[P'_Y]$ well-defined r.v.s g_+, g_- such that

$$\varphi'^+(A') = \int_{A'} g_+ dP'_Y, \quad \varphi'^-(A') = \int_{A'} g_- dP'_Y, \quad A' \in \mathcal{B}'_Y.$$

By setting $g = g_+ - g_-$, we have that g is $\mathcal{B}'_Y$-measurable, (a.s. $[P'_Y]$ finite), and a.s. $[P'_Y]$ well defined, and such that

$$\varphi'(A') = \int_{A'} g dP'_Y, \quad A' \in \mathcal{B}'_Y.$$

But

$$\int_{A'} g \, dP'_Y = \varphi'(A')[= \varphi'^+(A') - \varphi'^-(A') = \varphi^+(A) - \varphi^-(A)]$$

$$= \varphi(A) = \int_A X \, dP = \int_A \mathcal{E}^Y X \, dP_Y.$$

Thus,

$$\int_{A'} g \, dP'_Y = \int_{A'} g(y) dP'_Y = \int_A \mathcal{E}^Y X \, dP_Y.$$

On the other hand,

$$\int_{A'} g(y) dP'_Y = \int_A g(y) \, dP_Y,$$

by Lemma 2. Thus,

$$\int_A \mathcal{E}^Y X \, dP_Y = \int_A g(Y) \, dP_Y, \quad A \in \mathcal{B}_Y.$$

Since both $\mathcal{E}^Y X$ and $g(Y)$ are $\mathcal{B}_Y$-measurable, we have that $\mathcal{E}^Y X = g(Y)$ a.s., as was to be seen. ■

Remark 4. In the last theorem, we have shown that there exists a measurable function g on Ω' into $\Re$, (a.s. $[P'_Y]$ finite), and a.s. $[P'_Y]$ well defined such that $\mathcal{E}^Y X = \mathcal{E}(X \mid Y) = g(Y)$ a.s. Now, let y be in the range of Y, let $\{y\} \in \mathcal{B}'_Y$, and set $A_y = Y^{-1}(\{y\})$ (so that $A_y \in \mathcal{B}_Y$). Then for $\omega \in A_y$, $(\mathcal{E}^Y X)(\omega) = g(y)$ a.s. This is what we mean by writing

$$\mathcal{E}(X \mid Y = y) = g(y).$$

If $A \in X^{-1}$(Borel σ − field in $\Re$), it follows that $A = X^{-1}(B)$ for some $B \in$ (Borel σ-field in $\Re$). Then

$$\mathcal{E}^Y I_A = P^Y A = P^Y(X \in B), \quad (\text{since } A = (X \in B))$$

to be denoted by $P(X \in B \mid Y)$ and the notation $P(X \in B \mid Y = y)$ is used to denote the value of $P(X \in B \mid Y)$ at $\omega \in A_y$.

Following is an example where Theorem 11 is employed along with other results obtained in this chapter.

Example 3. Let $X_1, \ldots, X_n$ be i.i.d. r.v.s defined on the probability space $(\Omega, \mathcal{A}, P_\theta)$, where the parameter θ lies in the parameter space Θ, an open subset of $\Re$, and set $\mathbf{X} = (X_1, \ldots, X_n)$. Let $U = U(\mathbf{X})$ be a statistic (a measurable function of $\mathbf{X}$ not involving any unknown quantities), and let T be a sufficient statistic for θ (as defined in Example 1 of this chapter). The statistic U, as an estimator of θ, is called *unbiased*, if $\mathcal{E}_\theta U = \theta$ for all $\theta \in \Theta$. Consider the $\mathcal{E}_\theta(U|T)$. Then this conditional expectation has the following two properties: It is independent of θ (by sufficiency), and it depends on T only a.s. (by Theorem 11). Therefore, we may set $\mathcal{E}_\theta(U|T) = \phi(T)$. Then also, $\mathcal{E}_\theta \phi(T) = \mathcal{E}_\theta[\mathcal{E}_\theta(U|T)] = \mathcal{E}_\theta U = \theta$ for all θ, so that $\phi(T)$ is also an unbiased estimator of θ. Furthermore, $\sigma^2_\theta[\phi(T)] \le \sigma^2_\theta(U)$ for all θ, so that $\phi(T)$ is at least as good as U in terms of variances (which are assumed to be finite). The variances inequality is seen as follows. On $\Re$, consider the function $g(x) = (x - \theta)^2$, which is convex (because $g''(x) = 2 > 0$), and apply the Jensen inequality for conditional expectations (Theorem 8 in this chapter) to obtain

$$\begin{aligned}
[\phi(T) - \theta]^2 &= [\mathcal{E}_\theta(U|T) - \theta]^2 \\
&\overset{\text{a.s.}}{=} [\mathcal{E}_\theta(U - \theta|T)]^2 \\
&\le \mathcal{E}_\theta[(U - \theta)^2|T] \text{ a.s.}
\end{aligned}$$

Taking the $\mathcal{E}_\theta$-expectations of both sides, we get

$$\sigma^2_\theta[\phi(T)] = \mathcal{E}_\theta[\mathcal{E}_\theta(U|T) - \theta]^2 \le \mathcal{E}_\theta(U - \theta)^2 = \sigma^2_\theta(U)$$

for all θ. ■

Application 1. Let X and Y be discrete (real-valued) r.v.s defined on $(\Omega, \mathcal{A}, P)$, with Y taking on the values y_j with $P(Y = y_j) > 0$ for all $j \geq 1$. In the sequel, y will stand for such a value. Let $\mathcal{B}'_Y$ be the σ -field generated by $\{\{y_j\}, j \geq 1\}$, and let $\mathcal{B}_Y = Y^{-1}(\mathcal{B}'_Y)$. Then, for every $A \in \mathcal{B}_Y$, $A = Y^{-1}(B')$ for some $B' \in \mathcal{B}'_Y$. Let P_Y be the restriction of P to $\mathcal{B}_Y$, and let P'_Y be the probability distribution of Y; i.e., $P'_Y(B') = P_Y(A)$ for every $B' \in \mathcal{B}'_Y$, where $A = Y^{-1}(B')$. Let x be any one of the values $x_1, x_2, \ldots$ of X, set $B = \{x\}$, and let $C = X^{-1}(B) = X^{-1}(\{x\}) = (X = x)$. Define φ as follows:

$$\varphi(A) = \int_A I_C dP, \quad A \in \mathcal{B}_Y. \tag{9.10}$$

Then,

$$\begin{aligned}\varphi(A) &= \int_A I_C dP (= P(A \cap C)) = \int_A (\mathcal{E}^Y I_C) dP_Y \\ &= \int_A \mathcal{E}(I_C|Y) dP_Y = \int_A P(C|Y) dP_Y.\end{aligned} \tag{9.11}$$

Define φ' by

$$\varphi'(B') = \varphi(A), \quad B' \in \mathcal{B}'_Y, \quad A = Y^{-1}(B'). \tag{9.12}$$

Then φ' is a finite measure on $\mathcal{B}'_Y$ and P'_Y-continuous (since $P'_Y(B') = 0$ implies $P_Y(A) = 0$ or $P(A) = 0$, hence $\varphi(A) = 0$, so that $\varphi'(B') = 0$). Then, by Theorem 3 in Chapter 7, there exists a $\mathcal{B}'_Y$-measurable function g, nonnegative (a.s. $[P'_Y]$ finite), and a.s. $[P'_Y]$ well defined, such that

$$\varphi'(B') = \int_{B'} g(y) dP'_Y, \quad B' \in \mathcal{B}'_Y. \tag{9.13}$$

By Lemma 2,

$$\int_{B'} g(y) dP'_Y = \int_A g(Y) dP_Y \quad (A = Y^{-1}(B')), \tag{9.14}$$

and therefore, by means of (9.10)–(9.14),

$$\int_A g(Y) dP_Y = \int_A P(C|Y) dP_Y, \quad A \in \mathcal{B}_Y. \tag{9.15}$$

Since $g(Y)$ and $P(C|Y)$ are $\mathcal{B}_Y$-measurable, relation (9.15) implies that $P(C|Y) = P(X = x|Y) = g(Y)$ a.s., and, in particular,

$$P(C|Y)(\omega) = P(X = x|Y)(\omega) = g(y) \text{ a.s.} \quad \text{for } \omega \in A_y = Y^{-1}(\{y\}). \tag{9.16}$$

Next, for a fixed x denoting any one of the values $x_1, x_2, \ldots$ of X,

$$P(X = x|Y = y) = \frac{P(X = x, Y = y)}{P(Y = y)} \text{ is } \mathcal{B}'_Y \text{ -measurable (see also Exercise 10)}$$

and by (9.11),

$$\int_{B'} P(X = x|Y = y)dP'_Y = \sum_{y \in B'} \frac{P(X = x, Y = y)}{P(Y = y)} P(Y = y)$$
$$= \sum_{y \in B'} P(X = x, Y = y) = P(X = x, Y \in B')$$
$$= P(A \cap C) = \varphi(A)$$
$$= \varphi'(B') = \int_{B'} g(y)dP'_Y \quad \text{(by (9.12) and (9.13))}. \tag{9.17}$$

The $\mathcal{B}'_Y$-measurability of $P(X = x|Y = y)$ and of $g(y)$, and relation (9.17) imply that

$$P(X = x|Y = y) = g(y) \text{ a.s. } \left[P'_Y\right]. \tag{9.18}$$

Therefore, by (9.16) and (9.18),

$$P(X = x|Y = y) = P(X = x|Y)(\omega) \text{ a.s. } \quad \text{for } \omega \in A_y = Y^{-1}(\{y\}). \tag{9.19}$$

Relation (9.19) states that $P(X = x|Y = y)$ is a.s. equal to $P(C|Y)$ evaluated at each ω in A_y. This result demonstrates that, in the present setup, the general definition and the elementary definition of a conditional p.d.f. coincide (a.s.).

Exercises.

1. If X is an integrable r.v. and $\{A_n, n = 1, 2, \ldots\}$ are pairwise disjoint events, show that

$$\int_{\sum_{i=1}^{\infty} A_i} X\, dP = \sum_{i=1}^{\infty} \int_{A_i} X\, dP.$$

Hint: Split X into X^+ and X^- and use Corollary 1 (ii) to the Lebesgue Monotone Convergence Theorem (Theorem 1 in Chapter 5).

2. Let X be an integrable r.v. and, for $B \in \mathcal{A}$ with $P(B) > 0$, consider the conditional probability on $\mathcal{A}$, $P(\cdot|B)$. Then show that

$$\int_{\Omega} X\, dP(\cdot|B) = \frac{1}{P(B)} \int_B X\, dP.$$

Hint: Go through the familiar four steps that X is an indicator function, a simple r.v., a nonnegative r.v., any r.v.

3. Consider the probability space $(\Omega, \mathcal{A}, P) = ([0, 1], \mathcal{B}_{[0,1]}, \lambda)$, where λ is the Lebesgue measure, and let $\mathcal{F}$ be the σ-field generated by the class $\{[0, \frac{1}{4}], (\frac{1}{4}, \frac{2}{3}], (\frac{2}{3}, 1]\}$. Also, let X be the r.v. defined by: $X(\omega) = \omega^2$, $\omega \in \Omega$. Then show that

$$\mathcal{E}(X|\mathcal{F}) = \alpha_1 I_{[0,\frac{1}{4}]} + \alpha_2 I_{(\frac{1}{4},\frac{2}{3}]} + \alpha_3 I_{(\frac{2}{3},1]},$$

and compute $\alpha_1, \alpha_2, \alpha_3$.
Hint: Refer to the special cases discussed right after Definition 1.

4. Let X and Y be $\mathcal{B}$-measurable and integrable r.v.s. We further assume that $\int_B X\,dP \geq \int_B Y\,dP$ for every $B \in \mathcal{B}(\subseteq \mathcal{A})$. Then show that $X \geq Y$ a.s.
Hint: By setting $Z = Y - X$, we have $\int_B Z\,dP \leq 0$ for all $B \in \mathcal{B}$, and we wish to conclude that $Z \leq 0$ a.s. Set $C = (Z \leq 0)$ and $D = (Z > 0)(= C^c)$. Then it suffices to show that $P(D) = 0$. By taking $B = D$, we have $\int_D Z\,dP \leq 0$ and $\int_D Z\,dP = \int (ZI_D)dP \geq 0$ since $Z > 0$ on D, so that $\int_D Z\,dP = 0$. Thus, it suffices to show that, if for a r.v. Z with $D = (Z > 0)$ it holds that $\int_D Z\,dP = 0$, then $P(D) = 0$. This can be done through the four familiar steps.

5. **(i)** Let $X \geq 0$ a.s. on a set A with $P(A) > 0$, and suppose that $\int_A X\,dP = 0$. Then show that $X = 0$ a.s. on A.
(ii) Let $X \geq 0$, integrable, and $X = 0$ a.s. on a set $A \in \mathcal{B}(\subseteq A)$, with $P(A) > 0$. Then show that $\mathcal{E}^{\mathcal{B}}X = 0$ a.s. on A.

Hint: (i) With $C = (X > 0)$, we have $0 = \int_A X\,dP = \int_D X\,dP$, $D = A \cap C$. So, $\int_D X\,dP = 0$ and $X > 0$ on D. Show that $P(D) = 0$ which would be equivalent to saying that $X = 0$ a.s. on A. Do it by going through the four familiar steps.

(iii) Use the fact that $\int_B X\,dP = \int_B \mathcal{E}^{\mathcal{B}}X\,dP_{\mathcal{B}}$ for all $B \in \mathcal{B}$, replace B by $B \cap A$, and conclude that $I_A \mathcal{E}^{\mathcal{B}}X = 0$ a.s. This would imply that $\mathcal{E}^{\mathcal{B}}X = 0$ a.s. on A.

6. If $\mathcal{E}|X|^r < \infty$, then show that $|\mathcal{E}^{\mathcal{B}}X|^r \leq \mathcal{E}^{\mathcal{B}}|X|^r$ a.s., $r \geq 1$.
Hint: One way of establishing this inequality is to use the Jensen inequality (Theorem 8). For this purpose, take $g(x) = |x|^r$, $r \geq 1$, and observe that it is convex. (It is convex for $x \geq 0$ and symmetric about the y-axis, hence convex in $\Re$.).

7. If the r.v. X is integrable, then $\mathcal{E}^{\mathcal{B}}X$ is finite a.s.

8. Recall that a function $g : I(\text{open interval}) \subseteq \Re \to \Re$ is said to be *convex* if $g(\alpha_1 x_1 + \alpha_2 x_2) \leq \alpha_1 g(x_1) + \alpha_2 g(x_2)$ for all $\alpha_1, \alpha_2 \geq 0$ with $\alpha_1 + \alpha_2 = 1$, and all $x_1, x_2 \in I$.
Prove the following generalization: if g is as above, then:

$$g(\alpha_1 x_1 + \cdots + \alpha_n x_n) \leq \alpha_1 g(x_1) + \cdots + \alpha_n g(x_n) \qquad (*)$$

for any $n \geq 2$, any $\alpha_1, \ldots, \alpha_n \geq 0$ with $\alpha_1 + \cdots + \alpha_n = 1$, and all $x_1, \ldots, x_n \in I$.
Hint: Use the induction method. Inequality (*) is true for $n = 2$, assume it to be true for $n = k$ and establish it for $n = k + 1$. In the expression

$$g(\alpha_1 x_1 + \cdots + \alpha_{k+1} x_{k+1})$$

group the terms in two parts, one containing the first k terms and one containing the last term. In the first group, multiply and divide by $1 - \alpha_{k+1}$ (assuming, without loss of generality, that $\alpha_{k+1} < 1$), and use the induction hypothesis.

9. Fill in the details in proving Lemma 2.
Hint: Refer to the proof of Theorem 13 in Chapter 4.

10. Let X and Y be discrete r.v.s and recall (from the application following Theorem 11) that $\mathcal{B}'_Y$ is the σ-field of subsets of $\Re$ defined by $\mathcal{B}'_Y = \{B' \subseteq \Re;\ Y^{-1}(B') = A \text{ for some } A \in \mathcal{A}\}$. For $x, y \in \Re$ with $P(Y = y) > 0$, consider the conditional probability

$$P(X = x|Y = y) = P(X = x, Y = y)/P(Y = y)$$

and show that, for each fixed x, the function $P(X = x|Y = \cdot)$ is $\mathcal{B}'_Y$-measurable.

11. The Dominated Convergence Theorem in its classical form states: If $|X_n| \leq Y$ a.s., $n \geq 1$, $\mathcal{E}Y < \infty$, and either $X_n \underset{n\to\infty}{\overset{\text{a.s.}}{\to}} X$ or $X_n \underset{n\to\infty}{\overset{P}{\to}} X$, then $\mathcal{E}X_n \underset{n\to\infty}{\to} \mathcal{E}X$ finite.

In the framework of conditional expectations, we have shown that: If $|X_n| \leq Y$ a.s., $n \geq 1$, $\mathcal{E}Y < \infty$, and $X_n \underset{n\to\infty}{\overset{\text{a.s.}}{\to}} X$ with $\mathcal{E}X$ finite, then $\mathcal{E}^{\mathcal{B}}X_n \underset{n\to\infty}{\overset{\text{a.s.}}{\to}} \mathcal{E}^{\mathcal{B}}X$ for any σ-field $\mathcal{B} \subseteq \mathcal{A}$ (see Theorem 7 (iii$'$)).

By means of an example, show that the convergence $X_n \underset{n\to\infty}{\overset{\text{a.s.}}{\to}} X$ cannot be replaced by $X_n \underset{n\to\infty}{\overset{P}{\to}} X$ and still conclude that $\mathcal{E}^{\mathcal{B}}X_n \underset{n\to\infty}{\overset{\text{a.s.}}{\to}} \mathcal{E}^{\mathcal{B}}X$.

12. Let the r.v.s X and Y have the Bivariate Normal distribution with parameters μ_1, μ_2 in $\Re$, $0 < \sigma_1, \sigma_2 < \infty$, and $\rho \in [-1, 1]$, so that their joint probability density function (p.d.f.) is given by

$$p_{X,Y}(x, y) = \frac{1}{2\pi\sigma_1\sigma_2\sqrt{1-\rho^2}} e^{-q/2},$$

where

$$q = \frac{1}{1-\rho^2}\left[\left(\frac{x-\mu_1}{\sigma_1}\right)^2 - 2\rho\left(\frac{x-\mu_1}{\sigma_1}\right)\left(\frac{y-\mu_2}{\sigma_2}\right) + \left(\frac{y-\mu_2}{\sigma_2}\right)^2\right], \quad x, y \in \Re.$$

(i) Show that the exponent may be written thus:

$$\begin{aligned}(1-\rho^2)q &= \left[\left(\frac{y-\mu_2}{\sigma_2}\right) - \rho\left(\frac{x-\mu_1}{\sigma_1}\right)\right]^2 + (1-\rho^2)\left(\frac{x-\mu_1}{\sigma_1}\right)^2 \\ &= \left(\frac{y-b}{\sigma_2}\right)^2 + (1-\rho^2)\left(\frac{x-\mu_1}{\sigma_1}\right)^2,\end{aligned}$$

where $b = \mu_2 + \frac{\rho\sigma_2}{\sigma_1}(x - \mu_1)$.

(ii) From part (i), it follows that:

$$p_{X,Y}(x, y) = \frac{1}{\sqrt{2\pi}\sigma_1}\exp\left[-\frac{(x-\mu_1)^2}{2\sigma_1^2}\right] \times \frac{1}{\sqrt{2\pi}(\sigma_2\sqrt{1-\rho^2})}\exp\left[-\frac{(y-b)^2}{2(\sigma_2\sqrt{1-\rho^2})^2}\right].$$

From this expression, and without any actual integration, conclude that the r.v. X is distributed as $N(\mu_1, \sigma_1^2)$; i.e., $X \sim N(\mu_1, \sigma_1^2)$, and by symmetry, $Y \sim N(\mu_2, \sigma_2^2)$.

13. In reference to Exercise 12 (ii), and without any actual operations, conclude that the conditional distribution of the r.v. Y, give $X = x$, is $N(b, \sigma_2^2(1-\rho^2))$, where $b = \mu_2 + \frac{\rho\sigma_2}{\sigma_1}(x-\mu_1)$; and by symmetry, the conditional distribution of X, given $Y = y$, is $N(c, \sigma_1^2(1-\rho^2))$, where $c = \mu_1 + \frac{\rho\sigma_1}{\sigma_2}(y-\mu_2)$.

14. **(i)** In reference to Exercises 12 and 13, and by writing $\mathcal{E}(XY) = \mathcal{E}[\mathcal{E}(XY)|X]$, show that $\mathcal{E}(XY) = \mu_1\mu_2 + \rho\sigma_1\sigma_2$.

(ii) Use the definition of the covariance of two r.v.s X and Y (with finite second moments), $\text{Cov}(X, Y) = \mathcal{E}[(X - \mathcal{E}X)(Y - \mathcal{E}Y)] = \mathcal{E}(XY) - (\mathcal{E}X)(\mathcal{E}Y)$, in order to conclude that, in the present case, $\text{Cov}(X, Y) = \rho\sigma_1\sigma_2$.

(iii) From part (ii), conclude that, in the present case, the correlation coefficient of the r.v.s X and Y, $\rho(X, Y)$, is equal to ρ.

15. If the r.v.s X and Y have the Bivariate Normal distribution with parameters μ_1, μ_2 in $\Re$, $0 < \sigma_1, \sigma_2 < \infty$ and $\rho \in [-1, 1]$, set

$$U = \frac{X-\mu_1}{\sigma_1} \text{ and } V = \frac{Y-\mu_2}{\sigma_2}$$

and show that the r.v.s U and V have the Bivariate Normal distribution with parameters 0, 0, 1, 1 and ρ, by transforming the joint p.d.f. of X and Y into the joint p.d.f. of U and V.

16. Let X and Y be r.v.s defined on the probability space $(\Omega, \mathcal{A}, P)$, and suppose that $\mathcal{E}X^2 < \infty$. Then show that

(i) The conditional variance of X, given Y, is given by the formula

$$\begin{aligned}\text{Var}\,(X|Y) &= \mathcal{E}\{[X - \mathcal{E}(X|Y)]^2|Y\}\\ &= \mathcal{E}(X^2|Y) - [\mathcal{E}(X|Y)]^2 \text{ a.s.}\end{aligned}$$

(ii) $\text{Var}(X) = \mathcal{E}[\text{Var}\,(X|Y)] + \text{Var}\,[\mathcal{E}(X|Y)]$.

17. (*Wald*)

(i) Let $X_1, X_2, \ldots$ be r.v.s and let N be a r.v. taking the values $1, 2, \ldots$, all defined on the probability space $(\Omega, \mathcal{A}, P)$. Define the function X as

$$X(\omega) = X_1(\omega) + \cdots + X_{N(\omega)}(\omega),$$

and show that X is a r.v.

(ii) Now suppose that the X_i s are independent and identically distributed with $\mathcal{E}X_1 = \mu \in \Re$, that N is independent of the X_i s, and that $\mathcal{E}N < \infty$. Then show that $\mathcal{E}(X|N) = \mu N$, and therefore $\mathcal{E}X = \mu(\mathcal{E}N)$.

(iii) If in addition to the assumptions made in part (ii), it also holds that $\text{Var}(X_1) = \sigma^2 < \infty$ and $\text{Var}(N) < \infty$, then show that $\text{Var}(X|N) = \sigma^2 N$.

(iv) Use parts (ii) and (iii) here and part (ii) of Exercise 16 in order to conclude that $\text{Var}(X) = \sigma^2(\mathcal{E}N) + \mu^2 \text{Var}(N)$.

Hint: For parts (ii) and (iii), use the special case right after Definition 1, and part (i) of Exercise 16.

18. Let $\mathcal{B}$ be a sub-σ-field in $(\Omega, \mathcal{A}, P)$ which is equivalent to the trivial σ-field $\{\emptyset, \Omega\}$ (in the sense that, for every $B \in \mathcal{B}$, either $P(B) = 0$ or $P(B) = 1$), and let X be a $\mathcal{B}$-measurable (integrable) r.v. Then show that $X = \mathcal{E}X$ a.s.

19. In reference to Example 2, show that, for each $t = 0, 1, \ldots$, there are, indeed, $\binom{n+t-1}{t}$ $\mathbf{x}_i = (x_{i1}, \ldots, x_{in})$ where each $x_{i1}, \ldots, x_{in}$ ranges from 0 to t, and $x_{i1} + \cdots + x_{in} = t, i = 1, \ldots, \binom{n+t-1}{t}$.

CHAPTER

Independence

10

The concept of independence of two σ-fields and two r.v.s was introduced in Section 4 of the previous chapter, because it was needed in Lemma 1 there. What we do in this chapter is to elaborate to a considerable extent on the concept of independence and some of its consequences.

In Section 10.1, the relevant definitions are given, and a result regarding independence of functions of independent r.v.s is stated and proved. The highlight of the chapter is Theorem 1, which states that the factorization of the joint d.f. of n r.v.s to the individual d.f.s implies independence of the r.v.s involved.

Section 10.2 is devoted to establishing those auxiliary results, which are needed for the proof of Theorem 1.

In the final section of the chapter, the proof of Theorem 1 is given, as well as the proof of Lemma 1 in Chapter 9.

10.1 Independence of Events, σ-Fields, and Random Variables

Here, we recall the definition of independence of events, and then extend it to independence of classes of events (and in particular, fields or σ-fields), and independence of r.v.s. In all that follows, $(\Omega, \mathcal{A}, P)$ is the underlying probability space.

Definition 1. Two events A_1, A_2 are said to be *independent (stochastically, statistically, or in the probability sense)*, if

$$P(A_1 \cap A_2) = P(A_1)P(A_2). \tag{10.1}$$

For $n \geq 2$, the events $A_1, \ldots, A_n$ are said to be *independent*, if for any $k \geq 2$ and any (integers) $n_1, \ldots, n_k$ with $1 \leq n_1 < n_2 < \cdots < n_k \leq n$, it holds that

$$P(A_{n_1} \cap A_{n_2} \cap \cdots \cap A_{n_k}) = P(A_{n_1})P(A_{n_2}) \cdots P(A_{n_k}). \tag{10.2}$$

Any collection of events, $\{A_i, i \in I\}$, is said to be *independent*, if any finite subcollection is a set of independent events. ■

Definition 2. Two classes of events $\mathcal{C}_1$ and $\mathcal{C}_2$ with $\mathcal{C}_j \subseteq \mathcal{A}$, $j = 1, 2$, are said to be *independent*, if for all choices of events $A_j \in \mathcal{C}_j$, $j = 1, 2$, relation (10.1) holds. For

An Introduction to Measure-Theoretic Probability, Second Edition. http://dx.doi.org/10.1016/B978-0-12-800042-7.00010-4

$n \geq 2$, the classes of events $\mathcal{C}_j \subseteq \mathcal{A}$, $j = 1, \ldots, n$, are said to be *independent*, if for all choices of events $A_j \in \mathcal{C}_j$, $j = 1, \ldots, n$, the events $\{A_1, \ldots, A_n\}$ are independent; i.e., relation (10.2) holds. Any collection of classes of events $\mathcal{C}_j \subseteq \mathcal{A}$, $j \in I$, are said to be *independent*, if any finite subcollection out of these classes is a collection of independent classes.

In particular, if the classes $\mathcal{C}_j$ are fields $\mathcal{F}_j$ or σ-fields $\mathcal{A}_j$, then we talk about *independence* of fields or σ-fields, respectively. ■

Definition 3. Consider the r.v.s X_1 and X_2, and let $\mathcal{A}_j = X_j^{-1}(\mathcal{B})$, $j = 1, 2$, be the σ-fields induced by them in Ω ($\mathcal{B}$ being the Borel σ-field in $\Re$). Then we say that the r.v.s X_1 and X_2 are *independent*, if the σ-fields $\mathcal{A}_1$ and $\mathcal{A}_2$ are independent. The σ-field $\mathcal{A}_1$ and the r.v. X are said to be *independent* if the σ-fields $\mathcal{A}_1$ and $\mathcal{A}_2 = X^{-1}(\mathcal{B})$ are independent. For $n \geq 2$, the r.v.s X_j, $j = 1, \ldots, n$, are said to be *independent*, if the σ-fields $\mathcal{A}_j = X_j^{-1}(\mathcal{B})$, $j = 1, \ldots, n$, induced by them are independent. Finally, restricting ourselves to denumerably many r.v.s, we say that the r.v.s $\{X_n, n \geq 1\}$ are *independent*, if any finite collection of them is a set of independent r.v.s. ■

Remark 1. In order to establish independence of n fields $\mathcal{F}_j$ or σ-fields $\mathcal{A}_j$, $j = 1, \ldots, n$, it suffices to show that, for all choices of events $A_j \in \mathcal{F}_j$ or $A_j \in \mathcal{A}_j$, $j = 1, \ldots, n$, it holds that

$$P(A_1 \cap \cdots \cap A_n) = P(A_1) \cdots P(A_n). \tag{10.3}$$

In other words, it is not necessary to check (10.2) for all subcollections of k events out of the n events. This is so, because by taking $A_j = \Omega$ for $j \neq n_1, \ldots, n_k$, relation (10.3) reduces to (10.2).

Remark 2. From Definition 2, it is immediate that subclasses of independent classes are also independent.

We record below a simple but very useful result.

Proposition 1. Borel functions of independent r.v.s are independent r.v.s. That is, if the r.v.s $\{X_n, n \geq 1\}$ are independent and $Y_n = g_n(X_n)$, where $g_n : \Re \to \Re$ measurable, then the r.v.s $\{Y_n, n \geq 1\}$ are independent.

Proof. It follows immediately by Remark 2, because, if $\mathcal{A}_{X_n}$ are the σ-fields induced by the r.v.s X_n, and $\mathcal{A}_{Y_n}$ are the σ-fields induced by the r.v.s Y_n, then $\mathcal{A}_{Y_n} \subseteq \mathcal{A}_{X_n}$, $n \geq 1$. This is so because $Y_n^{-1}(B) = [g_n(X_n)]^{-1}(B) = X_n^{-1}[g_n^{-1}(B)] \in \mathcal{A}_{X_n}$ since $g_n^{-1}(B) \in \mathcal{B}$, $B \in \mathcal{B}$. ■

Now, consider the r.v.s X_j, $j = 1, \ldots, n$, and let $\mathcal{A}_j$, $j = 1, \ldots, n$ be the σ-fields induced by them. Then by Definition 3 and Remark 1, independence of the r.v.s $X_1, \ldots, X_n$ amounts to the validity of relation (10.3) for all choices of the events $A_j \in \mathcal{A}_j$, $j = 1, \ldots, n$. But $A_j \in \mathcal{A}_j$ means that $A_j = X_j^{-1}(B_j)$, some $B_j \in \mathcal{B}$. Therefore relation (10.3) becomes, equivalently,

$$P(X_1 \in B_1, \ldots, X_n \in B_n) = P(X_1 \in B_1) \ldots P(X_n \in B_n) \tag{10.4}$$

for all $B_j \in \mathcal{B}$, $j = 1, \ldots, n$. Thus, establishing independence for the r.v.s $X_1, \ldots, X_n$ amounts to checking relation (10.4). However, in lower level textbooks, it is claimed that the r.v.s $X_1, \ldots, X_n$ are independent, if

$$\begin{aligned} &P\left(X_1 \le x_1, \ldots, X_n \le x_n\right) = P(X_1 \le x_n) \ldots P(X_n \le x_n), \\ &\text{or } P\left(X_1 \in (-\infty, x_1], \ldots, X_n \in (-\infty, x_n]\right) \\ &= P\left(X_1 \in (-\infty, x_1]\right) \ldots P\left(X_n \in (-\infty, x_n]\right) \end{aligned} \tag{10.5}$$

for all $x_1, \ldots, x_n$ in $\Re$. Relation (10.5) is a very special case of (10.4), taken from it for $B_j = (-\infty, x_j]$, $j = 1, \ldots, n$. The question then arises as to whether this claim is valid, as it should be. The justification of this claim is the content of the next result.

Theorem 1. The r.v.s $X_1, \ldots, X_n$ are independent, if and only if relation (10.5) holds for all $x_1, \ldots, x_n$ in $\Re$. ■

As already stated, the validity of (10.5) is a special case of (10.4), taken from it for $B_j = (-\infty, x_j]$, $x_j \in \Re$; or to put it differently, by taking B_js from the class

$$\mathcal{C} = \{(-\infty, x];\ x \in \Re\}.$$

We show below that (10.5) also holds when this class is enlarged to include $\oslash$, $\Re$, and any interval in $\Re$ finite or not, and of any form. The proof of this is facilitated by the results established in the following section.

10.2 Some Auxiliary Results

In this section, three lemmas and one proposition are established, on which the proof of Theorem 1 is based.

Lemma 1. If relation (10.5) holds when the B_js in (10.4) are chosen from the class $\mathcal{C}$, then it holds if the B_js are chosen from the class $\mathcal{C}_0$:

$$\begin{aligned} \mathcal{C}_0 &= \{(-\infty, x], (-\infty, x), (x, \infty), [x, \infty), (x, y], \\ &\quad [x, y), (x, y), [x, y], \oslash, \Re;\ x, y \in \Re\} \\ &= \{\oslash, \Re, (x, \infty), (x, y], (-\infty, x], [x, y], \\ &\quad (x, y), [x, y);\ x, y \in \Re\} \ \text{(listed in a convenient order).} \end{aligned}$$

Proof. In (10.5), if at least one of the intervals $(-\infty, x_j]$, $j = 1, \ldots, n$, is replaced by $\oslash$, then it becomes identity, $0 = 0$. Thus, (10.5) holds with intervals $(-\infty, x_j]$, $j = 1, \ldots, n$, replaced by $\oslash$. Next,

$$\begin{aligned} &P(X_1 \in \Re, X_j \le x_j, j = 2, \ldots, n) \\ &= P\left(X_1 \in \bigcup_m (-\infty, y_m], X_j \le x_j, j = 2, \ldots, n\right) \text{ where } y_m \uparrow \infty, m \to \infty \end{aligned}$$

$$
\begin{aligned}
&= P\Big(\bigcup_m (X_1 \le y_m), X_j \le x_j, j = 2, \ldots, n\Big)\\
&= P\Big(\bigcup_m (X_1 \le y_m, X_j \le x_j, j = 2, \ldots, n)\Big)\\
&= P\Big(\lim_m (X_1 \le y_m, X_j \le x_j, j = 2, \ldots, n)\Big)\\
&= \lim_m P(X_1 \le y_m, X_j \le x_j, j = 2, \ldots, n)\\
&= \lim_m P(X_1 \le y_m) P(X_2 \le x_2) \cdots P(X_n \le x_n)\\
&= P(X_1 \in \Re) P(X_2 \le x_2) \cdots P(X_n \le x_n).
\end{aligned}
$$

So, the factorization in (10.5) holds if one of the intervals $(-\infty, x_j]$, $j = 1, \ldots, n$ (which without loss of generality may be taken to be the interval $(-\infty, x_1]$), is replaced by $\Re$. Assuming the factorization to be true when k of the preceding intervals $(2 \le k < n)$ are replaced by $\Re$, we show as before that it is also true if $k + 1$ intervals are replaced by $\Re$. So, by the induction hypothesis, any number of intervals may be replaced by $\Re$ and the factorization in (10.5) holds.

Next,

$$
\begin{aligned}
&P(X_1 > x, X_j \le x_j, j = 2, \ldots, n)\\
&= P(X_1 \in (\Re - (-\infty, x]), X_j \le x_j, j = 2, \ldots, n)\\
&= P((X_1 \in \Re, X_j \le x_j, j = 2, \ldots, n)\\
&\quad -(X_1 \le x, X_j \le x_j, j = 2, \ldots, n))\\
&= P(X_1 \in \Re, X_j \le x_j, j = 2, \ldots, n)\\
&\quad -P(X_1 \le x, X_j \le x_j, j = 2, \ldots, n)\\
&= P(X_1 \in \Re) P(X_2 \le x_2) \cdots P(X_n \le x_n)\\
&\quad -P(X_1 \le x) P(X_2 \le x_2) \cdots P(X_n \le x_n)\\
&= [P(X_1 \in \Re) - P(X_1 \le x)] P(X_2 \le x_2) \cdots P(X_n \le x_n)\\
&= P(X_1 > x) P(X_2 \le x_2) \cdots P(X_n \le x_n).
\end{aligned}
$$

As in the previous step, the factorization holds if any number of the intervals $(-\infty, x_j]$, $j = 1, \ldots, n$, is replaced by intervals of the form (x, ∞).

Next,

$$
\begin{aligned}
&P(x < X_1 \le y, X_j \le x_j, j = 2, \ldots, n)\\
&= P(((X_1 \le y) - (X_1 \le x)), X_j \le x_j, j = 2, \ldots, n)\\
&= P(X_1 \le y, X_j \le x_j, j = 2, \ldots, n)\\
&\quad -P(X_1 \le x, X_j \le x_j, j = 2, \ldots, n)\\
&= P(X_1 \le y) P(X_2 \le x_2) \cdots P(X_n \le x_n)\\
&\quad -P(X_1 \le x) P(X_2 \le x_2) \cdots P(X_n \le x_n)
\end{aligned}
$$

$$= [P(X_1 \le y) - P(X_1 \le x)]P(X_2 \le x_2) \cdots P(X_n \le x_n)$$
$$= P(x < X_1 \le y)P(X_2 \le x_2) \cdots P(X_n \le x_n),$$

and the same is true, if any number of the $(-\infty, x_j]$, $j = 1, \ldots, n$, is replaced by intervals of the form $(x, y]$.

Next,

$$\begin{aligned}
&P(X_1 < x, X_j \le x_j, j = 2, \ldots, n)\\
&= P\Big(X_1 \in \bigcup_m (-\infty, y_m], X_j \le x_j, j = 2, \ldots, n\Big) \text{ where } y_m \uparrow x, m \to \infty\\
&= P\left(\bigcup_m (X_1 \le y_m), X_j \le x_j, j = 2, \ldots, n\right)\\
&= P\left(\bigcup_m (X_1 \le y_m, X_j \le x_j, j = 2, \ldots, n)\right)\\
&= P\left(\lim_m (X_1 \le y_m, X_j \le x_j, j = 2, \ldots, n)\right)\\
&= \lim_m P(X_1 \le y_m, X_j \le x_j, j = 2, \ldots, n)\\
&= \lim_m P(X_1 \le y_m)P(X_2 \le x_2) \cdots P(X_n \le x_n)\\
&= P(X_1 < x)P(X_2 \le x_2) \cdots P(X_n \le x_n).
\end{aligned}$$

Then, arguing as in the previous steps, we conclude that (10.5) holds if any number of the $(-\infty, x_j]$, $j = 1, \ldots, n$, is replaced by intervals of the form $(-\infty, x)$.

Next,

$$\begin{aligned}
&P(x \le X_1 \le y, X_j \le x_j, j = 2, \ldots, n)\\
&= P(((X_1 \le y) - (X_1 < x)), X_j \le x_j, j = 2, \ldots, n)\\
&= P(X_1 \le y, X_j \le x_j, j = 2, \ldots, n)\\
&\quad - P(X_1 < x, X_j \le x_j, j = 2, \ldots, n)\\
&= P(X_1 \le y)P(X_2 \le x_2) \cdots P(X_n \le x_n)\\
&\quad - P(X_1 < x)P(X_2 \le x_2) \cdots P(X_n \le x_n)\\
&= [P(X_1 \le y) - P(X_1 < x)]P(X_2 \le x_2) \cdots P(X_n \le x_n)\\
&= P(x \le X_1 \le y)P(X_2 \le x_2) \cdots P(X_n \le x_n),
\end{aligned}$$

and arguing as before, we conclude that (10.5) holds if any number of the $(-\infty, x_j]$, $j = 1, \ldots, n$, is replaced by intervals of the form $[x, y]$.

Next,

$$\begin{aligned}
&P(X_1 \ge x, X_j \le x_j, j = 2, \ldots, n)\\
&= P\left(X_1 \in \bigcap_m (y_m, \infty), X_j \le x_j, j = 2, \ldots, n\right) \text{ where } y_m \uparrow x, m \to \infty
\end{aligned}$$

$$
\begin{aligned}
&= P\Big(\bigcap_m (X_1 > y_m), X_j \le x_j, j = 2, \ldots, n\Big) \\
&= P\Big(\bigcap_m (X_1 > y_m, X_j \le x_j, j = 2, \ldots, n)\Big) \\
&= P\Big(\lim_m (X_1 > y_m, X_j \le x_j, j = 2, \ldots, n)\Big) \\
&= \lim_m P(X_1 > y_m, X_j \le x_j, j = 2, \ldots, n) \\
&= \lim_m P(X_1 > y_m) P(X_2 \le x_2) \cdots P(X_n \le x_n) \\
&= P(X_1 \ge x) P(X_2 \le x_2) \cdots P(X_n \le x_n),
\end{aligned}
$$

and arguing as in previous steps, we conclude that (10.5) holds if any number of the $(-\infty, x_j]$, $j = 1, \ldots, n$, is replaced by intervals of the form $[x, \infty)$.

Next,

$$
\begin{aligned}
&P(x < X_1 < y, X_j \le x_j, j = 2, \ldots, n) \\
&= P\Big(X_1 \in \bigcup_m (x, y_m], X_j \le x_j, j = 2, \ldots, n\Big) \quad \text{where } y_m \uparrow y, m \to \infty \\
&= P\Big(\bigcup_m (x < X_1 \le y_m), X_j \le x_j, j = 2, \ldots, n\Big) \\
&= P\Big(\bigcup_m (x < X_1 \le y_m, X_j \le x_j, j = 2, \ldots, n)\Big) \\
&= P\Big(\lim_m (x < X_1 \le y_m, X_j \le x_j, j = 2, \ldots, n)\Big) \\
&= \lim_m P(x < X_1 \le y_m, X_j \le x_j, j = 2, \ldots, n) \\
&= \lim_m P(x < X_1 \le y_m) P(X_2 \le x_2) \cdots P(X_n \le x_n) \\
&= P(x < X_1 < y) P(X_2 \le x_2) \cdots P(X_n \le x_n),
\end{aligned}
$$

and as in previous steps, it follows that (10.5) holds when any number of the $(-\infty, x_j]$, $j = 1, \ldots, n$, is replaced by intervals of the form (x, y).

Finally,

$$
\begin{aligned}
&P(x \le X_1 < y, X_j \le x_j, j = 2, \ldots, n) \\
&= P\Big(X_1 \in \bigcap_m (y_m, y), X_j \le x_j, j = 2, \ldots, n\Big) \quad \text{where } y_m \uparrow x, m \to \infty \\
&= P\Big(\bigcap_m (y_m < X_1 < y), X_j \le x_j, j = 2, \ldots, n\Big) \\
&= P\Big(\bigcap_m (y_m < X_1 < y, X_j \le x_j, j = 2, \ldots, n)\Big)
\end{aligned}
$$

$$= P\left(\lim_m (y_m < X_1 < y, X_j \le x_j, j = 2, \ldots, n)\right)$$
$$= \lim_m P(y_m < X_1 < y, X_j \le x_j, j = 2, \ldots, n)$$
$$= \lim_m P(y_m < X_1 < y)P(X_2 \le x_2) \cdots P(X_n \le x_n)$$
$$= P(x \le X_1 < y)P(X_2 \le x_2) \cdots P(X_n \le x_n),$$

and as before, we conclude that (10.5) holds when any number of the intervals $(-\infty, x_j]$, $j = 1, \ldots, n$, is replaced by intervals of the form $[x, y)$. Finally, combining the 10 conclusions just reached, we have that the factorization (10.5) holds when the B_js in (10.4) are chosen in any way from the class $\mathcal{C}_0$. This completes the proof of the lemma. ■

Lemma 2. Let $\mathcal{F}$ be the field (by Exercise 7(ii) in Chapter 1) of all finite sums of members of $\mathcal{C}_0$. Then (10.5) holds, if any number of intervals $(-\infty, x_j]$, $j = 1, \ldots, n$, are replaced by elements of $\mathcal{F}$.

Proof. Let $I \in \mathcal{F}$. Then $I = \sum_{i=1}^m I_i$ with $I_i \in \mathcal{C}_0$, and

$$P(X_1 \in I, X_j \le x_j, j = 2, \ldots, n)$$
$$= P\left(X_1 \in \sum_i I_i, X_j \le x_j, j = 2, \ldots, n\right)$$
$$= P\left(\sum_i (X_1 \in I_i), X_j \le x_j, j = 2, \ldots, n\right)$$
$$= P\left(\sum_i (X_1 \in I_i, X_j \le x_j, j = 2, \ldots, n)\right).$$

However,

$$P\left(\sum_i (X_1 \in I_i, X_j \le x_j, j = 2, \ldots, n)\right)$$
$$= \sum_i P\left(X_1 \in I_i, X_j \le x_j, j = 2, \ldots, n\right).$$

Therefore

$$P(X_1 \in I, X_j \le x_j, j = 2, \ldots, n)$$
$$= \sum_i P(X_1 \in I_i)P(X_2 \le x_2) \cdots P(X_n \le x_n) \text{ (by Lemma 1)}$$
$$= P(X_2 \le x_2) \cdots P(X_n \le x_n) \sum_i P(X_1 \in I_i)$$
$$= P(X_2 \le x_2) \cdots P(X_n \le x_n) P\left(X_1 \in \sum_i I_i\right)$$

$$= P(X_2 \leq x_2) \cdots P(X_n \leq x_n) P(X_1 \in I)$$
$$= P(X_1 \in I) P(X_2 \leq x_2) \cdots P(X_n \leq x_n).$$

Thus, the factorization in (10.5) holds if one of the intervals $(-\infty, x_j]$, $j = 1, \ldots, n$ (which without loss of generality may be taken to be the interval $(-\infty, x_1]$), is replaced by a member I of $\mathcal{F}$. Assuming it to be true when k of the intervals $(2 \leq k < n)$ are replaced by members of $\mathcal{F}$, we show as before that the factorization also holds if $k+1$ intervals are replaced by Is. So, by the induction hypothesis, any number of intervals may be replaced by Is and the factorization holds. This completes the proof. ■

With $\mathcal{F}$ as in Lemma 2, we know that $\mathcal{B} = \sigma(\mathcal{F})$; i.e., $\mathcal{B}$ is generated by $\mathcal{F}$. Let $\mathcal{F}_j = X_j^{-1}(\mathcal{F})$ and $\mathcal{A}_j = X_j^{-1}(\mathcal{B})$, $j = 1, \ldots, n$. Then by Theorem 12 in Chapter 1, $\mathcal{A}_j = \sigma(\mathcal{F}_j)$, $j = 1, \ldots, n$. A reformulation of Lemma 2 in terms of elements in the $\mathcal{F}_j$ s is as follows: for all $F_j \in \mathcal{F}_j$, $j = 1, \ldots, n$,

$$P(F_1 \cap \cdots \cap F_n) = P(F_1) \cdots P(F_n). \tag{10.6}$$

It is our intention to show that (10.6) holds when the F_js are arbitrary members of the respective $\mathcal{A}_j$s. To this effect, let $\mathcal{M}_1$ be defined as follows:

$$\mathcal{M}_1 = \{B \in \mathcal{A}_1;\ P(B \cap F_2 \cap \cdots \cap F_n) = P(B)P(F_2) \cdots P(F_n)$$
$$\text{with } F_j \in \mathcal{F}_j, j = 2, \ldots, n\}. \tag{10.7}$$

Then, by (10.6) and (10.7), $\mathcal{F}_1 \subseteq \mathcal{M}_1 \subseteq \mathcal{A}_1$, and we shall show that $\mathcal{M}_1$ is a monotone class. To this end, let $\{A_m\}$, $m \geq 1$, be a monotone sequence of elements of $\mathcal{M}_1$ with limit A. We shall show that $A \in \mathcal{M}_1$, which amounts to showing that

$$P(A \cap F_2 \cap \cdots \cap F_n) = P(A)P(F_2) \cdots P(F_n) \tag{10.8}$$

for all $F_j \in \mathcal{F}_j$, $j = 2, \ldots, n$. Indeed, as $m \to \infty$,

$$\begin{aligned} P(A \cap F_2 \cap \cdots \cap F_n) &= P\left(\left(\lim_m A_m\right) \cap F_2 \cap \cdots \cap F_n\right) \\ &= P\left(\left(\bigcup_m A_m\right) \cap F_2 \cap \cdots \cap F_n\right), \quad \text{if } A_m \uparrow \\ &= P\left(\bigcup_m (A_m \cap F_2 \cap \cdots \cap F_n)\right) \\ &= P\left(\lim_m (A_m \cap F_2 \cap \cdots \cap F_n)\right) \\ &= \lim_m P(A_m \cap F_2 \cap \cdots \cap F_n) \\ &= \lim_m P(A_m)P(F_2) \cdots P(F_n) \\ &= P(A)P(F_2) \cdots P(F_n), \end{aligned}$$

and the same is true, if $A_m \downarrow$, by replacing union by intersection. This justifies the assertion made. Next, define $\mathcal{M}_2$ by

$$\begin{aligned} \mathcal{M}_2 = \{B \in \mathcal{A}_2;\ &P(A_1 \cap B \cap F_3 \cap \cdots \cap F_n) \\ &= P(A_1)P(B)P(F_3) \cdots P(F_n) \\ &\text{with } A_1 \in \mathcal{M}_1 \text{ and } F_j \in \mathcal{F}_j, j = 3, \ldots, n\}. \end{aligned} \tag{10.9}$$

Then, by (10.8) and (10.9), $\mathcal{F}_2 \subseteq \mathcal{M}_2 \subseteq \mathcal{A}_2$, and we shall show that $\mathcal{M}_2$ is a monotone class. This is done as in the previous step by letting $\{A_m\}$, $m \geq 1$, be a monotone sequence of elements of $\mathcal{M}_2$ with limit A, and showing that

$$P(A_1 \cap A \cap F_3 \cap \cdots \cap F_n) = P(A_1)P(A)P(F_3)\cdots P(F_n) \tag{10.10}$$

for all $A_1 \in \mathcal{M}_1$ and all $F_j \in \mathcal{F}_j$, $j = 3, \ldots, n$.

Continuing on like this, let $\mathcal{M}_n$ be the class defined by

$$\begin{aligned}\mathcal{M}_n = \{B \in \mathcal{A}_n;\ & P(A_1 \cap \cdots \cap A_{n-1} \cap B) \\ & = P(A_1)\cdots P(A_{n-1})P(B) \\ & \text{with } A_j \in \mathcal{M}_j,\ j = 1, \ldots, n-1\}.\end{aligned} \tag{10.11}$$

Then as before, $\mathcal{F}_n \subseteq \mathcal{M}_n \subseteq \mathcal{A}_n$, and $\mathcal{M}_n$ is a monotone class, because if $\{A_m\}$, $m \geq 1$, is a monotone sequence of elements of $\mathcal{M}_n$ with limit A, then it is shown as in the previous steps that

$$P(A_1 \cap \cdots \cap A_{n-1} \cap A) = P(A_1)\cdots P(A_{n-1})P(A) \tag{10.12}$$

for all $A_j \in \mathcal{M}_j$, $j = 1, \ldots, n-1$.

Gathering together the results just obtained, we then have the following

Lemma 3. For $j = 1, \ldots, n$, the classes $\mathcal{M}_j$ defined by (10.7), (10.9), and (10.11) are monotone classes with the property that $\mathcal{F}_j \subseteq \mathcal{M}_j \subseteq \mathcal{A}_j$, and by (10.12),

$$P\left(\bigcap_{j=1}^{n} A_j\right) = \prod_{j=1}^{n} P(A_j), \text{ for all } A_j \in \mathcal{M}_j,\ j = 1, \ldots, n. \tag{10.13}$$

The following proposition is an immediate consequence of what has just been discussed.

Proposition 2. If the fields $\mathcal{F}_j$, $j = 1, \ldots, n$ are independent, then so are the σ-fields generated by them, $\mathcal{A}_j = \sigma(\mathcal{F}_j)$, $j = 1, \ldots, n$.

Proof. In Lemma 3, it was proved that relation (10.6) implies relation (10.13) for all $\mathcal{A}_j \in \mathcal{M}_j$, where $\mathcal{M}_j$ are monotone classes with $\mathcal{F}_j \subseteq \mathcal{M}_j \subseteq \mathcal{A}_j$, $j = 1, \ldots, n$. But, by Theorem 6 in Chapter 1, $\mathcal{M}_j = \mathcal{A}_j$, $j = 1, \ldots, n$. This completes the proof. ■

10.3 Proof of Theorem 1 and of Lemma 1 in Chapter 9

We are now ready to prove Theorem 1 here and Lemma 1 in the previous chapter.

Proof of Theorem 1. If the r.v.s are independent, then

$$P\left(\bigcap_{j=1}^{n} A_j\right) = \prod_{j=1}^{n} P(A_j), \text{ for all } A_j \in \mathcal{A}_j,\ j = 1, \ldots, n,$$

and in particular, this is true for $A_j = X_j^{-1}((-\infty, x_j]),\ j = 1, \ldots, n$, which gives (10.5). In the other way around, if (10.5) holds, then so does (10.13) by way of Lemma 3 and Proposition 2, for all $A_j \in \mathcal{A}_j,\ j = 1, \ldots, n$, which establishes independence of $X_j,\ j = 1, \ldots, n$. ■

Remark 3. Relation (10.5) can also be written as follows in terms of d.f.s:

$$F_{X_1,\ldots,X_n}(x_1, \ldots, x_n) = F_{X_1}(x_1) \cdots F_{X_n}(x_n)$$

for all $x_1, \ldots, x_n$ in $\Re$.

This section is concluded with the proof of Lemma 1 in Chapter 9.

Proof of Lemma 1 in Chapter 9. The proof follows the four familiar steps. It is established successively for indicators, simple r.v.s, nonnegative r.v.s, and any r.v.s. So, let $X = I_A, Y = I_B, A, B \in \mathcal{A}$. Then the σ-fields $\mathcal{B}_X = \{\varnothing, A, A^c, \Omega\}$ and $\mathcal{B}_Y = \{\varnothing, B, B^c, \Omega\}$ are independent because of the independence of the r.v.s X and Y. Next, $XY = I_{A \cap B}$, so that $\mathcal{E}(XY) = P(A \cap B) = P(A)P(B) = (\mathcal{E}I_A)(\mathcal{E}I_B) = (\mathcal{E}X)(\mathcal{E}Y)$. Now let $X = \sum_{i=1}^m \alpha_i I_{A_i}$ and $Y = \sum_{j=1}^n \beta_j I_{B_j}$, so that $XY = \sum_i \sum_j \gamma_{ij} I_{A_i \cap B_j}$, where $\gamma_{ij} = \alpha_i \beta_j$, if $A_i \cap B_j \neq \varnothing$, and whatever; e.g., 0, if $A_i \cap B_j = \varnothing$, and

$$\begin{aligned}
\mathcal{E}(XY) = \sum_i \sum_j \gamma_{ij} P(A_i \cap B_j) &= \sum_i \sum_j \alpha_i \beta_j P(A_i \cap B_j) \\
&= \sum_i \sum_j \alpha_i \beta_j P(A_i) P(B_j) \\
&= \left[\sum_i \alpha_i P(A_i)\right] \times \left[\sum_j \beta_j P(B_j)\right] = (\mathcal{E}X)(\mathcal{E}Y).
\end{aligned}$$

Next, let X and Y be ≥ 0 r.v.s and let $n \to \infty$ in the remainder of this proof. Then there exist simple r.v.s X_n and Y_n such that $0 \leq X_n \uparrow X$ and $0 \leq Y_n \uparrow Y$, so that $0 \leq X_n Y_n \uparrow XY$. By the Lebesgue Monotone Convergence Theorem, we have then: $\mathcal{E}X_n \to \mathcal{E}X, \mathcal{E}Y_n \to \mathcal{E}Y$, so that $(\mathcal{E}X_n)(\mathcal{E}Y_n) \to (\mathcal{E}X)(\mathcal{E}Y)$, and $\mathcal{E}(X_n Y_n) \to \mathcal{E}(XY)$. However, $\mathcal{E}(X_n Y_n) = (\mathcal{E}X_n)(\mathcal{E}Y_n)$ by the previous step. Thus, $\mathcal{E}(XY) = (\mathcal{E}X)(\mathcal{E}Y) < \infty$. Finally, for any r.v.s X and Y, we have

$$XY = (X^+ - X^-)(Y^+ - Y^-) = X^+Y^+ - X^+Y^- - X^-Y^+ + X^-Y^-,$$

and all expectations $\mathcal{E}X^+, \mathcal{E}X^-, \mathcal{E}Y^+, \mathcal{E}Y^-$ are finite by the assumption that $\mathcal{E}|X| < \infty$ and $\mathcal{E}|Y| < \infty$. Then, by the previous step, all expectations $\mathcal{E}(X^+Y^+), \mathcal{E}(X^+Y^-), \mathcal{E}(X^-Y^+), \mathcal{E}(X^-Y^-)$ are finite, and therefore

$$\begin{aligned}
\mathcal{E}(XY) &= \mathcal{E}(X^+Y^+) - \mathcal{E}(X^+Y^-) - \mathcal{E}(X^-Y^+) + \mathcal{E}(X^-Y^-) \\
&= (\mathcal{E}X^+)(\mathcal{E}Y^+) - (\mathcal{E}X^+)(\mathcal{E}Y^-) \\
&\quad -(\mathcal{E}X^-)(\mathcal{E}Y^+) + (\mathcal{E}X^-)(\mathcal{E}Y^-) \\
&= (\mathcal{E}X^+)(\mathcal{E}Y^+ - \mathcal{E}Y^-) - (\mathcal{E}X^-)(\mathcal{E}Y^+ - \mathcal{E}Y^-) \\
&= (\mathcal{E}Y^+ - \mathcal{E}Y^-)(\mathcal{E}X^+ - \mathcal{E}X^-) \\
&= (\mathcal{E}X)(\mathcal{E}Y) \text{ finite.}
\end{aligned}$$

■

Exercises.

1. If the events $A_1, \ldots, A_n$ are independent, then so are the events $A'_1, \ldots, A'_n$ where A'_i is either A_i or $A_i^c, i = 1, \ldots, n$.
Hint: The proof is done by double induction. See also Theorem 6 in Chapter 2 in Roussas (1997).
2. Consider the measurable space $(\Omega, \mathcal{A})$, let $\mathcal{F}_i, i = 1, 2$, be fields with $\mathcal{F}_i \subseteq \mathcal{A}, i = 1, 2$, and define $\mathcal{F}$ by
$$\mathcal{F} = \{\text{all finite unions of } A\text{s with } A \in \mathcal{A};\ A = A_1 \cup A_2 \text{ or}$$
$$A = A_1 \cap A_2,\ A_i \in \mathcal{F}_i, i = 1, 2\}.$$
Then show that $\mathcal{F}$ is a field.
3. For any real numbers $p_1, \ldots, p_n$ such that $0 \le p_i \le 1, i = 1, \ldots, n$, show that
$$1 - \exp\left(-\sum_{i=m}^{n} p_i\right) \le 1 - \prod_{i=m}^{n}(1 - p_i) \le \sum_{i=m}^{n} p_i, \quad m = 1, \ldots, n.$$
Hint: For the left-hand side, use the inequality $e^x \ge 1 + x, x \in \Re$, and for the right-hand side employ the induction method.
4. Let $A_n, n = 1, 2, \ldots$, be independent events in the probability space $(\Omega, \mathcal{A}, P)$. Then show that
   (i) $\sum_{n=1}^{\infty} P(A_n) = \infty$ if and only if $P(\limsup_{n\to\infty} A_n) = 1$.
   (ii) $\sum_{n=1}^{\infty} P(A_n) < \infty$ if and only if $P(\limsup_{n\to\infty} A_n) = 0$.

   Hint: For part (i), use the left-hand side inequality in Exercise 3 here, and Exercise 3 in Chapter 3. For part (ii), use part (i) and Exercise 3 in Chapter 3 again.
   Remark: This result is referred to as the *Borel Zero-One* Criterion.
5. Let $A_n, n = 1, 2, \ldots$, be independent events in the probability space $(\Omega, \mathcal{A}, P)$, and suppose that $\lim_{n\to\infty} A_n$ exists, call it A. Then show that $P(A) = 0$ or $P(A) = 1$.
Hint: Use Exercise 4.
6. Let $X_n, n \ge 1$, be independent r.v.s distributed as $B(1, p)$, and set $\bar{X}_n = n^{-1}\sum_{i=1}^{n} X_i$. Then show that $\bar{X}_n \underset{n\to\infty}{\overset{P}{\to}} p$ and $\bar{X}_{k^2} \underset{k\to\infty}{\overset{\text{a.s.}}{\to}} p$.
Hint: For the second conclusion, use Exercise 4(i) here and Theorem 4 in Chapter 3.
7. **(i)** If $X_n, n = 1, 2, \ldots$, are independent r.v.s defined on the probability space $(\Omega, \mathcal{A}, P)$, show that $X_n \underset{n\to\infty}{\overset{\text{a.s.}}{\to}} 0$ if and only if $\sum_{n=1}^{\infty} P(|X_n| \ge \varepsilon) < \infty$ for every $\varepsilon > 0$.
   (ii) Reconcile this result with the result in Exercise 4 in Chapter 3.

   Hint: Use Exercise 4(ii) here and Theorem 4 in Chapter 3.
8. Refer to Exercise 11 in Chapter 9 and construct a concrete example, by means of independent r.v.s (e.g., Binomially distributed r.v.s), to demonstrate the correctness of the assertion made there.

9. The r.v.s $X_1, \ldots, X_k$ are independent if and only if in relation (10.4) all B_js are replaced by intervals $(a_j, b_j]$ with a_j, b_j in $\Re$ and $a_j < b_j$, $j = 1, \ldots, k$.
Hint: Work as in Lemma 1 in order to show that

$$P(X \leq x_1, a_j < X_j \leq b_j, j = 2, \ldots, k) = P(X \leq x_1) \prod_{j=2}^{k} P(a_j < X_j \leq b_j),$$

and complete the factorization by replacing one of the remaining X_j, $j = 2, \ldots, k$, at a time.

10. Let X and Y be independent r.v.s and suppose that $\mathcal{E}X$ exists. For every $B \in \mathcal{B}$, let $A = Y^{-1}(B)$ and show that $\int_A X dP = (\mathcal{E}X)P(A)$.

11. Show that a r.v. X is independent of itself if and only if $P(X = c) = 1$ for some (finite) constant c.

12. **(i)** For two r.v.s with finite second moments, it follows that, if X and Y are independent, then they are uncorrelated (i.e., $\rho(X, Y) = 0$, or equivalently, $Cov(X, Y) = 0$). Justify this statement.
(ii) For the case that the r.v.s X and Y have the Bivariate Normal distribution, use Exercise 14(iii) in Chapter 9 in order to show that, if X and Y are uncorrelated, then they are independent.

13. If the r.v.s X_1 and X_2 are independent, then show that

$$\mathcal{E}e^{it(X_1+X_2)} = \mathcal{E}e^{itX_1} \times \mathcal{E}e^{itX_2}, \quad t \in \Re,$$

and by induction,

$$\mathcal{E}e^{it(X_1+\cdots+X_k)} = \mathcal{E}e^{itX_1} \times \cdots \times \mathcal{E}e^{itX_k}, \quad t \in \Re,$$

for the independent r.v.s $X_1, \ldots, X_k$.
Hint: Write $e^{it(X_1+X_2)} = e^{itX_1} \times e^{itX_2} = [\cos(tX_1) + i\sin(tX_1)] \times [\cos(tX_2) + i\sin(tX_2)]$ and use Proposition 1.

14. **(i)** If the r.v.s X and Y are independent, distributed as $N(0, \sigma^2)$, and $U = X + Y$, $V = X - Y$, then show that U and V are independent, distributed as $N(0, 2\sigma^2)$, by transforming the joint p.d.f. of X and Y into the joint p.d.f. of U and V.
(ii) If the r.v.s X and Y are independent, distributed as $N(\mu_1, \sigma^2)$ and $N(\mu_2, \sigma^2)$, respectively, use part (i) in order to show that U and V are independent, distributed as $N(\mu_1 + \mu_2, 2\sigma^2)$ and $N(\mu_1 - \mu_2, 2\sigma^2)$, respectively.

15. Consider the probability space $(\Omega, \mathcal{A}, P)$, and let $A_1, \ldots, A_n$ be independent events with $P(A_k) = p$, $k = 1, \ldots, n$. Next, define the function $X : \Omega \to \Re$ as follows:

$$X(\omega) = \text{the number of } A_1, \ldots, A_n \text{ containing } \omega.$$

Then show that

(i) X is a r.v.
(ii) The distribution of X is $B(n, p)$ (i.e., Binomial with parameters n and p).

16. Consider the probability space $(\Omega, \mathcal{A}, P)$, let $\{A_n\}, n \geq 1$, be a sequence of events, and set $X_n = I_{A_n}$. Then show that the events $\{A_1, A_2, \ldots\}$ are independent if and only if the r.v.s $X_1, X_2, \ldots$ are independent.

17. If $X \sim B(n, p)$, compute the probability that P(number of Hs = number of $Ts + r$), $r = 0, 1, \ldots, n$.

18. Let $X_1, \ldots, X_n$ be independent identically distributed (i.i.d.) r.v.s defined on the probability space $(\Omega, \mathcal{A}, P)$ and having d.f. F. Let F_n be the *empirical* d.f. defined in terms of the X_is; i.e.,

$$F_n(x, \omega) = \frac{1}{n}[\text{number of } X_1(\omega), \ldots, X_n(\omega) \leq x].$$

Then show that $\sup\{|F_n(x, \cdot) - F(x)|; x \in \Re\} \stackrel{def}{=} D_n(\cdot)$ is a r.v. That is, although $D_n(\cdot)$ is arrived at through noncountable operations, it is still a r.v.
Hint: Define D_n^+ and D_n^- by: $D_n^+ = D_n^+(\cdot) \stackrel{def}{=} \sup_{x\in\Re}[F_n(x, \cdot) - F(x)]$, $D_n^- = D_n^-(\cdot) \stackrel{def}{=} \sup_{x\in\Re}[F(x) - F_n(x, \cdot)]$, so that $D_n = \max\{D_n^+, D_n^-\}$. Next, show that $D_n^+ = \max\{\max_{1\leq i\leq n}[\frac{i}{n} - F(y_i)], 0\}$, and $D_n^- = \max\{\max_{1\leq i\leq n}[F(y_i - 0) - \frac{i-1}{n}], 0\}$, where $y_i = x_{(i)}, i = 1, \ldots, n$, and $x_{(1)} \leq x_{(2)} \leq \cdots \leq x_{(n)}$ are the ordered x_i s, $x_i = X_i(\omega)$, and ω is an arbitrary but fixed $\omega \in \Omega$.

19. (*Glivenko–Cantelli*). Refer to Exercise 18 and show that

$$\sup\{|F_n(x, \omega) - F(x)|; x \in \Re\} \xrightarrow[n\to\infty]{\text{a.s.}} 0.$$

Hint: For $0 < p < 1$, define x_p by: $x_p = \inf\{x \in \Re; F(x) \geq p\}$, so that $F(x) \geq p$ for $x \geq x_p$, and $F(x) < p$ for $x < x_p$, which implies $F(x_p - 0) \leq p$. Next, replace p by i/k ($k \geq 2$ integer), $i = 0, 1, \ldots, k$, to get the points x_{ki}, with $-\infty \leq x_{k0} < x_{k1}$, and $x_{k,k-1} < x_{kk} \leq \infty$. Then, for $x \in [\frac{i}{k}, \frac{i+1}{k})$, $i = 0, 1, \ldots, k - 1$, it holds that

$$\frac{i}{k} \leq F(x_{ki}) \leq F(x) \leq F(x_{k,i+1} - 0) \leq \frac{i+1}{k},$$

so that $F(x_{k,i+1} - 0) - F(x_{ki}) \leq \frac{1}{k}$. Use this result and the nondecreasing property of F and F_n to obtain, for $x \in \Re$ and $i = 0, 1, \ldots, k$:

$$\begin{aligned} F_n(x) - F(x) &\leq [F_n(x_{k,i+1} - 0) - F(x_{k,i+1} - 0)] + \frac{1}{k}, \\ F_n(x) - F(x) &\geq [F_n(x_{ki}) - F(x_{ki})] - \frac{1}{k}, \end{aligned}$$

so that

$$\begin{aligned} |F_n(x) - F(x)| \leq \max\{&|F_n(x_{k,i+1} - 0) - F(x_{k,i+1} - 0)|, \\ &|F_n(x_{ki}) - F(x_{ki})|; i = 0, 1, \ldots, k - 1\} + \frac{1}{k}. \end{aligned}$$

Finally, take the sup over $x \in \Re$ (which leaves the right-hand side intact), and use the SLLN to each one of the (finitely many) terms on the right-hand side to arrive at the asserted conclusion.

This page is intentionally left blank

CHAPTER

Topics from the Theory of Characteristic Functions

11

This chapter is a rather extensive one consisting of nine sections. The main theme of the chapter is the introduction of the concept of a characteristic function (ch.f.) and the study of some of its properties, as well as some of the ways it is used for probabilistic purposes. It is to be emphasized that ch.f.s are not studied, to the extent they are, for their own sake; rather, they are looked upon as a powerful tool for the purpose of obtaining certain probability results.

A brief description of the sections is as follows. In the first section, the concept of the ch.f. of a d.f. is defined and some of its basic properties are established. In the following section, the so-called inversion formula is proven in several forms. The significance of this formula is that it allows recovery of the distribution by means of its ch.f. The application of this inversion formula is illustrated by way of two simple examples.

One of the basic convergences in probability is convergence in distribution, which, in practice, is not easy to check directly. The so-called Paul Lévy Continuity Theorem, which is the main result in Section 11.3, replaces convergence in distribution by convergence of ch.f.s; this latter convergence is much easier to handle. Convergence in distribution in higher than one-dimensional spaces is in essence replaced by convergence in distribution in the real line. This is done by way of the Cramér-Wold device, which is discussed in Section 11.4.

The convolution of two d.f.s, its interpretation, and several related results are discussed in Section 11.5, whereas some technical properties of ch.f.s are studied in the following section. An application of some of these results yields the Weak Law of Large Numbers and the Central Limit Theorem; this is done in Section 11.7. The basic result discussed in the next section is that, under certain regularity conditions, the moments of a distribution uniquely determine the distribution. For its rigorous justification certain concepts and results from complex analysis are required, which are dealt with in the final section of the chapter.

11.1 Definition of the Characteristic Function of a Distribution and Basic Properties

In all that follows, d.f.s are nonnegative, nondecreasing, right-continuous functions with finite variations; it is not assumed that the variations are necessarily bounded by 1 unless otherwise stated (see also Exercises 4 and 5 in Chapter 8).

An Introduction to Measure-Theoretic Probability, Second Edition. http://dx.doi.org/10.1016/B978-0-12-800042-7.00011-6

Definition 1. The characteristic function f of a d.f. F (in the sense of Definition 1 in Chapter 8; see also Remark 5 there) is, in general, a complex-valued function defined on $\Re$ by

$$f(t) = \int_{\Re} e^{itx} dF(x) = \int_{\Re} \cos tx dF(x) + i \int_{\Re} \sin tx dF(x). \tag{11.1}$$

The integration in (11.1) is to be understood either in the sense of Riemann–Stieltjes, or as integration with respect to the measure induced by F (see also Appendix B). The integral is well defined for all $t \in \Re$, since $\cos tx$ and $\sin tx$ are F-integrable. If F is the d.f. of a r.v. X, then (11.1) may be rewritten as

$$f_X(t) = \mathcal{E}e^{itX} = \mathcal{E}\cos tX + i\mathcal{E}\sin tX. \quad \blacksquare \tag{11.2}$$

Some basic properties of a ch.f. are gathered next in the form of a theorem.

Theorem 1.

(i) $|f(t)| \leq Var F, t \in \Re$, and $f(0) = Var F$. In particular, if $f(0) = 1$ and $0 \leq F(x) \leq 1$, then f is the ch.f. of a r.v.

(ii) f is uniformly continuous in $\Re$.

(iii) If f is the ch.f. of a r.v. X, then $f_{\alpha X+\beta}(t) = e^{i\beta t} f_X(\alpha t), t \in \Re$, where α and β are constants.

(iv) If f is the ch.f. of a r.v. X, then $f_{-X}(t) = \overline{f_X(t)}, t \in \Re$, where, for $z = x + iy (x, y \in \Re), \bar{z} = x - iy$.

(v) If for some positive integer n the nth moment $\mathcal{E}X^n$ is finite, then $\frac{d^n}{dt^n} f_X(t)|_{t=0} = i^n \mathcal{E}X^n$. $\blacksquare$

Remark 1. In the proof of the theorems, as well as in other cases, the following property is used:

$$\left| \int_{\Re} [g(x) + ih(x)] d\mu \right| \leq \int_{\Re} |g(x) + ih(x)| \, d\mu = \int_{\Re} [g^2(x) + h^2(x)]^{1/2} d\mu,$$

where g and h are real-valued functions, and $\int_{\Re} [g(x) + ih(x)] d\mu = \int_{\Re} g(x) d\mu + i \int_{\Re} h(x) d\mu$. Its justification is left as an exercise (see Exercise 1).

Proof of Theorem 1. For convenience omit $\Re$ in the integration. Then

(i) $|f(t)| = |\int e^{itx} dF(x)| \leq \int |e^{itx}| dF(x) = Var F$, and $f(0) = \int dF(x) = Var F$. If $f(0) = 1$, then $Var F = 1$, which together with $0 \leq F(x) \leq 1, x \in \Re$, implies $F(-\infty) = 0$, $F(\infty) = 1$, so that F is the d.f. of a r.v.

(ii) $|f(t+h) - f(t)| = |\int e^{i(t+h)x} dF(x) - \int e^{itx} dF(x)| = |\int [e^{i(t+h)x} - e^{itx}] dF(x)| = |\int [e^{itx}(e^{ihx} - 1)] dF(x)| \leq \int |e^{itx}(e^{ihx} - 1)| dF(x) = \int |e^{ihx} - 1| dF(x)$. Now $|e^{ihx} - 1| \leq 2$, which is independent of h and F-integrable. Furthermore, $e^{ihx} - 1 \to 0$ as $h \to 0$. Therefore the Dominated Convergence

Theorem applies and gives

$$\int |e^{ihx} - 1| dF(x) \to 0 \text{ as } h \to 0.$$

So $|f(t+h) - f(t)|$ is bounded by a quantity that is independent of t and $\to 0$ as $h \to 0$. This establishes uniform continuity for f.

(iii) $f_{\alpha X+\beta}(t) = \mathcal{E}e^{it(\alpha X+\beta)} = \mathcal{E}[e^{i\beta t}e^{i(\alpha t)X}] = e^{i\beta t}\mathcal{E}[e^{i(\alpha t)X}] = e^{i\beta t} f_X(\alpha t)$.

(iv) $f_{-X}(t) = \mathcal{E}e^{it(-X)} = \mathcal{E}e^{i(-t)X} = \mathcal{E}[\cos(-tX) + i\sin(-tX)] = \mathcal{E}[\cos(tX) - i\sin(tX)] = \mathcal{E}\cos tX - i\mathcal{E}\sin tX = \overline{\mathcal{E}\cos tX + i\mathcal{E}\sin tX} = \overline{f_X(t)}$.

(v) Consider, e.g., the interval $[-r, r]$ for some $r > 0$. Then, for $t \in [-r, r]$, $\frac{\partial}{\partial t}e^{itX} = iXe^{itX}$ exists, and $|\frac{\partial}{\partial t}e^{itX}| \leq |X|$, independent of t and integrable. Then, by Theorem 5 in Chapter 5,

$$\frac{d}{dt}f(t) = \frac{d}{dt}\int e^{itX}dF(x) = \int \left(\frac{\partial}{\partial t}e^{itX}\right) dF(x)$$
$$= i\int (Xe^{itX})dF(x),$$

and, in particular, $\frac{d}{dt}f(t)|_{t=0} = i\int (Xe^{itX})|_{t=0}dF(x) = i\mathcal{E}X$. The same applies for any k, $1 \leq k \leq n$, since $\frac{d^k}{dt^k}e^{itX} = i^kX^ke^{itX}$ exists, and $|\frac{d^k}{dt^k}e^{itX}| \leq |X^k|$, independent of t and integrable. In particular, $\frac{d^k}{dt^k}f(t)|_{t=0} = i^k\int X^k dF(x) = i^k\mathcal{E}X^k$. ■

11.2 The Inversion Formula

By means of relation (11.1), the d.f. F defines its ch.f. f. The converse is also true; i.e., if it is given that f is the ch.f. of a d.f. F, then F can be recovered by means of the so-called inversion formula. More precisely, we have the following result.

Theorem 2. Let F be a d.f. (not necessarily of a r.v.) and let f be its ch.f. Then

(i) For any $a, b \in \Re (a < b)$ and $T > 0$:

$$\frac{F(b) + F(b-)}{2} - \frac{F(a) + F(a-)}{2} = \lim_{T\to\infty}\frac{1}{2\pi}\int_{-T}^{T}\frac{e^{-ita} - e^{-itb}}{it}f(t)dt. \quad (11.3)$$

(ii) If $a, b \in C(F)$ and $T > 0$:

$$F(b) - F(a) = \lim_{T\to\infty}\frac{1}{2\pi}\int_{-T}^{T}\frac{e^{-ita} - e^{-itb}}{it}f(t)dt. \quad (11.4)$$

(iii) If X is a discrete r.v. taking on the value x_j with probability $P(X = x_j) = p(x_j)$, $j \geq 1$, then

$$p(x_j) = \lim_{T\to\infty} \frac{1}{2T} \int_{-T}^{T} e^{-itx_j} f(t)dt, \quad j \geq 1, T > 0. \qquad \blacksquare$$

In the course of the proof of the theorem, the following two facts will be needed, which are recorded here for easy reference.

Fact 1. The Dirichlet integrals $\int_a^b \frac{\sin x}{x} dx$ are bounded uniformly in $a, b \in \Re$.

Fact 2. $\int_{-\infty}^{0} \frac{\sin x}{x} dx = \int_0^{\infty} \frac{\sin x}{x} dx = \frac{\pi}{2}$. (See, e.g., integral 417 in Tallarida (1999).)
Also, the following remarks are in order.

Remark 2.

(i) As it will be seen in the proof of the theorem, the integrals on the right-hand sides of relations (11.3) and (11.4) are real.

(ii) It is to be noticed that in (11.3) and (11.4), we consider the Cauchy principal values of the integrals (i.e., integrals over intervals symmetric with respect to the origin); the integrals taken over arbitrary limits may fail to exist.

(iii) It is to be pointed out here that, whereas a d.f. F determines the corresponding ch.f. uniquely, the converse is not true. That is, a ch.f. determines a class of d.f.s through the difference $F(b) - F(a)$, $a, b \in C(F)$ (Theorem 2(ii)) rather than a unique d.f. Any two such d.f.s differ by a constant. Indeed, let F and G be two such d.f.s. Then, for x and a in $C(F) \cap C(G)$ with $a < x$, we have

$$F(x) - F(a) = G(x) - G(a) (= \text{the right-hand side in (11.4)}).$$

Letting $a \downarrow -\infty$ through $C(F) \cap C(G)$, we get $F(x) - F(-\infty) = G(x) - G(-\infty)$ or $F(x) - G(x) = F(-\infty) - G(-\infty)$, a constant for $x \in \Re$. Nevertheless, all these d.f.s define the same measure (Theorem 7 in Chapter 4). In particular, if f is the ch.f. of a r.v., then the corresponding measure is a probability measure (Theorem 1(i) here).

Proof of Theorem 2.

(i) Set

$$g(t) = \frac{e^{-ita} - e^{-itb}}{it}, \quad J(T) = \frac{1}{2\pi} \int_{-T}^{T} g(t) f(t)dt. \tag{11.5}$$

Expanding e^{-ita} and e^{-itb} around 0, up to terms of second order, dividing by $t(\neq 0)$, and taking the limit as $t \to 0$, we obtain $\lim_{t\to 0} g(t) = b - a$. Then define by continuity, $g(0) = b - a$. (See also Exercise 2.) Since

$$\frac{e^{-ita} - e^{-itb}}{it} = \int_a^b e^{-ity} dy \qquad \text{(both for } t \neq 0 \text{ and } t = 0\text{), and}$$

$$f(t) = \int_{-\infty}^{\infty} e^{itx} dF(x),$$

we have

$$J(T)=\frac{1}{2\pi}\int_{-T}^{T}\left(\int_a^b e^{-ity}dy\right)\left[\int_{-\infty}^{\infty}e^{itx}dF(x)\right]dt$$
$$=\frac{1}{2\pi}\int_{-T}^{T}\int_{-\infty}^{\infty}\int_a^b e^{it(x-y)}dydF(x)dt$$
$$=\frac{1}{2\pi}\int_{-\infty}^{\infty}\int_a^b\int_{-T}^{T}e^{it(x-y)}dtdydF(x), \tag{11.6}$$

where the change of the order of the integration is allowed by the Fubini Theorem (Theorem 12 in Chapter 5), since $|e^{it(x-y)}|=1$ is integrable with respect to the d.f. F (over $(-\infty,\infty)$), the Lebesgue measure (over (a,b)), and the Lebesgue measure again (over $(-T,T)$). Next, for $x\neq y$:

$$\int_{-T}^{T}e^{it(x-y)}dt=\frac{1}{i(x-y)}\int_{-T}^{T}de^{it(x-y)}$$
$$=\frac{1}{i(x-y)}[e^{iT(x-y)}-e^{-iT(x-y)}].$$

On the other hand, from $e^{iv}=\cos v+i\sin v$, $e^{-iv}=\cos v-i\sin v$, we get: $e^{iv}-e^{-iv}=2i\sin v$ or $\frac{1}{2i}(e^{iv}-e^{-iv})=\sin v$, so that

$$\frac{1}{i(x-y)}[e^{iT(x-y)}-e^{-iT(x-y)}]=2T\frac{\sin[T(x-y)]}{T(x-y)}.$$

Since $\frac{\sin x}{x}\to 1$ as $x\to 0$, $\frac{\sin x}{x}$ is defined to be 1 at $x=0$ by continuity. But for $x=y$: $\int_{-T}^{T}e^{it(x-y)}dt=\int_{-T}^{T}dt=2T$. Thus, for all x,y, relation (11.6) becomes

$$J(T)=\frac{T}{\pi}\int_{-\infty}^{\infty}\int_a^b\frac{\sin[T(x-y)]}{T(x-y)}dydF(x). \tag{11.7}$$

Next, consider the integral $\frac{T}{\pi}\int_a^b\frac{\sin[T(x-y)]}{T(x-y)}dy$ and set $T(x-y)=-u$, so that $dy=\frac{du}{T}$, $u:T(a-x),T(b-x)$, to get

$$\frac{T}{\pi}\int_a^b\frac{\sin[T(x-y)]}{T(x-y)}dy=\frac{1}{\pi}\int_{T(a-x)}^{T(b-x)}\frac{\sin u}{u}du\overset{def}{=}K_T(x), \tag{11.8}$$

so that by (11.7) and (11.8), we have

$$J(T)=\int_{-\infty}^{\infty}K_T(x)dF(x). \tag{11.9}$$

By Fact 1, $K_T(x)$ is bounded uniformly in T, a, b, and x, and its bound (a finite constant) is F-integrable. Next, by Fact 2,

$$\lim_{T\to\infty}K_T(x)=\left\{\begin{array}{l}0\text{ if }x<a\text{ or }x>b\\ 1\text{ if }a<x<b\\ \frac{1}{2}\text{ if }x=a\text{ or }x=b\end{array}\right\}\overset{def}{=}K(x). \tag{11.10}$$

By the boundedness of $K_T(x)$ by a constant (independent of T), F-integrable, and relation (11.10), the Dominated Convergence Theorem applies, and therefore (11.9) gives

$$
\begin{aligned}
\lim_{T\to\infty} J(T) &= \lim_{T\to\infty}\int_{-\infty}^{\infty} K_T(x)dF(x) = \int_{-\infty}^{\infty} K(x)dF(x)\\
&= \int_{-\infty}^{a-} K(x)dF(x) + \int_{a}^{a} K(x)dF(x) + \int_{a+}^{b-} K(x)dF(x)\\
&\quad + \int_{b}^{b} K(x)dF(x) + \int_{b+}^{\infty} K(x)dF(x)\\
&= 0 + \frac{1}{2}[F(a) - F(a-)] + 1\times[F(b-) - F(a)]\\
&\quad + \frac{1}{2}[F(b) - F(b-)] + 0\\
&= \frac{1}{2}[F(b) + F(b-)] - \frac{1}{2}[F(a) + F(a-)].
\end{aligned}
\tag{11.11}
$$

Then relations (11.5) and (11.11) complete the proof of part (i).

(ii) It follows from part (i), since $F(a) = F(a-)$ and $F(b) = F(b-)$.

(iii) We have

$$
\begin{aligned}
\int_{-T}^{T} e^{-itx_j} f(t)dt &= \int_{-T}^{T} e^{-itx_j}\left[\sum_k e^{itx_k} p(x_k)\right]dt\\
&= \int_{-T}^{T}\left[\sum_k e^{it(x_k-x_j)} p(x_k)\right]dt\\
&= \sum_k p(x_k)\int_{-T}^{T} e^{it(x_k-x_j)}dt,
\end{aligned}
$$

by the Fubini Theorem, because the integrand is integrable with respect to the counting measure (or the probability measure induced by $\{p(x_k)\}$; $k \geq 1$), and the Lebesgue measure over $[-T, T]$, and this is $= p(x_j)\times 2T + \sum_{k\neq j}\int_{-\infty}^{\infty} e^{it(x_k-x_j)}dt$. But for $x_k \neq x_j$,

$$
\begin{aligned}
\int_{-T}^{T} e^{it(x_k-x_j)}dt &= \int_{-T}^{T}\cos t(x_k - x_j)dt + i\int_{-T}^{T}\sin t(x_k - x_j)dt\\
&= \int_{-T}^{T}\cos t(x_k - x_j)dt\\
&= \frac{1}{(x_k - x_j)}\int_{-T}^{T} d\sin t(x_k - x_j)\\
&= \frac{\sin T(x_k - x_j) - \sin[-T(x_k - x_j)]}{x_k - x_j} = \frac{2\sin T(x_k - x_j)}{x_k - x_j}.
\end{aligned}
$$

Therefore, the original integral becomes

$$\begin{aligned}\frac{1}{2T}\int_{-T}^{T} e^{-itx_j} f(t)dt &= \frac{1}{2T} p(x_j) \times 2T \\ &+ \frac{1}{2T} \sum_{x_k \neq x_j} p(x_k) \frac{2\sin T(x_k - x_j)}{x_k - x_j} \\ &= p(x_j) + \sum_{x_k \neq x_j} p(x_k) \frac{\sin T(x_k - x_j)}{T(x_k - x_j)}.\end{aligned}$$

Set $T(x_k - x_j) = x$ and expand $\sin x$ around 0 up to terms of order one to obtain

$$\begin{aligned}\sin x &= \sin 0 + (x - 0)\cos x|_{x=x^*} \\ &\quad (\text{for some } x^* = \theta x,\ |\theta| \leq 1) \\ &= x\cos x^*, \quad \text{so that} \\ \frac{\sin T(x_k - x_j)}{T(x_k - x_j)} &= \frac{T(x_k - x_j)\cos x^*}{T(x_k - x_j)} = \cos x^*, \quad \text{and} \\ \left|\frac{\sin T(x_k - x_j)}{T(x_k - x_j)}\right| &= |\cos x^*| \leq 1\end{aligned}$$

independent of T and integrable (with respect to the measure $\{p(x_k), x_k \neq x_j\}$). Therefore, by the Dominated Convergence Theorem,

$$\begin{aligned}&\lim_{T\to\infty} \sum_{x_k \neq x_j} p(x_k) \frac{\sin T(x_k - x_j)}{T(x_k - x_j)} \\ &= \sum_{x_k \neq x_j} p(x_k) \lim_{T\to\infty} \frac{\sin T(x_k - x_j)}{T(x_k - x_j)} = 0 \\ &\left(\text{since } \left|\frac{\sin T(x_k - x_j)}{T(x_k - x_j)}\right| \leq \frac{1}{|T(x_k - x_j)|} \underset{T\to\infty}{\longrightarrow} 0\right)\end{aligned}$$

and hence $\lim_{T\to\infty} \frac{1}{2T}\int_{-T}^{T} e^{-itx_j} f(t)dt = p(x_j)$. ■

To this theorem, there are the following three corollaries.

Corollary 1. Let $a = x - h$ and $b = x + h$ $(h > 0)$ be continuity points of F. Then

$$F(x+h) - F(x-h) = \lim_{T\to\infty} \frac{1}{\pi} \int_{-T}^{T} \frac{\sin(th)}{t} e^{-itx} f(t)dt.$$

Proof. For this choice of a and b, we have

$$\begin{aligned}e^{-ita} - e^{-itb} &= e^{-it(x-h)} - e^{-it(x+h)} \\ &= e^{-itx}(e^{ith} - e^{-ith}) = e^{-itx} \times 2i\sin(th),\end{aligned}$$

so that
$$\frac{1}{2\pi}\int_{-T}^{T}\frac{e^{-ita}-e^{-itb}}{it}f(t)dt=\frac{1}{\pi}\int_{-T}^{T}\frac{\sin(th)}{t}e^{-itx}f(t)dt.$$
The result follows from part (ii). ■

Corollary 2. The d.f. F is differentiable and its derivative at a, $F'(a)=p(a)$, is given by
$$p(a)=\lim_{h\to 0}\lim_{T\to\infty}\frac{1}{2\pi}\int_{-T}^{T}\frac{1-e^{-ith}}{ith}e^{-ita}f(t)dt \tag{11.12}$$
if and only if the right-hand side in (11.12) exists.

Proof. In part (ii) of the theorem, set $b=a+h$ (with both a and $a+h$ continuity points of F, $h>0$). Then
$$e^{-ita}-e^{-itb}=e^{-ita}-e^{-it(a+h)}=e^{-ita}(1-e^{-ith}),$$
and hence
$$\frac{1}{h}[F(a+h)-F(a)]=\lim_{T\to\infty}\frac{1}{2\pi}\int_{-T}^{T}\frac{1-e^{-ith}}{ith}e^{-ita}f(t)dt. \tag{11.13}$$
Suppose first that $p(a)$ exists. Then, taking the limits in (11.13), as $h\to 0$, we have the desired result. Next, if the limit (as $h\to 0$) on the right-hand side of (11.13) exists, then $p(a)$ exists and is taken as stated. Similarly for $h<0$. ■

Corollary 3. If $\int_{-\infty}^{\infty}|f(t)|dt<\infty$, then the derivative $F'(x)=p(x)$ exists, is bounded and continuous in $\Re$, and is given by
$$p(x)=\frac{1}{2\pi}\int_{-\infty}^{\infty}e^{-itx}f(t)dt. \tag{11.14}$$

Proof. From the expansion $e^{-ith}=1-ithe^{z}$, with $z=-ith\theta$, θ real, $|\theta|\le 1$, we have $\frac{1-e^{-ith}}{ith}=e^{z}$, $(t\neq 0)$, and $\frac{1-e^{-ith}}{ith}$ is defined to be 1 for $t=0$ by continuity. Also, $\left|\frac{1-e^{-ith}}{ith}\right|\le|e^{z}|=1$. Thus
$$\int_{-T}^{T}\frac{1-e^{-ith}}{ith}e^{-ita}f(t)dt=\int_{-\infty}^{\infty}\frac{1-e^{-ith}}{ith}e^{-ita}f(t)I_{[-T,T]}(t)dt$$
with $\left|\frac{1-e^{-ith}}{ith}e^{-ita}f(t)I_{[-T,T]}(t)\right|\le|f(t)|$ independent of T and (Lebesgue-) integrable, and
$$\lim_{T\to\infty}\left\{\frac{1-e^{-ith}}{ith}e^{-ita}f(t)I_{[-T,T]}(t)\right\}=\frac{1-e^{-ith}}{ith}e^{-ita}f(t).$$
Then the Dominated Convergence Theorem applies and gives
$$\lim_{T\to\infty}\int_{-T}^{T}\frac{1-e^{-ith}}{ith}e^{-ita}f(t)dt=\int_{-\infty}^{\infty}\frac{1-e^{-ith}}{ith}e^{-ita}f(t)dt. \tag{11.15}$$

In the integrand on the right-hand side of (11.15), look at h as an index, and observe that $\left|\frac{1-e^{-ith}}{ith}e^{-ita}f(t)\right| \leq |f(t)|$ independent of h, (Lebesgue-) integrable, and $\frac{1-e^{-ith}}{ith}e^{-ita}f(t) \underset{h\to 0}{\to} e^{-ita}f(t)$. Therefore, the Dominated Convergence Theorem again gives

$$\lim_{h\to 0}\int_{-\infty}^{\infty}\frac{1-e^{-ith}}{ith}e^{-ita}f(t)dt = \int_{-\infty}^{\infty}e^{-ita}f(t)dt. \tag{11.16}$$

Thus, from (11.15) and (11.16),

$$\lim_{h\to 0}\lim_{T\to\infty}\frac{1}{2\pi}\int_{-T}^{T}\frac{1-e^{-ith}}{ith}e^{-ita}f(t)dt = \frac{1}{2\pi}\int_{-\infty}^{\infty}e^{-ita}f(t)dt. \tag{11.17}$$

Since the left-hand side of (11.17) is equal to $p(a)$ by (11.12), we have $p(a) = \frac{1}{2\pi}\int_{-\infty}^{\infty}e^{-ita}f(t)dt$. Replacing a by x, we have $p(x) = \frac{1}{2\pi}\int_{-\infty}^{\infty}e^{-itx}f(t)dt$. The boundedness of $p(x)$ follows by $p(x) = \left|\frac{1}{2\pi}\int_{-\infty}^{\infty}e^{-itx}f(t)dt\right| \leq \frac{1}{2\pi}\int_{-\infty}^{\infty}|f(t)|\,dt < \infty$, whereas continuity follows thus: $\left|\frac{1}{2\pi}e^{-itx}f(t)\right| \leq \frac{|f(t)|}{2\pi}$ independent of x, (Lebesgue-) integrable, and $\frac{1}{2\pi}e^{-itx}f(t) \underset{x\to x_0}{\to} \frac{1}{2\pi}e^{-itx_0}f(t)$. Then the Dominated Convergence Theorem completes the proof. ■

Here are two examples that illustrate how the inversion formula applies.

Example 1. Let the r.v. X be distributed as $B(n, p)$, so that its ch.f. is given by

$$f(t) = \sum_{k=0}^{n}e^{itk}\binom{n}{k}p^k q^{n-k} = \sum_{k=0}^{n}\binom{n}{k}(pe^{it})^k q^{n-k}$$
$$= (pe^{it}+q)^n,\ t \in \Re\ (q = 1-p).$$

Apply Theorem 2(iii) to recover $p(x)$, $x = 0, \ldots, n$. We have

$$\frac{1}{2T}\int_{-T}^{T}e^{-itx}f(t)dt = \frac{1}{2T}\int_{-T}^{T}e^{-itx}(pe^{it}+q)^n dt$$
$$= \frac{1}{2T}\int_{-T}^{T}\left[e^{-itx}\sum_{k=0}^{n}\binom{n}{k}(pe^{it})^k q^{n-k}\right]dt$$
$$= \frac{1}{2T}\int_{-T}^{T}\left[\sum_{k=0}^{n}\binom{n}{k}p^k q^{n-k}e^{i(k-x)t}\right]dt$$
$$= \frac{1}{2T}\sum_{k=0}^{n}\binom{n}{k}p^k q^{n-k}\int_{-T}^{T}e^{i(k-x)t}dt$$
$$= \frac{1}{2T}\sum_{k=0,k\neq x}^{n}\binom{n}{k}p^k q^{n-k}\int_{-T}^{T}e^{i(k-x)t}dt$$
$$+\frac{1}{2T}\binom{n}{x}p^x q^{n-x}\times 2T$$

$$= p(x) + \sum_{k=0,k\neq x}^{n} \binom{n}{k} p^k q^{n-k} \frac{e^{i(k-x)T} - e^{-i(k-x)T}}{2iT(k-x)}$$

$$= p(x) + \sum_{k=0,k\neq x}^{n} \binom{n}{k} p^k q^{n-k} \frac{\sin(k-x)T}{(k-x)T},$$

and by taking the limit, as $T \to \infty$, the second term on the right-hand side above tends to 0 (since $|\frac{\sin x}{x}| \leq \frac{1}{|x|} \to 0$ as $|x| \to \infty$). It follows that $\lim_{T\to\infty} \frac{1}{2T} \int_{-T}^{T} e^{-itx} f(t)dt = p(x)$ as in Theorem 2(iii). ■

Example 2. Let the r.v. X be distributed as $N(0, 1)$, so that its ch.f. is $f(t) = e^{-t^2/2}$. Since $\int_{\Re} |f(t)|dt = \int_{-\infty}^{\infty} e^{-t^2/2}dt < \infty$, Corollary 3 applies. We have then

$$\frac{1}{2\pi}\int_{-\infty}^{\infty} e^{-itx} f(t)dt = \frac{1}{2\pi}\int_{-\infty}^{\infty} e^{-itx} e^{-t^2/2}dt$$

$$= \frac{1}{2\pi}\int_{-\infty}^{\infty} e^{-(t^2+2itx)/2}dt$$

$$= \frac{1}{2\pi}\int_{-\infty}^{\infty} e^{-[t^2+2itx+(ix)^2]/2} e^{(ix)^2/2}dt$$

$$= \frac{e^{-x^2/2}}{2\pi}\int_{-\infty}^{\infty} e^{-(t+ix)^2/2}dt$$

$$= \frac{e^{-x^2/2}}{\sqrt{2\pi}}\int_{-\infty}^{\infty} \frac{1}{\sqrt{2\pi}} e^{-u^2/2}du$$

$$= \frac{e^{-x^2/2}}{\sqrt{2\pi}} \times 1 = \frac{1}{\sqrt{2\pi}} e^{-x^2/2} = p(x).$$ ■

11.3 Convergence in Distribution and Convergence of Characteristic Functions—The Paul Lévy Continuity Theorem

In this section, two versions of the Paul Lévy continuity theorem are stated and proved, after a number of auxiliary results have been established. The significance of this theorem is that convergence in distribution of a sequence of r.v.s to a r.v. is reduced to convergence of ch.f.s. Convergence in distribution is not easy to deal with, whereas convergence of ch.f.s is amenable to a large body of analytical facts and techniques.

Below, we define the so-called integral ch.f. of a d.f. F, which is needed as a tool in the sequel. To this end, let F be a d.f. (not necessarily of a r.v.) with a ch.f. f. In terms of f, we define the function $\hat{f}$ as follows:

$$\hat{f}(t) = \int_0^t f(v)dv, \quad t \in \Re. \tag{11.18}$$

Then

$$\hat{f}(t) = \int_0^t \left[\int_{\Re} e^{ivx} dF(x)\right] dv = \int_0^t \int_{\Re} e^{ivx} dF(x) dv = \int_{\Re} \int_0^t e^{ivx} dv\, dF(x)$$

(by the Fubini Theorem, since the integrand is integrable). Now, for $x \neq 0$, we have

$$\int_0^t e^{ivx} dv = \frac{1}{ix} \int_0^t de^{ivx} = \frac{1}{ix} e^{ivx}|_0^t = \frac{e^{itx} - 1}{ix},$$

while for $x = 0$, we have $\int_0^t dv = t$. Expanding $e^{(ix)t}$ around $t = 0$, we get

$$e^{itx} = 1 + itx + \frac{(itx)^2}{2} \times e^{it^* x}, \quad \text{for some } t^* \text{ with } |t^*| \leq 1,$$

so that

$$\frac{e^{itx} - 1}{ix} = t + (ix)\frac{t^2}{2} e^{it^* x}.$$

Thus, if $\frac{e^{itx}-1}{ix}$ is defined by continuity at $x = 0$ (which is what we usually do), then $\frac{e^{itx}-1}{ix}|_{x=0} = t$, and this is what we get as a value of $\int_0^t e^{ivx} dv$ for $x = 0$. So, for any x, we have then

$$\hat{f}(t) = \int_{\Re} \frac{e^{itx} - 1}{ix} dF(x). \tag{11.19}$$

Definition 2. The function $\hat{f}$ as defined in (11.18) or (11.19) is called the *integral ch.f.* of the d.f. F. ■

Remark 3. There is a one-to-one correspondence between f and $\hat{f}$. In fact, f uniquely determines $\hat{f}$ by means of (11.18). On the other hand (i.e., if we are told that $\hat{f}$ is the integral ch.f. of a ch.f. f and we want to recover f), since f is continuous, we have $\hat{f}'(t) = f(t)$, so that $\hat{f}$ uniquely determines f.

The simplest (and most commonly used) version of the continuity theorem is the following.

Theorem 3 (Paul Lévy Continuity Theorem). For $n \geq 1$, let F_n and F be d.f.s of r.v.s with respective ch.f.s f_n and f. Then, as $n \to \infty$,

(i) $F_n \Rightarrow F$ (or, equivalently, $F_n \xrightarrow{c} F$) implies $f_n \to f$ on $\Re$.

(ii) $f_n \to f$ on $\Re$ implies $F_n \Rightarrow F$ (or, equivalently, $F_n \xrightarrow{c} F$). ■

For the proof of part (ii), we need the following two auxiliary results.

Lemma 1. Let f_n and f be as in Theorem 3, and let $\hat{f}_n$ and $\hat{f}$ be the respective integral ch.f.s. Then, as $n \to \infty$, $f_n \to f$ on $\Re$ implies $\hat{f}_n \to \hat{f}$ on $\Re$.

Proof. For $t > 0$ (and similarly for $t < 0$), $|f_n(v)I_{[0,t]}(v)| \leq I_{[0,t]}(v)$ independent of n, (Lebesgue-) integrable, and $f_n(v)I_{[0,t]}(v) \underset{n\to\infty}{\to} f(v)I_{[0,t]}(v)$. Then, by the Dominated Convergence Theorem,

$$\int_0^t f_n(v)dv = \int_{\Re} f_n(v)I_{[0,t]}(v)dv \underset{n\to\infty}{\to} \int_{\Re} f(v)I_{[0,t]}(v)dv = \int_0^t f(v)dv,$$

or $\hat{f}_n \to \hat{f}$ on $\Re$. ∎

Lemma 2. Let F_n, F and f_n, f be as in Theorem 3, and let $\hat{f}_n$ and $\hat{f}$ be the respective integral ch.f.s of f_n and f. Then, as $n \to \infty$, $\hat{f}_n \to \hat{f}$ on $\Re$ implies $F_n \Rightarrow F$ (or, equivalently, $F_n \overset{c}{\to} F$).

In the course of the proof of this lemma, as well as elsewhere, the following elementary fact is employed, which is stated here as a remark.

Remark 4. If $\{z_n\}, n \geq 1$, is a sequence of (real or complex) numbers, then $z_n \to z_0$ as $n \to \infty$, if and only if for any subsequence $\{m\} \subseteq \{n\}$ there exists a further subsequence $\{r\} \subseteq \{m\}$ such that $z_r \to z_0$ as $r \to \infty$.

Proof of Lemma 2. In this proof, all limits are taken as $\{n\}$ or subsequencies thereof converge to ∞. To show that $F_n(x) \to F(x), x \in C(F)$. By Remark 4, it suffices to show that for every $\{m\} \subseteq \{n\}$ there exists $\{r\} \subseteq \{m\}$ such that $F_r(x) \to F(x), x \in C(F)$. Since $\{F_m\}$ (evaluated at $x \in C(F)$) is bounded (by 1), there exists $\{r\} \subseteq \{m\}$ such that $F_r \Rightarrow F_0$, some d.f. on $\Re$. (This is so by Theorem 5 in Chapter 8.) Clearly, $0 \leq F_0(x) \leq 1, x \in \Re$, and let f_0 and $\hat{f}_0$ be the ch.f. and the integral ch.f. of F_0, respectively. We have $\hat{f}_r(t) = \int_{\Re}\left[(e^{itx}-1)/ix\right]dF_r(x)$ with $(e^{itx}-1)/ix$ continuous (in x) over $\Re$ (for each arbitrary but fixed $t \in \Re$) and $|(e^{itx}-1)/ix| \leq 2/|x| \to 0$ as $|x| \to \infty$. Also, $F_r \Rightarrow F_0$. Then (by Theorem 7 in Chapter 8),

$$\int_{\Re} \frac{e^{itx}-1}{ix} dF_r(x) \to \int_{\Re} \frac{e^{itx}-1}{ix} dF_0(x),$$

or $\hat{f}_r \to \hat{f}_0$ on $\Re$.

However, $\hat{f}_n \to \hat{f}$ on $\Re$, so that $\hat{f}_r \to \hat{f}$ and $\hat{f}_0 = \hat{f}$ on $\Re$. It follows that $f_0 = f$ on $\Re$, and therefore $F_0 - F = c$, some constant c (see also Remark 2(iii)). We shall show that $c = 0$, which will establish the assertion that $F_n \Rightarrow F$. Indeed, from $f_0 = f$ on $\Re$, we have $f_0(0) = f(0)$, or $Var F_0 = Var F = 1$. So, $0 \leq F_0(x) \leq 1, x \in \Re$, and $Var F_0 = 1$. Then $F_0(-\infty) = 0$ and $F_0(\infty) = 1$. Finally, from $F_0(x) - F(x) = c$, we get, as $x \to -\infty$, $F_0(-\infty) - F(-\infty) = c$, or $0 - 0 = c$; i.e., $c = 0$ and $F_0 = F$. ∎

Proof of Theorem 3.

(i) For each arbitrary $t \in \Re$, e^{itx} is continuous (in x) and bounded (by 1). Since $F_n \overset{c}{\to} F$, Theorem 8 in Chapter 8 applies and gives that

$$f_n(t) = \int_{\Re} e^{itx} dF_n(x) \to \int_{\Re} e^{itx} dF(x) = f(t),$$

or $f_n \to f$ on $\Re$.

(ii) It follows from Lemmas 1 and 2.
(Or, in more detail, from $f_n \to f$ or $\Re$, we have

$$\hat{f}_n(t) = \int_0^t f_n(v)dv = \int_{\Re} \left[f_n(v) I_{[0,t]}(v) \right] dv$$

with $|f_n(v)I_{[0,t](v)}| \leq I_{[0,t]}(v)$ (since $|f_n(v)| \leq Var F_n = 1$) independent of n and (Lebesgue-) integrable. Also, $f_n(v)I_{[0,t]}(v) \to f(v)I_{[0,t]}(v)$. Hence

$$\hat{f}_n(t) = \int_{\Re} \left[f_n(v) I_{[0,t]}(v) \right] dv \to \int_{\Re} \left[f(v) I_{[0,t]}(v) \right] dv = \hat{f}(t);$$

i.e.; $\hat{f}_n \to \hat{f}$ on $\Re$. Next, in order to show that $F_n \Rightarrow F$, suffices to show that for every $\{F_m\} \subseteq \{F_n\}$, there is $\{F_r\} \subseteq \{F_m\}$ such that $F_r \Rightarrow F$. To this end, let $\{F_m\} \subseteq \{F_n\}$. Then there is $\{F_r\} \subseteq \{F_m\}$ such that $F_r \to F_0$, some d.f. This is so by Theorem 6 in Chapter 8. Let f_0 and $\hat{f}_0$ be the ch.f. and the integral ch.f., respectively, of F_0. Then $F_r \Rightarrow F_0$ implies $\hat{f}_r \to \hat{f}_0$ on $\Re$, by part (i). However, $\hat{f}_n \to \hat{f}$, by what it was just proved. Hence $\hat{f}_r \to \hat{f}$, and therefore $\hat{f} = \hat{f}_0$, which implies $f = f_0$. But then $F_0 - F = c$, a constant. To show that $c = 0$. Indeed, $0 \leq F_0 \leq 1$ (since $0 \leq F_n \leq 1$ for all n), and $1 = f(0) = f_0(0)$ implies that $Var F_0 = 1$. Then $F_0(-\infty) = 0$. Since $F_0(x) - F(x) = c, x \in \Re$, letting $x \downarrow -\infty$, we get $0 - 0 = c$, so that $F_0 = F$, and therefore $F_n \Rightarrow F$.) ■

A version of Theorem 3, whose part (ii) is seemingly less restrictive than part (ii) of Theorem 3, is the following.

Theorem 3* (Paul Lévy Continuity Theorem). For $n \geq 1$, let F_n be d.f.s of r.v.s with respective ch.f.s f_n. Then

(i) If $F_n \Rightarrow F$, a d.f. of a r.v. with ch.f. f, it follows that $f_n \to f$ on $\Re$, as $n \to \infty$.
(ii) Let $f_n \underset{n\to\infty}{\to} g$, some function on $\Re$ continuous at the origin. Then $F_n \underset{n\to\infty}{\Longrightarrow} F$, where F is a (uniquely determined) d.f. of a r.v. ■

Remark 5. Clearly, part (i) is the same in both Theorems 3 and 3*. In part (ii) of Theorem 3*, it is not required that the limit g of $\{f_n\}$ be a ch.f. at all (even less the ch.f. of a r.v.), but it turns out that g is, indeed, the ch.f. of a r.v. and that $F_n \underset{n\to\infty}{\Longrightarrow} F$, the uniquely determined d.f. corresponding to f.

For the proof of part (ii) of Theorem 3*, we need the following auxiliary results.

Lemma 3. Let f_n and g be as in Theorem 3*, and let $\hat{f}_n$ be the integral ch.f. corresponding to f_n. Then, as $n \to \infty$, $f_n \to g$ on $\Re$ implies that $\hat{f}_n \to \hat{g}$ on $\Re$, where $\hat{g}(t) = \int_0^t g(v)dv$ (where the integral is to be understood in the Lebesgue sense).

Proof. It is the same as that of Lemma 1, where f and $\hat{f}$ are replaced by g and $\hat{g}$, respectively. ■

Before the second auxiliary result is formulated, we have to introduce the concept of weak (and complete) convergence of a sequence of d.f.s to a d.f. up to an additive constant. At this point, recall that, if $F_n, n \geq 1$, and F are d.f.s (not necessarily of r.v.s), then weak convergence $F_n \Rightarrow F$ (as $n \to \infty$) means $F_n(x) \to F(x), x \in C(F)$. In this section, the need for a modification of this convergence arises. This is due to the fact that there is no one-to-one correspondence between d.f.s and the corresponding ch.f.s (although there is such one-to-one correspondence between the measures induced by the d.f.s and the respective ch.f.s). From relation (11.1), it follows that it is, actually, an entire class of d.f.s that induce the same ch.f.; namely, a d.f. F and any d.f. $F+c, c \in \Re$, constant, for which $F(x)+c \geq 0$, and any two members of this class differ by a constant. On the other hand, by means of the inversion formula, a ch.f. defines a class of d.f.s any two of which differ by a constant. So, for a d.f. F, all d.f.s $F+c, c \in \Re$ such that $F(x)+c \geq 0, x \in \Re$, have the same ch.f., and a given ch.f. f determines a class of d.f.s of the form just described. These observations lead to the following definition.

Definition 3. For $n \geq 1$, let F_n and F be d.f.s (not necessarily of r.v.s). We say that $\{F_n\}$ *converges weakly to* F *up to an additive constant*, and we write $F_n \underset{n\to\infty}{\Longrightarrow} F$ *uac*, if for every $\{n_1\} \subseteq \{n\}$ and $\{n_2\} \subseteq \{n\}$ with $F_{n_1} \underset{n\to\infty}{\Longrightarrow} F_1$ and $F_{n_2} \underset{n\to\infty}{\Longrightarrow} F_2$, d.f.s, it holds that $F_1 - F_2 = c$, some constant. Also, we say that $\{F_n\}$ *converges completely to* F *uac*, and write $F_n \overset{c}{\underset{n\to\infty}{\to}} F$ *uac*, if $F_n \underset{n\to\infty}{\Longrightarrow} F$ *uac* and $Var\,F_n \underset{n\to\infty}{\longrightarrow} Var\,F$. (The convergence $F_n \overset{c}{\underset{n\to\infty}{\longrightarrow}} F$, trivially, implies $F_n \overset{c}{\underset{n\to\infty}{\longrightarrow}} F$ *uac*.) ■

Remark 6.

- **(i)** If F_0 is the limiting d.f. of a (weakly) converging subsequence of $\{F_n\}, n \geq 1$, then any other limiting d.f. F is of the form $F = F_0 + c$, some $c \in \Re$.
- **(ii)** Let F_0 and F be as in part (i). Then, if $F(-\infty) = F_0(-\infty) = 0$, it follows that $F(x) = F_0(x), x \in \Re$. Consequently, if F_0 is the d.f. of a r.v., then so is F. Indeed, $F(x) = F_0(x) + c$, and as $x \to -\infty$, $F(-\infty) = F_0(-\infty) + c$, or $0 = 0 + c$, and $c = 0$.
- **(iii)** Clearly, all the limiting d.f.s F determine the same ch.f. f.
- **(iv)** If $F_n \underset{n\to\infty}{\Longrightarrow} F$, then, trivially, $F_n \underset{n\to\infty}{\Longrightarrow} F$ *uac*, and if $F_n \overset{c}{\underset{n\to\infty}{\longrightarrow}} F$, then $F_n \overset{c}{\underset{n\to\infty}{\longrightarrow}} F$ *uac*.

Lemma 4. For $n \geq 1$, let F_n be d.f.s of r.v.s with respective ch.f.s and integral ch.f.s f_n and $\hat{f}_n$. Let $\hat{g}$ be as in Lemma 3 (i.e., $\hat{g}(t) = \int_0^t g(v)dv, t \in \Re$, with g defined on $\Re$ and being continuous at the origin), so that $\hat{f}_n \to \hat{g}$ in $\Re$. Then $F_n \underset{n\to\infty}{\Longrightarrow} F$ *uac*, some d.f. F with ch.f. f, and $\hat{f} = \hat{g}$ on $\Re$.

Proof. In the proof, all limits are taken as $\{n\}$ or subsequences thereof tend to ∞. Let $\{n_1\}$ and $\{n_2\}$ be any subsequences of $\{n\}$ such that $F_{n_1} \Rightarrow F_1$ and $F_{n_2} \Rightarrow F_2$, some d.f.s. (Such subsequences exist, by Theorem 5 in Chapter 8, since the F_ns are bounded (by 1).) Let f_1, f_2, and $\hat{f}_1, \hat{f}_2$ be the respective ch.f.s and integral ch.f.s of F_1 and F_2. As in the proof of Lemma 2, $\hat{f}_{n_i} \to \hat{f}_i$ on $\Re, i = 1, 2$, and since

$\hat{f}_n \to \hat{g}$ on $\Re$, we conclude that $\hat{f}_1 = \hat{f}_2 (= \hat{g})$ on $\Re$, and hence $f_1 = f_2$ on $\Re$. Thus, all (weakly) convergent subsequences of $\{F_n\}$ determine the same ch.f., call it f. Therefore $F_1 - F_2 = c$, so that $F_n \Rightarrow F$ *uac*, where the d.f. F is in the class of d.f.s determined by f. Finally, $\hat{f} = \hat{f}_1 = \hat{f}_2 = \hat{g}$ from above. ∎

Proposition 1. For $n \geq 1$, let F_n be d.f.s of r.v.s with respective ch.f.s and integral ch.f.s f_n and $\hat{f}_n$, let g be a function on $\Re$ continuous at the origin, and let $\hat{g}(t) = \int_0^t g(v)dv, t \in \Re$ (where the integral is to be understood in the Lebesgue sense). Suppose that $f_n \underset{n\to\infty}{\longrightarrow} g$ on $\Re$. Then $F_n \underset{n\to\infty}{\Longrightarrow} F$, a (uniquely determined) d.f. of a r.v.

Proof. With all limits taken as $n \to \infty$, we have that $f_n \to g$ on $\Re$ implies $\hat{f}_n \to \hat{g}$ on $\Re$ (by Lemma 3), and this, in turn, implies (by Lemma 4) that $F_n \Rightarrow F$ *uac*, some d.f. with ch.f. f and $\hat{f} = \hat{g}$. From $\hat{f} = \hat{g}$ on $\Re$, we have

$$\int_0^t f(v)dv = \int_0^t g(v)dv, t \in \Re, \quad \text{or} \quad \frac{1}{t}\int_0^t f(v)dv = \frac{1}{t}\int_0^t g(v)dv, t \neq 0.$$

Taking the limits, as $t \to 0$, we get $f(0) = g(0)$ (by continuity at 0; see also Exercise 3). From $f_n \to g$ on $\Re$, we have $1 = f_n(0) \to g(0)$, so that $g(0) = f(0) = 1$. The d.f.s F_1 and F_2 in Lemma 4 take values in $[0, 1]$, as they are limits of such sequences. So, $0 \leq F_i(x) \leq 1, x \in \Re$, and $Var F_i = f(0) = 1, i = 1, 2$. It follows $F_i(-\infty) = 0$, so that $F_1(x) - F_2(x) = c$ yields, as $x \to -\infty, 0 - 0 = c$. Then the limiting d.f. F in Lemma 4 is uniquely determined, and is the d.f. of a r.v. ∎

Proof of Theorem 3*. All we have to do is to justify part (ii). However, this is the conclusion of Proposition 1. ∎

Convergence in distribution is preserved under continuity, as the following example shows.

Example 3. Let $X_1, X_2, \ldots,$ and X be r.v.s such that $X_n \underset{n\to\infty}{\overset{d}{\longrightarrow}} X$, and let $g : \Re \to \Re$ be continuous. Then $g(X_n) \underset{n\to\infty}{\overset{d}{\longrightarrow}} g(X)$.

Indeed, by Theorem 3, it suffices to show that $f_{g(X_n)}(t) \underset{n\to\infty}{\longrightarrow} f_{g(X)}(t)$. However,

$$\begin{aligned} f_{g(X_n)}(t) &= \mathcal{E}e^{itg(X_n)} = \mathcal{E}\cos[tg(X_n)] + i\mathcal{E}\sin[tg(X_n)] \\ &= \int_\Re \cos[tg(x)]dF_{X_n}(x) + i\int_\Re \sin[tg(x)]dF_{X_n}(x) \\ &\underset{n\to\infty}{\longrightarrow} \int_\Re \cos[tg(x)]dF_X(x) + i\int_\Re \sin[tg(x)]dF_X(x) \\ &\quad \text{(by Theorem 8 in Chapter 8,} \\ &\quad \text{since the integrands are bounded and continuous on } \Re) \\ &= \mathcal{E}e^{itg(X)} = f_{g(X)}(t). \end{aligned}$$

The foregoing convergence is valid, because $F_{X_n} \underset{n\to\infty}{\overset{c}{\longrightarrow}} F_X$, and $\cos[tg(x)]$ and $\sin[tg(x)]$ are bounded and continuous on $\Re$, so that Theorem 8 in Chapter 8 applies.

As a simple application, we have that $X_n \xrightarrow[n\to\infty]{d} Z \sim N(0,1)$ implies $X_n^2 \xrightarrow[n\to\infty]{d} Z^2$ whose distribution is the so-called chi-square with one degree of freedom, $Z^2 \sim \chi_1^2$.

It is of some importance to observe that suitable versions of the results incorporated in the Lemmas 1, 2, 3, 4, and Proposition 1 also hold under less restrictive conditions. This is the content of the following proposition.

Proposition 2. With $n \geq 1$, let F_n be (uniformly) bounded d.f.s (not necessarily of r.v.s) with ch.f.s f_n and integral ch.f.s $\hat{f}_n$. Then

(i) If $F_n \underset{n\to\infty}{\Longrightarrow} F$ *uac*, some d.f. F with ch.f. f and integral ch.f. $\hat{f}$, it follows that $\hat{f}_n \underset{n\to\infty}{\to} \hat{f}$ on $\mathfrak{R}$.

(ii) If $\hat{f}_n \underset{n\to\infty}{\to} \hat{g}$, some function on $\mathfrak{R}$, it follows that there exists a d.f. F (not necessarily of a r.v.) with ch.f. f and integral ch.f. $\hat{f}$ such that $F_n \underset{n\to\infty}{\Longrightarrow} F uac$ and $\hat{f} = \hat{g}$ on $\mathfrak{R}$.

Proof. With all limits taken as $\{n\}$ or subsequences thereof tend to ∞, we have

(i) In order to show that $\hat{f}_n \to \hat{f}$ on $\mathfrak{R}$, it suffices to show (by Remark 4) that for every $\{m\} \subseteq \{n\}$ there exists $\{r\} \subseteq \{m\}$ such that $\hat{f}_r \to \hat{f}$ on $\mathfrak{R}$. Indeed, by looking at $\{F_m\}$, there exists a subsequence $\{F_r\} \subseteq \{F_m\}$ (by the uniform boundedness of the F_ns and Theorem 5 in Chapter 8) such that $F_r \Rightarrow F$, some d.f. with ch.f. f and integral ch.f. $\hat{f}$. Then

$$\hat{f}_r(t) = \int_{\mathfrak{R}} \frac{e^{itx}-1}{ix} dF_r(x) \to \int_{\mathfrak{R}} \frac{e^{itx}-1}{ix} dF(x) = \hat{f}(r),$$

as in the proof of Lemma 2, or $\hat{f}_r \to \hat{f}$ on $\mathfrak{R}$. It follows that $\hat{f}_n \to \hat{f}$ on $\mathfrak{R}$.

(ii) For any $\{n_1\} \subseteq \{n\}$ and $\{n_2\} \subseteq \{n\}$ with $F_{n_1} \Rightarrow F_1$ and $F_{n_2} \Rightarrow F_2$, some d.f.s F_1 and F_2, one has (by part (i)) that $\hat{f}_{n_1} \to \hat{f}_1$ and $\hat{f}_{n_2} \to \hat{f}_2$ on $\mathfrak{R}$, where $\hat{f}_1$ and $\hat{f}_2$ are the integral ch.f.s corresponding to F_1 and F_2. Since also $\hat{f}_{n_1} \to \hat{g}$ and $\hat{f}_{n_2} \to \hat{g}$ on $\mathfrak{R}$, it follows that $\hat{f}_1 = \hat{f}_2 (= \hat{g})$, and hence $f_1 = f_2$ for the respective ch.f.s. Then $F_1 - F_2 = c$, some constant c, and hence $F_n \Rightarrow F$ *uac*, some d.f. F with ch.f. $f (= f_1 = f_2)$. ■

Corollary. For $n \geq 1$, let F_n, f_n and $\hat{f}_n$ be as in the proposition, and suppose that $f_n \underset{n\to\infty}{\to} g$ a.e., (with respect to Lebesgue measure) on $\mathfrak{R}$. Then $F_n \underset{n\to\infty}{\Longrightarrow} F uac$, some d.f. F (not necessarily of the r.v.) with ch.f. f, and $f = g$ a.e.,

Proof. For $n \geq 1$ and $t \in \mathfrak{R}$, $|f_n(t)|$ are uniformly bounded by a constant, and $f_n \to g$ a.e., where here and in the sequel all limits are taken as $n \to \infty$. Then

$$\int_0^t f_n(v)dv \to \int_0^t g(v)dv, \quad \text{or} \quad \hat{f}_n(t) \to \hat{g}(t) \overset{\text{def}}{=} \int_0^t g(v)dv, t \in \mathfrak{R}.$$

Then, by part (ii) of the proposition, $F_n \Rightarrow F$ *uac*, some d.f. F with ch.f. f and integral ch.f. $\hat{f}$, and $\hat{f} = \hat{g}$ on $\Re$. From

$$\hat{f}(t) = \int_0^t f(v)dv \quad \text{and} \quad \hat{g}(t) = \int_0^t g(v)dv,$$

it follows that $\hat{f}' = f$ on $\Re$, and $\hat{g}' = g$ a.e., (see, e.g., Theorem 10 on page 107 of Royden (1988)). However, $\hat{f} = \hat{g}$ on $\Re$. Hence $f = g$ a.e. ■

Proposition 3. With $n \geq 1$, let F_n be (uniformly) bounded d.f.s (not necessarily of r.v.s) with ch.f.s f_n and integral ch.f.s $\hat{f}_n$. Then

(i) If $F_n \underset{n\to\infty}{\overset{c}{\to}} F$ *uac*, some d.f. F with ch.f. f, it follows that $f_n \underset{n\to\infty}{\to} f$ on $\Re$.

(ii) If $f_n \underset{n\to\infty}{\to} g$, some function on $\Re$ continuous function at the origin, it follows that there exists a d.f. F (not necessarily of a r.v.) with ch.f. f such that $F_n \underset{n\to\infty}{\overset{c}{\to}} F$ *uac* and $f = g$ on $\Re$.

Proof. With all limits taken as $\{n\}$ or subsequences thereof tending to ∞, we have

(i) It suffices to prove that, for every $\{m\} \subseteq \{n\}$, there exists $\{r\} \subseteq \{m\}$ such that $f_r \to f$ on $\Re$. Looking at $\{F_m\}$, there exists $\{F_r\} \subseteq \{F_m\}$ such that $F_r \Rightarrow F$, some d.f. F with ch.f. f. Since for each arbitrary and fixed $t \in \Re$, e^{itx} is bounded and continuous in $\Re$ (as a function of x), it follows (by Theorem 8 in Chapter 8) that

$$f_r(t) = \int_{\Re} e^{itx} dF_r(x) \to \int_{\Re} e^{itx} dF(x) = f(t), t \in \Re,$$

so that $f_n \to f$ on $\Re$.

(ii) For $n \geq 1$ and $t \in \Re$, $|f_n(t)|$ are uniformly bounded by a constant, and $f_n \to g$ on $\Re$. Then

$$\hat{f}_n(t) = \int_0^t f_n(v)dv \to \int_0^t g(v)dv \overset{\text{def}}{=} \hat{g}(t), t \in \Re.$$

Therefore, by part (ii) of Proposition 2, it follows that there exists a d.f. F (not necessarily of a r.v.) with ch.f. f and integral ch.f. $\hat{f}$ such that $F_n \Rightarrow F$ *uac* and $\hat{f} = \hat{g}$ on $\Re$. That is,

$$\int_0^t f(v)dv = \int_0^t g(v)dv, \ t \in \Re, \text{ or } \frac{1}{t}\int_0^t f(v)dv = \frac{1}{t}\int_0^t g(v)dv, t \neq 0.$$

By taking the limits as $t \to 0$, we have then (see also Exercise 3) $f(0) = g(0)$. Since $f_n \to g$ on $\Re$, we have $f_n(0) \to g(0)$, or $Var F_n = f_n(0) \to g(0) = f(0) = Var F$. Therefore $F_n \overset{c}{\to} F$ *uac*. ■

11.4 Convergence in Distribution in the Multidimensional Case—The Cramér–Wold Device

For $k \geq 2$, one may define a k-dimensional d.f. F and establish properties similar to those stated and proved in Chapter 8 for a one-dimensional d.f. Also, one may define its ch.f. and establish properties and results analogous to the ones shown in the first two sections of this chapter. However, we will only restrict ourselves to the definition of the d.f. of a k-dimensional random vector and its ch.f., and state two theorems analogous to Theorems 2 and 3.

Definition 4.

(i) The *d.f. of a k-dimensional random vector* $\mathbf{X} = (X_1, \ldots, X_k)$, or the *joint d.f.* of the r.v.s $X_1, \ldots, X_k$, is defined by

$$F_{\mathbf{X}}(\mathbf{x}) = F_{X_1,\ldots,X_k}(x_1, \ldots, x_k)$$
$$= P(X_1 \leq x_1, \ldots, X_k \leq x_k), \quad x_1, \ldots, x_k \in \Re.$$

(ii) The *ch.f. of the random vector* $\mathbf{X}$, or the *joint ch.f.* of the r.v.s $X_1, \ldots, X_k$, is defined by

$$f_{\mathbf{X}}(\mathbf{t}) = \mathcal{E}e^{\mathbf{tX}'} = \mathcal{E}e^{t_1 X_1 + \cdots + t_k X_k},$$

$t_1, \ldots, t_k \in \Re$, where "$'$" denotes transpose.

(iii) For $n \geq 1$, let F_n be the d.f. of the k-dimensional random vector $\mathbf{X}_n = (X_{1n}, \ldots, X_{kn})$, and let F be the d.f. of the random vector $\mathbf{X} = (X_1, \ldots, X_k)$. Then $\mathbf{X}_n \underset{n\to\infty}{\overset{d}{\to}} \mathbf{X}$ or $F_n \underset{n\to\infty}{\Longrightarrow} F$, if $F_{\mathbf{X}_n}(\mathbf{x}) \underset{n\to\infty}{\to} F_{\mathbf{X}}(\mathbf{x})$ for all continuity points $\mathbf{x}$ of $F_{\mathbf{X}}$. ■

A version of Theorem 2 for the k-dimensional case reads as follows.

Theorem 2′. Let $\mathbf{X}$ be a k-dimensional random vector, $\mathbf{X} = (X_1, \ldots, X_k)$, with d.f. F and ch.f. f. Then, for continuity points $\mathbf{a} = (a_1, \ldots, a_k)$ and $\mathbf{b} = (b_1, \ldots, b_k)$ of F, it holds that

$$P(a_j < X_j \leq b_j, \quad j = 1, \ldots, k)$$
$$= \lim \left(\frac{1}{2\pi}\right)^k \int_{-T_k}^{T_k} \cdots \int_{-T_1}^{T_1} \prod_{j=1}^{k} \frac{e^{-it_j a_j} - e^{-it_j b_j}}{it_j} f(t_1, \ldots, t_k) \times$$
$$dt_1 \cdots dt_k, \text{ as } (0 <) T_j \to \infty, j = 1, \ldots, k.$$ ■

Also, a version of Theorem 3 is as follows.

Theorem 3′. For $n \geq 1$, let F_n be d.f.s of k-dimensional random vectors with ch.f.s f_n. Suppose $F_n \underset{n\to\infty}{\Longrightarrow} F$, a d.f. of a k-dimensional random vector with ch.f. f. Then $f_n \underset{n\to\infty}{\to} f$ on $\Re^k$.

Conversely, if $f_n \underset{n\to\infty}{\to} f$, on $\Re^k$, then $F_n \underset{n\to\infty}{\Longrightarrow} F$. (Thus, if $F_n, n \geq 1$, and F are d.f.s of k-dimensional random vectors with respective ch.f.s f_n and f, then in order to show that $F_n \underset{n\to\infty}{\Rightarrow} F$, it suffices to show that $f_n \underset{n\to\infty}{\to} f$ on $\Re^k$.) ■

However, a certain device, stated as the Cramér–Wold Theorem next, although it makes use of Theorem 3′, reduces the actual proof of weak convergence for the k-dimensional case to the one-dimensional case.

Theorem 4 (Cramér–Wold). Let $\mathbf{X}_n, \mathbf{X}$ be k-dimensional random vectors with respective d.f.s and ch.f.s F_n, F and f_n, f. Then $\mathbf{X}_n \underset{n\to\infty}{\overset{d}{\to}} \mathbf{X}$ (i.e., $F_n \underset{n\to\infty}{\Longrightarrow} F$) if and only if, for any $c_j \in \Re$, $j = 1, \ldots, k$,

$$\sum_{j=1}^{k} c_j X_{jn} \underset{n\to\infty}{\overset{d}{\to}} \sum_{j=1}^{k} c_j X_j, \quad \text{or} \quad \mathbf{c}\,\mathbf{X}'_n \underset{n\to\infty}{\overset{d}{\to}} \mathbf{c}\,\mathbf{X}',$$

where $\mathbf{X}_n = (X_{1n}, \ldots, X_{kn})$, $\mathbf{X} = (X_1, \ldots, X_k)$, and $\mathbf{c} = (c_1, \ldots, c_k)$. ■

Proof. With all limits taken as $n \to \infty$, let $\mathbf{X}_n \overset{d}{\to} \mathbf{X}$, equivalently, $F_n \Rightarrow F$. Then, by Theorem 3′, $f_n \to f$ on $\Re^k$, or

$$\mathcal{E}e^{it_1X_{1n}+\cdots+it_kX_{kn}} \to \mathcal{E}e^{it_1X_1+\cdots+it_kX_k}, \quad t_j \in \Re^k, \ j = 1, \ldots, k.$$

For any $c_j \in \Re$, take $t_j = c_j t$, $j = 1, \ldots, k$, any $t \in \Re$. Then the preceding relation is rewritten thus:

$$\mathcal{E}e^{i(c_1X_{1n})t+\cdots+i(c_kX_{kn})t} \to \mathcal{E}e^{i(c_1X_1)t+\cdots+i(c_kX_k)t},$$

or

$$\mathcal{E}e^{i(c_1X_{1n}+\cdots+c_kX_{kn})t} \to \mathcal{E}e^{i(c_1X_1+\cdots+c_kX_k)t},$$

and then, by Theorem 3, $\sum_{j=1}^{k} c_j X_{jn} \overset{d}{\to} \sum_{j=1}^{k} c_j X_j$.

Next, let $\sum_{j=1}^{k} c_j X_{jn} \overset{d}{\to} \sum_{j=1}^{k} c_j X_j$ for any $c_j \in \Re$, $j = 1, \ldots, k$. Then, by Theorem 3,

$$\mathcal{E}e^{i(c_1X_{1n}+\cdots+c_kX_{kn})t} \to \mathcal{E}e^{i(c_1X_1+\cdots+c_kX_k)t}, \quad t \in \Re.$$

Since for $t = 0$, both sides of this equation are equal to 1, suppose that $t \neq 0$, and for any $t_j \in \Re$, take $c_j = \frac{t_j}{t}$, $j = 1, \ldots, k$. Since $c_j t = t_j$, $j = 1, \ldots, k$, the last expression becomes

$$\mathcal{E}e^{it_1X_{1n}+\cdots+it_kX_{kn}} \to \mathcal{E}e^{it_1X_1+\cdots+it_kX_k},$$

and this implies that $\mathbf{X}_n \overset{d}{\to} \mathbf{X}$ by Theorem 3′. ■

11.5 Convolution of Distribution Functions and Related Results

The d.f.s to be considered in this section are as described in Section 11.1 of this chapter, unless otherwise specified. To this end, let F_1 and F_2 be two d.f.s, and define the function J by

$$J(x) = \int_{\Re} F_1(x - y)\,dF_2(y), \quad x \in \Re. \tag{11.20}$$

Definition 5. The function J defined by (11.20) is called the *convolution* (or *composition*) of the d.f.s F_1 and F_2 and is denoted by $J = F_1 * F_2$. ■

Theorem 5. Let J be defined by (11.20). Then J is a d.f., and, in particular, it is a d.f. of a r.v. if F_1 and F_2 are d.f.s of r.v.s. ■

Proof. Clearly, $J(x) \geq 0$ on $\Re$, and

$$J(x) = \int_{\Re} F_1(x-y)dF_2(y) \leq C_1 \int_{\Re} dF_2(y) = C_1 V_2, \tag{11.21}$$

where C_1 is a bound for F_1 and V_2 is the variation of F_2. It is nondecreasing, since for $x_1 > x_2$,

$$J(x_1) - J(x_2) = \int_{\Re} [F_1(x_1 - y) - F_1(x_2 - y)]dF_2(y) \geq 0,$$

as

$$x_1 > x_2 \text{ implies } F_1(x_1 - y) \geq F_1(x_2 - y).$$

Hence $J(x_1) \geq J(x_2)$. Next, J is continuous from the right. In fact, let $x_0 \in \Re$ and let $x \downarrow x_0$. Then $F_1(x-y) \downarrow F_1(x_0 - y)$ while $F_1(x-y) \leq C_1$, independent of x, and integrable. Hence, the Dominated Convergence Theorem gives

$$J(x) = \int_{\Re} F_1(x-y)dF_2(y) \to \int_{\Re} F_1(x_0 - y)dF_2(y) = J(x_0).$$

Finally, if F_1 and F_2 are d.f.s of r.v.s, then $J(x) \leq 1$, by (11.21), and $J(-\infty) = 0$, $J(\infty) = 1$. In fact, as $x \to -\infty$, then $F_1(x-y) \to 0$, $y \in \Re$. Since again $F_1(x-y) \leq 1$, integrable, we get, by the Dominated Convergence Theorem,

$$\int F_1(x-y)dF_2(y) \to 0; \quad \text{i.e., } J(x) \underset{x \to -\infty}{\to} 0, \text{ or } \; J(-\infty) = 0.$$

Next,

$$x \to \infty \text{ implies } F_1(x-y) \to 1, \; y \in \Re.$$

As before,

$$\int F_1(x-y)dF_2(y) \to \int 1 dF_2(y) = 1; \quad \text{i.e., } J(x) \underset{x \to \infty}{\to} 1,$$

or $J(\infty) = 1$. ■

Remark 7. If $J^*(x) = \int_{\Re} F_2(x-y)dF_1(y)$, then $J^*(x)$ is also a d.f., denoted by $F_2 * F_1$.

Now let f, f_1, f_2 be the ch.f.s corresponding to F, F_1, F_2. Then the following theorem is true.

Theorem 6. If $F = F_1 * F_2$, then $f = f_1 \times f_2$. Conversely, if f_1 and f_2 are the ch.f.s of the d.f.s F_1 and F_2, respectively, and if we set $f = f_1 \times f_2$, then f is the ch.f. of the d.f. F, where $F = F^* + c$, for some constant c, and $F^* = F_1 * F_2$. ■

Proof. Let $F = F_1 * F_2$. Then

$$f(t) = \int_{\Re} e^{itx} dF(x) = \lim_{\alpha \to -\infty, \beta \to \infty} \int_{(\alpha,\beta]} e^{itx} dF(x).$$

For $\alpha < \beta$, look at $(\alpha, \beta]$ and consider the following partition for each n:

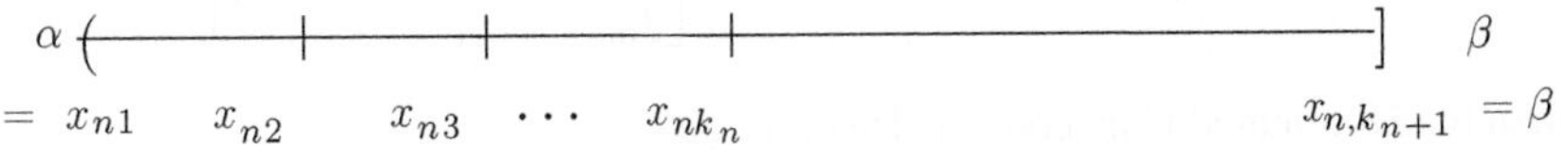

where the partitioning points are chosen so that

$$\max_{j=1,\ldots,k_n} (x_{n,j+1} - x_{nj}) \underset{n\to\infty}{\to} 0.$$

Then

$$\begin{aligned}
\int_{(\alpha,\beta]} e^{itx} dF(x) &= \lim_{n\to\infty} \sum_{j=1}^{k_n} e^{itx_{nj}} [F(x_{n,j+1}) - F(x_{nj})] \\
&= \lim_{n\to\infty} \sum_{j=1}^{k_n} e^{itx_{nj}} \left[\int_{\Re} F_1(x_{n,j+1} - y) dF_2(y) \right. \\
&\quad \left. - \int_{\Re} F_1(x_{nj} - y) dF_2(y) \right] \\
&= \lim_{n\to\infty} \int_{\Re} \sum_{j=1}^{k_n} e^{itx_{nj}} [F_1(x_{n,j+1} - y) \\
&\quad - F_1(x_{nj} - y)] dF_2(y) \\
&= \lim_{n\to\infty} \int_{\Re} \left\{ \sum_{j=1}^{k_n} e^{it(x_{nj}-y)} [F_1(x_{n,j+1} - y) \right. \\
&\quad \left. - F_1(x_{nj} - y)] e^{ity} \right\} dF_2(y). \qquad (11.22)
\end{aligned}$$

But $\sum_{j=1}^{k_n} e^{it(x_{nj}-y)} [F_1(x_{n,j+1} - y) - F_1(x_{nj} - y)]$ are partial sums tending to the integral of e^{itx} over the interval $(\alpha - y, \beta - y]$ with respect to F_1. Furthermore, these partial sums are bounded in absolute value by

$$\begin{aligned}
\sum_{j=1}^{k_n} [F_1(x_{n,j+1} - y) - F_1(x_{nj} - y)] &= F_1(x_{n,k_n+1} - y) \\
&\quad - F_1(x_{n1} - y) \le V_1,
\end{aligned}$$

the variation of F_1, independent of n, and F_2-integrable, whereas

$$\lim_{n\to\infty}\sum_{j=1}^{k_n} e^{it(x_{nj}-y)}[F_1(x_{n,j+1}-y) - F_1(x_{nj}-y)]e^{ity}$$

$$= \left[\int_{(\alpha-y,\beta-y]} e^{itx}\,dF_1(x)\right] e^{ity}.$$

Then the Dominated Convergence Theorem gives

$$\lim_{n\to\infty}\int_{\Re}\left\{\sum_{j=1}^{k_n} e^{it(x_{nj}-y)}[F_1(x_{n,j+1}-y) - F_1(x_{nj}-y)]e^{ity}\right\}dF_2(y)$$

$$= \int_{\Re}\left\{\lim_{n\to\infty}\sum_{j=1}^{k_n} e^{it(x_{nj}-y)}[F_1(x_{n,j+1}-y) - F_1(x_{nj}-y)]e^{ity}\right\}dF_2(y)$$

$$= \int_{\Re}\left[\int_{(\alpha-y,\beta-y]} e^{itx}\,dF_1(x)\right] e^{ity}\,dF_2(y),$$

or, by (11.22),

$$\int_{(\alpha,\beta]} e^{itx}\,dF(x) = \int_{\Re}\left[\int_{(\alpha-y,\beta-y]} e^{itx}\,dF_1(x)\right] e^{ity}\,dF_2(y). \tag{11.23}$$

Next,

$$\left|\left[\int_{(\alpha-y,\beta-y]} e^{itx}\,dF_1(x)\right] e^{ity}\right| \le V_1, \quad \text{independent of } \alpha,\, \beta,$$

and F_2-integrable, whereas

$$\lim_{\alpha\to-\infty,\beta\to\infty}\left[\int_{(\alpha-y,\beta-y]} e^{itx}\,dF_1(x)\right] e^{ity}$$

$$= \left[\int_{\Re} e^{itx}\,dF_1(x)\right] e^{ity} = f_1(t)e^{ity}.$$

Therefore, by the Dominated Convergence Theorem again, and (11.23),

$$\lim_{\alpha\to-\infty,\beta\to\infty}\int_{(\alpha,\beta]} e^{itx}\,dF(x)$$

$$= \lim_{\alpha \to -\infty, \beta \to \infty} \int_{\Re} \left[\int_{(\alpha - y, \beta - y]} e^{itx} dF_1(x) \right] e^{ity} dF_2(y)$$

$$= \int_{\Re} \left\{ \lim_{\alpha \to -\infty, \beta \to \infty} \left[\int_{(\alpha - y, \beta - y]} e^{itx} dF_1(x) \right] e^{ity} \right\} dF_2(y)$$

$$= \int_{\Re} f_1(t) e^{ity} dF_2(y) = f_1(t) \int_{\Re} e^{ity} dF_2(y) = f_1(t) f_2(t). \quad (11.24)$$

Since the left-hand side of (11.24) is $\int_{\Re} e^{itx} dF(x) = f(t)$, we have then $f(t) = f_1(t) f_2(t)$. Thus, if $F = F_1 * F_2$, then $f = f_1 \times f_2$.

From $F^* = F_1 * F_2$ and the direct part, we have $f^* = f_1 \times f_2$, where f^* is the ch.f. of F^*. Also, $f = f_1 \times f_2$. Thus, $f = f^*$. Hence f is a ch.f. and the corresponding d.f. is $F = F^* + c$, for some constant c. ■

To this theorem, there are the following three corollaries.

Corollary 1. The product of two ch.f.s is a ch.f.

Proof. Let f_1 and f_2 be two ch.f.s. For $j = 1, 2$, the ch.f. f_j determines a class $\mathcal{C}_j$ of d.f.s, any two of which differ by a constant. Let $F_j \in \mathcal{C}_j$, $j = 1, 2$, be any two d.f.s, and let $J = F_1 * F_2$. Then, by Theorem 5, J is a d.f., and let f be its ch.f. Then, by Theorem 6, $f = f_1 \times f_2$, so that the product $f_1 \times f_2$ is a ch.f. ■

Corollary 2. For any two d.f.s F_1 and F_2, we have $F_1 * F_2 = F_2 * F_1 uac$, and $F_1 * F_2 = F_2 * F_1$, if F_1 and F_2 are d.f.s of r.v.s.

Proof. Let $F_1 * F_2 = J$ and $F_2 * F_1 = J^*$ with respective ch.f.s f and f^*. Then, by Theorem 6, $f = f_1 \times f_2$ and $f^* = f_2 \times f_1$. Since $f_1 \times f_2 = f_2 \times f_1$, we have $f = f^*$, so that $J - J^* = c$, for some constant c, which proves the first assertion. For the second assertion, we have that if both F_1 and F_2 are d.f.s of r.v.s, then J and J^* are d.f.s of r.v.s by Theorem 5. Since $f = f^*$, by the first part here, it follows that $J = J^*$ (see Remark 6(ii)). ■

Corollary 3. If F_1 and F_2 are, respectively, the d.f.s of the independent r.v.s X_1 and X_2, then $F = F_1 * F_2 (= F_2 * F_1)$ is the d.f. of the r.v. $X_1 + X_2$.

Proof. By Lemma 1 in Chapter 10,

$$f_{X_1+X_2}(t) = \mathcal{E} e^{it(X_1+X_2)} = \mathcal{E}(e^{itX_1} \times e^{itX_2})$$
$$= (\mathcal{E} e^{itX_1})(\mathcal{E} e^{itX_2}) = f_{X_1}(t) f_{X_2}(t) = f(t),$$

where f is the ch.f. of $F_1 * F_2$ (by Corollary 1). Since $F_1 * F_2$ is a d.f. of a r.v., it follows that it is the d.f. of $X_1 + X_2$. ■

This section is concluded with the definition of symmetry of a r.v., and some results related to it.

Definition 6. The r.v. X is said to be *symmetric* about zero, if the r.v.s X and $-X$ have the same distribution; i.e.,

$$P(X \leq x) = P(-X \leq x) = P(X \geq -x), \quad x \in \Re. \quad ■$$

Theorem 7. We have

(i) For any ch.f. f, $f(-t) = \overline{f(t)}$, $t \in \Re$.
(ii) If f_X is the ch.f. of the r.v. X, then $\overline{f_X}$ is the ch.f. of the r.v. $-X$.
(iii) X is symmetric about zero if and only if its ch.f. f_X is real. ∎

Proof.

(i) Let F be any d.f. corresponding to f. Then

$$\begin{aligned} f(-t) &= \int_{\Re} e^{-itx} dF(x) \\ &= \int_{\Re} [\cos(-tx) + i\sin(-tx)] dF(x) \\ &= \int_{\Re} [\cos(tx) - i\sin(tx)] dF(x) \\ &= \overline{\int_{\Re} [\cos(tx) + i\sin(tx)] dF(x)} \\ &= \overline{\int_{\Re} e^{itx} dF(x)} = \overline{f(t)}. \quad \text{(See also Theorem 1(iv).)} \end{aligned}$$

(ii) By Theorem 1(iii),

$$f_{\alpha X+\beta}(t) = e^{i\beta t} f_X(\alpha t).$$

For $\alpha = -1, \beta = 0$, this becomes $f_{-X}(t) = f_X(-t)$. But $f_X(-t) = \overline{f_X(t)}$. Thus $\overline{f_X(t)}$ is the ch.f. of $-X$. (Or, by part (i), $f_{-X}(t) = \mathcal{E}e^{it(-X)} = \mathcal{E}e^{i(-t)X} = f_X(-t) = \overline{f_X(t)}$.)

(iii) Let X be symmetric about zero with d.f. F_X. Then

$$f_X(t) = \int_{\Re} e^{itx} dF_X(x) = \int_{\Re} e^{itx} dF_{-X}(x) = f_{-X}(t) = \overline{f_X(t)}$$

by part (ii); i.e., $f_X = \overline{f_X}$, so that f_X is real. Next, let f_X be real. Then $f_X = \overline{f_X}$. But $\overline{f_X} = f_{-X}$ by part (ii). Thus $f_X = f_{-X}$ or F_X and F_{-X} are the same; hence, X is symmetric about zero. ∎

11.6 Some Further Properties of Characteristic Functions

In this section, two main results (Theorems 8 and 9) are established pertaining to ch.f.s of r.v.s. Theorem 8 (and its corollary) may also be established for certain ch.f.s that are not necessarily ch.f.s of r.v.s.

Theorem 8. For $n = 1, 2, \ldots$, let f_n, f be ch.f.s of r.v.s Then, if $f_n \underset{n\to\infty}{\to} f$ on $\Re$, it follows that the convergence is uniform in closed intervals in $\Re$. ∎

Proof. It suffices to prove that $f_n(t) \to f(t)$ uniform in $t \in [-T, T]$, $T > 0$. Let F_n, F be the d.f.s corresponding to f_n, f. Then we have

$$\begin{aligned}|f_n(t) - f(t)| &= \left|\int_{\Re} e^{itx} dF_n(x) - \int_{\Re} e^{itx} dF(x)\right| \\ &\leq \left|\int_{(\alpha,\beta]} e^{itx} dF_n(x) - \int_{(\alpha,\beta]} e^{itx} dF(x)\right| \\ &+ \left|\int_{\Re-(\alpha,\beta]} e^{itx} dF_n(x)\right| + \left|\int_{\Re-(\alpha,\beta]} e^{itx} dF(x)\right|. \quad (11.25)\end{aligned}$$

But

$$\left|\int_{\Re-(\alpha,\beta]} e^{itx} dF(x)\right| \leq \int_{\Re-(\alpha,\beta]} dF(x) = 1 - [F(\beta) - F(\alpha)],$$

$$\left|\int_{\Re-(\alpha,\beta]} e^{itx} dF_n(x)\right| \leq \int_{\Re-(\alpha,\beta]} dF_n(x) = 1 - [F_n(\beta) - F_n(\alpha)].$$

Thus, (11.25) becomes

$$\begin{aligned}|f_n(t) - f(t)| &\leq \left|\int_{(\alpha,\beta]} e^{itx} dF_n(x) - \int_{(\alpha,\beta]} e^{itx} dF(x)\right| \\ &+\{1 - [F(\beta) - F(\alpha)]\} + \{1 - [F_n(\beta) - F_n(\alpha)]\}. \quad (11.26)\end{aligned}$$

Pick α, β to be continuity points F and such that

$$1 - [F(\beta) - F(\alpha)] < \frac{\varepsilon}{7}. \quad (11.27)$$

Now, with $n \to \infty$,

$$\begin{aligned}&f_n \to f \text{ implies } F_n \Rightarrow F (\text{by Theorem 3}) \text{ so that } F_n(\alpha) \to F(\alpha), \\ &F_n(\beta) \to F(\beta) \text{ and } 1 - [F_n(\beta) - F_n(\alpha)] < 1 - [F(\beta) - F(\alpha)] + \frac{\varepsilon}{7}\end{aligned}$$

for $n > n_1 = n_1(\varepsilon)$, and the last expression is $< \frac{2\varepsilon}{7}$ by means of (11.27). Then (11.26) becomes

$$\begin{aligned}|f_n(t) - f(t)| &\leq \left|\int_{(\alpha,\beta]} e^{itx} dF_n(x) - \int_{(\alpha,\beta]} e^{itx} dF(x)\right| + \frac{3\varepsilon}{7}, \\ n &\geq n_1. \quad (11.28)\end{aligned}$$

The proof of the theorem would be completed (by means of (11.28)) if we knew that

$$\int_{(\alpha,\beta]} e^{itx} dF_n(x) \underset{n\to\infty}{\to} \int_{(\alpha,\beta]} e^{itx}\, dF(x) \quad uniformly \text{ in } t \in [-T, T].$$

This convergence is true for each t (by the Helly–Bray Lemma, Theorem 6 in Chapter 8), and the uniformity in $t \in [-T, T]$ is the content of the following result.

Statement. Under the assumptions of Theorem 8 and with α and β being continuity points of F, it holds

$$\int_{(\alpha,\beta]} e^{itx} dF_n(x) \underset{n\to\infty}{\to} \int_{(\alpha,\beta]} e^{itx} dF(x) \quad uniformly \text{ in } t \in [-T, T].$$

Proof. Pick points

$$\alpha = x_1 < x_2 < \cdots < x_N < x_{N+1} = \beta$$

to be continuity points of F and such that

$$\max_{k=1,\ldots,N} (x_{k+1} - x_k) \leq \frac{\varepsilon}{7T},$$

and on $(\alpha, \beta]$, define the function g_t as follows: $g_t(x) = e^{itx_k}$ if $x \in (x_k, x_{k+1}]$, $k = 1, \ldots, N$. Pick $n \geq n_2 = n_2(\varepsilon, N)$, so that

$$|F_n(x_k) - F(x_k)| < \frac{\varepsilon}{7(N+1)}, \quad k = 1, \ldots, N+1. \tag{11.29}$$

Next,

$$\begin{aligned} &\left|\int_{(\alpha,\beta]} e^{itx} dF_n(x) - \int_{(\alpha,\beta]} e^{itx} dF(x)\right| \\ &\leq \int_{(\alpha,\beta]} |e^{itx} - g_t(x)| dF_n(x) \\ &+ \int_{(\alpha,\beta]} |e^{itx} - g_t(x)| dF(x) + \left|\int_{(\alpha,\beta]} g_t(x) dF_n(x) \right.\\ &\left. - \int_{(\alpha,\beta]} g_t(x) dF(x)\right|. \end{aligned} \tag{11.30}$$

But for $x \in (\alpha, \beta]$, $|e^{itx} - g_t(x)| = |e^{itx} - e^{itx_k}|$ (for some x_k) and this equals

$$\left|e^{itx_k}\right| \left|e^{it(x-x_k)} - 1\right| = \left|e^{it(x-x_k)} - 1\right| \leq |t(x - x_k)|,$$

since $\left|e^{ix} - 1\right| \leq |x|$, $x \in \Re$ (see also Exercise 4). Then the preceding expression is $\leq T\,|x - x_k| < T \times \frac{\varepsilon}{7T} = \frac{\varepsilon}{7}$.

Thus, (11.30) becomes

$$\begin{aligned} &\left|\int_{(\alpha,\beta]} e^{itx} dF_n(x) - \int_{(\alpha,\beta]} e^{itx} dF(x)\right| < \frac{2\varepsilon}{7} \\ &+ \left|\int_{(\alpha,\beta]} g_t(x) dF_n(x) - \int_{(\alpha,\beta]} g_t(x) dF(x)\right|. \end{aligned} \tag{11.31}$$

But, by the definition of $g_t(x)$,

$$\begin{aligned}
&\left|\int_{(\alpha,\beta]} g_t(x)dF_n(x) - \int_{(\alpha,\beta]} g_t(x)dF(x)\right| \\
&= \left|\sum_{k=1}^{N} e^{itx_k}[F_n(x_{k+1}) - F_n(x_k)] - \sum_{k=1}^{N} e^{itx_k}[F(x_{k+1}) - F(x_k)]\right| \\
&\le \sum_{k=1}^{N} |[F_n(x_{k+1}) - F_n(x_k)] - [F(x_{k+1}) - F(x_k)]| \\
&= \sum_{k=1}^{N} |[F_n(x_{k+1}) - F(x_{k+1})] - [F_n(x_k) - F(x_k)]| \\
&\le 2\sum_{k=1}^{N+1} |F_n(x_k) - F(x_k)| \\
&\le 2(N+1) \times \frac{\varepsilon}{7(N+1)} = \frac{2\varepsilon}{7} \text{ by (11.29).}
\end{aligned}$$

Thus, (11.31) yields, for $n \ge n_2 = n_2(\varepsilon, N)$,

$$\left|\int_{(\alpha,\beta]} e^{itx} dF_n(x) - \int_{(\alpha,\beta]} e^{itx} dF(x)\right| < \frac{4\varepsilon}{7}, \tag{11.32}$$

as was to be seen.

Completion of the Proof of Theorem 8. For $t \in [-T, T]$, (11.28) becomes, by means of (11.32), and for $n \ge n(\varepsilon) = \max\{n_1(\varepsilon), n_2(\varepsilon, N)\}$,

$$|f_n(t) - f(t)| \le \frac{3\varepsilon}{7} + \frac{4\varepsilon}{7} = \varepsilon. \qquad \blacksquare$$

Next, we recall the definition of continuous convergence and derive a simple result to be employed in the corollary following.

Definition 7. For $n \ge 1$, let g_n, g be functions defined on $A \subseteq \Re$ into $\Re$. Then, as $n \to \infty$, we say that $g_n \to g$ *continuously* in A if $g_n(t_n) \to g(t)$ whenever $t_n \to t, t_n, t \in A$. ■

Proposition 4. Let $g_n \underset{n\to\infty}{\to} g$ uniformly in $A \subseteq \Re$ and let g be continuous. Then $g_n \underset{n\to\infty}{\to} g$ continuously in A.

Proof. We have $|g_n(t_n) - g(t)| \le |g_n(t_n) - g(t_n)| + |g(t_n) - g(t)|$, and $g_n(t_n) - g(t_n) \underset{n\to\infty}{\to} 0$ by uniform convergence of g_n and $g(t_n) - g(t) \underset{n\to\infty}{\to} 0$ by continuity of g. ■

Corollary to Theorem 8. If $f_n \underset{n\to\infty}{\to} f$ and $t_n \underset{n\to\infty}{\to} t, t_n, t \in \Re$, then $f_n(t_n) \underset{n\to\infty}{\to} f(t)$ (i.e., f_n converges continuously in $\Re$).

Proof. With $n \to \infty$, let $t_n \to t$, $t_n, t \in [-T, T]$, $n \geq 1$, for some $T > 0$. Then $f_n \to f$ uniformly in $[-T, T]$ and f is continuous. Hence, by Proposition 4, the convergence is continuous in $[-T, T]$. In particular, $f_n(t_n) \to f(t)$. ∎

Lemma 5. Let X be a r.v. with ch.f. f. Then if $\mathcal{E}|X|^n < \infty$ for some positive integer n, it follows that $f^{(n)}(t) = \frac{d^n}{dt^n} f(t)$ is continuous in $\Re$.

Proof. As was seen in the proof of Theorem 1(v),

$$f^{(n)}(t) = \int_{\Re} i^n e^{itX} X^n dF(x).$$

Next,

$$i^n e^{itX} X^n \underset{t \to t_0}{\to} i^n e^{it_0 X} X^n \text{ and } \left| i^n e^{itX} X^n \right| \leq |X|^n,$$

independent of t and integrable. Hence the Dominated Convergence Theorem completes the proof. ∎

Theorem 9. Let X be a r.v. such that $\mathcal{E}|X|^n < \infty$ for some positive integer n, and let f be its ch.f. Set $m^{(k)} = \mathcal{E}X^k$, $k = 0, 1, \ldots, n$. Then one has

$$f(t) = \sum_{k=0}^{n-1} \frac{m^{(k)}}{k!}(it)^k + \rho_n(t), \quad t \in \Re,$$

where

(i) $\rho_n(t) = t^n \int_0^1 \frac{(1-x)^{n-1}}{(n-1)!} f^{(n)}(tx)dx$ (where the integral exists because of Lemma 5), or

(ii) $\rho_n(t) = \frac{m^{(n)}}{n!}(it)^n + o(t^n)$, or

(iii) $\rho_n(t) = \theta \frac{\mu^{(n)}}{n!}|t|^n$, where $\theta = \theta(n, t)$ is such that $|\theta| \leq 1$, and $\mu^{(n)} = \mathcal{E}|X|^n$. ∎

Proof.

(i) In the first place, $f^{(n)}(u)$ exists and is continuous (by Theorem 1(v) and Lemma 5), so that the integral indicated exists in the Riemann sense. Next, the result is, clearly, true for $t = 0$ $(0^0 = 1)$. Thus we may assume in the sequel that $t \neq 0$, if need be. Now

$$\begin{aligned}
\rho_n(t) &= t^n \int_0^1 \frac{(1-x)^{n-1}}{(n-1)!} f^{(n)}(tx)dx \\
&= \frac{1}{(n-1)!} \int_0^1 (t - tx)^{n-1} f^{(n)}(tx)d(tx) \\
&= \frac{1}{(n-1)!} \int_0^t (t - u)^{n-1} f^{(n)}(u)du \\
&\text{(by setting } tx = u)
\end{aligned}$$

$$
\begin{aligned}
&= \frac{1}{(n-1)!}\int_0^t (t-u)^{n-1} df^{(n-1)}(u) \\
&= \frac{1}{(n-1)!}(t-u)^{n-1} f^{(n-1)}(u)|_0^t + \frac{(n-1)}{(n-1)!} \\
&\quad \times \int_0^t (t-u)^{n-2} f^{(n-1)}(u)du \\
&= -\frac{t^{n-1}}{(n-1)!} f^{(n-1)}(0) \\
&\quad + \frac{1}{(n-2)!}\int_0^t (t-u)^{n-2} f^{(n-1)}(u)du;
\end{aligned}
$$

i.e., integrating by parts, we get

$\rho_n(t) = -\frac{t^{n-1}}{(n-1)!} f^{(n-1)}(0) + \frac{1}{(n-2)!}\int_0^t (t-u)^{n-2} f^{(n-1)}(u)du.$

Also,

$$
\begin{aligned}
\tfrac{1}{(n-2)!}\int_0^t (t-u)^{n-2} f^{(n-1)}(u)du &= \tfrac{1}{(n-2)!}\int_0^t (t-u)^{n-2} df^{(n-2)}(u) \\
&= \tfrac{1}{(n-2)!}(t-u)^{n-2} f^{(n-2)}(u)|_0^t \\
&\quad + \tfrac{(n-2)}{(n-2)!}\int_0^t (t-u)^{n-3} f^{(n-2)}(u)du \\
&= -\frac{t^{n-2}}{(n-2)!} f^{(n-2)}(0) \\
&\quad + \tfrac{1}{(n-3)!}\int_0^t (t-u)^{n-3} f^{(n-2)}(u)du,
\end{aligned}
$$

so that

$$
\begin{aligned}
\rho_n(t) &= -\frac{t^{n-1}}{(n-1)!} f^{(n-1)}(0) - \frac{t^{n-2}}{(n-2)!} f^{(n-2)}(0) \\
&\quad + \frac{1}{(n-3)!}\int_0^t (t-u)^{n-3} f^{(n-2)}(u)du.
\end{aligned}
$$

Proceeding in this manner, the $(n-1)$th integration by parts yields

$$
-\frac{t}{1!} f^{(1)}(0) + \int_0^t f^{(1)}(u)du = -\frac{t}{1!} f^{(1)}(0) + f(t) - 1,
$$

so that

$$
\begin{aligned}
\rho_n(t) &= -1 - \frac{t}{1!} f^{(1)}(0) - \cdots - \frac{t^{n-2}}{(n-2)!} f^{(n-2)}(0) \\
&\quad - \frac{t^{n-1}}{(n-1)!} f^{(n-1)}(0) + f(t) \\
&= -\sum_{k=0}^{n-1} \frac{t^k}{k!} f^{(k)}(0) + f(t)
\end{aligned}
$$

$$= -\sum_{k=0}^{n-1} \frac{t^k}{k!} i^k m^{(k)} + f(t)$$

$$\text{(since } f^{(k)}(0) = i^k m^{(k)})$$

$$= -\sum_{k=0}^{n-1} \frac{m^{(k)}}{k!} (it)^k + f(t), \text{ so that}$$

$$f(t) = \sum_{k=0}^{n-1} \frac{m^{(k)}}{k!} (it)^k + \rho_n(t), \quad t \in \Re.$$

(ii) By part (i), $\rho_n(t) = \frac{t^n}{(n-1)!} \int_0^1 (1-x)^{n-1} f^{(n)}(tx)dx$, so that, for $t \neq 0$,

$$\frac{(n-1)!}{t^n} \rho_n(t) = \int_0^1 (1-x)^{n-1} f^{(n)}(tx)dx$$

$$= \int_0^1 (1-x)^{n-1} \left[\int_{\Re} (iu)^n e^{itxu} dF(u) \right] dx$$

$$\left(\text{since } \frac{d^n}{d\lambda^n} f(\lambda) = \frac{d^n}{d\lambda^n} \int_{\Re} e^{i\lambda u} dF(u) = \int_{\Re} \left(\frac{\partial^n}{\partial \lambda^n} e^{i\lambda u} \right) dF(u) = \int_{\Re} (iu)^n e^{i\lambda u} dF(u) \right)$$

$$= i^n \int_0^1 \int_{\Re} (1-x)^{n-1} u^n e^{itxu} dF(u) dx$$

$$= i^n \int_{\Re} u^n \left[\int_0^1 (1-x)^{n-1} e^{itxu} dx \right] dF(u) \tag{11.33}$$

(by the Fubini Theorem, which applies since $|(1-x)^{n-1} u^n e^{itxu}| \leq |u|^n$ is Lebesgue $\times$ F-integrable over $[0, 1] \times \Re$).

Now, $|(1-x)^{n-1} e^{itxu}| = |1-x|^{n-1} \leq 1$ (over $[0, 1]$) independent of t and Lebesgue-integrable over $[0, 1]$. Furthermore, $(1-x)^{n-1} e^{itxu} \underset{t \to 0}{\to} (1-x)^{n-1}$, so that the Dominated Convergence Theorem yields

$$\int_0^1 (1-x)^{n-1} e^{itxu} dx \underset{t \to 0}{\to} \int_0^1 (1-x)^{n-1} dx = \frac{1}{n}.$$

Next,

$$\left| u^n \int_0^1 (1-x)^{n-1} e^{itxu} dx \right| \leq |u|^n \text{ independent of } t, \quad \text{and } F$$
$-$integrable.

Furthermore, $u^n \int_0^1 (1-x)^{n-1} e^{itxu} dx \underset{t\to 0}{\to} \frac{u^n}{n}$. Therefore, by the Dominated Convergence Theorem,

$$\int_{\Re} u^n \left[\int_0^1 (1-x)^{n-1} e^{itxu} dx\right] dF(u) \underset{t\to 0}{\to} \int_{\Re} \frac{u^n}{n} dF(u)$$
$$= \frac{1}{n} m^{(n)}.$$

It follows that $\int_{\Re} u^n [\int_0^1 (1-x)^{n-1} e^{itxu} dx] dF(u) = \frac{1}{n} m^{(n)} + o(1)$ (where $o(1) \to 0$ as $t \to 0$), and then by (11.33),

$$\frac{(n-1)!}{t^n} \rho_n(t) = \frac{i^n}{n} m^{(n)} + o(1) \text{ or } \rho_n(t)$$
$$= \frac{m^{(n)}}{n!} (it)^n + t^n o(1) = \frac{m^{(n)}}{n!} (it)^n + o(t^n),$$

as was to be seen.

(iii) Again, as in part (ii) (see first and last lines on the right-hand side of relation (11.33)),

$$\left|\int_0^1 (1-x)^{n-1} f^{(n)}(tx) dx\right| = \left|i^n \int_{\Re} u^n \left[\int_0^1 (1-x)^{n-1} e^{itxu} dx\right] dF(u)\right|$$
$$\leq \int_{\Re} |u|^n \left[\int_0^1 (1-x)^{n-1} dx\right] dF(u)$$
$$= \frac{1}{n} \int_{\Re} |u|^n dF(u) = \frac{1}{n} \mu^{(n)},$$

so that

$$|\rho_n(t)| = \left|\frac{t^n}{(n-1)!} \int_0^1 (1-x)^{n-1} f^n(tx) dx\right| \leq \frac{|t|^n}{n!} \mu^{(n)}.$$

Then there exists $\theta = \theta(n, t)$ with $|\theta| \leq 1$ such that $\rho_n(t) = \theta \frac{|t|^n}{n!} \times \mu^{(n)} = \theta \frac{\mu^{(n)}}{n!} |t|^n$. ∎

11.7 Applications to the Weak Law of Large Numbers and the Central Limit Theorem

The following two results are applications of two theorems in the previous sections, Theorems 3 and 9.

Application 1 (Weak Law of Large Numbers, WLLN). Let $X_1, \ldots, X_n$ be i.i.d. r.v.s with $\mathcal{E}X_1 = m^{(1)}$ finite. Then

$$\bar{X}_n = \frac{X_1 + \cdots + X_n}{n} \underset{n\to\infty}{\overset{P}{\to}} m^{(1)}.$$

Discussion. By Theorem 9 (ii), applied for $n = 1$, we get

$$f(t) = 1 + \frac{m^{(1)}}{1!}(it) + o(t) \quad \text{(where } f \text{ is the ch.f. of the } X_i\text{s).}$$

Hence

$$f_{\bar{X}_n}(t) = f^n\left(\frac{t}{n}\right) = \left[1 + m^{(1)} i\frac{t}{n} + o\left(\frac{t}{n}\right)\right]^n$$
$$= \left[1 + m^{(1)}\frac{it}{n} + \frac{t}{n}o(1)\right]^n = \left[1 + \frac{im^{(1)}t + to(1)}{n}\right]^n,$$

where for a fixed t, $o(1) \to 0$ as $n \to \infty$, so that $im^{(1)}t + to(1) \underset{n\to\infty}{\to} im^{(1)}t$. This implies that

$$\left[1 + \frac{im^{(1)}t + to(1)}{n}\right]^n \underset{n\to\infty}{\to} e^{im^{(1)}t}, \tag{11.34}$$

so that $f_{\bar{X}_n}(t) \underset{n\to\infty}{\to} e^{im^{(1)}t}$ which is the ch.f. of the r.v. X that is equal to $m^{(1)}$ with probability one. Hence, by Theorem 3,

$$\bar{X}_n \underset{n\to\infty}{\overset{d}{\to}} m^{(1)} \text{ or, } \quad \text{equivalently } \bar{X}_n \underset{n\to\infty}{\overset{P}{\to}} m^{(1)}.$$ ■

Remark 8. In relation (11.34), we use the familiar result $(1 + \frac{c_n}{n})^n \to e^c$ when $c_n \to c$ as $n \to \infty$ (see also Exercise 5).

Application 2 (Central Limit Theorem, CLT). Let $X_1, \ldots, X_n$ be i.i.d. r.v.s with $\mathcal{E}X_1 = m^{(1)}$ finite and $\sigma^2(X_1) = \sigma^2 \in (0, \infty)$. Then

$$\frac{S_n - \mathcal{E}S_n}{\sigma(S_n)} = \frac{\sqrt{n}[\bar{X}_n - m^{(1)}]}{\sigma} \underset{n\to\infty}{\overset{d}{\to}} Z \sim N(0, 1),$$

where

$$S_n = \sum_{j=1}^{n} X_j \text{ and } \bar{X}_n = \frac{1}{n}\sum_{j=1}^{n} X_j.$$

Discussion. Set $Y_j = \frac{X_j - m^{(1)}}{\sigma}$. Then the r.v.s $Y_1, \ldots, Y_n$ are i.i.d. with $\mathcal{E}Y_1 = 0$ and $\sigma^2(Y_1) = \mathcal{E}Y_1^2 = 1$. Let f_1 be the ch.f. of the Y_js. Then by Theorem 9 (ii), applied for $n = 2$, we get

$$f_1(t) = 1 + \frac{\mathcal{E}Y_1}{1!}(it) + \frac{\mathcal{E}Y_1^2}{2!}(it)^2 + o(t^2)$$
$$= 1 - \frac{t^2}{2} + o(t^2) = 1 - \frac{t^2}{2} + t^2 o(1).$$

Hence

$$f_{\frac{S_n - \mathcal{E}S_n}{\sigma(S_n)}}(t) = f_{\sum_{j=1}^{n}\frac{X_j - m^{(1)}}{\sigma\sqrt{n}}}(t) = f_{\sum_{j=1}^{n} Y_j/\sqrt{n}}(t) = f_1^n\left(\frac{t}{\sqrt{n}}\right)$$

$$= \left[1 - \frac{t^2}{2n} + \frac{t^2}{n} o(1)\right]^n$$

$$= \left[1 + \frac{-\frac{t^2}{2} + t^2 o(1)}{n}\right]^n \underset{n\to\infty}{\to} e^{-\frac{t^2}{2}},$$

since

$$-\frac{t^2}{2} + t^2 o(1) \underset{n\to\infty}{\to} -\frac{t^2}{2} \text{ as } o(1) \underset{n\to\infty}{\to} 0 \text{ for a fixed } t.$$

The fact that $e^{-\frac{t^2}{2}}$ is the ch.f. of a r.v. $Z \sim N(0, 1)$ completes the proof. ■

11.8 The Moments of a Random Variable Determine its Distribution

This section consists of two main results, Theorems 10 and 11. In the first of these results, a condition is given under which a ch.f. expands into an infinite series, and in the second, conditions are stated under which the moments of a r.v. completely determine its distribution. In the proof of Theorem 11, one needs a number of concepts and results from complex analysis, which are presented in the next section.

Theorem 10. Let X be a r.v. such that $\mathcal{E}|X|^n < \infty$ for $n = 1, 2, \ldots$, and let f be its ch.f. Then for any $t \in \Re$ for which the series $\sum_{n=0}^{\infty} \frac{m^{(n)}}{n!}(it)^n$ converges, one has $f(t) = \sum_{n=0}^{\infty} \frac{m^{(n)}}{n!}(it)^n$. ■

Proof. Set

$$S_n(t) = \sum_{k=0}^{n} \frac{m^{(k)}}{k!}(it)^k,$$

and observe that the assertion is true for $t = 0$ ($0^0 = 1$). So, let $t_0 \neq 0$ be a point for which $S_n(t_0)$ converges. Then $\frac{m^{(k)}}{k!}(it_0)^k \underset{k\to\infty}{\to} 0$. In particular, $\frac{m^{(2r)}}{(2r)!}(it_0)^{2r} \underset{r\to\infty}{\to} 0$, and this is equivalent to

$$\frac{\mu^{(2r)}}{(2r)!}|t_0|^{2r} \underset{r\to\infty}{\to} 0, \tag{11.35}$$

since $\mu^{(2r)} = m^{(2r)}$. From Theorem 9 (iii), applied for $n = 2r$, we have

$$f(t_0) = \sum_{k=0}^{2r-1} \frac{m^{(k)}}{k!}(it_0)^k + \theta \frac{\mu^{(2r)}}{(2r)!}|t_0|^{2r} = S_{2r-1}(t_0) + \theta \frac{\mu^{(2r)}}{(2r)!}|t_0|^{2r},$$

so that

$$|f(t_0) - S_{2r-1}(t_0)| = \left|\theta \frac{\mu^{(2r)}}{(2r)!}|t_0|^{2r}\right| \leq \frac{\mu^{(2r)}}{(2r)!}|t_0|^{2r} \underset{r\to\infty}{\to} 0$$

on account of (11.35). Thus

$$S_{2r-1}(t_0) \underset{r\to\infty}{\to} f(t_0). \tag{11.36}$$

Next,

$$|S_{2r}(t_0) - S_{2r-1}(t_0)| = \left|\frac{m^{(2r)}}{(2r)!}(it_0)^{2r}\right| = \frac{\mu^{(2r)}}{(2r)!}|t_0|^{2r} \underset{r\to\infty}{\to} 0$$

by (11.35). Hence $S_{2r}(t_0) \underset{r\to\infty}{\to} f(t_0)$ and this together with (11.36) gives that $S_n(t_0) \underset{n\to\infty}{\to} f(t_0)$. ■

Remark 9. Recall that if $\sum_{n=0}^{\infty} c_n z^n$ is a power series where z is, in general, a complex variable, then the radius ρ of convergence of the series is given by

$$\rho = 1/\limsup_{n\to\infty} |c_n|^{1/n},$$

so that $\rho^{-1} = \limsup_{n\to\infty}|c_n|^{1/n}$, and the series converges for $|z| < \rho$.

Lemma 6. Let X be a r.v. such that $\mu^{(n)} = \mathcal{E}|X|^n < \infty$ for all $n = 1, 2, \ldots$ (so that $m^{(n)} = \mathcal{E}X^n$ are also finite). Then the series

$$S(t) = \sum_{n=0}^{\infty} \frac{m^{(n)}}{n!}(it)^n \text{ and } S^*(t) = \sum_{n=0}^{\infty} \frac{\mu^{(n)}}{n!}|t|^n$$

have the same radii of convergence.

Proof. Let ρ_1 and ρ_2 be the radii of convergence of the series $S(t)$ and $S^*(t)$, respectively. Then, since $\left|\frac{m^{(n)}}{n!}\right|^{1/n} \le \left[\frac{\mu^{(n)}}{n!}\right]^{1/n}$ (by the fact that $|\mathcal{E}X^n| \le \mathcal{E}|X|^n$), it follows that $\frac{1}{\rho_1} \le \frac{1}{\rho_2}$, or $\rho_1 \ge \rho_2$. Thus, it suffices to show that $\rho_1 \le \rho_2$, or equivalently, $\frac{1}{\rho_2} \le \frac{1}{\rho_1}$. To this end, we have $\left[\frac{\mu^{(2n)}}{(2n)!}\right]^{1/2n} = \left[\frac{m^{(2n)}}{(2n)!}\right]^{1/2n}$, so that

$$\limsup_{n\to\infty}\left[\frac{\mu^{(2n)}}{(2n)!}\right]^{1/2n} = \limsup_{n\to\infty}\left[\frac{m^{(2n)}}{(2n)!}\right]^{1/2n} \le \frac{1}{\rho_1}. \tag{11.37}$$

Next, $[\mu^{(2n-1)}]^{1/(2n-1)} \le [\mu^{(2n)}]^{1/2n}$ (by the fact that $\mathcal{E}^{1/r}|X|^r \uparrow$ in $r > 0$), so that

$$\left[\frac{\mu^{(2n-1)}}{(2n-1)!}\right]^{1/(2n-1)} \le \left[\frac{\mu^{(2n)}}{(2n)!}\right]^{1/(2n)} \times \frac{[(2n)!]^{1/2n}}{[(2n-1)!]^{1/(2n-1)}}. \tag{11.38}$$

At this point, assume for a moment that

$$\lim_{n\to\infty} \frac{[(2n)!]^{1/2n}}{[(2n-1)!]^{1/(2n-1)}} = 1. \tag{11.39}$$

Then (11.38) gives by means of (11.37)

$$\limsup_{n\to\infty}\left[\frac{\mu^{(2n-1)}}{(2n-1)!}\right]^{1/(2n-1)} \le \frac{1}{\rho_1}. \tag{11.40}$$

Since

$$\frac{1}{\rho_2} = \limsup_{n\to\infty}\left[\frac{\mu^{(n)}}{n!}\right]^{1/n} = \max\left\{\limsup_{n\to\infty}\left[\frac{\mu^{(2n)}}{(2n)!}\right]^{1/(2n)},\right.$$
$$\left.\limsup_{n\to\infty}\left[\frac{\mu^{(2n-1)}}{(2n-1)!}\right]^{1/(2n-1)}\right\},$$

we have, from (11.37) and (11.40), that $\frac{1}{\rho_2} \le \frac{1}{\rho_1}$, so that $\rho_1 = \rho_2$. ■

Remark 10. The proof of (11.39) is left as an exercise (see Exercise 7).

Corollary to Lemma 6. A sufficient condition for the series $S(t)$ to converge for some $t_0 \neq 0$ is that

$$\limsup_{n\to\infty}\left|\frac{m^{(n)}}{n!}\right|^{1/n} < \infty \text{ or } \limsup_{n\to\infty}\frac{[\mu^{(n)}]^{1/n}}{n} < \infty.$$

Proof. The first assertion is immediate and the second follows from the first by Stirling's formula (see Exercise 8). ■

Theorem 11. Let X be r.v. with d.f. F and ch.f. f, and suppose that $\mathcal{E}X^n = m^{(n)} \in \Re$, $n \ge 1$, and let $\mu^{(n)} = \mathcal{E}|X|^n$. Then

(i) If the series $S(t) = \sum_{n=0}^{\infty}\frac{m^{(n)}}{n!}(it)^n$ converges for some $t_0 \neq 0$, it follows that the distribution of X is uniquely determined (by the moments of X).

(ii) A sufficient condition for $S(t)$ to converge for some $t_0 \neq 0$ is that $\limsup_{n\to\infty}\left[\frac{\mu^{(n)}}{n!}\right]^{1/n} < \infty$. ■

Proof.

(i) Let F_0 be a d.f. of a r.v., potentially different from F, with corresponding ch.f. f_0, such that

$$\int_{\Re} x^n dF_0(x) = m^{(n)}, n \ge 0.$$

Then the series $S_0(t) = \sum_{n=0}^{\infty}\frac{m^{(n)}}{n!}(it)^n (= S(t))$ converges for $t = t_0 \neq 0$, and hence for all t with $|t| \le t_0$, supposing without loss of generality that $t_0 > 0$. For each such t, $S(t)$ represents $f(t)$, and $S_0(t)$ represents $f_0(t)$ on account of

Theorem 10. Then, by Fact 4 in the next section, $f(z)$ and $f_0(z)$ are defined and are analytic for $|Im(z)| < t_0$. Furthermore,

$$f^{(n)}(z) = \sum_{k=n}^{\infty} \frac{i^k m^{(k)}}{(k-n)!} z^{k-n} = f_0^{(n)}(z), \, |z| < t_0,$$

and in particular,

$$f(z) = \sum_{k=0}^{\infty} \frac{i^k m^{(k)}}{k!} z^k = f_0(z), \, |z| < t_0. \tag{11.41}$$

Thus, $f(z)$ and $f_0(z)$ are analytic for $|Im(z)| < t_0$, and $f(z) = f_0(z)$ for $|z| < t_0$, by means of (11.41). Then, by Fact 5, $f(z) = f_0(z)$ for $|Im(z)| < t_0$ (by continuous extension). In particular, $f(t) = f_0(t)$ for $t \in \Re$, so that $F = F_0$.

(ii) Immediate by the Corollary to Lemma 6. ■

The last theorem is illustrated by the following example.

Example 4. Let $Z \sim N(0, 1)$, so that $m^{(2k)} = \frac{(2k)!}{2^k k!}$, $m^{(2k+1)} = 0, k = 0, 1, \ldots$. Then

$$f(t) = \sum_{k=0}^{\infty} \frac{m^{(2k)}}{(2k)!} (it)^{2k} = \sum_{k=0}^{\infty} \frac{1}{2^k k!} (it)^{2k} = \sum_{k=0}^{\infty} \frac{1}{k!} \left(-\frac{t^2}{2}\right)^k = e^{-\frac{t^2}{2}},$$

as was expected. ■

This section is concluded with some comments and a result involving the expansion of a logarithmic function.

Recall that if z is a complex number, then the log z is any complex number w such that $e^w = z$. This relation defines a many-valued function, since if w is a solution of $e^w = z$, then so is $w + 2n\pi i, n = 0, \pm 1, \ldots$, because $e^{2n\pi i} = 1$. Now, every complex number z may be written as $z = |z|e^{i\theta}$ for some θ with $-\pi < \theta \leq \pi$. Then $w = \log|z| + i\theta$ ($|z| \neq 0$), $-\pi < \theta \leq \pi$, is a solution of $e^w = z$, since $e^w = e^{\log|z| + i\theta} = |z|e^{i\theta} = z$. This solution is called the *principal branch* of the logarithm of z and is usually denoted by $\log_p z$.

In all that follows, we shall work with $\log_p$ although we shall not indicate the p. The following result will prove useful in many situations.

Lemma 7. For any complex number z, one has

$$\log(1+z) = z[1 + \varepsilon(z)] = z(1 + \theta z), \quad \text{if } |z| \leq \frac{1}{2},$$

where

$$|\varepsilon(z)| \leq |z| \text{ and } |\theta| = |\theta(z)| \leq 1.$$

Proof. For $|z| < 1$, it is known that

$$\log(1+z) = z - \frac{z^2}{2} + \frac{z^3}{3} - \frac{z^4}{4} + \cdots$$

$$
\begin{aligned}
&= \sum_{n=1}^{\infty}(-1)^{n+1}\frac{z^n}{n} \\
&= z\left(1 - \frac{z}{2} + \frac{z^2}{3} - \frac{z^3}{4} + \cdots\right) \\
&= z[1+\varepsilon(z)], \quad (z \neq 0),
\end{aligned} \tag{11.42}
$$

where

$$\varepsilon(z) = -\frac{z}{2} + \frac{z^2}{3} - \frac{z^3}{4} + \cdots$$

Now

$$
\begin{aligned}
|\varepsilon(z)| &= \left|-\frac{z}{2} + \frac{z^2}{3} - \frac{z^3}{4} + \cdots\right| \\
&= \left|\frac{z}{2}\left(-1 + \frac{2}{3}z - \frac{2}{4}z^2 + \cdots\right)\right| \\
&\leq \frac{|z|}{2}\left(1 + \frac{2}{3}|z| + \frac{2}{4}|z|^2 + \cdots.\right) \\
&\leq \frac{|z|}{2}\left(1 + |z| + |z|^2 + \cdots\right) \\
&= \frac{|z|}{2}\frac{1}{1-|z|} \leq |z|,
\end{aligned} \tag{11.43}
$$

provided $|z| \leq \frac{1}{2}$. Thus, we have $|\varepsilon(z)| \leq |z|$ for $|z| \leq \frac{1}{2}$ by (11.43) and $\log(1+z) = z[1+\varepsilon(z)]$ by (11.42). Of course, the fact that $|z| \leq \frac{1}{2}$ implies $|\varepsilon(z)| \leq \frac{1}{2}$; it also implies that we can write $\varepsilon(z) = \theta z$ for some $\theta = \theta(z)$ with $|\theta| \leq 1$; (i.e., $\theta(z) = \frac{\varepsilon(z)}{z}$). Thus, $\log(1+z) = z(1+\theta z)$. ■

11.9 Some Basic Concepts and Results from Complex Analysis Employed in the Proof of Theorem 11

In what follows, $\mathbb{C}$ stands for the complex plane, and $Im(z)$ stands for the imaginary part of the complex number $z = x + iy$; i.e., $Im(z) = y$, $x, y \in \Re$.

Definition 8. A function $g : S \subseteq \mathbb{C} \to \mathbb{C}$ is said to be *differentiable at* $z_0 \in S$ with *derivative* $g'(z_0)$, if

$$\frac{g(z) - g(z_0)}{z - z_0} \to g'(z_0)$$

as z tends to z_0 in all possible ways; g is *differentiable in* S, if it is differentiable at each $z \in S$. ■

Definition 9. The function g is called *analytic* in S, if it is differentiable in S. If $S = \mathbb{C}$ and g is analytic, then it is called *entire*. ■

Fact 1. If g is analytic in S, then the derivatives of all orders $g^{(n)}$, $n \geq 1$, exist (and are given by a certain formula involving the Cauchy integral).

Fact 2. A function g represented by a power series

$$g(z) = \sum_{n=0}^{\infty} c_n z^n, \quad |z| \leq r \text{ (some } r > 0)$$

is analytic in $|z| < r$.

Fact 3. If g is analytic for $|z| \leq r$ (some $r > 0$) (or analytic on and inside a simple closed contour C), then, for every z with $|z| < r$ (or every z inside C), $g(z)$ can be represented by a power series. More specifically, for every a with $|a| < r$ (or every a inside C),

$$g(z) = \sum_{n=0}^{\infty} g^{(n)}(a)(z-a)^n, \quad |z-a| < \delta,$$

where δ is the distance of a from the nearest point of the circumference $|z| = r$ (or the distance of a from the nearest point of C). In particular, $g(z) = \sum_{n=0}^{\infty} g^{(n)}(0)z^n$, $|z| < r$.

Facts 2 and 3 justify the following definition.

Definition 10. The ch.f. f is said to be r-*analytic*, if

$$f(t) = \sum_{n=0}^{\infty} \frac{a_n}{n!} t^n, \quad |t| < r \ (r > 0), \quad a_n \in \mathbb{C},$$

and is called *entire*, if $r = \infty$. ■

Let f be a ch.f. with corresponding d.f. F, and for $z = x + iy$, $x, y, \in \Re$, define $f(z)$ by

$$\begin{aligned} f(z) &= \int_{\Re} e^{izu} dF(u) = \int_{\Re} e^{ixu} \times e^{-yu} dF(u) \\ &= \int_{\Re} e^{-yu} \cos xu dF(u) + i \int_{\Re} e^{-yu} \sin xu dF(u), \end{aligned}$$

provided (see also Proposition 5 below),

$$\int_{\Re} e^{-yu} dF(u) < \infty. \tag{11.44}$$

Fact 4. Let f be a ch.f. of a r.v. with corresponding d.f. F, and assume that the moments $m^{(n)} = \int_{\Re} x^n dF(x)$ are finite for all $n = 1, 2, \ldots$, and that $S(t_0)$ given in Lemma 6 converges for some $t_0 \neq 0$. Then $f(t)$ is r-analytic for $|t| < r$ (some $r > 0$). Also, $f(z)$ is defined and is analytic for z in the strip defined by $|Im(z)| = |y| < r$.

This is so by Proposition 5, which follows. Furthermore, for z with $|z| < r$, all derivatives $f^{(n)}(z), n \geq 1$, exist and

$$f^{(n)}(z) = \sum_{k=n}^{\infty} \frac{i^k m^{(k)}}{(k-n)!} z^{k-n},$$

and, in particular,

$$f(z) = \sum_{k=0}^{\infty} \frac{m^{(k)}}{k!} (iz)^k.$$

Remark 11. That $f(z)$ is well defined for z with $|Im(z)| = |y| < r$ is due to the fact that relation (11.44) is satisfied here, as Proposition 5 below shows.

Fact 5. Let $g, g_0 : S \to \mathbb{C}$ (where $\{z \in \mathbb{C}; |Im(z)| < r\} \subseteq S \subseteq \mathbb{C}$), and suppose that $g(z)$ and $g_0(z)$ are analytic for z with $|Im(z)| < r$ $(r > 0)$. Furthermore, assume that $g(z) = g_0(z)$ for $|z| < r$. Then $g(z) = g_0(z)$ for z with $|Im(z)| < r$.

Proposition 5. Assume that the series $S(t) = \sum_{n=0}^{\infty} \frac{m^{(n)}}{n!}(it)^n$ converges for some $t_0 \neq 0$ (as we do in the formulation of Theorem 11), and assume without loss of generality that $t_0 > 0$. Then relation (11.44) is satisfied, so that $f(z)$ is well defined for z with $|Im(z)| = |y| < t_0$.

Proof. In the first place, convergence of the series $S(t)$ for t with $|t| < t_0$ implies convergence of the series $S^*(t) = \sum_{n=0}^{\infty} \frac{\mu^{(n)}}{n!}|t|^n$; this is so, by Lemma 6. Next, for $0 < t < t_0$,

$$\begin{aligned}
\int_{\Re} e^{t|x|} dF(x) &= \int_{\Re} \left(\sum_{n=0}^{\infty} \frac{|x|^n t^n}{n!} \right) dF(x) \\
&= \sum_{n=0}^{\infty} \int_{\Re} \frac{|x|^n t^n}{n!} dF(x) \\
&\quad \text{(by Corollary 1(ii) to Theorem 1 in Chapter 5)} \\
&= \sum_{n=0}^{\infty} \frac{t^n}{n!} \int_{\Re} |x|^n dF(x) = \sum_{n=0}^{\infty} \frac{\mu^{(n)}}{n!} t^n < \infty,
\end{aligned}$$

as already pointed out. But

$$\begin{aligned}
\int_{\Re} e^{t|x|} dF(x) &= \int_{-\infty}^{0} e^{t|x|} dF(x) + \int_{0}^{\infty} e^{t|x|} dF(x) \\
&= \int_{-\infty}^{0} e^{-tx} dF(x) + \int_{0}^{\infty} e^{tx} dF(x),
\end{aligned}$$

so that

$$\int_{-\infty}^{0} e^{-tx} dF(x) < \infty, \quad \int_{0}^{\infty} e^{tx} dF(x) < \infty. \tag{11.45}$$

Next, for $0 < y < t_0$,

$$\int_{\Re} e^{-yu} dF(u) = \int_{-\infty}^{0} e^{-yu} dF(u) + \int_{0}^{\infty} e^{-yu} dF(u)$$
$$\leq \int_{-\infty}^{0} e^{-yu} dF(u) + Var F < \infty \text{ (by (11.45))},$$

whereas, for $-t_0 < y < 0$,

$$\int_{\Re} e^{-yu} dF(u) = \int_{-\infty}^{0} e^{-yu} dF(u) + \int_{0}^{\infty} e^{-yu} dF(u)$$
$$= \int_{-\infty}^{0} e^{(-y)u} dF(u) + \int_{0}^{\infty} e^{(-y)u} dF(u)$$
$$\leq Var F + \int_{0}^{\infty} e^{(-y)u} dF(u) < \infty \text{ (by (11.45))}.$$

Thus, $\int_{\Re} e^{-yu} dF(u) < \infty$ for $|y| < t_0$, which is (11.44). ■

Remark 12. Material pertaining to this section may be found in the classical reference Titchmarsch (1939).

Exercises.

1. Let g and h be real-valued functions defined on $\Re$ for which $\int_{\Re} g(x)d\mu$ and $\int_{\Re} h(x)d\mu$ are finite, where μ is a (σ-finite) measure in $\Re$. Then show that:

$$\left| \int_{\Re} [g(x) + ih(x)] d\mu \right| \leq \int_{\Re} |g(x) + ih(x)| \, d\mu.$$

In particular, $|\mathcal{E}Z| \leq \mathcal{E}|Z|$, where Z is a complex-valued r.v.; i.e., $Z = X + iY$ with X and Y real-valued r.v.s.
Hint: Use polar coordinates.

2. In reference to the proof of Theorem 2, show that $\lim_{t \to 0} g(t) = b - a$ as claimed there.
3. In reference to the proof of Proposition 1 (see also proof of Proposition 3), provide the details of the convergence

$$\frac{1}{t} \int_{0}^{t} g(v) dv \to g(0) \text{ as } t \downarrow 0.$$

4. Show that $|e^{ix} - 1| \leq |x|$ for all $x \in \Re$.
Hint: Write $e^{ix} = \cos x + i \sin x$, express $\cos x$ in terms of $\sin^2 \frac{x}{2}$, and use the inequality $|\sin t| \leq |t|$, $t \in \Re$ (which you also must prove).
5. Show that $(1 + \frac{c_n}{n})^n \underset{n \to \infty}{\to} e^c$, when $c_n \underset{n \to \infty}{\to} c$, where $c_n \in \mathbb{C}$, the complex plane, $n \geq 1$.

6. Let X be a r.v. having the Cauchy distribution with parameters $\mu = 0$ and $\sigma = 1$ (i.e., the p.d.f. of X is given by $p(x) = \frac{1}{\pi} \times \frac{1}{1+x^2}$, $x \in \Re$). Then show that:

(i) The $\mathcal{E}X$ does not exist.

(ii) The ch.f. $f_X(t) = e^{-|t|}$, $t \in \Re$.
Next, let $X_1, \ldots, X_n$ be independent r.v.s distributed as X and set $S_n = X_1 + \cdots + X_n$. Then

(iii) Identify the ch.f. $f_{S_n/n}(t)$.

(iv) Show that $\frac{S_n}{n} \underset{n\to\infty}{\overset{P}{\nrightarrow}} 0$ by showing that $\frac{S_n}{n} \underset{n\to\infty}{\overset{d}{\nrightarrow}} 0$.

(Although, by intuition, one would expect such a convergence, because of symmetry about 0 of the Cauchy distribution!)
Hint: For part (ii), use the result $\int_0^\infty \frac{\cos(tx)}{1+x^2}dx = \frac{\pi}{2}e^{-|t|}$ (see, e.g., integral 403 in Tallarida (1999); also see integral 635 in the same reference).

7. Show that: $\frac{[(2n)!]^{1/2n}}{[(2n-1)!]^{1/(2n-1)}} \underset{n\to\infty}{\to} 1$.
Hint: Use the Stirling formula, which states that $n!/\sqrt{2\pi} \times n^{n+\frac{1}{2}} \times e^{-n}$ tends to 1 as $n \to \infty$.

8. Establish the validity of the relations claimed in the Corollary to Lemma 6.
Hint: Use the Stirling formula cited in the hint of Exercise 7.

9. If $X_1, \ldots, X_n$ are i.i.d. r.v.s with $\mathcal{E}X_1 = \mu \in \Re$ and $\sigma^2(X_1) = \sigma^2 \in (0, \infty)$, then (by Application 2 to Theorem 9 in this chapter) it follows that $\frac{S_n - n\mu}{\sigma\sqrt{n}} \underset{n\to\infty}{\Rightarrow} Z \sim N(0, 1)$, where $S_n = \sum_{j=1}^n X_j$.
Show that $\left\{\frac{S_n - n\mu}{\sigma\sqrt{n}}\right\}$ does *not* converge in probability as $n \to \infty$.
Hint: Set $Y_n = (S_n - n\mu)/\sigma\sqrt{n}$ and show that, as $n \to \infty$, $\{Y_n\}$ does not converge mutually in probability by showing that $\{Y_{2n} - Y_n\}$ does not converge in probability to 0.

10. According to the WLLN (Application 1 to Theorem 9), if the r.v.s $X_1, \ldots, X_n$ are i.i.d. with finite $\mathcal{E}X_1$, then

$$\bar{X}_n = \frac{S_n}{n} \underset{n\to\infty}{\overset{P}{\longrightarrow}} \mathcal{E}X_1, \text{ where } S_n = \sum_{j=1}^n X_j.$$

The following example shows that it is possible for $\left\{\frac{S_n}{n}\right\}$ to converge in probability to a finite constant, as $n \to \infty$, even if the $\mathcal{E}X_1$ does not exist. To this effect, for $j = 1, 2, \ldots$, let X_j be i.i.d. r.v.s such that $P(X_j = -n) = P(X_j = n) = c/n^2 \log n$, $n \geq 3$, where $c = \frac{1}{2}\sum_{n=3}^\infty \frac{1}{n^2}\log n)$. Then show that $\mathcal{E}X_1$ does not exist, but $\frac{S_n}{n} \underset{n\to\infty}{\overset{P}{\longrightarrow}} 0$, where $S_n = \sum_{j=1}^n X_j$.
Hint: Show that $\mathcal{E}X_1$ does not exist by showing that $\mathcal{E}X_1^+ = \mathcal{E}X_1^- = \infty$. Next, set $X_{nj} = X_j$ if $|X_j| < n$, and $X_{nj} = 0$ otherwise, $j = 1, 2, \ldots, n \geq 3$,

and let $S_n^* = \sum_{j=1}^n X_{nj}$. Then show that (i) $\frac{S_n}{n} - \frac{S_n^*}{n} \xrightarrow[n\to\infty]{P} 0$ by showing that $P\left(\frac{S_n}{n} \neq \frac{S_n^*}{n}\right) \underset{n\to\infty}{\longrightarrow} 0$; (ii) $\mathcal{E}\left(\frac{S_n^*}{n}\right) = 0$; (iii) $Var\left(\frac{S_n^*}{n}\right) \underset{n\to\infty}{\longrightarrow} 0$; (iv) from (ii) and (iii), conclude that $\frac{S_n^*}{n} \xrightarrow[n\to\infty]{P} 0$; then (i) and (iv) complete the proof.

In all Exercises 11–16, i is to be treated as a real number, subject, of course, to the requirement that $i^2 = -1$.

11. If $X \sim B(n, p)$, show that

$$f_X(t) = (pe^{it} + q)^n, \quad q = 1 - p.$$

12. If $X \sim P(\lambda)$, show that

$$f_X(t) = e^{\lambda e^{it} - \lambda}.$$

13. **(i)** If $Z \sim N(0, 1)$, show that $f_Z(t) = e^{-t^2/2}$.

(ii) If $X \sim N(\mu, \sigma^2)$, use the fact that $Z = \frac{X-\mu}{\sigma} \sim N(0, 1)$ and part (i) in order to show that

$$f_X(t) = e^{i\mu t - \frac{\sigma^2 t^2}{2}}.$$

14. If X has the Gamma distribution with parameters α and β; i.e., if its p.d.f. is given by

$$p(x; \alpha, \beta) = \frac{1}{\Gamma(\alpha)\beta^\alpha} x^{\alpha-1} e^{-x/\beta}, \quad x > 0 \text{ (and 0 for } x \le 0),$$

where the Gamma function $\Gamma(\alpha)$ is given by $\Gamma(\alpha) = \int_0^\infty y^{\alpha-1} e^{-y} dy$ $(\alpha, \beta > 0)$, then show that

$$f_X(t) = \frac{1}{(1 - i\beta t)^\alpha}.$$

In particular, for $\alpha = 1$ and $\beta = 1/\lambda$, we get the ch.f. of the Negative Exponential distribution with parameter λ; i.e., $f_X(t) = 1/(1 - \frac{it}{\lambda})$; and for $\alpha = \frac{r}{2}$ ($r > 0$ integer) and $\beta = 2$, we get the ch.f. of the chi-square distribution with r degrees of freedom; i.e., $f_X(t) = 1/(1 - 2it)^{r/2}$.

15. If the r.v.s X and Y have the Bivariate Normal distribution with parameters μ_1, μ_2 in $\Re$, $0 < \sigma_1, \sigma_2 < \infty$, and $\rho \in [-1, 1]$, show that their joint ch.f. is given by

$$f_{X,Y}(t_1, t_2) = \exp\left[i\mu_1 t_1 + i\mu_2 t_2 - \frac{1}{2}\left(\sigma_1^2 t_1^2 + 2\rho\sigma_1\sigma_2 t_1 t_2 + \sigma_2^2 t_2^2\right)\right].$$

For this purpose, do the following:

(i) Assume first that $\mu_1 = \mu_2 = 0$ and $\sigma_1 = \sigma_2 = 1$, and use Exercises 12 (ii) in Chapter 9 and 13 (ii) in this chapter to show that:

$$f_{X,Y}(t_1, t_2) = \exp\left[-\frac{1}{2}\left(t_1^2 + 2\rho t_1 t_2 + t_2^2\right)\right].$$

(ii) For the general case, use the transformations $U = (X - \mu_1)/\sigma_1$, $V = (Y - \mu_2)/\sigma_2$ and verify that $\mathcal{E}U = \mathcal{E}V = 0$, $Var(U) = Var(V) = 1$, $\rho(U, V) = \rho(X, Y) = \rho$. Then use Exercise 15 in Chapter 9 and part (i) here to arrive at the desired expression for the ch.f. $f_{X,Y}$.

16. Let the r.v.s X and Y have the Bivariate Normal distribution with parameters $\mu_1, \mu_2 \in \Re$, $0 < \sigma_1, \sigma_2 < \infty$, and $\rho \in [-1, 1]$, and set $U = X+Y$, $V = X-Y$.

(i) Verify that $\mathcal{E}U = \mu_1 + \mu_2$, $Var(U) = \sigma_1^2 + \sigma_2^2 + 2\rho\sigma_1\sigma_2$, $\mathcal{E}V = \mu_1 - \mu_2$, $Var(V) = \sigma_1^2 + \sigma_2^2 - 2\rho\sigma_1\sigma_2$, and $Cov(U, V) = \sigma_1^2 - \sigma_2^2$ (by using Exercises 12 (ii) and 14 (ii) in Chapter 9).

(ii) Since

$$
\begin{aligned}
f_{U,V}(t_1, t_2) &= \mathcal{E}e^{it_1U+it_2V} = \mathcal{E}e^{it_1(X+Y)+it_2(X-Y)} \\
&= \mathcal{E}e^{i(t_1+t_2)X+i(t_1-t_2)Y} \\
&= f_{X,Y}(t_1+t_2, t_1-t_2),
\end{aligned}
$$

use Exercise 15 in order to conclude that

$$
\begin{aligned}
f_{U,V}(t_1, t_2) = \exp\{&i(\mu_1+\mu_2)t_1 + i(\mu_1-\mu_2)t_2 \\
&-\frac{1}{2}\Big[(\sigma_1^2+\sigma_2^2+2\rho\sigma_1\sigma_2)t_1^2 + 2(\sigma_1^2-\sigma_2^2)t_1t_2 \\
&+(\sigma_1^2+\sigma_2^2-2\rho\sigma_1\sigma_2)t_2^2\Big]\}.
\end{aligned}
$$

(iii) From part (ii) and Exercise 15, conclude that the r.v.s U and V have the Bivariate Normal distribution with parameters $\mu_1+\mu_2$, $\mu_1-\mu_2$, $\sigma_1^2+\sigma_2^2+2\rho\sigma_1\sigma_2 = \tau_1^2$, $\sigma_1^2+\sigma_2^2-2\rho\sigma_1\sigma_2 = \tau_2^2$, and $\rho(U, V) = (\sigma_1^2-\sigma_2^2)/\tau_1\tau_2$.

(iv) From part (iii) and Exercise 12 (ii) in Chapter 10, conclude that U and V are independent if and only if $\sigma_1 = \sigma_2$.

17. In this exercise, the r.v.s $X_1, \ldots, X_k$ are independent with distributions as indicated, and $X = X_1+\cdots+X_k$. Then use Exercise 13 in Chapter 10, the inversion formula (Theorem 2), and the appropriate ch.f.s in order to show that:

(i) If $X_j \sim B(n_j, p)$, $j = 1, \ldots, k$, then $X \sim B(n, p)$, where $n = n_1+\cdots+n_k$.

(ii) If $X_j \sim P(\lambda_j)$, $j = 1, \ldots, k$, then $X \sim P(\lambda)$, where $\lambda = \lambda_1+\cdots+\lambda_k$.

(iii) If $X_j \sim N(\mu_j, \sigma_j^2)$, $j = 1, \ldots, k$, then $X \sim N(\mu, \sigma^2)$, where $\mu = \mu_1+\cdots+\mu_k$ and $\sigma^2 = \sigma_1^2+\cdots+\sigma_k^2$. Also, $c_1X_1+\cdots+c_kX_k \sim N(c_1\mu_1+\cdots+c_k\mu_k, c_1^2\sigma_1^2+\cdots+c_k^2\sigma_k^2)$, where $c_1, \ldots, c_k$ are constants.

(iv) If $X_j \sim$ Gamma with parameters α_j and β, $j = 1, \ldots, k$, then $X \sim$ Gamma with parameters $\alpha = \alpha_1+\cdots+\alpha_k$ and β. In particular, if X_j has the Negative Exponential distribution with parameter λ, $j = 1, \ldots, k$, then $X \sim$ Gamma with parameters $\alpha = k$ and $\beta = 1/\lambda$, whereas, if $X_j \sim \chi^2_{r_j}$, $j = 1, \ldots, k$, then $X \sim \chi^2_r$, where $r = r_1+\cdots+r_k$.

18. Show that the r.v.s $X_1, \ldots, X_k$ are independent if and only if, for all $t_1, \ldots, t_k$ in $\Re$, $f_{X_1,\ldots,X_k}(t_1, \ldots, t_k) = f_{X_1}(t_1) \times \cdots \times f_{X_k}(t_k)$.
Hint: Use Proposition 1 in Chapter 10 (and work as in Exercise 13 in the same chapter), and Theorem 2′.

19. If f is a ch.f., then show that f is positive definite; i.e., for all t_k, t_l reals and all complex numbers z_k, z_l, we have

$$\sum_{k=1}^{n}\sum_{l=1}^{n} f(t_k - t_l) z_k \bar{z}_l \geq 0$$

for every integer $n \geq 1$.

20. For $n = 1, 2, \ldots$, let X_n be a r.v. distributed as Poisson with parameter n, $X_n \sim P(n)$ (so that $\mathcal{E}X_n = Var(X_n) = n$), and set $Y_n = (X_n - n)/\sqrt{n}$. Then show that $Y_n \xrightarrow[n\to\infty]{d} Z \sim N(0, 1)$.
Hint: Use the expansion: $e^{iz} = 1 + iz - \frac{z^2}{2} - \frac{iz^3}{6} e^{iz^*}$ for some complex number z^*.

21. Let F_1 and F_2 be d.f.s, and let G be their convolution, $G = F_1 * F_2$. Then:

(i) If F_1 is absolutely continuous with respect to Lebesgue measure with p.d.f. p_1, it follows that G is absolutely continuous with respect to Lebesgue measure with p.d.f. p given by

$$p(u) = \int_{-\infty}^{\infty} p_1(u - y) dF_2(y).$$

(ii) Furthermore, if F_2 is also absolutely continuous with respect to Lebesgue measure with p.d.f. p_2, then

$$p(u) = \int_{-\infty}^{\infty} p_1(u - y) p_2(y) dy.$$

22. If the r.v.s X and Y are i.i.d. with ch.f. f, then the ch.f. of the r.v. $X - Y$ is $|f(t)|^2$.

23. Let $X_1, \ldots, X_n$ be independent r.v.s, each of which is symmetric about 0. Then the r.v. $X_1 + \cdots + X_n$ is also symmetric about 0.

24. For $n \geq 1$, let g_n and g be functions defined on $E \subseteq \Re^k$ into $\Re$, and recall that $\{g_n\}$ is said to converge *continuously* to g on E, if for every $x \in E$, $g_n(x_n) \to g(x)$ whenever $x_n \to x$, as $n \to \infty$. Then show that if $\{g_n\}$ converges continuously to g on E, it follows that g is continuous on E.

25. For $n \geq 1$, let g_n, g and E be as in Exercise 24. Then:

(i) If $\{g_n\}$ converges to g uniformly on E and g is continuous on E, it follows that $\{g_n\}$ converges continuously to g on E.
(ii) If E is compact and $\{g_n\}$ converges continuously to g on E, then the convergence is uniform.

26. For $n \geq 1$, let X_n, Y_n and X, Y be r.v.s defined on the probability space $(\Omega, \mathcal{A}, P)$, let d_1, d_2 and c_n be constants with $0 \neq c_n \to \infty$ as $n \to \infty$, and suppose that

$c_n(X_n - d_1, Y_n - d_2) \xrightarrow{d} (X, Y)$ as $n \to \infty$. Let $g : \Re^2 \to \Re$ be differentiable, and suppose its (first-order) partial derivatives $\dot{g}_x, \dot{g}_y$ are continuous at (d_1, d_2). Then show that, as $n \to \infty$,

$$\begin{aligned} c_n[g(X_n, Y_n) - g(d_1, d_2)] &\xrightarrow{d} [\dot{g}_x(d_1, d_2) \quad \dot{g}_y(d_1, d_2)](X \quad Y)' \\ &= \dot{g}_x(d_1, d_2)X + \dot{g}_y(d_1, d_2)Y. \end{aligned}$$

This page is intentionally left blank

CHAPTER

The Central Limit Problem: The Centered Case

12

In this chapter, we discuss what is perhaps the most important problem in probability theory. Namely, conditions are sought under which partial sums of independent r.v.s converge in distribution to some limiting law. Also, the determination of the totality of such laws is sought.

A special but extremely important case is that where the limiting law is the Normal distribution, and a very special case of it is the Central Limit Theorem (CLT) in its simplest and popular form.

In order to cover as many cases of practical importance as feasible, the approach used here is that of employing a triangular array of r.v.s. The r.v.s within each row are assumed to be independent, but no such assumption is made for those in distinct rows. Of course, it is assumed that the number of r.v.s in the rows tends to infinity as the rank of rows increases.

A brief description of what is done in this chapter is as follows. After basic notation and assumptions are taken care of, necessary and sufficient conditions are given for the sequence of partial sums to converge in distribution to a $N(0, 1)$ distributed r.v. This is the extremely useful Normal Convergence Criterion due to Lindeberg and Feller, and stated as Theorem 1. Its proof is derived as a special case of Theorem 2 stated and proved in Section 12.1, and is deferred to Section 12.3.

In Section 12.2, the problem is cast in broad generality. Instead of seeking conditions under which the limiting law is normal, we investigate under what conditions there is convergence in distribution. Also, the totality of such limiting laws is identified. The culmination of these discussions is Theorem 2, which follows after a number of auxiliary results are established.

In Section 12.3, first the proof of Theorem 1 is derived as a special case of that of Theorem 2. Then several sets of sufficient conditions are given for convergence in distribution to the $N(0, 1)$ distribution. One such condition is expressed in terms of absolute moments and is the basis of the so-called Liapounov Theorem (Theorem 3). If the sum of variances of the r.v.s in each row is not equal to 1, which is part of the basic assumption made, then Theorem 1 assumes an easy modification, which is Theorem 4. The next modification of Theorem 1 is that where one is dealing with a single sequence of r.v.s, and leads to Theorem 5. Should the underlying r.v.s be also identically distributed, then one has the Corollary to Theorem 5, which is the CLT in its simplest and popular form.

An Introduction to Measure-Theoretic Probability, Second Edition. http://dx.doi.org/10.1016/B978-0-12-800042-7.00012-8

This chapter is concluded with a final section, Section 12.4, where technical results stated in Section 12.2 are proved.

In conclusion, it is to be pointed out that the term "centered" used in the title of this chapter simply indicates that all r.v.s involved have expectations zero. This is always achieved by subtracting off the expectations from the r.v.s, or by "centering" the r.v.s.

12.1 Convergence to the Normal Law (Central Limit Theorem, CLT)

In the present section, we shall concern ourselves with an array of r.v.s X_{nj}, $j = 1, \ldots, k_n$, where $k_n \underset{n\to\infty}{\to} \infty$, and for each n, the r.v.s $X_{n1}, \ldots, X_{nk_n}$ are independent, but not necessarily identically distributed.

$$\begin{cases} X_{11}, X_{12}, \ldots, X_{1k_1} & \text{independent} \\ X_{21}, X_{22}, \ldots, X_{2k_2} & \text{independent} \\ \cdots\cdots\cdots\cdots\cdots & \cdots\cdots\cdots \\ X_{n1}, X_{n2}, \ldots, X_{nk_n} & \text{independent} \\ \cdots\cdots\cdots\cdots\cdots & \cdots\cdots\cdots \end{cases}$$

However, r.v.s in different rows need not be independent. The probability spaces on which rows are defined may even be different. We shall assume throughout that $\mathcal{E}X_{nj}$, $j = 1, \ldots, k_n$, are finite for all n. Then we may consider the r.v.s $Y_{nj} = X_{nj} - \mathcal{E}X_{nj}$ with $\mathcal{E}Y_{nj} = 0$. Thus, without loss of generality, we shall assume that

$$\mathcal{E}X_{nj} = 0, \quad j = 1, \ldots, k_n, \quad n = 1, 2, \ldots.$$

To summarize:

For each n, X_{nj}, $j = 1, \ldots, k_n$, are independent r.v.s. (within each row), $k_n \underset{n\to\infty}{\to} \infty$, and $\mathcal{E}X_{nj} = 0$, $j = 1, \ldots, k_n, n = 1, 2, \ldots$.

In the present section, we shall also assume that

$$\sigma_{nj}^2 = \sigma^2(X_{nj}) = \mathcal{E}X_{nj}^2 < \infty, \quad j = 1, \ldots, k_n, \quad n = 1, 2, \ldots,$$

and we shall set

$$s_n^2 = \sum_{j=1}^{k_n} \sigma_{nj}^2 \text{ and } s_n = +\sqrt{s_n^2}.$$

Then the problem to concern ourselves with is this: under what condition is it true that

$$\mathcal{L}\left(\frac{S_n}{s_n}\right) \underset{n\to\infty}{\Rightarrow} N(0, 1), \text{ where } S_n = \sum_{j=1}^{k_n} X_{nj}?$$

Here and in the sequel, it is convenient to denote the distribution of a r.v. X by $\mathcal{L}(X)$, and also read it as the (distribution) law of X. Then by $\mathcal{L}(X_n) \underset{n\to\infty}{\Rightarrow} \mathcal{L}$ we understand the usual weak convergence of d.f.s, or convergence of r.v.s in distribution. That is, $F_{X_n} \underset{n\to\infty}{\Rightarrow} F_X$, or $X_n \underset{n\to\infty}{\overset{d}{\to}} X$.

Before we formulate the problem precisely and solve it, we notice that for $k_n = n$, the array of the r.v.s in question becomes a triangular array; i.e.,

$$\begin{cases} X_{11} \\ X_{21}, X_{22} \\ \cdots\cdots\cdots \\ X_{n1}, X_{n2}, \ldots, X_{nn} \\ \cdots\cdots\cdots\cdots\cdots\cdots \end{cases}$$

Furthermore, if $X_{nj} = X_j$, $j = 1, \ldots, n$, and they are also identically distributed, then we have $\mathcal{E}X_j = 0$, $\sigma^2(X_j) = \sigma^2$, $s_n^2 = n\sigma^2$, and

$$\begin{cases} X_1 \\ X_1, X_2 \\ \cdots\cdots\cdots \\ X_1, X_2, \ldots, X_n \\ \cdots\cdots\cdots\cdots\cdots \end{cases}$$

with $S_n = \sum_{j=1}^n X_{nj} = \sum_{j=1}^n X_j$.

Since $s_n = \sigma\sqrt{n}$, then the problem of finding conditions under which

$$\mathcal{L}\left(\frac{S_n}{s_n}\right) = \mathcal{L}\left(\frac{\sum_{j=1}^n X_j}{\sigma\sqrt{n}}\right) \underset{n\to\infty}{\Rightarrow} N(0,1)$$

becomes the classical CLT.

Going back to the problem we started out with, we have the r.v.s X_{nj}, $j = 1, \ldots, k_n \underset{n\to\infty}{\to} \infty$, independent in each row,

$$\mathcal{E}X_{nj} = 0, \quad \sigma_{nj}^2 = \sigma^2(X_{nj}) = \mathcal{E}X_{nj}^2 < \infty, \quad s_n^2 = \sum_{j=1}^{k_n} \sigma_{nj}^2,$$

$$S_n = \sum_{j=1}^{k_n} X_{nj}, \quad \frac{S_n}{s_n} = \sum_{j=1}^{k_n} \frac{X_{nj}}{s_n}.$$

Then

$$\mathcal{E}\left(\frac{X_{nj}}{s_n}\right) = 0 \text{ and } \sum_{j=1}^{k_n} \sigma^2\left(\frac{X_{nj}}{s_n}\right) = \frac{1}{s_n^2}\sum_{j=1}^{k_n} \sigma_{nj}^2 = 1.$$

Thus, replacing X_{nj} by $Z_{nj} = \frac{X_{nj}}{s_n}$, we have $\sum_{j=1}^n \sigma^2(Z_{nj}) = 1$ and therefore, without loss of generality, we may assume that $s_n^2 = 1$.

In order to summarize, we have then:

For each n, the r.v.s X_{nj}, $j = 1, \ldots, k_n \underset{n\to\infty}{\to} \infty$ are row-wise independent, $\mathcal{E}X_{nj} = 0, \sigma^2_{nj} = \sigma^2(X_{nj}) < \infty, s^2_n = \sum_{j=1}^{k_n} \sigma^2_{nj} = 1$.

Set $S_n = \sum_{j=1}^{k_n} X_{nj}$ and find conditions under which

$$\mathcal{L}(S_n) \underset{n\to\infty}{\Rightarrow} N(0, 1).$$

The condition to be imposed upon the r.v.s X_{nj} in order for $\mathcal{L}(S_n) \underset{n\to\infty}{\Rightarrow} N(0, 1)$ to hold true is of the following nature: it is assumed that the contribution of each one of the summands X_{nj} in the sum S_n is *asymptotic negligible* in the following sense.

Definition 1. The r.v.s X_{nj}, $j = 1, \ldots, k_n \underset{n\to\infty}{\to} \infty$, are said to be *uniformly asymptotic negligible* (*u.a.n.*) if, for every $\varepsilon > 0$,

$$\max_{1\le j\le k_n} P(|X_{nj}| > \varepsilon) \underset{n\to\infty}{\to} 0. \qquad \blacksquare$$

Actually, we shall make an assumption that will imply *u.a.n.* Namely,

$$\max_{1\le j\le k_n} \sigma^2_{nj} \underset{n\to\infty}{\to} 0.$$

That this assumption implies *u.a.n.* is seen thus

$$\max_{1\le j\le k_n} P(|X_{nj}| > \varepsilon) \le \max_{1\le j\le k_n} \frac{\mathcal{E}X^2_{nj}}{\varepsilon^2} = \frac{1}{\varepsilon^2} \max_{1\le j\le k_n} \sigma^2_{nj} \underset{n\to\infty}{\to} 0.$$

We are going to show the following result, which is the general form of the CLT, and from which other variants follow.

Theorem 1 (Normal Convergence Criterion or Lindeberg–Feller Theorem). For each n, let the r.v.s X_{nj}, $j = 1, \ldots, k_n \underset{n\to\infty}{\to} \infty$, be row-wise independent, assume that $\mathcal{E}X_{nj} = 0$ and $\sigma^2_{nj} = \sigma^2(X_{nj}) < \infty$, and set $s^2_n = \sum_{j=1}^{k_n} \sigma^2_{nj} = 1$ (by assumption). Then

$$\mathcal{L}\left(\sum_{j=1}^{k_n} X_{nj}\right) \underset{n\to\infty}{\Rightarrow} N(0, 1) \text{ and } \max_{1\le j\le k_n} \sigma^2_{nj} \underset{n\to\infty}{\to} 0,$$

if and only if, for every $\varepsilon > 0$,

$$g_n(\varepsilon) = \sum_{j=1}^{k_n} \int_{(|x|\ge\varepsilon)} x^2 \, dF_{nj} \underset{n\to\infty}{\to} 0,$$

where F_{nj} is the d.f. of the r.v. X_{nj}. $\blacksquare$

This theorem is a special case of a more general result to be obtained in the next section. The proof of Theorem 1 and of some variations of it are given in Section 12.3.

This section is concluded with three examples. The first illustrates the checking of the condition $g_n(\varepsilon) \underset{n\to\infty}{\to} 0$; the second presents a case where the condition $g_n(\varepsilon) \underset{n\to\infty}{\to} 0$ is not fulfilled; and the third is meant to shed some light on what lies behind the condition $g_n(\varepsilon) \underset{n\to\infty}{\to} 0$.

Example 1. Let $Y_1, \ldots, Y_n$ be i.i.d. r.v.s with $\mathcal{E}Y_1 = \mu \in \Re$ and $\sigma^2(Y_1) = \sigma^2 \in (0, \infty)$, and set $X_{nj} = \frac{Y_j - \mu}{\sigma\sqrt{n}}$, $j = 1, \ldots, n$. Then, for each n, $X_{n1}, \ldots, X_{nn}$ are i.i.d. with $\mathcal{E}X_{n1} = 0$, $\sigma^2(X_{n1}) = \sigma^2_{n1} = \sigma^2_{nj} = \frac{1}{n}$, so that $s_n^2 = \sum_{j=1}^n \sigma^2_{nj} = 1$ (and $\max_{1\le j\le n}\sigma^2_{nj} = \frac{1}{n} \underset{n\to\infty}{\to} 0$). Then by Theorem 1, $\mathcal{L}\left(\sum_{j=1}^n X_{nj}\right) \underset{n\to\infty}{\Rightarrow} N(0,1)$ if and only if $g_n(\varepsilon) = \sum_{j=1}^n \int_{(|x|\ge\varepsilon)} x^2 dF_{nj}(x) \underset{n\to\infty}{\to} 0$ for every $\varepsilon > 0$, where $F_{nj} = F_n$ is the d.f. of the X_{nj}s. Actually, the asymptotic normality has been established in Application 2 of Chapter 11. So, it must hold that $g_n(\varepsilon) \underset{n\to\infty}{\to} 0$ for every $\varepsilon > 0$. That this is, indeed, the case is seen as follows:

$g_n(\varepsilon) = n\int_{(|x|\ge\varepsilon)} x^2 dF_n(x) = n\int_{(|x|\ge\varepsilon)} x^2 dF(\mu + x\sigma\sqrt{n})$, since $F_n(x) = P(X_{n1} \le x) = P(Y_1 \le \mu + x\sigma\sqrt{n}) = F(\mu + x\sigma\sqrt{n})$, where F is the d.f. of the Y_js. By setting $\mu + x\sigma\sqrt{n} = y$, we get

$$
\begin{aligned}
g_n(\varepsilon) &= \frac{n}{\sigma^2 n}\int_{(|y-\mu|\ge\varepsilon\sigma\sqrt{n})} (y-\mu)^2\, dF(y) \\
&= \frac{1}{\sigma^2}\int_{\Re} (y-\mu)^2\, dF(y) - \frac{1}{\sigma^2}\int_{(|y-\mu|<\varepsilon\sigma\sqrt{n})} (y-\mu)^2\, dF(y) \\
&= 1 - \frac{1}{\sigma^2}\int_{(|y-\mu|<\varepsilon\sigma\sqrt{n})} (y-\mu)^2\, dF(y) \\
&= 1 - \frac{1}{\sigma^2}\int_{\Re}\left[(y-\mu)^2 I_{(|y-\mu|<\varepsilon\sigma\sqrt{n})}\right] dF(y).
\end{aligned}
$$

However, $(y-\mu)^2 I_{(|y-\mu|<\varepsilon\sigma\sqrt{n})} \le (y-\mu)^2$ independent of n and F-integrable. Also, $(y-\mu)^2 I_{(|y-\mu|<\varepsilon\sigma\sqrt{n})} \underset{n\to\infty}{\to} (y-\mu)^2$. Then, by the Dominated Convergence Theorem, $\int_{\Re}\left[(y-\mu)^2 I_{(|y-\mu|<\varepsilon\sigma\sqrt{n})}\right] dF(y) \underset{n\to\infty}{\to} \int_{\Re}(y-\mu)^2 dF(y) = \sigma^2$. It follows that $g_n(\varepsilon) \underset{n\to\infty}{\to} 0$ for every $\varepsilon > 0$. (See also the Corollary to Theorem 5 later.) ■

Of course, the example just discussed also applies to the special case where the Y_js have the Binomial distribution $B(1, p)$ $(0 < p < 1)$. Here is a version of this special case.

Example 2. For $j = 1, \ldots, n$, let Y_{nj} be independently distributed as $B(1, p_n)$ where $0 < p_n < 1$ with $p_n \underset{n\to\infty}{\to} 0$, so that $np_n \underset{n\to\infty}{\to} \lambda \in (0, \infty)$. Then $\mathcal{E}Y_{nj} = p_n$ and $\sigma^2(Y_{nj}) = p_n q_n$, $j = 1, \ldots, n$, where $q_n = 1 - p_n$. Set $X_{nj} = (Y_{nj} - p_n)/\sqrt{np_nq_n}$, so that $X_{n1}, \ldots, X_{nn}$ are independent, $\mathcal{E}X_{nj} = 0$, $\sigma^2(X_{nj}) = \sigma^2_{nj} = 1/n$, and $s_n^2 = \sum_{j=1}^n \sigma^2_{nj} = 1$ (and $\max_{1\le j\le n}\sigma^2_{nj} = \frac{1}{n} \underset{n\to\infty}{\to} 0$). Therefore, for each n, the r.v.s

$X_{n1}, \ldots, X_{nn}$ satisfy the conditions of Theorem 1, and therefore $\mathcal{L}(S_n) \underset{n\to\infty}{\Rightarrow} N(0,1)$ if and only if $g_n(\varepsilon) = \sum_{j=1}^{n} \int_{(|x|\geq\varepsilon)} x^2 dF_{nj}(x) = n\int_{(|x|\geq\varepsilon)} x^2 dF_n(x) \underset{n\to\infty}{\to} 0$ for every $\varepsilon > 0$, where $S_n = \sum_{j=1}^{n} X_{nj}$ and F_n is the d.f. of the X_{nj}s. It is shown here that for all (sufficiently small) $\varepsilon > 0$, $g_n(\varepsilon) \nrightarrow 0$, so that $\mathcal{L}(S_n) \underset{n\to\infty}{\not\Rightarrow} N(0,1)$. To this end, observe that the r.v. X_{n1} takes on the values $(1-p_n)/\sqrt{np_nq_n}$ and $-p_n/\sqrt{np_nq_n}$ with respective probabilities p_n and q_n. Hence

$$\begin{aligned} g_n(\varepsilon) &= n\int_{(|x|\geq\varepsilon)} x^2\,dF_n(x) \\ &= n\int_{(x\geq\varepsilon)} x^2\,dF_n(x) + n\int_{(x\leq-\varepsilon)} x^2\,dF_n(x) \\ &= n\int_{(x\geq\varepsilon)} x^2\,dF_n(x), \end{aligned}$$

because for sufficiently large n, $-\varepsilon < -p_n/\sqrt{np_nq_n}$, since $p_n/\sqrt{np_nq_n}$ tends to 0 as $n \to \infty$. Next, $\frac{1-p_n}{\sqrt{np_nq_n}} = \frac{q_n}{\sqrt{np_nq_n}} \underset{n\to\infty}{\to} \frac{1}{\sqrt{\lambda}} > 2\varepsilon$ (for sufficiently small ε), so that for sufficiently large n (and sufficiently small ε) $\frac{q_n}{\sqrt{np_nq_n}} > \varepsilon$.

It follows that

$$n\int_{(x\geq\varepsilon)} x^2\,dF_n(x) = n \times \frac{q_n^2}{np_nq_n} \times p_n = q_n \underset{n\to\infty}{\to} 1;$$

i.e., $g_n(\varepsilon) \nrightarrow 0$ for all (sufficiently small) $\varepsilon > 0$, as asserted.

It is to be observed, however, that $S_n = a_nT_n + b_n$, where $T_n = \sum_{j=1}^{n} Y_{nj} \sim B(n, p_n)$, $a_n = 1/\sqrt{np_nq_n}$, $b_n = -np_n/\sqrt{np_nq_n}$, and that $P(T_n = t) \underset{n\to\infty}{\to} P(T = t)$, where $T \sim P(\lambda)$, $t = 0, 1, \ldots$ (see, e.g., Theorem 1, Chapter 3, in Roussas (1997)). It follows (by Theorem 7 in Chapter 5) that $a_nT_n + b_n \underset{n\to\infty}{\overset{d}{\to}} \frac{1}{\sqrt{\lambda}}T - \sqrt{\lambda} \overset{\text{def}}{=} X$, where X takes on the values $\frac{x}{\sqrt{\lambda}} - \sqrt{\lambda}$ with respective probabilities $e^{-\lambda}\frac{\lambda^x}{x!}$, $x = 0, 1, \ldots$.

(Refer also to the Example 4 here, as well as Application 1 in Chapter 13.) ■

Example 3. For $j = 1, \ldots, n$, let X_{nj} be independent r.v.s distributed as $N(0, 1/n)$, so that $\mathcal{E}X_{nj} = 0, \sigma^2(X_{nj}) = \sigma_{nj}^2 = \frac{1}{n}, s_n^2 = 1$, and $\max_{1\leq j\leq n}\sigma_{nj}^2 = \frac{1}{n} \underset{n\to\infty}{\to} 0$. Thus, the contribution of each one of the r.v.s X_{nj}, $j = 1, \ldots, n$, to their sum $S_n = \sum_{j=1}^{n} X_{nj}$ is negligible. More precisely, the r.v.s X_{nj}, $j = 1, \ldots, n$, are *u.a.n.* Since $S_n \sim N(0,1)$, it follows that $g_n(\varepsilon) \underset{n\to\infty}{\to} 0$ for every $\varepsilon > 0$ (by Theorem 1).

In order to gain some insight as to why $g_n(\varepsilon)$ should converge to 0, and what this fact reflects, let us proceed with the computation of $g_n(\varepsilon)$. Here

$$g_n(\varepsilon) = n\int_{(|x|\geq\varepsilon)} x^2\,dF_n(x), \quad \text{where } F_n \text{ is the d.f. of } X_{n1}$$

$$= n\int_{(|x|\geq\varepsilon)} x^2 p_n(x)dx, \quad \text{where } p_n(x) = \frac{\sqrt{n}}{\sqrt{2\pi}}e^{-\frac{nx^2}{2}}.$$

But

$$\begin{aligned}\int_{(|x|\geq\varepsilon)} x^2 p_n(x)dx &= \frac{\sqrt{n}}{\sqrt{2\pi}}\int_{(|x|\geq\varepsilon)} x^2 e^{-\frac{nx^2}{2}}\,dx\\ &= \frac{2\sqrt{n}}{\sqrt{2\pi}}\int_{(x\geq\varepsilon)} x^2 e^{-\frac{nx^2}{2}}\,dx,\end{aligned}$$

and

$$\int_{(x\geq\varepsilon)} x^2 e^{-\frac{nx^2}{2}}\,dx = \frac{\varepsilon}{n}e^{-\frac{n\varepsilon^2}{2}} + \frac{1}{n}\int_{(x\geq\varepsilon)} e^{-\frac{nx^2}{2}}\,dx$$

(by integration by parts).

Therefore,

$$g_n(\varepsilon) = \frac{2\varepsilon}{\sqrt{2\pi}}\sqrt{n}e^{-\frac{n\varepsilon^2}{2}} + \frac{2\sqrt{n}}{\sqrt{2\pi}}\int_{(x\geq\varepsilon)} e^{-\frac{nx^2}{2}}\,dx.$$

However, $\sqrt{n}e^{-n\varepsilon^2/2} \underset{n\to\infty}{\to} 0$, and

$$\begin{aligned}\frac{2\sqrt{n}}{\sqrt{2\pi}}\int_{(x\geq\varepsilon)} e^{-\frac{nx^2}{2}}\,dx &= 2\int_{(x\geq\varepsilon)} \frac{1}{\sqrt{2\pi}\times(1/\sqrt{n})}e^{-x^2/2(1/\sqrt{n})^2}\,dx\\ &= 2P(X_{n1}\geq\varepsilon)\\ &= P(|X_{n1}|\geq\varepsilon)\\ &= \max_{1\leq j\leq n} P(|X_{nj}|\geq\varepsilon)\\ &\leq \frac{1}{n\varepsilon^2} \underset{n\to\infty}{\to} 0.\end{aligned}$$

So, $g_n(\varepsilon)$ consists essentially of the maximum probability of the individual X_{nj}s being outside $(-\varepsilon, \varepsilon)$, and this ought to be small, as the X_{nj}s are concentrated around 0. Another way of looking at $g_n(\varepsilon)$ is to intepret it as the sum of the segments of the truncated variances of the X_{nj}s taken outside $(-\varepsilon, \varepsilon)$. Again, this quantity must be small, as the variances of the X_{nj}s are close to 0. ■

12.2 Limiting Laws of $\mathcal{L}(S_n)$ Under Conditions (C)

For reference, we gather together the basic assumptions made so far.

$$(C)\begin{cases}\text{For each } n, \text{ the r.v.s } X_{nj}, \quad j = 1, \ldots, k_n \underset{n\to\infty}{\to} \infty,\\ \quad\text{are row-wise independent,}\\ \mathcal{E}X_{nj} = 0,\ \sigma^2_{nj} = \sigma^2(X_{nj}) < \infty \text{ and } s^2_n = \sum_{j=1}^{k_n}\sigma^2_{nj} = 1;\\ \quad\text{also, } \max_{1\leq j\leq k_n}\sigma^2_{nj} \underset{n\to\infty}{\to} 0.\end{cases} \tag{12.1}$$

The purpose of this section is the identification of the limiting laws of $\mathcal{L}(S_n)$ under conditions (C).

In all that follows, maxima, summations, and products are over $1 \leq j \leq k_n$, F_{nj} *is the d.f. of* X_{nj} *and* f_{nj} *is the ch.f. of* X_{nj}. *Also,* $S_n = \sum_{j=1}^{k_n} X_{nj}$, *and by* $\log z$, z *complex, we mean* $\log_p z$ *(the principal branch of the* $\log z$*). Limits and limit suprema are taken as* $n \to \infty$, *unless otherwise specified. Our problem is that of finding the limiting laws (in the weak sense) of* $\mathcal{L}(S_n)$.

The following lemma is established first.

Lemma 1. With the foregoing notation and under (C) in (12.1), one has

$$\sum_j \{\log f_{nj}(t) - [f_{nj}(t) - 1]\} \to 0, \quad t \in \Re.$$

Proof. By Theorem 9 (iii) in Chapter 11, applied with $n = 2$, one has

$$\begin{aligned} f_{nj}(t) &= 1 + \frac{\mathcal{E}X_{nj}}{1!}(it) + \theta_{nj}\frac{\mathcal{E}X_{nj}^2}{2!}t^2 \\ &= 1 + \theta_{nj}\frac{\sigma_{nj}^2}{2}t^2, \quad |\theta_{nj}| = |\theta_{nj}(t)| \leq 1. \end{aligned}$$

Hence

$$|f_{nj}(t) - 1| = \left|\theta_{nj}\frac{\sigma_{nj}^2}{2}t^2\right| \leq \frac{\sigma_{nj}^2}{2}t^2, \tag{12.2}$$

so that

$$\max_j |f_{nj}(t) - 1| \leq \frac{t^2}{2}\max_j \sigma_{nj}^2 \to 0 \text{ (for each } t).$$

Therefore, for each $t \in \Re$ and for all $n \geq n(t)$ (independent of j), one has

$$|f_{nj}(t) - 1| \leq \frac{1}{2} \text{ uniformly in } j = 1, \ldots, k_n. \tag{12.3}$$

Next, for t as above and $n \geq n(t)$, we have, by means of (12.3) and Lemma 7 in Chapter 11,

$$\log f_{nj}(t) = \log\{1 + [f_{nj}(t) - 1]\} = [f_{nj}(t) - 1] + \theta_{nj}^*[f_{nj}(t) - 1]^2$$

for some $\theta_{nj}^* = \theta_{nj}^*(t)$ with $|\theta_{nj}^*| \leq 1$. Thus,

$$\log f_{nj}(t) - [f_{nj}(t) - 1] = \theta_{nj}^*[f_{nj}(t) - 1]^2,$$

and

$$|\log f_{nj}(t) - [f_{nj}(t) - 1]| \leq |f_{nj}(t) - 1|^2, \quad j = 1, \ldots, k_n, \tag{12.4}$$

since $|z^2| = |z|^2$ for a complex number z (see also Exercise 10).

Hence, by employing (12.4) and (12.2), one has

$$\begin{aligned}\left|\sum_j \{\log f_{nj}(t) - [f_{nj}(t) - 1]\}\right| &\leq \sum_j |\log f_{nj}(t) - [f_{nj}(t) - 1]| \\ &\leq \sum_j |f_{nj}(t) - 1|^2 \\ &\leq [\max_j |f_{nj}(t) - 1|] \sum_j |f_{nj}(t) - 1| \\ &\leq [\max_j |f_{nj}(t) - 1|] \frac{t^2}{2} \sum_j \sigma_{nj}^2 \\ &= \frac{t^2}{2} [\max_j |f_{nj}(t) - 1|]. \end{aligned} \tag{12.5}$$

On the other hand, as was seen earlier,

$$\max_j |f_{nj}(t) - 1| \leq \frac{t^2}{2} \max_j \sigma_{nj}^2 \to 0. \tag{12.6}$$

By means of (12.6), relation (12.5) then gives

$$\sum_j \{\log f_{nj}(t) - [f_{nj}(t) - 1]\} \to 0, \text{ as was to be seen.}$$

■

Remark 1. If $s_n^2 \leq c$ rather than equal 1, then the right-hand side in (12.5) is bounded by $\frac{t^2}{2} c \max_j |f_{nj}(t) - 1| \leq \frac{t^2}{2} c \max_j \sigma_{nj}^2 \to 0$, so that the result is still true.

Corollary. Under (C) and with $s_n^2 = 1$ replaced by $s_n^2 \leq c$, one has

$$\log f_n(t) - \psi_n(t) \to 0, \quad t \in \Re,$$

where

$$f_n(t) = \prod_j f_{nj}(t) \text{ and } \psi_n(t) = \sum_j \int_{\Re} (e^{itx} - 1) dF_{nj}(x).$$

Proof. By independence, f_n is the ch.f. of S_n and by the definition of f_n,

$$\sum_j \log f_{nj}(t) = \log f_n(t), \tag{12.7}$$

$$\begin{aligned} f_{nj}(t) - 1 &= \int_{\Re} e^{itx} \, dF_{nj}(x) - 1 = \int_{\Re} e^{itx} \, dF_{nj}(x) - \int_{\Re} dF_{nj}(x) \\ &= \int_{\Re} (e^{itx} - 1) dF_{nj}(x), \end{aligned}$$

so that

$$\sum_j [f_{nj}(t) - 1] = \sum_j \int_{\Re} (e^{itx} - 1)dF_{nj}(x) = \psi_n(t). \tag{12.8}$$

Then the result follows from the lemma (along with Remark 1) and (12.7) and (12.8). ■

From the corollary, it follows that, if $\psi_n(t) \to \psi(t)$, where $e^{\psi(t)}$, $t \in \Re$, is a ch.f. of a r.v. (see Proposition 2(i) below), then $f_n(t) \to e^{\psi(t)}$, $t \in \Re$, and therefore $\mathcal{L}(S_n) \Rightarrow$ to the law corresponding to $e^{\psi(t)}$. The converse is also true by Theorem 3 in Chapter 11 and continuity of $\log_p z$. Therefore, in order to find the limiting laws of $\mathcal{L}(S_n)$ (in the weak convergence sense), it suffices to find the (pointwise) limits of $\psi_n(t)$, $t \in \Re$. This is the subject matter of this section and the content of Theorem 2 below.

To this end, we have

$$\psi_n(t) = \sum_j \int_{\Re} (e^{itx} - 1)dF_{nj}(x) = \sum_j \int_{\Re} (e^{itx} - 1 - itx)dF_{nj}(x),$$

since

$$\int_{\Re} itx\, dF_{nj}(x) = it \int_{\Re} x\, dF_{nj}(x) = it\mathcal{E}X_{nj} = 0.$$

Next, $\frac{e^{itx}-1-itx}{x^2} \underset{x\to 0}{\to} -\frac{t^2}{2}$, by expanding e^{itx} around $t = 0$ (see, e.g., discussion following relation (11.18) in Chapter 11). Then by defining $\frac{e^{itx}-1-itx}{x^2}$ to be equal to $-\frac{t^2}{2}$ for $x = 0$ (i.e., by continuity), we get

$$\begin{aligned}
\psi_n(t) &= \sum_j \int_{\Re} \frac{e^{itx} - 1 - itx}{x^2} x^2\, dF_{nj}(x) \\
&= \sum_j \int_{\Re} \frac{e^{itx} - 1 - itx}{x^2} d\left[\int_{-\infty}^{x} y^2\, dF_{nj}(y)\right] \\
&= \int_{\Re} \frac{e^{itx} - 1 - itx}{x^2} d\sum_j \int_{-\infty}^{x} y^2\, dF_{nj}(y) \\
&= \int_{\Re} \frac{e^{itx} - 1 - itx}{x^2} dK_n(x),
\end{aligned}$$

or

$$\psi_n(t) = \int_{\Re} \frac{e^{itx} - 1 - itx}{x^2} dK_n(x), \quad K_n(x) = \sum_j \int_{-\infty}^{x} y^2\, dF_{nj}(y). \tag{12.9}$$

In connection with the function K_n defined in (12.9), we have

Proposition 1. The function K_n defined in (12.9) is the d.f. of a r.v.

Proof. In fact, $0 \leq K_n \uparrow$, clearly, $K_n(-\infty) = 0$ also clearly, and $K_n(\infty) = \sum_j \int_{-\infty}^{\infty} y^2 dF_{nj}(y) = \sum_j \sigma_{nj}^2 = 1$. K_n is also right-continuous. For this, it suffices to show that $\int_{-\infty}^{x} y^2 dF_{nj}(y)$ is right-continuous. To this end, let $x \downarrow x_0$. Then

$$\int_{-\infty}^{x} y^2\, dF_{nj}(y) - \int_{-\infty}^{x_0} y^2\, dF_{nj}(y) = \int_{(x_0,x]} y^2\, dF_{nj}(y)$$
$$= \int_{\Re} y^2 I_{(x_0,x]}(y)\, dF_{nj}(y),$$

and

$$y^2 I_{(x_0,x]} \underset{x\downarrow x_0}{\rightarrow} y^2 I_{\oslash}(y) = 0, \text{ whereas } y^2 I_{(x_0,x]}(y) \leq y^2$$

independent of x and F_{nj} – integrable.

Thus, the Dominated Convergence Theorem implies that

$$\int_{\Re} y^2 I_{(x_0,x]}(y)\, dF_{nj}(y) \underset{x\downarrow x_0}{\rightarrow} 0,$$

or equivalently,

$$\int_{-\infty}^{x} y^2\, dF_{nj}(y) \underset{x\downarrow x_0}{\rightarrow} \int_{-\infty}^{x_0} y^2\, dF_{nj}(y).$$ ■

Next,

$$\frac{|e^{itx} - 1 - itx|}{x^2} \leq \frac{|e^{itx} - 1| + |tx|}{|x|^2} \leq \frac{2|tx|}{|x|^2} = \frac{2|t|}{|x|} \underset{0<|x|\rightarrow\infty}{\rightarrow} 0$$

(by Exercise 4 in Chapter 11).

Therefore $\frac{e^{itx}-1-itx}{x^2}$ is bounded and also continuous as a function of x for each fixed t.

It follows that for any d.f. K (with finite variation) the integral below, to be denoted by $\psi(t)$,

$$\psi(t) = \int_{\Re} \frac{e^{itx} - 1 - itx}{x^2}\, dK(x), \tag{12.10}$$

exists and is finite (in norm).

In the following, two propositions, Propositions 2 and 3 and two lemmas, Lemmas 2 and 3, are stated before the main result of this section, Theorem 2, can be stated and proved. The proofs of the propositions are deferred to Section 12.4. The lemmas are purely complex analysis results and are stated here for the sake of completeness only.

Proposition 2. Let K be a d.f. of bounded variation. Then the complex-valued function ψ given in (12.10) is well defined as explained above. Furthermore:

(i) The function $e^{\psi(t)} \overset{def}{=} f(t)$ is the ch.f. of a r.v., X, say.
(ii) $\mathcal{E}X = 0$ and $\sigma^2(X) = \mathcal{E}X^2 = Var\, K$.

Corollary 1. Let K be a d.f. with variation 1, and let the function ψ be defined by (12.10), so that $e^{\psi(t)}$ is the ch.f. of a r.v. (by part (i) of the proposition). Then under conditions (C) in (12.1), $e^{\psi(t)}$ is the ch.f. of the sum of n i.i.d. r.v.s. That is, for each n, there exist i.i.d. r.v.s X_{nj}, $j = 1, \ldots, n$, satisfying (C), such that

$\mathcal{L}(S_n) = \mathcal{L}$ (corresponding to $e^{\psi(t)}$), and therefore $f_n(t) \to e^{\psi(t)}$ on $\Re$, where $S_n = \sum_j X_{nj}$ and f_n is the ch.f. of S_n.

Corollary 2. If in (C), the condition $s_n^2 = \sum_{j=1}^{k_n} s_{nj}^2 = 1$ is replaced by $s_n^2 \leq c(< \infty)$, and K is any d.f. of bounded variation, then the proposition and Corollary 1 still hold true.

Corollary 3. Let K be a d.f. of bounded variation, and on $\Re$, define ψ by (12.10). Then:

(i) The second derivative of $\psi(t)$ exists and is given by

$$\psi''(t) = -\int_{\Re} e^{itx}\, dK(x), \quad t \in \Re.$$

(ii) Furthermore, if $K(-\infty) = 0$, there is only one such K that defines the ψ in (12.10).

Proposition 3. For each n, let K_n be a d.f. such that *Var* $K_n \leq c(< \infty)$, and let ψ_n be defined by (12.10); i.e.,

$$\psi_n(t) = \int_{\Re} \frac{e^{itx} - 1 - itx}{x^2}\, dK_n(x), \quad t \in \Re.$$

(i) Suppose that $K_n \Rightarrow K$, a d.f. (with variation $\leq c$, by Theorem 3 (ii) in Chapter 8). Then $\psi_n(t) \to \psi(t), t \in \Re$, where ψ is given in (12.10); i.e.,

$$\psi(t) = \int_{\Re} \frac{e^{itx} - 1 - itx}{x^2}\, dK(x).$$

(ii) Conversely, if $\psi_n(t) \to \psi(t)$, $t \in \Re$, for some ψ, then $K_n \Rightarrow K$ *uac* (see Definition 3 in Chapter 11), a d.f. (with variation $\leq c$) and ψ is given as above.

The following two lemmas are complex analysis results.

Lemma 2. Let g be a complex-valued function defined on $\Re$ such that $g(0) = 1$, g is continuous, and $g(t) \neq 0$, $t \in \Re$. Then there exists a unique (single-valued) function λ defined on $\Re$ into $\mathbb{C}$, such that $\lambda(0) = 0$, and continuous on $\Re$, for which

$$g(t) = e^{\lambda(t)}, \quad t \in \Re.$$

Lemma 3. For each n, let g_n, g be functions satisfying the assumptions of Lemma 2, and let λ_n, λ be the corresponding functions for which the conclusion of the lemma holds. Then

$$g_n \to g, \quad t \in \Re$$

implies

$$\lambda_n \to \lambda, \quad t \in \Re.$$

We may proceed with the formulation and proof of the main result in this section; a special case of it is Theorem 1 as already stated.

Theorem 2. Let the r.v.s X_{nj}, $j = 1, \ldots, k_n$, $n \geq 1$, satisfy conditions (C) in (12.1), and let F_{nj} and f_{nj} be their respective d.f.s and ch.f.s. Set $S_n = \sum_j X_{nj}$ and let K_n be as in (12.9); i.e.,

$$K_n(x) = \sum_j \int_{-\infty}^{x} y^2\, dF_{nj}(y). \tag{12.11}$$

Then

(i) If $\mathcal{L}(S_n) \Rightarrow \mathcal{L}$, a d.f. of a r.v. with ch.f. $f \neq 0$ on $\Re$, then $K_n \Rightarrow K$ *uac*, a d.f. with *Var* $K \leq 1$, and $f(t) = e^{\psi(t)}$, where

$$\psi(t) = \int_{\Re} \frac{e^{itx} - 1 - itx}{x^2}\, dK(x). \tag{12.12}$$

(ii) With K_n given in (12.11), suppose $K_n \Rightarrow K$, a d.f. (with *Var* $K \leq 1$, by Theorem 3 (ii) in Chapter 8), and let ψ be defined by (12.12), so that e^{ψ} is the ch.f. of a r.v. (by Proposition 2(i)). Then $\mathcal{L}(S_n) \Rightarrow \mathcal{L}$, where $\mathcal{L}$ is the d.f. of a r.v. corresponding to the ch.f. e^{ψ} of a r.v. ■

Proof.

(i) Let $\mathcal{L}(S_n) \Rightarrow \mathcal{L}$, a d.f. of a r.v. with ch.f. $f \neq 0$ on $\Re$. Then (by Theorem 3 in Chapter 11) $f_n(t) \to f(t)$ on $\Re$, where $f_n(t) = \prod_j f_{nj}(t)$. Under (C), the Corollary to Lemma 1 yields

$$\log f_n(t) - \psi_n(t) \to 0, \quad t \in \Re, \tag{12.13}$$

where $\psi_n(t)$ is given in (12.9) (see also (12.8)); i.e.,

$$\psi_n(t) = \int_{\Re} \frac{e^{itx} - 1 - itx}{x^2}\, dK_n(x). \tag{12.14}$$

From (12.13), we get by exponentiation

$$f_n(t)e^{-\psi_n(t)} \to 1, \quad t \in \Re. \tag{12.15}$$

Then $f^{-1} f_n e^{-\psi_n} \to f^{-1}$, or $e^{-\psi_n} \to f^{-1}$, or $e^{\psi_n} \to f$; i.e.,

$$e^{\psi_n(t)} \to f(t), \quad t \in \Re \tag{12.16}$$

The ch.f. f has the properties of $f(0) = 1$, as the ch.f. of a r.v., and continuity (by Theorem 1 (i), (ii) in Chapter 11), whereas $f(t) \neq 0$ on $\Re$, by assumption. Thus f satisfies the assumptions of Lemma 2. So, there exists a unique function $\psi : \Re \to \mathbb{C}$, continuous on $\Re$, $\psi(0) = 0$, and such that $f(t) = e^{\psi(t)}$ on $\Re$. Therefore, by (12.16),

$$e^{\psi_n(t)} \to e^{\psi(t)} \text{ on } \Re. \tag{12.17}$$

Set $g_n(t) = e^{\psi_n(t)}$ and $g(t) = e^{\psi(t)} (= f(t))$. From the derivations leading to relation (12.10), it is immediate that $\psi(0) = 0$, and $\psi_n(t)$ is continuous on $\Re$ for each $n \geq 1$. Indeed,

$$\psi_n(t) = \int_{\Re} \frac{e^{itx} - itx - 1}{x^2}\, dK_n(x) \quad \text{and}$$

$$\left|\frac{e^{itx}-itx-1}{x^2}\right| \leq \frac{|e^{itx}-1|+|tx|}{x^2} \leq \frac{|tx|+|tx|}{x^2} = \frac{2|t||x|}{x^2}$$
$$= \frac{2|t|}{|x|} \leq \frac{2|t_0|}{|x|}, \quad |t| \leq |t_0| \quad (\text{some } t_0 \in \Re)$$

independent of $t \in [-|t_0|, |t_0|]$, K_n-integrable, and

$$\frac{e^{itx}-itx-1}{x^2} \underset{t\to t_0}{\longrightarrow} \frac{e^{it_0x}-it_0x-1}{x^2}.$$

Then the Dominated Convergence Theorem implies that $\psi_n(t) \underset{t\to t_0}{\longrightarrow} \psi(t_0)$. Therefore, the function $g_n(n \geq 1)$ and g satisfy the conditions of Lemma 2. This fact along with convergence (12.17) and Lemma 3 yield $\psi_n(t) \to \psi(t)$ on $\Re$.

So, ψ_n is given by (12.14) and $\psi_n(t) \to \psi(t), t \in \Re$, for some ψ (as described earlier). Then, by Proposition 3 (ii), $K_n \Rightarrow K$ *uac*, a d.f. with variation ≤ 1 and ψ is given by (12.12).

(ii) By Proposition 1, K_n is a d.f. of a r.v. $(n \geq 1)$ and the assumption is that $K_n \Rightarrow K$ some d.f. (with $Var\, K \leq 1$). Therefore Proposition 3 (i) applies and yields that $\psi_n(t) \to \psi(t), t \in \Re$, where $\psi_n(t)$ $(n \geq 1)$ and ψ are given by (12.14) and (12.12), respectively; hence $e^{\psi_n(t)} \to e^{\psi(t)}$ on $\Re$. By this and (12.15) (which holds under conditions (C)), we have then $f_n(t) \to e^{\psi(t)}$ on $\Re$, where $e^{\psi(t)}$ is a ch.f. of a r.v., by Proposition 2(i). Thus $\mathcal{L}(S_n) \Rightarrow \mathcal{L}$, where $\mathcal{L}$ is the d.f. of a r.v. corresponding to e^{ψ}; this is so by the converse part of Theorem 3 in Chapter 11. ■

Corollary. In conditions (C) given in (12.1), replace $s_n^2 = 1$ by $s_n^2 \leq c$. Then one has $Var\, K_n \to Var\, K$, if and only if $s_n^2 \to \sigma^2(\mathcal{L})$ (the variance of the distribution $\mathcal{L}$).

Proof. From (12.11), we have

$$Var\, K_n = \sum_j \int_{\Re} y^2\, dF_{nj}(y) = \sum_j \sigma_{nj}^2 = s_n^2.$$

On the other hand, Proposition 2(ii) gives $Var\, K = \sigma^2(\mathcal{L})$ (equal to the second moment of $\mathcal{L}$, the d.f. of a r.v. corresponding to e^{ψ}, where ψ is determined by K through (12.12), since its first moment is 0). Hence the result. ■

12.3 Conditions for the Central Limit Theorem to Hold

In this section, we operate basically under conditions (C) given in (12.1), as has been the case in the previous section. We provide necessary and sufficient conditions for the CLT to hold; in other words, establish Theorem 1, and also give several sets of sufficient conditions for the validity of the CLT.

Here the d.f. K; is selected to accommodate the needs of Theorem 1; namely, K is chosen as follows:

$$K(x) = \begin{cases} 0 & x < 0 \\ 1 & x \geq 0. \end{cases} \tag{12.18}$$

So, K is a d.f. of a r.v. with a jump of length 1 at the origin (and it uniquely determines $\psi(t)$). Then the formula

$$\psi(t) = \int_{\Re} \frac{e^{itx} - 1 - itx}{x^2} dK(x)$$

gives

$$\psi(t) = \int_{\{0\}} -\frac{t^2}{2} dK(0) = -\frac{t^2}{2},$$

so that $e^{\psi(t)} = e^{-\frac{t^2}{2}}$, which is the ch.f. of $N(0, 1)$. Recall that

$$K_n(x) = \sum_j \int_{-\infty}^{x} y^2 dF_{nj}(y), \tag{12.19}$$

where F_{nj} is the d.f. of X_{nj}. Then, by Theorem 2 (ii), if $K_n \Rightarrow K$, it follows that $\mathcal{L}(S_n) \Rightarrow N(0, 1)$, whereas (by part (i) of Theorem 2) if $\mathcal{L}(S_n) \Rightarrow N(0, 1)$, then $K_n \Rightarrow K\ uac$. Therefore in order to ensure that $\mathcal{L}(S_n) \Rightarrow N(0, 1)$, it suffices to find conditions for which $K_n \Rightarrow K$, where K_n and K are given by (12.19) and (12.18), respectively. This issue will be settled in the following two lemmas, the combination of which will also provide the proof of Theorem 1.

Lemma 4. For each $n \geq 1$, let X_{nj}, $j = 1, \ldots, k_n \to \infty$, be row-wise independent r.v.s with $\mathcal{E}X_{nj} = 0$, $\sigma^2_{nj} = \mathcal{E}X^2_{nj} < \infty$, and $s^2_n = \sum_{j=1}^{k_n} \sigma^2_{nj} = 1$. Let F_{nj} be the d.f. of X_{nj} and let $S_n = \sum_{j=1}^{k_n} X_{nj}$. Finally, suppose that, for every $\varepsilon > 0$,

$$g_n(\varepsilon) = \sum_{j=1}^{k_n} \int_{(|x| \geq \varepsilon)} x^2\, dF_{nj}(x) \to 0. \tag{12.20}$$

Then

(i) $\max_{1 \leq j \leq k_n} \sigma^2_{nj} \to 0$ (so that conditions (C) in (12.1) are fulfilled).
(ii) $K_n \Rightarrow K$, where K_n and K are given in (12.19) and (12.18), respectively.
(iii) $\mathcal{L}(S_n) \Rightarrow N(0, 1)$.

Proof.

(i) Clearly,

$$\begin{aligned}\sigma^2_{nj} &= \int_{\Re} x^2\, dF_{nj}(x) = \int_{(|x| \geq \varepsilon)} x^2\, dF_{nj}(x) + \int_{(|x| < \varepsilon)} x^2\, dF_{nj}(x) \\ &\leq g_n(\varepsilon) + \int_{(|x| < \varepsilon)} x^2\, dF_{nj}(x) \leq g_n(\varepsilon) + \varepsilon^2,\end{aligned}$$

so that $\max_j \sigma^2_{nj} \leq g_n(\varepsilon) + \varepsilon^2$. Hence $\limsup \max_j\ \sigma^2_{nj} \leq \varepsilon^2$, by (12.20), and letting $\varepsilon \to 0$, we get the result.

(ii) We have to show that $K_n(-\varepsilon) \to 0$ and $K_n(\varepsilon) \to 1$ for every $\varepsilon > 0$. To this end,

$$\begin{aligned} K_n(-\varepsilon) &= \sum_j \int_{(-\infty,-\varepsilon]} x^2\, dF_{nj}(x) \\ &\le \sum_j \int_{(-\infty,-\varepsilon]} x^2\, dF_{nj}(x) \\ &\quad + \sum_j \int_{[\varepsilon,\infty)} x^2\, dF_{nj}(x) \\ &= \sum_j \int_{(|x|\ge\varepsilon)} x^2\, dF_{nj}(x) = g_n(\varepsilon) \to 0. \end{aligned}$$

Also,

$$\begin{aligned} K_n(\varepsilon) &= \sum_j \int_{(-\infty,\varepsilon]} x^2\, dF_{nj}(x) = \sum_j \int_{(-\infty,\infty)} x^2\, dF_{nj}(x) \\ &\quad - \sum_j \int_{(\varepsilon,\infty)} x^2\, dF_{nj}(x) \\ &= 1 - \sum_j \int_{(\varepsilon,\infty)} x^2\, dF_{nj}(x) \to 1, \end{aligned}$$

because $\sum_j \int_{(\varepsilon,\infty)} x^2 dF_{nj}(x) \le g_n(\varepsilon) \to 0$. Thus $K_n \Rightarrow K$.

(iii) As has already been mentioned, part (i) ensures that conditions (C) in (12.1) hold. Then this, part (ii), and Theorem 2 (ii) imply that $\mathcal{L}(S_n) \Rightarrow \mathcal{L}$, where $\mathcal{L}$ is the distribution corresponding to $e^{\psi(t)}$, which here is $e^{-\frac{t^2}{2}}$. In other words, $\mathcal{L}(S_n) \Rightarrow N(0,1)$. ■

Lemma 5. Assume conditions (C), let F_{nj} be the d.f. of X_{nj}, and set $S_n = \sum_{j=1}^{k_n} X_{nj}$. Then, if $\mathcal{L}(S_n) \Rightarrow N(0,1)$, it follows that the convergence in (12.20) holds for every $\varepsilon > 0$.

Proof. Let K_n and K be given by (12.19) and (12.18), respectively. Then, by Theorem 2 (i), $\mathcal{L}(S_n) \Rightarrow N(0,1)$ implies $K_n \Rightarrow K$ *uac*. Next,

$$\begin{aligned} g_n(\varepsilon) &= \sum_j \int_{(|x|\ge\varepsilon)} x^2\, dF_{nj}(x) \\ &= \sum_j \int_{(-\infty,-\varepsilon]} x^2\, dF_{nj}(x) + \sum_j \int_{[\varepsilon,\infty)} x^2\, dF_{nj}(x) \\ &\le \sum_j \int_{(-\infty,-\varepsilon]} x^2\, dF_{nj}(x) + \sum_j \int_{(\frac{\varepsilon}{2},\infty)} x^2\, dF_{nj}(x) \\ &= K_n(-\varepsilon) + 1 - K_n\left(\frac{\varepsilon}{2}\right), \end{aligned}$$

and we wish to show that $g_n(\varepsilon) \to 0$. It suffices to show that for every $\{m\} \subseteq \{n\}$ there exists $\{r\} \subseteq \{m\}$ such that $g_r(\varepsilon) \underset{r\to\infty}{\to} 0$. By considering $\{K_m\}$, there exists $\{r\} \subseteq \{m\}$ such that $K_r \underset{r\to\infty}{\Rightarrow} K^*$ a d.f. (by Theorem 5 in Chapter 8), and $K^* = K + c$, some constant c. Therefore

$$\begin{aligned} K_r(-\varepsilon) + 1 - K_r\left(\frac{\varepsilon}{2}\right) &\underset{r\to\infty}{\to} K^*(-\varepsilon) + 1 - K^*\left(\frac{\varepsilon}{2}\right) \\ &= K(-\varepsilon) + c + 1 - K\left(\frac{\varepsilon}{2}\right) - c \\ &= 0 + 1 - 1 = 0. \end{aligned}$$

Thus, $g_r(\varepsilon) \underset{r\to\infty}{\to} 0$. ■

We may now proceed with the proof of Theorem 1.

Proof of Theorem 1. Suppose that $g_n(\varepsilon) \to 0$ for every $\varepsilon > 0$. Then, by Lemma 4 (i), $\max_j \sigma_{nj}^2 \to 0$, whereas by Lemma 4 (iii), $\mathcal{L}(S_n) \Rightarrow N(0, 1)$. On the other hand, under conditions (C), if $\mathcal{L}(S_n) \Rightarrow N(0, 1)$, then $g_n(\varepsilon) \to 0$ for every $\varepsilon > 0$; this is so by Lemma 5. The proof is complete. ■

In reference to Example 2, we show that Theorem 1 does not provide the asymptotic distribution of $\mathcal{L}(S_n)$. However, the asymptotic distribution of $\mathcal{L}(S_n)$ may be obtained directly, as is done in the following example. As usual, all limits are taken as $n \to \infty$.

Example 4. In reference to Example 2, show directly, by means of ch.f.s, that $\mathcal{L}(S_n) \Rightarrow \mathcal{L}(cX + d)$, where the r.v. X is distributed as $P(\lambda)$, $X \sim P(\lambda)$, $c = 1/\sqrt{\lambda}$, and $d = -\sqrt{\lambda}$.

Indeed, recalling that $S_n = \sum_{j=1}^n (Y_{nj} - p_n)/\sqrt{np_nq_n}$, we have

$$\begin{aligned} f_{S_n}(t) &= f_{\sum_{j=1}^n (Y_{nj}-p_n)/\sqrt{np_nq_n}}(t) = f_{\sum_{j=1}^n (Y_{nj}-p_n)}(t/\sqrt{np_nq_n}) \\ &= \prod_{j=1}^n f_{Y_{nj}-p_n}(t/\sqrt{np_nq_n}) \\ &= \prod_{j=1}^n e^{-ip_nt/\sqrt{np_nq_n}} f_{Y_{nj}}(t/\sqrt{np_nq_n}) \\ &= e^{-inp_nt/\sqrt{np_nq_n}}[f_{Y_{n1}}(t/\sqrt{np_nq_n})]^n \\ &= e^{-inp_nt/\sqrt{np_nq_n}}(p_ne^{it/\sqrt{np_nq_n}} + q_n)^n \\ &\qquad \text{(since } f_{Y_{n1}}(t) = (p_ne^{it} + q_n)^n\text{)} \\ &= q_n^n e^{-inp_nt/\sqrt{np_nq_n}}\left(1 + \frac{np_ne^{it/\sqrt{np_nq_n}}/q_n}{n}\right)^n. \end{aligned}$$

However,

$$q_n^n = (1 - p_n)^n = \left(1 + \frac{-np_n}{n}\right)^n \to e^{-\lambda},$$

$$e^{-inp_n t/\sqrt{np_n q_n}} \underset{n\to\infty}{\to} e^{-i\lambda t/\sqrt{\lambda}} = e^{-it\sqrt{\lambda}},$$

and

$$\left(1 + \frac{np_n e^{it/\sqrt{np_n q_n}}/q_n}{n}\right)^n \to e^{\lambda e^{it/\sqrt{\lambda}}},$$

since $\frac{np_n e^{it/\sqrt{np_n q_n}}}{q_n} \to \lambda e^{it/\sqrt{\lambda}}$. It follows that $f_{S_n}(t) \to e^{-\lambda} \times e^{-it\sqrt{\lambda}} \times e^{\lambda e^{it/\sqrt{\lambda}}} = e^{-it\sqrt{\lambda}} \times e^{\lambda e^{it/\sqrt{\lambda}} - \lambda}$, which is the ch.f. of the r.v. $cX + d = \frac{1}{\sqrt{\lambda}}X - \sqrt{\lambda}$, where $X \sim P(\lambda)$ with ch.f. $e^{\lambda e^{it} - \lambda}$. ■

Observe that the result obtained here is consistent with that to be arrived at in Application 1 of Chapter 13, where it is shown that $\mathcal{L}\left(\sum_{j=1}^{n} Y_{nj}\right) \Rightarrow P(\lambda)$.

As a consequence of Theorem 1, we have the following results.

Theorem 3 (Liapounov Theorem). For each $n \geq 1$, and with $n \to \infty$ throughout, let the r.v.s X_{nj}, $j = 1, \ldots, k_n \to \infty$, be row-wise independent and such that $\mathcal{E}X_{nj} = 0$ and $\sigma_{nj}^2 = \sigma^2(X_{nj}) < \infty$, with $s_n^2 = \sum_j \sigma_{nj}^2 = 1$. Then, if for some $\delta > 0$,

$$\sum_j \mathcal{E}|X_{nj}|^{2+\delta} \to 0, \tag{12.21}$$

it follows that

$$\mathcal{L}(S_n) \Rightarrow N(0, 1) \text{ and } \max_j \sigma_{nj}^2 \to 0, \tag{12.22}$$

where $S_n = \sum_j X_{nj}$. ■

Proof. By Theorem 1, it suffices to show that, for every $\varepsilon > 0$,

$$g_n(\varepsilon) = \sum_j \int_{(|x|\geq\varepsilon)} x^2 \, dF_{nj}(x) \to 0,$$

where F_{nj} is the d.f. of X_{nj}. We have

$$\begin{aligned}\mathcal{E}|X_{nj}|^{2+\delta} &= \int_{(-\infty,\infty)} |x|^{2+\delta} \, dF_{nj}(x) \geq \int_{(|x|\geq\varepsilon)} |x|^{2+\delta} \, dF_{nj}(x) \\ &= \int_{(|x|\geq\varepsilon)} |x|^2 |x|^\delta \, dF_{nj}(x) \geq \varepsilon^\delta \int_{(|x|\geq\varepsilon)} x^2 \, dF_{nj}(x),\end{aligned}$$

so that

$$\sum_j \mathcal{E}|X_{nj}|^{2+\delta} \geq \varepsilon^\delta g_n(\varepsilon).$$

Therefore, relation (12.21) implies $g_n(\varepsilon) \to 0$, which ensures (12.22). ■

Remark 2. The theorem is true, in particular, if $\delta = 1$; i.e., if

$$\sum_j \mathcal{E}|X_{nj}|^3 \to 0.$$

Corollary. The conclusion of the theorem holds true if condition (12.21) is replaced by

$$|X_{nj}| \leq M_{nj}(< \infty) \text{ a.s., } \max_j M_{nj} \to 0.$$

Proof. We have

$$\sum_j |X_{nj}|^3 = \sum_j \left(X_{nj}^2 |X_{nj}|\right) \overset{\text{a.s.}}{\leq} \left(\max_j M_{nj}\right) \sum_j X_{nj}^2,$$

so that, by taking expectation of both sides, we obtain

$$\sum_j \mathcal{E}|X_{nj}|^3 \leq \left(\max_j M_{nj}\right) \sum_j \mathcal{E}X_{nj}^2 = \max_j M_{nj} \to 0,$$

since $\sum_j \mathcal{E}X_{nj}^2 = \sum_j \sigma^2(X_{nj}) = s_n^2 = 1$. Then Remark 2 applies and gives the result. ■

Now Theorem 1 can be generalized as follows.

Theorem 4. For each $n \geq 1$, let the r.v.s X_{nj}, $j = 1, \ldots, k_n \to \infty$, be row-wise independent and such that $\mathcal{E}X_{nj} = 0$ and $\sigma_{nj}^2 = \sigma^2(X_{nj}) < \infty$. Set $s_n^2 = \sum_{j=1}^{k_n} \sigma_{nj}^2$. Then

$$\mathcal{L}\left(\frac{S_n}{s_n}\right) \Rightarrow N(0, 1) \text{ and } \max \frac{\sigma_{nj}^2}{s_n^2} \to 0\,,$$

if and only if, for every $\varepsilon > 0$,

$$g_n(\varepsilon) = \frac{1}{s_n^2} \sum_j \int_{(|x| \geq \varepsilon s_n)} x^2\, dF_{nj}(x) \to 0,$$

where $S_n = \sum_j X_{nj}$ and F_{nj} is the d.f. of X_{nj}. ■

Proof. Set

$$Y_{nj} = \frac{X_{nj}}{s_n}, \quad j = 1, \ldots, k_n.$$

Then for each n, the r.v.s Y_{nj}, $j = 1, \ldots, k_n$, are independent (within each row) and such that

$$\mathcal{E}Y_{nj} = 0,\ \tau_{nj}^2 = \sigma^2(Y_{nj}) = \frac{\sigma_{nj}^2}{s_n^2} < \infty, \text{ and set } \tau_n^2 = \sum_{j=1}^{k_n} \tau_{nj}^2 = 1.$$

That is, the basic assumptions of Theorem 1 are satisfied for the r.v.s Y_{nj}, $j = 1, \ldots, k_n$. Thus, by setting

$$T_n = \sum_j Y_{nj} = \frac{S_n}{s_n} \text{ and } g_n(\varepsilon) = \sum_j \int_{(|x|\geq\varepsilon)} x^2\, dG_{nj}(x),$$

where G_{nj} is the d.f. of Y_{nj}, we obtain $\mathcal{L}(T_n) \Rightarrow N(0, 1)$ and $\max_j \tau_{nj}^2 \to 0$, if and only if $g_n(\varepsilon) \to 0$ for every $\varepsilon > 0$. However,

$$G_{nj}(x) = P(Y_{nj} \leq x) = P\left(\frac{X_{nj}}{s_n} \leq x\right) = P(X_{nj} \leq xs_n) = F_{nj}(xs_n),$$

so that

$$\begin{aligned} g_n(\varepsilon) &= \sum_j \int_{(|x|\geq\varepsilon)} x^2\, dG_{nj}(x) = \sum_j \int_{(|x|\geq\varepsilon)} x^2\, dF_{nj}(xs_n) \\ &= \frac{1}{s_n^2} \sum_j \int_{(|y|\geq\varepsilon s_n)} y^2\, dF_{nj}(y) \text{ (by setting } xs_n = y) \\ &= \frac{1}{s_n^2} \sum_j \int_{(|x|\geq\varepsilon s_n)} x^2\, dF_{nj}(x). \end{aligned}$$

So, $\mathcal{L}(T_n) \Rightarrow N(0, 1)$ and $\max_j \tau_{nj}^2 \to 0$, if and only if

$$\frac{1}{s_n^2} \sum_j \int_{(|x|\geq\varepsilon s_n)} x^2\, dF_{nj}(x) \to 0,$$

or

$$\mathcal{L}\left(\frac{S_n}{s_n}\right) \Rightarrow N(0, 1) \text{ and } \max \frac{\sigma_{nj}^2}{s_n^2} \to 0, \text{ if and only if}$$

$$g_n(\varepsilon) = \frac{1}{s_n^2} \sum_j \int_{(|x|\geq\varepsilon s_n)} x^2\, dF_{nj}(x) \to 0 \text{ for every } \varepsilon > 0. \qquad \blacksquare$$

Specializing the result obtained to a single sequence of r.v.s, we have the following theorem.

Theorem 5. Let the r.v.s X_j, $j = 1, \ldots, k_n$, be independent and such that $\mathcal{E}X_j = 0$ and $\sigma_j^2 = \sigma^2(X_j) < \infty$. Set $s_n^2 = \sum_{j=1}^{k_n} \sigma_j^2$. Then

$$\mathcal{L}\left(\frac{S_n}{s_n}\right) \Rightarrow N(0, 1) \text{ and } \max_j \frac{\sigma_j^2}{s_n^2} \to 0\,, \tag{12.23}$$

if and only if, for every $\varepsilon > 0$,

$$g_n(\varepsilon) = \frac{1}{s_n^2} \sum_j \int_{(|x|\geq\varepsilon s_n)} x^2\, dF_j(x) \to 0\,, \tag{12.24}$$

where $S_n = \sum_j X_j$ and F_j is the d.f. of X_j. $\blacksquare$

Proof. For each n, set $Y_{nj} = \frac{X_j}{s_n}$, $j = 1, \ldots, k_n$. Then Y_{nj}, $j = 1, \ldots, k_n$, are independent,

$$\mathcal{E}Y_{nj} = 0, \quad \tau_{nj}^2 = \sigma^2(Y_{nj}) = \frac{\sigma_j^2}{s_n^2} < \infty, \text{ and } \tau_n^2 = \sum_{j=1}^{k_n} \tau_{nj}^2 = 1.$$

With $T_n = \sum_j Y_{nj} = \frac{S_n}{s_n}$ and $G_{nj} = F_j$ being the d.f. of Y_{nj}, $j = 1, \ldots, k_n$, Theorem 1 applies to the present Y_{nj}s and gives the result, since

$$\begin{aligned} g_n(\varepsilon) &= \sum_j \int_{(|x|\geq\varepsilon)} x^2\, dG_{nj}(x) = \sum_j \int_{(|x|\geq\varepsilon)} x^2\, dF_j(xs_n) \\ &= \frac{1}{s_n^2}\sum_j \int_{(|y|\geq\varepsilon s_n)} y^2\, dF_j(y) \text{ (by setting } xs_n = y) \\ &= \frac{1}{s_n^2}\sum_j \int_{(|x|\geq\varepsilon s_n)} x^2\, dF_j(x). \end{aligned}$$

■

Corollary. Let the r.v.s X_j, $j = 1, \ldots, n$, be i.i.d. with $\mathcal{E}X_j = 0$ and $\sigma^2 = \sigma^2(X_j) < \infty$. Then relation (12.24) is satisfied, and so is (12.23), which here becomes

$$\mathcal{L}\left(\frac{S_n}{s_n}\right) = \mathcal{L}\left(\frac{S_n}{\sigma\sqrt{n}}\right) \Rightarrow N(0, 1) \text{ (and } \frac{\sigma^2}{n\sigma^2} = \frac{1}{n} \to 0).$$

Proof. We have $s_n^2 = n\sigma^2$, so that

$$g_n(\varepsilon) = \frac{1}{n\sigma^2} n \int_{(|x|\geq\varepsilon\sigma\sqrt{n})} x^2\, dF(x) = \frac{1}{\sigma^2}\int_{(|x|\geq\varepsilon\sigma\sqrt{n})} x^2\, dF(x),$$

where F is the common d.f. of the X_js. Now, by the fact that $\int_{\Re} x^2 dF(x) < \infty$, and the relations

$$x^2 I_{(|x|\geq\varepsilon\sigma\sqrt{n})}(x) \leq x^2, \quad n \geq 1, \quad x^2 I_{(|x|\geq\varepsilon\sigma\sqrt{n})}(x) \to 0,$$

the Dominated Convergence Theorem applies and gives $\int_{(|x|\geq\varepsilon\sigma\sqrt{n})} x^2 dF(x) \to 0$. Hence the result. ■

Remark 3. The corollary just established provides the CLT in its most common form.

This section is concluded with an example providing a couple of useful results.

Example 5. Let $X_1, \ldots, X_n$ be i.i.d. r.v.s with $\mathcal{E}X_1 = \mu \in \Re$ and $Var(X_1) = \sigma^2 \in (0, \infty)$, and let $\bar{X}_n$ be the sample mean of the X_js (which is used for estimating μ), $\bar{X}_n = \frac{1}{n}\sum_{j=1}^n X_j$. Then $\mathcal{E}\bar{X}_n = \mu$, $Var(\bar{X}_n) = \frac{\sigma^2}{n}$, and by the CLT, $\frac{\sqrt{n}(\bar{X}_n - \mu)}{\sigma} \underset{n\to\infty}{\overset{d}{\longrightarrow}} Z \sim N(0, 1)$, or $\sqrt{n}(\bar{X}_n - \mu) \underset{n\to\infty}{\overset{d}{\to}} Y \sim N(0, \sigma^2)$.

When μ is known, the sample variance S_n^2 of the X_js (which is used for estimating σ^2) is $S_n^2 = \frac{1}{n}\sum_{j=1}^{n}(X_j - \mu)^2$, and $\frac{\sqrt{n}(\bar{X}_n-\mu)}{S_n} \xrightarrow[n\to\infty]{d} Y \sim N(0,1)$. Setting $Y_j = (X_j - \mu)^2$, the r.v.s $Y_1, \ldots, Y_n$ are i.i.d. with $\mathcal{E}Y_1 = Var(X_1) = \sigma^2$ and $Var(Y_1) = \mathcal{E}Y_1^2 - (\mathcal{E}Y_1)^2 = \mathcal{E}(X_1-\mu)^4 - \sigma^4 = \mu_4 - \sigma^4$, where μ_4 is the fourth central moment of X_1, assuming, of course, that $\mathcal{E}X_1^4 < \infty$. So, $\mathcal{E}S_n^2 = \sigma^2$ and $Var(S_n^2) = (\mu_4 - \sigma^4)/n$. By the CLT again,

$$\frac{\sqrt{n}(\bar{Y}_n - \sigma^2)}{\sqrt{\mu_4 - \sigma^4}} = \frac{\sqrt{n}(S_n^2 - \sigma^2)}{\sqrt{\mu_4 - \sigma^4}} \xrightarrow[n\to\infty]{d} Z \sim N(0,1),$$

or $\sqrt{n}(S_n^2 - \sigma^2) \xrightarrow[n\to\infty]{d} Y \sim N(0, \mu_4 - \sigma^4)$.

If μ is unknown, the sample variance used for estimating σ^2 is $\bar{S}_n^2 = \sum_{j=1}^{n}(X_j - \bar{X}_n)^2/(n-1)$. Then, as is easily seen, $\sum_{j=1}^{n}(X_j - \bar{X}_n)^2 = \sum_{j=1}^{n}(X_j - \mu)^2 - n(\bar{X}_n - \mu)^2$, and hence $\mathcal{E}\bar{S}_n^2 = \frac{1}{n-1}(n\sigma^2 - n \times \frac{\sigma^2}{n}) = \sigma^2$. Furthermore,

$$\bar{S}_n^2 = \frac{n}{n-1}S_n^2 - \frac{n}{n-1}(\bar{X}_n - \mu)^2,$$

so that

$$\sqrt{n}(\bar{S}_n^2 - \sigma^2) = \frac{n}{n-1}\sqrt{n}(S_n^2 - \sigma^2) + \frac{\sigma^2\sqrt{n}}{n-1} - \frac{\sqrt{n}}{n-1}[\sqrt{n}(\bar{X}_n - \mu)]^2.$$

As $n \to \infty$, $\frac{n}{n-1}\sqrt{n}(S_n^2 - \sigma^2) \xrightarrow[n\to\infty]{d} Y \sim N(0, \mu_4 - \sigma^4)$, by the result obtained earlier and Theorem 7 (ii) in Chapter 5, $\frac{\sigma^2\sqrt{n}}{n-1} \longrightarrow 0$, and $\frac{\sqrt{n}}{n-1}[\sqrt{n}(\bar{X}_n - \mu)]^2 \longrightarrow 0$, by the fact that $\sqrt{n}(\bar{X}_n - \mu) \xrightarrow{d} Y \sim N(0, \sigma^2)$, Example 3 in Chapter 11, and Theorem 7 (ii) in Chapter 5 again. Then, by the theorem just cited, it follows that $\sqrt{n}(\bar{S}_n^2 - \sigma^2) \xrightarrow{d} Y \sim N(0, \mu_4 - \sigma^4)$. ∎

12.4 Proof of Results in Section 12.2

In this section, we provide the proofs of Proposition 2 and its three corollaries stated in Section 12.2, as well as the proof of Proposition 3, also stated in Section 12.2. For the proof of Proposition 2, the following lemma is needed which is also of independent interest. To this effect, recall that from Theorem 1 and Lemma 5 of Chapter 11, we have that if $\mathcal{E}|X^n| < \infty$, then

$$\frac{d^n}{dt^n}f(t)|_{t=0} = i^n \mathcal{E}X^n$$

and $f^n(t)$ is continuous on $\Re$. The following result is also true (and is used in the proof of Proposition 2).

Lemma 6. Let X be a r.v. with d.f. F and ch.f. f. Then, if $f^{(2n)}(0)$ exists and is finite (in norm), it follows that $\mathcal{E}|X^k| < \infty$ for all $k \leq 2n$ and $\mathcal{E}X^k = \frac{1}{i^k} f^{(k)}(0)$.

Proof. For a complex-valued function g defined on $\Re$, set $\Delta g(u) = g(u+h) - g(u-h)$, $h \in \Re$, and define $\Delta^{(n)} g(u)$ recursively. Then it can be seen (see Exercise 7) that

$$\Delta^{(2n)} f(0) = \sum_{r=0}^{2n} (-1)^r \binom{2n}{r} f[(2n-2r)h], \tag{12.25}$$

and

$$\frac{\Delta^{(2n)} f(0)}{(2h)^{2n}} = f^{(2n)}(0) + \frac{1}{2^{2n}} \frac{o(h^{2n})}{h^{2n}} = f^{(2n)}(0) + \frac{1}{2^{2n}} o(1)$$
$$\text{with } o(1) \underset{h\to 0}{\longrightarrow} 0. \tag{12.26}$$

By replacing f by what it is equal to, (12.25) becomes as follows:

$$\begin{aligned}
\Delta^{(2n)} f(0) &= \sum_{r=0}^{2n} (-1)^r \binom{2n}{r} \int_{\Re} e^{i[(2n-2r)h]x}\, dF(x) \\
&= \int_{\Re} \left[\sum_{r=0}^{2n} (-1)^r \binom{2n}{r} (e^{-ihx})^r (e^{ihx})^{2n-r} \right] dF(x) \\
&= \int_{\Re} \left(e^{-ihx} - e^{ihx} \right)^{2n} dF(x) \\
&= \int_{\Re} \left(e^{ihx} - e^{-ihx} \right)^{2n} dF(x) \\
&= \int_{\Re} [2i \sin(hx)]^{2n} dF(x) = i^{2n} 2^{2n} \int_{\Re} [\sin(hx)]^{2n}\, dF(x) \\
&= (-1)^n 2^{2n} \int_{\Re} [\sin(hx)]^{2n}\, dF(x).
\end{aligned}$$

Thus

$$\frac{\Delta^{(2n)} f(0)}{(2h)^{2n}} = (-1)^n \int_{\Re} \left[\frac{\sin(hx)}{hx} \right]^{2n} x^{2n}\, dF(x). \tag{12.27}$$

Letting $h \to 0$ and utilizing (12.26), (12.27) becomes

$$f(0)^{(2n)} = (-1)^n \lim_{h\to 0} \int_{\Re} \left[\frac{\sin(hx)}{hx} \right]^{2n} x^{2n}\, dF(x). \tag{12.28}$$

Now $0 \leq \left[\frac{\sin(hx)}{hx} \right]^{2n} x^{2n}$ (here is where the even-order derivative is employed!). Thus, by the Fatou–Lebesgue Theorem (Theorem 2 in Chapter 5),

$$\lim_{h\to 0} \int_{\Re} \left[\frac{\sin(hx)}{hx} \right]^{2n} x^{2n}\, dF(x) = \liminf_{h\to 0} \int_{\Re} \left[\frac{\sin(hx)}{hx} \right]^{2n} x^{2n}\, dF(x)$$

$$\geq \int_{\Re} \liminf_{h\to 0} \left[\frac{\sin(hx)}{hx}\right]^{2n} x^{2n}\, dF(x)$$
$$= \int_{\Re} \lim_{h\to 0} \left[\frac{\sin(hx)}{hx}\right]^{2n} x^{2n}\, dF(x)$$
$$= \int_{\Re} x^{2n}\, dF(x) = \mathcal{E}X^{2n} = (-1)^n f^{(2n)}(0),$$

since $\frac{\sin t}{t} \underset{t\to 0}{\to} 1$. From this result and (12.28), it follows that $\mathcal{E}X^{2n} < \infty$ and hence $\mathcal{E}|X|^k < \infty$ for all $k \leq 2n$. That $\mathcal{E}X^k = \frac{1}{i^k} f^{(k)}(0)$ follows from Theorem 1(v) in Chapter 11. ■

Proof of Proposition 2. Recall that

$$\psi(t) = \int_{\Re} \frac{e^{itx} - 1 - itx}{x^2}\, dK(x), \tag{12.29}$$

where K is a d.f. of bounded variation.

(i) As was pointed out just prior to the formulation of the proposition, for each $t \in \Re$, the function $\frac{e^{itx}-1-itx}{x^2}$ is well defined for all $x \in \Re$ and also continuous and bounded. Thus, the integral in (12.29) is well defined (and finite in norm) as a Riemann–Stieltjes integral and is taken as the limit of Riemann–Stieltjes sums corresponding to any arbitrarily chosen collection of division points and any points in the corresponding intervals. More precisely, for $a_n < 0 < b_n$ with $a_n \to -\infty$ and $b_n \to \infty$ where, which here and in the sequel all limits are taken as $n \to \infty$, consider the interval $(a_n, b_n]$ and divide it into subintervals by the points

$$a_n = x_{n0} < x_{n1} < \cdots < x_{n,r_n-1} < x_{nr_n} = b_n \quad (r_n \to \infty),$$

which are chosen to be $\neq 0$ and such that

$$\max_{1\leq k\leq r_n} (x_{nk} - x_{n,k-1}) \to 0.$$

For each subdivision, we have the following corresponding Riemann–Stieltjes sum

$$T_n(t) \overset{def}{=} \sum_{k=1}^{r_n} \frac{e^{itx_{nk}}-1-itx_{nk}}{x_{nk}^2} K(x_{n,k-1}, x_{nk}]$$
$$= \sum_k \left\{ i\frac{-K(x_{n,k-1},x_{nk}]}{x_{nk}} t + \frac{K(x_{n,k-1},x_{nk}]}{x_{nk}^2}(e^{itx_{nk}} - 1)\right\}, \tag{12.30}$$

where $K(x_{n,k-1}, x_{nk}]$ stands for the variation of K over the interval $(x_{n,k-1}, x_{nk}]$. By setting

$$\alpha_{nk} = -\frac{K(x_{n,k-1}, x_{nk}]}{x_{nk}}, \quad \beta_{nk} = x_{nk}, \text{ and}$$
$$\lambda_{nk} = \frac{K(x_{n,k-1}, x_{nk}]}{x_{nk}^2} (\geq 0),$$

relation (12.30) becomes as follows:

$$T_n(t) = \sum_k \left[i\alpha_{nk}t + \lambda_{nk}(e^{i\beta_{nk}t} - 1)\right]. \quad (12.31)$$

Now, for each n, let $Y_{nk},\ k = 1, \ldots, r_n$, be independent (within each row) r.v.s such that Y_{nk} is $P(\lambda_{nk})$, the Poisson distribution with parameter λ_{nk}, so that its ch.f. is given by

$$f_{Y_{nk}}(t) = e^{\lambda_{nk}(e^{it}-1)}.$$

Set $Z_{nk} = \alpha_{nk} + \beta_{nk}Y_{nk}$. Then the r.v.s $Z_{nk},\ k = 1, \ldots, r_n$, are independent (within each row) and the ch.f. of the r.v. Z_{nk} is given by

$$f_{Z_{nk}}(t) = e^{i\alpha_{nk}t}e^{\lambda_{nk}(e^{i\beta_{nk}t}-1)} = e^{i\alpha_{nk}t+\lambda_{nk}(e^{i\beta_{nk}t}-1)}.$$

Thus

$$\exp T_n(t) = \exp\left\{\sum_k \left[i\alpha_{nk}t + \lambda_{nk}(e^{i\beta_{nk}t} - 1)\right]\right\}$$
$$= \prod_k e^{i\alpha_{nk}t+\lambda_{nk}(e^{i\beta_{nk}t}-1)}$$

is a ch.f. (that of the r.v. $\sum_k Z_{nk}$). But

$$T_n(t) \to \int_{\Re} \frac{e^{itx} - 1 - itx}{x^2} dK(x) = \psi(t),$$

so that

$$f_n(t) \stackrel{def}{=} e^{T_n(t)} \to e^{\psi(t)} \stackrel{def}{=} f(t). \quad (12.32)$$

Define $g(t; x)$ by: $g(t; x) = \frac{e^{itx}-1-itx}{x^2}$ for $x \neq 0$, and $g(t; x) = -\frac{t^2}{2}$ for $x = 0$. Restrict $|t| \leq 1$ and look at $g(t; x)$ as a function of x. Then, for $x \neq 0$, $|g(t; x)| = \left|\frac{e^{itx}-1-itx}{x^2}\right| \leq \frac{|e^{itx}-1|+|tx|}{x^2} \leq \frac{2|tx|}{x^2} \leq \frac{2}{|x|} \to 0$ as $|x| \to \infty$, and, for $x = 0$, $|g(x; t)| \leq 1$, so that $|g(\cdot\,; t)|$ is bounded by a bound independent of t ($|t| \leq 1$) and K-integrable. Also, the integrand tends to 0, as $t \to 0$, for all $x \in \Re$.

Then, as $t \to 0$, the Dominated Convergence Theorem yields $\psi(t) \to \psi(0) = 0$, so that ψ is continuous at 0 and then so is $f(t)$. Thus, we have that f_n are ch.f.s of r.v.s and, by (12.32), $f_n \to f$ in $\Re$ with f being continuous at the origin. Then the converse part of the Paul Lévy Continuity Theorem (Theorem 3 in Chapter 11) implies that f is the ch.f. of a uniquely determined d.f. of a r.v., call it X. This establishes part (i).

(ii) Restrict attention to $|t| \leq 1$, and observe that, for $x \neq 0$, $|\frac{\partial}{\partial t}g(t; x)| = \left|i\frac{e^{itx}-1}{x}\right| \leq \frac{|tx|}{|x|} = |t| \leq 1$ independent of t and K-integrable. Since

$$\psi(t) = \int_{\Re} g(t; x)dK(x) = \int_{\{0\}} g(t; x)dK(x) + \int_{\Re-\{0\}} g(t; x)dK(x)$$

$$= -\frac{v_0}{2}t^2 + \int_{\Re-\{0\}} \frac{e^{itx} - 1 - itx}{x^2} dK(x),$$
$$v_0 = K(0) - K(0-),$$

Theorem 5 in Chapter 5 applies and yields

$$\begin{aligned}\psi'(t) &= -v_0 t + \frac{d}{dt}\int_{\Re-\{0\}} \frac{e^{itx} - 1 - itx}{x^2} dK(x)\\ &= -v_0 t + \int_{\Re-\{0\}} \frac{\partial}{\partial t}\left(\frac{e^{itx} - 1 - itx}{x^2}\right) dK(x)\\ &= -v_0 t + \int_{\Re-\{0\}} \left(i\frac{e^{itx} - 1}{x}\right) dK(x). \qquad (12.33)\end{aligned}$$

In particular,

$$\psi'(0) = 0. \qquad (12.34)$$

Next, for $x \neq 0$, $|\frac{\partial^2}{\partial t^2} g(t; x)| = \left|\frac{\partial}{\partial t}\left(i\frac{e^{itx}-1}{x}\right)\right| = |-e^{itx}| = 1$, again independent of t and K-integrable. Then by the theorem just cited and (12.33),

$$\begin{aligned}\psi''(t) &= -v_0 + \frac{d}{dt}\int_{\Re-\{0\}} \left(i\frac{e^{itx} - 1}{x}\right) dK(x)\\ &= -v_0 + \int_{\Re-\{0\}} \frac{\partial}{\partial t}\left(i\frac{e^{itx} - 1}{x}\right) dK(x)\\ &= -v_0 - \int_{\Re-\{0\}} e^{itx}\, dK(x) = -\int_{\Re} e^{itx}\, dK(x). \qquad (12.35)\end{aligned}$$

In particular,

$$\psi''(0) = -Var\, K. \qquad (12.36)$$

Now, $e^{\psi(t)} = f(t)$ is the ch.f. of the r.v. X, whose second-order derivative at 0 is

$$\begin{aligned}\left(e^{\psi(t)}\right)'' |_{t=0} &= \left(\psi'(t)e^{\psi(t)}\right)' |_{t=0}\\ &= \left(\psi''(t)e^{\psi(t)} + [\psi'(t)]^2 e^{\psi(t)}\right)' |_{t=0}\\ &= \psi''(0) = -Var\, K \text{ finite},\end{aligned}$$

by (12.34) and (12.36) and the fact that $\psi(0) = 0$. Then Lemma 6 applies and gives

$$\mathcal{E}X = \frac{1}{i}\left(e^{\psi(t)}\right)' |_{t=0} = \psi'(t)e^{\psi(t)}|_{t=0} = 0,$$

and

$$\mathcal{E}X^2 = \frac{1}{i^2}\left(e^{\psi(t)}\right)'' |_{t=0} = -\psi''(0) = Var\, K \,(\text{by } (12.36)).$$

The proof of the proposition is completed. ■

Proof of Corollary 1. From (12.29), we get

$$\frac{\psi(t)}{n} = \int_{\Re} \frac{e^{itx} - 1 - itx}{x^2} d\frac{K(x)}{n},$$

where, for each n, $\frac{K}{n}$ is a d.f. of bounded variation. Then, by Proposition 2(i), $e^{\frac{\psi}{n}}$ is the ch.f. of a r.v. Let X_{nj}, $j = 1, \ldots, n$, be row-wise i.i.d. r.v.s with ch.f. $e^{\frac{\psi}{n}}$. Then the ch.f., f_n, of $S_n = \sum_{j=1}^n X_{nj}$ is $f_n = \prod_{j=1}^n e^{\frac{\psi}{n}} = e^{\psi}$, so that $\mathcal{L}(S_n) = \mathcal{L}$(corresponding to e^{ψ}). It remains for us to check conditions (C). By Proposition 2(ii), applied with $e^{\frac{\psi}{n}}$ rather than e^{ψ}, we have $\frac{d}{dt}e^{\psi(t)/n}|_{t=0} = 0$, $\frac{d^2}{dt^2}e^{\psi(t)/n}|_{t=0} = -Var\frac{K}{n}$, by (12.34) and (12.36), so that the first and second moments of the distribution corresponding to the ch.f. $e^{\frac{\psi}{n}}$ are 0 and $Var\frac{K}{n}$, respectively. Thus, $\mathcal{E}X_{nj} = 0$, and $\sigma_{nj}^2 = \sigma^2(X_{nj}) = Var\frac{K}{n} = \frac{1}{n}Var\,K = \frac{1}{n}$, $j = 1, \ldots, n$, so that $\max_j \sigma_{nj}^2 = \frac{1}{n} \to 0$, and $s_{nj}^2 = \sum_j \sigma_{nj}^2 = 1$. ■

Proof of Corollary 2. Actually, the only point so far where the condition $s_n^2 = 1$ was used was in the proof of Lemma 1, which, however, also holds true under the condition $s_n^2 \leq c$ (see Remark 1). So, if K is a d.f. with $Var\,K = v$, then $\frac{K}{nv}$ is also a d.f. with $Var\frac{K}{nv} = \frac{1}{n}$. Hence $e^{\frac{\psi}{n}}$ is the ch.f. of row-wise i.i.d. r.v.s X_{nj}, $j = 1, \ldots, n$, with $\mathcal{E}X_{nj} = 0$ and $\sigma^2(X_{nj}) = Var\frac{K}{n} = \frac{v}{n}$. It follows that $\sum_j \sigma_{nj}^2 = v$ and $\max_j \sigma_{nj}^2 = \frac{v}{n} \to 0$. ■

Proof of Corollary 3.

(i) For an arbitrary $t \in \Re$, consider, e.g., the interval $[t-1, t+1]$ and apply Theorem 5 in Chapter 5 to obtain, as in relation (12.35),

$$\psi''(t) = -\int_{\Re} e^{itx}\, dK(x).$$

(ii) Let K^* be another d.f. as described in part (ii), and such that

$$\int_{\Re} \frac{e^{itx} - 1 - itx}{x^2} dK^*(x) = \psi(t), \quad t \in \Re.$$

Then, as in part (i),

$$\psi''(t) = -\int_{\Re} e^{itx}\, dK(x) = -\int_{\Re} e^{itx}\, dK^*(x),$$

or

$$-\psi''(t) = \int_{\Re} e^{itx}\, dK(x) = \int_{\Re} e^{itx}\, dK^*(x). \tag{12.37}$$

In particular,

$$-\psi''(0) = Var\,K = Var\,K^* \stackrel{def}{=} v_0, \tag{12.38}$$

so that, by (12.37),

$$-\frac{\psi''(t)}{v_0} = \int_{\Re} e^{itx}\, d\frac{K(x)}{v_0} = \int_{\Re} e^{itx}\, d\frac{K^*(x)}{v_0}. \tag{12.39}$$

Thus, the d.f.s $\frac{K}{v_0}$ and $\frac{K^*}{v_0}$ have variation 1 and $\frac{K}{v_0}(-\infty) = \frac{K^*}{v_0}(-\infty) = 0$. Since, by (12.39), they have the same ch.f., it follows that $\frac{K}{v_0} = \frac{K^*}{v_0}$ in $\Re$, or $K = K^*$ in $\Re$. ■

Proof of Proposition 3.

(i) As in the proof of Proposition 2(i), let $g(t;x) = (e^{itx} - 1 - itx)/x^2$ for $x \neq 0$, and (see discussion following relation (12.32)) $g(t;x) = -t^2/2$ for $x = 0$. Then for $x \neq 0$, $|g(t;x)| \leq \frac{2|t|}{|x|} \to 0$ as $x \to \infty$, so that, for each $t \in \Re$, $g(t;x)$ is bounded and continuous in $\Re$ as a function of x. Thus, if $K_n \Rightarrow K$, a d.f. (with $Var\,K \leq c$), then the Helly–Bray Extended Lemma (Theorem 7 in Chapter 8; see also Remark 4 in the same chapter) applies and yields

$$\psi_n(t) = \int_{\Re} g(t;x)\, dK_n(x) \to \int_{\Re} g(t;x)\, dK(x) = \psi(t), \quad t \in \Re. \tag{12.40}$$

(ii) Consider the sequence $\{K_n\}$ and apply the Weak Compactness Theorem (Theorem 5 in Chapter 8) to obtain $\{K_m\} \subseteq \{K_n\}$ such that $K_m \underset{m\to\infty}{\Rightarrow} K^*$, some d.f. (with $Var\,K^*$ necessarily $\leq c$). Define ψ^* by

$$\psi^*(t) = \int_{\Re} \frac{e^{itx} - 1 - itx}{x^2}\, dK^*(x), \quad t \in \Re.$$

Then, by part (i), $\psi_m \underset{m\to\infty}{\to} \psi^*$ in $\Re$. Since also $\psi_m \underset{m\to\infty}{\to} \psi$ in $\Re$, it follows that $\psi^* = \psi$ in $\Re$. Next, let $\{K_r\} \subseteq \{K_n\}$ distinct from $\{K_m\}$ such that $K_r \underset{r\to\infty}{\Rightarrow} K^{**}$, some d.f. (of variation necessarily $\leq c$), and define ψ^{**} by

$$\psi^{**}(t) = \int_{\Re} \frac{e^{itx} - 1 - itx}{x^2}\, dK^{**}(x), \quad t \in \Re.$$

Then, as before, $\psi_r \underset{r\to\infty}{\to} \psi^{**}$ in $\Re$, whereas $\psi_r \underset{r\to\infty}{\to} \psi$ in $\Re$ also, so that $\psi^{**} = \psi$ in $\Re$. It follows that $\psi^{**} = \psi^*(= \psi)$ in $\Re$. On the other hand, by relation (12.37),

$$-\psi''(t) = \int_{\Re} e^{itx}\, dK^*(x) = \int_{\Re} e^{itx}\, dK^{**}(x),$$

so that the d.f.s K^* and K^{**} have the same ch.f. Then, by the inversion formula (Theorem 2 in Chapter 11) it follows that $K^* = K^{**} + C$, for a constant C. Hence (by Definition 3 in Chapter 11) $K_n \Rightarrow K$ *uac*, some d.f. K (with $Var\,K \leq c$), and all of them determine the same ψ. ■

Exercises.

1. If the independent r.v.s $X_j,\ j \geq 1$, are distributed as $U(-j, j)$, then show that the Lindeberg condition (see relation (12.24)) holds, so that $\mathcal{L}(\frac{S_n}{s_n}) \underset{n\to\infty}{\Rightarrow} N(0,1)$, where $S_n = \sum_{j=1}^{n} X_j$ and $s_n^2 = \sum_{j=1}^{n} \sigma_j^2$, $\sigma_j^2 = Var(X_j)$.
Hint: Recall that $X \sim U(\alpha, \beta)$ means that the r.v. X has the uniform distribution with parameters α and β ($\alpha < \beta$), its probability density function is $p(x) = \frac{1}{\beta-\alpha} I_{[\alpha,\beta]}(x)$, $\mathcal{E}X = \frac{\alpha+\beta}{2}$ and $Var(X) = \frac{(\alpha-\beta)^2}{12}$. Finally, recall that $\sum_{j=1}^{n} j^2 = n(n+1)(2n+1)/6$.
2. If the independent r.v.s $X_j,\ j \geq 2$, are distributed as follows:

$$P(X_j = -j^{\alpha}) = P(X_j = j^{\alpha}) = \frac{1}{j^{\beta}}, \quad P(X_j = 0) = 1 - \frac{2}{j^{\beta}} \quad (\alpha, \beta > 0),$$

show that the restriction $\beta < 1$ ensures that the Lindeberg condition (relation (12.24)) holds.
Hint: Show that the restriction $\beta < 1$ ensures that the set of js with $j = 1, \ldots, n$, and $|\pm j^{\alpha}| \geq \varepsilon s_n$, is empty for all $\varepsilon > 0$, for large n. For an arbitrary, but fixed $\beta (< 1)$, it is to be understood that $j \geq j_0$, where $j_0 = 2^{1/\beta}$, if $2^{1/\beta}$ is an integer, or $j_0 = [2^{1/\beta}] + 1$ otherwise. This ensures that $1 - \frac{2}{j^{\beta}}$ is nonnegative.
3. Let the r.v.s $X_j,\ j \geq 1$, be distributed as follows:

$$P(X_j = \pm j^{\alpha}) = \frac{1}{6j^{2(\alpha-1)}}, \quad P(X_j = 0) = 1 - \frac{1}{3j^{2(\alpha-1)}}, \quad \alpha > 1.$$

Show that the Lindeberg condition (relation (12.24)) holds, if and only if $\alpha < \frac{3}{2}$. Conclude that $\mathcal{L}(\frac{S_n}{s_n}) \underset{n\to\infty}{\Rightarrow} N(0,1)$, where $S_n = \sum_{j=1}^{n} X_j$ and $s_n^2 = \sum_{j=1}^{n} \sigma_j^2$, $\sigma_j^2 = Var(X_j)$.
Hint: For $\alpha < \frac{3}{2}$, show that $j^{2\alpha} < \varepsilon^2 s_n^2$, $j = 1, \ldots, n$, which is implied by $n^{2\alpha} < \varepsilon^2 s_n^2$ for large n, so that $g_n(\varepsilon) = 0$. Next, $g_n(\varepsilon) \geq 1 - \frac{\varepsilon^2}{18}(1-\frac{1}{k})(2-\frac{1}{k})\frac{k^{2\alpha}}{\varepsilon^2 s_n^2} k^{3-2\alpha}$, where $k = [(\varepsilon s_n)^{1/\alpha}]$, and conclude that the expression on the right-hand side does not converge to 0 for $\alpha \geq \frac{3}{2}$.
4. For $j = 1, 2, \ldots, n$, let X_j be independent r.v.s defined as follows:

$$X_j = \begin{cases} \pm j^2 & \text{with probability } 1/12j^2 \text{ each} \\ \pm j & \text{with probability } 1/12 \text{ each} \\ 0 & \text{with probability } 1 - 1/6 - 1/6j^2. \end{cases}$$

Then show that the Lindeberg condition (condition (12.24)) does *not* hold.

Hint: Recall that $\sum_{j=1}^{n} j^2 = n(n+1)(2n+1)/6$, and show that, for every $\varepsilon > 0$ and large n: $s_n^{-2}\sum_{j=1}^{n} j^2 I_{(|\pm j| \geq \varepsilon s_n)} = o(1)$, $s_n^{-2}\sum_{j=1}^{k} j^2 = o(1)$, $k = [(\varepsilon s_n)^{1/2}]$, in order to conclude that $g_n(\varepsilon) \underset{n\to\infty}{\nrightarrow} 0$.

5. Let $X_n, n \geq 1$, be independent r.v.s such that $|X_n| \leq M_n$ a.s. with $M_n = o(s_n)$ where $s_n^2 = \sum_{j=1}^{n} \sigma_j^2 \underset{n\to\infty}{\to} \infty$ and $\sigma_j^2 = Var(X_j)$.
Set $S_n = \sum_{j=1}^{n} X_j$ and show that

$$\mathcal{L}\left(\frac{S_n - \mathcal{E}S_n}{s_n}\right) \underset{n\to\infty}{\Rightarrow} N(0,1).$$

Hint: From the assumption $M_n = o(s_n)$, it follows that $\frac{M_n}{s_n} \underset{n\to\infty}{\to} 0$, so that $M_n < \varepsilon s_n, n > n_0 (= n(\varepsilon)), \varepsilon > 0$. Write

$$\frac{S_n - \mathcal{E}S_n}{s_n} = \frac{1}{s_n}\sum_{j=1}^{n_0}(X_j - \mathcal{E}X_j) + \frac{1}{s_n}\sum_{j=n_0+1}^{n}(X_j - \mathcal{E}X_j)\ (n > n_0),$$

and since the first term tends to 0, as $n \to \infty$, work with the second term only. To this end, set $Y_{nj} = (X_j - \mathcal{E}X_j)/\tau_n$, where $\tau_n^2 = s_n^2 - s_{n_0}^2$, and show that the r.v.s Y_{nj}, $j = n_0 + 1, \ldots, n$, satisfy the Liapounov condition (for $\delta = 1$) (see Theorem 3).

6. Let $X_{nj}, j = 1, \ldots, k_n \to \infty$ be row-wise independent r.v.s with $\mathcal{E}X_{nj} = 0$, $\sigma^2(X_{nj}) = \sigma_{nj}^2 < \infty$, $s_n^2 = \sum_{j=1}^{k_n}\sigma_{nj}^2 = 1$, and $\max_{1\leq j\leq k_n}\sigma_{nj}^2 \underset{n\to\infty}{\to} 0$. Then, with $S_n = \sum_{j=1}^{k_n} X_{nj}$, and under the assumption that $\mathcal{L}(S_n) \Rightarrow N(0,1)$, show that $\sum_{j=1}^{k_n} X_{nj}^2 \underset{n\to\infty}{\overset{P}{\to}} 1$.
Hint: By Theorem 1, it follows that, for every $\varepsilon > 0$, $g_n(\varepsilon) = \sum_j \int_{(|x|\geq\varepsilon s_n)} x^2 dF_{nj} \underset{n\to\infty}{\to} 0$, where F_{nj} is the d.f. of X_{nj}. One way of proceeding is to show that $\sum_j \sigma^2\left(|X_{nj}| I_{(|X_{nj}|\geq 1)}\right) \underset{n\to\infty}{\to} 0$, $\sum_j P(|X_{nj}| \geq \varepsilon) \underset{n\to\infty}{\to} 0$, $\sum_j \sigma^2\left(X_{nj}^2 I_{(X_{nj}^2<1)}\right) \underset{n\to\infty}{\to} 0$, $\sum_j X_{nj}^2 I_{(X_{nj}^2<1)} - \sum_j \mathcal{E}[X_{nj}^2 I_{(|X_{nj}^2|<1)}] \underset{n\to\infty}{\overset{P}{\to}} 0$, and, finally, $P(\sum_j X_{nj}^2 I_{(X_{nj}^2<1)} \neq \sum_j X_{nj}^2) \underset{n\to\infty}{\longrightarrow} 0$.

7. For a function $g : \Re \to \mathbb{C}$, set $\Delta g(u) = g(u+h) - g(u-h)$, $h \in \Re$, and define $\Delta^{(n)} g(u)$ recursively. Then

 (i) Show that

$$\Delta^{(m)} g(u) = \sum_{r=0}^{m} (-1)^r \binom{m}{r} g(u + (m-2r)h).$$

(ii) From part (i), and by expanding $f(t)$ around 0 up to terms of order $2n$, obtain

$$\frac{\Delta^{(2n)} f(0)}{(2h)^{2n}} = f^{(2n)}(0) + \frac{o(h^{2n})}{h^{2n}}$$

$$\left(\text{where } \frac{o(t)}{t} \to 0 \text{ as } t \to 0\right),$$

so that $f^{(2n)}(0) = \lim \frac{\Delta^{(2n)} f(0)}{(2h)^{2n}}$ as $h \to 0$.

Hint: Part (i) is proved by induction in m. In the process of doing so, the relation $\binom{m}{r+1} + \binom{m}{r} = \binom{m+1}{r+1}$ will be needed. In proving part (ii), the following relation is required (which you may use here without proof; see, however, the next exercise):

$$\sum_{r=0}^{2n} (-1)^r \binom{2n}{r} (2n - 2r)^k = \begin{cases} 0 & k < 2n \\ 2^{2n}(2n)! & k = 2n. \end{cases}$$

8. For $m = 1, 2, \ldots$ and k integer, show that

$$\sum_{r=0}^{m} (-1)^r \binom{m}{r} (m - 2r)^k = \begin{cases} 0 & \text{for } (0 \le)k < m \\ 2^m \times m! & k = m. \end{cases} \tag{12.41}$$

Hint: The proof is by induction. The result is easily checked for $m = 1, 2, 3$. Then assume (12.41) and show that

$$\sum_{r=0}^{m} (-1)^r \binom{m+1}{r} (m + 1 - 2r)^k = \begin{cases} 0 & \text{for } k < m + 1 \\ 2^{m+1} \times (m+1)! & k = m + 1. \end{cases} \tag{12.42}$$

First, prove (12.42) for $k < m + 1$. For this purpose, use the identity $\binom{m+1}{r} = \binom{m}{r} + \binom{m}{r-1}$, and in the process of the subsequent parts of the proof, use the expansion $(x+y)^k = \sum_{l=0}^{k} \binom{k}{l} x^l y^{k-l}$, along with (12.41). In establishing (12.42) for $k = m + 1$, write $(m + 1 - 2r)^{m+1} = (m + 1 - 2r)^m (m + 1 - 2r)$, use the fact that $r\binom{m+1}{r} = (m + 1)\binom{m}{r-1}$, repeatedly use the expansion just mentioned, and, of course, also employ relation (12.41).

9. For $j = 1, \ldots, k_n \underset{n\to\infty}{\to} \infty$, let X_{nj} be row-wise independent r.v.s with respective d.f.s F_{nj}. Then show that, for every $\varepsilon > 0$, the convergence

$$P\left[\max_{1 \le j \le k_n} |X_{nj}| \ge \varepsilon\right] \underset{n\to\infty}{\to} 0 \tag{12.43}$$

is equivalent to

$$\sum_{j=1}^{k_n} \int_{(|x| \geq \varepsilon)} dF_{nj}(x) \underset{n\to\infty}{\to} 0 \tag{12.44}$$

and implies the convergence

$$\max_{1\leq j\leq k_n} P(|X_{nj}| \geq \varepsilon) \underset{n\to\infty}{\to} 0. \tag{12.45}$$

Hint: Use Exercise 3 in Chapter 10.

10. Show that, for every complex number z, it holds $|z^2| = |z|^2$.

11. In the proof of Proposition 2 (see discussion following relation (12.36)), Lemma 6 is used for $n = 1$ to ensure finiteness of $\mathcal{E}X^2$, which, according to the lemma, follows from the existence and finiteness of $f^{(2)}(0)$. Prove this fact directly; i.e., show that the existence and finiteness of $f^{(2)}(0)$ implies finiteness of $\mathcal{E}X^2$.

CHAPTER

The Central Limit Problem: The Noncentered Case 13

This chapter is a continuation of the previous chapter. In the basic conditions (C), summarized in (12.1) of Chapter 12, the assumption that the r.v.s involved have expectations zero was included. In the present chapter, this assumption is dropped (hence the term "noncentered"). As it will be seen, however, this has ramifications that will be addressed in this chapter.

In Section 13.1, conditions (C) are restated for convenience, and then they are modified to conditions (C') listed in (13.1) by replacing the requirement $s_n^2 = 1$ with the less stringent requirement $s_n^2 \leq c(< \infty)$. It is then stated that the basic Theorem 2 in Chapter 12 still holds true under (C'). This statement is justified by carefully walking through all preceding results in Chapter 12. Then conditions (C') are modified even further by dropping the requirement that $\mathcal{E}X_{nj} = 0$. This brings us to conditions (C'') listed in (13.2). These are the conditions under which the present chapter operates. The next step is to relate quantities, such as d.f. and ch.f., corresponding to the centered r.v.s $X_{nj} - \mathcal{E}X_{nj}$ and to the noncentered r.v.s X_{nj}. This process and some observations bring us to relation (13.11), which is the starting point for the following section.

In Section 13.2, the quantity $\bar{\psi}$ plays the role of ψ of Chapter 12, and a new ψ is introduced as in relation (13.12). Then this section is devoted to stating and proving four propositions analogous to Proposition 2, its corollaries, and Proposition 3, in the previous chapter. The culmination of this section is Theorem 2′, analogous to Theorem 2 of Chapter 12.

In Section 13.3, two important special cases of Theorem 2′ are discussed. In the first one, the limiting law is the $N(\mu, \sigma^2)$ distribution, and in the second the limiting law is the Poisson distribution with parameter λ, $P(\lambda)$. The section (and the chapter) is concluded with an application to the second special case just described. This application produces the familiar result that, under certain conditions, binomial probabilities are approximated by Poisson probabilities.

13.1 Notation and Preliminary Discussion

Recalling the basic setting from the previous chapter, we are dealing with the triangular arrays of r.v.s that are subject to conditions (C) in relation (12.1) of that chapter. We

An Introduction to Measure-Theoretic Probability, Second Edition. http://dx.doi.org/10.1016/B978-0-12-800042-7.00013-X

reproduce them here for easy reference.

$$(C)\begin{cases}\text{For each } n \geq 1, \quad \text{the r.v.s } X_{nj}, \quad j = 1, \ldots, k_n \underset{n\to\infty}{\to} \infty, \\ \quad \text{are independent within each row,} \\ \mathcal{E}X_{nj} = 0, \quad \sigma^2_{nj} = \sigma^2(X_{nj}) < \infty \text{ and } s^2_n = \sum_{j=1}^{k_n} \sigma^2_{nj} = 1; \\ \quad \text{also,} \quad \max_{1\leq j\leq k_n} \sigma^2_{nj} \underset{n\to\infty}{\to} 0.\end{cases}$$

As has already been mentioned in the introductory part, the central purpose here is to drop the assumption that $\mathcal{E}X_{nj} = 0$, and study its ramifications. First, however, an observation is in order, and that is that the fundamental Theorem 2 in Chapter 12 holds true under conditions (C) by suppressing $s^2_n = 1$ and replacing it by $s^2_n \leq c(< \infty)$ for all n. Again, for convenient reference, restate the resulting assumptions and denote them by (C'). That is,

$$(C')\begin{cases}\text{For each } n, \quad \text{the r.v.s } X_{nj}, \quad j = 1, \ldots, k_n \underset{n\to\infty}{\to} \infty, \\ \quad \text{are row-wise independent,} \\ \mathcal{E}X_{nj} = 0, \quad \sigma^2_{nj} = \sigma^2(X_{nj}) < \infty \text{ and } s^2_n \leq c(< \infty); \\ \quad \text{also,} \quad \max_{1\leq j\leq k_n} \sigma^2_{nj} \underset{n\to\infty}{\to} 0.\end{cases} \tag{13.1}$$

The assertion then is that

Statement. Theorem 2 in Chapter 12 holds true under conditions (C') listed in (13.1).

Justification. All one has to do is to go carefully through the proofs of the theorem and of other results used, and see what bearing on them the condition $s^2_n = 1$ has. *All references made here will be to Chapter 12*. First, Lemma 1 and its corollary hold under the condition $s^2_n \leq c$ as was pointed out in Remark 1. In Proposition 1, it is to be observed that now $K_n, n \geq 1$, are d.f.s with $Var\,K_n \leq c$, and $K_n(-\infty) = 0$, but not necessarily d.f.s of r.v.s. Lemma 6 and Proposition 2 are not related to conditions (C). Corollary 1 to Proposition 2 holds true under the condition $s^2_n \leq c$, because of Corollary 2 to Proposition 2. Corollary 3 makes use of relation (12.35) pertaining to Proposition 2, which, however, is not related to conditions (C). Thus, Corollary 3 to Proposition 2 is not related to conditions (C) either. In Proposition 3, part (i) holds for d.f.s K_n with $Var\,K_n \leq c$ (which amounts to $s^2_n \leq c$) and $K_n(-\infty) = 0$. Part (ii) makes use of K_ns as just described and of Corollary 3 to Proposition 2, which is unrelated to conditions (C). In other words, Proposition 3 still holds, if $s^2_n \leq c$.

Finally, regarding Theorem 2, the proof of part (i) makes use of (12.13), which also holds true for $s^2_n \leq 1$ (because of Remark 1). Also, it uses Lemmas 2 and 3, which are unrelated to conditions (C), and of Proposition 3(ii), which holds under $s^2_n \leq c$. The proof of part (ii) makes use of Proposition 3(i) and (12.15), both holding under $s^2_n \leq c$. This completes the justification of the statement. ■

Now, starting with conditions (C') and dropping the requirement that $\mathcal{E}X_{nj} = 0$, we have conditions (C''):

$$(C'')\begin{cases} \text{For each } n, \quad \text{the r.v.s } X_{nj}, \quad j = 1, \ldots, k_n \underset{n\to\infty}{\to} \infty, \\ \quad \text{are row-wise independent,} \\ \mathcal{E}X_{nj} = \alpha_{nj}, \text{ say, finite,} \quad \sigma^2_{nj} = \sigma^2(X_{nj}) < \infty \text{ and} \\ s^2_n \le c(<\infty); \text{ also,} \quad \max_{1\le j\le k_n} \sigma^2_{nj} \underset{n\to\infty}{\to} 0. \end{cases} \tag{13.2}$$

To economize on the notation, we mention once and for all that, for a r.v. X with $\mathcal{E}X = \alpha$, finite, $\sigma^2(X) = \sigma^2 < \infty$, with d.f. F and ch.f. f, we set

$$\bar{X} = X - \alpha, \quad \text{so that } \mathcal{E}\bar{X} = 0 \text{ and } \sigma^2(\bar{X}) = \sigma^2, \tag{13.3}$$

and $\bar{F}$ and $\bar{f}$, respectively, for the d.f. and the ch.f. of $\bar{X}$.

For later reference, we mention here that

$$\bar{f}(t) = \mathcal{E}e^{it\bar{X}} = e^{-i\alpha t} f(t), \tag{13.4}$$

and

$$\bar{F}(x) = P(\bar{X} \le x) = P(X \le x + \alpha) = F(x + \alpha). \tag{13.5}$$

Also, all limits are taken as $n \to \infty$ unless otherwise specified, and all maxima, summations and products are over j ranging from 1 to k_n.

The ch.f.s $\bar{f}_{nj}$ and f_{nj} of the r.v.s $\bar{X}_{nj} = X_{nj} - \alpha_{nj}$ and X_{nj}, respectively, are on account of (13.4),

$$\bar{f}_{nj}(t) = e^{-i\alpha_{nj}t} f_{nj}(t), \tag{13.6}$$

so that

$$\bar{f}_n(t) = \prod_{j=1}^{k_n} \bar{f}_{nj}(t) = e^{-i\alpha_n t} f_n(t), \quad f_n(t) = \prod_{j=1}^{k_n} f_{nj}(t), \quad \alpha_n = \sum_{j=1}^{k_n} \alpha_{nj}. \tag{13.7}$$

The respective d.f.s $\bar{F}_{nj}$ and F_{nj} of the r.v.s $\bar{X}_{nj}$ and X_{nj} are related as follows, on account of (13.5):

$$\bar{F}_{nj}(x) = P(\bar{X}_{nj} \le x) = P(X_{nj} \le x + \alpha_{nj}) = F_{nj}(x + \alpha_{nj}). \tag{13.8}$$

Set

$$\bar{K}_n(x) = \sum_{j=1}^{k_n} \int_{-\infty}^{x} y^2 d\bar{F}_{nj}(y) = \sum_{j=1}^{k_n} \int_{-\infty}^{x} y^2 dF_{nj}(y + \alpha_{nj}), \tag{13.9}$$

and observe that $\bar{K}_n$ is a d.f. with $\bar{K}_n(-\infty) = 0$ and $Var\bar{K}_n \le c$. By means of $\bar{K}_n$, define the function $\bar{\psi}_n$ by

$$\bar{\psi}_n(t) = \int_{\Re} \frac{e^{itx} - 1 - itx}{x^2} d\bar{K}_n(x). \tag{13.10}$$

It is immediate that the r.v.s $\bar{X}_{nj}$, $j = 1, \ldots, k_n$, satisfy conditions (C'). Therefore Lemma 1 and its Corollary in Chapter 12 apply and give

$$\log \bar{f}_n(t) - \bar{\psi}_n(t) \to 0 \text{ on } \Re,$$

and hence, for all $t \in \Re$,

$$\bar{f}_n(t)e^{-\bar{\psi}_n(t)} \to 1, \quad \text{or } e^{-i\alpha_n t} f_n(t)e^{-\bar{\psi}_n(t)} \to 1, \quad \text{or } f_n(t)e^{-\psi_n(t)} \to 1, \qquad (13.11)$$

where $\psi_n(t) = i\alpha_n t + \bar{\psi}_n(t)$, $\alpha_n = \sum_{j=1}^{k_n} \alpha_{nj}$.

This will be the starting point of the next section where results analogous to Proposition 2, its Corollaries 1 and 3, and Proposition 3 will be formulated. Also, a result analogous to the fundamental Theorem 2 in Chapter 12 will be formulated and proved, which is, actually, the main purpose of this chapter.

13.2 Limiting Laws of $\mathcal{L}(S_n)$ Under Conditions (C'')

As has been the case throughout, S_n stands for partial sums; i.e., $S_n = \sum_{j=1}^{k_n} X_{nj}$, and the purpose of the present section is to determine conditions under which $\mathcal{L}(S_n)$ has a limiting distribution, and determine the class of all limiting laws of $\mathcal{L}(S_n)$. This is done in stages with the formulation and proof of two propositions and two corollaries (Proposition 2′, its two corollaries–Corollary 1′ and 3′—and Proposition 3′) the way it was done in Section 12.2 (of Chapter 12).

Proposition 2′. Let K be a d.f. of bounded variation and let α be a real number. Define ψ by

$$\psi(t) = i\alpha t + \bar{\psi}(t), \quad \text{where } \bar{\psi}(t) = \int_{\Re} \frac{e^{itx} - 1 - itx}{x^2} dK(x). \qquad (13.12)$$

Then $e^{\bar{\psi}(t)}$ is the ch.f. of a r.v. $\bar{X}$, say, and $e^{\psi(t)}$ is the ch.f. of the r.v. $X = \bar{X} + \alpha$. Furthermore, $\mathcal{E}\bar{X} = 0$, $\sigma^2(\bar{X}) = Var K$, so that $\mathcal{E}X = \alpha$ and $\sigma^2(X) = Var K$.

Proof. That $e^{\bar{\psi}(t)}$ is the ch.f. of a r.v. $\bar{X}$ follows from Proposition 2(i) in Chapter 12. That $e^{\psi(t)}$ is the ch.f. of the r.v. $X = \bar{X} + \alpha$ follows from the relation $e^{\psi(t)} = e^{i\alpha t}e^{\bar{\psi}(t)}$. Finally, that $\mathcal{E}\bar{X} = 0$ and $\sigma^2(\bar{X}) = Var K$ follow from Proposition 2(ii) in Chapter 12. ■

Corollary 1′. Let K be a d.f. with bounded variation, and let the function $\bar{\psi}$ be defined by (13.12), so that $e^{\bar{\psi}(t)}$ is the ch.f. of a r.v. (by part (i) of Proposition 2 in Chapter 12). Let α be a real number, and define ψ by (13.12). Then e^{ψ} is the ch.f. of a r.v. Finally, for each n, there exist i.i.d. r.v.s X_{nj}, $j = 1, \ldots, n$, satisfying conditions (C''), such that $\mathcal{L}(S_n) = \mathcal{L}$(corresponding to $e^{\psi(t)}$), where $S_n = \sum_j X_{nj}$.

Proof. As already mentioned, $e^{\bar{\psi}}$ is the ch.f. of a r.v. $\bar{X}$ which, by Proposition 2(ii) in Chapter 12, has $\mathcal{E}\bar{X} = 0$ and $\sigma^2(\bar{X}) = Var K$. Thus e^{ψ} is the ch.f. of the r.v.

$X = \bar{X} + \alpha$ and $\mathcal{E}X = \alpha$ and $\sigma^2(X) = Var K$. This is so, because by (13.12),

$$e^{\psi(t)} = e^{i\alpha t + \bar{\psi}(t)} = e^{i\alpha t} e^{\bar{\psi}(t)}, \quad t \in \Re,$$

and the right-hand side is the ch.f. of $X + \alpha$, since α and X are independent. Next, let $\bar{X}_{nj}$, $j = 1, \ldots, n$, be i.i.d. r.v.s with ch.f. $e^{\frac{\bar{\psi}(t)}{n}}$, and let $X_{nj} = \bar{X}_{nj} + \frac{\alpha}{n}$. Then the ch.f. of X_{nj} is $e^{\frac{i\alpha t}{n} + \frac{\bar{\psi}(t)}{n}}$, so that the ch.f. of $S_n = \sum_j X_{nj}$ is $e^{i\alpha t + \bar{\psi}(t)} = e^{\psi(t)}$. That is, $\mathcal{L}(S_n) = \mathcal{L}$ (corresponding to $e^{\psi(t)}$). Next, $\mathcal{E}\bar{X}_{nj} = 0$ and $\sigma^2(\bar{X}_{nj}) = \frac{1}{n} Var K$ by Proposition 2(ii) in Chapter 12, so that $\mathcal{E}X_{nj} = \frac{\alpha}{n}$ and $\sigma^2(X_{nj}) = \frac{1}{n} Var K$. It follows that $s_n^2 = Var K$ and $\max_j \sigma_{nj}^2 = \frac{1}{n} Var K \to 0$. The conditions (C'') are fulfilled, and the proof is completed. ■

Corollary 3′. Let K be a d.f. of bounded variation with $K(-\infty) = 0$, and let α be a real number. Define ψ as in (13.12). Then the pair (α, K) uniquely determines ψ, and vice versa. That is, if K^* is another d.f. as before and $\alpha^* \in \Re$ determining the same ψ, then $K = K^*$ and $\alpha = \alpha^*$.

Proof. Indeed, let K^* be a d.f. of bounded variation with $K^*(-\infty) = 0$ and let $\alpha^* \in \Re$, such that

$$i\alpha t + \bar{\psi}(t) = \psi(t) = \psi^*(t) = i\alpha^* t + \bar{\psi}^*(t), \quad t \in \Re, \tag{13.13}$$

where

$$\bar{\psi}^*(t) = \int_{\Re} \frac{e^{itx} - 1 - itx}{x^2} dK^*(x), \quad \text{and } \bar{\psi} \text{ is as in (13.12).}$$

Use (13.12), the defining relation for $\bar{\psi}^*$, and take the second-order derivative with respect to t as in relations (12.37) and (12.38) (of Chapter 12) in order to get $Var K = -\psi''(0) = -\psi^{*\prime\prime}(0) = Var K^* \overset{def}{=} v_0$, or by (12.39) in Chapter 12,

$$-\frac{\psi''(t)}{v_0} = \int_{\Re} e^{itx} d\frac{K}{v_0}(x) = \int_{\Re} e^{itx} d\frac{K^*}{v_0}(x).$$

Then, as in Corollary 3(ii) to Proposition 2 in Chapter 12, it follows that $K^* = K$ on $\Re$. Therefore, $\bar{\psi}^* = \bar{\psi}$ in $\Re$ and hence $\alpha^* = \alpha$, by (13.13). ■

Proposition 3′. For each n, let $\bar{K}_n$ be a d.f. such that $Var \bar{K}_n \leq c (< \infty)$, and let $\alpha_n \in \Re$. Define $\bar{\psi}_n$ and ψ_n by

$$\bar{\psi}_n(t) = \int_{\Re} \frac{e^{itx} - 1 - itx}{x^2} d\bar{K}_n(x), \quad \psi_n(t) = i\alpha_n t + \bar{\psi}_n(t), \ t \in \Re. \tag{13.14}$$

(i) Suppose that $\bar{K}_n \Rightarrow K$, some d.f. (with $Var K$ necessarily $\leq c$) and $\alpha_n \to \alpha \in \Re$. Then $\psi_n(t) \to \psi(t)$, $t \in \Re$, where ψ is defined by

$$\psi(t) = i\alpha t + \bar{\psi}(t) \text{ and } \bar{\psi}(t) = \int_{\Re} \frac{e^{itx} - 1 - itx}{x^2} dK(x). \tag{13.15}$$

(ii) Let $\psi_n(t)$ be defined by (13.14) and suppose that $\psi_n(t) \to \psi(t), t \in \Re$, some function on $\Re$. Then $\bar{K}_n \Rightarrow K$ uac, some d.f. (of $Var K \leq c$), and $\alpha_n \to \alpha \in \Re$. Furthermore, ψ corresponds to the pair (α, K) and is defined as in (13.15), where K is any member of the family of limiting distribution of $\{\bar{K}_n\}$.

Proof.

(i) The assumption $\bar{K}_n \Rightarrow K$ implies that $\bar{\psi}_n(t) \to \bar{\psi}(t), t \in \Re$, where $\bar{\psi}$ is defined by (13.15). This follows by Proposition 3(i) in Chapter 12. Then that $\psi_n \to \psi$ follows from the fact that $\alpha_n \to \alpha$ and relation (13.14).

(ii) In order to show that $\alpha_n \to \alpha \in \Re$, it suffices to show that for every $\{m\} \subseteq \{n\}$ there exists $\{r\} \subseteq \{m\}$ such that $\alpha_r \underset{r\to\infty}{\to} \alpha$. By considering $\{m\} \subseteq \{n\}$, there exists $\{r\} \subseteq \{m\}$ such that $\bar{K}_r \underset{r\to\infty}{\Rightarrow} K^*$, some d.f. (with $Var K^*$ necessarily $\leq c$), and define $\bar{\psi}^*$ by

$$\bar{\psi}^*(t) = \int_{\Re} \frac{e^{itx} - 1 - itx}{it} dK^*(x), \quad t \in \Re. \tag{13.16}$$

Then $\bar{\psi}_r(t) \underset{r\to\infty}{\to} \bar{\psi}^*(t)$, by Proposition 3(i) in Chapter 12. Also, $\psi_r(t) \underset{r\to\infty}{\to} \psi(t)$, by assumption. By (13.14), we have $\psi_r(t) = i\alpha_r t + \bar{\psi}_r(t)$, so that for $t \neq 0$,

$$\alpha_r = \frac{\psi_r(t) - \bar{\psi}_r(t)}{it} \underset{r\to\infty}{\to} \frac{\psi(t) - \bar{\psi}^*(t)}{it} \overset{def}{=} \alpha^* \in \Re,$$
$$\text{independent of } t. \tag{13.17}$$

Next, consider $\{s\} \subseteq \{n\}$ distinct from $\{m\}$. Then there exists $\{\nu\} \subseteq \{s\}$ such that $\bar{K}_\nu \underset{\nu\to\infty}{\Rightarrow} K^{**}$, some d.f. (with $Var K^{**}$ necessarily $\leq c$), and define $\bar{\psi}^{**}$ as in (13.16); i.e.,

$$\bar{\psi}^{**}(t) = \int_{\Re} \frac{e^{itx} - 1 - itx}{it} dK^{**}(x), \quad t \in \Re. \tag{13.18}$$

As before, $\bar{\psi}_\nu(t) \underset{\nu\to\infty}{\to} \bar{\psi}^{**}(t)$ and $\psi_\nu(t) \underset{\nu\to\infty}{\to} \psi(t)$, and for $t \neq 0$,

$$\alpha_\nu = \frac{\psi_\nu(t) - \bar{\psi}_\nu(t)}{it} \underset{\nu\to\infty}{\to} \frac{\psi(t) - \bar{\psi}^{**}(t)}{it} \overset{def}{=} \alpha^{**} \in \Re,$$
$$\text{independent of } t. \tag{13.19}$$

From (13.17) and (13.19), we have

$$\psi(t) = i\alpha^* t + \bar{\psi}^*(t) = i\alpha^{**} t + \bar{\psi}^{**}(t),$$

so that

$$\psi''(t) = \bar{\psi}^{*''}(t) = \bar{\psi}^{**''}(t).$$

On the other hand, from (13.15), (13.16), and by (12.37) (in Chapter 12),

$$\bar{\psi}''(t) = \bar{\psi}^{*\prime\prime}(t) = -\int_{\Re} e^{itx} dK^*(x) = -\int_{\Re} e^{itx} dK^{**}(x)$$
$$= \bar{\psi}^{**\prime\prime}(t),$$

so that the d.f.s K^* and K^{**} have the same ch.f. It follows that $K^* = K^{**} + c$, for some constant c, and hence K^* and K^{**} determine the same $\bar{\psi}$. Thus, $\bar{\psi}^*(t) = \bar{\psi}^{**}(t), t \in \Re$, and therefore (13.17) and (13.18) imply that $\alpha^* = \alpha^{**}$, call it α. Also, setting K for any limiting d.f. of $\{\bar{K}_n\}$, we have then shown that $\alpha_n \to \alpha$ and $K_n \Rightarrow K\ uac$, as was to be seen. ∎

We may now proceed with the formulation and proof of the main result in this section.

Theorem 2′. Let the r.v.s $X_{nj}, j = 1, \ldots, k_n, n \geq 1$, satisfy conditions ($C''$) in (13.2), let F_{nj} and f_{nj} be their respective d.f.s and ch.f.s, and let $\bar{F}_{nj}$ and $\bar{f}_{nj}$ be the respective d.f.s and ch.f.s of $\bar{X}_{nj} = X_{nj} - \alpha_{nj}, \mathcal{E}X_{nj} = \alpha_{n_j} \in \Re$. Let $S_n = \sum_j X_{nj}$ and let $\bar{K}_n$ be defined by

$$\bar{K}_n(x) = \sum_j \int_{-\infty}^{x} y^2 d\bar{F}_{nj}(y) = \sum_j \int_{-\infty}^{x} y^2 dF_{nj}(y + \alpha_{nj}),$$
$$\alpha_n = \sum_j \alpha_{nj}, \tag{13.20}$$

(where each $\bar{K}_n$ is a d.f. with $Var\bar{K}_n \leq c$ and $\bar{K}_n(-\infty) = 0$).

(i) Let $\mathcal{L}(S_n) \Rightarrow \mathcal{L}$, a d.f. of a r.v. with ch.f. $f \neq 0$ on $\Re$. Then $\bar{K}_n \Rightarrow K\ uac$, a d.f. (with $VarK \leq c$, by Theorem 3(ii) in Chapter 8), and $\alpha_n \to \alpha \in \Re$. Furthermore, $f(t) = e^{\psi(t)}$, where ψ is defined by

$$\psi(t) = i\alpha t + \bar{\psi}(t), \quad \alpha \in \Re, \text{ and}$$
$$\bar{\psi}(t) = \int_{\Re} \frac{e^{itx} - 1 - itx}{x^2} dK(x), \quad t \in \Re. \tag{13.21}$$

(ii) With $\bar{K}_n$ given in (13.20), suppose that $\bar{K}_n \Rightarrow K$, a d.f. (with $VarK \leq c$, by Theorem 3(ii) in Chapter 8), and $\alpha_n \to \alpha \in \Re$. Then $\mathcal{L}(S_n) \Rightarrow \mathcal{L}$, where $\mathcal{L}$ is the d.f. of a r.v. with ch.f. f given by $f(t) = e^{\psi(t)}$ (by Proposition 2′) and ψ is defined in (13.21).

Proof.

In the first place, by relation (13.11), we have

$$\bar{f}_n(t)e^{-\bar{\psi}_n(t)} \to 1 \text{ on } \Re, \tag{13.22}$$

where

$$\bar{f}_n(t) = e^{-i\alpha_n t} f_n(t), \quad f_n(t) = \prod_j f_{nj}(t),$$
$$\bar{\psi}_n(t) = \int_{\Re} \frac{e^{itx} - 1 - itx}{x^2} d\bar{K}_n(x), \tag{13.23}$$

and $\bar{K}_n$ and α_n are given in (13.20).

From (13.22) and (13.23) (see also (13.11)), we have

$$f_n(t)e^{-\psi_n(t)} \to 1 \text{ on } \Re, \quad \psi_n(t) = i\alpha_n t + \bar{\psi}_n(t). \tag{13.24}$$

We now proceed with the proof of the theorem.

(i) The assumption $\mathcal{L}(S_n) \Rightarrow \mathcal{L}$, a d.f. of a r.v., implies $f_n(t) \to f(t)$ on $\Re$ (by Theorem 3 in Chapter 11), whereas $f(t) \neq 0$ on $\Re$ implies that

$$e^{\psi_n(t)} \to f(t) \text{ on } \Re, \tag{13.25}$$

by (13.24) and the fact that $f^{-1}(t) f_n(t) \to 1$. Now, $f(0) = 1$ and f is continuous on $\Re$ as the ch.f. of a r.v., whereas $f(t) \neq 0$ on $\Re$, by assumption. Then, by Lemma 2 in Chapter 12, $f(t) = e^{\psi(t)}$, where ψ is a uniquely defined function on $\Re$ into $\mathbb{C}$ with $\psi(0) = 0$ and continuous on $\Re$. Next, $\psi_n(0) = 0$ and the ψ_ns are continuous in $\Re$, as has been seen. Furthermore, (13.25) becomes

$$e^{\psi_n(t)} \to e^{\psi(t)} \text{ on } \Re. \tag{13.26}$$

The conditions of Lemma 3 in Chapter 12 are satisfied, and the convergence (13.26) implies that $\psi_n(t) \to \psi(t)$ on $\Re$. On account of this and relations (13.23) and (13.24), Proposition 3′(ii) applies and gives $\bar{K}_n \Rightarrow K \; uac$, some d.f. (of $Var K \leq c$), and $\alpha_n \to \alpha \in \Re$. Furthermore, ψ corresponds to the pair (α, K), where K is any member of the family of limiting distributions of $\{\bar{K}_n\}$; i.e., ψ is given by (13.21).

(ii) Now suppose that $\bar{K}_n \Rightarrow K$, a d.f. as described in the theorem, and $\alpha_n \to \alpha \in \Re$. By means of K and α, define ψ as in (13.21). Then, with ψ_n defined by (13.24), Proposition 3′(i) applies and gives $\psi_n(t) \to \psi(t)$ on $\Re$, so that $e^{\psi_n(t)} \to e^{\psi(t)}$ on $\Re$. From this last convergence and relation (13.24), we obtain $f_n(t) \to e^{\psi(t)}$ on $\Re$ and $e^{\psi(t)}$ is continuous at the origin. Then, by the converse part of Theorem 3 in Chapter 11, it follows that $\mathcal{L}(S_n) \Rightarrow \mathcal{L}$, a d.f. of a r.v. with ch.f. f such that $f(t) = e^{\psi(t)}$ on $\Re$. This completes the proof. ∎

13.3 Two Special Cases of the Limiting Laws of $\mathcal{L}(S_n)$

In this section, two important special cases of Theorem 2′ are considered that result from two specific choices of the d.f. K. In one of those cases, the limiting law is Normal, and in the other the limiting law is the Poisson distribution.

For the first case, choose the d.f. K as follows:

$$K(x) = \begin{cases} 0 & \text{for } x < 0, \\ \sigma^2 & \text{for } x \geq 0. \end{cases} \tag{13.27}$$

Then the corresponding $\bar{\psi}$, defined in (13.21), becomes

$$\bar{\psi}(t) = -\frac{\sigma^2 t^2}{2}$$

(by recalling that the integrand is equal to $-\frac{t^2}{2}$ for $x = 0$). In the following, we will consider row-wise independent r.v.s X_{nj} with $\mathcal{E}X_{nj} = \alpha_{nj}$ and $\sigma^2(X_{nj}) = \sigma^2_{nj}$ finite with $\alpha_n = \sum_j \alpha_{nj} \to \mu \in \Re$ and $s_n^2 = \sum_j \sigma^2_{nj} \to \sigma^2 < \infty$. Define ψ, as in (13.21), by

$$\psi(t) = i\mu t + \bar{\psi}(t) = i\mu t - \frac{\sigma^2 t^2}{2}, \tag{13.28}$$

and observe that $e^{\psi(t)}$ is the ch.f. of the $N(\mu, \sigma^2)$ distribution. At this point, observe that, by Corollary 3′ and with K as in (13.27), the pair (μ, K) uniquely determines ψ. Then the following result holds.

Theorem 1. For each $n \geq 1$, let the r.v.s X_{nj}, $j = 1, \ldots, k_n \to \infty$, be row-wise independent with $\mathcal{E}X_{nj} = \alpha_{nj}$ and $\sigma^2(X_{nj}) = \sigma^2_{nj}$ finite, and such that

$$\alpha_n = \sum_{j=1}^{k_n} \alpha_{nj} \to \mu \in \Re, \quad s_n^2 = \sum_{j=1}^{k_n} \sigma^2_{nj} \to \sigma^2 < \infty.$$

Set $S_n = \sum_{j=1}^{k_n} X_{nj}$. Then

$$\mathcal{L}(S_n) \Rightarrow N(\mu, \sigma^2), \quad \text{and} \quad \max_{1 \leq j \leq k_n} \sigma^2_{nj} \to 0, \tag{13.29}$$

if and only if, for every $\varepsilon > 0$,

$$h_n(\varepsilon) = \sum_{j=1}^{k_n} \int_{(|x| \geq \varepsilon)} x^2 dF_{nj}(x + \alpha_{nj}) \to 0. \quad \blacksquare \tag{13.30}$$

The proof of Theorem 1 is facilitated by the following lemmas.

Lemma 1. For each $n \geq 1$, let X_{nj}, $j = 1, \ldots, k_n \to \infty$, be row-wise independent r.v.s with $\mathcal{E}X_{nj} = \alpha_{nj}$ and $\sigma^2(X_{nj}) = \sigma^2_{nj}$ finite, and such that

$$\alpha_n = \sum_{j=1}^{k_n} \alpha_{nj} \to \mu \in \Re, \quad s_n^2 = \sum_{j=1}^{k_n} \sigma^2_{nj} \to \sigma^2 < \infty.$$

Let F_{nj} be the d.f. of X_{nj}, and let $S_n = \sum_{j=1}^{k_n} X_{nj}$. Finally, suppose that, for every $\varepsilon > 0$,

$$h_n(\varepsilon) = \sum_{j=1}^{k_n} \int_{(|x|\geq\varepsilon)} x^2 dF_{nj}(x + \alpha_{nj}) \to 0. \tag{13.31}$$

Then

(i) $\max_{1\leq j\leq k_n} \sigma_{nj}^2 \to 0$ (so that conditions (C'') in (13.2) are satisfied).

(ii) With $\bar{K}_n$ defined by

$$\bar{K}_n(y) = \sum_{j=1}^{k_n} \int_{-\infty}^{x} y^2 dF_{nj}(y + \alpha_{nj}), \tag{13.32}$$

it holds $\bar{K}_n \Rightarrow K$, where K is given in (13.27).

(iii) $\mathcal{L}(S_n) \Rightarrow N(\mu, \sigma^2)$.

Proof.

(i) Clearly,

$$\begin{aligned}
\sigma_{nj}^2 &= \int_{\Re} (x - \alpha_{nj})^2 dF_{nj}(x) = \int_{\Re} y^2 dF_{nj}(y + \alpha_{nj}) \\
&\quad \text{(by setting } x - \alpha_{nj} = y; \quad \text{see also Appendix B for related properties)} \\
&= \int_{(|x|\geq\varepsilon)} x^2 dF_{nj}(x + \alpha_{nj}) \\
&\quad + \int_{(|x|<\varepsilon)} x^2 dF_{nj}(x + \alpha_{nj}) \\
&\leq h_n(\varepsilon) + \int_{(|x|<\varepsilon)} x^2 dF_{nj}(x + \alpha_{nj}) \\
&\leq h_n(\varepsilon) + \varepsilon^2.
\end{aligned}$$

Hence $\limsup \max_j \sigma_{nj}^2 \leq \varepsilon^2$, by (13.31), and letting $\varepsilon \to 0$, we get the result.

(ii) As in the proof of Lemma 4(ii) in Chapter 12, we have, for every $\varepsilon > 0$,

$$\begin{aligned}
\bar{K}_n(-\varepsilon) &= \sum_j \int_{(-\infty,-\varepsilon]} x^2 dF_{nj}(x + \alpha_{nj}) \\
&\leq \sum_j \int_{(|x|\geq\varepsilon)} x^2 dF_{nj}(x + \alpha_{nj}) = h_n(\varepsilon) \to 0,
\end{aligned}$$

and

$$\bar{K}_n(\varepsilon) = \sum_j \int_{(-\infty,\varepsilon]} x^2 dF_{nj}(x + \alpha_{nj})$$

$$= s_n^2 - \sum_j \int_{(\varepsilon,\infty)} x^2 dF_{nj}(x + \alpha_{nj}) \to \sigma^2,$$

because $\sum_j \int_{(\varepsilon,\infty)} x^2 dF_{nj}(x+\alpha_{nj}) \leq \sum_j \int_{[\varepsilon,\infty)} x^2 dF_{nj}(x+\alpha_{nj}) \leq h_n(\varepsilon) \to 0$. Thus, $\bar{K}_n \Rightarrow K$.

(iii) By part (ii), $\bar{K}_n \Rightarrow K$, and by assumption, $\alpha_n \to \mu$. Then, by Theorem 2′(ii), $\mathcal{L}(S_n) \Rightarrow \mathcal{L}$, where the d.f. $\mathcal{L}$ has ch.f. $e^{\psi(t)}$ with $\psi(t) = i\mu t + \bar{\psi}(t) = i\mu t - \frac{\sigma^2 t^2}{2}$. It follows that $\mathcal{L}$ is the $N(\mu, \sigma^2)$, and the proof is completed. ■

Lemma 2. For each $n \geq 1$, let X_{nj}, $j = 1, \ldots, k_n \to \infty$, be row-wise independent r.v.s with $\mathcal{E}X_{nj} = \alpha_{nj}$ and $\sigma^2(X_{nj}) = \sigma_{nj}^2$ finite, and such that

$$\alpha_n = \sum_{j=1}^{k_n} \alpha_{nj} \to \mu \in \Re, \quad s_n^2 = \sum_{j=1}^{k_n} \sigma_{nj}^2 \to \sigma^2 < \infty.$$

Set $S_n = \sum_{j=1}^{k_n} X_{nj}$ and suppose that

$$\mathcal{L}(S_n) \Rightarrow N(\mu, \sigma^2) \text{ and } \max_{1 \leq j \leq k_n} \sigma_{nj}^2 \to 0.$$

Then the convergence in (13.30) holds for every $\varepsilon > 0$.

Proof. The assumption $\mathcal{L}(S_n) \Rightarrow N(\mu, \sigma^2)$ implies that $\bar{K}_n \Rightarrow K$ *uac*, where K and $\bar{K}_n$ are given by (13.27) and (13.32), respectively. This is so by Theorem 2′ (i). Next, as in the proof of Lemma 5 in Chapter 12,

$$\begin{aligned} h_n(\varepsilon) &= \sum_j \int_{(|x| \geq \varepsilon)} x^2 dF_{nj}(x + \alpha_{nj}) \\ &\leq \bar{K}_n(-\varepsilon) + s_n^2 - \bar{K}_n\left(\frac{\varepsilon}{2}\right), \end{aligned}$$

and we wish to show that $h_n(\varepsilon) \to 0$. To this end, for every $\{m\} \subseteq \{n\}$ there exists $\{r\} \subseteq \{m\}$ such that $\bar{K}_r \underset{r\to\infty}{\Rightarrow} K^*$ a d.f., and $K^* = K + c$, for some constant c. It follows that

$$\begin{aligned} \bar{K}_r(-\varepsilon) + s_r^2 - \bar{K}_r\left(\frac{\varepsilon}{2}\right) &\underset{r\to\infty}{\to} K^*(-\varepsilon) + \sigma^2 - K^*\left(\frac{\varepsilon}{2}\right) \\ &= K(-\varepsilon) + c + \sigma^2 - K\left(\frac{\varepsilon}{2}\right) - c \\ &= 0 + \sigma^2 - \sigma^2 = 0, \end{aligned}$$

so that $h_r(\varepsilon) \underset{r\to\infty}{\to} 0$. Then $h_n(\varepsilon) \to 0$. ■

Proof of Theorem 1. It follows from Lemmas 1 and 2. ■

For the second special case of Theorem 2′, choose the d.f. K as follows:

$$K(x) = \begin{cases} 0 & \text{for } x < 1, \\ \lambda & \text{for } x \geq 1 \ (\lambda > 0). \end{cases} \tag{13.33}$$

Then, the corresponding $\bar{\psi}$ is given by

$$\bar{\psi}(t) = \int_{\Re} \frac{e^{itx} - 1 - itx}{x^2} dK(x) = (e^{it} - 1 - it)\lambda.$$

In (13.21), take $\alpha = \lambda$. Then the ψ that corresponds to (λ, K) is

$$\psi(t) = it\lambda + (e^{it} - 1 - it)\lambda = \lambda(e^{it} - 1), \tag{13.34}$$

where we observe that $e^{\psi(t)}$ is the ch.f. of $P(\lambda)$, the Poisson distribution with parameter λ.

Then, we have the following theorem.

Theorem 2. For each $n \geq 1$, let the r.v.s X_{nj}, $j = 1, \ldots, k_n \to \infty$, be row-wise independent with $\mathcal{E}X_{nj} = \alpha_{nj}$ and $\sigma^2(X_{nj}) = \sigma^2_{nj}$ finite, and such that $\alpha_n = \sum_{j=1}^{k_n} \alpha_{nj} \to \lambda$, $s_n^2 = \sum_{j=1}^{k_n} \sigma^2_{nj} \to \lambda$ $(0 < \lambda < \infty)$. Set $S_n = \sum_{j=1}^{k_n} X_{nj}$. Then

$$\mathcal{L}(S_n) \Rightarrow P(\lambda), \quad \text{and} \quad \max_{1 \leq j \leq k_n} \sigma^2_{nj} \to 0, \tag{13.35}$$

if and only if, for every $\varepsilon > 0$,

$$h_n(\varepsilon) = \sum_{j=1}^{k_n} \int_{(|x-1| \geq \varepsilon)} x^2 dF_{nj}(x + \alpha_{nj}) \to 0. \quad \blacksquare \tag{13.36}$$

The proof of Theorem 2 is facilitated by the following lemmas.

Lemma 3. For each $n \geq 1$, let X_{nj}, $j = 1, \ldots, k_n \to \infty$, be row-wise independent r.v.s with $\mathcal{E}X_{nj} = \alpha_{nj}$ and $\sigma^2(X_{nj}) = \sigma^2_{nj}$ finite, and such that

$$\alpha_n = \sum_{j=1}^{k_n} \alpha_{nj} \to \lambda \in \Re, \quad s_n^2 = \sum_{j=1}^{k_n} \sigma^2_{nj} \to \lambda.$$

Let F_{nj} be the d.f. of X_{nj}, and set $S_n = \sum_{j=1}^{k_n} X_{nj}$. Finally, suppose that, for every $\varepsilon > 0$,

$$h_n(\varepsilon) = \sum_{j=1}^{k_n} \int_{(|x-1| \geq \varepsilon)} x^2 dF_{nj}(x + \alpha_{nj}) \to 0. \tag{13.37}$$

Then,

(i) $\max_{1 \leq j \leq k_n} \sigma^2_{nj} \to 0$, (so that conditions (C'') in (13.2) are satisfied).

(ii) With $\bar{K}_n$ and K defined by (13.32) and (13.33), respectively, it holds $\bar{K}_n \Rightarrow K$.
(iii) $\mathcal{L}(S_n) \Rightarrow \mathcal{L}$, where $\mathcal{L}$ is the Poisson distribution with parameter λ.

Proof.

(i) Clearly,

$$\begin{aligned}\sigma_{nj}^2 &= \int_{\Re}(x-\alpha_{nj})^2 dF_{nj}(x) = \int_{\Re} y^2 dF_{nj}(y+\alpha_{nj}) \text{ (by setting } x-\alpha_{nj}=y)\\ &= \int_{(|x-1|\geq\varepsilon)} x^2 dF_{nj}(x+\alpha_{nj}) + \int_{(|x-1|<\varepsilon)} x^2 dF_{nj}(x+\alpha_{nj})\\ &\leq h_n(\varepsilon)+\varepsilon^2.\end{aligned}$$

Hence $\limsup \max_j \sigma_{nj}^2 \leq \varepsilon^2$, by (13.37), and letting $\varepsilon \to 0$, we get the result.

(ii) We have to show that, for every $\varepsilon > 0$, $\bar{K}_n(1-\varepsilon) \to 0$ and $\bar{K}_n(1+\varepsilon) \to \lambda$. Indeed,

$$\bar{K}_n(1-\varepsilon) = \sum_j \int_{(-\infty,1-\varepsilon]} x^2 dF_{nj}(x+\alpha_{nj}) \leq h_n(\varepsilon) \to 0,$$

whereas

$$\begin{aligned}0 &\leq s_n^2 - \sum_j \int_{(-\infty,1+\varepsilon]} x^2 dF_{nj}(x+\alpha_{nj})\\ &= \sum_j \int_{(1+\varepsilon,\infty)} x^2 dF_{nj}(x+\alpha_{nj})\\ &\leq \sum_j \int_{[1+\varepsilon,\infty)} x^2 dF_{nj}(x+\alpha_{nj}) \leq h_n(\varepsilon) \to 0.\end{aligned}$$

Therefore, since $s_n^2 \to \lambda$, it follows that

$$\bar{K}_n(1+\varepsilon) = \sum_j \int_{(-\infty,1+\varepsilon]} x^2 dF_{nj}(x+\alpha_{nj}) \to \lambda.$$

(iii) By part (i), $\bar{K}_n \Rightarrow K$, and by assumption $\alpha_n \to \lambda$. Then Theorem 2′(ii) applies and gives that $\mathcal{L}(S_n) \Rightarrow \mathcal{L}$, where the d.f. $\mathcal{L}$ has ch.f. $e^{\psi(t)}$ where $\psi(t) = i\lambda t + \bar{\psi}(t) = i\lambda t + (e^{it}-1-it)\lambda = \lambda(e^{it}-1)$, which is the ch.f. of the Poisson distribution with parameter λ. ■

Lemma 4. For each $n \geq 1$, let X_{nj}, $j = 1, \ldots, k_n \to \infty$, be row-wise independent r.v.s with $\mathcal{E}X_{nj} = \alpha_{nj}$ and $\sigma^2(X_{nj}) = \sigma_{nj}^2$ finite, and such that

$$\alpha_n = \sum_{j=1}^{k_n} \alpha_{nj} \to \lambda \in \Re, \quad s_n^2 = \sum_{j=1}^{k_n} \sigma_{nj}^2 \to \lambda.$$

Set $S_n = \sum_{j=1}^{k_n} \alpha_{nj}$ and suppose that

$$\mathcal{L}(S_n) \Rightarrow P(\lambda), \quad \text{and} \quad \max_{1 \le j \le k_n} \sigma_{nj}^2 \to 0.$$

Thus the convergence in (13.36) holds for every $\varepsilon > 0$.

Proof. Since $\alpha_n \to \lambda$ and $\mathcal{L}(S_n) \Rightarrow P(\lambda)$, Theorem 2′(i) implies that $\bar{K}_n \Rightarrow K$ uac. Next,

$$\begin{aligned}
h_n(\varepsilon) &= \sum_j \int_{(|x-1| \ge \varepsilon)} x^2 dF_{nj}(x + \alpha_{nj}) \\
&= \sum_j \int_{(-\infty, 1-\varepsilon]} x^2 dF_{nj}(x + \alpha_{nj}) \\
&\quad + \sum_j \int_{[1+\varepsilon, \infty)} x^2 dF_{nj}(x + \alpha_{nj}) \\
&\le \sum_j \int_{(-\infty, 1-\varepsilon]} x^2 dF_{nj}(x + \alpha_{nj}) \\
&\quad + \sum_j \int_{(1+\frac{\varepsilon}{2}, \infty)} x^2 dF_{nj}(x + \alpha_{nj}) \\
&= \bar{K}_n(1-\varepsilon) + s_n^2 - \bar{K}_n\left(1 + \frac{\varepsilon}{2}\right).
\end{aligned}$$

Thus, as in the proof of Lemma 2, passing to a subsequence $\{r\} \subseteq \{n\}$ for which $\bar{K}_r \underset{r\to\infty}{\Rightarrow} K^*$ a d.f. with $K^* = K + c$, for some constant c, we have

$$\begin{aligned}
\bar{K}_r(1-\varepsilon) + s_r^2 - \bar{K}_r\left(1 + \frac{\varepsilon}{2}\right) &\underset{r\to\infty}{\to} K^*(1-\varepsilon) + \lambda - K^*\left(1 + \frac{\varepsilon}{2}\right) \\
&= K(1-\varepsilon) + c + \lambda \\
&\quad - K\left(1 + \frac{\varepsilon}{2}\right) - c \\
&= 0 + \lambda - \lambda = 0,
\end{aligned}$$

so that $h_r(\varepsilon) \underset{r\to\infty}{\to} 0$, and then $h_n(\varepsilon) \to 0$. ■

Proof of Theorem 2. It follows from Lemmas 3 and 4. ■

As an application to Theorem 2, we present the familiar result, according to which Binomial probabilities are approximated by Poisson probabilities.

Application 1 (Approximation of a Binomial distribution by a Poisson distribution). For $n \ge 1$, let $X_{nj}, j = 1, \ldots, n$, be row-wise independent r.v.s distributed as $B(1, p_n)$ (Bernoulli with parameter p_n), such that $p_n \to 0$ and $np_n \to \lambda \in (0, \infty)$. Set $S_n = \sum_{j=1}^{n} X_{nj}$ (so that S_n is distributed as $B(n, p_n)$, Binomial with parameters n and p_n). Then $\mathcal{L}(S_n) \Rightarrow \mathcal{L}(X)$, where X is distributed as $P(\lambda)$ (Poisson with parameter λ).

First, we verify that the conditions of Theorem 2 are satisfied here. Indeed, $\mathcal{E}X_{nj} = p_n$ and $\alpha_n = \sum_{j=1}^{n} \alpha_{nj} = np_n \to \lambda$, by assumption, with $\lambda \in (0, \infty)$. Next, $\sigma_{nj}^2 = \sigma^2(X_{nj}) = p_n q_n$, where $q_n = 1 - p_n$. Also, $s_n^2 = \sum_{j=1}^{n} \sigma_{nj}^2 = np_n q_n \to \lambda$, since $np_n \to \lambda$ and $q_n \to 1$.

Then, in order to show that $\mathcal{L}(S_n) \Rightarrow \mathcal{L}(X)$ (and $\max_j \sigma_{nj}^2 \to 0$), it suffices to show that, for every $\varepsilon > 0$,

$$h_n(\varepsilon) = \sum_j \int_{(|x-1| \geq \varepsilon)} x^2 dF_{nj}(x + p_n) \to 0,$$

where $F_{nj} = F_n$ is the d.f. of X_{nj}. Since $F_n(x + \alpha_{nj}) = F_n(x + p_n)$, and F_n is given by

$$F_n(x) = \begin{cases} 0 & \text{for } x < 0, \\ q_n & \text{for } 0 \leq x < 1, \\ 1 & \text{for } x \geq 1, \end{cases}$$

we have

$$F_n(x + p_n) = \begin{cases} 0 & \text{for } x < -p_n, \\ q_n & \text{for } -p_n \leq x < 1 - p_n = q_n, \\ 1 & \text{for } x \geq 1 - p_n = q_n, \end{cases}$$

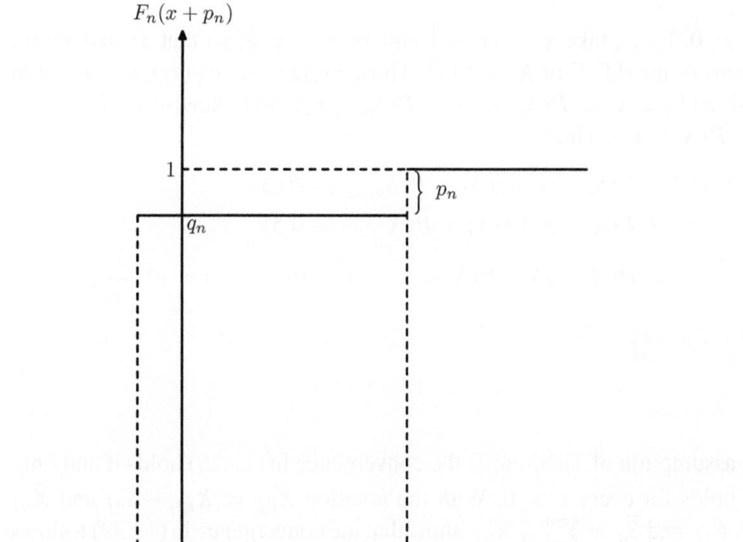

(see figure). We have

$$
\begin{aligned}
h_n(\varepsilon) &= \sum_j \int_{(|x-1|\geq\varepsilon)} x^2 dF_{nj}(x+\alpha_{nj}) \\
&= n\int_{(|x-1|\geq\varepsilon)} x^2 dF_n(x+p_n) \\
&= n\int_{(-\infty,1-\varepsilon]} x^2 dF_n(x+p_n) \\
&\quad + n\int_{[1+\varepsilon,\infty)} x^2 dF_n(x+p_n) \\
&= n\int_{(-\infty,1-\varepsilon]} x^2 dF_n(x+p_n) \\
&\quad \text{(since } F_n(x+p_n) = 1 \text{ for } x \geq 1+\varepsilon) \\
&= np_n^2 q_n (\text{since } q_n > 1-\varepsilon \text{ for sufficiently large } n, \text{ so that} \\
&\qquad \text{the only value taken by } x \text{ with positive probability is } -p_n) \\
&= np_n \times p_n q_n \leq np_n \times p_n \to \lambda \times 0 = 0.
\end{aligned}
$$

Thus, $h_n(\varepsilon) \to 0$, and therefore $\mathcal{L}(S_n) \Rightarrow P(\lambda)$. (Also, $\max_{1\leq j\leq n} \sigma_{nj}^2 = p_n q_n \leq p_n \to 0$.)

As a consequence of it, we have

$$
\binom{n}{x} p_n^x q_n^{n-x} \to e^{-\lambda}\frac{\lambda^x}{x!}, \quad x = 0, 1, \ldots
$$

In fact, for $x = 0, 1, \ldots$, take $x_1 = x - \frac{1}{2}$ and $x_2 = x + \frac{1}{2}$, so that x_1 and x_2 are continuity points of the d.f. F of $X \sim P(\lambda)$. Thus, $F_{S_n}(x_i) \to F_X(x_i), i = 1, 2$. On the other hand, $P(S_n = x) = P(S_n \leq x_2) - P(S_n \leq x_1)$, and likewise $P(X = x) = P(X \leq x_2) - P(X \leq x_1)$. Thus,

$$
\begin{aligned}
P(S_n = x) &= P(S_n \leq x + 0.5) - P(S_n \leq x - 0.5) \\
&\to P(X \leq x + 0.5) - P(X \leq x - 0.5) \\
&= P(X \leq x) - P(X \leq x-1) = P(X = x) = e^{-\lambda}\frac{\lambda^x}{x!},
\end{aligned}
$$

or $\binom{n}{x} p_n^x q_n^{n-x} \to e^{-\lambda}\frac{\lambda^x}{x!}$.

Exercises.

1. Under the assumption of Theorem 1, the convergence in (13.29) holds if and only if (13.30) holds for every $\varepsilon > 0$. With the notation $\bar{X}_{nj} = X_{nj} - \alpha_{nj}$ and $\bar{X}_{nj}$ having d.f. $\bar{F}_{nj}$, and $\bar{S}_n = \sum_{j=1}^n \bar{X}_{nj}$, show that the convergence in (13.29) follows from the condition

$$
\sum_{j=1}^{k_n} \int_{(|x|\geq\varepsilon s_n)} x^2 d\bar{F}_{nj}(x) \underset{n\to\infty}{\to} 0 \text{ for every } \varepsilon > 0.
$$

2. In Theorem 1 of this chapter, as well as Theorem 1 of Chapter 12, under some basic assumptions on the r.v.s X_{nj}, $j = 1, \ldots, k_n$, given in the theorems cited, a necessary and sufficient condition is given (namely, $g_n(\varepsilon) \underset{n\to\infty}{\longrightarrow} 0$ for every $\varepsilon > 0$) so that

$$\mathcal{L}(S_n) \underset{n\to\infty}{\Rightarrow} N(\mu, \sigma^2), \quad \mu \in \Re, \quad \sigma^2 \in (0, \infty),$$

where $S_n = \sum_{j=1}^{k_n} X_{nj}$. Thus,

$$P(S_n \leq x) = F_{S_n}(x) \underset{n\to\infty}{\longrightarrow} \Phi(x; \mu, \sigma^2), \quad x \in \Re,$$

where $\Phi(x; \mu, \sigma^2)$ is the d.f. of the $N(\mu, \sigma^2)$ distribution. Actually, the preceding convergence is *uniform* in $x \in \Re$. That is, for every $\varepsilon > 0$ there exists $N(\varepsilon) > 0$, *independent* of $x \in \Re$, such that $n \geq N(\varepsilon)$ implies

$$|F_{S_n}(x) - \Phi(x; \mu, \sigma^2)| < \varepsilon \text{ for all } x \in \Re. \qquad (*)$$

This result is a consequence of the following fact (known as *Pólya's Lemma*). Namely, if F and $\{F_n\}$ are d.f.s of r.v.s such that $F_n(x) \underset{n\to\infty}{\longrightarrow} F(x), x \in \Re$, and F is continuous, then the convergence is uniform in $x \in \Re$. Establish this fact.

Hint: Refer to Exercise 9 in Chapter 8. See also Lemma 1, page 206, in Roussas (1997).

Remark: The uniformity asserted in (*) is what legitimizes as simple an approximation as that in the Binomial case. Thus, if $X_1, \ldots, X_n$ are independent, distributed as $B(1, p)$, and if a and b are integers with $0 \leq a < b \leq n$, then, for sufficiently large n, $P(a < S_n \leq b) \simeq \Phi\left(\frac{b-np}{\sqrt{npq}}\right) - \Phi\left(\frac{a-np}{\sqrt{npq}}\right)$, where $S_n = \sum_{j=1}^{n} X_j$, $q = 1 - p$, and $\Phi(x) = \Phi(x; 0, 1)$. This is so, because $P(S_n \leq b) = P\left(\sum_{j=1}^{n} Z_j \leq \frac{b-np}{\sqrt{npq}}\right)$, where $Z_j = \frac{1}{\sqrt{n}} \frac{X_j - p}{\sqrt{pq}}$, $j = 1, \ldots, n$, and $\left|P\left(\sum_{j=1}^{n} Z_j \leq \frac{b-np}{\sqrt{npq}}\right) - \Phi\left(\frac{b-np}{\sqrt{npq}}\right)\right| < \varepsilon$ for $n \geq N(\varepsilon)$ independently of $\frac{b-np}{\sqrt{npq}}$; and likewise for $P(S_n \leq a)$.

This page is intentionally left blank

CHAPTER

14 Topics from Sequences of Independent Random Variables

This chapter consists of a selection of topics from sequences of r.v.s. The r.v.s considered are most often independent, although not necessarily identically distributed. The central result is the statement and proof of the Strong Law of Large Numbers (SLLN), Theorem 7.

It is important to notice that all results prior to Theorem 7, consisting of six theorems and two lemmas, are used directly or indirectly in the proof of the SLLN. This may serve as a tribute to its depth and complexity.

The chapter opens with the Kolmogorov inequalities, the detailed discussion of which takes up the entire first section. Kolmogorov inequalities provide upper and lower bounds for the probability of the maximum of the absolute value of partial sums of independent r.v.s. They are instrumental, albeit indirectly, in the proof of the SLLN.

Section 14.2 is devoted to discussing four theorems. Theorem 2 gives a sufficient condition, in terms of variances, for the series $\sum_{n\geq 1}(X_n - \mathcal{E}X_n)$ to converge a.s. Theorems 3 and 4 taken together state that, if $\sum_{n\geq 1} P(A_n) < \infty$, then $P(\limsup_{n\to\infty} A_n) = 0$, and if the former sum is ∞, then the latter probability is 1. The first conclusion (Borel–Cantelli Lemma) holds for any events, whereas the second conclusion (Borel Zero–One Criterion) requires independence. Theorem 5 discusses some technical consequences of the assumption that $\sum_{n\geq 1} P(X_n \neq X_{n'}) < \infty$ for two sequences of r.v.s. The following two lemmas, Lemma 3 (Toeplitz) and Lemma 4 (Kronecker), are analytic results and lend themselves here and in many other situations.

In Section 14.3, a precursor to the SLLN, Theorem 6, is discussed, as well as a lemma providing bounds for the $\mathcal{E}|X|$ of a r.v. X in terms of sums of certain probabilities, and, finally, the SLLN itself is proved. In the following section, it is shown that the SLLN essentially holds, even if the expectation of the underlying r.v.s is infinite. If, however, this expectation does not exist, then the averages S_n/n are unbounded with probability one.

In the final section, two main results are discussed. One is the Kolmogorov Zero–One Law, which states that tail events defined on a sequence of independent r.v.s have probability either 0 or 1. This conclusion has important ramifications regarding the limits of sequences, series, etc., of independent r.v.s. The other result gives necessary and sufficient conditions for the a.s. convergence of a series of independent r.v.s, and because of its form, it is referred to as the three series criterion. These main results

An Introduction to Measure-Theoretic Probability, Second Edition. http://dx.doi.org/10.1016/B978-0-12-800042-7.00014-1

are obtained after the concept of the tail σ-field is introduced and a related result is established.

14.1 Kolmogorov Inequalities

In this chapter, we consider certain topics related to sequences of independent r.v.s with a view of establishing the Strong Law of Large Numbers (SLLN) (Theorem 7). We start with Kolmogorov inequalities, which provide upper and lower bounds for probabilities of the maximum of partial sums of r.v.s centered at their expectations.

Theorem 1 (Kolmogorov inequalities). Let X_j, $j = 1, \ldots, n$, be independent r.v.s (not necessarily i.d.), and let $\mathcal{E}X_j$ be finite and $|X_j| \leq c$, where c is finite or ∞. Then, for every $\varepsilon > 0$, one has

$$1 - \frac{(\varepsilon + 2c)^2}{\sum_{j=1}^{n} \sigma_j^2} \leq P\left(\max_{1 \leq k \leq n} |S_k - \mathcal{E}S_k| \geq \varepsilon\right) \leq \frac{\sum_{j=1}^{n} \sigma_j^2}{\varepsilon^2}, \tag{14.1}$$

where $S_n = \sum_{j=1}^{n} X_j$ and $\sigma_j^2 = \sigma^2(X_j)$. ■

Remark 1. We observe that $\sigma_j^2 = \infty$ if and only if $\mathcal{E}X_j^2 = \infty$. Thus, if $\sigma_j^2 = \infty$ for some j, then $c = \infty$. Then the right-hand side of (14.1) becomes ∞ and the left-hand side becomes 0, interpreting $\frac{\infty}{\infty}$ as 1. Thus (14.1) is trivially true, and we may therefore assume that $\sigma_j^2 < \infty$ for all j. Next, if $c = \infty$, the left-hand side of (14.1), where c appears, is $-\infty$ and therefore we may assume that $c < \infty$, since the proof of the right-hand side of the inequality does not depend on the finiteness or not of c, as will be seen (see Lemma 1). Under this assumption, we have $|X_j| \leq c$, so that $|\mathcal{E}X_j| \leq c$ and therefore $|X_j - \mathcal{E}X_j| \leq 2c$. Thus, if we set $Y_j = X_j - \mathcal{E}X_j$, $j = 1, \ldots, n$, then the Y_js are independent, $\mathcal{E}Y_j = 0$, and $|Y_j| \leq c^*(= 2c) < \infty$. Therefore, in carrying out the proof of the theorem, we may assume that $\mathcal{E}X_j = 0$ and $|X_j| \leq 2c < \infty$ (or work with the Y_js).

The proof of the theorem is rather long, in particular the proof of the left-hand-side inequality. It might then be appropriate to split it in two parts in the form of lemmas.

Lemma 1. Under the assumptions of Theorem 1 and the additional nonrestrictive assumption that $\mathcal{E}X_j = 0$ (but without assuming finiteness of the constant c), the right-hand side of (14.1) holds.

Proof. For $\varepsilon > 0$, we set

$$A_k = \left(\max_{1 \leq j \leq k} |S_j| \geq \varepsilon\right), \quad k = 1, \ldots, n, \quad A_0 = \oslash, \tag{14.2}$$

and

$$B_k = \left(\max_{1 \leq j \leq k-1} |S_j| < \varepsilon \text{ and } |S_k| \geq \varepsilon\right), \quad k = 2, \ldots, n,$$
$$B_1 = (|S_1| \geq \varepsilon) = A_1, \quad S_0 = 0. \tag{14.3}$$

Then it is clear that $B_k \cap B_l = \oslash$, for $k \neq l$, and

$$A_j = \sum_{k=1}^{j} B_k, \quad j = 1, \ldots, n. \tag{14.4}$$

Indeed, if $C_j = (|S_j| \geq \varepsilon)$, $j = 1, \ldots, n$, then

$$\begin{aligned} A_n = \bigcup_{j=1}^{n} C_j &= C_1 + (C_1^c \cap C_2) + \cdots + (C_1^c \cap \cdots C_{n-1}^c \cap C_n) \\ &= B_1 + B_2 + \cdots + B_n. \end{aligned}$$

Next,

$$\begin{aligned} \int_{B_k} S_n^2 dP &= \int S_n^2 I_{B_k} dP = \mathcal{E}(S_n^2 I_{B_k}) = \mathcal{E}\left\{I_{B_k}[S_k + (S_n - S_k)]^2\right\} \\ &= \mathcal{E}[(I_{B_k} S_k) + I_{B_k}(S_n - S_k)]^2 \\ &= \mathcal{E}(I_{B_k} S_k^2) + 2\mathcal{E}[I_{B_k} S_k (S_n - S_k)] + \mathcal{E}[I_{B_k}(S_n - S_k)^2]. \quad (14.5) \end{aligned}$$

Clearly, $B_k \in \sigma(S_1, \ldots, S_k)$, so that I_{B_k} is $\sigma(S_1, \ldots, S_k)$-measurable. It follows that $I_{B_k} S_k$ is $\sigma(S_1, \ldots, S_k)$-measurable. Also, $I_{B_k} S_k$ and $S_n - S_k$ are independent, since the former is defined in terms of the r.v.s X_j, $j = 1, \ldots, k$, whereas the latter is defined in terms of the r.v.s X_j, $j = k+1, \ldots, n$. It follows that

$$\mathcal{E}[I_{B_k} S_k (S_n - S_k)] = \mathcal{E}(I_{B_k} S_k)\mathcal{E}(S_n - S_k) = 0,$$

so that (14.5) becomes

$$\begin{aligned} \int_{B_k} S_n^2 dP &= \mathcal{E}(I_{B_k} S_k^2) + \mathcal{E}[I_{B_k}(S_n - S_k)^2] \geq \mathcal{E}(I_{B_k} S_k^2) \\ &= \int_{B_k} S_k^2 dP \geq \varepsilon^2 P(B_k) \text{ by means of (14.3); i.e.,} \\ &\int_{B_k} S_n^2 dP \geq \varepsilon^2 P(B_k). \end{aligned}$$

Adding over $k = 1, \ldots, n$, and taking into consideration relation (14.4), we get then

$$\int_{A_n} S_n^2 dP \geq \varepsilon^2 P(A_n). \tag{14.6}$$

But

$$\int_{A_n} S_n^2 dP \leq \int S_n^2 dP = \mathcal{E} S_n^2 = \sum_j \sigma_j^2 \text{ by independence.}$$

Therefore

$$\sum_j \sigma_j^2 \geq \varepsilon^2 P(A_n) \text{ and hence } P(A_n) \leq \frac{\sum_j \sigma_j^2}{\varepsilon^2},$$

which is the right-hand side of the required inequality. ∎

Lemma 2. Under the assumptions of Theorem 1 (including boundedness, $|X_j| \leq c(< \infty)$, $j = 1, \ldots, n$), the left-hand side of (14.1) holds.

Proof. From (14.2) and (14.3), it follows that

$$\begin{aligned} &B_k \subseteq A_k,\ A_{k-1} \cap B_k = \oslash \text{ and, by (14.4)},\\ &A_k = (B_1 + \cdots + B_{k-1}) + B_k \\ &\quad = A_{k-1} + B_k, \end{aligned} \tag{14.7}$$

so that $A_{k-1} = A_k - B_k$, and $I_{A_{k-1}} = I_{A_k} - I_{B_k}$.

Thus

$$S_{k-1}I_{A_{k-1}} + X_k I_{A_{k-1}} = S_k I_{A_{k-1}} = S_k I_{A_k} - S_k I_{B_k},$$

by means of (14.7). Therefore, squaring out and taking the expectations, we obtain

$$\mathcal{E}(S_{k-1}^2 I_{A_{k-1}}) + \mathcal{E}(X_k^2 I_{A_{k-1}}) = \mathcal{E}(S_k^2 I_{A_k}) - \mathcal{E}(S_k^2 I_{B_k}), \tag{14.8}$$

because

$$\begin{aligned} \mathcal{E}(S_{k-1}I_{A_{k-1}} X_k I_{A_{k-1}}) &= \mathcal{E}[(S_{k-1}I_{A_{k-1}})X_k] \\ &= \mathcal{E}(S_{k-1}I_{A_{k-1}})\mathcal{E}X_k = 0, \end{aligned}$$

and

$$\mathcal{E}(S_k I_{A_k} S_k I_{B_k}) = \mathcal{E}(S_k^2 I_{A_k} I_{B_k}) = \mathcal{E}(S_k^2 I_{B_k}) \text{ since } B_k \subseteq A_k.$$

Now $I_{A_{k-1}}$ is defined in terms of $X_1, \ldots, X_{k-1}$ and hence it is independent of X_k. Therefore

$$\mathcal{E}(X_k^2 I_{A_{k-1}}) = \mathcal{E}X_k^2 \mathcal{E}I_{A_{k-1}} = \mathcal{E}X_k^2 P(A_{k-1}) = \sigma_k^2 P(A_{k-1}).$$

Then (14.8) becomes

$$\mathcal{E}(S_{k-1}^2 I_{A_{k-1}}) + \sigma_k^2 P(A_{k-1}) = \mathcal{E}(S_k^2 I_{A_k}) - \mathcal{E}(S_k^2 I_{B_k}). \tag{14.9}$$

But $S_k I_{B_k} = S_{k-1} I_{B_k} + X_k I_{B_k}$, and hence

$$|S_k I_{B_k}| \leq |S_{k-1} I_{B_k}| + |X_k I_{B_k}| < \varepsilon I_{B_k} + 2c I_{B_k} = (\varepsilon + 2c) I_{B_k}, \tag{14.10}$$

since $|S_{k-1}| < \varepsilon$ on B_k and $|X_k| \leq 2c$. Squaring out both sides of (14.10) and taking expectations, we get

$$\mathcal{E}(S_k^2 I_{B_k}) \leq (\varepsilon + 2c)^2 P(B_k). \tag{14.11}$$

Next, by means of (14.11), relation (14.9) becomes

$$\begin{aligned} \sigma_k^2 P(A_{k-1}) &= \mathcal{E}(S_k^2 I_{A_k}) - \mathcal{E}(S_k^2 I_{B_k}) - \mathcal{E}(S_{k-1}^2 I_{A_{k-1}}) \\ &\geq \mathcal{E}(S_k^2 I_{A_k}) - \mathcal{E}(S_{k-1}^2 I_{A_{k-1}}) - (\varepsilon + 2c)^2 P(B_k). \end{aligned} \tag{14.12}$$

But $A_{k-1} \subseteq A_n$, by virtue of (14.4), and therefore (14.12) becomes

$$\sigma_k^2 P(A_n) \geq \mathcal{E}(S_k^2 I_{A_k}) - \mathcal{E}(S_{k-1}^2 I_{A_{k-1}}) - (\varepsilon + 2c)^2 P(B_k). \tag{14.13}$$

Summing over $k = 1, \ldots, n$, on both sides of (14.13), we obtain, by means of (14.4),

$$\begin{aligned}
\left(\sum_k \sigma_k^2\right) P(A_n) &\geq [\mathcal{E}(S_1^2 I_{A_1}) - \mathcal{E}(S_0^2 I_{A_0}) + \mathcal{E}(S_2^2 I_{A_2}) - \mathcal{E}(S_1^2 I_{A_1}) \\
&\quad + \mathcal{E}(S_3^2 I_{A_3}) - \mathcal{E}(S_2^2 I_{A_2}) + \cdots + \mathcal{E}(S_n^2 I_{A_n}) \\
&\quad - \mathcal{E}(S_{n-1}^2 I_{A_{n-1}})] - (\varepsilon + 2c)^2 P(A_n) \\
&= [\mathcal{E}(S_n^2 I_{A_n}) - \mathcal{E}(S_0^2 I_{A_0})] - (\varepsilon + 2c)^2 P(A_n) \\
&= \mathcal{E}(S_n^2 I_{A_n}) - (\varepsilon + 2c)^2 P(A_n) \text{ (since } S_0 = 0) \\
&= \mathcal{E}\left[S_n^2 (1 - I_{A_n^c})\right] - (\varepsilon + 2c)^2 P(A_n) \\
&= \mathcal{E} S_n^2 - \mathcal{E}(S_n^2 I_{A_n^c}) - (\varepsilon + 2c)^2 P(A_n) \\
&= \left(\sum_k \sigma_k^2\right) - \mathcal{E}(S_n^2 I_{A_n^c}) - (\varepsilon + 2c)^2 P(A_n); \quad \text{i.e.,}
\end{aligned}$$

$$\left(\sum_k \sigma_k^2\right) P(A_n) \geq \left(\sum_k \sigma_k^2\right) - \mathcal{E}(S_n^2 I_{A_n^c}) - (\varepsilon + 2c)^2 P(A_n). \tag{14.14}$$

But $|S_n| < \varepsilon$ on A_n^c, since $A_n^c = \cap_{j=1}^n C_j^c = \cap_{j=1}^n (|S_j| < \varepsilon)$. Therefore (14.14) becomes

$$\begin{aligned}
\left(\sum_k \sigma_k^2\right) P(A_n) &\geq \left(\sum_k \sigma_k^2\right) - \varepsilon^2 P(A_n^c) - (\varepsilon + 2c)^2 P(A_n) \\
&= \left(\sum_k \sigma_k^2\right) - \varepsilon^2 P(A_n^c) - (\varepsilon + 2c)^2 [1 - P(A_n^c)] \\
&= \left(\sum_k \sigma_k^2\right) - (\varepsilon + 2c)^2 \\
&\quad + [(\varepsilon + 2c)^2 - \varepsilon^2] P(A_n^c) \\
&\geq \left(\sum_k \sigma_k^2\right) - (\varepsilon + 2c)^2; \quad \text{i.e.,}
\end{aligned}$$

$$\left(\sum_k \sigma_k^2\right) P(A_n) \geq \left(\sum_k \sigma_k^2\right) - (\varepsilon + 2c)^2.$$

Hence

$$P(A_n) \geq 1 - \frac{(\varepsilon + 2c)^2}{\sum_k \sigma_k^2},$$

as was to be shown. ■

Proof of Theorem 1. It is the combination of Lemmas 1 and 2. ■

Remark 2. In Kolmogorov inequalities, it is clear that, in forming partial sums, we can sum between any two positive integers $m+1$ and $m+r$ with $m+r \leq n$. Then, the inequalities become

$$1 - \frac{(\varepsilon + 2c)^2}{\sum_{j=m+1}^{m+r} \sigma_j^2} \leq P\left(\max_{1 \leq k \leq r} \left|\sum_{j=m+1}^{m+k} X_j - \mathcal{E} \sum_{j=m+1}^{m+k} X_j\right| \geq \varepsilon\right)$$

$$= P\left(\max_{1 \leq k \leq r} \left|\sum_{j=m+1}^{m+k} (X_j - \mathcal{E}X_j)\right| \geq \varepsilon\right) \leq \frac{\sum_{j=m+1}^{m+r} \sigma_j^2}{\varepsilon^2}.$$

This actually amounts to reindexing the r.v.s. Indeed, by setting $Y_j = X_{j+m}$ and $\tau_j^2 = \sigma^2(Y_j)$, $j = 1, \ldots, r$, we have that the Y_js are independent and $|Y_j| \leq c$. Then inequalities (14.1) hold; i.e.,

$$1 - \frac{(\varepsilon + 2c)^2}{\sum_{j=1}^{r} \tau_j^2} \leq P\left(\max_{1 \leq k \leq r} \left|\sum_{j=1}^{k} Y_j - \mathcal{E} \sum_{j=1}^{k} Y_j\right| \geq \varepsilon\right) \leq \frac{\sum_{j=1}^{r} \tau_j^2}{\varepsilon^2}.$$

But $\sum_{j=1}^{k} Y_j - \mathcal{E}\sum_{j=1}^{k} Y_j = \sum_{j=m+1}^{m+k} X_j - \mathcal{E}\sum_{j=m+1}^{m+k} X_j$ and $\sum_{j=1}^{r} \tau_j^2 = \sum_{j=m+1}^{m+r} \sigma_j^2$, and then the remark is valid.

14.2 More Important Results Toward Proving the Strong Law of Large Numbers

In this section, a number of results are established, which will lead to the proof of the SLLN. These results (six in number) are important in their own right and are used in many other situations in probability and mathematical statistics.

The next theorem states, in effect, that, if the variances of r.v.s do not vary wildly, then the series consisting of the r.v.s converges a.s. More precisely, we have

Theorem 2. For $n \geq 1$, let X_n be independent r.v.s such that $\mathcal{E}X_n$ is finite and $\sigma_n^2 = \sigma^2(X_n) < \infty$. Then, if $\sum_{n=1}^{\infty} \sigma_n^2 < \infty$, it follows that $\sum_{n=1}^{\infty}(X_n - \mathcal{E}X_n)$ converges a.s. ■

Proof. Set $T_k = \sum_{j=1}^{k}(X_j - \mathcal{E}X_j)$ and apply the right-hand side of the relation in Remark 2 in order to obtain

$$P\left(\max_{1\le k\le r} |T_{m+k} - T_m| \ge \varepsilon\right) \le \frac{\sum\limits_{j=m+1}^{m+r} \sigma_j^2}{\varepsilon^2}. \tag{14.15}$$

However,

$$\left(\max_{1\le k\le r} |T_{m+k} - T_m| \ge \varepsilon\right) = \big(|T_{m+k} - T_m| \ge \varepsilon \text{ for at least one } k \text{ with } 1 \le k \le r\big) = \bigcup_{k=1}^{r} \left(|T_{m+k} - T_m| \ge \varepsilon\right).$$

Thus, relation (14.15) becomes

$$P\left(\bigcup_{k=1}^{r} \left(|T_{m+k} - T_m| \ge \varepsilon\right)\right) \le \frac{\sum\limits_{j=m+1}^{m+r} \sigma_j^2}{\varepsilon^2}. \tag{14.16}$$

In (14.16), let $r \to \infty$ and use continuity from below of the probability measure P to obtain

$$P\left(\bigcup_{k=1}^{r} \left(|T_{m+k} - T_m| \ge \varepsilon\right)\right) \le \frac{\sum\limits_{j=m+1}^{\infty} \sigma_j^2}{\varepsilon^2}. \tag{14.17}$$

By letting $m \to \infty$ in (14.17), and using the fact that $\sum_{n\ge1} \sigma_n^2 < \infty$, we get

$$\lim_{m\to\infty} P\left(\bigcup_{k=1}^{\infty} \left(|T_{m+k} - T_m| \ge \varepsilon\right)\right) = 0 \text{ for every } \varepsilon > 0. \tag{14.18}$$

However, (14.18) is a necessary and sufficient condition for mutual a.s. convergence of $\{T_m\}$, by Theorem 4 in Chapter 3, and hence for a.s. convergence of $T_m = \sum_{j=1}^{m}(X_j - \mathcal{E}X_j)$, as was to be seen. ∎

The following result states that, if the probabilities of a sequence of events are small, then the $\limsup_{n\to\infty}$ of the sequence has probability zero.

Theorem 3 (Borel–Cantelli Lemma). Let $\{A_n\}$, $n \ge 1$, be an arbitrary sequence of events. Then, if $\sum_n P(A_n) < \infty$, it follows that $P(\limsup_{n\to\infty} A_n) = 0$. ∎

Proof. With n and m tending to ∞, we have

$$
\begin{aligned}
P(\limsup_n A_n) &= P\left(\bigcap_{n=1}^{\infty}\bigcup_{j=n}^{\infty} A_j\right) = P\left(\lim_n \bigcup_{j=n}^{\infty} A_j\right) \\
&= \lim_n P\left(\bigcup_{j=n}^{\infty} A_j\right) = \lim_n P\left(\lim_m \bigcup_{j=n}^{m} A_j\right) \\
&= \lim_n \lim_m P\left(\bigcup_{j=n}^{m} A_j\right). \qquad (14.19)
\end{aligned}
$$

However,

$$
\lim_m P\left(\bigcup_{j=n}^{m} A_j\right) \leq \lim_m \sum_{j=n}^{m} P(A_j) = \sum_{j=n}^{\infty} P(A_j),
$$

so that

$$
\lim_n \lim_m P\left(\bigcup_{j=n}^{m} A_j\right) \leq \lim_n \sum_{j=n}^{\infty} P(A_j) = 0, \qquad (14.20)
$$

since $\sum_{j=1}^{\infty} P(A_j) < \infty$. Relations (14.19) and (14.20) complete the proof. ■

The Borel Zero–One Criterion discussed later restates in its part (i) the Borel–Cantelli theorem for independent events. The important conclusion is part (ii), where it is stated what happens to the probability of the $\limsup_{n\to\infty}$ of the sequence when the individual probabilities are large.

Theorem 4 (Borel Zero–One Criterion). Let the events A_n, $n \geq 1$, be independent. Then

(i) If $\sum_n P(A_n) < \infty$, it follows that $P(\limsup_{n\to\infty} A_n) = 0$, and
(ii) If $\sum_n P(A_n) = \infty$, it follows that $P(\limsup_{n\to\infty} A_n) = 1$. ■

Proof.

(i) This is a special case of Theorem 3, which holds regardless of the independence or not of the events involved.
(ii) For the proof of (ii), we need the following inequality

$$
1 - e^{-\sum_{j=1}^{n} x_j} \leq 1 - \prod_{j=1}^{n}(1 - x_j) \text{ for } \; 0 \leq x_j \leq 1, \, j = 1, \ldots, n. \qquad (14.21)
$$

(For its justification, see Exercise 3 in Chapter 10.) Next, with n and m tending to ∞ we have, by (14.19),

$$
P(\limsup_n A_n) = \lim_n \lim_m P(\cup_{j=n}^{m} A_j)
$$

$$= \lim_n \lim_m [1 - P(\cup_{j=n}^m A_j)^c]$$
$$= \lim_n \lim_m [1 - P(\cap_{j=n}^m A_j^c)]$$
$$= \lim_n \lim_m \left[1 - \prod_{j=n}^m P(A_j^c)\right] \text{ (by independence)}$$
$$= \lim_n \lim_m \left\{1 - \prod_{j=n}^m [1 - P(A_j)]\right\}; \quad \text{i.e.,}$$
$$P(\limsup_n A_n) = \lim_n \lim_m \left\{1 - \prod_{j=n}^m [1 - P(A_j)]\right\}. \tag{14.22}$$

Applying (14.21) with

$$x_1 = P(A_n),\ x_2 = P(A_{n+1}), \ldots, x_{m-n+1} = P(A_m),$$

we obtain

$$1 - \prod_{j=n}^m [1 - P(A_j)] \geq 1 - e^{-\sum_{j=n}^m P(A_j)} \underset{m\to\infty}{\to} 1 - e^{-\sum_{j=n}^\infty P(A_j)}$$
$$= 1 - e^{-\infty} = 1.$$

Therefore

$$\lim_n \lim_m \left\{1 - \prod_{j=n}^m [1 - P(A_j)]\right\} \geq 1, \quad \text{so that, by (14.22),}$$
$$P(\limsup_n A_n) \geq 1, \quad \text{hence} \quad P(\limsup_n A_n) = 1. \qquad ■$$

Remark 3. The significance of the Borel Zero–One Criterion is that, when dealing with independent events, there is no room for the value of the probability of their $\limsup_{n\to\infty}$ other than either 0 or 1; no values strictly between 0 and 1 can occur.

The theorem just proved has the following three corollaries.

Corollary 1. Suppose that the events $A_n, n \geq 1$, are independent and $A_n \underset{n\to\infty}{\to} A$ (in the sense that

$$\liminf_{n\to\infty} A_n = \bigcup_{n=1}^\infty \bigcap_{j=n}^\infty A_j = \bigcap_{n=1}^\infty \bigcup_{j=n}^\infty A_j = \limsup_{n\to\infty} A_n = A).$$

Then $P(A)$ is either 0 or 1.

Proof. We have $A = \lim_n A_n$ and by the theorem, $P(A)$ is either 0 or 1 according to whether $\sum_n P(A_n) < \infty$ or $\sum_n P(A_n) = \infty$. ■

Corollary 2. For independent events $A_n, n \geq 1$, we have

$$P(\liminf_{n\to\infty} A_n) = \begin{cases} 1 \text{ if } \sum_n P(A_n^c) < \infty, \\ 0 \text{ if } \sum_n P(A_n^c) = \infty. \end{cases}$$

Proof. In fact,

$$\begin{aligned} P(\liminf_{n\to\infty} A_n) &= P\Big(\bigcup_{n=1}^{\infty}\bigcap_{j=n}^{\infty} A_j\Big) = 1 - P\left(\bigcup_{n=1}^{\infty}\bigcap_{j=n}^{\infty} A_j\right)^c \\ &= 1 - P\left(\bigcap_{n=1}^{\infty}\bigcup_{j=n}^{\infty} A_j^c\right) \\ &= 1 - P(\limsup_{n\to\infty} A_n^c) \\ &= \begin{cases} 1 - 0 = 1 \text{ if } \sum_n P(A_n^c) < \infty, \\ 1 - 1 = 0 \text{ if } \sum_n P(A_n^c) = \infty. \end{cases} \end{aligned}$$ ■

Corollary 3. Consider the independent r.v.s $X_n, n \geq 1$, and suppose that $X_n \underset{n\to\infty}{\overset{\text{a.s.}}{\to}} c$, constant. Then for every $\delta > 0$, one has

$$\sum_{n=1}^{\infty} P(|X_n - c| \geq \delta) < \infty.$$

Proof. Set $A_n(\delta) = A_n = (|X_n - c| \geq \delta)$. Then the events $A_n, n \geq 1$, are independent. Next, let N^c be the set of points for which $X_n(\omega) \underset{n\to\infty}{\to} c$. Then $P(N^c) = 1$, and for every $\omega \in N^c$ there are finitely many ns for which $\omega \in A_n$, because otherwise $X_n(\omega) \underset{n\to\infty}{\nrightarrow} c$. Since $\limsup_{n\to\infty} A_n$ consists of points belonging to infinitely many A_ns (by Exercise 30(ii) in Chapter 1), it follows that $\limsup_{n\to\infty} A_n \subseteq N$, so that $P(\limsup_{n\to\infty} A_n) = 0$. Then $\Sigma_{n=1}^{\infty} P(A_n) < \infty$, because otherwise $P(\limsup_{n\to\infty} A_n) = 1$. Hence $\Sigma_{n=1}^{\infty} P(|X_n - c| \geq \delta) < \infty$ for every $\delta > 0$. ■

In the result discussed next, we consider two sequences of r.v.s, and the probabilities of the events over which respective r.v.s differ from each other. If these probabilities are small, then one can reach a number of conclusions, which is the content of the next theorem.

Theorem 5. For $n \geq 1$, consider the r.v.s X_n and X'_n, and set

$$A_n = (X_n \neq X'_n), \quad A = (X_n \neq X'_n \text{ i.o.}) = \limsup_{n\to\infty} A_n$$

(where i.o. is read "infinitely often"; see also Exercise 30 in Chapter 1). Then, if $\Sigma_{n=1}^{\infty} P(A_n) < \infty$, it follows that

(i) $P(A) = 0$.

(ii) Let B and B' be the sets of convergence of $\sum_{n=1}^{\infty} X_n$ and $\sum_{n=1}^{\infty} X'_n$, respectively. Then $B \cap A^c = B' \cap A^c$ and $P(B \cap A) = P(B' \cap A) = P(B \cap B' \cap A) = 0$

(i.e., the set of convergence of the two series is essentially the same, although the limits may be distinct).

(iii) The sets of convergence of the sequences $\{\frac{S_n}{b_n}\}$ and $\{\frac{S'_n}{b_n}\}$, as $n \to \infty$, differ only by a null set and the limits of the sequences are the same, where

$$S_n = \sum_{j=1}^{n} X_j, \quad S'_n = \sum_{j=1}^{n} X'_j \quad \text{and} \quad 0 \neq b_n \uparrow \infty \text{ as } n \to \infty.$$

Proof.

(i) Throughout the proof all limits are taken as $n \to \infty$. Then, with $A_n = (X_n \neq X'_n)$, $n \geq 1$, and $A = (X_n \neq X'_n \text{ i.o.}) = \limsup_n A_n$, the result follows from Theorem 3. (The last equality follows from Exercise 30(ii) in Chapter 2.)

(ii) From $A = \limsup_n A_n = \bigcap_{n=1}^{\infty} \bigcup_{j=n}^{\infty} A_j$, we have $A^c = \bigcup_{n=1}^{\infty} \bigcap_{j=n}^{\infty} A_j^c = \liminf_n A_n^c$ (see Exercise 30(i) in Chapter 2). Then, for $\omega \in A^c$ or, equivalently, $\omega \in \liminf_n A_n^c$, we have $\omega \in \bigcap_{j=n_0+1}^{\infty} A_j^c$ for some $n_0 = n_0(\omega)$ and then $\omega \in A_n^c$ for all $n > n_0$, hence $X_n(\omega) = X'_n(\omega), n > n_0$. Since $\sum_n X_n(\omega) = \sum_{n=1}^{n_0} X_n(\omega) + \sum_{n>n_0} X_n(\omega)$, $\sum_n X'_n(\omega) = \sum_{n=1}^{n_0} X'_n(\omega) + \sum_{n>n_0} X_n(\omega)$, it follows that $\sum_n X_n(\omega) \to \sum_{n=1}^{n_0} X_n(\omega) + X(\omega)$, if and only if $\sum_n X'_n(\omega) \to \sum_{n=1}^{n_0} X'_n(\omega) + X'(\omega)$, where $X(\omega) = \sum_{n>n_0} X_n(\omega)$ and $X'(\omega) = \sum_{n>n_0} X'_n(\omega)$. Thus, $B \cap A^c \subseteq B' \cap A^c$ and $B' \cap A^c \subseteq B \cap A^c$, so that $B \cap A^c = B' \cap A^c$, and $P(B \cap B' \cap A) \leq P(B \cap A) = P(B' \cap A) \leq P(A) = 0$.

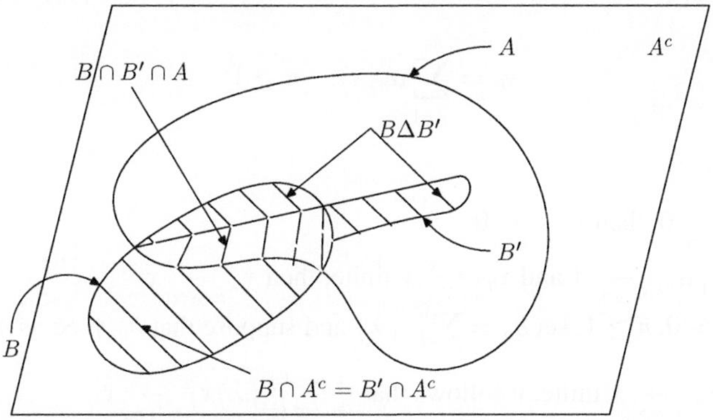

(iii) For each $\omega \in A^c$, there exists $n(\omega) = n_0$ such that $X_n(\omega) = X'_n(\omega)$ for all $n > n_0$. For such ns and $0 \neq b_n \uparrow \infty$, we write

$$\frac{S_n(\omega)}{b_n} = \frac{\sum_{j=1}^{n} X_j(\omega)}{b_n} = \frac{\sum_{j=1}^{n_0} X_j(\omega)}{b_n} + \frac{\sum_{j=n_0+1}^{n} X_j(\omega)}{b_n}$$

and

$$\frac{S'_n(\omega)}{b_n} = \frac{\sum_{j=1}^{n} X'_j(\omega)}{b_n} = \frac{\sum_{j=1}^{n_0} X'_j(\omega)}{b_n} + \frac{\sum_{j=n_0+1}^{n} X'_j(\omega)}{b_n},$$

so that

$$\frac{S_n(\omega)}{b_n} - \frac{S'_n(\omega)}{b_n} = \frac{1}{b_n}\sum_{j=1}^{n_0}\left[X_j(\omega) - X'_j(\omega)\right].$$

As $\sum_{j=1}^{n_0}\left[X_j(\omega) - X'_j(\omega)\right]$ is independent of n and $\frac{1}{b_n} \to 0$, it follows that $\frac{S_n(\omega)}{b_n} - \frac{S'_n(\omega)}{b_n} \to 0$. Therefore $\{\frac{S_n}{b_n}\}$ and $\{\frac{S'_n}{b_n}\}$ converge on the same subset of A^c, where $P(A^c) = 1$, and to the same limit. That is, if C and C' are the sets of convergence of $\{\frac{S_n}{b_n}\}$ and $\{\frac{S'_n}{b_n}\}$, respectively, then $C \cap A^c = C' \cap A^c$ as in part (ii), and $P(C \cap A) = P(C' \cap A) = P(C \cap C' \cap A) \leq P(A) = 0$. ■

This section is closed with two lemmas, the Toeplitz lemma and the Kronecker lemma, which are analytic rather than probabilistic. They are used decisively toward the proof of the SLLN, as well as in many other instances in probability and mathematical statistics.

Lemma 3 (Toeplitz Lemma). Let $\alpha_{nj}, j = 1, \ldots, k_n \underset{n\to\infty}{\to} \infty, n \geq 1$, be (real) numbers such that

$$\alpha_{nj} \underset{n\to\infty}{\to} 0 \text{ for each } j, \tag{14.23}$$

and

$$\sum_{j=1}^{k_n} |\alpha_{nj}| \leq c(< \infty) \text{ for all } n. \tag{14.24}$$

Let $\{x_n\}$ be a sequence of (real) numbers and define the sequence $\{y_n\}$ as follows:

$$y_n = \sum_{j=1}^{k_n} \alpha_{nj} x_j, \quad n \geq 1.$$

Then one has

(i) If $x_n \underset{n\to\infty}{\to} 0$, then $y_n \underset{n\to\infty}{\to} 0$.

(ii) If $\sum_{j=1}^{k_n} \alpha_{nj} \underset{n\to\infty}{\to} 1$ and $x_n \underset{n\to\infty}{\to} x$ finite, then $y_n \underset{n\to\infty}{\to} x$.

(iii) For $\lambda_n > 0, n \geq 1$, set $b_n = \sum_{j=1}^{n} \lambda_j$ and suppose that $b_n \uparrow \infty$ as $n \to \infty$.

Then, if $x_n \underset{n\to\infty}{\to} x$ finite, it follows that $\frac{1}{b_n}\sum_{j=1}^{k_n} \lambda_j x_j \underset{n\to\infty}{\to} x$.

Proof. Throughout the proof all limits are taken as $n \to \infty$.

(i) Since $x_n \to 0$, it follows that for every $\varepsilon > 0$, there exists an integer $n(\varepsilon) = n_0 > 0$ such that for $n > n_0$, $|x_n| < \frac{\varepsilon}{c}$. Thus for all sufficiently large n (so that $k_n > n_0$),

$$|y_n| = \left|\sum_{j=1}^{k_n} \alpha_{nj} x_j\right| = \left|\sum_{j=1}^{n_0} \alpha_{nj} x_j + \sum_{j=n_0+1}^{k_n} \alpha_{nj} x_j\right|$$

$$\leq \left|\sum_{j=1}^{n_0} \alpha_{nj} x_j\right|$$

$$+\frac{\varepsilon}{c} \sum_{j=n_0+1}^{k_n} |\alpha_{nj}|$$

$$\leq \left|\sum_{j=1}^{n_0} \alpha_{nj} x_j\right| + \varepsilon \text{ (by (14.24)), and this is}$$

$$\leq \left(\max_{1\leq j\leq n_0} |x_j|\right) \sum_{j=1}^{n_0} |\alpha_{nj}| + \varepsilon.$$

By taking now the limits and utilizing (14.23), we obtain

$$\limsup_n |y_n| \leq \varepsilon \text{ for every } \varepsilon > 0, \quad \text{so that } y_n \to 0.$$

(ii) We have

$$y_n = \sum_{j=1}^{k_n} \alpha_{nj} x_j = \left(\sum_j \alpha_{nj}\right) x + \sum_j \alpha_{nj}(x_j - x).$$

Now

$$\left(\sum_j \alpha_{nj}\right) x \to x \text{ since } \sum_j \alpha_{nj} \to 1, \quad \text{and } \sum_j \alpha_{nj}(x_j - x) \to 0$$

by (i) with x_n replaced by $x_n - x$, so that $x_n - x \to 0$. Thus $y_n \to x$.

(iii) Set

$$\alpha_{nj} = \frac{\lambda_j}{b_n}, \quad j = 1, \ldots, n.$$

Then $\sum_j \alpha_{nj} = 1$ and $\alpha_{nj} \underset{n\to\infty}{\to} 0$ for each fixed j, so that the assumptions in (ii) are satisfied. Therefore

$$\sum_j \alpha_{nj} x_j = \sum_j \frac{\lambda_j}{b_n} x_j = \frac{1}{b_n} \sum_j \lambda_j x_j \to x \text{ by (ii).} \qquad \blacksquare$$

Lemma 4 (Kronecker Lemma). Consider the sequence $\{x_n\}$, $n \geq 1$, of (real) numbers, and suppose that $\sum_{j=1}^n x_j \underset{n\to\infty}{\to} s$ finite. Let also $0 \neq b_n \uparrow \infty$ as $n \to \infty$. Then $\frac{1}{b_n}\sum_{j=1}^n b_j x_j \underset{n\to\infty}{\to} 0$.

Proof. Let $b_0 = 0$, and set $s_{n+1} = \sum_{j=1}^n x_j$ and $s_1 = 0$. Then one has

$$\frac{1}{b_n}\sum_{j=1}^n b_j x_j = \frac{1}{b_n}\sum_{j=1}^n b_j(s_{j+1} - s_j)$$

$$= \frac{1}{b_n}\left(\sum_{j=1}^{n} b_j s_{j+1} - \sum_{j=1}^{n} b_j s_j\right). \qquad (14.25)$$

But

$$\sum_{j=1}^{n} b_j s_{j+1} = \sum_{j=1}^{n-1} b_j s_{j+1} + b_n s_{n+1},$$

and if we set $j+1=r$, so that $r=2,\ldots,n$, we obtain

$$\sum_{j=1}^{n} b_j s_{j+1} = \sum_{r=2}^{n} b_{r-1} s_r + b_n s_{n+1} = \sum_{r=1}^{n} b_{r-1} s_r + b_n s_{n+1},$$

since $b_0 = 0$. Therefore (14.25) becomes

$$\frac{1}{b_n}\sum_{j=1}^{n} b_j x_j = \frac{1}{b_n}\left(\sum_{j=1}^{n} b_{j-1} s_j + b_n s_{n+1} - \sum_{j=1}^{n} b_j s_j\right)$$

$$= \frac{1}{b_n}\left[b_n s_{n+1} - \sum_{j=1}^{n}(b_j - b_{j-1}) s_j\right]$$

$$= s_{n+1} - \frac{1}{b_n}\sum_{j=1}^{n}(b_j - b_{j-1}) s_j. \qquad (14.26)$$

Thus, by setting $\lambda_j = b_j - b_{j-1} > 0$, $j \geq 1$, we have $b_n = \sum_{j=1}^{n} \lambda_j$ with $b_n \uparrow \infty$ as $n \to \infty$. Since $s_n \underset{n\to\infty}{\to} s$ finite, part (iii) of Lemma 3 applies and gives that

$$\frac{1}{b_n}\sum_{j=1}^{n} \lambda_j s_j = \frac{1}{b_n}\sum_{j=1}^{n}(b_j - b_{j-1}) s_j \underset{n\to\infty}{\to} s.$$

This result, together with (14.26), implies that $\frac{1}{b_n}\sum_{j=1}^{n} b_j x_j \underset{n\to\infty}{\to} s - s = 0$. ■

14.3 Statement and Proof of the Strong Law of Large Numbers

In this section, the SLLN is stated and proved. Before this is done, another theorem, Theorem 6, is established that is some kind of precursor to the SLLN. Also, an additional result is required for the proof of the SLLN, which is of independent interest; it is stated as a lemma.

Theorem 6 (Kolmogorov). For $n \geq 1$, consider the independent r.v.s X_n and suppose that $\mathcal{E}X_n$ is finite and $\sigma_n^2 = \sigma^2(X_n) < \infty$. Then, if $\Sigma_{n=1}^{\infty} \frac{\sigma_n^2}{b_n^2} < \infty$, for

$0 < b_n \uparrow \infty$ as $n \to \infty$, it follows that

$$\frac{S_n - \mathcal{E}S_n}{b_n} \underset{n\to\infty}{\overset{\text{a.s.}}{\to}} 0, \quad \text{where } S_n = \sum_{j=1}^{n} X_j. \qquad \blacksquare$$

Proof. We have

$$\sum_n \frac{\sigma_n^2}{b_n^2} = \sum_n \sigma^2\left(\frac{X_n}{b_n}\right) < \infty.$$

Therefore

$$\sum_n \left(\frac{X_n}{b_n} - \mathcal{E}\frac{X_n}{b_n}\right) = \sum_n \frac{X_n - \mathcal{E}X_n}{b_n}$$

converges a.s., by Theorem 2. Thus, for ω in the set of convergence of this series, we have that $\sum_n \frac{X_n(\omega) - \mathcal{E}X_n}{b_n}$ converges. For an arbitrary, but fixed such an ω, set

$$x_n = \frac{X_n(\omega) - \mathcal{E}X_n}{b_n}, \quad n \geq 1.$$

We have then that $\sum_{j=1}^{n} x_j \underset{n\to\infty}{\to} s(= s(\omega))$ finite. Then the Kronecker lemma applies and gives that

$$\frac{1}{b_n}\sum_{j=1}^{n} b_j x_j \underset{n\to\infty}{\to} 0, \text{ or equivalently,}$$

$$\frac{1}{b_n}\sum_{j=1}^{n} b_j \frac{X_j(\omega) - \mathcal{E}X_j}{b_j} = \frac{1}{b_n}\sum_{j=1}^{n}[X_j(\omega) - \mathcal{E}X_j]$$

$$= \frac{S_n(\omega) - \mathcal{E}S_n}{b_n} \underset{n\to\infty}{\to} 0.$$

Thus $\frac{S_n - \mathcal{E}S_n}{b_n} \underset{n\to\infty}{\to} 0$ a.s. $\blacksquare$

Remark 4. For $b_n = n$, we get $\frac{S_n - \mathcal{E}S_n}{n} \underset{n\to\infty}{\overset{\text{a.s.}}{\to}} 0$, provided $\sum_n \frac{\sigma_n^2}{n^2} < \infty$ (which happens, e.g., if $\sigma_n^2 = \sigma^2$ for all n), and if $\mathcal{E}X_n = \mu \in \Re$ for all n, then $\frac{S_n}{n} \underset{n\to\infty}{\overset{\text{a.s.}}{\to}} \mu$. However, this result is true without assuming finiteness of σ_n^2.

The following lemma will facilitate the proof of the SLLN in addition to being interesting in its own right.

Lemma 5. Let X be a r.v. and set

$$A_n = (|X| \geq n), \quad n \geq 1. \tag{14.27}$$

Then one has

$$\sum_{j=1}^{\infty} P(A_j) \leq \mathcal{E}|X| \leq 1 + \sum_{j=1}^{\infty} P(A_j)$$

(so that $\mathcal{E}|X| < \infty$ if and only if $\sum_{j=1}^{\infty} P(A_j) < \infty$).

Proof. Let $A_n, n \geq 1$, be defined by (14.27) and set $A_0 = \Omega$. Then as $n \to \infty$,

$$A_n \downarrow \oslash,\ B_n \stackrel{def}{=} A_{n-1} - A_n = (n - 1 \leq |X| < n),$$
$$B_i \cap B_j = \oslash, \quad i \neq j, \quad \sum_{j=1}^{\infty} B_j = \Omega. \tag{14.28}$$

Next,

$$(n-1)[P(A_{n-1}) - P(A_n)] = (n-1)P(B_n) \leq \int_{B_n} |X| dP$$
$$= \mathcal{E}(|X| I_{B_n}) \leq n P(B_n) = n[P(A_{n-1}) - P(A_n)], \tag{14.29}$$

and

$$n P(A_n) \leq \int_{A_n} |X| dP = \mathcal{E}(|X| I_{A_n}). \tag{14.30}$$

Now, for each $r \geq 1$, we have

$$\sum_{j=1}^{r} P(A_j) = 0 \times [P(A_0) - P(A_1)] +$$
$$1 \times [P(A_1) - P(A_2)] +$$
$$2 \times [P(A_2) - P(A_3)] +$$
$$\cdots\cdots\cdots\cdots\cdots\cdots\cdots +$$
$$(r-1) \times [P(A_{r-1}) - P(A_r)] +$$
$$r \times P(A_r)$$
$$= \sum_{j=1}^{r} (j-1)[P(A_{j-1}) - P(A_j)] + r P(A_r),$$

which, by means of (14.29) and (14.30) suitably applied, becomes

$$\leq \sum_{j=1}^{r} \mathcal{E}(|X| I_{B_j}) + \mathcal{E}(|X| I_{A_r})$$
$$= \int_{\Sigma_{j=1}^{r} B_j} |X| dP + \int_{A_r} |X| dP = \int_{(0 \leq |X| < r)} |X| dP$$
$$+ \int_{(|X| \geq r)} |X| dP$$
$$= \int_{\Omega} |X| dP = \mathcal{E}|X|,$$

since, clearly,

$$\sum_{j=1}^{r} B_j = \sum_{j=1}^{r} (j - 1 \leq |X| < j) = (0 \leq |X| < r); \tag{14.31}$$

i.e., $\sum_{j=1}^{r} P(A_j) \leq \mathcal{E}|X|$ for all $r \geq 1$, and hence

$$\sum_{j=1}^{\infty} P(A_j) \leq \mathcal{E}|X|. \tag{14.32}$$

Next, by the fact that $P(A_0) = P(\Omega) = 1$, one has

$$\begin{aligned}
1 + \sum_{j=1}^{r} P(A_j) &= 1 \times [P(A_0) - P(A_1)] + \\
&\quad 2 \times [P(A_1) - P(A_2)] + \\
&\quad 3 \times [P(A_2) - P(A_3)] + \\
&\quad \cdots\cdots\cdots\cdots\cdots\cdots\cdots + \\
&\quad r \times [P(A_{r-1}) - P(A_r)] + \\
&\quad (r+1) \times P(A_r) \\
&= \sum_{j=1}^{r} j[P(A_{j-1}) - P(A_j)] + \\
&\quad (r+1)P(A_r) \\
&\geq \sum_{j=1}^{r} j[P(A_{j-1}) - P(A_j)],
\end{aligned}$$

which, by means of (14.29), becomes

$$\begin{aligned}
&\geq \sum_{j=1}^{r} \mathcal{E}(|X|I_{B_j}) = \int_{(0 \leq |X| < r)} |X| dP \text{ (by (14.31))} \\
&= \int_{A_r^c} |X| dP \text{ (by (14.27))} \\
&= \mathcal{E}(|X|I_{A_r^c}); \quad \text{i.e.,}
\end{aligned}$$

$$\mathcal{E}(|X|I_{A_r^c}) \leq 1 + \sum_{j=1}^{r} P(A_j). \tag{14.33}$$

But $A_r \downarrow \oslash$ implies that $A_r^c \uparrow \Omega$, so that $I_{A_r^c} \underset{r \to \infty}{\to} 1$. Thus, $|X|I_{A_r^c} \uparrow |X|$ as $r \to \infty$ and therefore the Monotone Convergence Theorem gives that $\mathcal{E}(|X|I_{A_r^c}) \uparrow \mathcal{E}|X|$ as $r \to \infty$. Therefore (14.33) becomes, as $r \to \infty$,

$$\mathcal{E}|X| \leq 1 + \sum_{j=1}^{\infty} P(A_j). \tag{14.34}$$

Combining (14.32) and (14.34), one has then

$$\sum_{j=1}^{\infty} P(A_j) \leq \mathcal{E}|X| \leq 1 + \sum_{j=1}^{\infty} P(A_j). \qquad \blacksquare \tag{14.35}$$

Finally, we are in a position to state and prove the SLLN.

Theorem 7 (Strong Law of Large Numbers, SLLN; Kolmogorov). For $n \geq 1$, let X_n be i.i.d. r.v.s, let X be a r.v. distributed as the X_ns, and set $S_n = \sum_{j=1}^{n} X_j$. Then if $\mathcal{E}|X| < \infty$, it follows that $\frac{S_n}{n} \underset{n\to\infty}{\overset{\text{a.s.}}{\longrightarrow}} \mathcal{E}X$. On the other hand, if $\frac{S_n}{n} \underset{n\to\infty}{\overset{\text{a.s.}}{\longrightarrow}} c$, a finite constant, it follows that $\mathcal{E}|X| < \infty$ and $c = \mathcal{E}X$.

Proof. Throughout the proof all limits are taken as $n \to \infty$. Let $A_n, n \geq 1$, be defined by (14.27) in terms of the r.v. X. Then one has relation (14.35).

First, suppose that $\mathcal{E}|X| < \infty$, and we will show that $\frac{S_n}{n} \underset{\text{a.s.}}{\to} \mathcal{E}X$. To this end, define the truncated r.v.s $\bar{X}_n$ as follows:

$$\bar{X}_n = \begin{cases} X_n \text{ if } |X_n| < n \\ 0 \text{ if } |X_n| \geq n \end{cases}, \quad n \geq 1, \tag{14.36}$$

(so that $\bar{X}_n = X_n I_{A_n^c}$). Also, set $\bar{S}_n = \sum_{j=1}^{n} \bar{X}_j$. Then

$$\sum_n P\left(X_n \neq \bar{X}_n\right) = \sum_n P\left(|X_n| \geq n\right) = \sum_n P(|X| \geq n) = \sum_n P\left(A_n\right) \leq \mathcal{E}|X|$$
$$< \infty \quad (\text{by} \quad (14.35)).$$

Then, by Theorem 5 (iii), $\{\frac{S_n}{n}\}$ and $\{\frac{\bar{S}_n}{n}\}$ converge on events that differ with probability 0 and to the same limit. So, if $\frac{\bar{S}_n}{n} \overset{\text{a.s.}}{\to} \mathcal{E}X$, then $\frac{S_n}{n} \overset{\text{a.s.}}{\to} \mathcal{E}X$. Thus, it suffices to show that

$$\frac{\bar{S}_n}{n} \overset{\text{a.s.}}{\to} \mathcal{E}X. \tag{14.37}$$

We have, by means of (14.36),

$$\mathcal{E}\bar{X}_n = \mathcal{E}\left(X_n I_{(|X_n|<n)}\right) = \mathcal{E}\left(X I_{A_n^c}\right), \quad \text{and this converges to } \mathcal{E}X, \tag{14.38}$$

(see also Exercise 5) since $|X I_{A_n^c}| \leq |X|$, independent of n, integrable and $X I_{A_n^c} \to X$, so that the Dominated Convergence Theorem applies. Next, $\frac{\mathcal{E}\bar{S}_n}{n} = \frac{1}{n}\sum_{j=1}^{n} \mathcal{E}\bar{X}_j$, which is of the form $\frac{1}{b_n}\sum_{j=1}^{n} \lambda_j x_j$ with $x_j = \mathcal{E}\bar{X}_j, \lambda_j = 1, b_n = n$. Since also $x_n \to \mathcal{E}X$, finite (by (14.38)), the third part of the Toeplitz Lemma (Lemma 3) applies and gives that

$$\frac{\mathcal{E}\bar{S}_n}{n} \to \mathcal{E}X. \tag{14.39}$$

By means of (14.39), the convergence in (14.37) would be true, if we would show that

$$\frac{\bar{S}_n - \mathcal{E}\bar{S}_n}{n} \xrightarrow{\text{a.s.}} 0. \tag{14.40}$$

This would be true (by Theorem 6 applied with $b_n = n$), if we would show that $\sum_n \sigma^2\left(\frac{\bar{X}_n}{n}\right) < \infty$. To this end, consider $(n-1 \le |X| < n) = B_n = A_{n-1} - A_n = A_{n-1} \cap A_n^c, n \ge 1$. By the fact that $A_n^c = (|X| < n)$, and $B_m = (m-1 \le |X| < m)$, it is clear that

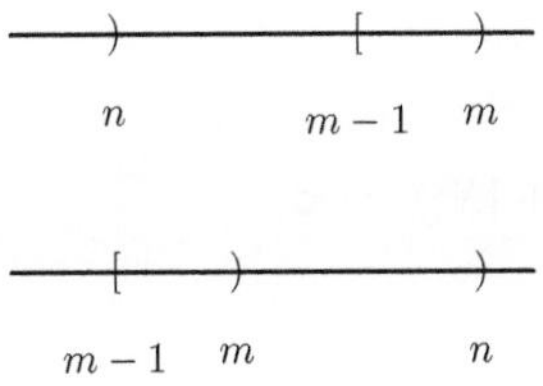

$$A_n^c \cap B_m = \begin{cases} \emptyset & \text{if } m > n \\ B_m & \text{if } m \le n. \end{cases} \tag{14.41}$$

Thus,

$$\begin{aligned} I_{B_m}\left(\sum_{n=1}^{\infty} \frac{X^2}{n^2} I_{A_n^c}\right) &= \sum_{n=1}^{\infty} \frac{X^2}{n^2} I_{A_n^c \cap B_m} \\ &= \sum_{n=m}^{\infty} \frac{X^2}{n^2} I_{B_m} \text{ (by means of (14.41))}, \\ &= I_{B_m} X^2 \left[\frac{1}{m^2} + \frac{1}{(m+1)^2} + \cdots\right]. \end{aligned}$$

If $\omega \in B_m$, then $m - 1 \le |X_m(\omega)| < m$ by the definition of B_m. Thus, in any case, the preceding expression is bounded by

$$\begin{aligned} &I_{B_m} m^2 \left[\frac{1}{m^2} + \frac{1}{(m+1)^2} + \cdots\right] \\ &= I_{B_m}\left\{1 + m^2\left[\frac{1}{(m+1)^2} + \frac{1}{(m+2)^2} + \cdots\right]\right\}. \end{aligned}$$

Clearly (see also the picture at the end of this proof),

$$\frac{1}{(m+1)^2} + \frac{1}{(m+2)^2} + \cdots < \int_m^{\infty} \frac{dx}{x^2} = \frac{1}{m}.$$

Therefore

$$\begin{aligned} &I_{B_m}\left(\sum_{n=1}^{\infty} \frac{X^2}{n^2} I_{A_n^c}\right) < (1+m) I_{B_m} = [2 + (m-1)] I_{B_m} \\ &\le (2 + |X|) I_{B_m} \end{aligned}$$

by the definition of B_m in (14.28). So

$$I_{B_m}\left(\sum_{n=1}^{\infty}\frac{X^2}{n^2}I_{A_n^c}\right) < (2+|X|)I_{B_m}.$$

Hence

$$\int_{B_m}\left(\sum_{n=1}^{\infty}\frac{X^2}{n^2}I_{A_n^c}\right)dP \le \int_{B_m}(2+|X|)dP.$$

Summing over $m = 1, 2, \ldots$, and taking into consideration relation (14.28), we then obtain (by Corollary 1 to Theorem 1 in Chapter 5)

$$\mathcal{E}\left(\sum_{n=1}^{\infty}\frac{X^2}{n^2}I_{A_n^c}\right) \le \mathcal{E}(2+|X|) < \infty. \tag{14.42}$$

Next,

$$\sum_n \sigma^2\left(\frac{\bar{X}_n}{n}\right) = \sum_n \frac{\sigma^2(\bar{X}_n)}{n^2} \le \sum_n \frac{\mathcal{E}\bar{X}_n^2}{n^2} = \sum_n \frac{\mathcal{E}(X^2 I_{A_n^c})}{n^2} \quad \text{(by} \quad (14.36))$$

$$= \mathcal{E}\left(\sum_n \frac{X^2}{n^2}I_{A_n^c}\right) < \infty \quad \text{(by the corollary just cited, and (14.42))}.$$

That is,

$$\sum_n \frac{\sigma^2(\bar{X}_n)}{n^2} < \infty.$$

Then Theorem 6 applies with $b_n = n$ and gives that $\frac{\bar{S}_n - \mathcal{E}\bar{S}_n}{n} \xrightarrow{\text{a.s.}} 0$, which is (14.40). To summarize:

$$\mathcal{E}|X| < \infty \text{ implies } \frac{S_n}{n} \xrightarrow{\text{a.s.}} \mathcal{E}X. \tag{14.43}$$

Next, suppose that $\frac{S_n}{n} \xrightarrow{\text{a.s.}} c$, finite, and we will show that $\mathcal{E}|X| < \infty$ and $c = \mathcal{E}X$. By setting $S_0 = 0$, we have

$$\frac{X_n}{n} = \frac{S_n - S_{n-1}}{n} = \frac{S_n}{n} - \frac{n-1}{n} \times \frac{S_{n-1}}{n-1} \xrightarrow{\text{a.s.}} 0.$$

Therefore, by taking $\delta = 1$ in Corollary 3 of the Borel Zero–One Criterion (Theorem 4), we obtain

$$\sum_n P\left(\left|\frac{X_n}{n}\right| \ge 1\right) < \infty, \quad \text{or} \sum_n P\left(|X_n| \ge n\right) = \sum_n P(|X| \ge n) < \infty,$$

or equivalently, $\sum_n P(A_n) < \infty$. Then inequality (14.35) implies that $\mathcal{E}|X| < \infty$. That is,

$$\frac{S_n}{n} \xrightarrow{\text{a.s.}} c, \quad \text{finite, implies } \mathcal{E}|X| < \infty. \tag{14.44}$$

But then $\frac{S_n}{n} \xrightarrow{\text{a.s.}} \mathcal{E}X$ by (14.43). From (14.43) and (14.36), we have that $c = \mathcal{E}X$ and $\mathcal{E}|X| < \infty$.

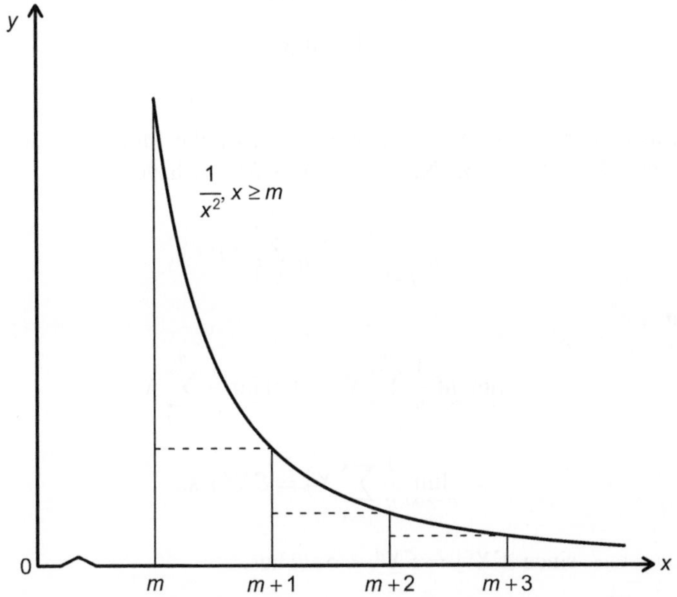

$\frac{1}{(m+1)^2} + \frac{1}{(m+2)^2} + \cdots$ is the area of the orthogonals between m and ∞ which is smaller than the area under the curve $y = \frac{1}{x^2}$ between m and ∞, which is equal to $\int_m^\infty \frac{dx}{x^2} = \frac{1}{m}$. ■

14.4 A Version of the Strong Law of Large Numbers for Random Variables with Infinite Expectation

In Theorem 7, the a.s. limit of the average S_n/n was finite. In this section, a certain version of the theorem is established. Specifically, it is shown that, if the common expectation of the X_js exists, but it is either ∞ or $-\infty$, the average S_n/n still converges (actually, diverges) a.s. to the expectation. However, if the common expectation does not exist, then the average S_n/n is unbounded with probability one.

We start with the following lemma.

Lemma 6.

(i) Let X be a nonnegative r.v. with $\mathcal{E}X = \infty$ and let $X_n, n \geq 1$, be independent r.v.s distributed as X. Then $\frac{1}{n}\sum_{j=1}^n X_j \underset{n\to\infty}{\overset{\text{a.s.}}{\to}} \infty$.

(ii) Let X be a nonpositive r.v. with $\mathcal{E}X = -\infty$ and let $X_n, n \geq 1$, be independent r.v.s distributed as X. Then $\frac{1}{n}\sum_{j=1}^n X_j \underset{n\to\infty}{\overset{\text{a.s.}}{\to}} -\infty$.

Proof.

(i) For $c > 0$, define the truncated r.v. X^c as follows:

$$X^c = \begin{cases} X \text{ if } X \leq c, \\ c \text{ if } X > c. \end{cases}$$

Then, as $c \uparrow \infty, 0 \leq X^c \uparrow X$ and hence, by the Monotone Convergence Theorem, $\mathcal{E}X^c \uparrow \mathcal{E}X = \infty$. Next, since $X \geq X^c$, we have

$$\frac{1}{n}\sum_{j=1}^{n} X_j \geq \frac{1}{n}\sum_{j=1}^{n} X_j^c,$$

so that

$$\liminf_{n\to\infty} \frac{1}{n}\sum_{j=1}^{n} X_j \geq \liminf_{n\to\infty} \frac{1}{n}\sum_{j=1}^{n} X_j^c$$
$$= \lim_{n\to\infty} \frac{1}{n}\sum_{j=1}^{n} X_j^c = \mathcal{E}X^c \text{ a.s.},$$

by Theorem 7, since $|\mathcal{E}X^c| = \mathcal{E}X^c < \infty$. Thus,

$$\liminf_{n\to\infty} \frac{1}{n}\sum_{j=1}^{n} X_j \geq \mathcal{E}X^c \text{ a.s.}$$

for every $c > 0$. Letting $c \uparrow \infty$ and utilizing the fact that $\mathcal{E}X^c \uparrow \mathcal{E}X = \infty$ (by Theorem 1 in Chapter 5), we then obtain that

$$\liminf_{n\to\infty} \frac{1}{n}\sum_{j=1}^{n} X_j \geq \infty \text{ a.s.},$$

so that $\frac{1}{n}\sum_{j=1}^{n} X_j \underset{n\to\infty}{\overset{\text{a.s.}}{\to}} \infty$.

(ii) By letting $Y = -X$, we have that Y satisfies the assumptions in (i). Thus

$$\frac{1}{n}\sum_{j=1}^{n}(-X_j) \underset{n\to\infty}{\overset{\text{a.s.}}{\to}} \mathcal{E}(-X) = \infty,$$

or equivalently,

$$\frac{1}{n}\sum_{j=1}^{n} X_j \underset{n\to\infty}{\overset{\text{a.s.}}{\to}} -\infty,$$

as was to be seen. ■

The next theorem is a straightforward application of the foregoing lemma and Theorem 7.

Theorem 8. For $n \geq 1$, let X_n be i.i.d. r.v.s such that $\mathcal{E}X_1$ exists but $\mathcal{E}|X_1| = \infty$. Then the SLLN still holds; i.e., $\frac{1}{n}\sum_{j=1}^{n} X_j \underset{n\to\infty}{\overset{\text{a.s.}}{\to}} \mathcal{E}X_1 (= \pm\infty)$. ■

Proof. We have $X_j = X_j^+ - X_j^-$, so that

$$\frac{1}{n}\sum_{j=1}^{n} X_j = \left(\frac{1}{n}\sum_{j=1}^{n} X_j^+\right) - \left(\frac{1}{n}\sum_{j=1}^{n} X_j^-\right) \underset{n\to\infty}{\overset{\text{a.s.}}{\to}} \mathcal{E}X_1^+ - \mathcal{E}X_1^-$$

by Theorem 7, applied to that one of $\mathcal{E}X_1^+, \mathcal{E}X_1^-$ which is $< \infty$, and Lemma 6, applied to the other one. Since $\mathcal{E}X_1^+ - \mathcal{E}X_1^- = \mathcal{E}X_1$, we obtain

$$\frac{1}{n}\sum_{j=1}^{n} X_j \underset{n\to\infty}{\overset{\text{a.s.}}{\to}} \mathcal{E}X_1,$$

as was to be seen. ■

Finally, consider a r.v. X for which $\mathcal{E}X$ does not exist (e.g., X may be a Cauchy distributed r.v.). If the r.v.s $X_n, n \geq 1$, are independent and distributed as X, one would like to make some kind of a statement for $\frac{1}{n}\sum_{j=1}^{n} X_j$. To this effect, one has the following result.

Theorem 9. For $n \geq 1$, let X_n be i.i.d. r.v.s and suppose that $\mathcal{E}X_1$ does not exist. Set $S_n = \sum_{j=1}^{n} X_j$. Then the sequence $\{\frac{S_n}{n}\}$ is unbounded with probability 1. (That is, for every $M > 0$, it holds $P\left(\left|\frac{S_n}{n}\right| > M \text{ i.o.}\right) = P\left[\limsup_{n\to\infty}\left(\left|\frac{S_n}{n}\right| > M\right)\right] = 1$.) ■

Proof. From $\frac{X_n}{n} = \frac{S_n}{n} - \frac{n-1}{n} \times \frac{S_{n-1}}{n-1}$ (and with $S_0 = 0$), it follows that $|\frac{X_n}{n}| \leq |\frac{S_n}{n}| + |\frac{S_{n-1}}{n-1}|$ and therefore, if $\{\frac{X_n}{n}\}$ is unbounded, then so is $\{\frac{S_n}{n}\}$. Thus, it suffices to show that $\{\frac{X_n}{n}\}$ is unbounded with probability 1. For $M > 0$, set

$$A_n(M) = \left(\frac{|X_n|}{n} > M\right).$$

Then $\{\frac{X_n}{n}\}$ is unbounded with probability 1, if and only if $P[\limsup_{n\to\infty} A_n(M)] = 1$ for every $M > 0$. From Theorem 4 (Borel Zero–One Criterion), it follows that $P[\lim_{n\to\infty}\sup A_n(M)] = 1$ for every $M > 0$, if and only if $\sum_{n=1}^{\infty} P[A_n(M)] = \infty$ for every $M > 0$. Now

$$\sum_{n=1}^{\infty} P(A_n(M)) = \sum_{n=1}^{\infty} P(|X_n| > nM) = \sum_{n-1}^{\infty} P(|X| > nM)$$

(where the r.v. X is distributed as the X_ns)

$$= \sum_{n=1}^{\infty}\sum_{k=n}^{\infty} P\left(kM < |X| \leq (k+1)M\right)$$

$$
\begin{aligned}
&= \sum_{k=1}^{\infty} kP\left[kM < |X| \le (k+1)M\right] \\
&= \sum_{k=1}^{\infty} k \int_{[kM<|X|\le(k+1)M]} dP \\
&= \sum_{k=1}^{\infty} k \int_{(kM<|x|\le(k+1)M)} dP_X \\
&\quad \text{(where } P_X \text{ is the probability distribution of } X\text{)} \\
&= \sum_{k=1}^{\infty} \int_{(kM<|x|\le(k+1)M)} k\,dP_X.
\end{aligned}
$$

But $k \ge 1$ implies $2k \ge k+1$ and hence $2kM \ge (k+1)M$. On the other hand, on the set $(kM < |x| \le (k+1)M)$, $|x| \le (k+1)M$, so that $|x| \le 2kM$ or $\frac{|x|}{2M} \le k$. Therefore the last expression is

$$
\begin{aligned}
&\ge \frac{1}{2M} \sum_{k=1}^{\infty} \int_{(kM<|x|\le(k+1)M)} |x|\,dP_X \\
&= \frac{1}{2M} \int_{(|x|>M)} |x|\,dP_X = \frac{1}{2M} \int_{(|X|>M)} |X|\,dP,
\end{aligned}
$$

and this is $= \infty$. This is so, because if it were $< \infty$, then

$$
\begin{aligned}
\mathcal{E}|X_1| &= \int_{\Omega} |X_1|\,dP = \int_{(|X_1|>M)} |X_1|\,dP + \int_{(|X_1|\le M)} |X_1|\,dP \\
&= \int_{(|x|>M)} |x|\,dP_{X_1} + \int_{(|x|\le M)} |x|\,dP_{X_1} < \infty,
\end{aligned}
$$

which would imply that both $\mathcal{E}X_1^+$ and $\mathcal{E}X_1^-$ are $< \infty$, a contradiction to the nonexistence of the $\mathcal{E}X_1$. Thus, $\sum_{n=1}^{\infty} P(A_n(M)) = \infty$ for every $M > 0$ and the desired result follows. ∎

To Theorems 7, 8, and 9, we then have the following.

Corollary. For $n \ge 1$, let X_n be i.i.d. r.v.s and set $S_n = \sum_{j=1}^{n} X_j$. Then, if $\mathcal{E}X_1$ is finite, one has that $\frac{S_n}{n} \underset{n\to\infty}{\overset{\text{a.s.}}{\to}} \mathcal{E}X_1$, whereas if $\mathcal{E}|X_1| = \infty$, one has that $\{\frac{S_n}{n}\}$ is unbounded with probability 1.

Proof. The first conclusion is one direction of Theorem 7. As for the second conclusion, if $\mathcal{E}X_1$ exists and $\mathcal{E}|X_1| = \infty$, then $\frac{S_n}{n} \underset{n\to\infty}{\overset{\text{a.s.}}{\to}} \mathcal{E}X_1$, by Theorem 8, and $\mathcal{E}X_1$ is either ∞ or $-\infty$. Thus, $\frac{S_n}{n} \underset{n\to\infty}{\overset{\text{a.s.}}{\to}} \pm\infty$, so that $P\left(\frac{|S_n|}{n} \ge M \text{ i.o.}\right) = 1$, and the conclusion then follows. Finally, the only other case that $\mathcal{E}|X_1| = \infty$ may happen is when $\mathcal{E}X_1$ does not exist ($\mathcal{E}X_1^+ = \mathcal{E}X_1^- = \infty$, so that $\mathcal{E}|X_1| = \infty$), but then Theorem 9 applies and yields the result. ∎

14.5 Some Further Results on Sequences of Independent Random Variables

In this section, we discuss two main results, Theorems 10 and 11, the first specifying the tail σ-field defined on a sequence of r.v.s, and the second providing necessary and sufficient conditions for the convergence of a series of r.v.s. When talking about convergence here, it is to be understood that the limit, be it a number or a (nondegenerate) r.v., is finite. In these results, the concept of a tail σ-fields is used. For its definition, and an associated result, see Definition 1 and Proposition 1 later.

First, we introduce the concept of the tail σ-field associated with a sequence of r.v.s, and discuss some operations on the r.v.s involved, which produce r.v.s measurable with respect to the tail σ-field.

To this end, let $X_n, n = 1, 2, \ldots,$ be r.v.s defined on the probability space $(\Omega, \mathcal{A}, P)$; these r.v.s need not be independent. For each $n \geq 1$ and $k \geq 0$, let $\mathcal{A}_{n,n+k}$ be the σ-field induced by the $k+1$ r.v.s $X_n, \ldots, X_{n+k}$; i.e., $\mathcal{A}_{n,n+k} = \sigma(X_n, \ldots, X_{n+k})$. Clearly, for fixed n, the σ-fields $\mathcal{A}_{n,n+k}$ are nondecreasing in k. In fact, the sets B^{k+1} in $\mathcal{B}^{k+1}$ can be viewed as sets B^{k+2} in $\mathcal{B}^{k+2}$ of the form $B^{k+2} = B^{k+1} \times \Re \in \mathcal{B}^{k+2}$, and, clearly, $\{B^{k+2} \in \mathcal{B}^{k+2}; B^{k+2} = B^{k+1} \times \Re \text{ for } B^{k+1} \in \mathcal{B}^{k+1}\} \subseteq \mathcal{B}^{k+2}$. Hence

$$\begin{aligned}
\mathcal{A}_{n,n+k} &= (X_n, \ldots, X_{n+k})^{-1}(\mathcal{B}^{k+1}) \\
&= (X_n, \ldots, X_{n+k}, X_{n+k+1})^{-1}(\mathcal{B}^{k+1} \times \Re) \\
&\subseteq (X_n, \ldots, X_{n+k}, X_{n+k+1})^{-1}(\mathcal{B}^{k+2}) \\
&= \mathcal{A}_{n,n+k+1}.
\end{aligned}$$

Next, set $\mathcal{F}_n = \cup_{k \geq 0} \mathcal{A}_{n,n+k}$. Then $\mathcal{F}_n$ is a field (by Exercise 10 in Chapter 1) (whereas it need not be a σ-field), and let $\mathcal{A}_n$ be the σ-field generated by $\mathcal{F}_n$; i.e., $\mathcal{A}_n = \sigma(\mathcal{F}_n)$. Clearly, $\mathcal{A}_n$ is the smallest σ-field with respect to which all r.v.s $X_n, X_{n+1}, \ldots$ are measurable. The σ-fields $\mathcal{A}_{n+1}, \mathcal{A}_{n+2}, \ldots$ are constructed likewise. Evidently, $\mathcal{F}_n \supseteq \mathcal{F}_{n+1} \supseteq \cdots$, and this implies that $\mathcal{A}_n \supseteq \mathcal{A}_{n+1} \supseteq \cdots$. Set $\mathcal{T} = \cap_{n \geq 1} \mathcal{A}_n$, so that the σ-field $\mathcal{T} \subseteq \mathcal{A}_n$, for all $n \geq 1$.

Definition 1. $\mathcal{T}$, as just defined, is a σ-field called the *tail σ-field* (of the sequence of r.v.s $\{X_n\}, n \geq 1$). The members of $\mathcal{T}$ are called *tail events*, and r.v.s measurable with respect to $\mathcal{T}$ are called *tail r.v.s*. ■

Here are some $\mathcal{T}$-measurable r.v.s resulting upon operating on a given sequence of any r.v.s $X_n, n \geq 1$.

Proposition 1. Let $\{X_n\}, n \geq 1$, be r.v.s. Then

(i) $\liminf_{n\to\infty} X_n$, $\limsup_{n\to\infty} X_n$, and $\lim_{n\to\infty} X_n$, if it exists, are all $\mathcal{T}$-measurable.

(ii) The set of convergence of the series $\sum_{n\geq 1} X_n$ is $\mathcal{T}$-measurable.

Proof.

(i) $\liminf_{n\to\infty} X_n = \lim_{n\to\infty}(\inf_{m\geq n} X_m) = \lim_{n\to\infty} Y_n$, where $Y_n = \inf_{m \geq n} X_m$. But Y_n is $\mathcal{A}_n$-measurable since it is defined in terms of the r.v.s $X_m, m \geq n$.

Likewise, Y_{n+1} is $\mathcal{A}_{n+1}$-measurable and hence $\mathcal{A}_n$-measurable since $\mathcal{A}_{n+1} \subseteq \mathcal{A}_n$. Thus $Y_n, Y_{n+1}, \ldots$ are all $\mathcal{A}_n$-measurable and hence so is their limit, Y, say. So, Y is $\mathcal{A}_n$-measurable for all $n \geq 1$, and therefore it is $\mathcal{T}$-measurable. In other words, $\liminf_{n\to\infty} X_n$ is $\mathcal{T}$-measurable. The $\mathcal{T}$-measurability of $\limsup_{n\to\infty} X_n$ follows similarly, since $\limsup_{n\to\infty} X_n = \lim_{n\to\infty}(\sup_{m\geq n} X_m)$. Alternatively, $\limsup_{n\to\infty} X_n = -\liminf_{n\to\infty}(-X_n)$, and the previous proof applies. Finally, the $\lim_{n\to\infty} X_n$, if it exists, is $\mathcal{T}$-measurable, since it is equal to $\limsup_{n\to\infty} X_n$ $(= \liminf_{n\to\infty} X_n)$.

(ii) Set $S_{1,n} = \sum_{j=1}^n X_j, n \geq 1$. Then "the set of convergence of $S_{1,n}$" $= (S_{1,n}$ converges, as $n \to \infty) = \bigcap_{m\geq 1}\bigcup_{n\geq 1}\bigcap_{\nu\geq 1}\left(|S_{1,n+\nu} - S_{1,n}| < \frac{1}{m}\right)$ (by Theorem 3 in Chapter 3). However, the r.v.s $S_{1,n}$ and $S_{1,n+\nu}, n \geq 1, \nu \geq 1$ are all $\mathcal{A}_1$-measurable, so that the set $(S_{1,n}$ converges, as $n \to \infty)$ is also $\mathcal{A}_1$-measurable. Next, $(S_{1,n}$ converges, as $n \to \infty) = (S_{k,n}$ converges, as $n \to \infty)$, for each $k \geq 1$, where $S_{k,n} = \sum_{j=k}^n X_j$, and the set $(S_{k,n}$ converges, as $n \to \infty)$ is $\mathcal{A}_k$-measurable. Thus, $(S_{1,n}$ converges, as $n \to \infty)$ is $\mathcal{A}_k$-measurable, for every $k \geq 1$, or "the set of convergence of $\sum_{n\geq 1} X_n$" is $\mathcal{A}_k$-measurable for all $k \geq 1$, and hence $\mathcal{T}$-measurable. ■

We now proceed with the statement and proof of the so-called Kolmogorov Zero–One Law.

Theorem 10 (Kolmogorov Zero–One Law). If the r.v.s $X_n, n \geq 1$, are independent, then the tail σ-field $\mathcal{T}$ is equivalent to the trivial σ-field $\{\oslash, \Omega\}$ in the sense that, for every $A \in \mathcal{T}$, we have $P(A) = 0$ or $P(A) = 1$. ■

Proof. The independence of the r.v.s $X_n, n \geq 1$, implies that the σ-fields $\sigma(X_1, \ldots, X_n)$ and $\sigma(X_{n+1}, X_{n+2}, \ldots)$ are independent for all n. It follows that $\sigma(X_1, \ldots, X_n)$ and $\mathcal{T}$ are independent, since $\mathcal{T}(= \bigcap_{n\geq 1} \mathcal{A}_n) \subseteq \sigma(X_{n+1}, X_{n+2}, \ldots)$ for all $n \geq 0$ (and $\mathcal{T} \subseteq \sigma(X_n, X_{n+1}, \ldots)$ for all $n \geq 1$). Therefore the field $\mathcal{F} = \bigcup_{n\geq 1}\sigma(X_1, \ldots, X_n)$ and $\mathcal{T}$ are independent. But then the σ-fields $\sigma(\mathcal{F}) = \sigma(X_1, X_2, \ldots)$ and $\mathcal{T}$ are independent. This is so by Proposition 2 in Chapter 10. Since $\mathcal{T} \subseteq \sigma(X_1, X_2, \ldots)$, as already mentioned, it follows that $\mathcal{T}$ is independent of itself. Then for $A \in \mathcal{T}$, we have $P(A \cap A) = P(A)P(A)$, which can happen only if $P(A) = 0$ or $P(A) = 1$. ■

To this theorem, there is the following important corollary, according to which sequences and series of independent r.v.s either converge a.s. or diverge a.s.; i.e., they cannot converge (or diverge) on a set A with $0 < P(A) < 1$.

Corollary. If the r.v.s $X_n, n \geq 1$, are independent, then the $\underline{X} \overset{\text{def}}{=} \liminf_{n\to\infty} X_n$, $\bar{X} \overset{\text{def}}{=} \limsup_{n\to\infty} X_n$, are (possibly extended) degenerate r.v.s with probability 1; i.e., they are constants (included $\pm\infty$) with probability 1. Moreover, the limits as $n \to \infty$, if they exist, of $\{X_n\}$, $\{S_n\}$, and $\{\frac{S_n}{b_n}\}$, where $S_n = \sum_{j=1}^n X_j$ and $0 < b_n \uparrow \infty$, are constants, as above, with probability 1.

Proof. In Proposition 1, it was proved that $\liminf_{n\to\infty} X_n$, $\limsup_{n\to\infty} X_n$ and the $\lim X_n$, if it exists, are all $\mathcal{T}$-measurable r.v.s. The same was proved to be true for the set of convergence of the series $\sum_{n\geq 1} X_n$, or of the sequence $\{S_n\}$, from which the same follows for the sequence $\left\{\frac{S_n}{b_n}\right\}$. The proof will be completed by showing that, if a (possibly extended) r.v. X is $\mathcal{T}$-measurable, then $P(X = c) = 1$. Indeed, if $P(X \leq x) = 1$ for every x, then, by letting $x \downarrow -\infty$, we get $P(X = -\infty) = 1$. Likewise, if $P(X > x) = 1$ for all x, then, by letting $x \uparrow \infty$, we get $P(X = \infty) = 1$. The remaining case is that there exist a, b in $\Re$ with $a < b$ such that $P(X < a) = 0$ and $P(X > b) = 0$. This implies that $P(a \leq X \leq b) = 1$. Then, by considering the intervals $\left[a, \frac{a+b}{2}\right)$ and $\left[\frac{a+b}{2}, b\right]$, we have that either $P\left(a \leq X < \frac{a+b}{2}\right) = 1$ or $P\left(\frac{a+b}{2} \leq X \leq b\right) = 1$. Call A_1 that interval for which $P(X \in A_1) = 1$. Split A_1 into halves, and let A_2 be that half for which $P(X \in A_2) = 1$. Continuing on like this, we have a sequence $A_n \downarrow$ with $P(X \in A_n) = 1, n \geq 1$. Clearly, as $n \to \infty$, $A_n \downarrow \{x_0\}$, some point in $[a, b]$, and then $P(X = x_0) = 1$. ∎

In establishing Theorem 11 later, we need two auxiliary results, which are also of independent interest.

Lemma 7. Let $X_n, n \geq 1$, be independent r.v.s with finite expectations, and let $\sigma_n^2 = \sigma^2(X_n)$. Then

(i) If $\sum_{n\geq 1} \sigma_n^2 < \infty$, then $\sum_{n\geq 1}(X_n - \mathcal{E}X_n)$ converges a.s.

(ii) If $\sum_{n\geq 1} \sigma_n^2 = \infty$ and $|X_n| \leq c(< \infty)$ a.s., $n \geq 1$, then $\sum_{n\geq 1}(X_n - \mathcal{E}X_n)$, with probability 1, does not converge.

(iii) If $|X_n| \leq c(< \infty)$ a.s., $n \geq 1$, then $\sum_{n\geq 1}(X_n - \mathcal{E}X_n)$ converges a.s., if and only if $\sum_{n\geq 1} \sigma_n^2 < \infty$.

Proof.

(i) See the proof of Theorem 2.

(ii) As in part (i), set $T_k = \sum_{j=1}^{k}(X_j - \mathcal{E}X_j)$ and apply the left-hand side of the relation in Remark 2 to obtain

$$1 - \frac{(\varepsilon + 2c)^2}{\sum_{j=m+1}^{m+r} \sigma_j^2} \leq P\left(\max_{1\leq k\leq r} |T_{m+k} - T_m| \geq \varepsilon\right). \tag{14.45}$$

Since

$$\begin{aligned}\left(\max_{1\leq k\leq r} |T_{m+k} - T_m| \geq \varepsilon\right) &= \big(|T_{m+k} - T_m| \geq \varepsilon \\ &\qquad \text{for at least one } k \text{ with } 1 \leq k \leq r\big) \\ &= \bigcup_{k=1}^{r} \left(|T_{m+k} - T_m| \geq \varepsilon\right),\end{aligned}$$

relation (14.45) becomes

$$1-\frac{(\varepsilon+2c)^2}{\sum_{j=m+1}^{m+r}\sigma_j^2}\le P\left(\bigcup_{k=1}^{r}(|T_{m+k}-T_m|\ge\varepsilon)\right). \tag{14.46}$$

In (14.46), let $r\to\infty$ and use the assumption that $\sum_{n\ge1}\sigma_n^2=\infty$ to obtain

$$1\le P\left(\bigcup_{k=1}^{\infty}(|T_{m+k}-T_m|\ge\varepsilon)\right).$$

Hence

$$\lim_{m\to\infty}P\left(\bigcup_{k=1}^{\infty}(|T_{m+k}-T_m|\ge\varepsilon)\right)=1 \text{ for every } \varepsilon>0. \tag{14.47}$$

However, (14.47) is a necessary and sufficient condition for mutual a.s. nonconvergence of $\{T_m\}$, by Theorem 4 in Chapter 3, and hence a.s. nonconvergence of $\{T_m\}$.

(iii) Follow from parts (i) and (ii). ■

Lemma 8. For the independent r.v.s $X_n, n\ge1$, suppose that $|X_n|\le c(<\infty)$ a.s., $n\ge1$, and that $\sum_{n\ge1}X_n$ converges a.s. Then both $\sum_{n\ge1}\sigma_n^2$ and $\sum_{n\ge1}\mathcal{E}X_n$ converge.

Proof. For each $n\ge1$, let Y_n be a r.v. having the same distribution as X_n and such that the r.v.s $X_1, Y_1, X_2, Y_2, \ldots$ are independent. Set $Z_n=X_n-Y_n$, so that, for $n\ge1$, the r.v.s Z_n are independent, $|Z_n|\le 2c$ a.s., $\mathcal{E}Z_n=0$, and $\sigma^2(Z_n)=2\sigma^2(X_n)=2\sigma_n^2$. Furthermore, the a.s. convergence of $\sum_{n\ge1}X_n$ implies a.s. convergence for $\sum_{n\ge1}Y_n$ as well as a.s. convergence for $\sum_{n\ge1}Z_n$, since $\sum_{n\ge1}Z_n=\sum_{n\ge1}X_n-\sum_{n\ge1}Y_n$. By the direct part of Lemma 7 (iii), it follows that $\sum_{n\ge1}\sigma^2(Z_n)<\infty$, and hence $\sum_{n\ge1}\sigma^2(X_n)<\infty$. Then, by Lemma 7 (i), $\sum_{n\ge1}(X_n-\mathcal{E}X_n)$ converges a.s. Since $\sum_{n\ge1}\mathcal{E}X_n=\sum_{n\ge1}X_n-\sum_{n\ge1}(X_n-\mathcal{E}X_n)$, it follows that $\sum_{n\ge1}\mathcal{E}X_n$ converges. ■

The following result provides necessary and sufficient conditions for a.s. convergence of a series of independent r.v.s.

Theorem 11 (Three Series Criterion; Kolmogorov). Let $X_n, n\ge1$, be independent r.v.s, and for some $c>0$, let X_n^c be the truncation of X_n at c; i.e.,

$$X_n^c=\begin{cases}X_n \text{ if } |X_n|<c\\ 0 \text{ if } |X_n|\ge c\end{cases}, \quad n\ge1.$$

Consider the three series:

$$\text{(i)}\sum_{n\ge1}P(|X_n|\ge c),\quad \text{(ii)}\sum_{n\ge1}\sigma^2(X_n^c),\quad \text{(iii)}\sum_{n\ge1}\mathcal{E}X_n^c.$$

Then

(a) If for some $c > 0$, all three series (i)–(iii) converge, then the series $\sum_{n\geq 1} X_n$ converges a.s.

(b) If $\sum_{n\geq 1} X_n$ converges a.s., then all three series (i)–(iii) converge for every $c > 0$. ■

Proof.

(a) Clearly, $(|X_n| \geq c) = (X_n \neq X_n^c)$, and by (i), $\sum_{n\geq 1} P(X_n \neq X_n^c) < \infty$. Therefore, by Theorem 5 (ii), the series $\sum_{n\geq 1} X_n$ and $\sum_{n\geq 1} X_n^c$ converge (essentially) on the same set. On account of (ii), Theorem 2 applies and gives that $\sum_{n\geq 1}(X_n^c - \mathcal{E}X_n^c)$ converges a.s. Then, by (iii), $\sum_{n\geq 1} X_n^c$ converges a.s., and hence so does $\sum_{n\geq 1} X_n$.

(b) Now, a.s. convergence of $\sum_{n\geq 1} X_n$ implies $X_n \underset{n\to\infty}{\overset{\text{a.s.}}{\to}} 0$. Then, by Corollary 3 to Theorem 4, it follows that, for every $c > 0$, $\sum_{n\geq 1} P(|X_n| \geq c) < \infty$. Thus the series in (i) converges. Next, as noted in part (a), the series $\sum_{n\geq 1} X_n$ and $\sum_{n\geq 1} X_n^c$ converge (essentially) on the same set. Since $\sum_{n\geq 1} X_n$ converges a.s., so does $\sum_{n\geq 1} X_n^c$. However, $|X_n^c| < c$, $n \geq 1$. Then Lemma 8 applies and yields that $\sum_{n\geq 1} \sigma^2(X_n^c) < \infty$ and $\sum_{n\geq 1} \mathcal{E}X_n^c$ converges. So, the series in (ii) and (iii) converge, and the proof is completed. ■

Exercises 1.

1. In reference to Exercise 6 in Chapter 11, explain why $\frac{S_n}{n} \underset{n\to\infty}{\overset{P}{\nrightarrow}} 0$.
2. For a r.v. X, show that $\mathcal{E}|X| < \infty$ if and only if $\sum_{n=1}^{\infty} P(|X| \geq nc) < \infty$ for some fixed constant $c > 0$.
3. Let $X_1, X_2, \ldots$ be independent r.v.s, and let the r.v. X be $\mathcal{A}_n = \sigma(X_n, X_{n+1}, \ldots)$-measurable for every $n \geq 1$. Then show that $P(X = c) = 1$ for some (finite) constant c.
4. For $n = 1, 2, \ldots$, suppose that the r.v.s X_n are independent and that $\sum_{n=1}^{\infty} \mathcal{E}|X_n| < \infty$. Then show that $\sum_{n=1}^{\infty} |X_n|$ converges a.s.
Hint: Use the Markov inequality, the special case of Theorem 6 in Chapter 6, and Theorem 11 here.
5. For $n \geq 1$, let X_n and X be r.v.s defined on the measure space $(\Omega, \mathcal{A}, \mu)$. Then, by Theorem 4 in Chapter 3, $X_n \underset{n\to\infty}{\to} X$ a.e. if and only if $\mu(\cap_{n=1}^{\infty} \cup_{\nu=0}^{\infty}(|X_{n+\nu} - X| \geq \frac{1}{k}) = 0$ for each arbitrary but fixed $k = 1, 2, \ldots$ Replace μ by a probability measure P, and show that $X_n \underset{n\to\infty}{\to} 0$ a.s. if and only if $P(\limsup_{n\to\infty} A_n) = 0$, where $A_n = (|X_n| \geq \frac{1}{k})$ for each arbitrary but fixed $k = 1, 2, \ldots$
6. Consider any events (independent or not) A_n, $n \geq 1$, and suppose that $\sum_{n=1}^{\infty} P(A_n) < \infty$. Then $P(\limsup_{n\to\infty} A_n) = 0$. This is so by Theorem 3 in this chapter (see also Theorem 4(i)). Regarding the converse of this statement, we have that, if $P(\limsup_{n\to\infty} A_n) = 0$ and the events A_n, $n \geq 1$, are independent, then $\sum_{n=1}^{\infty} P(A_n) < \infty$. Justify this assertion.

7. Regarding the converse stated in Exercise 6, if the events $A_n, n \geq 1$, are *not* independent, then $P(\limsup_{n\to\infty} A_n) = 0$ need *not* imply that $\sum_{n=1}^{\infty} P(A_n) < \infty$. Give one or two concrete examples to demonstrate this assertion.
Hint: Take $(\Omega, \mathcal{A}, P) = ((0, 1), \mathcal{B}_{(0,1)}, \lambda)$, λ being the Lebesgue measure. Then

(a) Take $X_n = I_{(0, \frac{1}{n})}, n \geq 1$, and show that $X_n \underset{n\to\infty}{\to} 0$, so that $X_n \underset{n\to\infty}{\to} 0$ a.s. Then, by Exercise 5, $P(\limsup_{n\to\infty} A_n) = 0$, where $A_n = (|X_n| \geq 1/k)$ for any arbitrary but fixed $k = 1, 2, \ldots$ Also, show that $\sum_{n=1}^{\infty} P(A_n) = \infty$.

(b) Take $X_n = I_{(0, \frac{1}{n^2})}, n \geq 1$, and show that $X_n \underset{n\to\infty}{\to} 0$, so that $X_n \underset{n\to\infty}{\to} 0$ a.s. Again, $P(\limsup_{n\to\infty} A_n) = 0$, as in (a). Also, show that $\sum_{n=1}^{\infty} P(A_n) < \infty$.

CHAPTER

Topics from Ergodic Theory 15

The ultimate purpose of this chapter is the formulation and proof of the Ergodic Theorem (see Theorem 1 and its Corollaries 1 and 4, as well as Theorem 3 and its Corollaries 1 and 2). To this effect, several concepts must be introduced first, and a substantial number of results must also be established; most of them are of independent interest.

The chapter is organized in six sections. In Section 15.1, the basic concept of a discrete parameter (stochastic) process is introduced as well as the special case of the coordinate process. Also, the concept of (strict) stationarity is introduced, and some characterizations of it are discussed.

In Section 15.2, the concept of a measurable measure-preserving transformation is introduced, as well as the special case of the shift transformation. Stationary processes are then defined by means of a measurable measure-preserving transformation and a r.v. It is shown that the coordinate process is defined by way of the shift transformation, and that the coordinate process is stationary if and only if the shift transformation is measure-preserving. Furthermore, the interplay between stationary processes and the coordinate process is studied.

The concepts of invariant and of almost sure invariant sets under a transformation are taken up in Section 15.3. It is shown, among other things, that invariant and almost invariant sets form σ-fields and one is the completion of the other. Also, the concept of ergodicity of a transformation is introduced here.

In Section 15.4, the concept of invariance, relative to a transformation, is extended to a r.v., and it is shown that a r.v. is invariant if and only if it is measurable with respect to the σ-field of invariant sets. Also, it is shown that a transformation is ergodic if and only if every r.v., invariant relative to the underlying transformation, is equal to a constant with probability 1.

In the subsequent section, the Ergodic Theorem as well as the so-called Maximal Ergodic Theorem are formulated and proved, and their forms are also noted under ergodicity of the underlying transformation.

In Section 15.6, a process **X** is considered, without being stipulated that it is defined by means of a measurable measure-preserving transformation, and the concepts of invariance of a set and of a r.v., relative to **X**, are defined. The invariant sets form a σ-field, and a r.v. is invariant if and only if it is measurable with respect to the σ-field of invariant sets. Ergodicity of **X** is also defined here. The Ergodic Theorem

An Introduction to Measure-Theoretic Probability, Second Edition. http://dx.doi.org/10.1016/B978-0-12-800042-7.00015-3

is reformulated and proved, and its form is noted under ergodicity of $\mathbf{X}$. The section is concluded with the derivation of processes by means of a given process $\mathbf{X}$, and the study of properties inherited from $\mathbf{X}$, such as stationarity, invariance, and ergodicity.

15.1 Stochastic Process, the Coordinate Process, Stationary Process, and Related Results

As has already been mentioned, the ultimate purpose of this chapter is to present a proof of the Ergodic Theorem. To this end, one will have first to build up the necessary mathematical machinery, and prove a series of auxiliary results. To start with, let

$$\mathbf{X} = (X_1, X_2, \ldots) \quad \text{and} \quad \mathbf{X}' = (X_1', X_2', \ldots)$$

be two (discrete time parameter) *stochastic processes* or just *processes* (i.e., two infinite sequences of r.v.s) defined on the probability spaces $(\Omega, \mathcal{A}, P)$ and $(\Omega', \mathcal{A}', P')$, respectively, and taking values in $(\Re^\infty, \mathcal{B}^\infty)$, the infinite cartesian product of Borel real lines $(\Re, \mathcal{B})$.

Definition 1. The processes $\mathbf{X}$ and $\mathbf{X}'$ just described have the same distribution, if for every $B \in \mathcal{B}^\infty$, one has

$$P\left(\mathbf{X} \in B\right) = P'\left(\mathbf{X}' \in B\right). \qquad \blacksquare$$

Clearly, two processes with the same distribution are indistinguishable from the probabilistic point of view. Thus, what counts is the distribution of a process rather than the probability space on which it is defined. It would then be desirable to replace the probability space of a given process by another one that would be easier to deal with. This can actually be done, as will be shown in the sequel.

Definition 2. Consider the measurable space $(\Re^\infty, \mathcal{B}^\infty)$, where the points of $\Re^\infty$ are denoted by $\mathbf{x} = (x_1, x_2, \ldots)$. Then, for each $n \geq 1$, define on $\Re^\infty$ the real-valued function

$$\hat{X}_n(\mathbf{x}) = x_n.$$

Then $\hat{X}_n$ is a r.v. and the process $\hat{\mathbf{X}} = (\hat{X}_1, \hat{X}_2, \ldots)$ is called the *coordinate process*. $\blacksquare$

That $\hat{X}_n$ is a r.v. is immediate, since for every set A in $\mathcal{B}$, $\hat{X}_n^{-1}(A)$ is the cylinder $\underbrace{\Re \times \cdots \times \Re}_{n-1} \times A \times \Re \times \cdots$ which is in $\mathcal{B}^\infty$. We then have the following result.

Proposition 1. Let $\mathbf{X} = (X_1, X_2, \ldots)$ be a process defined on the probability space $(\Omega, \mathcal{A}, P)$ and taking values in $(\Re^\infty, \mathcal{B}^\infty)$. Then there is a probability measure $\hat{P}$ on $\mathcal{B}^\infty$ such that the given process and the coordinate process $\hat{\mathbf{X}} = (\hat{X}_1, \hat{X}_2, \ldots)$ have the same distribution, under P and $\hat{P}$, respectively.

Proof. The probability measure $\hat{P}$ in question is simply the distribution of $\mathbf{X}$ under P; i.e., for $B \in \mathcal{B}^\infty$, $\hat{P}(B) = P(\mathbf{X} \in B)$. Since, clearly, for every $B \in \mathcal{B}^\infty$, $(\hat{\mathbf{X}} \in B) = B$, we have $P(\mathbf{X} \in B) = \hat{P}(\hat{\mathbf{X}} \in B)$. $\blacksquare$

In view of this proposition, all definitions and results involving the distribution of a process may be given in terms of the coordinate process by employing the appropriate probability measure. Thus, the process $\mathbf{X} = (X_1, X_2, \ldots)$, defined on $(\Omega, \mathcal{A}, P)$ into $(\Re^\infty, \mathcal{B}^\infty)$, may be thought of as being the coordinate process and then $(\Omega, \mathcal{A}, P) = (\Re^\infty, \mathcal{B}^\infty, \hat{P})$.

We now give the following definition.

Definition 3. The process $\mathbf{X} = (X_1, X_2, \ldots)$ is said to be (*strictly*) *stationary*, if for every $m \geq 1$, every $1 \leq n_1 < n_2 < \cdots < n_m$, with $n_1, n_2, \ldots, n_m$ integers and every $k \geq 1$, one has

$$P\left[\left(X_{n_1}, X_{n_2}, \ldots, X_{n_m}\right) \in B\right] = P\left[\left(X_{n_1+k}, X_{n_2+k}, \ldots, X_{n_m+k}\right) \in B_{(k)}\right], \qquad (15.1)$$

where B is any cylinder with base a Borel set in $\prod_{j=1}^m \mathcal{B}_{n_j}$ and all of whose sides are equal to $\Re$, and $B_{(k)}$ is the cylinder with the same base as that of B but located in $\prod_{j=1}^m \mathcal{B}_{n_j+k}$ and all of whose sides are equal to $\Re$. ∎

In the sequel, we will write B, rather than $B_{(k)}$, but the preceding explanation should be kept in mind. We shall also write B for the cylinder with base the Borel set B in the σ-field $\prod_{j=1}^m \mathcal{B}_{n_j}$.

Since the distribution of the random vector $(X_{n_1}, X_{n_2}, \ldots, X_{n_m})$ is determined by the joint d.f.of $X_{n_1}, X_{n_2}, \ldots, X_{n_m}$, relation (15.1) is equivalent to the following one,

$$F_{X_{n_1}, X_{n_2}, \ldots, X_{n_m}}(x_{n_1}, x_{n_2}, \ldots, x_{n_m}) = F_{X_{n_1+k}, X_{n_2+k}, \ldots, X_{n_m+k}}(x_{n_1}, x_{n_2}, \ldots, x_{n_m}), \qquad (15.2)$$

for all $x_{n_1}, x_{n_2}, \ldots, x_{n_m} \in \Re$.

From (15.1) (or (15.2)), it follows, in particular, that the r.v.s $X_n, n \geq 1$, are identically distributed. Interpreting n as time, the concept of stationarity is then clear: *No matter what times they are associated with, the joint distribution of any finite number of r.v.s from a stationary process is the same, provided their relative distance (in time) remains the same. In particular, no matter at what time one starts making observations, the distribution of the outcomes is the same.* From (15.1), it also follows that, for every $k \geq 1$, and any $B \in \mathcal{B}^\infty$,

$$P\left[(X_1, X_2, \ldots) \in B\right] = P\left[(X_{k+1}, X_{k+2}, \ldots) \in B\right]. \qquad (15.3)$$

This is so because all finite dimensional distributions of the processes $\mathbf{X} = (X_1, X_2, \ldots)$ and $\mathbf{X}^{(k)} = (X_{k+1}, X_{k+2}, \ldots)$ are the same (by (15.1)), and then so are the induced infinite dimensional measures in (15.3) (by the Carathéodory Extension Theorem and the Kolmogorov Consistency Theorem; for the latter, see, e.g., Loève (1963), page 93).

Actually, stationarity is characterized by the joint distributions of all finitely many consecutive r.v.s. More precisely, one has

Proposition 2. The process $\mathbf{X} = (X_1, X_2, \ldots)$ is stationary, if and only if, for every $m, k \geq 1$, and every $B \in \mathcal{B}^m$, one has

$$P\left[\left(X_1, X_2, \ldots, X_m\right) \in B\right] = P\left[\left(X_{k+1}, X_{k+2}, \ldots, X_{k+m}\right) \in B\right]. \tag{15.4}$$

Proof. One direction is obvious from (15.1) by taking $n_j = j$, $j = 1, \ldots, m$. So, it suffices to show the other direction. To this end, let $1 \leq n_1 < n_2 < \cdots < n_m$. One has, by (15.4),

$$\begin{aligned} &F_{X_1, X_2, \ldots, X_{n_m}}(x_1, x_2, \ldots, x_{n_m}) \\ &= F_{X_{k+1}, X_{k+2}, \ldots, X_{k+n_m}}(x_1, x_2, \ldots, x_{n_m}) \end{aligned} \tag{15.5}$$

for all $x_1, x_2, \ldots, x_{n_m} \in \Re$. In (15.5), replace the xs that are $\neq$ from x_{n_j}, $j = 1, \ldots, m$, by ∞(in the sense of letting them $\to \infty$). Then one has

$$\begin{aligned} &F_{X_{n_1}, X_{n_2}, \ldots, X_{n_m}}(x_{n_1}, x_{n_2}, \ldots, x_{n_m}) \\ &= F_{X_{n_1+k}, X_{n_2+k}, \ldots, X_{n_m+k}}(x_{n_1}, x_{n_2}, \ldots, x_{n_m}), \end{aligned}$$

which is (15.2). But (15.2) implies (15.1). ■

We also have the following result.

Proposition 3. For $n \geq 1$, let X_n be independent r.v.s. Then $\mathbf{X} = (X_1, X_2, \ldots)$ is stationary, if and only if the X_ns are identically distributed.

Proof. If $\mathbf{X}$ is stationary, the X_n s are always identically distributed. For the converse, we have

$$\begin{aligned} F_{X_1, \ldots, X_n}(x_1, \ldots, x_n) &= F_{X_1}(x_1) \cdots F_{X_n}(x_n) \\ &= F_{X_{k+1}}(x_1) \cdots F_{X_{k+n}}(x_n) \\ &= F_{X_{k+1}, \ldots, X_{k+n}}(x_1, \ldots, x_n) \end{aligned}$$

for every $k \geq 1$, and then Proposition 2 completes the proof. ■

Given a stationary process, one may derive any number of other stationary processes. This is so because of the following result.

Proposition 4. Let $\mathbf{X} = (X_1, X_2, \ldots)$ be a stationary process defined on $(\Omega, \mathcal{A}, P)$ and taking values in $(\Re^\infty, \mathcal{B}^\infty)$, and let φ_m be defined as follows:

$$\varphi_m : (\Re^m, \mathcal{B}^m) \to (\Re, \mathcal{B}), \text{ measurable}, \quad 1 \leq m \leq \infty.$$

Set

$$Y_n = \varphi_m\left(X_n, X_{n+1}, \ldots, X_{n+m-1}\right), \quad n \geq 1. \tag{15.6}$$

Then the process $\mathbf{Y} = (Y_1, Y_2, \ldots)$ is stationary.

Proof. By Proposition 2, it suffices to prove that, for every $n, k \geq 1$, and every $B \in \mathcal{B}^n$, one has

$$P\left[(Y_1, Y_2, \ldots, Y_n) \in B\right] = P\left[(Y_{k+1}, Y_{k+2}, \ldots, Y_{k+n}) \in B\right]. \tag{15.7}$$

Set $y_j = \varphi_m(x_j, x_{j+1}, \ldots, x_{j+m-1})$, $j \geq 1$, and let $A = \{(x_1, x_2, \ldots, x_{n+m-1}) \in \Re^{n+m-1};\ (y_1, y_2, \ldots, y_n) \in B\}$. Then one has

$$\begin{aligned}
P\left[(Y_1, Y_2, \ldots, Y_n) \in B\right] &= P\left[(X_1, X_2, \ldots, X_{n+m-1}) \in A\right] \\
&= P\left[(X_{k+1}, X_{k+2}, \ldots, X_{k+n+m-1}) \in A\right] \\
&\quad \text{(by stationarity of } \mathbf{X}) \\
&= P\left[(Y_{k+1}, Y_{k+2}, \ldots, Y_{k+n}) \in B\right] \\
&\quad \text{which is (15.7).}
\end{aligned}$$

■

Corollary. If the r.v.s X_n, $n \geq 1$, are i.i.d., then the process $\mathbf{Y}$, as defined by (15.6), is stationary.

Proof. It follows from Propositions 3 and 4. ■

15.2 Measure-Preserving Transformations, the Shift Transformation, and Related Results

Consider the probability space $(\Omega, \mathcal{A}, P)$, and let

$$T : (\Omega, \mathcal{A}) \to (\Omega, \mathcal{A}), \quad T^{-1}(\mathcal{A}) \subseteq \mathcal{A};$$

i.e., T is a measurable transformation on Ω into itself. T will be always assumed to be measurable, whether explicitly stated or not.

Definition 4. The transformation T is said to be *measure-preserving*, if

$$P\left(T^{-1}A\right) = P\left(A\right) \quad \text{for every} \quad A \in \mathcal{A}.$$

■

Remark 1. Actually, it suffices to show that $P\left(T^{-1}A\right) = P(A)$ for A belonging to a measure-determining class, such as the field $\mathcal{F}$ generating $\mathcal{A}$.

More precisely,

Proposition 5. Let $\mathcal{F}$ be a field such that $\sigma(\mathcal{F}) = \mathcal{A}$, and assume that, for every $A \in \mathcal{F}$, $P(T^{-1}A) = P(A)$. Then, for every $A \in \mathcal{A}$, $P(T^{-1}A) = P(A)$.

Proof. Let $\mathcal{M} = \{A \in \mathcal{A};\ P(T^{-1}A) = P(A)\}$. Then $\mathcal{M}$ is a monotone class. Indeed, $\mathcal{M} \subseteq \mathcal{F}$ by assumptions. Thus, $\mathcal{M} \neq \emptyset$. Next, let $A_n \in \mathcal{M}$, $n \geq 1$, be monotone. To show that $\lim A_n \in \mathcal{M}$, or equivalently, $P(T^{-1}(\lim A_n)) = P(\lim A_n)$ where here and in the sequel the limits are taken as $n \to \infty$. To this end, we have, for $A_n \uparrow$:

$$\begin{aligned}
P(T^{-1}(\lim A_n)) &= P\left(T^{-1}\left(\bigcup_{n\geq 1} A_n\right)\right) = P\left(\bigcup_{n\geq 1} T^{-1}A_n\right) \\
&= P(\lim T^{-1}A_n) = \lim P(T^{-1}A_n) = \lim P(A_n) = P(\lim A_n).
\end{aligned}$$

For $A_n \downarrow$:

$$P(T^{-1}(\lim A_n)) = P\left(T^{-1}\left(\bigcap_{n\geq 1} A_n\right)\right) = P\left(\bigcap_{n\geq 1} T^{-1}A_n\right)$$
$$= P(\lim T^{-1}A_n) = \lim P(T^{-1}A_n) = \lim P(A_n) = P(\lim A_n).$$

So, $\mathcal{M}$ is monotone, contains $\mathcal{F}$ with $\sigma(\mathcal{F}) = \mathcal{A}$, and is contained in $\mathcal{A}$. Hence $\mathcal{M} = \mathcal{A}$ (by Theorem 6 in Chapter 1). ∎

Remark 2. If T is measure-preserving, then $P\left(T^{-n}A\right) = P(A), n \geq 1$, where $T^{-n}A = T^{-1}(T^{-(n-1)}A), n \geq 1$, and $T^{-0} = I$, the identity in Ω. In fact, for $n = 1$ the statement is true. Let us assume it to be true for some $n > 1$. Then $P[T^{-(n+1)}A] = P[T^{-1}(T^{-n}A)] = P(T^{-n}A)$ (by the measure-preserving property of T) $= P(A)$ (by the induction hypothesis).

In terms of a measure-preserving transformation T, one may define any number of stationary processes. In fact, let X be a r.v. defined on $(\Omega, \mathcal{A}, P)$, and define the r.v.s $X_n, n \geq 1$, as follows:

$$X_n(\omega) = X\left(T^{n-1}\omega\right), \quad \omega \in \Omega, \tag{15.8}$$

where

$$T^2\omega = T(T\omega), T^k\omega = T(T^{k-1}\omega), \quad \text{and } T^0 = I, \quad \text{the identity in } \Omega. \tag{15.9}$$

Then one has the following result.

Proposition 6. Let T be a (measurable) measure-preserving transformation defined on $(\Omega, \mathcal{A}, P)$ into itself, and let X be a r.v. defined on this probability space. Let $X_n, n \geq 1$, be defined by (15.8) and (15.9). Then the process $\mathbf{X} = (X_1, X_2, \ldots)$ is a stationary process.

Proof. In the first place, the measurability of T implies the measurability of T^2, since for $A \in \mathcal{A}$, $(T^2)^{-1}(A) = T^{-1}[T^{-1}(A)]$, and by induction, T^n is measurable for all $n \geq 1$. Thus, $X_n, n \geq 1$, are r.v.s. It remains to prove stationarity. For $B \in \mathcal{B}^n$, we set

$$A = \left\{\omega \in \Omega;\ \left(X_1(\omega), \ldots, X_n(\omega)\right) \in B\right\} \quad \text{and}$$
$$A' = \left\{\omega \in \Omega;\ \left(X_{k+1}(\omega), \ldots, X_{k+n}(\omega)\right) \in B\right\}$$

and we wish to show that $P(A) = P(A')$. This will follow, by the measure-preserving property of T, by showing that

Lemma 1. With A and A' as just defined, we have $A' = T^{-k}A$.

Proof. First, $T^{-k}A \subseteq A'$. Indeed, $\omega \in T^{-k}A$ implies $T^k\omega \overset{def}{=} \omega' \in A$. Then

$$\begin{aligned} B &\ni (X_1(\omega'), \ldots, X_n(\omega')) \\ &= (X(T^0\omega'), \ldots, X(T^{n-1}\omega')) \\ &= (X(T^0T^k\omega), \ldots, X(T^{n-1}T^k\omega)) = (X(T^k\omega), \ldots, X(T^{k+n-1}\omega)) \\ &= (X_{k+1}(\omega), \ldots, X_{k+n}(\omega)), \quad \text{so that } \omega \in A'; \quad \text{i.e., } T^{-k}A \subseteq A'. \end{aligned}$$

Next, $A' \subseteq T^{-k}A$. That is, for $\omega \in A'$, to show $\omega \in T^{-k}A$ or $T^k\omega \stackrel{def}{=} \omega' \in A$. We have

$$
\begin{aligned}
(X_1(\omega'), \ldots, X_n(\omega')) &= (X(T^0\omega'), \ldots, X(T^{n-1}\omega')) \\
&= (X(T^0T^k\omega), \ldots, X(T^{n-1}T^k\omega)) = (X(T^k\omega), \ldots, X(T^{k+n-1}\omega)) \\
&= (X_{k+1}(\omega), \ldots, X_{k+n}(\omega))
\end{aligned}
$$

and this does belong in B since $\omega \in A'$. Thus, $(X_1(\omega'), \ldots, X_n(\omega')) \in B$, which implies that $\omega' \in A$. Hence $A' \subseteq T^{-k}A$, and therefore $A' = T^{-k}A$. ■

A certain transformation to be introduced next is of special interest.

Definition 5. The transformation S, defined as

$$S : \Re^\infty \to \Re^\infty, \text{ so that } S(x_1, x_2, \ldots) = (x_2, x_3, \ldots),$$

is called the *shift* transformation. ■

We then have the following result.

Proposition 7. The shift transformation S defined on $(\Re^\infty, \mathcal{B}^\infty)$ is measurable.

Proof. In order to prove measurability for S, it suffices to show that

$$S^{-1}\left(I_1 \times \cdots \times I_n \times \Re \times \cdots\right) \in \mathcal{B}^\infty \text{ for any } n \geq 1,$$

and any intervals I_j, $j = 1, \ldots, n$, in $\Re$. We have

$$
\begin{aligned}
S^{-1}\left(I_1 \times \cdots \times I_n \times \Re \times \cdots\right) &= \left\{\mathbf{x} = (x_1, x_2, \ldots) \in \Re^\infty;\quad S\left(\mathbf{x}\right)\right. \\
&= \left.(x_2, x_3, \ldots) \in I_1 \times \cdots \times I_n \times \Re \times \cdots\right\} \\
&= \left\{\mathbf{x} = (x_1, x_2, \ldots) \in \Re^\infty;\quad x_2 \in I_1, \ldots, x_{n+1} \in I_n, x_j \in \Re, j \geq n+2\right\} \\
&= \Re \times I_1 \times \cdots \times I_n \times \Re \times \cdots, \qquad (15.10)
\end{aligned}
$$

which is in $\mathcal{B}^\infty$. ■

In terms of the shift transformation S, the coordinate process $\hat{\mathbf{X}} = (\hat{X}_1, \hat{X}_2, \ldots)$ can be expressed as

$$\hat{X}_n(\mathbf{x}) = \hat{X}_1\left(S^{n-1}\mathbf{x}\right), n \geq 1 \quad \left(S^0 \text{ is the identity in } \Re^\infty\right). \qquad (15.11)$$

We now have the following result.

Proposition 8. If the coordinate process $\hat{\mathbf{X}} = (\hat{X}_1, \hat{X}_2, \ldots)$, defined on $(\Re^\infty, \mathcal{B}^\infty, \hat{P})$, is stationary, then the shift transformation S is measure-preserving, and vice versa.

Proof. Since the class of cylinders $\left\{I_1 \times \cdots \times I_n \times \Re \times \cdots;\ I_j, j = 1, \ldots, n,\right.$ intervals in $\left.\Re\right\}$ determines the measure $\hat{P}$ on $\mathcal{B}^\infty$, by Remark 1 and Proposition 5, it suffices to show that S is measure-preserving over this class of sets alone. We have

$$\hat{P}\left(I_1 \times \cdots \times I_n \times \Re \times \cdots\right) = \hat{P}\left[\left(\hat{X}_1, \ldots, \hat{X}_n\right) \in \left(I_1 \times \cdots \times I_n\right)\right]$$
$$= \hat{P}\left(\hat{X}_1 \in I_1, \ldots, \hat{X}_n \in I_n\right) = \hat{P}\left(\hat{X}_2 \in I_1, \ldots, \hat{X}_{n+1} \in I_n\right)$$
$$\text{(by stationarity of } \hat{\mathbf{X}}\text{)},$$
$$= \hat{P}\left(\Re \times I_1 \times \cdots \times I_n \times \Re \times \cdots\right)$$

which, by (15.10), completes one direction of the proof. The converse is true by Proposition 6. ■

Now Propositions 7 and 8 state, in effect, that every stationary process is *essentially* generated by a measure-preserving transformation. More precisely, we have

Proposition 9. Let $\mathbf{X} = \left(X_1, X_2, \ldots\right)$ be a stationary process defined on the probability space $(\Omega, \mathcal{A}, P)$, and let $\hat{P}$ be the distribution of $\mathbf{X}$, under P. Then the coordinate process $\hat{\mathbf{X}}$ is defined in terms of the shift transformation S, has the same distribution $\mathbf{X}$, and is stationary; S is measure-preserving.

Proof. It follows from (15.11), Propositions 8, and 1. ■

Thus, from a probability point of view, if $\mathbf{X} = \left(X_1, X_2, \ldots\right)$ is a stationary process defined on the probability space $(\Omega, \mathcal{A}, P)$, we may assume that there is a (measurable) measure-preserving transformation T on $(\Omega, \mathcal{A}, P)$ into itself such that $X_n = X(T^{n-1}), n \geq 1$, for some r.v. X, in the sense that $X_n(\omega) = X[T^{n-1}(\omega)], \omega \in \Omega, n \geq 1$.

15.3 Invariant and Almost Sure Invariant Sets Relative to a Transformation, and Related Results

The basic concepts in this section are those of an invariant and of an almost invariant set given below.

Definition 6. Let T be a (measurable) transformation on $(\Omega, \mathcal{A}, P)$ into itself. A set $A \in \mathcal{A}$ is said to be *invariant* (under T), if $T^{-1}A = A$. A set $A \in \mathcal{A}$ is said to be *a.s. invariant*, if $P(A \,\triangle\, T^{-1}A) = 0$, or equivalently, $P(A \cap T^{-1}A) = P(A) = P(T^{-1}A)$. ■

Remark 3. It is easily seen that, if A is invariant, then $P(T^{-n}A) = P(A)$, and if T is a.s. invariant, then $P(A \,\triangle\, T^{-n}A) = 0, n \geq 0$. That is,

Proposition 10. If A is a.s. invariant, then $P(A \,\triangle\, T^{-n}A) = 0$, so that $A \overset{\text{a.s.}}{=} T^{-n}A$ for $n \geq 0$.

Proof. In the first place, we observe that $A \overset{\text{a.s.}}{=} B$ and $B \overset{\text{a.s.}}{=} C$ imply $A \overset{\text{a.s.}}{=} C$. Indeed, $A \overset{\text{a.s.}}{=} B$ means that $P(A \,\triangle\, B) = 0$ or $P(A \cap B^c) = P(A^c \cap B) = 0$, and $B \overset{\text{a.s.}}{=} C$

means that $P(B \bigtriangleup C) = 0$ or $P(B \cap C^c) = P(B^c \cap C) = 0$. Then

$$\begin{aligned}
P(A \cap C^c) &= P[(A \cap C^c) \cap B] + P[(A \cap C^c) \cap B^c] \\
&= P[A \cap (B \cap C^c)] + P[(A \cap B^c) \cap C^c] \\
&\leq P(B \cap C^c) + P(A \cap B^c) = 0, \\
P(A^c \cap C) &= P[(A^c \cap C) \cap B] + P[(A^c \cap C) \cap B^c] \\
&= P[(A^c \cap B) \cap C] + P[(B^c \cap C) \cap A^c] \\
&\leq P(A^c \cap B) + P(B^c \cap C) = 0.
\end{aligned}$$

Thus, $P(A \cap C^c) = P(A^c \cap C) = 0$ and hence $P(A \bigtriangleup C) = 0$ or $A \overset{\text{a.s.}}{=} C$.

Next, by the a.s. invariance of A, we have,

$P(A) = P(T^{-1}A) = P(A \cap T^{-1}A)$, and by the measure-preserving property of T,

$$P(T^{-1}A) = P(T^{-2}A) = P[T^{-1}(A \cap T^{-1}A)] = P[T^{-1}A \cap T^{-2}A].$$

Thus $P(T^{-1}A) = P(T^{-2}A) = P[T^{-1}A \cap T^{-2}A]$, hence $P(T^{-1}A \bigtriangleup T^{-2}A) = 0$ or $T^{-1}A \overset{\text{a.s.}}{=} T^{-2}A$.

Since also $A \overset{\text{a.s.}}{=} T^{-1}A$, we have, by the introductory observation, $A \overset{\text{a.s.}}{=} T^{-2}A$. Likewise, $T^{-2}A \overset{\text{a.s.}}{=} T^{-3}A \overset{\text{a.s.}}{=} \cdots \overset{\text{a.s.}}{=} T^{-n}A$, so that $A \overset{\text{a.s.}}{=} T^{-n}A$, $n \geq 0$. ∎

Proposition 11. Let T be a (measurable) measure-preserving transformation on $(\Omega, \mathcal{A}, P)$ into itself. Then one has

(i) If $A = B$ a.s. and B is a.s. invariant, then so is A.
(ii) If B is a.s. invariant, then $P(B^c \cap T^{-n}B) = 0$ and $P(B \cap T^{-n}B) = P(B)$, $n \geq 0$.

Proof. First, show that $A \overset{\text{a.s.}}{=} B$ implies $T^{-1}A \overset{\text{a.s.}}{=} T^{-1}B$. Indeed,

$$\begin{aligned}
P(T^{-1}A) &= P(A)(\text{by the measure-preserving property of T}) \\
&= P(B)(\text{since } A \overset{\text{a.s.}}{=} B) \\
&= P(T^{-1}B)(\text{by the measure-preserving property of T}),
\end{aligned}$$

and

$$\begin{aligned}
P(A) = P(B) &= P(A \cap B)(\text{since } A \overset{\text{a.s.}}{=} B) \\
&= P[T^{-1}(A \cap B)] \\
&\qquad (\text{by the measure-preserving property of T}) \\
&= P(T^{-1}A \cap T^{-1}B).
\end{aligned}$$

So, $P(T^{-1}A) = P(T^{-1}B) = P(T^{-1}A \cap T^{-1}B)$, or $P(T^{-1}A \bigtriangleup T^{-1}B) = 0$, so that $T^{-1}A \overset{\text{a.s.}}{=} T^{-1}B$.

We now proceed with the proof of the proposition.

(i) $A \overset{\text{a.s.}}{=} B$ implies $T^{-1}A \overset{\text{a.s.}}{=} T^{-1}B$, by the introductory observation. But $B \overset{\text{a.s.}}{=} T^{-1}B$ by Proposition 10. Thus $A \overset{\text{a.s.}}{=} B \overset{\text{a.s.}}{=} T^{-1}B \overset{\text{a.s.}}{=} T^{-1}A$, so that $A \overset{\text{a.s.}}{=} T^{-1}A$, and A is a.s. invariant.

(ii) By Proposition 10, we have $B \overset{\text{a.s.}}{=} T^{-n}B$, since B is a.s. invariant. Hence $P(B \cap T^{-n}B) = P(B)$ and $P(B^c \cap T^{-n}B) = 0$. ■

We also have

Proposition 12. The class $\mathcal{J}$ of invariant sets in $\mathcal{A}$ under a (measurable) transformation T is a σ-field.

Proof. We have that $T^{-1}\Omega = \Omega$, so that $\Omega \in \mathcal{J}$ and so $\mathcal{J}$ is not empty. Next,

$$T^{-1}A^c = \left(T^{-1}A\right)^c = A^c, \quad \text{if } A \in \mathcal{J}, \text{ so that } A^c \in \mathcal{J}.$$

Finally,

$$T^{-1}\left(\bigcup_j A_j\right) = \bigcup_j T^{-1}A_j = \bigcup_j A_j, \quad \text{if } A_j \in \mathcal{J},$$

$$\text{so that } \bigcup_j A_j \in \mathcal{J}.$$ ■

Proposition 13. Let $\mathcal{J}'$ be the class of all a.s. invariant sets in $(\Omega, \mathcal{A}, P)$ under the (measurable) measure-preserving transformation T. Then, for every $A' \in \mathcal{J}'$, there exists a set $A \in \mathcal{J}$ such that $A = A'$ a.s.; i.e., $P(A \,\triangle\, A') = 0$.

Proof. Let $C \in \mathcal{A}$ and set $B = \bigcup_{n=0}^{\infty} T^{-n}C$. Then

$$T^{-n}B \subseteq B, \quad n \geq 0. \tag{15.12}$$

In fact,

$$T^{-n}B = T^{-n}\left(\bigcup_{j=0}^{\infty} T^{-j}C\right) = \bigcup_{j=0}^{\infty} T^{-n-j}C = \bigcup_{j=n}^{\infty} T^{-j}C$$
$$\subseteq \bigcup_{j=0}^{\infty} T^{-j}C = B.$$

Next, set

$$D = \bigcap_{n=0}^{\infty} T^{-n}B. \tag{15.13}$$

Then D is invariant. In fact,

$$T^{-1}D = T^{-1}\left(\bigcap_{n=0}^{\infty} T^{-n}B\right) = \bigcap_{n=0}^{\infty} T^{-n-1}B = \bigcap_{n=1}^{\infty} T^{-n}B$$

$$
\begin{aligned}
&= \left(\bigcap_{n=1}^{\infty} T^{-n}B\right) \bigcap B + \left(\bigcap_{n=1}^{\infty} T^{-n}B\right) \bigcap B^c \\
&= \left(\bigcap_{n=1}^{\infty} T^{-n}B\right) \bigcap B \\
&= \bigcap_{n=0}^{\infty} T^{-n}B = D,
\end{aligned}
$$

because $B^c \cap T^{-1}B = \oslash$, since $T^{-1}B \subseteq B$, by (15.12). (Actually, this may be considered as a way of constructing invariant sets.)

Suppose now that $C \in \mathcal{J}'$. Then we will show that $D = C$ a.s. by showing that $B = C$ a.s. and $D = B$ a.s. In fact, since $C \subseteq B$, it suffices to show that $P(B \cap C^c) = 0$. We have

$$
\begin{aligned}
P(B \cap C^c) &= P\left[\left(\bigcup_{n=0}^{\infty} T^{-n}C\right) \cap C^c\right] = P\left[\bigcup_{n=0}^{\infty} (C^c \cap T^{-n}C)\right] \\
&\leq \sum_{n=0}^{\infty} P\left(C^c \cap T^{-n}C\right).
\end{aligned}
$$

But

$$
P\left(C^c \cap T^{-n}C\right) = 0, \quad \text{by Proposition 11(ii).}
$$

Thus, $B = C$ a.s., and hence B is a.s. invariant by Proposition 11(i). Now from (15.13), we have that $D \subseteq B$, so that

$$
\begin{aligned}
D \,\triangle\, B = B \cap D^c &= B \cap \left(\bigcap_{n=0}^{\infty} T^{-n}B\right)^c \\
&= B \cap \left[\bigcup_{n=0}^{\infty} (T^{-n}B)^c\right] \\
&= B \cap \left(\bigcup_{n=0}^{\infty} T^{-n}B^c\right) \\
&= \bigcup_{n=0}^{\infty} (B \cap T^{-n}B^c) = \bigcup_{n=1}^{\infty} (B \cap T^{-n}B^c).
\end{aligned}
$$

But

$$
\begin{aligned}
B \cap T^{-n}B^c &= B \cap T^{-n}(\Omega - B) = B \cap (T^{-n}\Omega - T^{-n}B) \\
&= B \cap (\Omega - T^{-n}B) \\
&= B - B \cap T^{-n}B \quad \text{and} \quad P(B \cap T^{-n}B) = P(B),
\end{aligned}
$$

by the fact that B is a.s.invariant, and Proposition 11 (ii). Thus, $P(B \cap T^{-n}B^c) = 0$, and $P(D \triangle B) = 0$, so that $D = B$ a.s. and hence $D = C$ a.s. To summarize, C is a.s. invariant, D is invariant, and $C = D$ a.s. Hence, if we set $A' = C$, then $A = D$. ■

Proposition 14. Let $\mathcal{J}$ be as in Proposition 12, and define the class $\mathcal{J}''$ as follows:

$$\mathcal{J}'' = \{A \in \mathcal{A};\quad P(A \triangle B) = 0 \text{ for some } B \in \mathcal{J}\}. \tag{15.14}$$

Then

(i) Every $A \in \mathcal{J}''$ is a.s. invariant.
(ii) $\mathcal{J}''$ is a σ-field.

Proof.

(i) First, $\mathcal{J}'' \neq \emptyset$ since, clearly, $\mathcal{J}' \subseteq \mathcal{J}''$ by Proposition 13. Next, for $A \in \mathcal{J}''$, one has that there exists $B \in \mathcal{J}$ such that $P(A \triangle B) = 0$. Thus, $A = B$ a.s. Since B is trivially a.s. invariant, it follows, by Proposition 11(i), that A is a.s. invariant.
(ii) If $A \in \mathcal{J}''$, then for some $B \in \mathcal{J}, 0 = P(A \triangle B) = P(A \cap B^c) + P(A^c \cap B) = P[A^c \cap (B^c)^c] + P[(A^c)^c \cap B^c] = P(A^c \triangle B^c)$ and $B^c \in \mathcal{J}$, so that $A^c \in \mathcal{J}''$. Next, let $A_j \in \mathcal{J}''$, $j = 1, 2, \ldots$ Then

$$0 = P(A_j \triangle B_j) = P(A_j \cap B_j^c) + P(A_j^c \cap B_j) \text{ for some } B_j \in \mathcal{J}.$$

We then set $B = \bigcup_i B_i$ and get

$$\begin{aligned}
P\left(\bigcup_j A_j \triangle B\right) &= P\left(\bigcup_j A_j \cap B^c\right) + P\left[\left(\bigcup_j A_j\right)^c \cap B\right] \\
&= P\left(\bigcup_j A_j \cap \bigcap_i B_i^c\right) + P\left(\bigcap_j A_j^c \cap \bigcup_i B_i\right) \\
&\leq P\left[\bigcup_j \left(A_j \cap B_j^c\right)\right] + P\left[\bigcup_i \left(A_i^c \cap B_i\right)\right] \\
&\leq \sum_j P\left(A_j \cap B_j^c\right) + \sum_i P\left(A_i^c \cap B_i\right) = 0.
\end{aligned}$$

Thus, $\bigcup_j A_j \in \mathcal{J}''$. ■

Definition 7. The σ-field $\mathcal{J}''$, as defined by (15.14), is called the *completion* of $\mathcal{J}$ with respect to $\mathcal{A}$ and P. (See also Exercise 17 in Chapter 2.) ■

Proposition 15. Let $\mathcal{J}'$ be the class of all a.s. invariant sets in $\mathcal{A}$ under the (measurable) measure-preserving transformation T. Then $\mathcal{J}'$ is a σ-field and, indeed, $\mathcal{J}' = \mathcal{J}''$.

Proof. By Proposition 13, $\mathcal{J}' \subseteq \mathcal{J}''$, whereas, by Proposition 14(i), $\mathcal{J}'' \subseteq \mathcal{J}'$. Thus, $\mathcal{J}' = \mathcal{J}''$. ■

Definition 8. Let T be a (measurable) transformation on $(\Omega, \mathcal{A}, P)$ into itself. Then T is said to be *ergodic*, if for every $A \in \mathcal{J}$, one has $P(A) = 0$ or 1. ■

15.4 Measure-Preserving Ergodic Transformations, Invariant Random Variables Relative to a Transformation, and Related Results

The concept of invariance is also defined for r.v.s.

Definition 9. Let T be a (measurable) transformation on $(\Omega, \mathcal{A}, P)$ into itself, and let X be a r.v. on the same space. We say that X is *invariant* (relative to T), if $X(\omega) = X(T\omega), \omega \in \Omega$. ■

Remark 4. For an invariant r.v. X, it is immediate that $X(\omega) = X(T^n\omega), n \geq 0$ (and for $n = -1, -2, \ldots$, if T is one-to-one onto). Indeed, let $\omega' = T^{-1}\omega$. Then

$$X(T^{-1}\omega) = X(\omega') = X(T\omega') = X(T(T^{-1}\omega)) = X(\omega),$$

and likewise, if $T^{-n}\omega = \omega'$, then

$$\begin{aligned} X(T^{-n}\omega) &= X(\omega') = X(T\omega') = X(T(T^{-n}\omega)) \\ &= X(T^{-(n-1)}\omega) = X(\omega), \end{aligned}$$

by the induction hypothesis.

The fact that $X(\omega) = X(T^n\omega), \omega \in \Omega, n = 0, \pm 1, \ldots$, means that $X(\omega)$ remains *constant* on the *orbit* $T^n\omega, n = 0, \pm 1, \ldots$

Now the question arises as to what r.v.s are invariant. The answer to this is given by the following result.

Proposition 16. Let T be a (measurable) transformation on $(\Omega, \mathcal{A}, P)$ into itself, and let X be a r.v. on the same space. Then X is *invariant* (relative to T), if and only if X is $\mathcal{J}$-measurable, where $\mathcal{J}$ is the σ-field of (T-) invariant sets in $\mathcal{A}$.

Proof. Let X be invariant, and set $A(x) = \{\omega \in \Omega;\ X(\omega) \leq x\}, x \in \Re$.

Then

$$\begin{aligned} T^{-1}A(x) &= \{\omega \in \Omega;\ T\omega \in A(x)\} = \{\omega \in \Omega;\ X(T\omega) \leq x\} \\ &= \{\omega \in \Omega;\ X(\omega) \leq x\} = A(x), \end{aligned}$$

since $X(T\omega) = X(\omega)$. Thus, $T^{-1}A(x) = A(x), x \in \Re$, and this establishes $\mathcal{J}$-measurability for X.

Next, let X be $\mathcal{J}$-measurable. We shall show that X is invariant. Since every X is the pointwise limit of simple r.v.s, it suffices to show the result for the case that X is only an indicator function. So let $X = I_A$ with $A \in \mathcal{J}$. Then

$$X(T\omega) = I_A(T\omega) = I_{T^{-1}A}(\omega) = I_A(\omega) = X(\omega),$$
$$\text{since } \; T^{-1}A = A. \quad \text{(See also Exercise 4.)}$$

■

Ergodicity and invariant r.v.s are related as follows.

Proposition 17. Let T be a (measurable) transformation on $(\Omega, \mathcal{A}, P)$ into itself. Then T is ergodic, if and only if every real-valued r.v. invariant relative to T, defined on the same probability space, is a.s. equal to a finite constant.

Proof. Suppose that every invariant r.v. is a.s. equal to a constant, and let A be an arbitrary set in $\mathcal{J}$. If we set $X = I_A$, then X is $\mathcal{J}$-measurable, and therefore invariant, by Proposition 16. Since X is equal to 1 or 0 with probability $P(A)$ or $P(A^c)$, respectively, we have that

$$P(A) = 1 \quad \text{or} \quad P(A^c) = 1. \quad \text{So } P(A) = 0 \quad \text{or} \quad P(A) = 1.$$

Next, suppose that T is ergodic, so that $P(A) = 0$ or 1 for every $A \in \mathcal{J}$. We shall prove that every invariant r.v. X is a.s. equal to a constant. Since X is invariant, it is $\mathcal{J}$-measurable and hence $P(X < x) = 0$ or 1 for every $x \in \Re$. On the other hand, $P(X < x) \to 1$ as $x \to \infty$. Hence $P(X < x) = 1$ for all $x \geq x_1$, some sufficiently large x_1. Set

$$x_0 = \inf\{x \in \Re; \quad P(X < x) = 1\};$$

x_0 is finite; i.e., $x_0 > -\infty$, because otherwise $P(X < x) = 1$ for all x would imply $1 = P(X < x) \downarrow P(X = -\infty)$ as $x \downarrow -\infty$, a contradiction.

Then

$$P\left(x_0 - \varepsilon < X < x_0 + \varepsilon\right) = P\left(X < x_0 + \varepsilon\right)$$
$$-P\left(X \leq x_0 - \varepsilon\right) = 1 - 0 = 1$$

for every $\varepsilon > 0$. Letting $\varepsilon \downarrow 0$ and observing that

$$\left(x_0 - \varepsilon < X < x_0 + \varepsilon\right) \downarrow \left(X = x_0\right),$$

we get $P(X = x_0) = 1$. ■

15.5 The Ergodic Theorem, Preliminary Results

We are now in a position to formulate and prove the Ergodic Theorem.

Theorem 1 (Ergodic Theorem). Let T be a (measurable) measure-preserving transformation on $(\Omega, \mathcal{A}, P)$ into itself, and let X be a r.v. on the same space such that $\mathcal{E}|X| < \infty$. Define $X_n, n \geq 1$, as

$$X_n(\omega) = X(T^{n-1}\omega), \quad \omega \in \Omega,$$

and let $\mathcal{J}$ be the invariant σ-field (relative to T). Then

$$\frac{1}{n}\sum_{j=1}^{n} X_j(\omega) = \frac{1}{n}\sum_{j=1}^{n} X(T^{j-1}\omega) \underset{n\to\infty}{\overset{\text{a.s.}}{\to}} \mathcal{E}\left(X|\mathcal{J}\right). \quad \blacksquare \tag{15.15}$$

For the proof of Theorem 1, we need the following results.

Lemma 2. Let T be a (measurable) measure-preserving transformation on $(\Omega, \mathcal{A}, P)$ into itself, and let X be a r.v. on the same space for which $\mathcal{E}X$ exists. Then $\mathcal{E}X(\omega) = \mathcal{E}X(T\omega)$.

Proof. Since any r.v. is split into two nonnegative r.v.s, which are nondecreasing pointwise limits of nonnegative simple r.v.s, it suffices to show the result for the case that X is an indicator r.v. So, let $X = I_A$, $A \in \mathcal{A}$. Then $\mathcal{E}X(\omega) = P(A)$ and $\mathcal{E}X(T\omega) = \mathcal{E}I_A(T\omega) = \mathcal{E}I_{T^{-1}A}(\omega) = P(T^{-1}A) = P(A)$ by the measure-preserving property of T; i.e., $\mathcal{E}X(\omega) = \mathcal{E}X(T\omega)$. (See also Exercise 5.) ∎

Alternative Simpler Proof. Let $A = X^{-1}(B)$, $B \in \mathcal{B}$. Then $(XT)^{-1}(B) = T^{-1}A$, and $P(A) = P(T^{-1}A)$. Thus, X and XT have the same distribution under P. But

$$\int_\Omega XdP = \int_\Re xdQ_X \quad \text{and} \quad \int_\Omega (XT)dP = \int_\Re xdQ_{XT}.$$

Since $F_X \equiv F_{XT}$, we obtain $\int_\Omega XdP = \int_\Omega (XT)dP$. ∎

Theorem 2 (Maximal Ergodic Theorem). Let T, X and X_n, $n \geq 1$, be as in Theorem 1, and, for $\omega \in \Omega$, set

$$S_k(\omega) = \sum_{j=1}^{k} X_j(\omega) = \sum_{j=1}^{k} X(T^{j-1}\omega), \quad k \geq 1, \tag{15.16}$$

and

$$M_n(\omega) = \max\left[0, S_1(\omega), \ldots, S_n(\omega)\right]. \tag{15.17}$$

Then

$$\int_{(M_n>0)} XdP \geq 0. \quad \blacksquare \tag{15.18}$$

Proof. For $1 \leq k \leq n$, we have, by replacing in (15.16) and (15.17) ω by $T\omega$,

$$M_n(T\omega) \geq S_k(T\omega),$$

so that, by (15.16),

$$X(\omega) + M_n(T\omega) \geq X(\omega) + S_k(T\omega) = S_{k+1}(\omega),$$

since $S_k(T\omega) = \sum_{j=1}^{k} X(T^{j-1}T\omega) = S_{k+1}(\omega) - X(\omega) = \sum_{j=1}^{k} X(T^j\omega) = X_2(\omega) + \cdots + X_{k+1}(\omega)$.

Thus,

$$X(\omega) \geq S_{k+1}(\omega) - M_n(T\omega), \quad k = 1, \ldots, n. \tag{15.19}$$

But

$$S_1(\omega) = X(\omega) \quad \text{and} \quad M_n(T\omega) \geq 0,$$

so that

$$X(\omega) \geq S_1(\omega) - M_n(T\omega). \tag{15.20}$$

From (15.19) and (15.20), it follows that

$$X(\omega) \geq \max\left[S_1(\omega), \ldots, S_n(\omega)\right] - M_n(T\omega).$$

Therefore

$$\begin{aligned}\int_{(M_n>0)} X(\omega)dP &\geq \int_{(M_n>0)} \left\{\max\left[S_1(\omega), \ldots, S_n(\omega)\right]\right.\\ &\quad \left. - M_n(T\omega)\right\} dP\\ &= \int_{(M_n>0)} \left[M_n(\omega) - M_n(T\omega)\right] dP, \quad \text{by (15.17)}.\end{aligned}$$

But

$$\begin{aligned}\int_{(M_n>0)} M_n(\omega)dP &= \int_{(M_n\geq 0)} M_n(\omega)dP\\ &= \int M_n(\omega)dP, \quad \text{since } M_n \geq 0.\end{aligned}$$

Therefore

$$\begin{aligned}\int_{(M_n>0)} \left[M_n(\omega) - M_n(T\omega)\right] dP &= \int M_n(\omega)dP\\ &\quad - \int_{(M_n>0)} M_n(T\omega)dP\\ &\geq \int M_n(\omega)dP\\ &\quad - \int M_n(T\omega)dP = 0,\end{aligned}$$

by Lemma 2. It follows that $\int_{(M_n>0)} XdP \geq 0$, which is (15.18). ∎

Lemma 3. Let $T, X, X_n, n \geq 1$, and $\mathcal{J}$ be as in Theorem 1, and suppose that $\mathcal{E}\left(X|\mathcal{J}\right) = 0$ a.s. Then relation (15.15) holds true.

Proof. Let $S_n, n \geq 1$, be defined by (15.16), and let $\bar{X} = \limsup \frac{S_n}{n}$, where here and in the sequel all limits are taken as $n \to \infty$.

Then

$$\begin{aligned}\bar{X}(T\omega) &= \limsup \frac{S_n(T\omega)}{n} = \limsup \frac{S_{n+1}(\omega) - X(\omega)}{n}\\ &= \limsup \left[-\frac{X(\omega)}{n} + \frac{n+1}{n} \times \frac{S_{n+1}(\omega)}{n+1}\right] = \limsup \frac{S_{n+1}(\omega)}{n+1} = \bar{X}(\omega);\end{aligned}$$

i.e., $\bar{X}(T\omega) = \bar{X}(\omega)$, which means that $\bar{X}$ is invariant. Then $\bar{X}$ is measurable with respect to $\mathcal{J}$, and therefore, if for $\varepsilon > 0$ we set $D = (\bar{X} > \varepsilon)$, it follows that D is invariant.

Now define the r.v. X^* as follows:

$$X^*(\omega) = \left[X(\omega) - \varepsilon\right] I_D(\omega). \tag{15.21}$$

Let $X_n^* = X^*(T^{n-1})$, $n \geq 1$, and let S_k^* and M_n^* be the quantities defined by (15.16) and (15.17), respectively, when X is replaced by X^*. Then Theorem 2 implies that

$$\int_{(M_n^*>0)} X^* dP \geq 0. \tag{15.22}$$

Define the sets A_n, $n \geq 1$, by

$$A_n = \left(M_n^* > 0\right).$$

Then, clearly,

$$A_n = \left(M_n^* > 0\right) = \left(\max_{1\leq k\leq n} S_k^* > 0\right). \tag{15.23}$$

Also, let the set A be defined by

$$A = \left(\sup_{k\geq 1} S_k^* > 0\right). \tag{15.24}$$

Then, from (15.23) and (15.24), it clearly follows that

$$A_n \uparrow A. \tag{15.25}$$

From the fact that D is invariant, we have that $T^{-j}D = D$, $j \geq 0$, so that

$$\begin{aligned} X^*(T^j\omega) &= \left[X(T^j\omega) - \varepsilon\right] I_D(T^j\omega) = \left[X(T^j\omega) - \varepsilon\right] I_{T^{-j}D}(\omega) \\ &= \left[X(T^j\omega) - \varepsilon\right] I_D(\omega); \quad \text{i.e.,} \\ X^*(T^j\omega) &= \left[X(T^j\omega) - \varepsilon\right] I_D(\omega). \end{aligned}$$

Therefore

$$\frac{S_k^*(\omega)}{k} = \left[\frac{S_k(\omega)}{k} - \varepsilon\right] I_D(\omega),$$

and

$$\begin{aligned} A = \left(\sup_{k\geq 1} S_k^* > 0\right) &= \left(\sup_{k\geq 1} \frac{S_k^*}{k} > 0\right) \\ &= \left[\sup_{k\geq 1}\left(\frac{S_k}{k} - \varepsilon\right) I_D > 0\right] \\ &= \left(\sup_{k\geq 1} \frac{S_k}{k} > \varepsilon\right) \cap D, \quad \text{so that } A \subseteq D. \end{aligned}$$

Now, from the definition of lim sup, it follows that

$$\limsup \frac{S_k}{k} \leq \sup_{k\geq 1} \frac{S_k}{k}; \quad \text{i.e.,} \ \bar{X} \leq \sup_{k\geq 1} \frac{S_k}{k}.$$

Since on D, $\bar{X} > \varepsilon$, it follows that on D, $\sup_{k\geq 1} \frac{S_k}{k} > \varepsilon$, so that

$$D \subseteq \left(\sup_{k\geq 1} \frac{S_k}{k} > \varepsilon\right) \cap D = \left(\sup_{k\geq 1} \frac{S_k^*}{k} > 0\right)$$
$$= \left(\sup_{k\geq 1} S_k^* > 0\right) = A.$$

Thus $A = D$.

Therefore (15.25) gives $A_n \uparrow D$. So we have $X^* I_{A_n} \to X^* I_D$, $\left|X^* I_{A_n}\right| \leq |X^*| \leq |X| + \varepsilon$, independent of n, and $\mathcal{E}\left(|X| + \varepsilon\right) < \infty$. Hence the Dominated Convergence Theorem applies and gives that $\int_{A_n} X^* dP \to \int_D X^* dP$. But $\int_{A_n} X^* dP = \int_{(M_n^*>0)} X^* dP \geq 0$,by (15.23), so that $\int_D X^* dP \geq 0$.

Next, one has, by means of (15.21),

$$\begin{aligned}\int_D X^* dP &= \int_D (X - \varepsilon) I_D dP \\ &= \int_D (X - \varepsilon) dP = \int_D X dP - \varepsilon P(D) \\ &= \int_D \mathcal{E}(X|\mathcal{J}) dP - \varepsilon P(D) \\ &= 0 - \varepsilon P(D) = -\varepsilon P(D).\end{aligned}$$

Therefore $-\varepsilon P(D) \geq 0$, and hence $P(D) = 0$. That is to say, $P(\bar{X} > \varepsilon) = 0$, for every $\varepsilon > 0$, so that $\bar{X} \leq 0$ a.s. So the conclusion so far has been that

$$\limsup \frac{S_n}{n} \leq 0 \quad \text{a.s.} \tag{15.26}$$

Next, replacing the r.v. X by $-X$, the corresponding S_n is $-S_n$, and the same arguments above yield

$$\text{a.s.} \quad 0 \geq \limsup\left(-\frac{S_n}{n}\right) = -\liminf \frac{S_n}{n} \ \text{or} \ \liminf \frac{S_n}{n} \geq 0 \ \text{a.s.}$$

This result, together with (15.26), then gives that

$$\frac{S_n}{n} \xrightarrow{\text{a.s.}} 0.$$

■

Proof of Theorem 1. Let $Y = X - \mathcal{E}\left(X|\mathcal{J}\right)$. Then $\mathcal{E}\left(Y|\mathcal{J}\right) = 0$ a.s., and of course, $\mathcal{E}\left(X|\mathcal{J}\right)$ is $\mathcal{J}$-measurable, hence invariant, by Proposition 16. Then, by Remark 4,

$$\mathcal{E}\left(X|\mathcal{J}\right)(T^n\omega) = \mathcal{E}\left(X|\mathcal{J}\right)(\omega), \quad n \geq 0, \quad \omega \in \Omega.$$

It follows that $\frac{1}{n}\sum_{j=1}^{n} Y\left(T^{j-1}\omega\right) = \frac{1}{n}\sum_{j=1}^{n} X\left(T^{j-1}\omega\right) - \mathcal{E}\left(X|\mathcal{J}\right)$, whereas

$$\frac{1}{n}\sum_{j=1}^{n} Y\left(T^{j-1}\omega\right) \overset{\text{a.s.}}{\to} 0,$$

by Lemma 3. Thus

$$\frac{1}{n}\sum_{j=1}^{n} X\left(T^{j-1}\omega\right) \overset{\text{a.s.}}{\to} \mathcal{E}\left(X|\mathcal{J}\right).$$

■

Corollary 1. If T is also ergodic, then the right-hand side of (15.15) is equal to $\mathcal{E}X$ a.s.

Proof. $\mathcal{E}\left(X|\mathcal{J}\right)$ is $\mathcal{J}$-measurable, and since T is ergodic, it follows that $\mathcal{E}\left(X|\mathcal{J}\right) = c$ a.s. by Proposition 17. Then $\mathcal{E}\left[\mathcal{E}\left(X|\mathcal{J}\right)\right] = \mathcal{E}X = c$. Thus $\mathcal{E}\left(X|\mathcal{J}\right) = \mathcal{E}X$ a.s. ■

Corollary 2. If T is also ergodic, then for every $A \in \mathcal{A}$, one has

$$\frac{1}{n}\sum_{j=1}^{n} I_A\left(T^{j-1}\omega\right) \underset{n\to\infty}{\overset{\text{a.s.}}{\to}} P(A). \tag{15.27}$$

Proof. It follows by Corollary 1, and the fact that a.s. $\mathcal{E}\left(I_A|\mathcal{J}\right) = \mathcal{E}I_A = P(A)$. ■

Remark 5. Relation (15.27) says, in effect, that, if T is ergodic, then, for almost all $\omega \in \Omega$, the proportion of the points $\omega, T^1\omega, T^2\omega, \ldots$, as $n \to \infty$, which lie in any set $A \in \mathcal{A}$, is equal to $P(A)$.

Definition 10. Consider the measurable space $(\Omega, \mathcal{A})$, and let P_1, P_2 be two probability measures on $\mathcal{A}$. We say that P_1 and P_2 are *orthogonal*, and we write $P_1 \perp P_2$, if there exists a set $A \in \mathcal{A}$, such that $P_1(A) = 1$ and $P_2(A) = 0$ (so that $P_2(A^c) = 1$). ■

Corollary 3. Let $T : (\Omega, A) \to (\Omega, \mathcal{A})$ be a (measurable) measure-preserving transformation with respect to two probability measures P_1 and P_2 on $\mathcal{A}$, and suppose that T is also ergodic with respect to both P_1 and P_2. Then either $P_1 \equiv P_2$ or $P_1 \perp P_2$.

Proof. It suffices to prove that, if $P_1 \neq P_2$, then $P_1 \perp P_2$. To this end, let $A \in \mathcal{A}$ be such that $P_1(A) \neq P_2(A)$. Then, by Corollary 1,

$$\frac{1}{n}\sum_{j=1}^{n} I_A\left(T^{j-1}\omega\right) \underset{n\to\infty}{\to} P_1(A)$$

except on a set $B_1 \in \mathcal{A}$ such that $P_1(B_1) = 0$. Also,

$$\frac{1}{n}\sum_{j=1}^{n} I_A\left(T^{j-1}\omega\right) \underset{n\to\infty}{\to} P_2(A)$$

except on a set $B_2 \in \mathcal{A}$ such that $P_2(B_2) = 0$. Since $P_1(A) \neq P_2(A)$, we have that $B_1^c \cap B_2^c = \oslash$, since, for $\omega \in B_1^c$, $\frac{1}{n}\sum_{j=1}^{n} I_A(T^{j-1}\omega) \underset{n\to\infty}{\to} P_1(A)$ and, for $\omega \in B_2^c$, $\frac{1}{n}\sum_{j=1}^{n} I_A(T^{j-1}\omega) \underset{n\to\infty}{\to} P_2(A)$, and $P_1(A) \neq P_2(A)$, so that $B_1 \supseteq B_2^c$, and therefore $P_2(B_1) \geq P_2(B_2^c) = 1$; i.e., $P_1(B_1) = 0$ or $P_1(B_1^c) = 1$ and $P_2(B_1) = 1$, so that $P_1 \perp P_2$. ■

Corollary 4. Under the same assumptions as those of Theorem 1, one has

$$\frac{1}{n}\sum_{j=1}^{n} X_j(\omega) = \frac{1}{n}\sum_{j=1}^{n} X\left(T^{j-1}\omega\right) \underset{n\to\infty}{\overset{(1)}{\to}} \mathcal{E}\left(X|\mathcal{J}\right).$$

Proof. As it follows from the proof of Theorem 1, we may assume, without loss of generality, that $\mathcal{E}\left(X|\mathcal{J}\right) = 0$ a.s. Then we set

$$Y_n(\omega) = \frac{1}{n}\sum_{j=1}^{n} X_j(\omega) = \frac{1}{n}\sum_{j=1}^{n} X\left(T^{j-1}\omega\right), \quad \omega \in \Omega,\ n \geq 1,$$

and we shall show that, as $n \to \infty$ (here as well as throughout the proof),

$$\mathcal{E}\left|Y_n\right| \to 0. \tag{15.28}$$

By Theorem 1, we have that $Y_n \overset{\text{a.s.}}{\to} 0$. Then Egorov's Theorem (see Exercise 11 in Chapter 3) implies that, for every $\varepsilon > 0$ there exists $A = A(\varepsilon) \in \mathcal{A}$ such that $P(A) \leq \varepsilon$ and $Y_n \to 0$ uniformly on A^c. Thus, for all sufficiently large n independent of $\omega \in A^c$, one has

$$\begin{aligned}\mathcal{E}\left|Y_n\right| &= \int_A \left|Y_n\right| dP + \int_{A^c} \left|Y_n\right| dP \leq \int_A \left|Y_n\right| dP + \varepsilon P(A^c)\\ &\leq \frac{1}{n}\sum_{j=1}^{n}\int_A \left|X_j\right| dP + \varepsilon P(A^c) = \int_A |X| dP + \varepsilon P(A^c),\end{aligned}$$

and therefore

$$\limsup \mathcal{E}\left|Y_n\right| \leq \int_A |X|\, dP + \varepsilon P(A^c). \tag{15.29}$$

But

$$\int_A |X|\,dP = \int_{A\cap(|X|>k)} |X|\,dP + \int_{A\cap(|X|\le k)} |X|\,dP$$
$$\le \int_{A\cap(|X|>k)} |X|\,dP + kP(A)$$
$$\le \int_{(|X|>k)} |X|\,dP + \varepsilon k,$$

and therefore (15.29) becomes

$$\limsup \mathcal{E}\,|Y_n| \le \int_{(|X|>k)} |X|\,dP + \varepsilon k + \varepsilon P(A^c).$$

Letting $\varepsilon \to 0$, we obtain

$$\limsup \mathcal{E}\,|Y_n| \le \int_{(|X|>k)} |X|\,dP.$$

Now letting $k \to \infty$ and utilizing the fact that $\mathcal{E}\,|X| < \infty$, we get $\limsup \mathcal{E}\,|Y_n| = 0$. Thus, $\mathcal{E}\,|Y_n| \to 0$, which is (15.28). ■

Now, let $\mathbf{X} = \left(X_1, X_2, \ldots\right)$ be a stationary process defined on $(\Omega, \mathcal{A}, P)$ into $(\Re^\infty, \mathcal{B}^\infty)$, and let $\hat{P}$ be the distribution of $\mathbf{X}$ under P. Then the coordinate process $\hat{\mathbf{X}} = (\hat{X}_1, \hat{X}_2, \ldots)$ defined on $(\Re^\infty, \mathcal{B}^\infty, \hat{P})$ is also stationary (by Proposition 1); it assumes the representation

$$\hat{X}_n(\mathbf{x}) = \hat{X}_1\left(S^{n-1}(\mathbf{x})\right) \quad \text{(by (15.11))},$$

where S is the shift transformation on $\Re^\infty$(see Definition 5); and the shift transformation S is measurable and measure-preserving (by Proposition 8). The processes $\mathbf{X}$ and $\hat{\mathbf{X}}$ are equivalent probabilistically, in the sense that they have the same distribution (under P and $\hat{P}$, respectively). And, whereas $\mathbf{X}$ may not have a representation in terms of a measure-preserving transformation T on Ω into itself, $\hat{\mathbf{X}}$ does. It follows, in particular, that

$$\mathcal{E}X_1 = \int X_1 dP = \int \hat{X}_1 d\hat{P} = \mathcal{E}\hat{X}_1,$$

provided these expectations exist. Thus, if $\mathcal{E}\,|X_1| < \infty$, and the shift transformation is also ergodic, then we have that

$$\frac{1}{n}\sum_{j=1}^{n} \hat{X}_j \to \mathcal{E}\hat{X}_1 = \mathcal{E}X_1 \quad \text{a.s. and in the first mean}$$

(by Corollaries 1 and 3). Hence

$$\frac{1}{n}\sum_{j=1}^{n} X_j \to \mathcal{E}X_1 \text{ a.s. and in the first mean.}$$

15.6 Invariant Sets and Random Variables Relative to a Process, Formulation of the Ergodic Theorem in Terms of Stationary Processes, Ergodic Processes

Now, it would be desirable to have a definition of invariance and ergodicity, and a formulation of the Ergodic Theorem in terms of the original process without passing to the equivalent coordinate process and without the assumption that **X** is generated by a transformation T. This is actually done here. The appropriate definition of invariance of a set in $\mathcal{A}$ is the following one.

Definition 11. A set $A \in \mathcal{A}$ is said to be *invariant* (relative to **X**), if there exists a set $B \in \mathcal{B}^\infty$ such that $A = \mathbf{X}^{-1}B$ and such that

$$A = \left(X_n, X_{n+1}, \ldots\right)^{-1} B \quad \text{for all} \quad n \geq 1. \qquad \blacksquare$$

Proposition 18. The class $\mathcal{J}$ of invariant sets in $\mathcal{A}$ is a σ-field.

Proof. $\mathcal{J}$ is nonempty since $\Omega \in \mathcal{J}$, and closure of $\mathcal{J}$ under countable unions and complementation is immediate. (See also Exercise 6.) ■

Remark 6. Consider the coordinate process $\hat{\mathbf{X}}$ and the shift transformation S. Then the σ-field of invariant sets under S, $\mathcal{J}_S$, say, is defined by

$$\mathcal{J}_S = \left\{B \in \mathcal{B}^\infty; \quad S^{-1}B = B\right\}.$$

Because then

$$S^{-n}B = B, \quad \text{and} \quad S^{-n}B = \underbrace{\Re \times \cdots \times \Re}_{n} \times B, \ n \geq 0,$$

we have that

$$\mathcal{J}_S = \left\{B \in \mathcal{B}^\infty; \quad B = \underbrace{\Re \times \cdots \times \Re}_{n} \times B, \ n \geq 0\right\}.$$

Next, let $\hat{\mathcal{J}}$ be the σ-field of invariant sets relative to $\hat{\mathbf{X}}$; i.e.,

$$\hat{\mathcal{J}} = \left\{B \in \mathcal{B}^\infty; \quad \left(\hat{X}_n, \hat{X}_{n+1}, \ldots\right)^{-1} B = B, \ n \geq 0\right\}.$$

But, clearly,

$$\left(\hat{X}_n, \hat{X}_{n+1}, \ldots\right)^{-1} B = \underbrace{\Re \times \cdots \times \Re}_{n-1} \times B, \ n \geq 1.$$

Thus,

$$\hat{\mathcal{J}} = \left\{B \in \mathcal{B}^\infty; \quad B = \underbrace{\Re \times \cdots \times \Re}_{n} \times B, \ n \geq 0\right\}, \text{ so that } \mathcal{J}_S = \hat{\mathcal{J}}.$$

Also,

Definition 12. A r.v. Y on $(\Omega, \mathcal{A})$ is said to be *invariant* (relative to $\mathbf{X}$), if there exists a measurable mapping φ on $(\Re^\infty, \mathcal{B}^\infty)$ into $(\Re, \mathcal{B})$ such that

$$Y = \varphi\left(X_n, X_{n+1}, \ldots\right) \quad \text{for all} \quad n \geq 1. \qquad \blacksquare$$

The following definition and result will be needed later. Let $Z_1 : (\Omega, \mathcal{A}) \to (\Omega_1, \mathcal{A}'_1)$, measurable, $Z_2 : (\Omega, \mathcal{A}) \to (\Omega_2, \mathcal{A}'_2)$, measurable. Then

Definition 13. We say that Z_2 is a function of Z_1, if for all $\omega, \omega' \in \Omega$, $\omega \neq \omega'$ for which $Z_1(\omega) = Z_1(\omega')$, it follows that $Z_2(\omega) = Z_2(\omega')$. ■

Lemma 4. Let Z_1, Z_2 be defined as before, and suppose that the σ-field induced by Z_2, $\mathcal{A}_2 \subseteq \mathcal{A}_1$, the σ-field induced by Z_1. Then, if $\mathcal{A}'_2$ contains the singletons of points in the range of Z_2, it follows that

(i) Z_2 is a function of Z_1.

(ii) There exists a unique function Z on $(Z_1(\Omega), \mathcal{A}'_1 \cap Z_1(\Omega)$ to $(\Omega_2, \mathcal{A}'_2))$ defined by

$$Z(\omega_1) = Z_2(\omega; Z_1(\omega) = \omega_1), \quad \text{so that } Z_2 = Z(Z_1).$$

(iii) Z is $\mathcal{A}'_1 \cap Z_1(\Omega)$-measurable.

The proof of this lemma is left as an exercise (Exercise 7).

Proposition 19. Let Y be a r.v. defined on $(\Omega, \mathcal{A})$. Then Y is *invariant* (relative to $\mathbf{X}$), if and only if Y is $\mathcal{J}$-measurable, where $\mathcal{J}$ is the σ-field of invariant sets (relative to $\mathbf{X}$).

Proof. Let Y be invariant. Then there exists a measurable mapping φ on $(\Re^\infty, \mathcal{B}^\infty)$ into $(\Re, \mathcal{B})$ such that

$$Y = \varphi\left(X_n, X_{n+1}, \ldots\right) \quad \text{for all} \quad n \geq 1.$$

Let $C \in \mathcal{B}$. Then $A = Y^{-1}C = \left(X_n, X_{n+1}, \ldots\right)^{-1}(\varphi^{-1}C) = \left(X_n, X_{n+1}, \ldots\right)^{-1}B$, where $B \in \mathcal{B}^\infty$ and $A \in \mathcal{A}$. Thus, for every $A \in Y^{-1}\left(\mathcal{B}\right)$, there exists $B \in \mathcal{B}^\infty$ such that

$$A = \left(X_n, X_{n+1}, \ldots\right)^{-1} B \quad \text{for all} \quad n \geq 1, \text{ so that } A \in \mathcal{J};$$
$$\text{i.e., } Y \text{ is } \mathcal{J}\text{-measurable.}$$

Next, let Y be $\mathcal{J}$-measurable. Then $Y^{-1}\left(\mathcal{B}\right) \subseteq \mathcal{J}$. Make the following identification:

$$(\Omega_1, \mathcal{A}_1) = (\Re^\infty, \mathcal{B}^\infty), (\Omega_2, \mathcal{A}_2) = (\Re, \mathcal{B}), \quad Z_1 = \left(X_1, X_2, \ldots\right), \; Z_2 = Y.$$

From the definition of $\mathcal{J}$, it follows that

$$\mathcal{J} \subseteq \left(X_n, X_{n+1}, \ldots\right)^{-1} \mathcal{B}^\infty \quad \text{for all} \quad n \geq 1.$$

Thus,

$$Y^{-1}(\mathcal{B}) \subseteq (X_1, X_2, \ldots)^{-1} \mathcal{B}^{\infty}.$$

Since $\mathcal{B}$ contains all singletons in $\Re$, it follows that the Definition 13 applies and gives the existence of a measurable mapping φ defined on the range of $(X_1, X_2, \ldots)$ into $(\Re, \mathcal{B})$ such that

$$Y = \varphi(X_1, X_2, \ldots).$$

By stationarity of $\mathbf{X}$, it follows that the range of $(X_1, X_2, \ldots)$ is the same as that of $(X_n, X_{n+1}, \ldots)$ for all $n \geq 1$. Thus, $\varphi(X_n, X_{n+1}, \ldots)$ is well defined. In order to complete the proof, it suffices to show that

$$\varphi(X_1, X_2, \ldots) = \varphi(X_n, X_{n+1}, \ldots) \quad \text{for all} \quad n \geq 1.$$

To see this, let

$$B_x = \varphi^{-1}\{x\}, \quad x \in \Re, \text{ and let } A_x = (X_1, X_2, \ldots)^{-1} B_x.$$

Then $A_x = Y^{-1}\{x\}$, so that $A_x \in \mathcal{J}$. But then $A_x = (X_n, X_{n+1}, \ldots)^{-1} B_x$ for all $n \geq 1$, so that for $\omega \in A_x$, $(X_n(\omega), X_{n+1}(\omega), \ldots) \in B_x$ for all $n \geq 1$, and hence $\varphi\left[(X_n(\omega), X_{n+1}(\omega), \ldots)\right] = x$ for all $n \geq 1$. This completes the proof. ■

Now the Ergodic Theorem becomes as follows.

Theorem 3 (Ergodic Theorem). Let $\mathbf{X} = (X_1, X_2, \ldots)$ be a stationary process defined on $(\Omega, \mathcal{A}, P)$, and let $\mathcal{J}$ be the σ-field of invariant events (relative to $\mathbf{X}$). Then, if $|\mathcal{E}X_1| < \infty$, one has

$$\frac{1}{n}\sum_{j=1}^{n} X_j \underset{n\to\infty}{\overset{\text{a.s.}}{\to}} \mathcal{E}(X_1|\mathcal{J}). \quad ■ \tag{15.30}$$

Proof. Let $\hat{P}$ be the distribution of $\mathbf{X}$ under P, and let $\hat{\mathbf{X}}$ be the coordinate process. Then Theorem 1 applies to the process $\hat{\mathbf{X}}$ and gives that

$$\frac{1}{n}\sum_{j=1}^{n} \hat{X}_j \underset{n\to\infty}{\to} \hat{X} = \mathcal{E}\left(\hat{X}_1|\mathcal{J}_S\right) \text{ a.s.} \left[\hat{P}\right].$$

Let $B_0 \in \mathcal{B}^{\infty}$ be the exceptional set, and let $A_0 = \mathbf{X}^{-1}B_0$. Then $P(A_0) = \hat{P}(B_0) = 0$, and on A_0^c, $\frac{1}{n}\sum_{j=1}^{n} X_j$ converges to X, say. We shall show that $X = \mathcal{E}(X_1|\mathcal{J})$ a.s. We show that X is invariant, and hence $\mathcal{J}$-measurable, by Proposition 19. To this end, set

$$\varphi_{n,k} = \frac{X_n + \cdots + X_{n+k-1}}{k}, \quad n \geq 1. \quad \text{Then } X = \lim_{k\to\infty} \varphi_{1,k}$$

on A_0^c.

But

$$\lim_{k\to\infty} \varphi_{n,k} = \lim_{k\to\infty}\left(\frac{X_1+\cdots+X_{n+k-1}}{n+k-1}\times\frac{n+k-1}{k}-\frac{X_1+\cdots+X_{n-1}}{k}\right)$$
$$= \lim_{k\to\infty}\frac{X_1+\cdots+X_{n+k-1}}{n+k-1} = X \text{ on } A_0^c.$$

Thus, if we set $\lim_{k\to\infty}\varphi_{1,k} = \varphi(X_1, X_2, \ldots)$ on A_0^c, then $\lim_{k\to\infty}\varphi_{n,k} = \varphi(X_n, X_{n+1}, \ldots)$ on A_0^c, and both are equal to X; i.e., $(X =)\varphi(X_n, X_{n+1}, \ldots) = \varphi(X_1, X_2, \ldots), n \geq 1$ on A_0^c, and hence X is invariant.

From the fact that $X = \varphi(X_n, X_{n+1}, \ldots), n \geq 1$, and the convergences

$$\frac{1}{n}\sum_{j=1}^{n} X_j \underset{n\to\infty}{\to} X \text{ on } A_0^c, \quad \frac{1}{n}\sum_{j=1}^{n}\hat{X}_j \underset{n\to\infty}{\to} \hat{X} \text{ on } B_0^c,$$

where

$$A_0^c = \mathbf{X}^{-1}B_0^c \text{ with } P(A_0^c) = \hat{P}(B_0) = 0,$$

we get that

$$\hat{X} = \varphi\left(\hat{X}_1, \hat{X}_2, \ldots\right) \text{ a.s.} \left[\hat{P}\right]. \tag{15.31}$$

Thus, for every $A = \mathbf{X}^{-1}B,\ B \in \mathcal{B}^\infty$, we get

$$\int_A X dP = \int_A \varphi(X_1, X_2, \ldots) dP = \int_B \varphi\left(\hat{X}_1, \hat{X}_2, \ldots\right) d\hat{P} = \int_B \hat{X} d\hat{P}$$

is finite and

$$\int_A\left(\frac{1}{n}\sum_{j=1}^{n}X_j\right)dP = \int_B\left(\frac{1}{n}\sum_{j=1}^{n}\hat{X}_j\right)d\hat{P} \underset{n\to\infty}{\to} \int_B \hat{X} d\hat{P} = \int_A X dP,$$

by Corollary 4 (in Section 15.5), because

$$\left|\int_A\left[\left(\frac{1}{n}\sum_{j=1}^{n}\hat{X}_j\right) - \hat{X}\right]dP\right| \leq \int_A\left|\frac{1}{n}\sum_{j=1}^{n}\hat{X}_j - \hat{X}\right|dP$$
$$\leq \mathcal{E}\left|\frac{1}{n}\sum_{j=1}^{n}\hat{X}_j - \hat{X}\right| \underset{n\to\infty}{\to} 0,$$

so that

$$\int_A\left(\frac{1}{n}\sum_{j=1}^{n}X_j\right)dP = \frac{1}{n}\sum_{j=1}^{n}\int_A X_j dP \underset{n\to\infty}{\to} \int_A X dP. \tag{15.32}$$

This is true, in particular, for $A \in \mathcal{J}$. But then there exists $B \in \mathcal{B}^{\infty}$ such that $A = \left(X_k, X_{k+1}, \ldots\right)^{-1} B, k \geq 1$. By stationarity, we then obtain

$$\int_A X_k dP = \int_{(X_k, X_{k+1}, \ldots)^{-1}B} X_k dP = \int_{(X_1, X_2, \ldots)^{-1}B} X_1 dP = \int_A X_1 dP.$$

Therefore (15.32) implies that $\int_A X_1 dP = \int_A X dP$ for every $A \in \mathcal{J}$. It follows that $X = \mathcal{E}\left(X_1 | \mathcal{J}\right)$ a.s. ■

Corollary 1. The convergence in (15.30) holds true in the first mean.

Proof. We have $X = \varphi\left(X_1, X_2, \ldots\right)$, and $X = \mathcal{E}\left(X_1 | \mathcal{J}\right)$ a.s., so that $\mathcal{E}\left(X_1 | \mathcal{J}\right) = \varphi\left(X_1, X_2, \ldots\right)$ a.s. Thus,

$$\begin{aligned}
\mathcal{E}\left|\frac{1}{n}\sum_{j=1}^{n} X_j - \mathcal{E}\left(X_1 | \mathcal{J}\right)\right| &= \mathcal{E}\left|\frac{1}{n}\sum_{j=1}^{n} X_j - \varphi\left(X_1, X_2, \ldots\right)\right| \\
&= \int\left|\frac{1}{n}\sum_{j=1}^{n} X_j - \varphi\left(X_1, X_2, \ldots\right)\right| dP = \int\left|\frac{1}{n}\sum_{j=1}^{n} \hat{X}_j - \varphi\left(\hat{X}_1, \hat{X}_2, \ldots\right)\right| d\hat{P} \\
&= \int\left|\frac{1}{n}\sum_{j=1}^{n} \hat{X}_j - \hat{X}\right| d\hat{P}
\end{aligned}$$

(by (15.31)), and this tends to 0 as $n \to \infty$ by Corollary 4 (in Section 15.5). ■

Definition 14. The stationary process $\mathbf{X} = \left(X_1, X_2, \ldots\right)$ is said to be *ergodic*, if $P(A) = 0$ or 1 for every $A \in \mathcal{J}$ ■.

We then have the following corollary to Theorem 3.

Corollary 2. If $\mathbf{X} = \left(X_1, X_2, \ldots\right)$ of Theorem 3 is also ergodic, then (15.30) holds true with the right-hand side being replaced by $\mathcal{E}X_1$.

Proof. For $A \in \mathcal{J}, \int_A \mathcal{E}(X_1|\mathcal{J})dP = \int_A X_1 dP$. For $P(A) = 1, \int_A \mathcal{E}(X_1|\mathcal{J})dP = \int_\Omega \mathcal{E}(X_1|\mathcal{J})dP = \mathcal{E}[\mathcal{E}(X_1|\mathcal{J})] = \mathcal{E}X_1 = \int_\Omega \mathcal{E}X_1 dP = \int_A \mathcal{E}X_1 dP$; and for $P(A) = 0, \int_A \mathcal{E}(X_1|\mathcal{J})dP = 0 = \int_A \mathcal{E}X_1 dP$. Hence $\mathcal{E}(X_1|\mathcal{J}) = \mathcal{E}X_1$ a.s. ■

In Proposition 4, it was seen that functions defined on a stationary process also produce stationary processes. The same is true with regards to ergodicity. More precisely, one has

Proposition 20. Let $\mathbf{X} = \left(X_1, X_2, \ldots\right)$ be a stationary process defined on $(\Omega, \mathcal{A}, P)$ and φ be defined as follows:

$$\varphi : (\Re^m, \mathcal{B}^m) \to (\Re, \mathcal{B}), \quad \text{measurable, } 1 \leq m \leq \infty.$$

Set $Y_n = \varphi\left(X_n, X_{n+1}, \ldots, X_{n+m-1}\right), n \geq 1$, and $\mathbf{Y} = \left(Y_1, Y_2, \ldots\right)$.

Then

(i) The process $\mathbf{Y}$ is stationary.
(ii) If $\mathcal{J}_X$ and $\mathcal{J}_Y$ are the invariant σ-fields associated with the processes $\mathbf{X}$ and $\mathbf{Y}$, respectively, it follows that $\mathcal{J}_Y \subseteq \mathcal{J}_X$.
(iii) If $\mathbf{X}$ is ergodic, so is $\mathbf{Y}$.

Proof.

(i) It follows from Proposition 4.
(ii) Let $A \in \mathcal{J}_Y$. Then there exists $B \in \mathcal{B}^\infty$ such that $A = (Y_1, Y_2, \ldots)^{-1} B$, and actually,

$$A = (Y_n, Y_{n+1}, \ldots)^{-1} B, \quad n \geq 1.$$

Set

$$\varphi_j = \varphi(x_j, x_{j+1}, \ldots, x_{j+m-1}), \quad j \geq 1, \text{ and let } C = \{(x_n, x_{n+1}, \ldots);$$
$$(\varphi_n, \varphi_{n+1}, \ldots) \in B\} \in \mathcal{B}^\infty.$$

For $n \geq 1$,

$$\begin{aligned} A = (Y_n, Y_{n+1}, \ldots)^{-1} B &= (\varphi(X_n, \ldots, X_{n+m-1}), \\ &\quad \varphi(X_{n+1}, \ldots, X_{(n+1)+m-1}), \ldots)^{-1} B \\ &= (\varphi(X_n, \ldots, X_{n+m-1}), \varphi(X_{n+1}, \ldots, X_{(n+1)+m-1}), \ldots) \in B \\ &= (X_n, X_{n+1}, \ldots) \in C = (X_n, X_{n+1}, \ldots)^{-1} C. \end{aligned}$$

Thus, A is invariant with respect to $\mathbf{X}$, and hence $A \in \mathcal{J}_{\mathbf{X}}$.
(iii) It is immediate from (ii) and ergodicity of $\mathbf{X}$. ■

Exercises.

1. Let $(\Omega, \mathcal{A}, P) = ([0, 1), \mathcal{B}_{[0,1)}, \lambda)$ where λ is the Lebesgue measure, and let the transformation T be defined by

$$T(x) = x + \frac{1}{2}, \quad x \in \left[0, \frac{1}{2}\right), \quad T(x) = x - \frac{1}{2}, \quad x \in \left[\frac{1}{2}, 1\right).$$

Then show that T is measurable and measure-preserving.
2. Let $(\Omega, \mathcal{A}) = ([0, 1), \mathcal{B}_{[0,1)})$, and let the transformation T be defined by

$$T(x) = cx, \quad x \in [0, 1), \text{ where } c \text{ is a constant in } (0, 1).$$

Then show that there is no probability measure P on $\mathcal{B}_{[0,1)}$ such that $P(\{x\}) = 0, x \in [0, 1)$, and for which the transformation T is measure-preserving.
3. Refer to Exercise 1 and examine the transformation T from the ergodicity viewpoint.
4. Complete the proof of the converse part of Proposition 16.

5. Complete the proof of Lemma 2.

6. Work out the details of the proof of Proposition 18.

7. Complete the proof of Lemma 4.

Hint: (i) For $\omega, \omega' \in \Omega$ with $\omega \neq \omega'$ and $Z_1(\omega) = Z_1(\omega')$, suppose that $Z_2(\omega) \neq Z_2(\omega')$. Use the assumptions that $\mathcal{A}_2'$ contains the singletons of points in $Z_2(\Omega)$, and that $\mathcal{A}_2 \subseteq \mathcal{A}_1$, in order to arrive at a contradiction.

(ii) By part (i), Z is well defined, and $Z_2 = Z(Z_1)$. If also $Z_2 = Z'(Z_1)$, then show that $Z(\omega_1) = Z'(\omega_1)$ for all $\omega_1 \in Z_1(\Omega)$.

(iii) For $D \in \mathcal{A}_2'$, we have $\mathcal{A}_1 \subseteq \mathcal{A}_2 \ni A = Z_2^{-1}(D) = Z_1^{-1}(B)$ with $B = Z^{-1}(D) \subseteq Z_1(\Omega)$, and $A = Z_1^{-1}(C)$ for some $C \in \mathcal{A}_1'$. Conclude that $B = C \cap Z_1(\Omega)$.

8. If $X_n, n \geq 1$, are i.i.d. r.v.s defined on the probability space $(\Omega, \mathcal{A}, P)$, then

(i) Show that the process $\mathbf{X} = (X_1, X_2, \ldots)$ is ergodic.

(ii) Derive the Strong Law of Large Numbers under the assumption that $\mathcal{E}X_1$ is finite.

Hint: (i) By Proposition 3, it follows that $\mathbf{X} = (X_1, X_2, \ldots)$ is stationary. Let $\mathcal{J}$ be the σ-field of invariant sets relative to $\mathbf{X}$ (see Definition 11), and let $\mathcal{T}$ be the tail σ-field defined on $\mathbf{X} = (X_1, X_2, \ldots)$ (see Definition 1 and preceding discussion in Chapter 14). Then $\mathcal{J} \subseteq \mathcal{T}$. Conclude the discussion by using Theorem 10 in Chapter 14.

(ii) Refer to Corollary 2 of Theorem 3.

CHAPTER

16 Two Cases of Statistical Inference: Estimation of a Real-Valued Parameter, Nonparametric Estimation of a Probability Density Function

The objective of this chapter is to present some cases, where probabilistic results obtained in previous chapters are used for statistical inference purposes. This should enhance the interest of statisticians in the probability part of the book.

There are two instances of statistical inference considered here. One is estimation of a (real-valued) parameter by means of the so-called Maximum Likelihood Estimator (or Estimate) (MLE), and the other is the nonparametric estimation of an unknown probability density function (p.d.f.) by means of the kernel method of estimation. The former problem is discussed in Sections 16.1,16.2,16.3, and the latter in Sections 16.5 and 16.6.

16.1 Construction of an Estimate of a Real-Valued Parameter

As a brief introduction, let $X_1, \ldots, X_n$ be independent identically distributed (i.i.d.) observations (real-valued r.v.s) defined on the probability space $(\Omega, \mathcal{A}, P_\theta)$, where here θ is a real-valued parameter taking values in the parameter space Θ, an open subset of the real line $\Re$. The problem is to construct on estimator (or estimate) of θ; i.e., a (measurable) function of the X_is taking values in Θ. There are several ways of going about it, and one of the most popular is that of using a MLE (should such an estimate exist). That is, form the so-called likelihood function of the X_is—which in the present set-up is their joint p.d.f.—and then attempt to maximize it with respect to the parameter θ over the parameter space Θ. Should a unique maximizer exist, to be denoted by $\hat{\theta}_n = \hat{\theta}(X_1, \ldots, X_n)$, then $\hat{\theta}_n$ is proclaimed to be the MLE of θ. There are several reasons as to why such an estimate is desirable, but we are not going to elaborate on it here. Instead, we are going to give a set of conditions under which one may construct a sequence of roots, $\tilde{\theta}_n = \tilde{\theta}(X_1, \ldots, X_n)$, of the likelihood functions, which is a strongly consistent estimate of θ; i.e., $\tilde{\theta}_n \underset{n\to\infty}{\to} \theta$ a.s. $[P_\theta], \theta \in \Theta$.

An Introduction to Measure-Theoretic Probability, Second Edition. http://dx.doi.org/10.1016/B978-0-12-800042-7.00016-5

Under an enhanced set of assumptions, such an estimate is also asymptotically normal, when suitably normalized. These results also hold true for the MLE, should it exist.

16.2 Construction of a Strongly Consistent Estimate of a Real-Valued Parameter

Let $X_1, \ldots, X_n$ be i.i.d. r.v.s defined on the probability space $(\Omega, \mathcal{A}, P_\theta)$, where the probability measure P_θ depends on a real-valued parameter θ lying in the parameter space Θ, an open subset of $\Re$. Let X be a r.v. distributed as the X_is, and suppose it has a p.d.f. $f(\cdot;\theta)$ (a Radon-Nikodym derivative with respect to Lebesgue measure; see Remark 1, Chapter 7). We intend to give sufficient conditions under which a strongly consistent estimate of θ may be constructed, based on the likelihood function. Below, we list perhaps the most economical conditions under which this happens.

Assumptions.

(A1) Assume Θ to be an open interval in $\Re$.
(A2) Suppose that the positivity set of $f(\cdot;\theta)$ is independent of θ; i.e., $S = \{x \in \Re;\ f(x;\theta) > 0\}$ is independent of θ.
(A3) Suppose that the p.d.f. $f(\cdot;\theta)$ is *identifiable*; i.e., for every θ_1 and θ_2 in Θ with $\theta_1 \neq \theta_2$, there exists a Borel set $B(\subseteq S)$ with $P_\theta(X \in B) > 0$ for all $\theta \in \Theta$ such that $f(x;\theta_1) \neq f(x;\theta_2)$ for $x \in B$.
(A4) Suppose that the derivative $\frac{\partial}{\partial\theta} f(x;\theta)$ exists for all $\theta \in \Theta$ and all $x \in S$.
(A5) Let θ_0 be the unknown but true value of the parameter θ, and for any $\theta \in \Theta$, set

$$K(\theta_0, \theta) = \mathcal{E}_{\theta_0} \log \frac{f(X;\theta_0)}{f(X;\theta)} \tag{16.1}$$

(where log stands throughout for the natural logarithm). Then assume that $K(\theta_0, \theta) < \infty$.

In the proof of the consistency theorem stated below, the following result is used which is stated here as a lemma.

Lemma 1. Let $K(\theta_0, \theta)$ be defined as in (16.1). Then, under assumptions (A2) and (A3), it follows that $K(\theta_0, \theta) > 0$.

Proof. (*Outline*). Set $y = -\log x (x > 0)$. Then $\frac{d^2y}{dx^2} = x^{-2} > 0$, so that this function is convex. Then apply Jensen's inequality (see Theorem 15, Chapter 6) with $g(X) = -\log\,[f(X;\theta_1)/f(X;\theta_0)]$ to obtain

$$-\log\ \mathcal{E}_{\theta_0}[f(X;\theta_1)/f(X;\theta_0)] \leq \mathcal{E}_{\theta_0}\{-\log[f(X;\theta_1)/f(X;\theta_0)]\} = K(\theta_0, \theta_1).$$

However,

$$\mathcal{E}_{\theta_0}[f(X;\theta_1)/f(X;\theta_0)] = \int_S \left[\frac{f(x;\theta_1)}{f(x;\theta_0)} f(x;\theta_0)\right] dx$$

$$= \int_S f(x;\theta_1)dx = \int_{\Re} f(x;\theta_1)dx = 1,$$

so that $-\log\, 1 = 0 \le K(\theta_0,\theta_1)$. By (A3), it is seen that $K(\theta_0,\theta_1) > 0$. ■

In all that follows, we work on the set S (see Assumption (A2)), θ_0 stands for the true (but unknown) value of the parameter θ, $\mathbf{X}_n$ stands for the random vector $(X_1,\ldots,X_n)$, and δ is an arbitrary positive number. The likelihood function is $L_n(\theta|\mathbf{X}_n) = \prod_{i=1}^n f(X_i;\theta)$, and the log-likelihood function is $\ell_n(\theta|\mathbf{X}_n) = \log\, L_n(\theta|\mathbf{X}_n)$.

We may now state the consistency result alluded to earlier.

Theorem 1. Under assumptions (A1)–(A5), we can construct a sequence $\{\widetilde{\theta}_n\}$, $n \ge 1$, of estimates of θ_0, where $\widetilde{\theta}_n$ is a root of the log-likelihood equation $\frac{\partial}{\partial\theta}\ell_n(\widetilde{\theta}_n|\mathbf{X}_n) = 0$ on a set of P_{θ_0}-probability 1, such that $\widetilde{\theta}_n \xrightarrow[(P_{\theta_0})]{a.s.} \theta_0$ as $n\to\infty$ (i.e., $\widetilde{\theta}_n \underset{n\to\infty}{\longrightarrow} \theta_0$ on a set of P_{θ_0}-probability 1).

Proof. The proof consists of two parts. First, it is shown that, for $\delta > 0$ sufficiently small, there exists a positive integer n_δ such that, for $n \ge n_\delta$ and on a set of P_{θ_0}-probability 1,

$$\frac{\partial}{\partial\theta}\ell_n(\theta_\delta|\mathbf{X}_n) = 0, \quad |\theta_\delta - \theta_0| < \delta,$$

where θ_δ is a (local) maximum of the log-likelihood function $\ell_n(\theta|\mathbf{X}_n)$. Next, suitably exploiting this result, we proceed with the construction of the $\widetilde{\theta}_n$'s as described. To this end, set $I_\delta = (\theta_0-\delta, \theta_0+\delta)$, $\bar{I}_\delta = [\theta_0-\delta, \theta_0+\delta]$, and take δ small enough, so that $\bar{I}_\delta \subset \Theta$. Also, set

$$J_{+\delta} = -K(\theta_0,\theta_0{+}\delta) = \mathcal{E}_{\theta_0}\log\frac{f(X;\theta_0+\delta)}{f(X;\theta_0)} = \int_S\left[\log\frac{f(x;\theta_0+\delta)}{f(x;\theta_0)}\right]f(x;\theta_0)\,dx, \tag{16.2}$$

so that $J_{+\delta} \le 0$ (by Lemma 1), and indeed, $J_{+\delta} < 0$ on account of (A3). Then

$$\frac{1}{n}\left[\ell_n(\theta_0+\delta|\mathbf{X}_n) - \ell_n(\theta_0|\mathbf{X}_n)\right] = \frac{1}{n}\log\prod_{i=1}^n\frac{f(X_i;\theta_0+\delta)}{f(X_i;\theta_0)}$$

$$= \frac{1}{n}\sum_{i=1}^n\log\frac{f(X_i;\theta_0+\delta)}{f(X_i;\theta_0)} \underset{n\to\infty}{\longrightarrow} J_{+\delta} \text{ on a set } N_{+\delta}^c,$$

say, with $P_{\theta_0}(N_{+\delta}^c) = 1$ (by the SLLN, see Theorem 4 in Chapter 14). Thus, for every $\omega \in N_{+\delta}^c$, there exists $n_{+\delta} = n_{+\delta}(\omega)$ positive integer such that

$$\left|\frac{1}{n}\left[\ell_n(\theta_0+\delta|\mathbf{X}_n) - \ell_n(\theta_0|\mathbf{X}_n)\right] - J_{+\delta}\right| < -\frac{1}{2}J_{+\delta} \text{ for } n \ge n_{+\delta}. \tag{16.3}$$

From this point on, we work with $\omega \in N_{+\delta}^c$ and $n \ge n_{+\delta}$. The inequality in (16.3) implies

$$\frac{1}{n}\left[\ell_n(\theta_0+\delta|\mathbf{X}_n) - \ell_n(\theta_0|\mathbf{X}_n)\right] < \frac{1}{2}J_{+\delta},$$

or

$$\ell_n(\theta_0 + \delta|\mathbf{X}_n) < \ell_n(\theta_0|\mathbf{X}_n) + \frac{n}{2}J_{+\delta},$$

and hence

$$\ell_n(\theta_0 + \delta|\mathbf{X}_n) < \ell_n(\theta_0|\mathbf{X}_n) \quad (\text{since } J_{+\delta} < 0). \tag{16.4}$$

At this point, refer to relation (16.2), set $J_{-\delta} = -K(\theta_0, \theta_0 - \delta)$, and work as above in order to conclude that

$$\ell_n(\theta_0 - \delta|\mathbf{X}_n) < \ell_n(\theta_0|\mathbf{X}_n) \tag{16.5}$$

for $\omega \in N^c_{-\delta}$ with $P_{\theta_0}(N^c_{-\delta}) = 1$ and $n \geq n_{-\delta} = n_{-\delta}(\omega)$ positive integer. Set $N^c_\delta = N^c_{+\delta} \cap N^c_{-\delta}$ (so that $P_{\theta_0}(N^c_\delta) = 1$), and $n_\delta = \max(n_{+\delta}, n_{-\delta})$. Then, for $\omega \in N^c_\delta$ and $n \geq n_\delta$, both relations (16.4) and (16.5) hold. Now, look at $\ell_n(\theta|\mathbf{X}_n)$ as a function of $\theta \in \bar{I}_\delta$ (always with $\omega \in N^c_\delta$ and $n \geq n_\delta$). By (A4), $\ell_n(\theta|\mathbf{X}_n)$ is continuous in θ (by differentiability). Then $\ell_n(\theta|\mathbf{X}_n)$ attains (at least) one (local) maximum; i.e., there is $\theta_\delta = \theta_\delta(\mathbf{X}_n)$ with $\theta_\delta \in I_\delta$ (open interval, not closed $\bar{I}_\delta$), because of inequalities (16.4) and (16.5). So, $\ell_n(\theta_\delta|\mathbf{X}_n)$ is a (local) maximum of $\ell_n(\theta|\mathbf{X}_n), \theta \in I_\delta$, and of course, $|\theta_\delta - \theta_0| < \delta$. On the other hand, any local maxima are roots of the log-likelihood equation $\frac{\partial}{\partial\theta}\ell_n(\theta|\mathbf{X}_n) = 0$ (because of the differentiability of $\ell_n(\theta|\mathbf{X}_n)$). Thus we have

$$\frac{\partial}{\partial\theta}\ell_n(\theta_\delta|\mathbf{X}_n) = 0, \quad |\theta_\delta - \theta| < \delta \tag{16.6}$$

(on the set N^c_δ with $P_{\theta_0}(N^c_\delta) = 1$, and for $n \geq n_\delta$).

We now embark on the construction of the desired sequence $\{\tilde{\theta}_n\}, n \geq 1$. To this end, for each $k = 1, 2, \ldots$, select $(0 <)\delta_k < \delta$ with $\delta_k \downarrow 0$, and consider the respective intervals $I_{\delta_k}, \bar{I}_{\delta_k}$, as well as the sets $N^c_{\delta_k}$ with $P_{\theta_0}(N^c_{\delta_k}) = 1$ for all k. Set $N^c = \cap_{k=1}^\infty N^c_{\delta_k}$, so that $P_{\theta_0}(N^c) = 1$ (see also Exercise 4, Chapter 2), and from this point on, work with $\omega \in N^c$. On the basis of the arguments in the first part of this proof, for each $k \geq 1$, there exists $n_{\delta_k}(\omega) = n_{\delta_k} = n_k$ positive integer, such that for $n \geq n_k$, there exists (at least) one (local) maximum $\theta_{\delta_k}(\mathbf{X}_n) = \theta_{\delta_k}$ in the interval $(\theta_0 - \delta_k, \theta_0 + \delta_k)$ with the properties (see also relations in (16.6)),

$$\frac{\partial}{\partial\theta}\ell_n\left(\theta_{\delta_k}|\mathbf{X}_n\right) = 0, \quad \left|\theta_{\delta_k} - \theta_0\right| < \delta_k. \tag{16.7}$$

Consider the subsequence $\{n_k\} \subseteq \{n\}$, and without loss of generality, we may assume that $n_1 < n_2 < \cdots \uparrow \infty$. Next, for each $n = 1, 2, \ldots$, apply the following rule: Look at $n_i < n_{i+1}, i \geq 1$, and for all those ns with $n_i \leq n < n_{i+1}$, take $\tilde{\theta}_n = \theta_{\delta_i}$. Then

$$\frac{\partial}{\partial\theta}\ell_n\left(\tilde{\theta}_n|\mathbf{X}_n\right) = \frac{\partial}{\partial\theta}\ell_n\left(\theta_{\delta_i}|\mathbf{X}_n\right) = 0, \quad \left|\tilde{\theta}_n - \theta_0\right| = \left|\theta_{\delta_i} - \theta_0\right| < \delta_i. \tag{16.8}$$

Summarizing the results and taking into consideration relation (16.8), we have then that, for $\omega \in N^c$ with $P_{\theta_0}(N^c) = 1$, there exists a sequence of roots, $\{\tilde{\theta}_n\}, n \geq 1$, of the log-likelihood equation such that $\tilde{\theta}_n \to \theta_0$ as $n \to \infty$. In other words, $\tilde{\theta}_n \to \theta_0$ a.s. (with respect to P_{θ_0}-probability), as was to be seen. ■

To the above theorem, we have the following corollary.

Corollary. Suppose that for all sufficiently large n the MLE $\widehat{\theta}_n$ is the unique root of the log-likelihood equation. Then, under assumptions (A1)–(A5), $\widehat{\theta}_n \underset{(P_{\theta_0})}{\overset{a.s.}{\longrightarrow}} \theta_0$ as $n \to \infty$.

Proof. All arguments employed in the proof of the theorem apply when $\widetilde{\theta}_n$ is replaced by $\widehat{\theta}_n$. In particular, relation (16.8) becomes, on a set of P_{θ_0}-probability 1,

$$\frac{\partial}{\partial \theta} \ell_n \left(\widehat{\theta}_n | \mathbf{X}_n \right) = \frac{\partial}{\partial \theta} \ell_n \left(\theta_{\delta_i} | \mathbf{X}_n \right) = 0, \quad \left| \widehat{\theta}_n - \theta_0 \right| = \left| \theta_{\delta_i} - \theta_0 \right| < \delta_i.$$

Given that $\delta_i \to 0$ as $i \to \infty$, the proof follows. ■

Remark 1. If the parameter space Θ is r-dimensional ($r \geq 1$), then in assumption (A4) the derivative is replaced by the partial derivatives. The proof of Lemma 1 is independent of the dimensionality of Θ. The absolute value in Θ is replaced by the usual Euclidean distance, denote it by $|\cdot|$, and all arguments in the proof go through. Then the conclusion $\widetilde{\theta}_n \underset{(P_{\theta_0})}{\overset{a.s.}{\longrightarrow}} \theta_0$, equivalently, $|\widetilde{\theta}_n - \theta_0| \underset{(P_{\theta_0})}{\overset{a.s.}{\longrightarrow}} 0$, becomes $|\widetilde{\theta}_n - \theta_0| \underset{(P_{\theta_0})}{\overset{a.s.}{\longrightarrow}} 0$ in the new setting, as $n \to \infty$. In other words, the conclusion of the theorem still holds true.

16.3 Some Preliminary Results

The lemma to be established below is employed in the proof of a certain uniform version of the SLLN, as well as in the proof of the asymptotic normality of the MLE. The lemma is formulated in a way to cover the case of random vectors, and the case that the parameter θ is multidimensional.

Lemma 2. Let T and $\mathcal{X}$ be subsets of finite-dimensional Euclidean spaces, let φ be a real-valued function defined on the product space $\mathcal{X} \times T$, and let X be a random vector defined on the probability space $(\Omega, \mathcal{A}, P)$ and taking values in $\mathcal{X}$. Make the following assumptions:

- **(i)** T is compact.
- **(ii)** For every $x \in \mathcal{X}, \varphi(x, t)$ is continuous in t (and measurable in x for every $t \in T$).
- **(iii)** $|\varphi(x, t)| \leq h(x), t \in T$, for some nonnegative (measurable) function h defined on $\mathcal{X}$ with $\mathcal{E}h(X) < \infty$.

Let $X_1, \ldots, X_n$ be independent random vectors distributed as X, and let $\mu(t) = \mathcal{E}\varphi(X, t)$. Then, with probability 1,

$$\limsup_{n \to \infty} \left\{ \sup_{t \in T} \left[\frac{1}{n} \sum_{j=1}^{n} \varphi(X_j, t) \right] \right\} \leq \sup_{t \in T} \mu(t).$$

Proof. For some (small) $\rho > 0$, set $\psi(x, t, \rho) = \sup_{|t'-t|<\rho} \varphi(x, t')$, where $|\cdot|$ is the usual distance in the Euclidean space $\mathcal{X}$. Then $\psi(x, t, \rho) \downarrow \varphi(x, t)$ as $\rho \downarrow 0$, by the assumed continuity in t of $\varphi(x,t)$. All arguments below hold for all elements of T. Thus, $\varphi(x,t') \leq h(x)$ implies $\sup_{|t'-t|<\rho} \varphi(x,t') \leq h(x)$ or $\psi(x,t,\rho) \leq h(x)$, equivalently, $-h(x) \leq -\psi(x,t,\rho)$ and $-\psi(x,t,\rho)\uparrow -\varphi(x,t)$ as $\rho\downarrow 0$. Replacing x by X, we have then $-h(X) \leq -\psi(X,t,\rho)$ and $-\psi(X,t,\rho)\uparrow -\varphi(X,t)$ as $\rho\downarrow 0$. Then $\mathcal{E}[-\psi(X,t,\rho)]\uparrow\mathcal{E}[-\varphi(X,t)]$, or equivalently, $\mathcal{E}\psi(X,t,\rho)\downarrow\mathcal{E}\varphi(X,t)$ as $\rho\downarrow 0$. (This is so by Corollary 2, Chapter 5) Or, $\mathcal{E}\psi(X,t,\rho)\downarrow\mu(t)$ as $\rho\downarrow 0$. Thus, for an arbitrary (small) $\varepsilon > 0$ and each $t \in T$, there exists a $\rho_t > 0$ sufficiently small, such that

$$\mathcal{E}\,\psi(X,t,\rho_t) < \mu(t) + \varepsilon.$$

The collection of (balls) $\{S_t = \{t';\, |t'-t| < \rho_t\}, t \in T\}$, clearly, cover T. Then, by assumption (i), there is a finite number (of these balls) $S_{t_i}, i = 1,\ldots,m$ whose union is T. (This is so by the Heine–Borel covering theorem.) Now, it is clear that for each $t \in T$, there is at least one $i = 1,\ldots,m$, such that $t \in S_{t_i}$. Then, from the definition of $\psi(x,t,\rho)$, it follows that $\varphi(x,t) \leq \psi(x,t_i,\rho_{t_i}), t \in S_{t_i}$. Replacing x by X_j, we get

$$\varphi(X_j,t) \leq \psi(X_j,t_i,\rho_{t_i}), \quad t \in S_{t_i}, \quad j = 1,\ldots,n,$$

and hence

$$\frac{1}{n}\sum_{j=1}^{n}\varphi(X_j,t) \leq \frac{1}{n}\sum_{j=1}^{n}\psi(X_j,t_i,\rho_{t_i}), \quad t \in S_{t_i}.$$

It follows that

$$\sup_{t\in T}\left[\frac{1}{n}\sum_{j=1}^{n}\varphi(X_j,t)\right] \leq \sup_{t\in T}\left[\frac{1}{n}\sum_{j=1}^{n}\psi(X_j,t_i,\rho_{t_i})\right]$$
$$= \max_{1\leq i\leq m}\left[\frac{1}{n}\sum_{j=1}^{n}\psi(X_j,t_i,\rho_{t_i})\right].$$

However, by the SLLN,

$$\frac{1}{n}\sum_{j=1}^{n}\psi(X_j,t_i,\rho_{t_i}) \underset{n\to\infty}{\overset{a.s.}{\longrightarrow}} \mathcal{E}\psi(X,t_i,\rho_{t_i}), \quad i = 1,\ldots,m,$$

and hence

$$\max_{1\leq i\leq m}\left[\frac{1}{n}\sum_{j=1}^{n}\psi(X_j,t_i,\rho_{t_i})\right] \underset{n\to\infty}{\overset{a.s.}{\longrightarrow}} \max_{1\leq i\leq m}\mathcal{E}\psi(X,t_i,\rho_{t_i}).$$

On the other hand, for sufficiently large n,

$$\mathcal{E}\,\psi(X,t_i,\rho_{t_i}) < \mu(t_i) + \varepsilon, \quad i = 1,\ldots,m,$$

so that

$$\max_{1\leq i\leq m} \mathcal{E}\psi(X, t_i, \rho_{t_i}) \leq \max_{1\leq i\leq m} [\mu(t_i) + \varepsilon]$$
$$= \max_{1\leq i\leq m} \mu(t_i) + \varepsilon \leq \sup_{t\in T} \mu(t) + \varepsilon.$$

Hence, for sufficiently large n,

$$\max_{1\leq i\leq m} \left[\frac{1}{n}\sum_{j=1}^{n} \psi(X_j, t_i, \rho_{t_i})\right] \leq \sup_{t\in T} \mu(t) + \varepsilon \quad a.s.,$$

and then

$$\sup_{t\in T} \left[\frac{1}{n}\sum_{j=1}^{n} \varphi(X_j, t)\right] \leq \sup_{t\in T} \mu(t) + \varepsilon \quad a.s.$$

Taking the limsup as $n\to\infty$, we get

$$\text{limsup} \left\{ \sup_{t\in T} \left[\frac{1}{n}\sum_{j=1}^{n} \varphi(X_j, t)\right]\right\} \leq \sup_{t\in T} \mu(t) + \varepsilon \quad a.s.$$

Finally, letting $\varepsilon\to 0$, we obtain

$$\text{limsup} \left\{ \sup_{t\in T} \left[\frac{1}{n}\sum_{j=1}^{n} \varphi(X_j, t)\right]\right\} \leq \sup_{t\in T} \mu(t) \quad a.s. \qquad \blacksquare$$

The result below is a uniform version of the SLLN, which, of course, is interesting on its own right; it will also be employed in establishing asymptotic normality of the strongly consistent estimates of Theorem 1 and of the MLE, should it exist.

Proposition 1. In the notation of Lemma 2 and under the assumptions (i)–(iii) made there, it holds

$$\lim_{n\to\infty} \left[\sup_{t\in T} \left|\frac{1}{n}\sum_{j=1}^{n} \varphi(X_j, t) - \mu(t)\right|\right] = 0 \quad a.s.$$

(Thus, as $n\to\infty$, $\frac{1}{n}\sum_{j=1}^{n} \varphi(X_j, t) \xrightarrow{a.s.} \mu(t)$ uniformly in $t \in T$.)

Proof. In the first place, the function $\mu(t) = \mathcal{E}\varphi(X, t)$ is continuous. Indeed, $\varphi(X, t^*) \to \varphi(X, t^*)$ as $t^* \to t$ (by assumption (ii)), and $|\varphi(X, t^*)| \leq h(X)$, independent of t^*, with $\mathcal{E}h(X) < \infty$ (by assumption (iii)). It follows that

$$\mu(t^*) = \mathcal{E}\varphi(X, t^*) \underset{t^*\to t}{\longrightarrow} \mathcal{E}\varphi(X, t) = \mu(t).$$

(This is so by the Dominated Convergence Theorem, Theorem 3, Chapter 5.) Next, suppose that the proposition is true when $\mu(t) = 0$; i.e., we suppose that

$$\lim_{n\to\infty}\left[\sup_{t\in T}\left|\frac{1}{n}\sum_{j=1}^{n}\varphi(X_j,t)\right|\right] = 0 \quad a.s.,$$

and replace $\varphi(x,t)$ by $\psi(x,t) \stackrel{def}{=} \varphi(x,t) - \mu(t)$. Clearly, $\psi(x,t)$ satisfies assumption (ii), and

$$|\psi(x,t)| = |\varphi(x,t) - \mu(t)| \leq |\varphi(x,t)| + |\mu(t)| \leq h(x) + \mathcal{E}h(X),$$

since $|\varphi(X,t)| \leq h(X)$ is equivalent to $-h(X) \leq \varphi(X,t) \leq h(X)$ and hence $-\mathcal{E}h(X) \leq \mathcal{E}\varphi(X,t) \leq \mathcal{E}h(X)$, or $|\mathcal{E}\varphi(X,t)| \leq \mathcal{E}h(X)$, equivalently, $|\mu(t)| \leq \mathcal{E}h(X)$. So $\psi(x,t)$ also satisfies assumption (iii) (with $h(x)$ replaced by $h(x) + \mathcal{E}h(X)$). Then

$$\lim_{n\to\infty}\left[\sup_{t\in T}\left|\frac{1}{n}\sum_{j=1}^{n}\psi(X_j,t)\right|\right] = 0 \quad a.s.,$$

or

$$\lim_{n\to\infty}\left\{\sup_{t\in T}\left|\frac{1}{n}\sum_{j=1}^{n}\left[\varphi(X_j,t) - \mu(t)\right]\right|\right\}$$
$$= \lim_{n\to\infty}\left[\sup_{t\in T}\left|\frac{1}{n}\sum_{j=1}^{n}\varphi(X_j,t) - \mu(t)\right|\right] = 0 \quad a.s. \tag{16.9}$$

In the sequel, we work with $\psi(x,t)$ and apply Lemma 2 to get

$$\limsup_{n\to\infty}\left\{\sup_{t\in T}\left[\frac{1}{n}\sum_{j=1}^{n}\psi(X_j,t)\right]\right\} \leq 0 \quad a.s. \tag{16.10}$$

Clearly, the function $-\psi(x,t)$ also satisfies assumptions (ii)–(iii) (and $\mathcal{E}[-\psi(X,t)] = 0$). Then Lemma 2 yields

$$\limsup_{n\to\infty}\left\{\sup_{t\in T}\left[-\frac{1}{n}\sum_{j=1}^{n}\psi(X_j,t)\right]\right\} \leq 0 \quad a.s. \tag{16.11}$$

On the basis of the last two conclusions and Remark 2 below, we have

$$0 \leq \limsup_{n\to\infty}\left\{\sup_{t\in T}\left|\frac{1}{n}\sum_{j=1}^{n}\psi(X_j,t)\right|\right\}$$

$$\leq \limsup_{n\to\infty}\left\{\sup_{t\in T}\left[\frac{1}{n}\sum_{j=1}^{n}\psi(X_j,t)\right]\right\}+\limsup_{n\to\infty}\left\{\sup_{t\in T}\left[-\frac{1}{n}\sum_{j=1}^{n}\psi(X_j,t)\right]\right\}$$
$$\leq 0 \quad a.s. \tag{16.12}$$

Therefore the $\lim_{n\to\infty}\left[\sup_{t\in T}\left|\frac{1}{n}\sum_{j=1}^{n}\psi(X_j,t)\right|\right]$ exists with probability 1 and equals 0. Reverting to the function $\varphi(x,t)$, we obtain the desired result, as is explained in relation (16.9).

The completion of the proof requires justification of the second inequality on the right-hand side in relation (16.12). This is done below.

Remark 2. For any functions $g_n(t), t \in T, n \geq 1$, the following are true:

$$\sup_{t\in T}|g_n(t)| = \max\left\{\sup_{t\in T} g_n(t), \sup_{t\in T}[-g_n(t)]\right\},$$

and

$$\limsup_{n\to\infty}\left[\sup_{t\in T}|g_n(t)|\right] \leq \limsup_{n\to\infty}\left[\sup_{t\in T} g_n(t)\right] + \limsup_{n\to\infty}\left\{\sup_{t\in T}[-g_n(t)]\right\}.$$

The first assertion is immediate. As for the second, we have: Set

$$\alpha_n = \sup_{t\in T} g_n(t), \quad \beta_n = \sup_{t\in T}[-g_n(t)],$$

and $\alpha = \limsup_{n\to\infty} \alpha_n$, $\beta = \limsup_{n\to\infty} \beta_n$. Then, from the definition of the $\limsup_{n\to\infty}$, we have that, for every $\varepsilon > 0$ and all sufficiently large n, $\alpha_n \leq \alpha+\varepsilon$, $\beta_n \leq \beta+\varepsilon$. Hence $\max\{\alpha_n, \beta_n\} \leq (\alpha+\beta)+\varepsilon$, and therefore

$$\limsup_{n\to\infty}[\max\{\alpha_n,\beta_n\}] \leq (\alpha+\beta)+\varepsilon$$
$$= \left(\limsup_{n\to\infty}\alpha_n + \limsup_{n\to\infty}\beta_n\right)+\varepsilon.$$

Letting $\varepsilon\to 0$, we get the desired result. ■

16.4 Asymptotic Normality of the Strongly Consistent Estimate

The objective here is to establish asymptotic normality for the strongly consistent estimate constructed in Theorem 1 and of the MLE, should it exist, under suitable conditions. For this purpose, we review the relevant notation and list conditions under which this result may be established. To this end, let X be a r.v. defined on the probability space $(\Omega, \mathcal{A}, P_\theta)$, where the parameter θ belongs in the parameter space Θ.

Assumptions.

(B1) The parameter space Θ is an open subset of $\Re$.

(B2) The set $S = \{x \in \Re;\ f(x;\theta) > 0\}$ is independent of θ.

(B3) The p.d.f. $f(\cdot;\theta)$ is *identifiable*; i.e., for every θ_1 and θ_2 in Θ with $\theta_1 \neq \theta_2$, there exists a set $B(\subseteq S)$ with $P_\theta(X \in B) > 0$ for all $\theta \in \Theta$ such that $f(x;\theta_1) \neq f(x;\theta_2)$ for $x \in B$.

(B4) The derivative $\frac{\partial}{\partial\theta} f(x;\theta)$ exists for all θ and all $x \in S$ and

$$\frac{\partial}{\partial\theta}\int_S f(x;\theta)dx = \int_S \frac{\partial}{\partial\theta} f(x;\theta)dx.$$

(B5) Let θ_0 be the (unknown) true value of the parameter θ, and set

$$K(\theta_0,\theta) = \mathcal{E}_{\theta_0}\left[\log\frac{f(X;\theta_0)}{f(X;\theta)}\right], \quad \theta \in \Theta.$$

Then it is assumed that $0 < K(\theta_0,\theta) < \infty$.

(B6) The derivative $\frac{\partial^2}{\partial\theta^2} f(x;\theta)$ exists for all θ and all $x \in S$, is continuous in θ, and

$$\frac{\partial^2}{\partial\theta^2}\int_S f(x;\theta)dx = \int_S \frac{\partial^2}{\partial\theta^2} f(x;\theta)dx.$$

(B7) Set

$$I(\theta_0) = \mathcal{E}_{\theta_0}\left[\frac{\partial}{\partial\theta}\log f(X;\theta)\Big|_{\theta=\theta_0}\right]^2$$

for the Fisher information number. Then it is assumed that $0 < I(\theta_0) < \infty$.

(B8) For each $\theta \in \Theta$, there is a compact neighborhood $N(\theta)$ with θ belonging in the interior of $N(\theta)$, such that $\left|\frac{\partial^2}{\partial\theta^2}\log f(x;\theta')\right| \leq H(x)$ for all θ' in $N(\theta)$ and all $x \in S$, for some bounding (measurable) function H with $\mathcal{E}_\theta H(X) < \infty$.

Notice that (B1)–(B3) are the same as (A1)–(A3), and (B4) is a strengthening of (A4). It has already been seen (see Theorem 1) that, under assumptions (B1)–(B5), there is at least one root $\widetilde{\theta}_n$ of the likelihood equation, with probability 1, so that $\widetilde{\theta}_n \underset{(P_{\theta_0})}{\xrightarrow{a.s.}} \theta_0$ as $n\to\infty$. Here it will be shown that this sequence is also asymptotically normal. Namely,

Theorem 2. Let θ_0 be the (unknown) true value of the parameter θ, and let $\{\widetilde{\theta}_n\}$ be a sequence as in Theorem 1. Then, under assumptions (B1)–(B8),

$$\sqrt{n}\left(\widetilde{\theta}_n - \theta_0\right) \underset{(P_{\theta_0})}{\xrightarrow{d}} N(0, 1/I(\theta_0)), \quad \text{as } n\to\infty.$$

The following results, stated as lemmas, are used on many occasions, and either directly or indirectly in the proof of Theorem 2.

Lemma 3. Let X be a r.v. defined on the probability space $(\Omega, \mathcal{A}, P_\theta)$, $\theta \in \Theta$, with p.d.f. $f(\cdot;\theta)$. Then:

(i) Under assumptions (B1), (B2), and (B4),

$$\mathcal{E}_\theta \left[\frac{\partial}{\partial\theta} \log f(X;\theta)\right] = 0.$$

(ii) Under assumptions (B1), (B2), and (B6),

$$\begin{aligned} I(\theta) &= \int_{\Re} \left[\frac{\partial}{\partial\theta} \log f(x;\theta)\right]^2 f(x;\theta)\,dx = \mathcal{E}_\theta \left[\frac{\partial}{\partial\theta} \log f(X;\theta)\right]^2 \\ &= \mathrm{Var}_\theta \left[\frac{\partial}{\partial\theta} \log f(X;\theta)\right] \quad \text{(by part (i))} \\ &= -\mathcal{E}_\theta \left[\frac{\partial^2}{\partial\theta^2} \log f(X;\theta)\right]. \end{aligned}$$

(iii) Under assumptions (B1), (B2), (B4), and (B6),

$$\frac{1}{\sqrt{n}} \sum_{j=1}^{n} \frac{\partial}{\partial\theta} \log f(X_j;\theta) \underset{(P_{\theta_0})}{\overset{d}{\longrightarrow}} N(0,\ I(\theta)), \text{ as } n \to \infty,$$

where $X_1, \ldots, X_n$ are independent r.v.s distributed as the r.v. X.

Proof.

(i) We have

$$\begin{aligned} 1 &= \int_S f(x;\theta)\,dx, \quad \text{so that by differentiation,} \\ 0 &= \frac{d}{d\theta}\int_S f(x;\theta)\,dx = \int_S \frac{\partial}{\partial\theta} f(x;\theta)dx \text{ (by Theorem 4, Chapter 5)} \\ &= \int_S \left[\frac{\partial f(x;\theta)}{\partial\theta} \Big/ f(x;\theta)\right] f(x;\theta)dx \\ &= \int_S \left[\frac{\partial}{\partial\theta} \log f(x;\theta)\right] f(x;\theta)\,dx = \mathcal{E}_\theta \left[\frac{\partial}{\partial\theta} \log f(X;\theta)\right]; \quad \text{i.e.,} \end{aligned}$$

$\mathcal{E}_\theta \left[\frac{\partial}{\partial\theta} \log f(X;\theta)\right] = 0$, as was to be seen.

(ii) Once again,

$$\begin{aligned} 1 &= \int_S f(x;\theta)\,dx, \quad \text{and hence} \\ 0 &= \frac{d^2}{d\theta^2}\int_S f(x;\theta)\,dx = \int_S \frac{\partial^2}{\partial\theta^2} f(x;\theta)dx \text{ (by Theorem 4, Chapter 5)} \end{aligned}$$

$$
\begin{aligned}
&= \int_S \frac{\partial^2}{\partial\theta^2} f(x;\theta)dx - \int_S \left[\frac{\partial}{\partial\theta} \log\ f(x;\theta)\right]^2 f(x;\theta)\,dx \\
&\qquad + \int_S \left[\frac{\partial}{\partial\theta} \log\ f(x;\theta)\right]^2 f(x;\theta)\,dx \\
&= \int_S \left\{ \frac{\frac{\partial^2}{\partial\theta^2} f(x;\theta)}{f(x;\theta)} - \frac{\left[\frac{\partial}{\partial\theta} f(x;\theta)\right]^2}{f^2(x;\theta)} \right\} f(x;\theta)\,dx + I(\theta) \\
&= \int_S \frac{\left[\frac{\partial^2}{\partial\theta^2} f(x;\theta)\right] f(x;\theta) - \left[\frac{\partial}{\partial\theta} f(x;\theta)\right]^2}{f^2(x;\theta)} f(x;\theta)\,dx + I(\theta) \\
&= \int_S \left\{ \frac{\partial}{\partial\theta} \left[\frac{\frac{\partial}{\partial\theta} f(x;\theta)}{f(x;\theta)} \right] \right\} f(x;\theta)\,dx + I(\theta) \\
&= \int_S \left\{ \frac{\partial}{\partial\theta} \left[\frac{\partial}{\partial\theta} \log\ f(x;\theta) \right] \right\} f(x;\theta)\,dx + I(\theta) \\
&= \int_S \left[\frac{\partial^2}{\partial\theta^2} \log\ f(x;\theta) \right] f(x;\theta)\,dx + I(\theta) \\
&= \mathcal{E}_\theta \left[\frac{\partial^2}{\partial\theta^2} \log\ f(X;\theta) \right] + I(\theta), \quad \text{so that}
\end{aligned}
$$

$$
I(\theta) = -\mathcal{E}_\theta \left[\frac{\partial^2}{\partial\theta^2} \log\ f(X;\theta) \right], \quad \text{as was to be seen.}
$$

(iii) By parts (i) and (ii),

$$
\mathcal{E}_\theta \left[\frac{\partial}{\partial\theta} \log\ f(X;\theta) \right] = 0, \quad \mathrm{Var}_\theta \left[\frac{\partial}{\partial\theta} \log\ f(X;\theta) \right] = I(\theta).
$$

Then the CLT (Corollary to Theorem 2, Chapter 12) yields

$$
\frac{1}{\sqrt{nI(\theta)}} \sum_{j=1}^{n} \frac{\partial}{\partial\theta} \log\ f(X_j;\theta) \xrightarrow[n\to\infty]{d} N(0,1),
$$

and hence (by Theorem 4 (ii), Chapter 5),

$$
\frac{1}{\sqrt{n}} \sum_{j=1}^{n} \frac{\partial}{\partial\theta} \log\ f(X_j;\theta) \xrightarrow[n\to\infty]{d} N(0, I(\theta)). \qquad \blacksquare
$$

Lemma 4. Set

$$
\psi(X;\theta) = \frac{\partial}{\partial\theta} \log\ f(X;\theta), \quad \text{and} \quad \dot{\psi}(X;\theta) = \frac{\partial^2}{\partial\theta^2} \log\ f(X;\theta).
$$

Also, let

$$B_n = -\int_0^1 \frac{1}{n}\sum_{j=1}^{n} \dot{\psi}\ [X_j; \theta_0 + \lambda(\widetilde{\theta}_n - \theta_0)]\, d\lambda. \tag{16.13}$$

Then, under assumptions (B1)–(B3), (B4), (B6)–(B8), it holds $B_n \underset{(P_{\theta_0})}{\overset{a.s.}{\longrightarrow}} I(\theta_0)$, as $n \to \infty$.

Proof. By assumption (B8),

$$|\dot{\psi}(X;\theta)| \le H(X), \quad \theta \in N(\theta_0), \quad \mathcal{E}_{\theta_0} H(X) < \infty,$$

whereas by assumption (B6) (continuity of $\dot{\psi}(x;\theta)$ in θ), $\dot{\psi}(X;\theta) \to \dot{\psi}(X;\theta_0)$ as $\theta \to \theta_0$. Then

$$\mathcal{E}_{\theta_0}\dot{\psi}(X;\theta) \underset{\theta\to\theta_0}{\longrightarrow} \mathcal{E}_{\theta_0}\dot{\psi}(X;\theta_0)\ (= -I(\theta_0), \text{ by Lemma 3(ii)}). \tag{16.14}$$

(This is so by the Dominated Convergence Theorem, Theorem 3, Chapter 5.) For some $\rho > 0$, set $S_\rho = \{\theta \in \Theta; |\theta - \theta_0| \le \rho\}$, and in Lemma 2 identify T with S_ρ, $\varphi(x,t)$ with $\dot{\psi}(x;\theta)$, and $\mu(t)$ with $\mathcal{E}_{\theta_0}\dot{\psi}(X;\theta)$. Then assumptions(i)–(iii) in the lemma just cited are satisfied. Under these assumptions, Proposition 1 applies and yields

$$\sup_{\theta\in S_\rho}\left|\frac{1}{n}\sum_{j=1}^{n}\dot{\psi}(X_j;\theta) - \mathcal{E}_0\dot{\psi}(X;\theta)\right| \underset{(P_{\theta_0})}{\overset{a.s.}{\longrightarrow}} 0 \quad \text{as } n\to\infty.$$

This means that for every ω in an event E_1 with $P_{\theta_0}(E_1) = 1$ the last convergence holds (pointwise) when the r.v.s X_j are evaluated at ω; i.e.,

$$\sup_{\theta\in S_\rho}\left|\frac{1}{n}\sum_{j=1}^{n}\dot{\psi}[X_j(\omega);\theta] - \mathcal{E}_{\theta_0}\dot{\psi}(X;\theta)\right| \longrightarrow 0 \quad \text{as } n\to\infty.$$

Hence, for every $\varepsilon > 0$ there exists $N_1 = N_1(\varepsilon,\omega) > 0$ integer, such that

$$\sup_{\theta\in S_\rho}\left|\frac{1}{n}\sum_{j=1}^{n}\dot{\psi}[X_j(\omega);\theta] - \mathcal{E}_{\theta_0}\dot{\psi}(X;\theta)\right| < \frac{\varepsilon}{2}, \quad n \ge N_1. \tag{16.15}$$

On the other hand, by Theorem 1, $\widetilde{\theta}_n \underset{(P_{\theta_0})}{\overset{a.s.}{\longrightarrow}} \theta_0$ as $n\to\infty$, which implies that, for every ω in an event E_2 with $P_{\theta_0}(E_2) = 1$, $\widetilde{\theta}_n \underset{n\to\infty}{\longrightarrow} \theta_0$. Thus, for $\omega \in E_2$ and $\rho > 0$ as above there exists $N_2 = N_2(\rho,\omega) > 0$ integer such that

$$|\widetilde{\theta}_n(\omega) - \theta_0| < \rho, \quad n \ge N_2. \tag{16.16}$$

Set $N_0 = N_0(\varepsilon, \rho, \omega) = \max\{N_1, N_2\}$ and restrict ω in $E_1 \cap E_2$, call it E_0 (with $P_{\theta_0}(E_0) = 1$). Then both inequalities (16.15) and (16.16) hold, provided $n \geq N_0$; i.e.,

$$\sup_{\theta \in S_\rho} \left| \frac{1}{n} \sum_{j=1}^{n} \dot{\psi}[X_j(\omega); \theta] - \mathcal{E}_{\theta_0} \dot{\psi}(X; \theta) \right| < \frac{\varepsilon}{2} \quad \text{and} \quad |\widetilde{\theta}_n(\omega) - \theta_0| < \rho, \quad n \geq N_0. \tag{16.17}$$

From this point on, we work with ω in E_0 (with $P_{\theta_0}(E_0) = 1$) and $n \geq N_0$. Also, for notational convenience, omit the evaluation of r.v.s at ω. We have then

$$\begin{aligned}
|B_n(\theta_0) - I(\theta_0)| &= \left| - \int_0^1 \frac{1}{n} \sum_{j=1}^{n} \dot{\psi}\left[X_j; \theta_0 + \lambda(\widetilde{\theta}_n - \theta_0)\right] d\lambda - I(\theta_0) \right| \\
&= \left| \int_0^1 \frac{1}{n} \sum_{j=1}^{n} \dot{\psi}\left[X_j; \theta_0 + \lambda(\widetilde{\theta}_n - \theta_0)\right] d\lambda + I(\theta_0) \right| \\
&= \left| \int_0^1 \frac{1}{n} \sum_{j=1}^{n} \dot{\psi}\left[X_j; \theta_0 + \lambda(\widetilde{\theta}_n - \theta_0)\right] d\lambda - \mathcal{E}_{\theta_0} \dot{\psi}(X; \theta_0) \right| \\
&\leq \left| \int_0^1 \frac{1}{n} \sum_{j=1}^{n} \dot{\psi}\left[X_j; \theta_0 + \lambda(\widetilde{\theta}_n - \theta_0)\right] d\lambda - \mathcal{E}_{\theta_0} \dot{\psi}(X; \theta) \right| \\
&\quad + \left| \mathcal{E}_{\theta_0} \dot{\psi}(X; \theta) - \mathcal{E}_{\theta_0} \dot{\psi}(X; \theta_0) \right| \\
&= \left| \int_0^1 \left\{ \frac{1}{n} \sum_{j=1}^{n} \dot{\psi}\left[X_j; \theta_0 + \lambda(\widetilde{\theta}_n - \theta_0)\right] - \mathcal{E}_{\theta_0} \dot{\psi}(X; \theta) \right\} d\lambda \right| \\
&\quad + \left| \mathcal{E}_{\theta_0} \dot{\psi}(X; \theta) - \mathcal{E}_{\theta_0} \dot{\psi}(X; \theta_0) \right|.
\end{aligned} \tag{16.18}$$

But on E_0 and for $n \geq N_0$,

$$\begin{aligned}
\left| [\theta_0 + \lambda(\widetilde{\theta}_n - \theta_0)] - \theta_0 \right| &= |\lambda(\widetilde{\theta}_n - \theta_0)| \\
&\leq |\widetilde{\theta}_n - \theta_0| \quad (\text{since } 0 \leq \lambda \leq 1) \\
&< \rho \text{ (by (16.17))},
\end{aligned}$$

so that $\theta_0 + \lambda(\widetilde{\theta}_n - \theta_0)$ $(= \theta(\omega))$, call it θ^*, lies in S_ρ. Also, on the right-hand side of (16.18), replace the arbitrary θ by θ^*. We have then

$$\begin{aligned}
|B_n(\theta_0) - I(\theta_0)| \leq & \left| \int_0^1 \left[\frac{1}{n} \sum_{j=1}^{n} \dot{\psi}(X_j; \theta^*) - \mathcal{E}_{\theta_0} \dot{\psi}(X; \theta^*) \right] d\lambda \right| \\
& + \left| \mathcal{E}_{\theta_0} \dot{\psi}(X; \theta^*) - \mathcal{E}_{\theta_0} \dot{\psi}(X; \theta_0) \right|.
\end{aligned} \tag{16.19}$$

However, by (16.17),

$$\left|\frac{1}{n}\sum_{j=1}^{n}\dot{\psi}(X_j;\theta^*) - \mathcal{E}_{\theta_0}\dot{\psi}(X;\theta^*)\right| < \frac{\varepsilon}{2}$$

as long as θ^* stays in S_ρ (which it does) and for $n \geq N_0$, which implies that

$$\left|\int_0^1\left[\frac{1}{n}\sum_{j=1}^{n}\dot{\psi}(X_j;\theta^*) - \mathcal{E}_{\theta_0}\dot{\psi}(X;\theta^*)\right]d\lambda\right|$$

$$\leq \int_0^1\left|\left[\frac{1}{n}\sum_{j=1}^{n}\dot{\psi}(X_j;\theta^*) - \mathcal{E}_{\theta_0}\dot{\psi}(X;\theta^*)\right]d\lambda\right| < \frac{\varepsilon}{2}. \tag{16.20}$$

Furthermore, by (16.14), for $\varepsilon > 0$ there exists $\delta(\varepsilon) > 0$ such that

$$\left|\mathcal{E}_{\theta_0}\dot{\psi}(X;\theta^*) - \mathcal{E}_{\theta_0}\dot{\psi}(X;\theta_0)\right| < \frac{\varepsilon}{2}, \tag{16.21}$$

provided $|\theta^* - \theta_0| < \delta(\varepsilon)$. This will be, indeed, the case if ρ is chosen to be $< \delta(\varepsilon)$. So, for every $\varepsilon > 0$, choose $\rho < \delta(\varepsilon)$, so that (16.21) is satisfied. Inequality (16.20) is also satisfied for ω in E_0 (with $P_{\theta_0}(E_0) = 1$) and $n \geq N_0$. Combining relations (16.18), (16.20), and (16.21), we obtain $|B_n(\theta_0) - I(\theta_0)| < \varepsilon$ on the event E_0 (with $P_{\theta_0}(E_0) = 1$), provided $n \geq N_0$. This is equivalent to saying that $B_n(\theta_0) \underset{(P_{\theta_0})}{\longrightarrow} I(\theta_0)$ as $n \to \infty$. ■

The remark below will be used in the proof of Theorem 2.

Remark 3. When expanding a function according to Taylor's formula, the following form of a remainder often proves convenient. To this effect, let g be a real-valued function defined on $\Re$, and assume it has a continuous derivative to be denoted by $\dot{g}$. Then

$$g(x+t) = g(x) + t\int_0^1 \dot{g}(x+\lambda t)d\lambda.$$

A similar expression holds when we assume the existence of higher-order derivatives, as well as when g is defined on $\Re^k$ $(k \geq 2)$.

Such formulas are exhibited in relation (8.14.3), page 186, in Dieudonné (1960).

Proof of Theorem 2. For the independent r.v.s $X_1, \ldots, X_n$ with p.d.f. $f(\cdot;\theta)$, consider the likelihood function $L_n(\theta|\mathbf{X}_n) = \prod_{j=1}^{n} f(X_j;\theta)$, where $\mathbf{X}_n = (X_1, \ldots, X_n)$, and set

$$\ell_n(\theta)(= \ell_n(\theta|\mathbf{X}_n)) = \log\ L_n(\theta|\mathbf{X}_n) = \sum_{j=1}^{n}\log\ f(X_j;\theta).$$

Then, by the notation introduced in Lemma 4,

$$\dot{\ell}_n(\theta) = \frac{\partial}{\partial\theta}\ell_n(\theta) = \sum_{j=1}^{n}\psi(X_j;\theta),$$

and hence

$$\ddot{\ell}_n(\theta) = \frac{\partial}{\partial\theta}\dot{\ell}_n(\theta) = \sum_{j=1}^{n} \dot{\psi}(X_j;\theta).$$

Consider $\dot{\ell}_n(\theta)$ and, by using Remark 3 with $g(x)$ replaced by $\dot{\ell}_n(\theta)$, expand it to obtain

$$\begin{aligned}\dot{\ell}_n(\theta) &= \dot{\ell}_n(\theta_0) + (\theta-\theta_0)\int_0^1 \ddot{\ell}_n[\theta_0+\lambda(\theta-\theta_0)]\,d\lambda\\ &= \dot{\ell}_n(\theta_0) + (\theta-\theta_0)\int_0^1 \left\{\sum_{j=1}^{n} \dot{\psi}[X_j;\theta_0+\lambda(\theta-\theta_0)]\right\} d\lambda. \quad (16.22)\end{aligned}$$

In (16.22), replace θ by $\widetilde{\theta}_n$, recall that $\dot{\ell}_n(\widetilde{\theta}_n) = 0$ (with P_{θ_0}-probability 1), and divide both sides by $\sqrt{n}$ to get

$$\begin{aligned}\frac{1}{\sqrt{n}}\dot{\ell}_n(\theta_0) &= \sqrt{n}(\widetilde{\theta}_n-\theta_0)\int_0^1 \left\{-\frac{1}{n}\sum_{j=1}^{n} \dot{\psi}[X_j;\theta_0+\lambda(\widetilde{\theta}_n-\theta_0)]\right\} d\lambda\\ &= B_n(\theta_0)\times\sqrt{n}(\widetilde{\theta}_n-\theta_0) \quad \text{(by the definition of } B_n(\theta_0) \text{ in (16.13))}.\end{aligned}$$

By Lemma 4, $B_n(\theta_0) \xrightarrow[(P_{\theta_0})]{a.s.} I(\theta_0)$ as $n\to\infty$. Therefore, with P_{θ_0}-probability 1 and sufficiently large n, $B_n^{-1}(\theta_0)$ exists and, clearly, $B_n^{-1}(\theta_0) \xrightarrow[(P_{\theta_0})]{a.s.} I^{-1}(\theta_0)$. Also, by Lemma 3 (iii), $\frac{1}{\sqrt{n}}\dot{\ell}_n(\theta_0) \xrightarrow[(P_{\theta_0})]{d} N(0, I(\theta_0))$. Then Slutsky's theorem (Theorem 8(ii), Chapter 8) implies that

$$B_n^{-1}(\theta_0)\times\frac{1}{\sqrt{n}}\dot{\ell}_n(\theta_0) \xrightarrow[(P_{\theta_0})]{d} \frac{1}{I(\theta_0)}\times N(0, I(\theta_0)) = N(0, I^{-1}(\theta_0)).$$

From $\frac{1}{\sqrt{n}}\dot{\ell}_n(\theta_0) = B_n(\theta_0)\times\sqrt{n}(\widetilde{\theta}_n-\theta_0)$, we get $\sqrt{n}(\widetilde{\theta}_n-\theta_0) = B_n^{-1}(\theta_0)\times\frac{1}{\sqrt{n}}\dot{\ell}_n(\theta_0)$ (with P_{θ_0}-probability 1), so that, as $n\to\infty$,

$$\sqrt{n}(\widetilde{\theta}_n-\theta_0) \xrightarrow[(P_{\theta_0})]{d} N(0, I^{-1}(\theta_0)),$$

as was to be seen. ■

To this theorem, there is the following corollary.

Corollary. Suppose that for all sufficiently large n and with P_{θ_0}-probability 1 there exists a unique MLE $\widehat{\theta}_n$ of θ_0. Then, under assumptions (B1)–(B8) and as $n\to\infty$, we have

$$\sqrt{n}(\widehat{\theta}_n-\theta_0) \xrightarrow[(P_{\theta_0})]{d} N(0, I^{-1}(\theta_0)).$$

Proof. It is immediate, since all arguments used in the proofs apply when $\widetilde{\theta}_n$ is replaced by $\widehat{\theta}_n$ (see also Corollary to Theorem 1). ■

Remark 4. When the parameter space Θ is r-dimensional, assumptions (B1), (B4), (B6)–(B8) have got to be modified suitably. The proofs of Lemma 2 and Proposition 1 are independent of the dimensionality of T. However, the formulations of Theorem 2 and of Lemmas 3 and 4 must be modified, as well as their proofs. The conclusion of the theorem, nevertheless, remains valid, properly interpreted.

Most of the relevant derivations for the multidimensional parameter case, both for Theorems 1 and 2, can be found in Theorem 17, page 114, and Theorem 18, page 121, in Ferguson (1996).

In reference to Theorem 1, it should be pointed out that the assumptions made there are sufficient but not necessary—albeit economical—for the theorem to hold. The following example illustrates the point.

Example 1. Let $X \sim U(0, \theta)$. Then assumptions (A2) and (A4) are *not* satisfied. On the basis of a random sample of size n, $X_1, \ldots, X_n$, the MLE $\widehat{\theta}_n$ is given by $\widehat{\theta}_n = X_{(n)}$, and its p.d.f. is given by $g_n(t;\theta) = \frac{n}{\theta^n} t^{n-1}$, $0 < t < \theta$. Set $Y_n = n(\widehat{\theta}_n - \theta)$. Then its p.d.f. is $f_{Y_n}(y;\theta) = \frac{1}{\theta^n}\left(\theta + \frac{y}{n}\right)^{n-1}$, $-n\theta < y < 0$, and the p.d.f. of $Z_n = \widehat{\theta}_n - \theta$ is $f_{Z_n}(t;\theta) = \frac{n}{\theta^n}(\theta + t)^{n-1}$, $-\theta < t < 0$. It follows that, for every $\varepsilon > 0$,

$$\begin{aligned}
P_\theta(|\widehat{\theta}_n - \theta| > \varepsilon) &= P_\theta(\widehat{\theta}_n - \theta > \varepsilon \text{ or } \widehat{\theta}_n - \theta < -\varepsilon)\\
&= P_\theta(\widehat{\theta}_n - \theta > \varepsilon) + P_\theta(\widehat{\theta}_n - \theta < -\varepsilon)\\
&= P_\theta(\widehat{\theta}_n - \theta < -\varepsilon)\\
&= \frac{n}{\theta^n}\int_{-\theta}^{-\varepsilon} (\theta + t)^{n-1}\,dt = \frac{n}{\theta^n} \times \frac{1}{n}(\theta + t)^n \Big|_{-\theta}^{-\varepsilon}\\
&= \frac{1}{\theta^n}(\theta - \varepsilon)^n = \left(1 - \frac{\varepsilon}{\theta}\right)^n.
\end{aligned}$$

By choosing $(0 <)\varepsilon < \theta$, so that $0 < 1 - \frac{\varepsilon}{\theta} < 1$, we get $\sum_{n=1}^{\infty}\left(1 - \frac{\varepsilon}{\theta}\right)^n = \frac{1-\frac{\varepsilon}{\theta}}{1-(1-\frac{\varepsilon}{\theta})} = \frac{\theta-\varepsilon}{\varepsilon}$ (finite). It can be argued that this is a sufficient condition for $\widehat{\theta}_n \xrightarrow[(P_\theta)]{a.s.} \theta$. (This follows by combining the results in Exercise 4, Chapter 2, and Exercises 3, and 4 (i), Chapter 3.) So, in this case the conclusion of the theorem holds, while some of its assumptions fail to be satisfied.

The same example also makes a case for Theorem 2. We have, for $-n\theta < y < 0$,

$$\begin{aligned}
f_{Y_n}(y;\theta) &= \frac{1}{\theta^n}\left(\theta + \frac{y}{n}\right)^{n-1}\\
&= \frac{1}{\theta}\left(1 + \frac{\frac{n-1}{n} \times \frac{y}{\theta}}{n-1}\right)^{n-1} \underset{n\to\infty}{\longrightarrow} \frac{1}{\theta}e^{y/\theta}, \quad y < 0,
\end{aligned}$$

which is the p.d.f. of the Negative Exponential distribution with parameter $1/\theta$, call it $f_Y(\cdot;\theta)$. From the fact that $f_{Y_n}(y;\theta) \underset{n\to\infty}{\longrightarrow} f_Y(y;\theta)$, $y < 0$, and by means of the

Dominated Convergence Theorem (Theorem 3, Chapter 5), it follows that

$$F_{Y_n}(y) = \int_{-\infty}^{y} f_{Y_n}(t;\theta)\,dt \underset{n\to\infty}{\longrightarrow} \int_{-\infty}^{y} f_Y(t;\theta)\,dt, \quad y < 0;$$

i.e., $n(\widehat{\theta}_n - \theta)\xrightarrow[(P_\theta)]{d} Y \sim$ Negative Exponential with parameter $\frac{1}{\theta}$, and hence $\sqrt{n}(\widehat{\theta}_n - \theta)\xrightarrow[(P_\theta)]{d} 0$. Thus, in this case, the conclusion of Theorem 2 *fails* to hold.

16.5 Nonparametric Estimation of a Probability Density Function

The problem we are faced with here is the following: we are given n i.i.d. r.v.s $X_1, \ldots, X_n$ with p.d.f. f (of the continuous type), for which very little is known, and we are asked to construct a nonparametric estimate $\hat{f}_n(x)$ of $f(x)$, for each $x \in \Re$, based on the random sample $X_1, \ldots, X_n$. The approach to be used here is the so-called *kernel-estimation* approach. According to this method, we select a (known) p.d.f. (of the continuous type) to be denoted by K and to be termed a *kernel*, subject to some rather minor requirements. Also, we choose a sequence of positive numbers, denoted by $\{h_n\}$, which has the property that $h_n \to 0$ as $n \to \infty$ and satisfies some additional requirements. The numbers $h_n, n \geq 1$, are referred to as *bandwidths* for a reason to be seen below (see Example 2). Then, on the basis of the random sample $X_1, \ldots, X_n$, the kernel K, and the bandwidths $h_n, n \geq 1$, the proposed estimate of $f(x)$ is $\hat{f}_n(x)$ given by:

$$\hat{f}_n(x) = \frac{1}{nh_n}\sum_{i=1}^{n} K\left(\frac{x - X_i}{h_n}\right). \tag{16.23}$$

Remark 5. In the spirit of motivation for using the estimate in (16.23), observe first that $\frac{1}{h_n}K\left(\frac{x-y}{h_n}\right)$ is a p.d.f. as a function of y for fixed x. Indeed,

$$\int_{-\infty}^{\infty} \frac{1}{h_n}K\left(\frac{x-y}{h_n}\right)dy = \int_{\infty}^{-\infty} \frac{1}{h_n}K(t)(-h_n)dt \quad \left(\text{by setting} \frac{x-y}{h_n} = t\right)$$

$$= \int_{-\infty}^{\infty} K(t)dt = 1.$$

Next, evaluate $\frac{1}{h_n}K\left(\frac{x-y}{h_n}\right)$ at $y = X_i,\ i = 1, \ldots, n$, and then form the average of these values to produce $\hat{f}_n(x)$.

A further motivation for the proposed estimate is the following. Let F be the d.f. of the X_is, and let F_n be the empirical d.f. based on $X_1, \ldots, X_n$, so that, for fixed x, $F_n(y)$ takes on the value $\frac{1}{n}$ at each one of the points $y = X_i,\ i = 1, \ldots, n$. Then weigh $\frac{1}{h_n}K\left(\frac{x-y}{h_n}\right)$ by $\frac{1}{n}$ and sum up from 1 to n (which is the same as integrating $\frac{1}{h_n}K\left(\frac{x-y}{h_n}\right)$ with respect to F_n) obtain $\hat{f}_n(x)$ again.

Example 2. Construct the kernel estimate of $f(x)$, for each $x \in \Re$, by using the $U(-1, 1)$ kernel; i.e., by taking

$$K(x) = \frac{1}{2}, \quad \text{for } -1 \leq x \leq 1, \text{ and } 0, \text{ otherwise.}$$

Here, it is convenient to use the indicator notation; namely, $K(x) = I_{[-1,1]}(x)$ (where, it is recalled, $I_A(x) = 1$ if $x \in A$, and 0 if $x \in A^c$). Then the estimate (16.23) becomes as follows:

$$\hat{f}_n(x) = \frac{1}{nh_n} \sum_{i=1}^{n} I_{[-1,1]}\left(\frac{x - X_i}{h_n}\right), \quad x \in \Re. \tag{16.24}$$

So, $I_{[-1,1]}(\frac{x-X_i}{h_n}) = 1$, if and only if $x - h_n \leq X_i \leq x + h_n$; in other words, in forming $\hat{f}_n(x)$, we use only those observations X_i which lie in the window $[x - h_n, x + h_n]$. The breadth of this window is, clearly, determined by h_n, and this is the reason that h_n is referred to as the bandwidth.

Usually, the minimum of assumptions required of the kernel K and the bandwidth h_n, in order for us to be able to establish some desirable properties of the estimate $\hat{f}_n(x)$ given in (16.23), are the following:

$$\left.\begin{array}{r} K \text{ is a bounded p.d.f.; i.e., } \sup\{K(x);\ x \in \Re\} < \infty. \\ xK(x) \text{ tends to 0 as } x \to \pm\infty; \text{ i:e:, } |xK(x)| \underset{|x|\to\infty}{\longrightarrow} 0. \\ K \text{ is symmetric about 0; i.e., } K(-x) = K(x),\ x \in \Re. \end{array}\right\} \tag{16.25}$$

$$\left.\begin{array}{ll} \text{As } n\to\infty\text{: (i)} & (0 <)h_n \to 0 \\ \text{(ii)} & nh_n \to \infty \end{array}\right\}. \tag{16.26}$$

Remark 6. Observe that requirements (16.25) are met for the kernel used in (16.24). Furthermore, the convergences in (16.26) are satisfied if one takes, e.g., $h_n = n^{-\alpha}$ with $0 < \alpha < 1$. Below, we record three (asymptotic) results regarding the estimate $\hat{f}_n(x)$ given in (16.23).

Theorem 3. Under assumptions (16.25) and (16.26)(i), the estimate $\hat{f}_n(x)$ given in (16.23) is an *asymptotically unbiased* estimate of $f(x)$ for every $x \in \Re$ at which f is continuous; i.e.,

$$\mathcal{E}\hat{f}_n(x) \to f(x) \quad \text{as } n\to\infty.$$

Theorem 4. Under assumptions (16.25) and (16.26)(i)–(ii), the estimate $\hat{f}_n(x)$ given in (16.23) is a *consistent in quadratic mean* estimate of $f(x)$ for every $x \in \Re$ at which f is continuous; i.e.,

$$\mathcal{E}[\hat{f}_n(x) - f(x)]^2 \to 0 \quad \text{as } n\to\infty.$$

Theorem 5. Under assumptions (16.25) and (16.26)(i)–(ii), the estimate $\hat{f}_n(x)$ given in (16.23) is *asymptotically normal*, when properly normalized, for every $x \in \Re$

at which f is continuous; i.e.,

$$\frac{\hat{f}_n(x) - \mathcal{E}\hat{f}_n(x)}{\sigma[\hat{f}_n(x)]} \underset{n\to\infty}{\overset{d}{\longrightarrow}} Z \sim N(0, 1).$$

At this point, it is only fitting to mention that the concept of kernel estimation of a p.d.f. was introduced by Murray Rosenblatt (1956), and it was popularized in a fundamental paper by Parzen (1962). There, one can find the proofs of the above theorems, along with other results. See references for more details.

This section is concluded with a fundamental result (stated as Theorem A), which is needed in the proof of Theorems 3–5.

Theorem A (Bochner). Make the following assumptions:

- **(i)** The function $K : \Re\to\Re$ is such that $|K(y)| \le M(< \infty)$ for every y, $\int_{-\infty}^{\infty} |K(y)| dy < \infty$, and $|yK(y)|\to 0$ as $|y|\to\infty$.
- **(ii)** The function $g : \Re\to\Re$ is such that $\int_{-\infty}^{\infty} |g(y)|\, dy < \infty$.
- **(iii)** The (real) numbers h_n, $n = 1, 2, \ldots$, are such that $0 < h_n\to 0$ as $n\to\infty$.

For each $x \in \Re$, define g_n by:

$$g_n(x) = \frac{1}{h_n}\int_{-\infty}^{\infty} K\left(\frac{y}{h_n}\right) g(x-y)dy \left(= \frac{1}{h_n}\int_{-\infty}^{\infty} K\left(\frac{x-t}{h_n}\right) g(t)\, dt,\right.$$
$$\left.\text{by setting} x - y = t\right).$$

Then, for every continuity point x of g, it holds

$$g_n(x) \underset{n\to\infty}{\longrightarrow} g(x)\int_{-\infty}^{\infty} K(y)\, dy. \tag{16.27}$$

Proof. In the first place, set h instead of h_n throughout the proof, and observe that

$$\int_{-\infty}^{\infty} K(y)\, dy = \int_{-\infty}^{\infty} K\left(\frac{t}{h}\right)\frac{dt}{h} = \frac{1}{h}\int_{-\infty}^{\infty} K\left(\frac{y}{h}\right) dy \quad \left(\text{by setting } y = \frac{t}{h}\right).$$

Then

$$\begin{aligned}
&\left|g_n(x) - g(x)\int_{-\infty}^{\infty} K\,(y)\, dy\right| \\
&= \left|\frac{1}{h}\int_{-\infty}^{\infty} K\,\left(\frac{y}{h}\right) g(x-y)\, dy - g(x)\frac{1}{h}\int_{-\infty}^{\infty} K\,\left(\frac{y}{h}\right) dy\right| \\
&= \left|\int_{-\infty}^{\infty} \frac{1}{h} K\,\left(\frac{y}{h}\right) [g(x-y) - g(x)]\, dy\right| \\
&\le \int_{-\infty}^{\infty} |g(x-y) - g(x)| \frac{1}{h}\left|K\,\left(\frac{y}{h}\right)\right| dy
\end{aligned}$$

$$= \int_{(|y|\le\delta)} |g(x-y)-g(x)|\frac{1}{h}\left|K\left(\frac{y}{h}\right)\right|dy$$
$$+\int_{(|y|>\delta)} |g(x-y)-g(x)|\frac{1}{h}\left|K\left(\frac{y}{h}\right)\right|dy$$
$$(\text{for } \delta > 0)$$
$$\le \sup_{|y|\le\delta}|g(x-y)-g(x)|\int_{(|y|\le\delta)}\frac{1}{h}\left|K\left(\frac{y}{h}\right)\right|dy+|g(x)|\int_{(|y|>\delta)}\frac{1}{h}\left|K\left(\frac{y}{h}\right)\right|dy$$
$$+\int_{(|y|>\delta)}\frac{|g(x-y)|}{|y|}\times\frac{|y|}{h}\left|K\left(\frac{y}{h}\right)\right|dy$$
$$\le \sup_{|y|>\delta}|g(x-y)-g(x)|\int_{(|y|\le\delta)}\frac{1}{h}\left|K\left(\frac{y}{h}\right)\right|dy+|g(x)|\int_{(|y|>\delta)}\frac{1}{h}\left|K\left(\frac{y}{h}\right)\right|dy$$
$$+\frac{1}{\delta}\int_{(|y|>\delta)}|g(x-y)|\times\frac{|y|}{h}\left|K\left(\frac{y}{h}\right)\right|dy$$
$$\left(\text{since } |y|>\delta \text{ is equivalent to } \frac{1}{|y|}<\frac{1}{\delta}\right)$$
$$\le \sup_{|y|\le\delta}|g(x-y)-g(x)|\int_{-\infty}^{\infty}\frac{1}{h}\left|K\left(\frac{y}{h}\right)\right|dy+|g(x)|\int_{(|y|>\delta)}\frac{1}{h}\left|K\left(\frac{y}{h}\right)\right|dy$$
$$+\frac{1}{\delta}\int_{(|t|>\delta/h)}|g(x-ht)||tK(t)|h\,dt$$
$$\left(\text{by setting } \frac{y}{h}=t \text{ in the last integral above}\right)$$
$$= \sup_{|y|\le\delta}|g(x-y)-g(x)|\int_{-\infty}^{\infty}|K(t)|\,dt+|g(x)|\int_{(|t|>\delta/h)}|K(t)|\,dt$$
$$+\frac{1}{\delta}\sup_{|t|>\delta/h}|tK(t)|\int_{-\infty}^{\infty}|g(x-ht)|h\,dt$$
$$\left(\text{by setting } \frac{y}{h}=t \text{ in the first two integrals}\right)$$
$$= \max_{|y|\le\delta}|g(x-y)-g(x)|\int_{-\infty}^{\infty}|K(t)|\,dt+|g(x)|\int_{(|t|>\delta/h)}|K(t)|\,dt$$
$$+\frac{1}{\delta}\sup_{|t|>\delta/h}|tK(t)|\int_{-\infty}^{\infty}|g(y)|\,dy \quad \text{(because the sup can be replaced by max,}$$
by continuity of g at x, and by setting $x-ht=y$ in the last integral).

However, as $n\to\infty$ (which implies $h\to 0$), the following things happen:

$$\int_{(|t|>\delta/h)}|K(t)|\,dt\to 0, \text{ following from :}$$

$\int_{-\infty}^{\infty} |K(t)|\, dt < \infty$, $|K(t)|I_{(|t|>\delta/h)} \to 0$, $K(t)I_{(|t|>\delta/h)} \leq |K(t)|$ independent of n, and the Dominated Convergence Theorem (Theorem 3(ii), Chapter 5); and

$$\sup_{|t|>\delta/h} |tK(t)| \to 0, \text{ following from } |yK(y)| \to 0 \text{ as } |y| \to \infty.$$

Taking the limits, as $n \to \infty$, we obtain then

$$\limsup \left| g_n(x) - g(x) \int_{-\infty}^{\infty} K(y)\, dy \right| \leq \max_{|y|\leq\delta} |g(x-y) - g(x)| \int_{-\infty}^{\infty} |K(t)|\, dt.$$

Finally, letting $\delta \to 0$, and using continuity of g at x, we have

$$\lim_{n\to\infty} \left| g_n(x) - g(x) \int_{-\infty}^{\infty} K(y)\, dy \right| = 0,$$

and hence

$$g_n(x) \underset{n\to\infty}{\longrightarrow} g(x) \int_{-\infty}^{\infty} K(y)\, dy,$$

which is what relation (16.27) asserts. ■

Remark 7. Under the first assumption in (16.25) (i.e., $K(x) \leq M(< \infty), x \in \Re$), it holds $\int_{-\infty}^{\infty} K^r(x)\, dx < \infty$ for every $r > 1$. This is so, because $K^r(x) = K^{r-1}(x)K(x) \leq M^{r-1}K(x)$, so that $\int_{-\infty}^{\infty} K^r(x)\, dx \leq M^{r-1}$.

Corollary. Under the assumptions of Theorem A and with the r.v. X distributed as the X_js, it holds

$$\frac{1}{h_n}\mathcal{E}K^r\left(\frac{x-X}{h_n}\right) \underset{n\to\infty}{\longrightarrow} f(x) \int_{-\infty}^{\infty} K^r(x)dx,$$

for every $x \in \Re$ continuity point of f, $r \geq 1$.

Proof. Writing h instead of h_n, we have

$$\begin{aligned}
\frac{1}{h}\mathcal{E}K^r\left(\frac{x-X}{h}\right) &= \frac{1}{h}\int_{-\infty}^{\infty} K^r\left(\frac{x-y}{h}\right) f(y)\, dy \\
&= \frac{1}{h}\int_{-\infty}^{\infty} K^r\left(\frac{z}{h}\right) f(x-z)\, dy \quad \text{(by setting } x-y=z\text{)} \\
&\underset{n\to\infty}{\longrightarrow} f(x) \int_{-\infty}^{\infty} K^r(z)\, dz,
\end{aligned}$$

by Theorem A applied with K replaced by K^r and g by f. ■

16.6 Proof of Theorems 3–5

On the basis of Theorem A, its corollary, and standard probability arguments, we may now proceed with the proof of Theorems 3–5.

Proof of Theorem 3. By the fact that the X_js are i.i.d., we obtain

$$\mathcal{E}\,\widehat{f}_n(x) = \frac{1}{nh} \times n\mathcal{E}K\left(\frac{x-X}{h}\right) = \frac{1}{h}\mathcal{E}K\left(\frac{x-X}{h}\right) \underset{n\to\infty}{\longrightarrow} f(x),$$

by the Corollary to Theorem A applied with $r = 1$. ■

The following results are needed in the proof of Theorem 4 below.

Lemma 5. Under assumptions (16.25) and (16.26) (i) and (ii), and for every $x \in \Re$ continuity point of f, it holds:

(i) $\sigma^2[f_n(x)] \to 0$.

(ii) $(nh_n)\sigma^2[f_n(x)] \to f(x)\int_{-\infty}^{\infty} K^2(z)\,dz$.

Proof.

(i) Indeed,

$$\begin{aligned}
\sigma^2[\widehat{f}_n(x)] &= \mathrm{Var}\left[\frac{1}{nh}\sum_{j=1}^{n} K\left(\frac{x-X_j}{h}\right)\right] \\
&= \frac{1}{(nh)^2} \times n\,\mathrm{Var}\left[K\left(\frac{x-X}{h}\right)\right] \\
&= \frac{1}{nh^2}\left\{\mathcal{E}K^2\left(\frac{x-X}{h}\right) - \left[\mathcal{E}K\left(\frac{x-X}{h}\right)\right]^2\right\} \\
&= \frac{1}{nh} \times \frac{1}{h}\mathcal{E}K^2\left(\frac{x-X}{h}\right) - \frac{1}{n}\left[\frac{1}{h}\mathcal{E}K\left(\frac{x-X}{h}\right)\right]^2 \\
&\underset{n\to\infty}{\longrightarrow} 0 \times f(x)\int_{-\infty}^{\infty} K^2(z)dz - 0 \times f^2(x) = 0, \qquad (16.28)
\end{aligned}$$

by the Corollary to Theorem A, applied for $r = 1$ and $r = 2$, and by part (ii) in (16.26).

(ii) From part (i),

$$\begin{aligned}
(nh)\sigma^2\left[\widehat{f}_n(x)\right] &= \frac{1}{h}\mathcal{E}K^2\left(\frac{x-X}{h}\right) - h\left[\frac{1}{h}\mathcal{E}K\left(\frac{x-X}{h}\right)\right]^2 \\
&\underset{n\to\infty}{\longrightarrow} f(x)\int_{-\infty}^{\infty} K^2(z)\,dz - 0 \times f^2(x) \\
&= f(x)\int_{-\infty}^{\infty} K^2(z)\,dz.
\end{aligned}$$

■

Proof of Theorem 4. In the first place, for any r.v. U with $\mathcal{E}U^2 < \infty$ and any constant c, the following identity is immediate, by adding and subtracting $\mathcal{E}U$,

$$\mathcal{E}(U-c)^2 = \sigma^2(U) + (\mathcal{E}U - c)^2.$$

Applying this identity with $U = \widehat{f_n}(x)$ and $c = f(x)$, we get

$$\mathcal{E}\left[\widehat{f_n}(x) - f(x)\right]^2 = \sigma^2\left[\widehat{f_n}(x)\right] + \left[\mathcal{E}\widehat{f_n}(x) - f(x)\right]^2.$$

Then, for every $x \in \Re$ continuity point of f, it holds $\mathcal{E}\left[\widehat{f_n}(x) - f(x)\right]^2 \underset{n\to\infty}{\longrightarrow} 0$ on account of Theorem 3 and Lemma 1(i). ■

Corollary. Under the assumptions of Theorem 1 and for every $x \in \Re$ continuity point of f, it holds

$$\widehat{f_n}(x) \underset{n\to\infty}{\overset{P}{\longrightarrow}} f(x).$$

Proof. Indeed, for every $\varepsilon > 0$, the Tchebichev inequality yields

$$P\left[\left|\widehat{f_n}(x) - f(x)\right| > \varepsilon\right] \leq \varepsilon^{-2}\mathcal{E}\left|\widehat{f_n}(x) - f(x)\right|^2 \underset{n\to\infty}{\longrightarrow} 0. \quad ■$$

Proof of Theorem 5. For an arbitrary but fixed $x \in \Re$ (continuity point of f throughout the proof), and any n, set

$$X_{nj} = \left\{\frac{1}{h}K\left(\frac{x - X_j}{h}\right) - \mathcal{E}\left[\frac{1}{h}K\left(\frac{x - X}{h}\right)\right]\right\} \Big/ \sqrt{n}\left\{\text{Var}\left[\frac{1}{h}K\left(\frac{x - X}{h}\right)\right]\right\}^{1/2},$$

$j = 1, \ldots, n$, where the r.v. X is distributed as the X_js. Then, as is easily seen by means of (16.23) and the third line on the right-hand side of (16.28), we have

$$\sum_{j=1}^{n} X_{nj} = \frac{\hat{f}_n(x) - \mathcal{E}\hat{f}_n(x)}{\sigma\left[\hat{f}_n(x)\right]}.$$

Consider the triangular array of r.v.s X_{nj}, $j = 1, \ldots, n$, $n \geq 1$, and observe that, within each row, they are i.i.d. with expectation 0 and variance $\sigma^2(X_{nj}) = 1/n$, so that $\sum_{j=1}^{n}\sigma^2(X_{nj}) = 1$ (and $\max_{1\leq j\leq n}\sigma^2(X_{nj}) = \frac{1}{n} \underset{n\to\infty}{\longrightarrow} 0$). Thus, the conditions of Theorem 1, Chapter 12, hold. Therefore a necessary and sufficient condition that $\sum_{j=1}^{n} X_{nj} \underset{n\to\infty}{\overset{d}{\longrightarrow}} Z \sim N(0, 1)$ is that, for every $\varepsilon > 0$,

$$g_n(\varepsilon) = n\int_{(|x|\geq\varepsilon)} x^2\, dF_n(x) \underset{n\to\infty}{\longrightarrow} 0,$$

where F_n is the d.f. of the X_{nj}s, $j = 1, \ldots, n$. On the other hand, a sufficient condition for $g_n(\varepsilon) \underset{n\to\infty}{\longrightarrow} 0$ is that (see Theorem 3, Chapter 12)

$$\sum_{j=1}^{n}\mathcal{E}|X_{nj}|^3 \underset{n\to\infty}{\longrightarrow} 0.$$

Since for any r.v. $U \geq 0$, it is clearly true that $|U - \mathcal{E}U| \leq U + \mathcal{E}U$, we have

$$\mathcal{E}|U - \mathcal{E}U|^3 \leq \mathcal{E}(U + \mathcal{E}U)^3 = \mathcal{E}U^3 + 3(\mathcal{E}U^2)(\mathcal{E}U) + 3(\mathcal{E}U)^3 + (\mathcal{E}U)^3$$
$$= \mathcal{E}U^3 + 3(\mathcal{E}U^2)(\mathcal{E}U) + 4(\mathcal{E}U)^3,$$

we have, upon replacing U by $\frac{1}{h}K\left(\frac{x-X_j}{h}\right)$, and summing over j from 1 to n:

$$\sum_{j=1}^{n} \mathcal{E}\left|\frac{1}{h}K\left(\frac{x-X_j}{h}\right) - \mathcal{E}\left[\frac{1}{h}K\left(\frac{x-X}{h}\right)\right]\right|^3$$
$$\leq \sum_{j=1}^{n} \mathcal{E}\left\{\frac{1}{h}K\left(\frac{x-X_j}{h}\right) + \mathcal{E}\left[\frac{1}{h}K\left(\frac{x-X}{h}\right)\right]\right\}^3$$
$$= \sum_{j=1}^{n} \mathcal{E}\left[\frac{1}{h}K\left(\frac{x-X_j}{h}\right)\right]^3 + 3\sum_{j=1}^{n}\left\{\mathcal{E}\left[\frac{1}{h}K\left(\frac{x-X_j}{h}\right)\right]^2\right\}$$
$$\times\left\{\mathcal{E}\left[\frac{1}{h}K\left(\frac{x-X}{h}\right)\right]\right\} + 4\sum_{j=1}^{n}\left\{\mathcal{E}\left[\frac{1}{h}K\left(\frac{x-X_j}{h}\right)\right]\right\}^3$$
$$= n\mathcal{E}\left[\frac{1}{h}K\left(\frac{x-X}{h}\right)\right]^3 + 3n\left\{\mathcal{E}\left[\frac{1}{h}K\left(\frac{x-X}{h}\right)\right]^2\right\}$$
$$\times\left\{\mathcal{E}\left[\frac{1}{h}K\left(\frac{x-X}{h}\right)\right]\right\} + 4n\left\{\mathcal{E}\left[\frac{1}{h}K\left(\frac{x-X}{h}\right)\right]\right\}^3$$
$$= n\left\{\frac{1}{h^2} \times \frac{1}{h}\mathcal{E}K^3\left(\frac{x-X}{h}\right) + \frac{3}{h^2} \times \left[\frac{1}{h}\mathcal{E}K^2\left(\frac{x-X}{h}\right)\right] \times h\right.$$
$$\left.\times\left[\frac{1}{h}\mathcal{E}K\left(\frac{x-X}{h}\right)\right] + \frac{4h^2}{h^2} \times \left[\frac{1}{h}\mathcal{E}K\left(\frac{x-X}{h}\right)\right]^3\right\}$$
$$= \frac{n}{h^2}\left\{\frac{1}{h}\mathcal{E}K^3\left(\frac{x-X}{h}\right) + 3 \times \left[\frac{1}{h}\mathcal{E}K^2\left(\frac{x-X}{h}\right)\right] \times h \times \left[\frac{1}{h}\mathcal{E}K\left(\frac{x-X}{h}\right)\right]\right.$$
$$\left. + 4h^2 \times \left[\frac{1}{h}\mathcal{E}K\left(\frac{x-X}{h}\right)\right]^3\right\} \stackrel{def}{=} I_{1n},$$

or $I_{1n} = \frac{n}{h^2} \times J_{1n}$, where J_{1n} is the quantity within the curly brackets in the last expression above. However, by the Corollary to Theorem A (applied with $r = 3$, $r = 2$, and $r = 1$), we have, as $n \to \infty$,

$$\frac{1}{h}\mathcal{E}K^3\left(\frac{x-X}{h}\right) \to f(x)\int_{-\infty}^{\infty} K^3(x)\,dx,\ \frac{1}{h}\mathcal{E}K^2\left(\frac{x-X}{h}\right) \to f(x)\int_{-\infty}^{\infty} K^2(x)\,dx,$$

$$\frac{1}{h}\mathcal{E}K\left(\frac{x-X}{h}\right)\to f(x)\int_{-\infty}^{\infty}K(x)\,dx=f(x),\qquad \text{so that}$$

$$\begin{aligned}J_{1n}\to f(x)&\int_{-\infty}^{\infty}K^3(x)\,dx+3f(x)\int_{-\infty}^{\infty}K^2(x)\,dx\times 0\times f(x)\\&+0\times[f(x)]^3=f(x)\int_{-\infty}^{\infty}K^3(x)\,dx.\end{aligned}\tag{16.29}$$

Also, set

$$\begin{aligned}I_{2n}&=\left[\sqrt{n}\left\{\operatorname{Var}\left[\frac{1}{h}K\left(\frac{x-X}{h}\right)\right]\right\}^{1/2}\right]^3\\&=n\sqrt{n}\left\{\operatorname{Var}\left[\frac{1}{h}K\left(\frac{x-X}{h}\right)\right]\right\}^{3/2}\\&=n\sqrt{n}\left[\mathcal{E}\left[\frac{1}{h}K\left(\frac{x-X}{h}\right)\right]^2-\left\{\mathcal{E}\left[\frac{1}{h}K\left(\frac{x-X}{h}\right)\right]\right\}^2\right]^{3/2}\\&=n\sqrt{n}\left\{\frac{1}{h^2}\mathcal{E}K^2\left(\frac{x-X}{h}\right)-\left[\frac{1}{h}\mathcal{E}K\left(\frac{x-X}{h}\right)\right]^2\right\}^{3/2}\\&=\frac{n\sqrt{n}}{h^3}\left\{\mathcal{E}K^2\left(\frac{x-X}{h}\right)-\left[\mathcal{E}K\left(\frac{x-X}{h}\right)\right]^2\right\}^{3/2}\\&=\frac{n\sqrt{n}}{h^{3/2}}\left\{\frac{1}{h}\mathcal{E}K^2\left(\frac{x-X}{h}\right)-h\left[\frac{1}{h}\mathcal{E}K\left(\frac{x-X}{h}\right)\right]^2\right\}^{3/2}\\&=\frac{n\sqrt{n}}{h^{3/2}}J_{2n},\qquad\text{or}\end{aligned}$$

$I_{2n}=\frac{n\sqrt{n}}{h^{3/2}}J_{2n}$, where J_{2n} is the quantity within the curly brackets in the last expression above. However, by the Corollary to Theorem A (applied with $r=2$, and $r=1$), we have, as $n\to\infty$,

$$\frac{1}{h}\mathcal{E}K^2\left(\frac{x-X}{h}\right)\to f(x)\int_{-\infty}^{\infty}K^2(x)\,dx,\qquad\frac{1}{h}\mathcal{E}K\left(\frac{x-X}{h}\right)\to f(x),$$

so that

$$J_{2n}\to f(x)\int_{-\infty}^{\infty}K^2(x)dx+0\times f(x)=f(x)\int_{-\infty}^{\infty}K^2(x)dx.\tag{16.30}$$

Therefore

$$\sum_{j=1}^{n} \mathcal{E}|X_{nj}|^3 \leq \frac{I_{1n}}{I_{2n}} = \frac{nh^{-2}}{n\sqrt{n}h^{-3/2}} \times \frac{J_{1n}}{J_{2n}} = \frac{1}{\sqrt{nh}} \times \frac{J_{1n}}{J_{2n}} \longrightarrow 0,$$

by means of (16.26)(i), (ii), (16.29), and (16.30). This implies that $g_n(\varepsilon) \underset{n\to\infty}{\longrightarrow} 0$ for every $\varepsilon > 0$, and therefore $\sum_{j=1}^{n} X_{nj} \underset{n\to\infty}{\overset{d}{\longrightarrow}} Z \sim N(0, 1)$. ■

This page is intentionally left blank

Brief Review of Chapters 1–16

A

The purpose of this appendix is to present a brief summary of the content of each chapter in this book. Before embarking on the study of each chapter, it would be advisable that the reader review the respective summary in this appendix. In this way, one obtains ahead of time an integrated picture of the entire chapter. This brief review also serves as a guide to the reader as to where various kinds of material are to be found.

Chapter 1 Certain Classes of Sets, Measurability, and Pointwise Approximation

In this chapter, the concepts of a field, of a σ-field, and of a monotone class are introduced and their basic properties and relationships are studied (Definitions 1, 2, and 4, and Theorems 1–6). Measurable spaces and product measurable spaces are also introduced, as well as the concept of a measurable function, in general, and of a random vector and of a random variable (r.v.), in particular (Definitions 3, 5–11, and Theorems 7–16). The chapter is concluded with the fundamental result of approximating pointwise a r.v. by a sequence of simple r.v.s (Theorem 17 and its Corollary).

Chapter 2 Definition and Construction of a Measure and its Basic Properties

The basic concepts in this chapter are those of a measure, in general, and of a probability measure in particular (Definition 1). A number of basic results are also established (Theorems 1 and 2). The concept of the outer measure is introduced and is then used in going from a measure defined on a field to a measure defined on the σ-field generated by the field (Definitions 4 and 5, and Theorems 3 and 4); the Carathéodory Extension Theorem is instrumental here (Definition 6 and Theorem 5). The chapter is concluded with the relationship between a distribution function and the measure induced by it, and as a by-product of the discussion, we obtain the Lebesgue measure in the real line (Theorems 6 and 7).

An Introduction to Measure-Theoretic Probability, Second Edition. http://dx.doi.org/10.1016/B978-0-12-800042-7.00024-4

Chapter 3 Some Modes of Convergence of Sequences of Random Variables and their Relationships

In this chapter, we introduce two modes of convergence of a sequence of r.v.s: almost everywhere (a.e.) convergence, and convergence in measure, as well as their mutual versions (Definitions 1 and 2). It is shown that a.e. convergence and convergence in measure are equivalent to a.e. mutual convergence and mutual convergence in measure, respectively (Theorems 2 and 6). These convergences become almost sure (a.s.) convergence (a.s. mutual convergence) and convergence (mutual convergence) in probability when the underlying measure is a probability measure. It is also shown that for a finite measure, a.e. convergence implies convergence in measure (Corollary to Theorem 4). Furthermore, necessary and sufficient conditions are found for a.e. (and a.e. mutual) convergence (Theorem 4).

Chapter 4 The Integral of a Random Variable and its Basic Properties

This chapter is devoted to the step-by-step definition of the integral of a r.v. X over an abstract measure space $(\Omega, \mathcal{A}, \mu)$, $\int_\Omega X d\mu$ (Theorems 1–3, and Definitions 1 and 2), and the proof of the basic properties of the integral (Theorems 4–12). The integral of X becomes the expectation of X (expected value of X, mean value of X), $\mathcal{E}X$, when μ is a probability measure P. Finally, the (probability) distribution of a r.v. X under a probability measure P, P_X, is defined (Definition 4), and it is shown that, for $g : \Re \to \Re$ measurable, it holds that $\int_\Omega g(X)dP = \int_\Re g(x)dP_X$ (Theorem 13).

Chapter 5 Standard Convergence Theorems, The Fubini Theorem

In this chapter, one finds the standard convergence theorems, such as the Lebesgue Monotone Convergence Theorem, the Fatou–Lebesgue Theorem, and the Dominated Convergence Theorem (Theorems 1–3). All these theorems, provide, in effect, conditions under which the limit can be taken under the integral sign. As an application of such theorems, one also establishes conditions for interchanging the operations of differentiation and integration (Theorems 4 and 5). Next, convergence in distribution of a sequence of r.v.s is defined (Definition 2), and the fact that this convergence is implied by convergence in probability is stated (Theorem 6), along with the very convenient Slutsky Theorem (Theorem 7). The second part of the chapter is devoted to a detailed discussion of the Product Measure Theorem (Theorem 11) and of the Fubini Theorem (Theorem 12), which gives conditions for interchanging the order of integration.

Chapter 6 Standard Moment and Probability Inequalities, Convergence in the rth Mean and its Implications

The first part of this chapter is devoted to the discussion of the basic moment inequalities, such as the Hölder inequality (and its special case the Cauchy–Schwaz inequality) (Theorem 2), which provide bounds for the expectation of the product of two r.v.s by moments of the individual r.v.s; the Minkowski and the c_r-inequality (Theorems 3 and 4), which provide bounds of a moment of the sum of two r.v.s by moments of the individual r.v.s; and the Jensen inequality (Theorem 5), where a convex function is involved. Moments are also used for providing bounds for certain probabilities, with the Markov and the Tchebichev inequalities being taken as special cases (Theorem 6). In the second part of the chapter, convergence in the rth mean is introduced (Definitions 2 and 3), as well as the concepts of uniform integrability and uniform (absolute) continuity (Definitions 4 and 5), along with a related result (Theorem 11). Some implications of convergence in the rth mean are studied (Theorems 7, 8, and 12), sufficient conditions for convergence in the rth mean are given (Theorems 8 and 13, and Corollaries 1–3), and, finally, necessary and sufficient conditions for convergence in the rth mean are discussed (Theorems 9, 13, and 14; see also Theorem 15).

Chapter 7 The Hahn–Jordan Decomposition Theorem, The Lebesgue Decomposition Theorem, and the Radon–Nikodym Theorem

This chapter discusses three theorems, each of which is instrumental in establishing the next one. The first theorem is the Hahn–Jordan Decomposition Theorem (Theorem 1), which, in effect, says that a σ-additive set function ϕ defined on $(\Omega, \mathcal{A})$ is uniquely written as a signed measure; i.e., as the difference of two measures, $\phi = \phi^+ - \phi^-$. In the Lebesgue Decomposition Theorem (Theorem 2), one starts out with two σ-finite measures μ and ν defined on $(\Omega, \mathcal{A})$, and shows that ν is uniquely written as the sum of two measures, $\nu = \nu_c + \nu_s$, where ν_s is μ-singular and ν_c is μ-continuous. Furthermore $\nu_c(A) = \int_A X d\mu$, $A \in \mathcal{A}$, where the r.v. X is nonnegative, and a.e.$[\mu]$ finite and unique. A specialization of the Lebesgue Decomposition Theorem yields the Radon–Nikodym Theorem (Theorem 3), according to which, if ν is already μ-continuous and μ is λ-continuous, then $\nu(A) = \int_A X d\mu = \int_A X(\frac{d\mu}{d\lambda})d\lambda$, $A \in \mathcal{A}$.

Chapter 8 Distribution Functions and Their Basic Properties, Helly–Bray Type Results

This chapter deals with distribution functions (d.f.s) and sequences of d.f.s. Here a d.f. is a nonnegative, bounded, nondecreasing, and right-continuous function (Definition 1;

see also Remark 5). It is shown that a d.f. has several properties, some of which are the following. It is uniquely determined by a nonnegative, bounded, and nondecreasing function defined on a set dense in $\Re$ (Propositions 1 and 2); its discontinuities, if any, are jumps, which are countably many (Theorem 1); it is uniquely decomposed into the d.f.s, F_d, F_{cc}, and F_{cs}, say, where F_d is a step function (Definition 2), the measure induced by F_{cc} is λ-continuous, where λ is the Lebesgue measure, and the measure induced by F_{cs} is λ-singular (Theorem 2 and its Corollary). Next, a sequence $\{F_n\}$, $n \geq 1$, of d.f.s is said to converge weekly to a d.f. F, if $F_n(x) \underset{n\to\infty}{\longrightarrow} F(x)$, for all continuity points x of F (Definition 3; see also Theorem 4), and it is shown that, given an arbitrary (bounded) sequence of d.f.s, one can always extract a subsequence, which converges weakly to a d.f. (Theorem 5). Finally, the so-called Helly–Bray results are established, where conditions are given under which $\int_{(\alpha,\beta]} g(x)dF_n(x) \underset{n\to\infty}{\longrightarrow} \int_{(\alpha,\beta]} g(x)dF(x)$ (Theorem 6), and $\int_{\Re} g(x)dF_n(x) \underset{n\to\infty}{\longrightarrow} \int_{\Re} g(x)dF(x)$ (Theorems 7 and 8).

Chapter 9 Conditional Expectation and Conditional Probability, and Related Properties and Results

All r.v.s considered in this chapter are assumed to be integrable. For a r.v. X defined on $(\Omega, \mathcal{A}, P)$, its conditional expectation, given a σ-field $\mathcal{B} \subseteq \mathcal{A}$, is defined, $\mathcal{E}^{\mathcal{B}}X$, by exploiting the Radon–Nikodym Theorem (Definition 1). The conditional probability of $A \in \mathcal{A}$, given $\mathcal{B}$, is taken as a special case for $X = I_A$. The conditional expectation has numerous properties, some of which are established in this chapter. A group of properties of the conditional expectation encompass the following ones: the expectation of the conditional expectation is the expectation; the conditional expectation of a $\mathcal{B}$-measurable r.v. X is X a.s.; the conditional expectation of a constant c is c a.s.; the conditional expectation preserves order a.s.; the conditional expectation preserves linearity a.s.; a $\mathcal{B}$-measurable r.v. can be pulled out a.s. when taking the conditional expectation of the product of two r.v.s; when taking the conditional expectation successively with respect to two nested σ-field, one is left a.s. with the conditional expectation with respect to the smaller σ-field; and finally, when a r.v. and $\mathcal{B}$ are independent, then its conditional expectation is its expectation a.s. (Theorems 1, 2, 4, 9, and 10).

Another group of properties provides versions of the Lebesgue Monotone Convergence Theorem (Theorem 3), of the Fatou–Lebesgue Theorem, and of the Dominated Convergence Theorem for the conditional expectation (Theorem 7).

Still another group of properties establishes versions of the following inequalities for the conditional expectation: Hölder, Minkowski, c_r (Theorem 5), and Jensen inequalities (Theorem 8). It is also shown that convergence in the rth mean implies the same convergence for the conditional expectations (Theorem 6).

Finally, it is proved that the conditional expectation of the r.v. X, given another r.v. Y (i.e., given the σ-field induced by Y), is a.s. equal to a function of Y (Theorem 11).

Chapter 10 Independence

This brief chapter deals with the concept of independence. It starts with independence of a finite number of events (Definition 1), proceeds with independence of a finite number of classes of events (Definition 2), and concludes with the definition of independence of a finite number of r.v.s, which is reduced to that of the induced σ-fields (Definition 3). An easy result is established, according to which if $X_1, \ldots, X_n$ are independent and $Y_j = g_j(X_j)$ for some $g_j : \Re \to \Re$ measurable, $j = 1, \ldots, n$, then $Y_1, \ldots, Y_n$ are also independent (Proposition 1). However, the highlight of this chapter is the justification of the hard part of the characterization of independence of n r.v.s through d.f.s. Namely, $X_1, \ldots, X_n$ are independent if and only if $F_{X_1,\ldots,X_n}(x_1, \ldots, x_n) = F_{X_1}(x_1)\ldots F_{X_n}(x_n)$ for all $x_1, \ldots, x_n$ in $\Re$ (Theorem 1). In the process, it is also shown that σ-fields generated by independent fields are independent (Proposition 2). The chapter is concluded with the proof of the familiar result that, under independence, the expectation of the product of two r.v.s is equal to the product of their expectations (proof of Lemma 1 stated in Chapter 9).

Chapter 11 Topics from the Theory of Characteristic Functions

This is an extensive chapter on characteristic functions (ch.f.s) and their ramifications. The ch.f. of a general d.f. is defined, and we obtain as a special case the ch.f. of a d.f. of a r.v. (Definition 1). An installment of basic properties of a ch.f. are established, including its boundedness (in norm), its uniform continuity, and the relations of its derivatives to the moments of the respective r.v. (in case the ch.f. is that of a r.v.). Also, established is the fact that a r.v. X is symmetric about 0 (Definition 7), if and only if its ch.f. is real, as well as the fact that should the ch.f.s $f_n(t) \underset{n\to\infty}{\longrightarrow} f(t)$, some ch.f., then this convergence is uniform in closed intervals in $\Re$ (Theorems 1, 7, and 8).

A d.f. determines a ch.f. through the definition of a ch.f. The important fact is that the converse is also true. This deep result is referred to as the inversion formula and is established here in great detail (Theorem 2 and its Corollaries); also, it is illustrated by a couple of examples (Examples 1 and 2).

The next major undertaking is that of establishing the P. Lévy Continuity Theorem (Theorem 3; see also Theorem 3*). The most useful and involved part of it states, in effect, that weak convergence of d.f.s, which is hard to handle, is reduced to convergence of the respective ch.f.s. This latter result is an analytical result and much more tractable.

Next, a brief passage is made to the k-dimensional case, which includes formulation of the appropriate versions of the inversion formula (Theorem 2′), and of the Continuity Theorem (Theorem 3′). Also, it is shown how the Cramér-Wold device reduces, in effect, the k-dimensional case to the one-dimensional case (Theorem 4).

The convolution of two d.f.s is defined (Definition 6), and it is shown to be a symmetric relation. Also, the relation of the convolution of two d.f.s to their ch.f.s is discussed (Theorem 5, and Theorem 6 and its Corollaries 1 and 2). In case the d.f.s are those of r.v.s, it is shown that their convolution is simply the d.f. of the sum of two independent r.v.s (Corollary 3 to Theorem 6).

Returning to a ch.f., it is shown how a ch.f. can be expanded in a Taylor-type manner with the remainder given in three different forms (Theorem 9). As a simple application of this expansion, we obtain the Weak Law of Large Numbers (WLLN) (Application 1), and the Central Limit Theorem (CLT) (Application 2). The last topic discussed here refers to the connection between the moments of a r.v. and its distribution. It is shown that, under certain conditions, the moments of a r.v. completely determine its distribution through its ch.f. (Theorem 11). This result elevates the significance of moments.

In the last short section of this chapter, some basic concepts and results from complex analysis are reviewed, in order to provide the necessary support for some of the derivations.

Chapter 12 The Central Limit Problem: The Centered Case

Indisputably, the most important theorem in probability is the Central Limit Theorem (CLT). This is the subject matter of this chapter in an expanded form. The problem is cast as follows: A triangular array of independent (within each row) r.v.s is considered, and the sequence of the sums of the row r.v.s is investigated from the weak convergence viewpoint. The basic assumption is made that the contribution of each r.v. to the row sum is negligible, in a sense that is made precise in terms of probabilities and variances. Then, under some additional minor assumptions, a necessary and sufficient condition is stated for the row sums to converge in distribution to a $N(0, 1)$ distributed r.v. This result is the so-called Lindeberg–Feller Theorem (Theorem 1).

Actually, the scope of this chapter is much broader than just stated. Specifically, under a set of very general conditions (which are summarized under (C) in relation (12.1)), the first question posed is, what is the class of all possible limit laws of the row sums (in the weak convergence sense), and the second, under what conditions do the row sums converge weakly to a given member of this class? The answer to these questions is provided by Theorem 2, and they constitute what is known as the Central Limit Problem. The proof of Theorem 2 is long and is deferred to the last section of the chapter. As one would suspect, Theorem 1 should fall out of Theorem 2, which it does as is explicitly shown. There are another three theorems revolving around the CLT. Theorem 3 (Liapounov) gives sufficient conditions in terms of moments for the CLT to hold. Theorem 4 is a variation of Theorem 1, and Theorem 5 discusses the case where the triangular array of r.v.s collapses into a sequence of r.v.s.

Chapter 13 The Central Limit Problem: The Non-centered Case

The problem investigated in this chapter is pretty much the same as the problem considered in the previous chapter with the following two differences. The row r.v.s are not assumed to be centered at their expectations, and the sum of their variances is not assumed to be equal to 1, but rather to be bounded by a (finite) constant. This necessitates carrying along the sum of the expectations of the row r.v.s, as well as the sum of their variances. Then a version of Theorem 2 of the previous chapter is established (Theorem 2′ in this chapter). As special cases of Theorem 2′, we state necessary and sufficient conditions for the row sums to converge weakly to a $N(\mu, \sigma^2)$ distributed r.v. (Theorem 1), as well as necessary and sufficient conditions for the row sums to converge weakly to a Poisson, $P(\lambda)$, distributed r.v. (Theorem 2). As an application of Theorem 2, the familiar result is established, according to which, under suitable conditions, Binomial probabilities are approximated by (converge to) Poisson probabilities.

Chapter 14 Topics from Sequences of Independent Random Variables

The ultimate purpose of this chapter is the proof of the Kolmogorov Law of Large Numbers (LLN) stated in Theorem 7. In the process of doing so, a number of auxiliary results are established that, however, are of interest in their own right. The first such result is the Kolmogorov inequality (Theorem 1). This is a two-sided inequality, which is useful in many circumstances. An application of the right-hand side of Kolmogorov's inequality provides a sufficient condition for the a.s. convergence of the series $\sum_{n=1}^{\infty}(X_n - \mathcal{E}X_n)$ of independent r.v.s (Theorem 2). Next, for any events $A_n, n \geq 1$, the Borel–Cantelli Lemma (Theorem 3) states that if $\sum_{n=1}^{\infty} P(A_n) < \infty$, then $P(\overline{\lim}_{n\to\infty} A_n) = 0$. Should the events be independent, the relation $\sum_{n=1}^{\infty} P(A_n) = \infty$ implies that $P(\overline{\lim}_{n\to\infty} A_n) = 1$, which is one part of the Borel Zero–One Criterion (Theorem 4). An application of Theorem 3 yields the result that, for any two sequences $\{X_n\}$ and $\{X'_n\}, n \geq 1$, of r.v.s, the condition $\sum_{n=1}^{\infty} P(X_n \neq X'_n) < \infty$ implies that the series $\sum_{n=1}^{\infty} X_n$ and $\sum_{n=1}^{\infty} X'_n$ converge essentially on the same set; also, the sequences $\{S_n/n\}$ and $\{S'_n/n\}$ converge essentially on the same set and have the same limit as $n \to \infty$, where $S_n = \sum_{j=1}^{n} X_j$ and likewise for S'_n. This is so by Theorem 5. The next two results are two analytical lemmas (Toeplitz and Kronecker), followed by a weak form of an SLLN for independent r.v.s (Theorem 6). Finally, before the actual justification of the Kolmogorov SLLN is undertaken, an interesting lemma is established, according to which, for a r.v. X, $\sum_{j=1}^{\infty} P(|X| \geq j) \leq \mathcal{E}|X| \leq 1 + \sum_{j=1}^{\infty} P(|X| \geq j)$. This result provides, among other things, a necessary and sufficient condition for integrability of the r.v. X.

For independent identically distributed (i.i.d.) r.v.s with expectation either ∞ or $-\infty$, Theorem 8 shows that the averages S_n/n still converge (actually, diverge) a.s. to the expectation as $n \to \infty$. However, if the expectation does not exist, then the sequence $\{S_n/n\}$ is unbounded with probability 1; i.e., for every $M > 0$, $P[\overline{\lim}_{n\to\infty}(|S_n/n| > M)] = 1$. This is what Theorem 9 asserts.

For independent r.v.s, it is shown that their tail σ-field $\mathcal{T}$ (Definition 1) obeys the Kolmogorov Zero–One Law (Theorem 10); i.e., $\mathcal{T}$ is equivalent to the trivial σ-field

$\{\emptyset, \Omega\}$, or to put it differently, $P(A) = 0$ or 1 for every $A \in \mathcal{T}$. The chapter is concluded with the so-called Three Series Criterion (Theorem 11), which provides necessary and sufficient conditions for the a.s. convergence of the series $\sum_{n=1}^{\infty} X_n$ of independent r.v.s.

Chapter 15 Topics from Ergodic Theory

The ultimate purpose of this chapter is to state and prove the Ergodic Theorem in its a.s. form, as well as the first-mean form. To this end, we have to introduce the necessary concepts, and also establish a long list of auxiliary results that will allow us to prove the Ergodic Theorem.

We start out with a discrete time parameter stationary (stochastic) process (Definition 3), and then provide a characterization of such processes (Proposition 2). It is easily seen that independent r.v.s constitute a stationary process, if and only if they are i.i.d. (Proposition 3). Also, a general way of constructing a stationary process by means of a given stationary process is provided (Proposition 4). Since throughout the chapter we keep moving between an abstract probability space and the countably infinite product space of the Borel real line, we need the concept of the coordinate process, which is given in Definition 2.

The second installment of notation and results starts out with a measurable transformation T defined on $(\Omega, \mathcal{A}, P)$ into itself; it is defined when T is said to be measure-preserving (Definition 4), and it is shown that, if T is measure-preserving on a field $\mathcal{F}$, then it is measure-preserving on the σ-field $\sigma(\mathcal{F})$ (Proposition 5). By means of T and a r.v. X, one defines r.v.s as in relations (15.8) and (15.9), and then one shows that these r.v.s form a stationary sequence, if and only if T is measure-preserving (Proposition 6). For the reason cited in the previous paragraph, we also introduce here the shift transformation (Definition 5), and show that this transformation is measurable (Proposition 7) and that the coordinate process is stationary, if and only if the shift transformation is measure-preserving (Proposition 8).

For a measurable transformation T as given earlier, we define invariance and a.s. invariance of an event relative to T (Definition 6), and we show that the classes of invariant and a.s. events form σ-fields (Propositions 12 and 14). The transformation T is said to be ergodic, if the σ-field of invariant events (relative to T), $\mathcal{J}$, is equivalent to $\{\emptyset, \Omega\}$ (Definition 8).

We then proceed to define invariance of a r.v. X relative to a transformation T (Definition 9) and to show that X is invariant if and only if it is $\mathcal{J}$-measurable (Proposition 16). Then the transformation T is ergodic if and only if every $\mathcal{J}$-measurable real-valued r.v. is a.s. equal to a constant (Proposition 17).

At this point, we have all we need to prove the so-called Maximal Ergodic Theorem (Theorem 2), and then the Ergodic Theorem in its a.s. form (Theorem 1), as well as the theorem in its first mean form (Corollary 4). When the transformation T is ergodic, then the limit in the Ergodic Theorem is a.s. equal to the $\mathcal{E}X$ (Corollary 1).

In the final section of this chapter, we consider a stationary process **X**, but we do not assume that **X** is generated by a measure-preserving transformation T. Then some of the concepts defined earlier in the chapter are redefined in terms of **X**. Thus, invariance of an event A relative to **X** is defined (Definition 11), and it is shown that the class of invariant events, $\mathcal{J}$, is a σ-field (Proposition 19). Also, invariance of a r.v. Y relative to **X** is defined (Definition 12), and it is shown that Y is invariant if and only if it is $\mathcal{J}$-measurable (Proposition 20). Then for a stationary process, as considered here, the Ergodic Theorem is restated and proved in its a.s. form as well as in the first mean (Theorem 3 and Corollary 1). If the process **X** is ergodic; i.e., if $\mathcal{J}$ is equivalent to $\{\oslash, \Omega\}$, then the limit in the Ergodic Theorem is a.s. equal to the $\mathcal{E}X_1$ (Corollary 2). Finally, by means of the X_ns, we define the Y_n r.v.s, as in Proposition 21, and show that the resulting process **Y** inherits the properties of **X**. That is, if **X** is stationary, so is **Y**, and if **X** is ergodic, so is **Y**.

Chapter 16 Two Cases of Statistical Inference: Estimation of a Real-valued Parameter, Nonparametric Estimation of a Probability Density Function

The purpose of the addition of this chapter is to demonstrate as to how some of the theorems, corollaries, etc. discussed in the book apply in establishing statistical inference results. Two cases of statistical estimation are selected for this demonstration. The first is the maximum likelihood estimation of a real-valued parameter in an assumed (parametric) model, and the other is the nonparametric estimation of a probability density function via the kernel methodology.

Actually, in the former case, it is shown that there is a sequence of roots of the log-likelihood function which sequence is almost surely consistent, as an estimate of the underlying parameter, as well as asymptotically normal. In the latter case, the proposed estimate is shown to be asymptotically unbiased, consistent in quadratic mean and in the probability sense, and also asymptotically normal.

This page is intentionally left blank

APPENDIX

Brief Review of Riemann–Stieltjes Integral

B

In this part of the Appendix, a brief review is given for the Riemann–Stieltjes integral, and its relation to the Riemann integral, as well as the Lebesgue integral on the real line.

To this effect, let $g : [a, b] \to \Re$ $(-\infty < a < b < \infty)$, and let F be a d.f. A *partition* $\mathcal{P}$ of $[a, b]$ is a set of points $\{x_0, x_1, \ldots, x_n\}$ with $a = x_0 < x_1 < \cdots < x_n = b$, and another partition $\mathcal{Q}$ of $[a, b]$ is said to be *finer* than $\mathcal{P}$ if $\mathcal{P} \subset \mathcal{Q}$. In terms of $\mathcal{P}$ and $\boldsymbol{\xi} = \{\xi_1, \ldots, \xi_n\}$, define the Riemann–Stieltjes sum

$$S(\mathcal{P}, \boldsymbol{\xi}) = \sum_{j=1}^{n} g(\xi_j)\left[F(x_j) - F(x_{j-1})\right],$$

where ξ_j is an arbitrary point in $[x_{j-1}, x_j]$, $j = 1, \ldots, n$.

The function g is said to be *Riemann–Stieltjes integrable* with respect to F over the interval $[a, b]$ if there is a number γ and a partition $\mathcal{P}$ such that, for every $\varepsilon > 0$, every partition $\mathcal{Q}$ finer than $\mathcal{P}$ and any choice of $\xi_j \in [x_{j-1}, x_j]$, $j = 1, \ldots, n$, it holds $|S(\mathcal{Q}, \boldsymbol{\xi}) - \gamma| < \varepsilon$. Then γ is called the *Riemann–Stieltjes integral* of g (with respect to F over $[a, b]$) and is denoted by $\int_a^b g dF$ or $\int_a^b g(x) dF(x)$.

Conditions ensuring the existence of $\int_a^b g dF$ can be found, e.g., in Theorem 9–19, page 206, and Theorem 9–26, page 211, in Apostol (1958). It is to be mentioned here that throughout this appendix the term "existence" of an integral always includes its being finite.

Remark 1. If the integrator is a monotone function, which is the case in this book, the Riemann–Stieltjes integral may be approached by means of the so-called upper and lower Stieltjes sums, and upper and lower Stieltjes integrals. Necessary and sufficient conditions are also given for the existence of the Riemann–Stieltjes integral by means of the quantities just mention. For these, see, e.g., Definitions 9–14, 9–16, 9–18, pages 204–206, and Theorem 9–19, page 206, in Apostol (1958).

The Riemann–Stieltjes integral has properties similar to those of the Riemann integral. See, e.g., Theorems 9–2, 9–3, 9–4, 9–6, 9–7, pages 193–196, and Theorem 9–29, page 213, in Apostol (1958). The following result provides a relation between the Riemann–Stieltjes and the Riemann integral of g.

Proposition 1. Suppose that the Riemann–Stieltjes integral $\int_a^b g(x) dF(x)$ exists and that F has a continuous derivative F' on $[a, b]$. Then the Riemann integral

An Introduction to Measure-Theoretic Probability, Second Edition. http://dx.doi.org/10.1016/B978-0-12-800042-7.00025-6

$\int_a^b g(x)F'(x)dx$ exists and we have

$$\int_a^b g(x)dF(x) = \int_a^b g(x)F'(x)dx.$$

Proof. See, e.g., Theorem 9–8, page 197, in Apostol (1958). ■

The following version of the previous result provides a necessary and sufficient condition for the existence of $\int_a^b g(x)dF(x)$.

Proposition 2. Let g be a bounded function on $[a, b]$ and let F have a continuous derivative F' on $[a, b]$. Then gF' is Riemann integrable over $[a, b]$ if and only if g is Riemann–Stieltjes integrable with respect to F over $[a, b]$, and then

$$\int_a^b g(x)dF(x) = \int_a^b g(x)F'(x)dx.$$

Proof. See, e.g., Proposition 2, page 705, in Fristedt and Gray (1997). ■

For an arbitrary $a \in \Re$, the improper Riemann–Stieltjes integral $\int_a^\infty g(x)dF(x)$ is defined in a way similar to that of the improper Riemann integral $\int_a^\infty g(x)dx$. Specifically, let a be kept fixed and suppose that $\int_a^b g(x)dF(x)$ exists for every $b > a$. We also suppose that $\lim_{b\to\infty} \int_a^b g(x)dF(x)$ exists. Then this limit is denoted by $\int_a^\infty g(x)dF(x)$ and is called the *improper* Riemann–Stieltjes integral of g with respect to F over $[a, \infty)$.

The result below gives a necessary and sufficient condition for the existence of $\int_a^\infty g(x)dF(x)$.

Proposition 3. Let $g : [a, \infty)\ \Re$ be Riemann–Stieltjes integrable with respect to F on every interval $[c, d]$ with $a \le c < d$. Then $\int_a^\infty g(x)dF(x)$ exists, if and only if for every $\varepsilon > 0$ there exists $M = M(\varepsilon) > 0$ such that $|\int_c^d g(x)dF(x)| < \varepsilon$ for all $d > c \ge M$.

Proof. Assume that $\int_a^\infty fdF$ exists and denote it by I. This means that $I = \lim_{b\to\infty} \int_a^b fdF$. Then for every $\varepsilon > 0$, choose $M = M(\varepsilon)$ so that $|I - \int_a^b fdF| < \varepsilon$ for $b \ge M$. Next, for $d > c \ge M$,

$$|\int_c^d fdF| \le |I - \int_a^c fdF| + |I - \int_a^d fdF| \le \varepsilon + \varepsilon = 2\varepsilon,$$

so that the condition $|\int_c^d fdF| < 2\varepsilon$, $d > c \ge M$ is satisfied.

Next, assume that $|\int_c^d fdF| < \varepsilon$ for $d > c \ge M$, and let $(a \le)a_n \underset{n\to\infty}{\longrightarrow} \infty$. By setting $A_n = \int_a^{a_n} fdF$, it is easy to see that $\{A_n\}$, $n \ge 1$, is a Cauchy sequence. Therefore there is $A \in \Re$ such that $\lim A_n = A$ as $n \to \infty$. Likewise, for $(a \le)b_n \underset{n\to\infty}{\longrightarrow} \infty$, set $B_n = \int_a^{b_n} fdF$, so that $\{B_n\}$, $n \ge 1$, is also a Cauchy sequence. Hence there is $B \in \Re$ such that $\lim B_n = B$ as $n \to \infty$. But then

$$|A - B| \le |A - A_n| + |B - B_n| + |A_n - B_n| \tag{B.1}$$

$$= |A - A_n| + |B - B_n| + |\int_{a_n}^{b_n} fdF| \tag{B.2}$$

$$< \varepsilon + \varepsilon + \varepsilon = 3\varepsilon \tag{B.3}$$

for sufficiently large n, so that $a_n, b_n \geq M$. It follows that $A = B$; i.e., the $lim \int_a^{a_n} fdF$ is independent of the sequence $(a \leq)a_n \underset{n\to\infty}{\longrightarrow} \infty$, and hence the $\lim_{b\to\infty} \int_a^b fdF = \int_a^\infty fdF$ exists. ■

The *improper* integral $\int_{-\infty}^b fdF$ is defined similarly. If both $\int_{-\infty}^a fdF$ and $\int_a^\infty fdF$ exist for some $a \in \Re$, we say that the *improper* integral $\int_{-\infty}^\infty fdF$ exists and is defined by the sum

$$\int_{-\infty}^{\infty} fdF = \int_{-\infty}^{a} fdF + \int_{a}^{\infty} fdF.$$

The choice of a is clearly immaterial.

If $g : \Re \to \Re$ is a Borel function, so that $g(X)$ is a r.v. if X is a r.v., it has been seen (see Theorem 13 in Chapter 4) that

$$\mathcal{E}g(X) = \int_\Omega g(X)dP = \int_\Re g(x)dP_X$$

in the sense that, if one side exists so does the other and they are equal. Also, the further notation $\int_\Re g(x)dF_X$ was used to denote $\int_\Re g(x)dP_X$. Here P_X is the (probability) distribution of X, under P, and F_X is its d.f. However if $\int_R g(x)dF_X$ is employed as a Riemann–Stieltjes integral (as is the case in the Helly-Bray type results in Section 8.3 of Chapter 8), one should ensure conditions under which it is, indeed, true that $\int_\Re g(x)dP_X = \int_\Re g(x)dF_X$.

The results below answer this question.

Proposition 4. Suppose that $g : \Re \to \Re$ Borel is Riemann–Stieltjes integrable on every interval $[a, b]$ with $\infty < a < b < \infty$ with respect to the d.f. F_X. Then, if $\int_\Re g(x)dP_X$ exists, it follows that

$$\int_\Re g(x)dP_X = \int_\Re g(x)dF_X (= \int_{-\infty}^{\infty} g(x)dF_X).$$

Proof. See, e.g., Theorem 14, page 111, in Fristedt and Gray (1997). ■

The last proposition above requires the existence of the Riemann–Stieltjes integral $\int_a^b f(x)dF_X$ over every interval $[a, b]$ $(-\infty < a < b < \infty)$ for the conclusion to be reached. The result below provides conditions for this to be the case.

Proposition 5. If $g : [a, b] \to \Re$ is continuous, then the Riemann–Stieltjes integral $\int_a^b g(x)dF_X$ exists.

Proof. See, e.g., Theorem 9–26, page 211, in Apostol (1958). ■

Remark 2. Use of suitable combinations of results provided in this Appendix justify fully the arguments employed in discussing Theorems 6–8 in Chapter 8.

This page is intentionally left blank

APPENDIX

Notation and Abbreviations

C

Table C.1 Notation and Abbreviations

$\mathcal{F}, \mathcal{A}$	field, σ-field
$\mathcal{F}(C), \sigma(C)$	field generated by class C, σ-field generated by class C
$\mathcal{A}_A$	σ-field of members of $\mathcal{A}$ which are subsets of A
$(\Omega, \mathcal{A})$	measurable space
$\Re^k, \mathcal{B}^k, k \geq 1$	k-dimensional Euclidean space, Borel σ-field
$(\Re^1, \mathcal{B}^1) = (\Re, \mathcal{B})$	Borel real line
$\uparrow, \downarrow$	increasing (nondecreasing), decreasing (nonincreasing)
$P, (\Omega, \mathcal{A}, P)$	probability measure, probability space
I_A	indicator of the set A
$X^{-1}(B)$	inverse image of the set B under X
$\mathcal{A}_X$ or X^{-1}(σ-field)	σ-field induced by X
$(X \in B) = X^{-1}(B)$	the set of points in Ω for which X takes values in B
r.v., r. vector	random variable, random vector
$B(n, p)$	Binomial distribution (or r.v.) with parameters n, p
$P(\lambda)$	Poisson distribution (or r.v.) with parameter λ
$N(\mu, \sigma^2)$	Normal distribution (or r.v.) with parameters μ, σ^2
Φ	distribution function of $N(0, 1)$
$\mathcal{E}X$ or $\mu(X)$ or μ_X or just μ	expectation (mean value, mean) of X
$\mathcal{E}^{\mathcal{B}}X$	conditional expectation of r.v. X, given σ-field $\mathcal{B}$
$P^{\mathcal{B}}A$	conditional probability of event A, given σ-field $\mathcal{B}$
$\sigma^2(X)$ $(\sigma(X))$ or σ_X^2 (σ_X) or just σ^2 (σ)	variance (standard deviation) of X
Cov (X, Y), $\rho(X, Y)$	Covariance of X, Y, correlation coefficient of X, Y
φ_X or $\varphi_{X_1,\ldots,X_n}$ or just φ	characteristic function (ch.f.) of X or joint ch.f. of $X_1, \ldots, X_n$
d.f.	distribution function
$\xrightarrow{a.s.}$	almost sure (a.s.) convergence (convergence with probability 1)
$\xrightarrow{a.e.}$	almost everywhere (a.e.) convergence
$\xrightarrow{a.u.}$	almost uniform convergence
$\xrightarrow{c}$	complete convergence (of d.f.s)
$\xrightarrow{w}$	weak convergence (of d.f.s)

(Continued)

An Introduction to Measure-Theoretic Probability, Second Edition. http://dx.doi.org/10.1016/B978-0-12-800042-7.00026-8

Table C.1 Continued

$\xrightarrow{d}, \xrightarrow{\mu}, \xrightarrow{P}$	convergence in distribution, in measure, in probability
$\xrightarrow{q.m.}, \xrightarrow{(r)}$	convergence in quadratic mean, in the rth mean
$(\Omega, \mathcal{A}, P_\theta)$	probability space
θ or $\boldsymbol{\theta}$	parameter
Θ	parameter space
MLE	Maximum Likelihood Estimate
$\tilde{\theta}_n$ or $\hat{\theta}_n$	estimate of θ
$I(\theta)$	Fisher information
p.d.f	probability density function

Selected References

[1] Apostol, Tom M. (1958) *Mathematical Analysis*, second printing. Addison-Wesley Publishing Company, Reading, MA.

[2] Billingsley, Patrick (1995) *Probability and Measure*, 3rd ed. John Wiley & Sons, New York.

[3] Billingsley, Patrick (1999) *Convergence of Probability Measures*, 2nd ed. John Wiley & Sons, New York.

[4] Dieudonné, J. (1960) *Foundations of Modern Analysis*. Academic Press, Boston.

[5] Ferguson, Thomas S. (1996). *A Course in Large Sample Theory*. Chapman & Hall/CRC.

[6] Fristedt, Bert and Gray, Lawrence (1997) *A Modern Approach to Probability Theory*. Birkhäuser, Boston. Basel, Berlin.

[7] Hardy, G. H., Littlewood, J. E., and Pölya, G. (1967) *Inequalities*, 2nd ed. Cambridge University Press, Cambridge, U.K.

[8] Loève, Michel (1963) *Probability*, 3rd ed. Van Nostrand Company, Princeton, NJ.

[9] Munroe, M. E. (1953) *Introduction to Measure and Integration*. Addison-Wesley Publishing Company, Reading, MA.

[10] Neveu, J. (1965) *Foundations of the Calculus of Probability*. Holden-Day, San Francisco.

[11] Parzen, E. (1962) On estimation of a probability density function and mode. *Annals of Mathematical Statistics*, Vol. 33, pages 1065–1076.

[12] Rao, C. R. (1965) *Linear Statistical Inference and Its Applications*. John Wiley & Sons, New York.

[13] Rosenblatt, M. (1956) Remarks on some nonparametric estimates of a density function. *Annals of Mathematical Statistics*, Vol. 27, pages 832–837.

[14] Roussas, George G. (1997) A Course in Mathematical Statistics, 2nd ed. Academic Press, Boston.

[15] Royden, H.L. (1988) *Real Analysis*, 3rd ed. Macmillan, New York.

[16] Scheffé, Henry (1947) A Useful Convergence Theorem for Probability Distributions. *Annals of Mathematical Statistics*, Vol. 18, pages 434–438.

[17] Shiryaev, A. N. (1995) *Probability*, 2nd ed. Springer, New York.

[18] Tallarida, Ronald J. (1999) *Pocket Book of Integrals and Mathematical Formulas*, 3rd ed. Chapman & Hall/CRC.

An Introduction to Measure-Theoretic Probability, Second Edition. http://dx.doi.org/10.1016/B978-0-12-800042-7.00027-X

[19] Titchmarsch, E. C. (1939) *The Theory of Functions*, 2nd ed. Oxford University Press, Oxford, U.K.
[20] Vestrup, Eric M. (2003) *The Theory of Measure and Integration*. John Wiley & Sons, Inc., New York.

Revised Answers Manual to an Introduction to Measure-Theoretic Probability

George G. Roussas
University of California, Davis, United States

Chapter 1

Certain Classes of Sets, Measurability, Pointwise Approximation

1. **(i)** $x \in \underline{\lim}_{n\to\infty} A_n$ if and only if $x \in \cup_{n\geq 1}\cap_{j\geq n} A_j$, so that $x \in \cap_{j\geq n_0} A_j$ for some $n_0 \geq 1$, and then $x \in A_j$ for all $j \geq n_0$, or $x \in \cup_{j\geq n} A_j$ for all $n \geq 1$, so that $x \in \cap_{n\geq 1}\cup_{j\geq 1} A_j \overline{\lim}_{n\to\infty} A_n$.

(ii) $\left(\underline{\lim}_{n\to\infty} A_n\right)^c = \left(\cup_{n\geq 1}\cap_{j\geq n} A_j\right)^c = \cap_{n\geq 1}\cup_{j\geq n} A_j^c = \overline{\lim}_{n\to\infty} A_n^c$, $\left(\overline{\lim}_{n\to\infty} A_n\right)^c = \left(\cap_{n\geq 1}\cup_{j\geq n} A_j\right)^c = \cup_{n\geq 1}\cap_{j\geq n} A_j^c = \underline{\lim}_{n\to\infty} A_n^c$.
Let $\lim_{n\to\infty} A_n = A$. Then $\underline{\lim}_{n\to\infty} A_n^c = \left(\overline{\lim}_{n\to\infty} A_n\right)^c = \left(\lim_{n\to\infty} A_n\right)^c = A^c$, and $\overline{\lim}_{n\to\infty} A_n = \left(\underline{\lim}_{n\to\infty} A_n\right)^c = \left(\lim_{n\to\infty} A_n\right)^c = A^c$, so that $\lim_{n\to\infty} A_n^c$ exists and is A^c.

(iii) To show that $\underline{\lim}_{n\to\infty}(A_n \cap B_n) = (\underline{\lim}_{n\to\infty} A_n) \cap (\underline{\lim}_{n\to\infty} B_n)$. Equivalently,

$$\bigcup_{n=1}^{\infty}\bigcap_{j=n}^{\infty}(A_j \cap B_j) = \left(\bigcup_{n=1}^{\infty}\bigcap_{j=n}^{\infty} A_j\right) \cap \left(\bigcup_{n=1}^{\infty}\bigcap_{j=n}^{\infty} B_j\right).$$

Indeed, let x belong to the left-hand side. Then $x \in \cap_{j=n_0}^{\infty}(A_j \cap B_j)$ for some $n_0 \geq 1$, hence $x \in (A_j \cap B_j)$ for all $j \geq n_0$, and then $x \in A_j$ and $x \in B_j$ for all $j \geq n_0$. Hence $x \in \cap_{j=n_0}^{\infty} A_j$ and $x \in \cap_{j=n_0}^{\infty} B_j$, so that $x \in \cup_{n=1}^{\infty}\cap_{j=n}^{\infty} A_j$ and $x \in \cup_{n=1}^{\infty}\cap_{j=n}^{\infty} B_j$; i.e., x belongs to the right-hand side. Next, let x belong to the right-hand side. Then $x \in \cup_{n=1}^{\infty}\cap_{j=n}^{\infty} A_j$ and $x \in \cup_{n=1}^{\infty}\cap_{j=n}^{\infty} B_j$, so that $x \in \cap_{j=n_1}^{\infty} A_j$ and $x \in \cap_{j=n_2}^{\infty} B_j$ for some $n_1, n_2 \geq 1$. Then $x \in \cap_{j=n_0}^{\infty} A_j$ and $x \in \cap_{j=n_0}^{\infty} B_j$ where $n_0 = \max(n_1, n_2)$, and hence $x \in A_j$ and $x \in B_j$ for all $j \geq n_0$. Thus, $x \in (A_j \cap B_j)$ for all $j \geq n_0$, so that $x \in \cap_{j=n_0}^{\infty}(A_j \cap B_j)$ and hence $x \in \cup_{n=1}^{\infty}\cap_{j=n}^{\infty}(A_j \cap B_j)$; i.e., x belongs to the left-hand side.

Next, $\overline{\lim}_{n\to\infty}(A_n \cup B_n) = \overline{\lim}_{n\to\infty}(A_n^c \cap B_n^c)^c = [\underline{\lim}_{n\to\infty}(A_n^c \cap B_n^c)]^c$ (by part (ii)), and this equals to $[(\underline{\lim}_{n\to\infty} A_n^c) \cap (\underline{\lim}_{n\to\infty} B_n^c)]^c$ (by what we just proved), and this equals $[(\overline{\lim}_{n\to\infty} A_n)^c \cap (\overline{\lim}_{n\to\infty} B_n)^c]^c = (\overline{\lim}_{n\to\infty} A_n) \cup (\overline{\lim}_{n\to\infty} B_n)$, as was to be seen.

An Introduction to Measure-Theoretic Probability. http://dx.doi.org/10.1016/B978-0-12-800042-7.00028-1

(iv) To show that: $\overline{\lim}_{n\to\infty}(A_n \cap B_n) \subseteq (\overline{\lim}_{n\to\infty}A_n) \cap (\overline{\lim}_{n\to\infty}B_n)$ and $\underline{\lim}_{n\to\infty}(A_n \cup B_n) \supseteq (\underline{\lim}_{n\to\infty}A_n) \cup (\underline{\lim}_{n\to\infty}B_n)$.
Suffices to show: $\cap_{n=1}^{\infty}\cup_{j=n}^{\infty}(A_j \cap B_j) \subseteq \left(\cap_{n=1}^{\infty}\cup_{j=n}^{\infty}A_j\right) \cap \left(\cap_{n=1}^{\infty}\cup_{j=n}^{\infty}B_j\right)$.

Indeed, let x belong to the left-hand side. Then $x \in \cup_{j=n}^{\infty}(A_j \cap B_j)$ for all $n \geq 1$, so that $x \in (A_j \cap B_j)$ for some $j \geq n$ and all $n \geq 1$. Then $x \in A_j$ and $x \in B_j$ for some $j \geq n$ and all $n \geq 1$, hence $x \in \cup_{j=n}^{\infty}A_j$ and $x \in \cup_{j=n}^{\infty}B_j$ for all $n \geq 1$, so that $x \in \cap_{n=1}^{\infty}\cup_{j=n}^{\infty}A_j$ and $x \in \cap_{n=1}^{\infty}\cup_{j=n}^{\infty}B_j$, and hence $x \in \left(\cap_{n=1}^{\infty}\cup_{j=n}^{\infty}A_j\right) \cap \left(\cap_{n=1}^{\infty}\cup_{j=n}^{\infty}B_j\right)$; i.e., x belongs to the right-hand side. So, the above inclusion is correct.

Also, to show that : $\left(\cup_{n=1}^{\infty}\cap_{j=n}^{\infty}A_j\right) \cup \left(\cup_{n=1}^{\infty}\cap_{j=n}^{\infty}B_j\right) \subseteq \cup_{n=1}^{\infty}\cap_{j=n}^{\infty}(A_j \cup B_j)$.

Indeed, let x belong to the left-hand side. Then $x \in \cup_{n=1}^{\infty}\cap_{j=n}^{\infty}A_j$ or $x \in \cup_{n=1}^{\infty}\cap_{j=n}^{\infty}B_j$ or to both. Let $x \in \cup_{n=1}^{\infty}\cap_{j=n}^{\infty}A_j$. Then $x \in \cap_{j=n_0}^{\infty}A_j$ for some $n_0 \geq 1$, hence $x \in A_j$ for all $j \geq n_0$, and then $x \in (A_j \cup B_j)$ for all $j \geq n_0$, so that $x \in \cup_{n=1}^{\infty}\cap_{j=n}^{\infty}(A_j \cup B_j)$; i.e., x belongs to the right-hand side. Similarly if $x \in \cup_{n=1}^{\infty}\cap_{j=n}^{\infty}B_j$.

An alternative proof of the second part is as follows:

$$\begin{aligned}
\underline{\lim}(A_n \cup B_n) &= \bigcup_{n=1}^{\infty}\bigcap_{k=n}^{\infty}(A_k \cup B_k) = \left[\bigcap_{n=1}^{\infty}\bigcup_{k=n}^{\infty}(A_k^c \cap B_k^c)\right]^c \\
&= \left[\overline{\lim}(A_k^c \cap B_k^c)\right]^c \supseteq \left[\left(\overline{\lim}A_k^c\right) \cap \left(\overline{\lim}B_k^c\right)\right]^c \\
&\qquad \text{(by the previous part)} \\
&= \left(\bigcap_{n=1}^{\infty}\bigcup_{k=n}^{\infty}\right)^c \cup \left(\bigcap_{n=1}^{\infty}\bigcup_{k=n}^{\infty} B_k^c\right)^c \\
&= \left(\bigcup_{n=1}^{\infty}\bigcap_{k=n}^{\infty} A_k\right) \cup \left(\bigcup_{n=1}^{\infty}\bigcap_{k=n}^{\infty} B_k\right) = (\underline{\lim}A_n) \cup (\underline{\lim}B_n).
\end{aligned}$$

(v) That the inverse inclusions in part (iv) need not hold is demonstrated by the following
Counterexample:
Let $A_{2j-1} = A$, $A_{2j} = A_0$ and $B_{2j-1} = B$, $B_{2j} = B_0$, $j \geq 1$, for some events A, A_0, B and B_0. Then: $\underline{\lim}_{n\to\infty}A_n = A \cap A_0$, $\overline{\lim}_{n\to\infty}A_n = A \cup A_0$, $\underline{\lim}_{n\to\infty}B_n = B \cap B_0$, $\overline{\lim}_{n\to\infty}B_n = B \cup B_0$, $\overline{\lim}_{n\to\infty}(A_n \cap B_n) = (A \cap B) \cup (A_0 \cap B_0)$, $\underline{\lim}_{n\to\infty}(A_n \cup B_n) = (A \cup B) \cap (A_0 \cup B_0)$. Therefore $(A \cup B) \cap (A_0 \cup B_0)$ need not contain $(A \cup A_0) \cap (B \cup B_0)$, and $(A \cap A_0) \cup (B \cap B_0)$ need not contain $(A \cup B) \cap (A_0 \cup B_0)$.
As a concrete example, take $\Omega = \Re$, $A = (0, 1]$, $A_0 = [2, 3]$, $B = [1, 2]$, $B_0 = [3, 4]$. Then: $(A \cup B) \cap (A_0 \cup B_0) = (0, 2]$, $(A \cup A_0) \cap$

$(B \cup B_0) = ((0,1] \cup [2,3]) \cap ([1,2] \cup [3,4]) = \{1\} \cup \{3\} = \{1,3\} \not\supseteq (0,2]$, and $(A \cap A_0) \cup (B \cap B_0) = \oslash \cup \oslash = \oslash$, $(A \cup B) \cap (A_0 \cup B_0) = (0,2] \cap [2,4] = \{2\}$ not contained in $\oslash$.

(vi) If $\lim_{n\to\infty} A_n = A$ and $\lim_{n\to\infty} B_n = B$, then by parts (iii) and (iv): $\overline{\lim}_{n\to\infty}(A_n \cap B_n) \subseteq A \cap B$ and $\underline{\lim}_{n\to\infty}(A_n \cap B_n) = A \cap B$. Thus, $A \cap B = \underline{\lim}_{n\to\infty}(A_n \cap B_n) \subseteq \overline{\lim}_{n\to\infty}(A_n \cap B_n) \subseteq A \cap B$, so that $\lim_{n\to\infty}(A_n \cap B_n) = A \cap B$. Likewise: $A \cup B \subseteq \underline{\lim}_{n\to\infty}(A_n \cup B_n) \subseteq \overline{\lim}_{n\to\infty}(A_n \cup B_n) = A \cup B$, so that $\lim_{n\to\infty}(A_n \cup B_n) = A \cup B$.

(vii) Since $A_n \Delta B = (A_n - B) + (B - A_n) = (A_n \cap B^c) + (B \cap A_n^c)$, we have $\lim_{n\to\infty}(A_n \cap B^c) = (\lim_{n\to\infty} A_n) \cap B^c = A \cap B^c$ by part (vi), and $\lim_{n\to\infty}(B \cap A_n^c) = B \cap (\lim_{n\to\infty} A_n^c) = B \cap A^c$ by parts (vi) and (ii). Therefore, by part (vi) again, $\lim_{n\to\infty}(A_n \Delta B) = \lim_{n\to\infty}[(A_n \cap B^c) + (B \cap A_n^c)] = \lim_{n\to\infty}(A_n \cap B^c) + \lim_{n\to\infty}(B \cap A_n^c) = (A \cap B^c) + (B \cap A^c) = A \Delta B$.

(viii) $A_{2j-1} = B$, $A_{2j} = C$, $j \geq 1$. Then, as in part (v), $\underline{\lim}_{n\to\infty} A_n = B \cap C$ and $\overline{\lim}_{n\to\infty} A_n = B \cup C$. The $\lim_{n\to\infty} A_n$ exists if and only if $B \cap C = B \cup C$, or $B \cup C = (B \cap C^c) + (B^c \cap C) + (B \cap C) = B \cap C$. Then, by the pairwise disjointness of $B \cap C^c$, $B^c \cap C$ and $B \cap C$, we have $B \cap C^c = B^c \cap C = \oslash$. From $B \cap C^c = \oslash$, it follows that $B \subseteq C$, and from $B^c \cap C = \oslash$, it follows that $C \subseteq B$. Therefore $B = C$. Thus, $\lim_{n\to\infty} A_n$ exists if and only if $B = C$. #

2.

(i) All three sets $\underline{A}$, $\overline{A}$, and A (if it exists) are in $\mathcal{A}$, because they are expressed in terms of A_n, $n \geq 1$, by means of countable operations.

(ii) Let $A_n \uparrow$. Then $\underline{\lim}_{n\to\infty} A_n = \cup_{n=1}^{\infty} \cap_{j=n}^{\infty} A_j = \cup_{n=1}^{\infty} A_n$, and $\overline{\lim}_{n\to\infty} A_n = \cap_{n=1}^{\infty} \cup_{j=n}^{\infty} A_j = \cup_{j=n}^{\infty} A_j = \cup_{j=1}^{\infty} A_j = \cup_{n=1}^{\infty} A_n$, so that $\lim_{n\to\infty} A_n = \cup_{n=1}^{\infty} A_n$.

If $A_n \downarrow$, then $A_n^c \uparrow$ and hence $\cap_{n=1}^{\infty} \cup_{j=n}^{\infty} A_j^c = \cup_{n=1}^{\infty} \cap_{j=n}^{\infty} A_j^c = \cup_{n=1}^{\infty} A_n^c$, so that, by taking the complements, $\cup_{n=1}^{\infty} \cap_{j=n}^{\infty} A_j = \cap_{n=1}^{\infty} \cup_{j=n}^{\infty} A_j = \cap_{n=1}^{\infty} A_n$, so that $\lim_{n\to\infty} A_n = \cap_{n=1}^{\infty} A_n$. #

3.

(i) $\cap_{j\in I} \mathcal{F}_j \neq \oslash$ since, e.g., $\Omega \in \mathcal{F}_j$, $j \in I$. Next, if $A \in \cap_{j\in I} \mathcal{F}_j$ for all $j \in I$, and hence $A^c \in \mathcal{F}_j$ for all $j \in I$, so that $A^c \in \cap_{j\in I} \mathcal{F}_j$. Finally, if $A, B \in \cap_{j\in I} \mathcal{F}_j$, then $A, B \in \mathcal{F}_j$ for all $j \in I$, and hence $A \cup B \in \mathcal{F}_j$ for all $j \in I$, so that $A \cup B \in \cap_{j\in I} \mathcal{F}_j$.

(ii) If $A_i \in \cap_{j\in I} \mathcal{A}_j$, $i = 1, 2, \ldots$, then $A_i \in \mathcal{A}_j$, $i = 1, 2, \ldots$, for all $j \in I$, and hence $\cup_{i=1}^{\infty} A_i \in \mathcal{A}_j$ for all $j \in I$, so that $\cup_{i=1}^{\infty} A_i \in \cap_{j\in I} \mathcal{A}_j$. #

4. Let $\Omega = \Re$, $\mathcal{F} = \{A \subseteq \Re;$ either A or A^c is finite$\}$, and let $A_j = \{1, 2, \ldots, j\}$, $j \geq 1$. Then $\mathcal{F}$ is a field and $A_j \in \mathcal{F}$, $j \geq 1$, but $\cup_{j=1}^{\infty} A_j = \{1, 2, \ldots\} \notin \mathcal{F}$, because neither this set nor its complement is finite.

Also, if $B_j = \{j+1, j+2, \ldots\}$, then $B_j \in \mathcal{F}_j$ since B_j^c is finite, whereas $\cap_{j=1}^{\infty} B_j = \cap_{j=1}^{\infty} A_j^c = \left(\cup_{j=1}^{\infty} A_j\right)^c \notin \mathcal{F}$, as it has been seen already. #

5. Clearly, $\mathcal{C}$ is $\neq \oslash$, every member of $\mathcal{C}$ is a countable union of members of $\mathcal{P}$, and $\mathcal{C}$ is the smallest σ-field containing $\mathcal{P}$, if indeed, is a σ-field. If $B \in \mathcal{C}$, then $B = \cup_{i \in I} A_i$ for some $I \subseteq \mathbb{N} = \{1, 2, \ldots\}$, and then $B^c = \cup_{j \in J} A_j$, where $J = \mathbb{N} - I$, so that $B^c \in \mathcal{C}$. Finally, if $B_j \in \mathcal{C}$, $j = 1, 2, \ldots$, then $B_j = \cup_{i \in I_j} A_{ji}$, where $I_j \subseteq \mathbb{N}$ and $I_i \cap I_j = \oslash$. Then $\cup_{j=1}^{\infty} B_j = \cup_{j=1}^{\infty} \cup_{i \in I_j} A_{ji}$, the union of members of $\mathcal{P}$, so that $\cup_{j=1}^{\infty} B_j$ belongs in $\mathcal{C}$. #

6. Since $\mathcal{C}_j$ and $\mathcal{C}'_j \subseteq \mathcal{C}_0$, $j = 1, \ldots, 8$, it follows that $\sigma(\mathcal{C}_j)$ and $\sigma(\mathcal{C}'_j) \subseteq \sigma(\mathcal{C}_0) = \mathcal{B}$, so that it suffices to show that $\mathcal{B} \subseteq \sigma(\mathcal{C}_j)$ and $\mathcal{B} \subseteq \sigma(\mathcal{C}'_j)$, which are implied, respectively, by $\mathcal{C}_0 \subseteq \sigma(\mathcal{C}_j)$ and $\mathcal{C}_0 \subseteq \sigma(\mathcal{C}'_j)$, $j = 1, \ldots, 8$. As an example, consider the classes mentioned in the hint.
So, to show that $\mathcal{C}_0 \subseteq \sigma(\mathcal{C}_1)$. In all that follows, all limits are taken as $n \to \infty$. Indeed, for $y_n \downarrow y$, we have $(x, y_n) \in \mathcal{C}_1$ and $\cap_{n=1}^{\infty}(x, y_n) = (x, y] \in \sigma(\mathcal{C}_1)$. Likewise, for $x_n \uparrow x$, we have $(x_n, y) \in \mathcal{C}_1$ and $\cap_{n=1}^{\infty}(x_n, y) = [x, y) \in \sigma(\mathcal{C}_1)$. Next, with x_n and y_n as above, $(x_n, y_n) \in \mathcal{C}_1$ and $\cap_{n=1}^{\infty}(x_n, y_n) = [x, y] \in \sigma(\mathcal{C}_1)$. Also, for $x_n \downarrow -\infty$, we have $(x_n, a) \in \mathcal{C}_1$ and $\cap_{n=1}^{\infty}(x_n, a) = (-\infty, a) \in \sigma(\mathcal{C}_1)$, and likewise $(x_n, a] \in \mathcal{C}_1$ and $\cup_{n=1}^{\infty}(x_n, a] = (-\infty, a] \in \sigma(\mathcal{C}_1)$. Finally, $(b, \infty) = (-\infty, b]^c \in \sigma(\mathcal{C}_1)$, and $[b, \infty) = (-\infty, b)^c \in \sigma(\mathcal{C}_1)$. It follows that $\mathcal{C}_0 \subseteq \sigma(\mathcal{C}_1)$.

That $\mathcal{C}_0 \in \sigma(\mathcal{C}'_1)$ is seen as follows. For (x, y), there exist x_n and y_n rationals with $x_n \downarrow x$ and $y_n \uparrow y$, so that $(x, y) = \cup_{n=1}^{\infty} \in \sigma(\mathcal{C}'_j)$. Also, for $y_n \downarrow y$, we have $(x, y_n) \in \sigma(\mathcal{C}'_1)$, as was just proved, and then $\cap_{n=1}^{\infty}(x, y_n) = (x, y] \in \sigma(\mathcal{C}'_1)$. Likewise, with $x_n \uparrow x$, we have $(x_n, y) \in \sigma(\mathcal{C}'_1)$ and then $\cap_{n=1}^{\infty}(x_n, y) = [x, y) \in \sigma(\mathcal{C}'_1)$. Also, with $x_n \uparrow x$ and $y_n \downarrow y$, we have $(x_n, y_n) \in \sigma(\mathcal{C}'_1)$, and $\cap_{n=1}^{\infty}(x_n, y_n) = [x, y] \in \sigma(\mathcal{C}'_1)$. Likewise, with $x_n \downarrow -\infty$, we have $(x_n, a) \in \sigma(\mathcal{C}'_1)$ and $\cup_{n=1}^{\infty}(x_n, a) = (-\infty, a) \in \sigma(\mathcal{C}'_1)$, whereas $(x_n, a] \in \sigma(\mathcal{C}'_1)$, so that $\cup_{n=1}^{\infty}(x_n, a] = (-\infty, a] \in \sigma(\mathcal{C}'_1)$. Finally, $(b, \infty) = (-\infty, b]^c \in \sigma(\mathcal{C}'_1)$ since $(-\infty, b] \in \sigma(\mathcal{C}'_1)$, and $[b, \infty) = (-\infty, b)^c \in \sigma(\mathcal{C}'_1)$ since $(-\infty, b) \in \sigma(\mathcal{C}'_1)$. It follows that $\mathcal{C}_0 \subseteq \sigma(\mathcal{C}'_1)$.

A slightly alternative version of the proof follows. We will show (a) $\sigma(\mathcal{C}_1) = \mathcal{B}$ and (b) $\sigma(\mathcal{C}'_1) = \mathcal{B}$.

(a) $\sigma(\mathcal{C}_1) = \mathcal{B}$.
That $\sigma(\mathcal{C}_1) \subseteq \mathcal{B}$ is clear; to show $\mathcal{B} \subseteq \sigma(\mathcal{C}_1)$ it suffices to show that $\mathcal{C}_0 \subseteq \sigma(\mathcal{C}_1)$. To this end, we show that $(x, y] \in \sigma(\mathcal{C}_1)$. Indeed, $\left(x, y + \frac{1}{n}\right) \in \mathcal{C}_1$, so that $\bigcap_{n=1}^{\infty}\left(x, y + \frac{1}{n}\right) = (x, y] \in \sigma(\mathcal{C}_1)$. Next, $\left(x - \frac{1}{n}, y\right) \in \mathcal{C}_1$, so that $\bigcap_{n=1}^{\infty}\left(x - \frac{1}{n}, y\right) = [x, y) \in \sigma(\mathcal{C}_1)$. Also, $\left(x - \frac{1}{n}, y + \frac{1}{n}\right) \in \mathcal{C}_1$, so that $\bigcap_{n=1}^{\infty}\left(x - \frac{1}{n}, y + \frac{1}{n}\right) = [x, y] \in \sigma(\mathcal{C}_1)$. Next, $(-n, x) \in \mathcal{C}_1$, so that $\bigcup_{n=1}^{\infty}(-n, x) = (-\infty, x) \in \sigma(\mathcal{C}_1)$. Also, $\left(-\infty, x + \frac{1}{n}\right] \in \mathcal{C}_1$, so that $\bigcap_{n=1}^{\infty}\left(-\infty, x + \frac{1}{n}\right] = (-\infty, x] \in \sigma(\mathcal{C}_1)$. Likewise, $(x, n) \in \mathcal{C}_1$, so that $\bigcup_{n=1}^{\infty}(x, n) = (x, \infty) \in \sigma(\mathcal{C}_1)$; and $\left(x - \frac{1}{n}, \infty\right) \in \sigma(\mathcal{C}_1)$, so that $\bigcap_{n=1}^{\infty}\left(x - \frac{1}{n}, \infty\right) = [x, \infty) \in \sigma(\mathcal{C}_1)$. The proof is complete.

(b) $\sigma(\mathcal{C}'_1) = \mathcal{B}$.

Since, clearly, $\sigma(\mathcal{C}_1') \subseteq \sigma(\mathcal{C}_1)$, it suffices to show that $\sigma(\mathcal{C}_1) \subseteq \sigma(\mathcal{C}_1')$. For $x, y \in \Re$ with $x < y$, there exist $x_n \downarrow x$ and $y_n \uparrow y$ with x_n, y_n rational numbers and $x_n < y_n$ for each n. Since $(x_n, y_n) \in \mathcal{C}_1'$, it follows that $\bigcup_{n=1}^{\infty}(x_n, y_n) = (x, y) \in \sigma(\mathcal{C}_1')$. So $\mathcal{C}_1 \subseteq \sigma(\mathcal{C}_1')$, and hence $\sigma(\mathcal{C}_1) \subseteq \sigma(\mathcal{C}_1')$. The proof is complete. #

7. **(i)** Let $A \in \mathcal{C}$. Then there are the following possible cases:

(a) $A = \sum_{i=1}^{m} I_i$, $I_i = (\alpha_i, \beta_i]$, $i = 1, \ldots, m$.

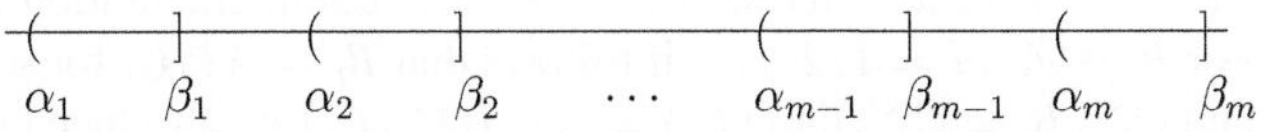

Then $A^c = (-\infty, \alpha_1] + (\beta_1, \alpha_2] + \ldots + (\beta_{m-1}, \alpha_m] + (\beta_m, \infty)$ and this is in $\mathcal{C}$.

(b) A consists only of intervals of the form $(-\infty, \alpha]$. Then there can be only one such interval; i.e., $A = (-\infty, \alpha]$ and hence $A^c = (\alpha, \infty)$ which is in $\mathcal{C}$.

(c) A consists only of intervals of the form (β, ∞). Then there can only be one such interval; i.e., $A = (\beta, \infty)$ so that $A^c = (-\infty, \beta]$ which is in $\mathcal{C}$.

(d) A consists only of intervals of the form $(-\infty, \alpha]$ and (β, ∞). Then A will be as follows: $A = (-\infty, \alpha] + (\beta, \infty)$ $(\alpha < \beta)$, so that $A^c = (\alpha, \infty) \cap (-\infty, \beta] = (\alpha, \beta]$ which is in $\mathcal{C}$.

(e) Finally, let A consist of intervals of all forms. Then A is as below:

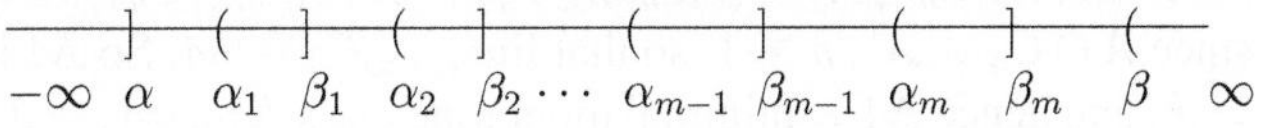

Then, clearly,

$$A^c = (\alpha, \alpha_1] + (\beta_1, \alpha_2] + \ldots + (\beta_{m-1}, \alpha_m] + (\beta_m, \beta]$$

which is in $\mathcal{C}$. So, $\mathcal{C}$ is closed under complementation. It is also closed under the union of two sets A and B in $\mathcal{C}$, because, clearly, the union of two such sets is also a member of $\mathcal{C}$. Thus, $\mathcal{C}$ is a field. Next, let $\mathcal{C}_2 = \{(\alpha, \beta];\ \alpha, \beta \in \Re, \alpha < \beta\}$. Then, by Exercise 6, $\sigma(\mathcal{C}_2) = \mathcal{B}$. Also, $\mathcal{C}_2 \subset \mathcal{C}$, so that $\mathcal{B} = \sigma(\mathcal{C}_2) \subseteq \sigma(\mathcal{C})$. Furthermore, $\mathcal{C} \subseteq \sigma(\mathcal{C}_0) = \mathcal{B}$ and hence $\sigma(\mathcal{C}) \subseteq \mathcal{B}$. It follows that $\sigma(\mathcal{C}) = \mathcal{B}$.

(ii) If $A \in \mathcal{C}$, then $A = \sum_{i=1}^{m} I_i$, where I_is are of the forms: (α, β), $(\alpha, \beta]$, $[\alpha, \beta)$, $[\alpha, \beta]$, $(-\infty, \alpha)$, $(-\infty, \alpha]$, (β, ∞), $[\beta, \infty)$. But $(\alpha, \beta)^c = (-\infty, \alpha] + [\beta, \infty)$, $(\alpha, \beta]^c = (-\infty, \alpha] + (\beta, \infty)$, $[\alpha, \beta)^c = (-\infty, \alpha) + (\beta, \infty)$, $[\alpha, \beta]^c = (-\infty, \alpha) + (\beta, \infty)$, $(-\infty, \alpha)^c = [\alpha, \infty)$, $(-\infty, \alpha]^c = (\alpha, \infty)$, $(\beta, \infty)^c = (-\infty, \beta]$, and $[\beta, \infty)^c = (-\infty, \beta)$. Then, considering all possibilities as in part (i), we conclude that $A^c \in \mathcal{C}$ in all cases. Next, for A as above and $B = \sum_{j=1}^{n} J_j$ with J_j being from among the above intervals, it follows that $A \cup B$ is a finite sum of intervals as above, and hence $A \cup B \in \mathcal{C}$. Thus, $\mathcal{C}$ is a field. Finally, from $\mathcal{C}_0 \subset \mathcal{C} \subset \mathcal{B}$, it follows that $\mathcal{B} = \sigma(\mathcal{C}_0) \subseteq \sigma(\mathcal{C}) \subseteq \mathcal{B}$, so that $\sigma(\mathcal{C}) = \mathcal{B}$. #

8. Clearly, $\mathcal{F}_A$ is $\neq \oslash$ since, for example, $A = A \cap \Omega$ and hence $A \in \mathcal{F}_A$. Next, for $B \in \mathcal{F}_A$, it follows that $B = A \cap C$, $C \in \mathcal{F}$, and B_A^c(=complement of B with respect to A)=$A \cap C^c \in \mathcal{F}_A$ since $C^c \in \mathcal{F}$. Finally, for $B_1, B_2 \in \mathcal{F}_A$, it follows that $B_i = A_i \cap C_i$, $C_i \in \mathcal{F}$, $i = 1, 2$, and then $B_1 \cup B_2 = A \cap (C_1 \cup C_2) \in \mathcal{F}_A$, since $C_1 \cup C_2 \in \mathcal{F}$. #

9. That $\mathcal{A}_A \neq \oslash$ and that it is closed under complementation is as in Exercise 8. For $B_i \in \mathcal{A}_A, i = 1, 2, \ldots$, it follows that $B_i = A \cap C_i$ for some $C_i \in \mathcal{A}, i \geq 1$, and $\cup_{i=1}^\infty B_i = \cup_{i=1}^\infty (A \cap C_i) = A \cap \left(\cup_{i=1}^\infty C_i\right) \in \mathcal{A}_A$ since $\cup_{i=1}^\infty C_i \in \mathcal{A}$. Thus, $\mathcal{A}_A$ is a σ-field. Since $\mathcal{F} \subseteq \mathcal{A}$, it follows that $\mathcal{F}_A \subseteq \mathcal{A}_A$ and hence $\sigma(\mathcal{F}_A) \subseteq \mathcal{A}_A$. Since for every $\mathcal{F} \subseteq \mathcal{A}_i, i \in I$, it follows $\mathcal{F}_A \subseteq \mathcal{A}_{i,A}, i \in I$, then $\sigma(\mathcal{F}_A) \subseteq \cap_{i \in I} \mathcal{A}_{i,A}$. Also, $\sigma(\mathcal{F}_A) = \cap_{j \in J} \mathcal{A}_j^*$ for all σ-fields of subsets of A with $\mathcal{A}_j^* \supseteq \mathcal{F}_A$. In order to show that $\sigma(\mathcal{F}_A) = \mathcal{A}_A$, it must be shown that for every σ-field $\mathcal{A}^*$ of subsets of A with $\mathcal{A}^* \supseteq \mathcal{F}_A$, we have $\mathcal{A}^* \supseteq \mathcal{A}_A$. That this is, indeed, the case is seen as follows. Define the class $\mathcal{M}$ by : $\mathcal{M} = \{C \in \mathcal{A};\ A \cap C \in \mathcal{A}^*\}$. Then, clearly, $\mathcal{F} \subseteq \mathcal{M} \subseteq \mathcal{A}$ and $\mathcal{M}_A (= \mathcal{M} \cap A) \subseteq \mathcal{A}^*$. This is so because, for $C \in \mathcal{F}$, it follows that $C \cap A \in \mathcal{F}_A$ and hence $C \cap A \in \mathcal{A}^*$ $(\supseteq \mathcal{F}_A)$. Also, with $\mathcal{M}_A = \{C \subseteq A; C = M \cap A, M \in \mathcal{M}\}$, it follows that $\mathcal{M}_A \subseteq \mathcal{A}^*$ from the definition of $\mathcal{M}$. We assert that $\mathcal{M}$ is a monotone class. Indeed, let $C_n \in \mathcal{M}$ with $C_n \uparrow$ or $C_n \downarrow$. Then, for the case that $C_n \uparrow$, $A \cap (\lim_{n\to\infty} C_n) = A \cap \left(\cup_{n=1}^\infty C_n\right) = \cup_{n=1}^\infty \left(A \cap C_n\right) \in \mathcal{A}^*$ since $A \cap C_n \in \mathcal{A}^*, n \geq 1$, so that $\lim_{n\to\infty} C_n \in \mathcal{M}$. Likewise, for $C_n \downarrow$, $A \cap (\lim_{n\to\infty} C_n) = A \cap (\cap_{n=1}^\infty C_n) = \cap_{n=1}^\infty (A \cap C_n) \in \mathcal{A}^*$ since $A \cap C_n \in \mathcal{A}^*, n \geq 1$, so that $\lim_{n\to\infty} C_n \in \mathcal{M}$. So $\mathcal{M}$ is a monotone class $\supseteq \mathcal{F}$, and hence $\mathcal{M} \supseteq$ minimal monotone class $\mathcal{M}_0$, say, $\supseteq \mathcal{F}$. Since $\mathcal{F}$ is a field, it follows that $\mathcal{M}_0$ is a σ-field and indeed $\mathcal{M}_0 = \mathcal{A}$ (by Theorem 6). Finally, $\mathcal{A} = \mathcal{M}_0 \subseteq \mathcal{M}$ implies $\mathcal{A}_A = \mathcal{M}_{0,A} \subseteq \mathcal{M}_A \subseteq \mathcal{A}^*$, as was to be seen. #

10. Set $\mathcal{F} = \cup_{n=1}^\infty \mathcal{A}_n$, and let $A \in \mathcal{F}$. Then $A \in \mathcal{A}_n$ for some n, so that $A^c \in \mathcal{A}_n$ and hence $A \in \mathcal{F}$. Next, let $A, B \in \mathcal{F}$. Then $A \in \mathcal{A}_{n_1}$, $B \in \mathcal{A}_{n_2}$ for some n_1 and n_2, and let $n_0 = \max(n_1, n_2)$. Then $A, B \in \mathcal{A}_{n_0}$, so that $A \cup B \in \mathcal{A}_{n_0}$ and $A \cup B \in \mathcal{F}$. Then, $A^c \in \mathcal{F}$ and $A \cup B \in \mathcal{F}$, so that $\mathcal{F}$ is a field.
It need not be a σ-field.
Counterexample: Let $\Omega = \Re$ and let $\mathcal{A}_n = \{A \subseteq [-n, n];$ either A or A^c is countable$\}$, $n \geq 1$. Then $\mathcal{A}_n$ is a σ-field (by Example 8) and $\mathcal{A}_n \uparrow$. However, $\mathcal{F}$ is not a σ-field because, if $A_n = \{$rationals in $[-n, n]\}$, $n \geq 1$, and if we set $A = \cup_{n=1}^\infty A_n$, then $A \notin \mathcal{F}$, because otherwise $A \in \mathcal{A}_n$ for some n, which cannot happen. #

11. Set $\mathcal{M} \cap_{j \in I} \mathcal{M}_j$ and let $A_n \in \mathcal{M}, n \geq 1$, where the A_ns form a monotone sequence. Then $A_n \in \mathcal{M}_j$ for each $j \in I$ and all $n \geq 1$, so that $\lim_{n\to\infty} A_n$ is also in $\mathcal{M}_j$. Since this is true for all $j \in I$, it follows that $\lim_{n\to\infty} A_n$ is in $\mathcal{M}$, and $\mathcal{M}$ is a monotone class. #

12. Let $\Omega = \{1, 2, \ldots\}$, $\mathcal{M} = \{\oslash, \{1, \ldots, n\}, \{n, n+1, \ldots\}, n \geq 1, \Omega\}$. Then $\mathcal{M}$ is a monotone class, but not a field, because, e.g., if $A = \{1, \ldots, n\}$ and $B = \{n-2, n-1, \ldots\}$ $(n \geq 3)$, then $A, B \in \mathcal{M}$, but $A \cap B = \{n-2, n-1, n\} \notin \mathcal{M}$.

As another example, let $\Omega = (0, 1)$ and $\mathcal{M} = \{(0, 1 - \frac{1}{n}], n \geq 1, \Omega\}$. Then $\mathcal{M}$ is a monotone class and $(0, \frac{1}{2}] \in \mathcal{M}$, but $(0, \frac{1}{2}]^c = (\frac{1}{2}, 1) \notin \mathcal{M}$.

Still as a third example, let $\Omega = \Re$ and let $\mathcal{M} = \{\emptyset, (0, n), (-n, 0), n \geq 1, (0, \infty), (-\infty, 0)\}$. Then $\mathcal{M}$ is a monotone class, but not a field since, for $A = (-1, 0)$ and $B = (0, 1)$, we have $A, B, \in \mathcal{M}$, but $A \cup B = (-1, 1) \notin \mathcal{M}$. #

13. **(i)** For $\omega = (\omega_1, \omega_2) \in E^c$, we have $\omega \notin E = A \times B$, so that either $\omega_1 \notin A$ or $\omega_2 \notin B$ or both. Let $\omega_1 \notin A$. Then $\omega_1 \in A^c$ and $(\omega_1, \omega_2) \in A^c \times \Omega_2$, whether or not $\omega_2 \in B$. Hence $E^c \subseteq (A \times B^c) + (A^c \times \Omega_2)$. If $\omega_1 \in A$, then $\omega_2 \notin B$, so that $(\omega_1, \omega_2) \in A \times B^c$ and $E^c \subseteq (A \times B^c) + (A^c \times \Omega_2)$. Next, if $(\omega_1, \omega_2) \in A \times B^c$, then $\omega_1 \in A$ and $\omega_2 \notin B$, so that $(\omega_1, \omega_2) \notin E$ and hence $(\omega_1, \omega_2) \in E^c$. If $(\omega_1, \omega_2) \in A^c \times \Omega_2$, then $\omega_1 \notin A$ and hence $(\omega_1, \omega_2) \notin A \times B = E$ whether or not $\omega_2 \in B$. Thus $(\omega_1, \omega_2) \in E^c$. In both cases, $(A \times B^c) + (A^c \times \Omega_2) \supseteq E^c$ and equality follows. The second equality is entirely symmetric.

(ii) Let $(\omega_1, \omega_2) \in E_1 \cap E_2$, so that $(\omega_1, \omega_2) \in E_1$ and $(\omega_1, \omega_2) \in E_2$ and hence $\omega_1 \in A_1, \omega_2 \in B_1$, and $\omega_1 \in A_2, \omega_2 \in B_2$. It follows that $\omega_1 \in A_1 \cap A_2, \omega_2 \in B_1 \cap B_2$ and hence $(\omega_1, \omega_2) \in (A_1 \cap A_2) \times (B_1 \cap B_2)$. Next, $(\omega_1, \omega_2) \in (A_1 \cap A_2) \times (B_1 \cap B_2)$, so that $\omega_1 \in A_1 \cap A_2$ and $\omega_2 \in B_1 \cap B_2$. Thus, $\omega_1 \in A_1, \omega_1 \in A_2$ and $\omega_2 \in B_1, \omega_2 \in B_2$, so that $(\omega_1, \omega_2) \in A_1 \cap B_1$ and $(\omega_1, \omega_2) \in A_2 \cap B_2$, or $(\omega_1, \omega_2) \in E_1 \cap E_2$, so that equality occurs. The second conclusion is immediate.

(iii) Indeed, $E_1 \cap F_1 = (A_1 \cap A_1') \times (B_1 \cap B_1')$ and $E_2 \cap F_2 = (A_2 \cap A_2') \times (B_2 \cap B_2')$, by part (ii), and the first equality follows. Next, again by part (ii), and replacing E_1 by $(A_1 \cap A_1') \times (B_1 \cap B_1')$ and E_2 by $(A_2 \cap A_2') \times (B_2 \cap B_2')$, we obtain the second equality. The third equality is immediate. Finally, the last conclusion is immediate. #

14. **(i)** Either by the inclusion process or as follows:

$$
\begin{aligned}
&(A_1 \times B_1) - (A_2 \times B_2) \\
&\quad = (A_1 \times B_1) \cap (A_2 \times B_2)^c \\
&\quad = (A_1 \times B_1) \cap [(A_2 \times B_2^c) + (A_2^c \times \Omega_2)] \text{ (by Lemma 2)} \\
&\quad = (A_1 \times B_1) \cap (A_2 \times B_2^c) + (A_1 \times B_1) \cap (A_2^c \times \Omega_2) \\
&\quad = (A_1 \cap A_2) \times (B_1 \cap B_2^c) + (A_1 \cap A_2^c) \times (B_1 \cap \Omega_2) \text{ (clearly)} \\
&\quad = (A_1 \cap A_2) \times (B_1 - B_2) + (A_1 - A_2) \times B_1.
\end{aligned}
$$

(ii) Let $A \times B = \emptyset$. Then $(x, y) \in A \times B$, so that $x \in A$ and $y \in B$. Also, $(x, y) \in \emptyset$ and this can happen only if at least one of A or B is $= \emptyset$. On the other hand, if at least one of A or B is $= \emptyset$, then, clearly, $A \times B = \emptyset$.

(iii) Let $A_1 \times B_1 \subseteq A_2 \times B_2$. Then $(x, y) \in A_1 \times B_1$, so that $x \in A_1$ and $y \in B_1$. Also, $(x, y) \in A_2 \times B_2$ implies $x \in A_2$ and $y \in B_2$. Thus, $A_1 \subseteq A_2$ and $B_1 \subseteq B_2$. Next, let $A_1 \subseteq A_2$ and $B_1 \subseteq B_2$. Then $A_1 \times B_1 \subseteq A_2 \times B_2$ since $(x, y) \in A_1 \times B_1$ if and only if $x \in A_1$ and $y \in B_1$. Hence, $x \in A_2$ and $y \in B_2$ or $(x, y) \in A_2 \times B_2$.

(iv) $A_1 \times B_1 \neq \oslash$ and $A_2 \times B_2 \neq \oslash$. Then $A_1 \times B_1 = A_2 \times B_2$ or $A_1 \times B_1 \subseteq A_2 \times B_2$ and then (by (iii)), $A_1 \subseteq A_2$ and $B_1 \subseteq B_2$. Also, $A_2 \times B_2 = A_1 \times B_1$ or $A_2 \times B_2 \subseteq A_1 \times B_1$, and then (by (iii) again), $A_2 \subseteq A_1$ and $B_2 \subseteq B_1$.

So, both $A_1 \subseteq A_2$ and $A_2 \subseteq A_1$, and therefore $A_1 = A_2$. Likewise, $B_1 \subseteq B_2$ and $B_2 \subseteq B_1$ so that $B_1 = B_2$.

(v)

$$A \times B = (A_1 \times B_1) + (A_2 \times B_2) \qquad (*)$$

From $\oslash = (A_1 \times B_1) \cap (A_2 \times B_2) = (A_1 \cap A_2) \times (B_1 \cap B_2)$ and part (ii), we have that at least one of $A_1 \cap A_2$, $B_1 \cap B_2$ is $\oslash$. Let $A_1 \cap A_2 = \oslash$. Then the claim is that $A = A_1 + A_2$. In fact, $(x, y) \in A \times B$ implies $x \in A$ (and $y \in B$). Also, (x, y) belonging to the right-hand side of (*) implies $(x, y) \in A_1 \times B_1$ or $(x, y) \in A_2 \times B_2$. Let $(x, y) \in A_1 \times B_1$. Then $x \in A_1$ (and $y \in B_1$), so that $A \subseteq A_2$. On the other hand, $(x, y) \in A_2 \times B_2$ implies $x \in A_2$ (and $y \in B_2$), so that $A \subseteq A_2$. Thus, $A \subseteq A_1 + A_2$. Next, let again (x, y) belong to the right-hand side of (*). Then $(x, y) \in A_1 \times B_1$ or $(x, y) \in A_2 \times B_2$. Now $(x, y) \in A_1 \times B_1$ implies that $x \in A_1$ (and $y \in B_1$). Also, (x, y) belonging to the left-hand side of (*) implies $(x, y) \in A \times B$, so that $x \in A$ (and $y \in B$). Hence $A_1 \subseteq A$. Likewise, $(x, y) \in A_2 \times B_2$ implies $A_2 \subseteq A$, so that $A_1 + A_2 \subseteq A$, and hence $A = A_1 + A_2$. Next, let $A = A_1 + A_2$. Then $A \times B = (A_1 + A_2) \times B = (A_1 \times B) + (A_2 \times B)$. Also, $A \times B = (A_1 \times B_1) + (A_2 \times B_2)$. Thus, $(A_1 \times B) + (A_2 \times B) = (A_1 \times B_1) + (A_2 \times B_2)$. (x, y) belonging to the left-hand side of (*) implies $(x, y) \in A_1 \times B$ or $(x, y) \in A_2 \times B$. $(x, y) \in A_1 \times B$ yields $y \in B$ (and $x \in A_1$). Same if $(x, y) \in A_2 \times B$. Also, (x, y) belonging to the right-hand side of (*) implies $(x, y) \in A_1 \times B_1$ or $(x, y) \in A_2 \times B_2$. For $(x, y) \in A_1 \times B_1$, we have $y \in B_1$ (and $x \in A_1$), so that $B \subseteq B_1$. For $(x, y) \in A_2 \times B_2$, we have $B \subseteq B_2$ likewise. Next, let again (x, y) belong to the right-hand side of (*). Then $(x, y) \in A_1 \times B_1$ or $(x, y) \in A_2 \times B_2$. For $(x, y) \in A_1 \times B_1$, we have $y \in B_1$ (and $x \in A_1$). Thus $B_1 \subseteq B$. For $(x, y) \in A_2 \times B_2$, we have $B_2 \subseteq B$. It follows that $B = B_1 = B_2$.

To summarize: $A_1 \cap A_2 = \oslash$ implies $A = A_1 + A_2$ and $B = B_1 = B_2$. Likewise, $B_1 \cap B_2 = \oslash$ implies $B = B_1 + B_2$ and $A = A_1 = A_2$. Furthermore, $A_1 \cap A_2 = \oslash$ and $B_1 \cap B_2 = \oslash$ cannot happen simultaneously. Indeed, $A_1 \cap A_2 = \oslash$ implies $A = A_1 + A_2$, and $B_1 \cap B_2 = \oslash$ implies $B = B_1 + B_2$. Then $A \times B = (A_1 + A_2) \times (B_1 + B_2) = (A_1 \times B_1) + (A_2 \times B_2) + (A_1 \times B_2) + (A_2 \times B_1)$. Also, $A \times B = (A_1 \times B_1) + (A_2 \times B_2)$, so that: $(A_1 \times B_1) + (A_2 \times B_2) + (A_1 \times B_2) + (A_2 \times B_1) = (A_1 \times B_1) + (A_2 \times B_2)$. Then $(A_1 \times B_2) + (A_2 \times B_1) = \oslash$ implies $(A_1 \times B_2) = (A_2 \times B_1) = \oslash$, so that at least one of $A_1, A_2, B_1, B_2 = \oslash$ (by part (ii)). However, this is not possible by the fact that $A_1 \times B_1 \neq \oslash$, $A_2 \times B_2 \neq \oslash$. #

15. **(i)** If either A or $B = \oslash$, then, clearly, $A \times B = \oslash$. Next, if $A \times B = \oslash$, and $A \neq \oslash$ and $B \neq \oslash$, then there exist $\omega_1 \in A$ and $\omega_2 \in B$, so that $(\omega_1, \omega_2) \in A \times B$, a contradiction.

(ii) Both directions of the first assertion are immediate. Without the assumption E_1 and $E_2 \neq \oslash$, the result need not be true. Indeed, let $\Omega_1 = \Omega_2$, $A_1 \neq \oslash$, $B_1 = A_2 = B_2 = \oslash$. Then $E_1 = E_2 = \oslash$, but $A_1 \not\subseteq A_2$. #

16. **(i)** If at least one of $A_1, \ldots, A_n$ is $= \oslash$, then, clearly, $A_1 \times \ldots \times A_n = \oslash$. Next, let $E = \oslash$ and suppose that $A_i \neq \oslash, i = 1, \ldots, n$. Then there exists $\omega_i \in A_i, i = 1, \ldots, n$, so that $(\omega_1, \ldots, \omega_n) \in E$, a contradiction.

(ii) Let $\omega = (\omega_1, \ldots, \omega_n) \in E \cap F$, or $(\omega_1, \ldots, \omega_n) \in (A_1 \times \ldots \times A_n) \cap (B_1 \times \ldots \times B_n)$. Then $(\omega_1, \ldots, \omega_n) \in A_1 \times \ldots \times A_n$ and $(\omega_1, \ldots, \omega_n) \in B_1 \times \ldots \times B_n$. It follows that $\omega_i \in A_i$ and $\omega_i \in B_i, i = 1, \ldots, n$, so that $\omega_i \in A_i \cap B_i, i = 1, \ldots, n$, and hence $(\omega_1, \ldots, \omega_n) \in (A_1 \cap B_1) \times \ldots \times (A_n \cap B_n)$. Next, let $(\omega_1, \ldots, \omega_n) \in (A_1 \cap B_1) \times \ldots \times (A_n \cap B_n)$. Then $\omega_i \in A_i \cap B_i, i = 1, \ldots, n$, so that $\omega_i \in A_i$ and $\omega_i \in B_i, i = 1, \ldots, n$. It follows that $(\omega_1, \ldots, \omega_n) \in A_1 \times \ldots \times A_n$ and $(\omega_1, \ldots, \omega_n) \in B_1 \times \ldots \times B_n$, so that $(\omega_1, \ldots, \omega_n) \in (A_1 \times \ldots \times A_n) \cap (B_1 \times \ldots \times B_n)$. #

17. We have $E = F + G$ and E, F, G are all $\neq \oslash$. This implies that A_i, B_i, and $C_i, i = 1, \ldots, n$ are all $\neq \oslash$; this is so by Exercise 16(i). Furthermore, by Exercise 16(ii):

$$F \cap G = (B_1 \times \ldots \times B_n) \cap (C_1 \times \ldots \times C_n) = (B_1 \cap C_1) \times \ldots \times (B_n \cap C_n),$$

whereas $F \cap G = \oslash$. It follows that $B_j \cap C_j = \oslash$ for at least one $j, 1 \leq j \leq n$. Without loss of generality, suppose that $B_1 \cap C_1 = \oslash$. Then we shall show that $A_1 = B_1 + C_1$ and $A_i = B_i = C_i, i = 2, \ldots, n$. To this end, let $\omega_j \in A_j$, $j = 1, \ldots, n$. Then $(\omega_1, \ldots, \omega_n) \in A_1 \times \ldots \times A_n$ or $(\omega_1, \ldots, \omega_n) \in E$ or $(\omega_1, \ldots, \omega_n) \in (F + G)$. Hence $(\omega_1, \ldots, \omega_n) \in F$ or $(\omega_1, \ldots, \omega_n) \in G$. Let $(\omega_1, \ldots, \omega_n) \in F$. Then $(\omega_1, \ldots, \omega_n) \in B_1 \times \ldots \times B_n$ and hence $\omega_1 \in B_1$ or $\omega_1 \in (B_1 \cup C_1)$, so that $A_1 \subseteq B_1 \cup C_1$. Likewise if $(\omega_1, \ldots, \omega_n) \in G$. Next, let $\omega_j \in B_j, j = 1, \ldots, n$. Then $(\omega_1, \ldots, \omega_n) \in B_1 \times \ldots \times B_n$ or $(\omega_1, \ldots, \omega_n) \in F$ or $(\omega_1, \ldots, \omega_n) \in E$ or $(\omega_1, \ldots, \omega_n) \in (A_1 \times \ldots \times A_n)$, hence $\omega_1 \in A_1$, which implies that $B_1 \subseteq A_1$. By taking $\omega_j \in C_j, j = 1, \ldots, n$ and arguing as before, we conclude that $C_1 \subseteq A_1$. From $B_1 \subseteq A_1$ and $C_1 \subseteq A_1$, we obtain $B_1 \cup C_1 \subseteq A_1$. Since also $A_1 \subseteq B_1 \cup C_1$, we get $A_1 = B_1 \cup C_1$. Since $B_1 \cap C_1 = \oslash$, we have then $A_1 = B_1 + C_1$.

It remains for us to show that $A_i = B_i = C_i, i = 2, \ldots, n$. Without loss of generality, it suffices to show that $A_2 = B_2 = C_2$, the remaining cases being treated symmetrically. As before, let $\omega_j \in A_j, j = 1, \ldots, n$. Then $(\omega_1, \ldots, \omega_n) \in (A_1 \times \ldots \times A_n)$ or $(\omega_1, \ldots, \omega_n) \in E$ or $(\omega_1, \ldots, \omega_n) \in (F+G)$. Hence either $(\omega_1, \ldots, \omega_n) \in F$ or $(\omega_1, \ldots, \omega_n) \in G$. Let $(\omega_1, \ldots, \omega_n) \in F$. Then $(\omega_1, \ldots, \omega_n) \in B_1 \times \ldots \times B_n$ and hence $\omega_2 \in B_2$, so that $A_2 \subseteq B_2$.

Likewise $A_2 \subseteq C_2$ if $(\omega_1, \ldots, \omega_n) \in G$. Next, let $(\omega_1, \ldots, \omega_n) \in B_1 \times \ldots \times B_n$ or $(\omega_1, \ldots, \omega_n) \in F$ or $(\omega_1, \ldots, \omega_n) \in (F + G)$ or $(\omega_1, \ldots, \omega_n) \in E$ or $(\omega_1, \ldots, \omega_n) \in (A_1 \times \ldots \times A_n)$ and hence $\omega_2 \in A_2$, so that $B_2 \subseteq A_2$. It follows that $A_2 = B_2$. We arrive at the same conclusion $A_2 = B_2$ if we take $(\omega_1, \ldots, \omega_n) \in G$. So, to sum it up, $A_1 = B_1 + C_1$, and $A_2 = B_2 = C_2$, and by symmetry, $A_i = B_i = C_i, i = 3, \ldots, n$.

A variation to the above proof is as follows.

Let $E = F + G$ or $A_1 \times \ldots \times A_n = (B_1 \times \ldots \times B_n) + (C_1 \times \ldots \times C_n)$, and let $(\omega_1, \ldots, \omega_n) \in E$. Then $(\omega_1, \ldots, \omega_n) \in A_1 \times \ldots \times A_n$, so that $\omega_i \in A_i, i = 1, \ldots, n$. Then $\omega_i \in B_i, i = 1, \ldots, n$ or $\omega_i \in C_i, i = 1, \ldots, n$ (but not both). So, $A_i = B_i \cup C_i, i = 1, \ldots, n$ and $A_j = B_j + C_j$ for at least one j. Consider the case $n = 2$, and without loss of generality suppose that $A_1 = B_1 + C_1$, $A_2 = B_2 \cup C_2$. Then, clearly:

$$\begin{aligned} A_1 \times A_2 &= (B_1 + C_1) \times (B_2 \cup C_2) \\ &= (B_1 \times B_2) \cup (C_1 \times C_2) \cup (B_1 \times C_2) \cup (C_1 \times B_2). \end{aligned}$$

However, $A_1 \times A_2 = (B_1 \times B_2) + (C_1 \times C_2)$, and this implies that $B_1 \times C_2 \subseteq B_1 \times B_2$ and $C_1 \times B_2 \subseteq B_1 \times C_2$, hence $C_2 \subseteq B_2$ and $B_2 \subseteq C_2$, so that $B_2 = C_2 (= A_2)$. Next, assume the assertion to be true for n and consider:

$$A_1 \times \ldots \times A_n \times A_{n+1} = (B_1 \times \ldots \times B_n \times B_{n+1}) + (C_1 \times \ldots \times C_n \times C_{n+1}),$$

or $A^n \times A_{n+1} = (B^n \times B_{n+1}) = (C^n \times C_{n+1})$, where $A^n = A_1 \times \ldots \times A_n$, $B^n = B_1 \times \ldots \times B_n$ and $C^n = C_1 \times \ldots \times C_n$. Apply the reasoning used in the case $n = 2$ by replacing A_1 by A^n and A_2 by A_{n+1} (so that B_1, B_2 and C_1, C_2 are replaced, respectively, by B^n, B_{n+1} and C^n, C_{n+1}) to get that:

$$A^n = B^n + C^n, \quad A_{n+1} = B_{n+1} \cup C_{n+1}.$$

The first union is a "+" by the induction hypothesis. The second union may or may not be a "+" as of now. Then:

$$\begin{aligned} A^n \times A_{n+1} &= (B^n \cup C^n) \times (B_{n+1} \cup C_{n+1}) \\ &= (B^n \times B_{n+1}) \cup (C^n \times C_{n+1}) \cup (B^n \times C_{n+1}) \cup (C^n \times B_{n+1}). \end{aligned}$$

However, $A^n \times A_{n+1} = (B^n \times B_{n+1}) + (C^n \times C_{n+1})$. Therefore $B^n \times C_{n+1} \subseteq B^n \times B_{n+1}$ and $C^n \times B_{n+1} \subseteq C^n \times C_{n+1}$, so that $C_{n+1} \subseteq B_{n+1}$ and $B_{n+1} \subseteq C_{n+1}$, and hence $B_{n+1} = C_{n+1}$. The proof is completed. #

18. The only properties of the σ-fields $\mathcal{A}_1$ and $\mathcal{A}_2$ used in the proof of Theorem 7 is that $\mathcal{A}_i, i = 1, 2$ are closed under the intersection of two sets in them and also closed under complementations. Since these properties hold also for the case that $\mathcal{A}_i, i = 1, 2$ are fields, $\mathcal{F}_i, i = 1, 2$, the proof is completed. #

19. $\mathcal{C}$ as defined here need not be a σ-field. Here is a *Counterexample*: $\Omega_1 = \Omega_2 = [0, 1]$. For $n \geq 2$, let $I_{n1} = [0, \frac{1}{n}]$, $I_{nj} = (\frac{j-1}{n}, \frac{j}{n}]$, $j = 2, \ldots, n$, and set $E_{nj} = I_{nj} \times I_{nj}$, $j = 1, \ldots, n$. Also, let

$Q_n = \sum_{j=1}^{n} E_{nj}, n \geq 2$. Then Q_n belongs to the field of all finite sums of rectangles. Furthermore, it is clear that $\cap_{n=2}^{\infty} Q_n = D$, where D is the main diagonal determined by the origin and the point (1,1). (See picture below.) However, D is not in the class of all countable sums of rectangles, since it cannot be written as such. D is written as $D = \cup_{x \in [0,1]}(x, x)$, an uncountable union.

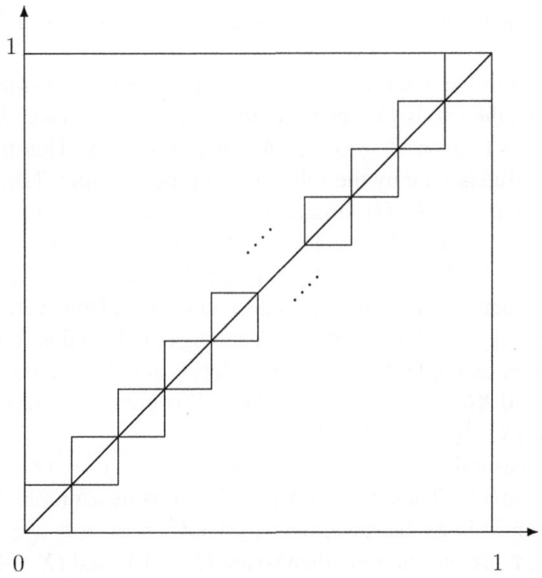

Note: In the picture, the first rectangle $E_{n1} = [0, \frac{1}{n}] \times [0, \frac{1}{n}]$, and the subsequent rectangles E_{nj} are: $E_{nj} = (\frac{j-1}{n}, \frac{j}{n}], j = 2, 3, \ldots, n$. #

20. That $\mathcal{C} \neq \oslash$ is obvious. For $A \in \mathcal{C}$, there exists $A' \in \mathcal{A}'$ such that $A = X^{-1}(A')$. Then $A^c = [X^{-1}(A')]^c = X^{-1}[(A')^c]$ with $(A')^c \in \mathcal{A}'$. Thus $A^c \in \mathcal{C}$. Finally, if $A_j \in \mathcal{C}, j = 1, 2, \ldots$, then $A_j = X^{-1}(A'_j)$ with $A'_j \in \mathcal{A}'$, and hence $\cup_{j=1}^{\infty} A_j = \cup_{j=1}^{\infty} X^{-1}(A'_j) = X^{-1}\left(\cup_{j=1}^{\infty} A'_j\right)$ with $\cup_{j=1}^{\infty} A'_j \in \mathcal{A}'$, so that $\cup_{j=1}^{\infty} A_j \in \mathcal{C}$, and $\mathcal{C}$ is a σ-field. #

21. That $\mathcal{C}' \neq \oslash$ is obvious. For $A' \in \mathcal{C}'$, there exists $A \in \mathcal{A}$ such that $A = X^{-1}(A')$. Then $X^{-1}[(A')^c] = [X^{-1}(A')]^c = A^c \in \mathcal{A}$, so that $(A')^c \in \mathcal{C}'$. Finally, for $A'_j \in \mathcal{C}', j = 1, 2, \ldots$, there exists $A_j \in \mathcal{A}$ such that $A_j = X^{-1}(A'_j)$ and $X^{-1}\left(\cup_{j=1}^{\infty} A'_j\right) = \cup_{j=1}^{\infty} X^{-1}(A'_j) = \cup_{j=1}^{\infty} A_j \in \mathcal{A}$, so that $\cup_{j=1}^{\infty} A'_j \in \mathcal{C}'$. It follows that $\mathcal{C}'$ is a σ-field. #

22. A simple example is the following. Let $\Omega = \{a, b, c, d\}$, $\mathcal{A} = \{\oslash, \{a\}, \{b, c, d\}, \Omega\}$, $X(a) = X(b) = 1$, $X(c) = 2$, $X(d) = 3$. Then $\Omega' = \{1, 2, 3\}$ and $X(\{a\}) = \{1\}$, $X(\{b, c, d\}) = \{1, 2, 3\}$, so that $\mathcal{C}' = \{\oslash, \{1\}, \{1, 2, 3\}\}$ which is not a σ-field. #

23. Let $X = \sum_{i=1}^{n} \alpha_i I_{A_i}$ and suppose that $A_i \in \mathcal{A}, i = 1, \ldots, n$. Then for any $B \in \mathcal{B}$, $X^{-1}(B) = \cup A_i$ where the union is taken over those is for which $\alpha_i \in B$.

Since this union is in $\mathcal{A}$, it follows that X is a r.v. Next, let X be a r.v. Then, by assuming without loss of generality that $\alpha_i \neq \alpha_j, i \neq j$, we have $X^{-1}(\{\alpha_i\}) = A_i \in \mathcal{A}$ since $\{\alpha_i\} \in \mathcal{B}, i = 1, \ldots, n$. Clearly, the same reasoning applies when $X = \sum_{i=1}^{\infty} \alpha_i I_{A_i}$. #

24. Let ω belong to the right-hand side. Then $X(\omega) < r$ and $Y(\omega) < x - r$ for some $r \in Q$, so that $X(\omega) + Y(\omega) < x$ and hence ω belongs to the left-hand side. Next, let ω belong to the left-hand side, so that $X(\omega) + Y(\omega) < x$ or $X(\omega) < x - Y(\omega)$. But then there exists $r \in Q$ such that $X(\omega) < r < x - Y(\omega)$ or $X(\omega) < r$ and $r < x - Y(\omega)$ or $X(\omega) < r$ and $Y(\omega) < x - r$, so that ω belongs to the right-hand side. #

25. If X is a r.v., then so is $|X|$, because for all $x \geq 0$, we have $|X|^{-1}((-\infty, x)) = (|X| < x) = (-x < X < x) \in \mathcal{A}$, since X is a r.v. That the converse is not necessarily true is seen by the following simple example. Take $\Omega = \{a, b, c, d\}$, $\mathcal{A} = \{\oslash, \{a, b\}, \{c, d\}, \Omega\}$, and define X by: $X(a) = -1, X(b) = 1, X(c) = -2, X(d) = 2$. Then $\Omega' = \{-2, -1, 1, 2\}$, and let $\mathcal{A}' = \mathcal{P}(\Omega')$. We have $|X|^{-1}(\{1\}) = \{a, b\}, |X|^{-1}(\{2\}) = \{c, d\}, |X|^{-1}(\{-2\}) = |X|^{-1}(\{-1\}) = \oslash$, and all these sets are in $\mathcal{A}$, so that $|X|$ is measurable. However, $X^{-1}(\{-1\}) = \{a\}$ and $X^{-1}(\{-2\}) = \{c\}$, none of which belongs in $\mathcal{A}$, so that X is not measurable.

As another example, let B be a non-Borel set in $\Re$, and define X by: $X(\omega) = 1, \omega \in B$, and $X(\omega) = -1, \omega \in B^c$. Then X is not $\mathcal{B}$-measurable as $X^{-1}(\{1\}) = B \notin \mathcal{B}$, but $|X|^{-1}(\{1\}) = \Re \in \mathcal{B}$. #

26. $X + Y$ is measurable by Exercise 24. Next, $(-Y \leq y) = (Y \geq -y) \in \mathcal{A}$, so that $-Y$ is measurable. Then $X + (-Y) = X - Y$ is measurable. Now, if Z is measurable, then so is Z^2 because, for $z \geq 0$, $(Z^2 \leq z) = (-\sqrt{z} \leq Z \leq \sqrt{z}) \in \mathcal{A}$. Thus, if X, Y are measurable, then so are $(X+Y)^2$ and $(X-Y)^2$, and therefore so is: $(X+Y)^2 - (X-Y)^2$. But $(X+Y)^2 - (X-Y)^2 = 4XY$. Thus, $4XY$ is measurable, and then so is, clearly, XY.
Finally, if $P(Y \neq 0) = 1$, then, for $y \neq 0$, $(\frac{1}{Y} \leq y) = (Y \geq \frac{1}{y}) \in \mathcal{A}$, so that $\frac{1}{Y}$ is measurable. Thus, X and Y are measurable, and $P(Y \neq 0) = 1$, so that X and $\frac{1}{Y}$ are measurable. Then $X \times \frac{1}{Y} = \frac{X}{Y}$ is measurable. #

27. Since $\sigma(\mathcal{T}_m) = \mathcal{B}^m$, it suffices to show (by Theorem 2) that $f^{-1}(\mathcal{T}_m) \subseteq \mathcal{B}^m$ for f to be measurable. By continuity of f, $f^{-1}(\mathcal{T}_m) \subseteq \mathcal{T}_n \subseteq \mathcal{B}^n$, since $\sigma(\mathcal{T}_n) = \mathcal{B}^n$. Thus, f is measurable. Then, for $B \in \mathcal{B}^m$, $[f(X)]^{-1} = X^{-1}[f^{-1}(B)] \in \mathcal{A}$, since $f^{-1}(B) \in \mathcal{B}^n$ and X is measurable. #

28. For any r.v. Z, it holds: $Z = Z^+ - Z^-$ and $|Z| = Z^+ + Z^-$. Hence $Z^+ = \frac{1}{2}(|Z| + Z)$, $Z^- = \frac{1}{2}(|Z| - Z)$.
Applying this to X, Y and $X + Y$, we get:

$$X^+ = \frac{1}{2}(|X| + X),\ Y^+ = \frac{1}{2}(|Y| + Y),\ (X+Y)^+ = \frac{1}{2}[|X+Y| + (X+Y)].$$

Hence

$$X^+ + Y^+ = \frac{1}{2}[(|X| + |Y|) + (X+Y)] \geq \frac{1}{2}[|X+Y| + (X+Y)] = (X+Y)^+.$$

Likewise,

$$X^- = \frac{1}{2}(|X| - X),\ Y^- = \frac{1}{2}(|Y| - Y),\ (X+Y)^- = \frac{1}{2}[|X+Y| - (X+Y)]$$

and hence

$$X^- + Y^- = \frac{1}{2}[(|X|+|Y|) - (X+Y)] \geq \frac{1}{2}[|X+Y| - (X+Y)] = (X+Y)^-.$$

Alternative proof:
Let $X+Y \leq 0$. Then $(X+Y)^+ = 0 = 0+0 \leq X^+ + Y^+$. Let $X+Y > 0$. Then $(X+Y)^+ = X+Y \leq X^+ + Y^+$, because $X = X^+ - X^- \leq X^+$ and $Y = Y^+ - Y^- \leq Y^+$. Thus, $(X+Y)^+ \leq X^+ + Y^+$. Again, let $X+Y < 0$. Then $(X+Y)^- = -(X+Y) = -X-Y \leq X^- + Y^-$, because $X = X^+ - X^-$ or $-X = X^- - X^+ \leq X^-$ and $Y = Y^+ - Y^-$ or $-Y = Y^- - Y^+ \leq Y^-$. Next, let $X+Y \geq 0$. Then $(X+Y)^- = 0 = 0+0 \leq X^- + Y^-$, so that $(X+Y)^- \leq X^- + Y^-$. So, again: $(X+Y)^+ \leq X^+ + Y^+$ and $(X+Y)^- \leq X^- + Y^-$. #

29. **(i)** From the definition of B_m, we have: $B_1 = A_1$, and for $m \geq 2$, $B_m = A_1^c \cap \ldots \cap A_{m-1}^c \cap A_m$.

(ii) For $i \neq j$ (e.g., $i < j$), B_i is either A_1 (for $i = 1$) or $B_i = A_1^c \cap \ldots \cap A_{i-1}^c \cap A_i$, whereas $B_j = A_1^c \cap \ldots \cap A_{j-1}^c \cap A_j$, and $B_i \cap B_j = \emptyset$, because B_i contains A_i and B_j contains A_i^c (since $i \leq j-1$).

(iii) Let $\omega = \sum_{m=1}^{\infty} B_m$. Then either $\omega \in B_1 = A_1$, and hence $\omega \in \cup_{n=1}^{\infty} A_n$, or $\omega \notin A_i, i = 1, \ldots, n-1$ and $\omega \in A_n$, so that $\omega \in \cup_{n=1}^{\infty} A_n$. Thus, $\sum_{m=1}^{\infty} B_m \subseteq \cup_{n=1}^{\infty} A_n$. Next, let $\omega \in \cup_{n=1}^{\infty} A_n$. Then either $\omega \in A_1 = B_1$, so that $\omega \in \sum_{m=1}^{\infty} B_m$, or $\omega \notin A_i, i = 1, \ldots, n-1$ and $\omega \in A_n$. Then $\omega \in B_n$, so that $\omega \in \sum_{m=1}^{\infty} B_m$. #

30. **(i)** We have $\underline{\lim}_{n\to\infty} A_n = \cup_{n=1}^{\infty} \cap_{k=n}^{\infty} A_k$, so that $\omega \in (\underline{\lim}_{n\to\infty} A_n)$ or $\omega \in \cup_{n=1}^{\infty} \cap_{k=n}^{\infty} A_k$, therefore $\omega \in \cap_{k=n_0}^{\infty} A_k$ for some n_0, and hence $\omega \in A_k$ for all $k \geq n_0$. Next, let $\omega \in A_n$ for all but finitely many ns; i.e., $\omega \in A_n$ for all $n \geq n_0$. Then $\omega \in \cap_{k=n_0}^{\infty} A_k$ and hence $\omega \in \cup_{n=1}^{\infty} \cap_{k=n}^{\infty} A_k$, which completes the proof.

(ii) Here $\overline{\lim}_{n\to\infty} A_n = \cap_{n=1}^{\infty} \cup_{k=n}^{\infty} A_k$, and hence $\omega \in (\overline{\lim}_{n\to\infty} A_n)$ or $\omega \in \cap_{n=1}^{\infty} \cup_{k=n}^{\infty} A_k$ implies that $\omega \in \cup_{k=n}^{\infty} A_k$ for $n \geq 1$. From $\omega \in \cup_{k=1}^{\infty} A_k$, let k_1 be the first k for which $\omega \in A_{k_1}$. Next, consider $\cup_{k=k_1+1}^{\infty} A_k$, and from $\omega \in \cup_{k=k_1+1}^{\infty} A_k$, let k_2 be the first k ($\geq k_1 + 1$) for which $\omega \in A_{k_2}$. Continuing like this, we get that ω belongs to infinitely many A_ns. In the other way around, if ω belongs to infinitely many A_ns, that means that there exist $1 < k_1 < k_2 < \ldots$ such that $\omega \in A_{k_j}, j = 1, 2, \ldots$ Then $\omega \in \cup_{k=k_j}^{\infty} A_k, j \geq 1$, and hence $\omega \in \cup_{k=n}^{\infty} A_k$ for $1 \leq n \leq k_1$ and $k_j < n < k_{j+1}, j \geq 1$. Thus, $\omega \in \cap_{n=1}^{\infty} \cup_{k=n}^{\infty} A_k$ and the result follows. #

31. From $A_k \subseteq B_k, k \geq 1$, we have $\cup_{k=n}^{\infty} A_k \subseteq \cup_{k=n}^{\infty} B_k, n \geq 1$, and hence $\cap_{n=1}^{\infty} \cup_{k=n}^{\infty} A_k \subseteq \cap_{n=1}^{\infty} \cup_{k=n}^{\infty} B_k$ or $\overline{\lim}_{n\to\infty} A_n \subseteq \overline{\lim}_{n\to\infty} B_n$ or (A_n i.o.) $\subseteq$ (B_n i.o.) (by Exercise 2). #

32. We have $\underline{\lim}_{n\to\infty} A_n = \cup_{n=1}^{\infty} \cap_{k=n}^{\infty} A_k$ and $\cap_{k=n}^{\infty} A_k = \cap_{k=n}^{\infty}\{r \in (1-\frac{1}{k+1}, 1+\frac{1}{k});\ r \in Q\} = \{1\}$ for all n, so that $\cup_{n=1}^{\infty}\cap_{k=n}^{\infty} A_k = \{1\}$; i.e., $\underline{\lim}_{n\to\infty} A_n = \{1\}$. Next, $\overline{\lim}_{n\to\infty} A_n = \cap_{n=1}^{\infty} \cup_{k=n}^{\infty} A_k$ and $\cup_{k=n}^{\infty} A_k = \cup_{k=n}^{\infty}\{r \in (1-\frac{1}{k+1}, 1+\frac{1}{k});$ $r \in Q\} = \{r \in (1-\frac{1}{n+1}, 1+\frac{1}{n});\ r \in Q\}$, so that $\cap_{n=1}^{\infty} \cup_{k=n}^{\infty} A_k = \cap_{n=1}^{\infty}\{r \in (1-\frac{1}{n+1}, 1+\frac{1}{n});\ r \in Q\} = \{1\}$. Thus, $\underline{\lim}_{n\to\infty} A_n = \overline{\lim}_{n\to\infty} A_n = \{1\} = \lim_{n\to\infty} A_n$. #

33. Here $\underline{\lim}_{n\to\infty} A_n = \cup_{n=1}^{\infty} \cap_{k=n}^{\infty} A_k$, and consider the $\cap_{k=n}^{\infty} A_k$ for n odd or even. Then

$$\bigcap_{k=2n-1}^{\infty} A_k = \Big(\bigcap_{\substack{k \text{ odd} \\ \geq 2n-1}} A_k\Big) \cap \Big(\bigcap_{\substack{k \text{ even} \\ \geq 2n}} A_k\Big),$$

and

$A_{2n-1} \cap A_{2n+1} \cap \ldots = [-1, \frac{1}{2n-1}] \cap [-1, \frac{1}{2n+1}] \cap \ldots = [-1, 0]$, $A_{2n} \cap A_{2n+2} \cap \ldots = [0, \frac{1}{2n}) \cap [0, \frac{1}{2n+2}) \cap \ldots = \{0\}$, so that $\cap_{k=2n-1}^{\infty} A_k = [-1, 0] \cap \{0\} = \{0\}$. Next,

$$\bigcap_{k=2n}^{\infty} A_k = \Big(\bigcap_{\substack{k \text{ even} \\ \geq 2n}} A_k\Big) \cap \Big(\bigcap_{\substack{k \text{ odd} \\ \geq 2n+1}} A_k\Big),$$

and

$A_{2n} \cap A_{2n+2} \cap \ldots = [0, \frac{1}{2n}) \cap [0, \frac{1}{2n+2}) \cap \ldots = \{0\}$, $A_{2n+1} \cap A_{2n+3} \cap \ldots = [-1, \frac{1}{2n+1}] \cap [-1, \frac{1}{2n+3}] \cap \ldots = [-1, 0]$, so that $\cap_{k=2n}^{\infty} A_k = \{0\} \cap [-1, 0] = \{0\}$. It follows that $\cup_{n=1}^{\infty} \cap_{k=n}^{\infty} A_n = \{0\} = \underline{\lim}_{n\to\infty} A_n$.

Next, $\overline{\lim}_{n\to\infty} A_n = \cap_{n=1}^{\infty} \cup_{k=n}^{\infty} A_k$, and consider the $\cup_{k=n}^{\infty} A_k$ for odd and even values of n. We have

$$\bigcup_{k=2n-1}^{\infty} A_k = \Big(\bigcup_{\substack{k \text{ odd} \\ \geq 2n-1}} A_k\Big) \cup \Big(\bigcup_{\substack{k \text{ even} \\ \geq 2n}} A_k\Big),$$

and

$A_{2n-1} \cup A_{2n+1} \cup \ldots = [-1, \frac{1}{2n-1}] \cup [-1, \frac{1}{2n+1}] \cup \ldots = [-1, \frac{1}{2n-1}]$, $A_{2n} \cup A_{2n+2} \cup \ldots = [0, \frac{1}{2n}) \cup [0, \frac{1}{2n+2}) \cup \ldots = [0, \frac{1}{2n})$, so that $\cup_{k=2n-1}^{\infty} A_k = [-1, \frac{1}{2n-1}] \cup [0, \frac{1}{2n}) = [-1, \frac{1}{2n-1}]$. Next,

$$\bigcup_{k=2n}^{\infty} A_k = \Big(\bigcup_{\substack{k \text{ even} \\ \geq 2n}} A_k\Big) \cup \Big(\bigcup_{\substack{k \text{ odd} \\ \geq 2n+1}} A_k\Big),$$

and

$A_{2n} \cup A_{2n+2} \cup \ldots = [0, \frac{1}{2n}) \cup [0, \frac{1}{2n+2}) \cup \ldots = [0, \frac{1}{2n})$, $A_{2n+1} \cup A_{2n+3} \cup \ldots = [-1, \frac{1}{2n+1}] \cup [-1, \frac{1}{2n+3}] \cup \ldots = [-1, \frac{1}{2n+1}]$, so that $\cup_{k=2n}^{\infty} A_k = [0, \frac{1}{2n}) \cup$

$[-1, \frac{1}{2n+1}] = [-1, \frac{1}{2n})$. It follows that

$$\bigcap_{n=1}^{\infty} \bigcup_{k=n}^{\infty} A_k = [-1, 1] \cap [-1, \tfrac{1}{2}) \cap [-1, \tfrac{1}{3}] \cap [-1, \tfrac{1}{4}) \cap \ldots$$
$$= [-1, 0] = \overline{\lim_{n\to\infty}} A_n.$$

So, $\underline{\lim}_{n\to\infty} A_n = \{0\}$ and $\overline{\lim}_{n\to\infty} A_n = [-1, 0]$, so that the $\lim_{n\to\infty} A_n$ does not exist. #

34. **(i)** We have:

$$\{[0, 1), [1, 2), \ldots, [n-1, n)\} \subset \{[0, 1), [1, 2), \ldots, [n-1, n), [n, n+1)\}$$

and hence $\mathcal{A}_n \subseteq \mathcal{A}_{n+1}$. That $\mathcal{A}_n \subset \mathcal{A}_{n+1}$ follows by the fact that, e.g., $[n, n+1)$ cannot belong in $\mathcal{A}_n$ since all members of $\mathcal{A}_n$ are $\subseteq [0, n)$.

(ii) Let $A_1 \in \mathcal{A}_1$, $A_2 \in \mathcal{A}_2$ but not in $\mathcal{A}_1, \ldots, A_n \in \mathcal{A}_n$ but not in $\mathcal{A}_{n-1}, \ldots,$ and set $A = \cup_{i=1}^{\infty} A_i$. Then $A \notin \cup_{n=1}^{\infty} \mathcal{A}_n$, because otherwise, $A \in \mathcal{A}_n$ for some n. However, this is not possible since $\cup_{i=n+1}^{\infty} A_i \notin \mathcal{A}_n$.

(iii) $\mathcal{A}_1 = \{\emptyset, [0, 1), [0, 1)^c = (-\infty, 0) \cup [1, \infty), \Re\}$, $\mathcal{A}_2 = \{\emptyset, [0, 1), [1, 2), (-\infty, 0) \cup [1, \infty), (-\infty, 1) \cup [2, \infty), [0, 2), (-\infty, 0) \cup [2, \infty), \Re\}$. #

35. **(i)** First, observe that all intersections $A_1' \cap \ldots \cap A_n'$ are pairwise disjoint, so that their unions are, actually, sums. Next, if A and B are in $\mathcal{C}$, it is clear that $A \cup B$ is a sum of intersections $A_1' \cap \ldots \cap A_n'$ (the sum of those intersections in A and those intersections in B), so that $A \cup B$ is in $\mathcal{C}$. Now, if $A \in \mathcal{C}$, then A^c is the sum of all those intersections $A_1' \cap \ldots \cap A_n'$ which are not part of A. Hence A^c is also in $\mathcal{C}$, and $\mathcal{C}$ is a field.

(ii) In forming $A_1' \cap \ldots \cap A_n'$, we have 2 choices at each one of the n steps. Thus, there are 2^n sets of the form $A_1' \cap \ldots \cap A_n'$. Next, in forming their sums, we select k of those members at a time, where $k = 0, 1, \ldots, 2^n$. Therefore the total number of sums is: $\binom{2^n}{0} + \binom{2^n}{1} + \ldots + \binom{2^n}{2^n} = 2^{2^n}$. #

36. **(i)** If $\omega \in A$, then $f(\omega) \in f(A)$ and $\omega \in f^{-1}[f(A)]$. For a concrete example, take $f : \Re \to [0, 1)$ where $f(x) = x^2$, and let $A = [0, 1)$. Then $f(A) = f([0, 1]) = [0, 1)$, and $f^{-1}([0, 1)) = (-1, 1)$. It follows that $f^{-1}[f(A)] = f^{-1}([0, 1)) = (-1, 1) \supset [0, 1) = A$.

(ii) Let $\omega' \in f[f^{-1}(B)]$ which implies that there exists $\omega \in f^{-1}(B)$ such that $f(\omega) = \omega'$. Also, $\omega \in f^{-1}(B)$ implies that $f(\omega) \in B$. Since also $f(\omega) = \omega'$, it follows that $\omega' \in B$. Thus $f[f^{-1}(B)] \subseteq B$.

For a concrete example, let $f : \Re \to \Re$ with $f(x) = c$. Take $B = (c-1, c+1)$, so that $f^{-1}[(c-1, c+1)] = \Re$ and $f(\Re) = \{c\} \subset (c-1, c+1)$. That is, $f[f^{-1}(B)] = \{c\} \subset (c-1, c+1) = B$. #

37. **(i)** Since $X^{-1}(\{-1\}) = A_1$, $X^{-1}(\{1\}) = A_1^c \cap A_2$, and $X^{-1}(\{0\}) = A_1^c \cap A_2^c$, and A_1, $A_1^c \cap A_2$, $A_1^c \cap A_2^c$ are in $\mathcal{A}$, X is a r.v.

(ii) We have $X^{-1}(\{-1\}) = \{a, b\}$, $X^{-1}(\{1\}) = \{c\}$, $X^{-1}(\{2\}) = \{d\}$, and neither $\{c\}$ nor $\{d\}$ are in $\mathcal{A}$. Then X is not $\mathcal{A}$-measurable.

(iii) We have $X^{-1}(\{-2\}) = \{-2\}$, $X^{-1}(\{-1\}) = \{-1\}$, $X^{-1}(\{0\}) = \{0\}$, $X^{-1}(\{1\}) = \{1\}$, $X^{-1}(\{2\}) = \{2\}$, so that $X^{-1}(\mathcal{B})$ is the field induced in Ω by the partition: $\{\{-2\}, \{-1\}, \{0\}, \{1\}, \{2\}\}$.
The values taken on by X^2 are 0, 1, 4, and $(X^2)^{-1}(\{0\}) = \{0\}$, $(X^2)^{-1}(\{1\}) = \{-1, 1\}$, $(X^2)^{-1}(\{4\}) = \{-2, 2\}$, so that the field induced by X^2 is the one generated by the sets $\{0\}, \{-1, 1\}, \{-2, 2\}$, and it is, clearly, strictly contained in the one induced by X. #

38. For a fixed k, let $\mathcal{A}_{k,n} = (X_k, \ldots, X_{k+n-1})^{-1}(\mathcal{B})$. Then the σ-fields $\mathcal{A}_{k,n}$, $n \geq 1$, form a nondecreasing sequence and therefore $\mathcal{F}_k = \cup_{n=1}^{\infty} \mathcal{A}_{k,n}$ is a field (but it may fail to be a σ-field; see Exercise 10 in this chapter) and $\mathcal{B}_k = \sigma(\mathcal{F}_k)$. Likewise, $\mathcal{B}_l = \sigma(\mathcal{F}_l)$ where $\mathcal{F}_l = \cup_{n=1}^{\infty} \mathcal{A}_{l,n}$.

However, $\cup_{n=k}^{\infty} \mathcal{A}_n \supseteq \cup_{n=l}^{\infty} \mathcal{A}_n$, so that $\mathcal{B}_k = \sigma(\cup_{n=k}^{\infty} \mathcal{A}_n) \supseteq \sigma(\cup_{n=l}^{\infty} \mathcal{A}_n) = \mathcal{B}_l$. This is so by the way the σ-fields $\mathcal{B}_k$ and $\mathcal{B}_l$ are generated (see Theorem 2(ii) in this chapter). #

39. Since S_k is a function of the X_js, $j = 1, \ldots, k$, $k = 1, \ldots, n$ it follows that $\sigma(S_k) \subseteq \sigma(X_1, \ldots, X_n)$, $k = 1, \ldots, n$. Hence $\cup_{k=1}^{n} \sigma(S_k) \subseteq \sigma(X_1, \ldots, X_n)$ and then $\sigma(\cup_{k=1}^{n} \sigma(S_k)) \subseteq \sigma(X_1, \ldots, X_n)$ or $\sigma(S_1, \ldots, S_n) \subseteq \sigma(X_1, \ldots, X_n)$. Next, $X_k = S_k - S_{k-1}$, $k = 1, \ldots, n$ $(S_0 = 0)$, so that X_k is a function of the S_js, $k = 1, \ldots, n$. Then, as above, $\sigma(X_1, \ldots, X_n) \subseteq \sigma(S_1, \ldots, S_n)$, and equality follows. #

40. Consider the function $f : \Re \to \Re$ defined by $y = f(x) = x + c$. Then, clearly, $f(B) = B_c$. The existing inverse of f, f^{-1}, is given by: $x = f^{-1}(y) = x - c$, and it is clear that $(f^{-1})(B_c) = B$. By setting $g = f^{-1}$, so that $g^{-1} = f$, we have that $g^{-1}(B)(= f(B)) = B_c$. So, g^{-1} is continuous and hence measurable, and $g^{-1}(B) = B_c$. Since B is measurable then so is B_c. #

41. **(i)** Clearly, $\mathcal{F} \neq \oslash$. Next, to show that $\mathcal{F}$ is closed under complementation. Indeed, if $A \in \mathcal{F}$, then

$$A = \bigcup_{i=1}^{n} A_i = \bigcup_{i=1}^{n} \bigcap_{j=1}^{m_i} A_i^j$$
$$= (A_1^1 \cap \ldots \cap A_1^{m_1}) \cup \ldots \cup (A_n^1 \cap \ldots \cap A_n^{m_n})$$

with all $A_1^1, \ldots, A_1^{m_1}, \ldots, A_n^1, \ldots, A_n^{m_n}$ in $\mathcal{F}_1$, so that

$$A^c = [A_1^1 \cap \ldots \cap A_1^{m_1}) \cup \ldots \cup (A_n^1 \cap \ldots \cap A_n^{m_n})]^c$$
$$= [(A_1^1)^c \cup \ldots \cup (A_1^{m_1})^c] \cap \ldots \cap [(A_n^1)^c \cup \ldots \cup (A_n^{m_n})^c]$$
$$= \bigcup_{i_1=1}^{m_1} \cdots \bigcup_{i_n=1}^{m_n} [(A_1^{i_1})^c \cap \ldots \cap (A_n^{i_n})^c].$$

The fact that $A_1^{i_1}, \ldots, A_n^{i_n}$ are in $\mathcal{F}_1$ implies that $(A_1^{i_1})^c, \ldots, (A_n^{i_n})^c$ are also in $\mathcal{F}_1$, as follows from the definition of $\mathcal{F}_1$. So, A^c is a finite union of a finite intersection of members of $\mathcal{F}_1$, and hence $A^c \in \mathcal{F}_3(= \mathcal{F})$,

by the definition of $\mathcal{F}_3$. Next, let $A, B \in \mathcal{F}$. To show that $A \cup B \in \mathcal{F}$. Indeed, $A, B \in \mathcal{F}$ implies that $A = A_1 \cup \ldots \cup A_m = (A_1^1 \cap \ldots \cap A_1^{k_1}) \cup \ldots \cup (A_m^1 \cap \ldots \cap A_m^{k_m})$ with $A_i^1, \ldots, A_i^{k_i}$ in $\mathcal{F}_1, i = 1, \ldots, m$, $B = B_1 \cup \ldots \cup B_n = (B_1^1 \cap \ldots \cap B_1^{l_1}) \cup \ldots \cup (B_n^1 \cap \ldots \cap B_n^{l_n})$ with $B_j^1, \ldots, B_j^{l_j}$ in $\mathcal{F}_1, j = 1, \ldots, n$,
so that

$$\begin{aligned} A \cup B &= [(A_1^1 \cap \ldots \cap A_1^{k_1}) \cup \ldots \cup (A_m^1 \cap \ldots \cap A_m^{k_m})] \cup \\ &\quad [(B_1^1 \cap \ldots \cap B_1^{l_1}) \cup \ldots \cup (B_n^1 \cap \ldots \cap B_n^{l_n})] \\ &= (A_1^1 \cap \ldots \cap A_1^{k_1}) \cup \ldots \cup (A_m^1 \cap \ldots \cap A_m^{k_m}) \cup \\ &\quad (B_1^1 \cap \ldots \cap B_1^{l_1}) \cup \ldots \cup (B_n^1 \cap \ldots \cap B_n^{l_n}), \end{aligned}$$

which is a finite union of finite intersections of members of $\mathcal{F}_1$. It follows that $A \cup B$ is in $\mathcal{F}_3 (= \mathcal{F})$, so that $\mathcal{F}$ is a field.

(ii) Trivially, $\mathcal{C} \subseteq \mathcal{F}$, so that $\mathcal{F}(\mathcal{C}) \subseteq \mathcal{F}$. To show that $\mathcal{F} \subseteq \mathcal{F}(\mathcal{C})$. Let $A \in \mathcal{F}$. Then, by part (i), $A = (A_1^1 \cap \ldots \cap A_1^{m_1}) \cup \ldots \cup (A_n^1 \cap \ldots \cap A_n^{m_n})$ with all $A_1^1, \ldots, A_1^{m_1}, \ldots, A_n^1, \ldots, A_n^{m_n}$ in $\mathcal{F}_1$.
Clearly, $\mathcal{F}_1 \subseteq \mathcal{F}(\mathcal{C})$ by the definition of $\mathcal{F}_1$. Thus, $A_i^1, \ldots, A_i^{m_i}$ are in $\mathcal{F}(\mathcal{C})$, for $i = 1, \ldots, n$, and then the intersections $A_i^1 \cap \ldots \cap A_i^{m_i}, i = 1, \ldots, n$ are in $\mathcal{F}(\mathcal{C})$, and therefore so is their union $(A_1^1 \cap \ldots \cap A_1^{m_1}) \cup \ldots \cup (A_n^1 \cap \ldots \cap A_n^{m_n})$. Since this union is A, it follows that $A \in \mathcal{F}(\mathcal{C})$. Thus, $\mathcal{F} \subseteq \mathcal{F}(\mathcal{C})$, and the proof is completed. #
Remark: In Exercise 41, in the proof that $A \in \mathcal{F}$ implies $A^c \in \mathcal{F}$, the following property was used (in a slightly different notation for simplification); namely, $(C_1^1 \cup \ldots \cup C_1^{m_1}) \cap \ldots \cap (C_n^1 \cup \ldots \cup C_n^{m_n}) = \cup_{i_1=1}^{m_1} \ldots \cup_{i_n=1}^{m_n} (C_1^{i_1} \cap \ldots \cap C_n^{i_n})$.
This is justified as follows: Let ω belong to the right-hand side. Then ω belongs to at leats one of the $m_1 \times \ldots \times m_n$ members of the union, for example, $\omega \in (C_1^{i_1'} \cap \ldots \cap C_n^{i_n'})$ for some $1 \le i_1' \le m_1, \ldots, 1 \le i_n' \le m_n$. But then $\omega \in (C_1^1 \cup \ldots \cup C_1^{m_1}), \ldots, \omega \in (C_n^1 \cup \ldots \cup C_n^{m_n})$, and therefore $\omega \in [C_1^1 \cup \ldots \cup C_1^{m_1}) \cap \ldots \cap (C_n^1 \cup \ldots \cup C_n^{m_n})]$, or ω belongs to the left-hand side. Next, let ω belong to the left-hand side. Then $\omega \in (C_1^1 \cup \ldots \cup C_1^{m_1}), \ldots, \omega \in (C_n^1 \cup \ldots \cup C_n^{m_n})$, so that $\omega \in C_1^{i_1'}, \ldots, \omega \in C_n^{i_n'}$ for some $1 \le i_1' \le m_1, \ldots, 1 \le i_n' \le m_n$. But then $\omega \in (C_1^{i_1'} \cap \ldots \cap C_n^{i_n'})$, and $C_1^{i_1'} \cap \ldots \cap C_n^{i_n'}$ is one of the $m_1 \times \ldots \times m_n$ members of the union on the right-hand side. It follows that ω belongs to the right-hand side, and the justification is completed. #

42. Let $A \in \mathcal{A}$. Then $A = \cup_{i=1}^{\infty} A_i, A_i = A_i^1 \cap A_i^2 \cap \ldots$ with $A_i^1, A_i^2, \ldots$ in $\mathcal{A}_1, i \ge 1$. Then

$$A^c = (\bigcup_{i=1}^{\infty} A_i)^c = \bigcap_{i=1}^{\infty} A_i^c = \bigcap_{i=1}^{\infty} (A_i^1 \cap A_i^2 \cap \ldots)^c$$

$$= \bigcap_{i=1}^{\infty} [(A_i^1)^c \cup (A_i^2)^c \cup \ldots]$$
$$= [(A_1^1)^c \cup (A_1^2)^c \cup \ldots] \cap [(A_2^1)^c \cup (A_2^2)^c \cup \ldots]$$
$$\cap \ldots \cap [(A_n^1)^c \cup (A_n^2)^c \cup \ldots] \cap \ldots,$$

and this is equal to

$$\cup [(A_1^{i_1})^c (\cap A_2^{i_2})^c \cap \ldots \cap (A_n^{i_n})^c \cap \ldots]$$

with $i_1, i_2, \ldots, i_n, \ldots$ integers ≥ 1, and the union extends over all choices of the sets $(A_1^{i_1})^c, (A_2^{i_2})^c, \ldots, (A_n^{i_n})^c, \ldots$ from the respective collections: $(A_i^1)^c$, $(A_i^2)^c, \ldots, (A_i^n)^c, \ldots$, $i = 1, 2, \ldots, n, \ldots$ However, these choices produce $\aleph_0 \times \aleph_0 \times \ldots \times \aleph_0 \times \ldots = \aleph_0^{\aleph_0} = \aleph$ where $\aleph_0$ and $\aleph$ are the cardinal numbers of a countable set and of the continuum, respectively. Thus, there are uncountable members in the union, and hence the union need not be in $\mathcal{A}$. In other word, A^c need not be in $\mathcal{A}$, so that $\mathcal{A}$ need not be a σ-field.
Remark: For the justification of the equality, asserted in the derivations related to A^c, refer to the remark following the proof of Exercise 41. #

Chapter 2

Definition and Construction of a Measure and its Basic Properties

1. If Ω is finite, then μ is ≥ 0, $\mu(\oslash) = 0$ and finitely additive (since there are only finitely many subsets of Ω). Thus, μ is a measure, and also finite. If Ω is denumerable, $\Omega = \{\omega_1, \omega_2, \ldots\}$, then $\mu \geq 0$, $\mu(\oslash) = 0$, and if $A_n, n \geq 1$, are $\neq \oslash$ and pairwise disjoint, then $\mu\left(\sum_{n=1}^{\infty} A_n\right) = \infty$ and $\sum_{n=1}^{\infty} \mu\left(A_n\right) = \infty$ since each term is ≥ 1. Thus, μ is a measure. It is σ-finite, since $\Omega = \sum_{n=1}^{\infty} \{\omega_n\}$ and $\mu(\{\omega_n\}) = 1$ (finite). #
2. **(i)** Let $A_i \in \mathcal{C}, i = 1, \ldots, n, A_i \cap A_j = \oslash, i \neq j$, and set $A = \sum_{i=1}^{n} A_i$, so that $A \in \mathcal{C}$. Then either A is finite or A^c is finite. If A is finite, then all $A_i, i = 1, \ldots, n$, are finite, and therefore $P(A) = 0 = 0 + \ldots + 0 = P(A_1) + \ldots + P(A_n)$. If A^c is finite, then A is not finite and hence at least one of $A_1, \ldots, A_n$ is not finite; call A_{i_0} such an event. We claim that A_{i_0} is unique. Indeed, if A_i and $A_j, i \neq j$, are not finite, then A_i^c and A_j^c are finite. Since $A_i \cap A_j = \oslash$, it follows that $A_i \subset A_j^c$ and hence A_i is finite, a contradiction. Then $\sum_{i=1}^{n} P(A_i) = P(A_{i_0}) = 1$ (since $P(A_i) = 0, i \neq i_0$, as being all finite), and $P(A) = 1$. Hence $P(A) = \sum_{i=1}^{n} P(A_i)$.
   (ii) Let $\Omega = \{\omega_1, \omega_2, \ldots\}$ and take $A_i = \{\omega_i\}$, so that $A_i \cap A_j = \oslash$, $i \neq j$, and $P(A_i) = 0$ for all i. However, $P\left(\sum_{i=1}^{\infty} A_i\right) (= P(\Omega)) = 1$ since $\sum_{i=1}^{\infty} A_i$ is infinite (and $\left(\sum_{i=1}^{n} A_i\right)^c = \oslash$ finite). Therefore $P\left(\sum_{i=1}^{\infty} A_i\right) = 1 \neq 0 = \sum_{i=1}^{\infty} P(A_i)$, and P is not σ-additive.

(iii) Let $A_n \in \mathcal{C}, n \geq 1, A_i \cap A_j = \oslash, i \neq j$, and set $A = \sum_{n=1}^{\infty} A_n$, so that $A \in \mathcal{C}$. Then either A is finite or A^c is finite. If A is finite, then all A_ns are finite (indeed, A is only the sum of finitely many of the A_ns) and hence $P(A_n) = 0$ for all n, and also $P(A) = 0$. Thus, $P(A) = \sum_{n=1}^{\infty} P(A_n)$ (actually, the σ-additivity here degenerates to finite additivity). If A^c is finite, then A is infinite. Since Ω is uncountable, it follows that at least one of the A_ns is infinite, because otherwise A would be countable (so that $A + A^c = \Omega$ is countable, a contradiction) ; call A_{n0} such an event. We claim that A_{n0} is unique. Indeed, if A_i and $A_j, i \neq j$, are infinite, then A_i^c and A_j^c are finite. Since $A_i \cap A_j = \oslash$, it follows that $A_i \subset A_j^c$ and hence A_i is finite, a contradiction. Then $\sum_{n=1}^{\infty} P(A_n) = P(A_{n0}) = 1$ (since $P(A_n) = 0, n \neq n_0$, as being all finite), and $P(A) = 1$. Hence $P(A) = \sum_{n=1}^{\infty} P(A_n)$.
Finally, it is clear that $P(A) \geq 0$, $P(\oslash) = 0$ and $P(\Omega) = 1$. These properties along with the σ-additivity just established make P a probability measure. #

3. Clearly, $P(A) \geq 0$, $P(\oslash) = 0$ and $P(\Omega) = 1$ since $\Omega^c = \oslash$ countable. It remains to establish σ-additivity. Let $A_n \in \mathcal{C}, n \geq 1, A_i \cap A_j = \oslash, i \neq j$, and set $A = \cup_{n=1}^{\infty} A_n$. Since $A \in \mathcal{C}$, it follows that either A is countable or A^c is countable. If A is countable, then all A_ns are countable, and hence $P(A) = 0$ and $P(A_n) = 0, n \geq 1$, so that $P(A) = \sum_{n=1}^{\infty} P(A_n)$. If A^c is countable, then A is uncountable, and therefore at least one of the A_ns is uncountable; call A_{n0} such an event. We claim that A_{n0} is unique. Indeed, if A_i and $A_j, i \neq j$, are uncountable, then A_i^c and A_j^c are countable. Since $A_i \cap A_j = \oslash$, it follows that $A_i \subset A_j^c$ and hence A_i is countable, a contradiction. Then $\sum_{n=1}^{\infty} P(A_n) = P(A_{n0}) = 1$ (since $P(A_i) = 0, i \neq n_0$, as being all countable), and $P(A) = 1$. Hence $P(A) = \sum_{n=1}^{\infty} P(A_n)$. #

4. $P(A_n) = 1$ if and only if $P(A_n^c) = 0$, which implies that $P\left(\cup_{n=1}^{\infty} A_n^c\right) \leq \sum_{n=1}^{\infty} P(A_n^c) = 0$; i.e., $P\left(\cup_{n=1}^{\infty} A_n^c\right) = 0$ or $P\left[\left(\cap_{n=1}^{\infty} A_n\right)^c\right] = 0$, and hence $P\left(\cap_{n=1}^{\infty} A_n\right) = 1$. #

5. For each $n \geq 2$, there are at most $n - 1$ events A_is for which $P(A_i) > \frac{1}{n}$, because otherwise, we could choose n events with $P(A_{i_j}) > \frac{1}{n}$, so that $\sum_{j=1}^{n} P(A_{i_j}) > 1$. However, $\sum_{j=1}^{n} A_{i_j} \subseteq \Omega$ and $\sum_{j=1}^{n} P(A_{i_j}) = P\left(\sum_{j=1}^{n} A_{i_j}\right)$ (by pairwise disjointness), and this is $\leq P(\Omega) = 1$, a contradiction. Thus, if $I_n = \{i \in I;\ P(A_i) > \frac{1}{n}\}$, then the cardinality of I_n is $\leq n-1$. Set $I_0 = \{i \in I;\ P(A_i) > 0\}$. Then, clearly, $I_0 = \cup_{n=2}^{\infty} I_n$, and since each I_n is finite, I_0 is countable. #

6. Clearly, $\mu(A) \geq 0$ and $\mu(\oslash) = 0$. To establish σ-additivity. To this end, let $A_n \in \mathcal{A}, A_i \cap A_j = \oslash, i \neq j$, and set $A = \sum_{n=1}^{\infty} A_n$. Then:

$$\mu(A) = \sum_{\omega_n \in A} p_n = \sum_{i=1}^{\infty} \sum_{\omega_n \in A_i} p_n = \sum_{i=1}^{\infty} \mu(A_i). \#$$

7. Let $\Omega_+ = \{\omega_n\text{s};\ p_n > 0\}$. Then the atoms are those A which are of the form: $A = \{\omega_n\} \cup N$, where $\oslash \subseteq N \subseteq \Omega - \Omega_+$. #

8. $\mu\left(\underline{\lim}_{n\to\infty} A_n\right) = \mu\left(\cup_{n=1}^{\infty} \cap_{i=n}^{\infty} A_i\right) = \mu\left(\lim_{n\to\infty} \cap_{i=n}^{\infty} A_i\right) = \lim_{n\to\infty} \mu\left(\cap_{i=n}^{\infty} A_i\right) \leq \underline{\lim}_{n\to\infty} \mu(A_n)$ since $\cap_{i=n}^{\infty} A_i \subseteq A_n$. Next, $\mu\left(\overline{\lim}_{n\to\infty} A_n\right) = \mu\left(\cap_{n=1}^{\infty} \cup_{i=n}^{\infty} A_i\right) = \mu\left(\lim_{n\to\infty} \cup_{i=n}^{\infty} A_i\right) = \lim_{n\to\infty} \mu\left(\cup_{i=n}^{\infty} A_i\right)$, provided $\mu\left(\cup_{i=n}^{\infty} A_i\right) < \infty$ for some n, and this is $\geq \overline{\lim}_{n\to\infty} \mu(A_n)$ since $\cup_{i=n}^{\infty} A_i \supseteq A_n$. #

9. μ^0 is an outer measure; i.e., $\mu^0(\oslash) = 0$, μ^0 is $\uparrow$, and μ^0 is sub-σ-additive, because: $\mu^0(\oslash) = I_{\oslash}(\omega) = 0$; $A \subseteq B$ implies $I_A(\omega_0) \leq I_B(\omega_0)$, so that $\mu^0(A) = I_A(\omega_0) \leq I_B(\omega_0) = \mu^0(B)$; clearly, $I_{\cup_{i=1}^{\infty} A_i}(\omega_0) \leq \sum_{i=1}^{\infty} I_{A_i}(\omega_0)$, so that $\mu^0\left(\cup_{i=1}^{\infty} A_i\right) = I_{\cup_{i=1}^{\infty} A_i}(\omega_0) \leq \sum_{i=1}^{\infty} I_{A_i}(\omega_0) = \sum_{i=1}^{\infty} \mu^0(A_i)$. #

10.

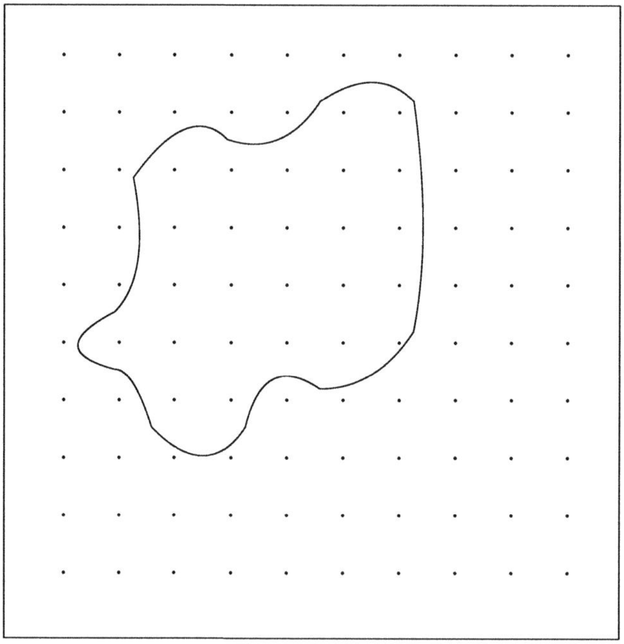

That $\mu^0(\oslash) = 0$ and $\uparrow$ are obvious. Denote by C_i the i-th column, $i = 1, \ldots, 10$, and let $A_n \subseteq \Omega, n \geq 1$. To show $\mu^0\left(\cup_{n\geq 1} A_n\right) \leq \sum_{n\geq 1} \mu^0(A_n)$. Set $A = \cup_{n\geq 1} A_n$ and suppose $\mu(A) = k$. Then there exist k columns $C_{i_1}, \ldots, C_{i_k}$ such that $C_{i_j} \cap A \neq \oslash$, $j = 1, \ldots, k$. This implies that there exists at least one $x_j \in C_{i_j} \cap A$ with $x_j \in C_{i_j}$ and $x_j \in A$, so that $x_j \in C_{i_j}$ and $x_j \in A_{n_j}$, $j = 1, \ldots, k$, where $n_1, \ldots, n_k$ are chosen from the set $\{1, 2, \ldots\}$ and need not be distinct. Then $\mu^0(A_{n_j}) \geq 1$, $j = 1, \ldots, k$, and therefore:

$$k \leq \sum_{j=1}^{k} \mu^0(A_{n_j}) \leq \sum_{n\geq 1} \mu^0(A_{n_j}) \text{ or } \mu^0\left(\underset{n\geq 1}{\cup} A_n\right) \leq \sum_{n\geq 1} \mu^0(A_n). \#$$

11.

(i) In the first place, it is clear that $\mu^*(\oslash) = 0$ and $\mu^*(\Omega) = 1$. Next, let $\oslash \subset A \subset \Omega$. The only covering of A by member of $\mathcal{F}$ is Ω, so that $\mu^*(A) = 1$. Thus, $\mu^*(A) = 0$ if $A = \oslash$ and $\mu^*(A) = 1$ if $A \neq \oslash$.

(ii) First, $\oslash$ and Ω are μ^*-measurable, and let $\oslash \subset A \subset \Omega$ (which implies $\oslash \subset A^c \subset \Omega$). Then A cannot be μ^*-measurable. Indeed, in the required equality $\mu^*(D) = \mu^*(A \cap D) + \mu^*(A^c \cap D)$, take $D = \Omega$. Then the left-hand side is $\mu^*(D) = \mu^*(\Omega) = 1$, and the right-hand side is $\mu^*(A \cap \Omega) + \mu^*(A^c \cap \Omega) = \mu^*(A) + \mu^*(A^c) = 1 + 1 = 2$, and the equality is violated. Hence $\mathcal{A}^* = \{\oslash, \Omega\}$. #

12.

(i) $\mathcal{C}$ is not a field because, e.g., $\{\omega_1, \omega_2\} \cup \{\omega_1, \omega_3\} = \{\omega_1, \omega_2, \omega_3\} \notin \mathcal{C}$.

(ii) Clearly, $\mu(A) \geq 0$ and $\mu(\oslash) = 0$. The only two disjoint sets whose sum is also in $\mathcal{C}$ are: $\{\omega_1, \omega_2\} + \{\omega_3, \omega_4\} = \Omega$, $\{\omega_1, \omega_3\} + \{\omega_2, \omega_4\} = \Omega$, and, by taking measures, we have: $3+3 = 6$, $3+3 = 6$, so that μ is a measure.

(iii) On $\mathcal{C}$: $\mu_1(\oslash) = \mu_2(\oslash) = 0$, $\mu_1(\Omega) = \mu_2(\Omega) = 6$, $\mu_1(\{\omega_1, \omega_2\}) = 3 = \mu_2(\{\omega_1, \omega_2\})$, $\mu_1(\{\omega_1, \omega_3\}) = 3 = \mu_2(\{\omega_1, \omega_2\})$, $\mu_1(\{\omega_2, \omega_4\}) = 3 = \mu_2(\{\omega_2, \omega_4\})$, $\mu_1(\{\omega_3, \omega_4\}) = 3 = \mu_2(\{\omega_3, \omega_4\})$, so that $\mu_1 = \mu_2$ on $\mathcal{C}$.

(iv) Write out the subsets of Ω and their coverages by unions of members of $\mathcal{C}$ with the smallest measures to get:

$$
\begin{aligned}
\omega_1 &: \{\omega_1, \omega_2\} \\
\omega_2 &: \{\omega_1, \omega_2\} \\
\omega_3 &: \{\omega_1, \omega_3\} \\
\omega_4 &: \{\omega_2, \omega_4\} \\
\{\omega_1, \omega_2\} &: \{\omega_1, \omega_2\} \\
\{\omega_1, \omega_3\} &: \{\omega_1, \omega_3\} \\
\{\omega_1, \omega_4\} &: \{\omega_1, \omega_2\} \cup \{\omega_2, \omega_4\} \cup, \{\omega_1, \omega_2\} \\
&\quad \cup \{\omega_3, \omega_4\}, \{\omega_1, \omega_3\} \cup \{\omega_2, \omega_4\}, \\
&\quad \{\omega_1, \omega_3\} \cup \{\omega_3, \omega_4\} \\
\{\omega_2, \omega_3\} &: \{\omega_1, \omega_2\} \cup \{\omega_1, \omega_3\} \cup, \{\omega_1, \omega_2\} \\
&\quad \cup \{\omega_3, \omega_4\}, \{\omega_1, \omega_3\} \cup \{\omega_2, \omega_4\}, \\
&\quad \{\omega_2, \omega_4\} \cup \{\omega_3, \omega_4\} \\
\{\omega_2, \omega_4\} &: \{\omega_2, \omega_4\} \\
\{\omega_3, \omega_4\} &: \{\omega_3, \omega_4\} \\
\{\omega_1, \omega_2, \omega_3\} &: \{\omega_1, \omega_2\} \cup \{\omega_1, \omega_3\} \\
\{\omega_1, \omega_2, \omega_4\} &: \{\omega_1, \omega_2\} \cup \{\omega_2, \omega_4\} \\
\{\omega_1, \omega_3, \omega_4\} &: \{\omega_1, \omega_3\} \cup \{\omega_2, \omega_4\} \\
\{\omega_2, \omega_3, \omega_4\} &: \{\omega_2, \omega_4\} \cup \{\omega_3, \omega_4\}. \text{ Then:}
\end{aligned}
$$

$$\begin{aligned}
&\mu^*(\{\omega_1\}) = \mu^*(\{\omega_2\}) = \mu^*(\{\omega_3\}) = \mu^*(\{\omega_4\}) = 3,\\
&\mu^*(\{\omega_1, \omega_2\}) = \mu^*(\{\omega_1, \omega_3\}) = \mu^*(\{\omega_2, \omega_4\}) = \mu^*(\{\omega_3, \omega_4\}) = 3,\\
&\mu^*(\{\omega_1, \omega_4\}) = \mu^*(\{\omega_2, \omega_3\}) = 6,\\
&\mu^*(\{\omega_1, \omega_2, \omega_3\}) = \mu^*(\{\omega_1, \omega_2, \omega_4\}) = \mu^*(\{\omega_1, \omega_3, \omega_4\})\\
&\qquad = \mu^*(\{\omega_2, \omega_3, \omega_4\}) = 6.
\end{aligned}$$

(v) By part (iv), $\mu^* \neq \mu_1 \neq \mu_2$ because, e.g., $\mu_1(\{\omega_1, \omega_4\}) = 2$, $\mu_2(\{\omega_1, \omega_4\}) = 4$ and $\mu^*(\{\omega_1, \omega_4\}) = 6$, all distinct. #

13. **(i)** Immediate.

(ii) The only partition of Ω with members in $\mathcal{C}$ is $\{A, A^c\}$ and $\mu(A) = \mu(A^c) = 0$ $(A^c = \{0, 2, 4, \ldots\})$.

(iii) On $\mathcal{C}$, $\mu_1(\oslash) = \mu_2(\oslash) = 0$, and $\mu_1(A) = \mu_1(A^c) = \mu_1(\Omega) = \infty = \mu_2(A) = \mu_2(A^c) = \mu_2(\Omega)$.

(iv) Let $\oslash \subset B \subset \Omega$. Then the only possible coverages of B by members of $\mathcal{C}$ are: A, A^c, Ω, all of which have μ-measure ∞. Thus, $\mu^*(B) = \infty$ for every B as above.

(v) Let $\oslash \subset B \subset \Omega$. Then if $D \subset \Omega$ is $= \oslash$, from $\oslash = (B \cap \oslash) + (B^c \cap \oslash)$, it follows that $0 = 0$, whereas for $D \neq \oslash$, the relation $D = (B \cap D) + (B^c \cap D)$ implies that at least one of $B \cap D$ and $B^c \cap D$ is $\neq \oslash$. Hence $\infty = \infty$ and the equality holds again. Since $\oslash$ and Ω are always μ^*-measurable, it follows that $\mathcal{A}^* = \mathcal{P}(\Omega)$. #

15. **(i)** To show that $A \Delta M = (A - N) \cup [N \cap (A \Delta M)]$, where $M \subseteq N$. We have

$$\begin{aligned}
A \Delta M &= (A \Delta M) \cap \Omega = (A \Delta M) \cap (N \cup N^c)\\
&= [(A \Delta M) \cap N] \cup [(A \Delta M) \cap N^c]\\
&= [N \cap (A \Delta M)] \cup \{[(A - M) \cup (M - A)] \cap N^c\}\\
&= [N \cap (A \Delta M)] \cup \{[(A \cap M^c) \cup (A^c \cap M)] \cap N^c\}\\
&= [N \cap (A \Delta M)] \cup (A \cap M^c \cap N^c) \cup (A^c \cap M \cap N^c)\\
&= [N \cap (A \Delta M)] \cup (A \cap M^c \cap N^c)\\
&\quad \text{(since } M \subseteq N \text{ implies } N^c \subseteq M^c \text{ and hence } M \cap N^c = \oslash)\\
&= [N \cap (A \Delta M)] \cup (A \cap N^c) \text{ (since } N^c \subseteq M^c)\\
&= (A - N) \cup [N \cap (A \Delta M)].
\end{aligned}$$

(ii)

$$\begin{aligned}
A \cup M &= [(A - N) \cup (A \cap N)] \cup M\\
&= (A - N) \cup [(A \cap N) \cup M]\\
&= (A - N) \cup [(A \cap N) \cup (M \cap N)] \text{ (since } M \subseteq N)\\
&= (A - N) + [(A \cup M) \cap N]\\
&= (A - N) \Delta [N \cap (A \cup M)]\\
&\quad \text{(since for } B \text{ and } C \text{ with } B \cap C = \oslash,\ B + C = B \Delta C).
\end{aligned}$$

(iii) Let $B \in \mathcal{A}^*$. Then $B = A\Delta M$ for some $A \in \mathcal{A}$ and $M \subseteq N, N \in \mathcal{A}$ with $\mu(N) = 0$. By part (i), $B = A\Delta M = (A - N) \cup [N \cap (A\Delta M)]$ with $(A-N) \in \mathcal{A}$ and $N \cap (A\Delta M) \subseteq N$. That is, B is of the form $A \cup M$ with A replaced by $A - N$ (a member of $\mathcal{A}$) and M replaced by $N \cap (A\Delta M)$ (which is a subset of N with $N \in \mathcal{A}$ and $\mu(N) = 0$). It follows that $B \in \bar{\mathcal{A}}$. Next, let $B \in \bar{\mathcal{A}}$. Then $B = A \cup M$ for some $A \in \mathcal{A}$ and some $M \subseteq N$ with $N \in \mathcal{A}$ and $\mu(N) = 0$. By part (ii), $B = A \cup M = (A - N)\Delta[N \cap (A \cup M)]$ with $(A - N) \in \mathcal{A}$ and $N \cap (A \cup M) \subseteq N$. That is, B is of the form $A \cup M$ with A replaced by $A - N$ (a member of $\mathcal{A}$) and M replaced by $N \cap (A \cup M)$ (which is a subset of $N \in \mathcal{A}$ and $\mu(N) = 0$). It follows that $B \in \mathcal{A}^*$. Therefore $\mathcal{A}^* = \bar{\mathcal{A}}$.

Note: Parts (i) and (ii) are also established by showing that each side is contained in the other. This is done as follows.

(i) Let ω belong to the left-hand side; i.e., $\omega \in A\Delta M$, so that $\omega \in A$ and $\omega \notin M$. That $\omega \notin M$ implies that either $\omega \notin N$ or $\omega \in N$. If $\omega \notin N$, then $\omega \in (A - N)$, so that ω belongs to the right-hand side. If $\omega \in N$, then $\omega \in [N \cap (A\Delta M)]$, so that ω belongs to the right-hand side again.
Next, let ω belong to the right-hand side. Then $\omega \in (A - N)$ or $\omega \in [N \cap (A\Delta M)]$. If $\omega \in (A - N)$, then $\omega \in A$ and $\omega \notin N$, so that $\omega \in A$ and $\omega \notin M$. It follows that ω belongs to the left-hand side. On the other hand, if $\omega \in [N \cap (A\Delta M)]$, then $\omega \in (A\Delta M)$, so that ω belongs to the left-hand side again.

(ii) Let ω belong to the left-hand side; i.e., $\omega \in A \cup M$, so that $\omega \in A$ and $\omega \in M$ or to both. Let $\omega \in A$. Also, either $\omega \in N$ or $\omega \notin N$. If $\omega \in N$, then $\omega \in [N \cap (A \cup M)]$, so that ω belongs to the right-hand side. If $\omega \notin N$, then $\omega \in (A - N)$, so that ω belongs to the right-hand side again. Finally, let $\omega \in M$. Then $\omega \in N$ and hence $\omega \in [N \cap (A \cup M)]$, so that ω belongs to the right-hand side.
Next, let ω belong to the right-hand side. Then either $\omega \in (A - N)$ or $\omega \in [N \cap (A \cup M)]$. Let $\omega \in (A - N)$. Then $\omega \in A$ and $(\omega \notin N)$, so that ω belongs to the left-hand side. If $\omega \in [N \cap (A \cup M)]$, then $\omega \in (A \cup M)$, so that ω belongs to the left-hand side. #

16. $\bar{\mathcal{A}}(= \mathcal{A}^*) \neq \oslash$ since, e.g., $\Omega = \Omega \cup \oslash$, $\Omega \in \mathcal{A}$, $\mu(\oslash) = 0$, so that $\Omega \in \bar{\mathcal{A}}$. Next, for $B \in \bar{\mathcal{A}}$ to show that $B^c \in \bar{\mathcal{A}}$. Now $B \in \bar{\mathcal{A}}$ implies $B = A \cup M$, $A \in \mathcal{A}$, $M \subseteq N \in \mathcal{A}$, $\mu(N) = 0$. Then

$$\begin{aligned}
B^c &= (A \cup M)^c = A^c \cap M^c \\
&= A^c \cap [M^c \cap (N \cup N^c)] \\
&= A^c \cap [(M^c \cap N) \cup (M^c \cap N^c)] \\
&= A^c \cap [(M^c \cap N) \cup N^c] \text{ (since } M \subseteq N \text{ implies } N^c \subseteq M^c) \\
&= (A^c \cap N^c) \cup (N \cap M^c \cap A^c)
\end{aligned}$$

with $A^c \cap N^c \in \mathcal{A}$ and $N \cap M^c \cap A^c \subseteq N$. That is, B^c is of the form $A \cup M$ with A (a member of $\mathcal{A}$) replaced by $A^c \cap N^c$ and M ($\subseteq N \in \mathcal{A}$ with $\mu(N) = 0$) replaced by $N \cap M^c \cap A^c$. It follows that $B^c \in \bar{\mathcal{A}}$. Finally, let $B_i \in \bar{\mathcal{A}}$, $i = 1, 2, \ldots$ Then $B_i = A_i \cup M_i$ with $A_i \in \mathcal{A}$ and $M_i \subseteq N_i \in \mathcal{A}$ with $\mu(N_i) = 0$, $i \geq 1$. Therefore

$$\bigcup_{i=1}^{\infty} B_i = \bigcup_{i=1}^{\infty} (A_i \cup M_i) = (\bigcup_{i=1}^{\infty} A_i) \cup (\bigcup_{i=1}^{\infty} M_i) \text{ with } \bigcup_{i=1}^{\infty} A_i \in \mathcal{A}$$

and $\cup_{i=1}^{\infty} M_i \subseteq \cup_{i=1}^{\infty} N_i$, a member of $\mathcal{A}$ with $\mu(\cup_{i=1}^{\infty} N_i) = 0$. It follows that $\cup_{i=1}^{\infty} B_i$ belongs in $\bar{\mathcal{A}}$, and $\bar{\mathcal{A}}$ is a σ-field. #

17.

(i) In the first place, the definition $\mu^*(A \Delta M) = \mu(A)$ implies $\mu^*(A \cup M) = \mu(A)$. Indeed, $A \cup M = (A - N)\Delta[N \cap (A \cup M)]$ with $(A - N) \in \mathcal{A}$ and $N \cap (A \cup M) \subseteq N \in \mathcal{A}$, $\mu(N) = 0$. Therefore $\mu^*(A \cup M) = \mu(A - N) = \mu(A \cap N^c) = \mu(A \cap N^c) + \mu(A \cap N) = \mu[(A \cap N^c) \cup (A \cap N)] = \mu(A)$. In the process of the proof, we also have seen that $\mu(A - N) = \mu(A)$.

(ii) As it was just seen, $\mu^*(A \cup M) = \mu(A - N) = \mu(A)$. We show that μ^* so defined on $\mathcal{A}^*$ is well-defined. That is, if $B = A_1 \cup M_1 = A_2 \cup M_2$, then $\mu(A_1) = \mu(A_2)$. Indeed,

$$A_1 = (A_1 \cap A_2) + (A_1 \cap A_2^c) = (A_1 \cap A_2)\Delta(A_1 \cap A_2^c).$$

Next, $A_1 \cap A_2^c \subseteq M_2$, because $x \in (A_1 \cap A_2^c)$ implies $x \in A_1$ and $x \notin A_2$, hence $x \in (A_1 \cup M_1)$ and $x \notin A_2$, so that $x \in B$ and $x \notin A_2$. This implies that $x \in (A_2 \cup M_2)$ and $x \notin A_2$, so that $x \in M_2$. Thus, $A_1 \cap A_2^c \subseteq M_2 \subseteq N_2$. From this and the fact that $B = (A_1 \cap A_2)\Delta(A_1 \cap A_2^c)$, it follows that $\mu^*(B) = \mu(A_1 \cap A_2)$ ($= \mu(A_1)$). Likewise, $A_2 = (A_1 \cap A_2)\Delta(A_1^c \cap A_2)$ with $A_1^c \cap A_2 \subseteq M_1 \subseteq N_1$, so that $\mu^*(B) = \mu(A_1 \cap A_2)$ ($= \mu(A_2)$). It follows that $\mu(A_1) = \mu(A_2)$ and μ^* is well-defined.

(iii) Clearly, $\mu^*(\emptyset) = \mu^*(\emptyset \Delta \emptyset) = \mu(\emptyset) = 0$, and $\mu^*(A \cup M) = \mu(A)$ (as was seen in part (i)) and this is ≥ 0. Finally, let $B_i \in \bar{\mathcal{A}}$, $i = 1, 2, \ldots$, $B_i \cap B_j = \emptyset$, $i \neq j$. Then $B_i = A_i \cup M_i$, $B_j = A_j \cup M_j$, and

$$\emptyset = B_i \cap B_j = (A_i \cap A_j) \cup (A_i \cap M_j) \cup (M_i \cap A_j) \cup (M_i \cap M_j),$$

so that $A_i \cap A_j = \emptyset$. Therefore

$$\mu^*\left(\sum_{i=1}^{\infty} B_i\right) = \mu^*\left[\bigcup_{i=1}^{\infty} (A_i \cup M_i)\right] = \mu^*\left[\left(\bigcup_{i=1}^{\infty} A_i\right) \cup \left(\bigcup_{i=1}^{\infty} M_i\right)\right]$$
$$= \mu\left(\bigcup_{i=1}^{\infty} A_i\right) = \mu\left(\sum_{i=1}^{\infty} A_i\right) = \sum_{i=1}^{\infty} \mu(A_i)$$

$$= \sum_{i=1}^{\infty} \mu^*(A_i \cup M_i)$$

$$= \sum_{i=1}^{\infty} \mu^*(B_i).$$

It follows that μ^* is a measure on $\bar{\mathcal{A}}(= \mathcal{A}^*)$. #

18. **(i)** Let $B \in \hat{\mathcal{C}}$ and suppose that $B = A$ for some $A \in \mathcal{A}$. Then $B = A \Delta \oslash$ with $\oslash \in \mathcal{A}$ and $\mu(\oslash) = 0$, so that $B \in \mathcal{A}^*$. If $B \subseteq N$ for some $N \in \mathcal{A}$ with $\mu(N) = 0$, we have $B = B \Delta \oslash$ with $\oslash \in \mathcal{A}$ and $B \subseteq N \in \mathcal{A}$ with $\mu(N) = 0$, so that $B \in \mathcal{A}^*$. Thus $\hat{\mathcal{C}} \subseteq \mathcal{A}^*$.

(ii) $\hat{\mathcal{C}} \subseteq \mathcal{A}^*$ implies that $\sigma(\hat{\mathcal{C}}) = \hat{\mathcal{A}} \subseteq \mathcal{A}^*$, so it suffices to show that $\mathcal{A}^* \subseteq \hat{\mathcal{A}}$. Let $B \in \mathcal{A}^*$, so that $B = A \Delta M$ with $A \in \mathcal{A}$ and $M \subseteq N, N \in \mathcal{A}$, $\mu(N) = 0$. Since $A = A \Delta \oslash$, it follows that $A \in \hat{\mathcal{C}}$ and hence $A \in \hat{\mathcal{A}}$. Also, $M = \oslash \Delta M$, so that $M \in \hat{\mathcal{C}}$ and hence $M \in \hat{\mathcal{A}}$. Thus, $A, M \in \hat{\mathcal{A}}$ and therefore $A \Delta M \in \hat{\mathcal{A}}$ or $B \in \hat{\mathcal{A}}$. #

19. Let $\Omega = \{\omega_1, \omega_2, \omega_3, \omega_4\}$, and let $\mathcal{A} = \{\oslash, \{\omega_1, \omega_2\}, \{\omega_3, \omega_4\}, \{\omega_1, \omega_2, \omega_3, \omega_4\}\}$. Then $\mathcal{A}$ is, trivially, a σ-field. On $\mathcal{A}$, define μ as follows: $\mu(\oslash) = 0 = \mu(\{\omega_1, \omega_2\})$, $\mu(\{\omega_3, \omega_4\}) = \mu(\{\omega_1, \omega_2, \omega_3, \omega_4\}) = 1$. Then, clearly, μ is a measure on $\mathcal{A}$. But $\{\omega_1\} \subset \{\omega_1, \omega_2\} \in \mathcal{A}$ with $\mu(\{\omega_1, \omega_2\}) = 0$ whereas $\{\omega_1\} \notin \mathcal{A}$. #

20. Recall that μ^0 is an outer measure on $\mathcal{P}(\Omega)$ if $\mu^0(\oslash) = 0$, μ^0 is $\uparrow$ and sub-σ-additive. Now, let $N \in \mathcal{A}^0$ with $\mu^0(N) = 0$, and let M be an arbitrary subset of N. To show that $M \in \mathcal{A}^0$. It suffices to show that $\mu^0(D) \geq \mu^0(M \cap D) + \mu^0(M^c \cap D)$ for every $D \subseteq \Omega$. We have: $M \subseteq N$, hence $M \cap D \subseteq N \cap D$ and $\mu^0(M \cap D) \leq \mu^0(N \cap D) = 0$, so that $\mu^0(M \cap D) = 0$. Next, $M^c \cap D \subseteq D$ and $\mu^0(M^c \cap D) \leq \mu^0(D)$, so that $\mu^0(D) \geq \mu^0(M \cap D) + \mu^0(M^c \cap D)$ for every $D \subseteq \Omega$. #

21. On $\mathcal{B}$, define μ in the following manner: $\mu(B)$ = number of integers in B. Then, clearly, μ is a measure satisfying the condition μ(finite interval) $< \infty$. Next, let $x_n \uparrow -2$, so that $\mu((x_n, 0]) = 3$ for all sufficiently large n, and hence $F_c(x_n) = c - 3$ for all sufficiently large n. But $F_c(-2) = c - \mu((-2, 0]) = c - 2$. Hence F_c is not left-continuous. #

22. Indeed, if μ were additive, then $c = \mu(\oslash) = \mu(\oslash \cup \oslash) = \mu(\oslash) + \mu(\oslash) = 2c$, so that $2 = 1$, a contradiction. #

23. For $n = 2$, let μ_1 and μ_2 be σ-finite, and let $\{A_1^1, A_2^1, \ldots\}$ and $\{A_1^2, A_2^2, \ldots\}$ be the associated partitions for which $\mu_1(A_i^1) < \infty$, $\mu_2(A_i^2) < \infty, i \geq 1$. Then $\{A_i^1 \cap A_j^2, i, j \geq 1\}$ is a partition of Ω and $\mu(A_i^1 \cap A_j^2) = \mu_1(A_i^1 \cap A_j^2) + \mu_2(A_i^1 \cap A_j^2) < \infty, i, j \geq 1$, so that μ is σ-finite.

Next, assume the assertion to be true for $n = k$ and we will establish it for $n = k + 1$. By setting $\mu_0 = \mu_1 + \ldots + \mu_k$, we have that both μ_0 and μ_{k+1} are σ-finite, and let $\{B_i, i \geq 1\}$ and $\{A_i^{k+1}, i \geq 1\}$ be the associated partitions for which $\mu_0(B_i) < \infty$, $\mu_{k+1}(A_i^{k+1}) < \infty, i \geq 1$. Then $\{B_i \cap A_j^{k+1}, i, j \geq 1\}$

is a partition of Ω, and $\mu_0(B_i \cap A_j^{k+1}) \leq \mu_0(B_i) < \infty$, $\mu_{k+1}(B_i \cap A_j^{k+1}) \leq \mu_{k+1}(A_j^{k+1}) < \infty$, $i, j \geq 1$. Thus,

$$(\mu_1 + \ldots + \mu_{k+1})(B_i \cap A_j^{k+1}) = (\mu_1 + \ldots + \mu_k)(B_i \cap A_j^{k+1}) + \mu_{k+1}(B_i \cap A_j^{k+1}) < \infty,\ i, j \geq 1, \text{ so that}$$

$\mu_1 + \ldots + \mu_{k+1}$ is σ-finite. #

24. **(i)** Clearly, $(A \cap B^c) \cup (A^c \cap B) = A \Delta B = (A \cup B) - (A \cap B)$. Hence

$$\begin{aligned} P[(A \cap B^c) \cup (A^c \cap B)] &= P[(A \cup B) - (A \cap B)] \\ &= P(A \cup B) - P(A \cap B) \text{ (since } A \cap B \subseteq A \cup B) \\ &= P(A) + P(B) - P(A \cap B) - P(A \cap B) \\ &= P(A) + P(B) - 2P(A \cap B). \end{aligned}$$

(ii) We will use the induction hypothesis.
For $n = 2$, we have:

$$P(A_1 \cup A_2) = P(A_1) + P(A_2) - P(A_1 \cap A_2),$$

so that

$$\begin{aligned} P(A_1 \cap A_2) &= P(A_1) + P(A_2) - P(A_1 \cup A_2) \\ &\geq P(A_1) + P(A_2) - 1. \end{aligned}$$

Next, assume it to be true for $n = k$ and establish it for $n = k + 1$. Indeed,

$$\begin{aligned} P(A_1 \cap \ldots \cap A_{k+1}) &= P[(A_1 \cap \ldots \cap A_k) \cap A_{k+1}] \\ &\geq P(A_1 \cap \ldots \cap A_k) + P(A_{k+1}) - 1 \\ &\geq \sum_{i=1}^{k} P(A_i) - (k-1) + P(A_{k+1}) - 1 \\ &= \sum_{i=1}^{k+1} P(A_i) - [(k+1) - 1]. \ \# \end{aligned}$$

25. $\underline{\lim}_{n\to\infty} A_n = \cup_{n=1}^{\infty} \cap_{k=n}^{\infty} A_k = \cup_{n=1}^{\infty} \{\omega_2\} = \{\omega_2\}$, $\overline{\lim}_{n\to\infty} A_n = \cap_{n=1}^{\infty} \cup_{k=n}^{\infty} A_k = \cap_{n=1}^{\infty} \{\omega_1, \omega_2, \omega_3\} = \{\omega_1, \omega_2, \omega_3\}$, so that $P(\underline{\lim}_{n\to\infty} A_n) = P(\{\omega_2\}) = \frac{1}{3}$, $P(\overline{\lim}_{n\to\infty} A_n) = P(\{\omega_1, \omega_2, \omega_3\}) = \frac{7}{10}$; also, $P(A_{2n-1}) = P(\{\omega_1, \omega_2\}) = \frac{1}{2}$, $P(A_{2n}) = P(\{\omega_2, \omega_3\}) = \frac{8}{5}$, so that $\underline{\lim}_{n\to\infty} P(A_n) = \frac{1}{2}$ and $\overline{\lim}_{n\to\infty} P(A_n) = \frac{8}{5}$. Observe that

$$P(\underline{\lim}_{n\to\infty} A_n) = \frac{1}{3} \neq \frac{1}{2} = \underline{\lim}_{n\to\infty} P(A_n),$$

and

$$P(\overline{\lim}_{n\to\infty} A_n) = \frac{7}{10} \neq \frac{8}{5} = \overline{\lim}_{n\to\infty} P(A_n). \ \#$$

26. **(i)** If $\{\omega_i\} \in \mathcal{A}$ for all ω_i, then, clearly, every subset of Ω is in $\mathcal{A}$, so that $\mathcal{A} = \mathcal{P}(\Omega)$. On the other hand, if $\mathcal{A} = \mathcal{P}(\Omega)$, then all subjects of Ω are in $\mathcal{A}$, and in particular, so are $\{\omega_i\}$ for all ω_is.

(ii) It is immediate. #

27. **(i)** That $\mu(A) \geq 0$ and $\mu(\oslash) = 0$ are immediate. Next, let $A_1, \ldots, A_n$ be pairwise disjoint. Then to show that $\mu(\sum_{i=1}^{n} A_i) = \sum_{i=1}^{n} \mu(A_i)$. If at least one of the A_is is infinite, then $\sum_{i=1}^{n} A_i$ is infinite, so that $\mu(\sum_{i=1}^{n} A_i) = \infty$. Also, at least one of the terms on the right-hand side is ∞, so that $\sum_{i=1}^{n} \mu(A_i) = \infty$. On the other hand, if all $A_1, \ldots, A_n$ are finite, then $\sum_{i=1}^{n} A_i$ is finite and hence $\mu(\sum_{i=1}^{n} A_i) = 0$. The right-hand side is also equal to 0 since each term is 0. Next, μ is not σ-additive, because if all A_is are finite, then $\sum_{i=1}^{\infty} A_i$ is infinite, so that $\mu(\sum_{i=1}^{\infty} A_i) = \infty$, whereas $\sum_{i=1}^{\infty} \mu(A_i) = \sum_{i=1}^{\infty} 0 = 0$.

(ii) Clearly, $\Omega = \cup_{n=1}^{\infty} A_n$, where $A_n = \{\omega_1, \ldots, \omega_n\}$, so that $A_n \subset A_{n+1}$, $n \geq 1$, and $\mu(A_n) = 0$ for all n. Since $\mu(A_n) = 0, n \geq 1$, it follows that $\mu(A_n^c) = \infty$ for all n. #

28. **(i)** We have to prove that $\mu^0(\oslash) = 0$, $\mu^0(A) \leq \mu^0(B)$ for $A \subset B$, and μ^0 is a sub-σ-additive. That $\mu^0(\oslash) = 0$ holds by the definition of μ^0. Next, suppose that $A \subset B$. There are three cases to consider. Let B be finite. Then A is finite, and $\mu^0(A) = \frac{a}{a+1} < \frac{b}{b+1} = \mu^0(B)$ since $a < b$. Let B be infinite but A be finite. Then $\mu^0(A) = \frac{a}{a+1} < 1 = \mu^0(B)$. Finally, let both A and B be infinite. Then $\mu^0(A) = 1 \leq 1 = \mu^0(B)$.
Now to establish sub-σ-additivity:

$$\mu^0(\bigcup_{n=1}^{\infty} A_n) \leq \sum_{n=1}^{\infty} \mu^0(A_n).$$

Suppose that at least one of the A_ns is infinite, e.g., A_{n_0}. Then the union $\cup_{n=1}^{\infty} A_n$ is infinite, and hence $\mu^0(\cup_{n=1}^{\infty} A_n) = 1$, whereas $\sum_{n=1}^{\infty} \mu^0(A_n) \geq 1$, since $\mu^0(A_{n_0}) = 1$ and $\mu^0(A_n) \geq 0, n \geq 1$. Next, let all A_n be finite and $\neq \oslash$. Then $\cup_{n=1}^{\infty} A_n$ is infinite, so that $\mu^0(\cup_{n=1}^{\infty} A_n) = 1$. As for the right-hand side, $\mu^0(A_n) = \frac{a_n}{a_n+1} \geq \frac{1}{2}$ for all n, so that $\sum_{n=1}^{\infty} \mu^0(A_n) = \infty$. Finally, suppose that only finitely many of the A_ns are finite, e.g., $A_{n_1}, \ldots, A_{n_k}$. Then, clearly, $\sup(A_{n_1} \cup \ldots \cup A_{n_k}) \leq \sup A_{n_1} + \ldots + \sup A_{n_k}$, so that $\mu^0(\cup_{n=1}^{\infty} A_n) \leq \sum_{n=1}^{\infty} \mu^0(A_n)$. Therefore μ^0 is an outer measure.

(ii) By Remark 6(i), A is μ^0-measurable if

$$\mu^0(D) \geq \mu^0(A \cap D) + \mu^0(A^c \cap D) \text{ for every } D \subseteq \Omega.$$

Also, by Remark 6(ii), $\oslash$ and Ω are μ^0-measurable, so to investigate the last inequality for $\oslash \subset A \subset \Omega$. Consider the following possible cases. Let both A and A^c be infinite, and take $D = \Omega$. Then $\mu^0(\Omega) = 1$, but

$\mu^0(A \cap \Omega) + \mu^0(A^c \cap \Omega) = \mu^0(A) + \mu^0(A^c) = 1 + 1 = 2$, so that the inequality is violated. Let A be infinite but A^c be finite, and take $D = \Omega$. Then $\mu^0(\Omega) = 1$, but $\mu^0(A \cap \Omega) + \mu^0(A^c \cap \Omega) = \mu^0(A) + \mu^0(A^c) = 1 + \frac{c}{c+1}$, $c = \sup A^c$. Again, the inequality is violated. Finally, let A be finite (so that A^c is infinite), and take $D = \Omega$. Once again, $\mu^0(\Omega) = 1$, and $\mu^0(A \cap \Omega) + \mu^0(A^c \cap \Omega) = \mu^0(A) + \mu^0(A^c) = \frac{a}{a+1} + 1$, $a = \sup A$. So, the inequality is violated. The conclusion then is that $\mathcal{A}_0 = \{\emptyset, \Omega\}$. #

29. It is immediate since:

$$P(-X \le -m) = P(X \ge m) \ge \frac{1}{2}, \text{ and}$$
$$P(-X \ge -m) = P(X \le m) \ge \frac{1}{2}. \#$$

30. By symmetry, we have

$$\begin{aligned} P(X \le x) &= P(-X \le x) = P(X \ge -x) \\ &= 1 - P(X < -x) \ge 1 - P(X \le -x). \end{aligned}$$

For $x = 0$, this becomes

$$P(X \le 0) \ge 1 - P(x \le 0), \text{ or } P(x \le 0) \ge \frac{1}{2}.$$

Again, by symmetry,

$$P(X \ge x) = P(-X \ge x) = P(X \le -x).$$

For $x = 0$, this relation becomes $P(X \ge 0) = P(X \le 0)$. But $P(X \le 0) \ge \frac{1}{2}$ as already shown. Thus, $P(X \ge 0) \ge \frac{1}{2}$, and 0 is a median for X. #

31. From $B \subseteq A \cup B$, we get $\mu^0(B) \le \mu^0(A \cup B)$. However, $\mu^0(A \cup B) \le \mu^0(A) + \mu^0(B) = \mu^0(B)$ (by the sub-additivity property of μ^0). Thus, $\mu^0(B) \le \mu^0(A \cup B) \le \mu^0(B)$, so that $\mu^0(A \cup B) = \mu^0(B)$. #

32. Let $N = (f \ne g)$, and let $B \in \mathcal{B}$. Then $f^{-1}(B) \in \mathcal{A}$, by assuming that, e.g., f is measurable. Also, $g^{-1}(B) = \{[g^{-1}(B)] \cap N\} \cup \{[g^{-1}(B)] \cap N^c\} = \{[g^{-1}(B)] \cap N\} \cup f^{-1}(B)$ (since $f = g$ on N^c). But $[g^{-1}(B)] \cap N \subseteq N$ with $\mu(N) = 0$. Thus, $[g^{-1}(B)] \cap N$ is in $\mathcal{A}$, and hence $g^{-1}(B)$ is in $\mathcal{A}$. It follows that g is measurable. #

33. Indeed, $B \in \mathcal{B}$, we have $f^{-1}(B) \subseteq A$ with $\mu[f^{-1}(B)] = 0$, so that $f^{-1}(B) \in \mathcal{A}$, and hence f is measurable. #

34. **(i)** We have to show that μ is nonnegative, $\mu(\emptyset) = 0$, and μ is σ-additive. Indeed, $\mu(A) = \mu_1(A) + \mu_2(A) \ge 0$; $\mu(\emptyset) = \mu_1(\emptyset) + \mu_2(\emptyset) = 0$; $\mu(\sum_{i=1}^{\infty} A_i) = \mu_1(\sum_{i=1}^{\infty} A_i) + \mu_2(\sum_{i=1}^{\infty} A_i) = \sum_{i=1}^{\infty} \mu_1(A_i) + \sum_{i=1}^{\infty} \mu_2(A_i) = \sum_{i=1}^{\infty} (\mu_1 + \mu_2)(A_i) = \sum_{i=1}^{\infty} \mu(A_i)$.

(ii) Suppose that, e.g., μ_1 is complete, or more properly, $\mathcal{A}$ is complete with respect to μ_1, which means that $\mathcal{A}$ contains all subsets of the μ_1-null sets. So, let $A \in \mathcal{A}$ with $\mu(A) = 0$. Then $\mu_1(A)(= \mu_2(A)) = 0$. Thus, for an arbitrary $B \subseteq A$, we have $\mu_1(B) \leq \mu_1(A) = 0$ and $B \in \mathcal{A}$. It follows that $\mu(B) \leq \mu(A) = 0$, so that μ is complete. #

35. **(i)** Unions of any two members of $\mathcal{C}_2$ produce elements in $\mathcal{C}_2$ except for two new elements; namely,

$$(A \cap B) \cup (A^c \cap B^c) \quad \text{and} \quad (A \cap B^c) \cup (A^c \cap B).$$

Beyond the obvious results, we have:

$$\begin{aligned}
&A \cup (A^c \cap B) = A \cup B,\ A \cup (A^c \cap B^c) = A \cup B^c;\\
&A^c \cup (A \cap B) = A^c \cup B,\ A^c \cup (A \cap B^c) = A^c \cup B^c;\\
&B \cup (A \cap B^c) = A \cup B,\ B \cup (A^c \cap B^c) = A^c \cup B;\\
&B^c \cup (A \cap B) = A \cup B^c,\ B^c \cup (A^c \cap B) = A^c \cup B^c;\\
&(A \cap B) \cup (A^c \cap B^c) \text{ new element},\\
&(A \cap B) \cup (A^c \cup B^c) = (A \cap B) \cup (A \cap B)^c = \Omega;\\
&(A \cap B^c) \cup (A^c \cap B) \text{ new element},\\
&(A \cap B^c) \cup (A^c \cup B) = \Omega;\\
&(A^c \cap B^c) \cup (A \cup B) = (A \cup B)^c \cup (A \cup B) = \Omega.
\end{aligned}$$

(ii) Closeness under complementation is immediate for all elements except, perhaps, for the last two, each of which is the complement of the other. Indeed,

$$\begin{aligned}
[(A \cap B) \cup (A^c \cap B^c)]^c &= (A^c \cup B^c) \cap (A \cup B)\\
&= [(A^c \cup B^c) \cap A] \cup [(A^c \cup B^c) \cap B]\\
&= (A \cap B^c) \cup (A^c \cap B).
\end{aligned}$$

In checking closeness under unions, it suffices to restrict ourselves to forming unions of two elements, one taken from each one of the classes:

$$\begin{aligned}
&\{(A \cap B) \cup (A^c \cap B^c), (A \cap B^c) \cup (A^c \cap B)\},\\
&\{A,\ A^c,\ B,\ B^c,\ A \cap B,\ A \cap B^c,\ A^c \cap B,\ A^c \cap B^c\},
\end{aligned}$$

as well as any two elements from the second class above. To this end, and except for the obvious results, we have:

$$\begin{aligned}
&A \cup [(A \cap B) \cup (A^c \cap B^c)] = A \cup (A^c \cap B^c) = A \cup B^c;\\
&A \cup [(A \cap B^c) \cup (A^c \cap B)] = A \cup (A^c \cap B) = A \cup B;\\
&A^c \cup [(A \cap B) \cup (A^c \cap B^c)] = A^c \cup (A \cap B) = A^c \cup B;\\
&A^c \cup [(A \cap B^c) \cup (A^c \cap B)] = A^c \cup (A \cap B^c) = A^c \cup B^c;\\
&B \cup [(A \cap B) \cup (A^c \cap B^c)] = B \cup (A^c \cap B^c) = A^c \cup B;
\end{aligned}$$

$$B \cup [(A \cap B^c) \cup (A^c \cap B)] = B \cup (A \cap B^c) = A \cup B;$$
$$B^c \cup [(A \cap B) \cup (A^c \cap B^c)] = B^c \cup (A \cap B) = A \cup B^c;$$
$$B^c \cup [(A \cap B^c) \cup (A^c \cap B)] = B^c \cup (A \cap B^c) = B^c;$$
$$(A \cap B) \cup (A^c \cup B^c) = (A \cap B) \cup (A \cap B)^c = \Omega,$$
$$(A \cap B) \cup [(A \cap B^c) \cup (A^c \cap B)] = A \cup (A^c \cap B) = A \cup B;$$
$$(A \cap B^c) \cup (A^c \cup B) = \Omega,$$
$$(A \cap B^c) \cup [(A \cap B) \cup (A^c \cap B^c)] = A \cup (A^c \cap B^c) = A \cup B^c;$$
$$(A^c \cap B) \cup (A \cup B^c) = \Omega,$$
$$(A^c \cap B) \cup [(A \cap B) \cup (A^c \cap B^c)] = A^c \cup (A \cap B) = A^c \cup B;$$
$$(A^c \cap B^c) \cup (A \cup B) = (A \cup B)^c \cup (A \cup B) = \Omega,$$
$$(A^c \cap B^c) \cup [(A \cap B^c) \cup (A^c \cap B)] = B^c \cup (A^c \cap B) = A^c \cup B^c.$$

Again, except for the obvious results, we have:

$$(A \cup B) \cup [(A \cap B) \cup (A^c \cap B^c)] = (A \cup B) \cup (A^c \cap B^c)$$
$$= (A \cup B) \cup (A \cup B)^c = \Omega;$$
$$(A \cup B^c) \cup [(A \cap B^c) \cup (A^c \cap B)] = (A \cup B^c) \cup (A^c \cap B) = \Omega;$$
$$(A^c \cup B) \cup [(A \cap B^c) \cup (A^c \cap B)] = (A^c \cup B) \cup (A \cap B^c) = \Omega;$$
$$(A^c \cup B^c) \cup [(A \cap B) \cup (A^c \cap B^c)] = (A^c \cup B^c) \cup (A \cap B)$$
$$= (A \cup B)^c \cup (A \cap B) = \Omega. \#$$

Chapter 3

Some Modes of Convergence of a Sequence of Random Variables and their Relationships

1. Indeed, $|X_n - X| = (X_n - X)^+ + (X_n - X)^-$, so that $(X_n - X)^+ \leq |X_n - X|$, $(X_n - X)^- \leq |X_n - X|$. Hence, for every $\varepsilon > 0$, $\mu[(X_n - X)^+ \geq \varepsilon] \leq \mu[|X_n - X| \geq \varepsilon] \underset{n\to\infty}{\longrightarrow} 0$, and likewise, $\mu[(X_n - X)^- \geq \varepsilon] \leq \mu[|X_n - X| \geq \varepsilon] \underset{n\to\infty}{\longrightarrow} 0$.
Next, recall that (Exercise 28, Chapter 1) that for any two r.v.s X and Y, $(X + Y)^+ \leq X^+ + Y^+$ and $(X + Y)^- \leq X^- + Y^-$. Hence
$$X_n^+ = ((X_n - X) + X)^+ \leq (X_n - X)^+ + X^+,$$
$$X^+ = ((X - X_n) + X_n)^+ \leq (X - X_n)^+ + X_n^+ = (X_n - X)^- + X_n^+,$$
because, as is easily seen, $(-Z)^+ = Z^-$. Then
$$-(X_n - X)^- \leq X_n^+ - X^+ \leq (X_n - X)^+,$$
or $|X_n^+ - X^+| \leq (X_n - X)^+ + (X_n - X)^- = |X_n - X|$, and therefore
$$\mu(|X_n^+ - X^+| \geq \varepsilon) \leq \mu(|X_n - X| \geq \varepsilon) \underset{n\to\infty}{\longrightarrow} 0,$$
so that $X_n^+ \xrightarrow[n\to\infty]{\mu} X^+$. Likewise, $X_n^- \xrightarrow[n\to\infty]{\mu} X^-$. #

2. **(i)** Let $(\Omega, \mathcal{A}, \mu) = (\Re, \mathcal{B}, \lambda)$, λ the Lebesgue measure, and for $n \geq 1$, let $X_n = I_{(n,\infty)}$ and $X = 0$. Then $X_n(\omega) \underset{n\to\infty}{\longrightarrow} 0$ for every $\omega \in \Re$ (since if $n_0 = n_0(\omega)$ is the smallest positive integer which is $\geq \omega$, then $X_n(\omega) = 0$ for all $n > n_0$), and in particular $X_n \overset{a.s.}{\underset{n\to\infty}{\longrightarrow}} 0$. However, for $0 < \varepsilon < 1$, $\{\omega \in \Re; |X_n(\omega) - X(\omega)| > \varepsilon\} = \{\omega \in \Re; X_n(\omega) = 1\} = (n, \infty)$ and $\lambda((n,\infty)) = \infty$, so that $X_n \overset{\lambda}{\underset{n\to\infty}{\nrightarrow}} 0$.

As another example, take $\Omega = \{1, 2, \ldots\}$, $\mathcal{A} = \mathcal{P}(\Omega)$, μ the counting measure, and let $A_n = \{1, \ldots, n\}$, $n \geq 1$. Take $X_n = I_{A_n}$ and $X = 1$. Then $X_n(\omega) \underset{n\to\infty}{\longrightarrow} X(\omega)$, $\omega \in \Omega$, and hence $X_n \overset{a.e.}{\underset{n\to\infty}{\longrightarrow}} X$. However, for $0 < \varepsilon < 1$, $\{\omega \in \Omega; |X_n(\omega) - X(\omega)| > \varepsilon\} = \{\omega \in \Omega; |X_n(\omega) - X(\omega)| = 1\} = \{\omega \in \Omega; X_n(\omega) = 0\} = A_n^c = \{n+1, n+2, \ldots\}$, and $\mu(A_n^c) = \infty$.

(ii) Let $\Omega = (0, 1]$, $\mathcal{A} = \mathcal{B}_\Omega$, and let λ be the Lebesgue measure. For $n = 1, 2, \ldots$, consider the partition of $(0, 1]$ by the 2^{n-1} intervals $\left(\frac{k-1}{2^{n-1}}, \frac{k}{2^{n-1}}\right]$, $k = 1, \ldots, 2^{n-1}$. Define the r.v.s Y_{nk} by: $Y_{nk}(\omega) = 1$ for $\omega \in \left(\frac{k-1}{2^{n-1}}, \frac{k}{2^{n-1}}\right]$ and $Y_{nk}(\omega) = 0$ otherwise, $k = 1, \ldots, 2^{n-1}$, $n \geq 1$, and set $\{X_1, X_2, \ldots\} = \{Y_{11}, Y_{21}, Y_{22}, Y_{31}, Y_{32}, Y_{33}, Y_{34}, \ldots\}$.

Then $X_n \overset{\lambda}{\underset{n\to\infty}{\longrightarrow}} 0$, but $\{X_n(\omega)\}$ does not convergence to 0 not even for a single $\omega \in \Omega$. This is so because for every $\omega \in (0, 1]$, ω belongs to infinitely many intervals $\left(\frac{k-1}{2^{n-1}}, \frac{k}{2^{n-1}}\right]$ and it does not belong to infinitely many such intervals. Consequently, $X_n(\omega) = 1$ for infinitely many n and $X_n(\omega) = 0$ for infinitely many n, so that $\liminf_{n\to\infty} X_n = 0 \neq 1 = \limsup_{n\to\infty} X_n$, and $\lim_{n\to\infty} X_n$ does not exist. #

3. We have

$$P\left(\limsup_{n\to\infty} A_n\right) = P\left(\bigcap_{n=1}^{\infty} \bigcup_{k=n}^{\infty} A_k\right) = P\left(\lim_{n\to\infty} \bigcup_{k=n}^{\infty} A_n\right)$$

$$= \lim_{n\to\infty} P\left(\bigcup_{k=n}^{\infty} A_n\right) \leq \lim_{n\to\infty} \sum_{k=n}^{\infty} P(A_k) = 0.$$

since $\sum_{n=1}^{\infty} P(A_n) < \infty$. Hence $P\left(\limsup_{n\to\infty} A_n\right) = 0$. #

4. **(i)** Set $A_n = (|X_n| \geq \frac{1}{k})$, $k = 1, 2, \ldots$ By assumption, $\sum_{n=1}^{\infty} P(A_n) < \infty$, so that $P\left(\limsup_{n\to\infty} A_n\right) = 0$, by Exercise 3. However,

$$\limsup_{n\to\infty} A_n = \bigcap_{n=1}^{\infty} \bigcup_{\nu=n}^{\infty} A_\nu = \bigcap_{n=1}^{\infty} \bigcup_{\nu=n}^{\infty} \left(|X_\nu| \geq \frac{1}{k}\right)$$

$$= \bigcap_{n=1}^{\infty} \bigcup_{\nu=1}^{\infty} \left(|X_{n+\nu}| \geq \frac{1}{k}\right) = B_k^c, \text{ say. Thus}$$

$0 = P\left(\limsup_{n\to\infty} A_n\right) = P(B_k^c)$, $k \geq 1$, so that $P(B_k) = 1$, $k \geq 1$. Then $P\left(\cap_{k=1}^{\infty} B_k\right) = 1$, by Exercise 4 in Chapter 2. But $\cap_{k=1}^{\infty} B_k =$

$\bigcap_{k=1}^{\infty} \bigcup_{n=1}^{\infty} \bigcap_{\nu=1}^{\infty} \left(|X_{n+\nu}| < \frac{1}{k}\right)$, and this is the event for which $X_n \underset{n\to\infty}{\longrightarrow} 0$, by Theorem 3. Thus, $P\left(X_n \underset{n\to\infty}{\longrightarrow} 0\right) = 1$ or $X_n \overset{a.s.}{\underset{n\to\infty}{\longrightarrow}} 0$.

(ii) Let Ω, $\mathcal{A}$ and P be as in the hint, and define X_n by $X_n(\omega) = 1$, if $\omega \in (0, \frac{1}{n})$, and 0 otherwise, $n = 1, 2, \ldots$ Then, for $\omega \in \Omega$, $X_n(\omega) = 0$, provided $n \geq n_0(\omega) = \frac{1}{\omega}$, so that $X_n(\omega) \underset{n\to\infty}{\longrightarrow} 0$ for every $\omega \in \Re$, and in particular, $X_n \overset{a.s.}{\underset{n\to\infty}{\longrightarrow}} 0$. However, for $0 < \varepsilon < 1$, $\sum_{n=1}^{\infty} P(|X_n| \geq \varepsilon) = \sum_{n=1}^{\infty} P(X_n = 1) = \sum_{n=1}^{\infty} \frac{1}{n} = \infty$. #

5. If $X_n \overset{P}{\underset{n\to\infty}{\longrightarrow}} X$, then for every subsequences $\{X'_n\} \subseteq \{X_n\}$ there is a further subsequence $\{X''_n\} \subseteq \{X'_n\}$ such that $X''_n \overset{a.s.}{\underset{n\to\infty}{\longrightarrow}} X'$, and $P(X' \neq X) = 0$. This is so by Theorem 5(ii). Next, suppose that for every subsequence $\{X'_n\} \subseteq \{X_n\}$ there is a further subsequence $\{X''_n\} \subseteq \{X'_n\}$ such that $X''_n \overset{a.s.}{\underset{n\to\infty}{\longrightarrow}} X'$, and that any two limiting r.v.s differ on a set of probability 0. If $\{X_n\}$ does not converge in probability to a r.v. X, then there must exist an $\varepsilon > 0$ such that $P(|X_n - X| \geq \varepsilon) \underset{n\to\infty}{\nrightarrow} 0$. Therefore there exists $\delta > 0$ for which there is no $N = N(\delta) > 0$ integer such that $P(|X_n - X| \geq \varepsilon) < \delta, n \geq N$. In other words, there exists $n_1 < n_2 < \ldots \uparrow \infty$ for which $P(|X_{n_k} - X| \geq \varepsilon) \geq \delta, k \geq 1$. Then, for every subsequence $\{X'_{n_k}\} \subseteq \{X_{n_k}\}$, $P(|X'_{n_k} - X| \geq \varepsilon) \geq \delta, k \geq 1$, so that no subsequence of $\{X_{n_k}\}$ converges $a.s.$ to X (or to another r.v. X' differing from X on a set of probability 0); a contradiction. Therefore $X_n \overset{P}{\underset{n\to\infty}{\longrightarrow}} X$. #

6. Let $(\Omega, \mathcal{A}, P) = ((0, 1], \mathcal{B}_{(0,1]}, \lambda)$, where λ is the Lebesgue measure, and define X_n by $X_n = I_{(\frac{1}{n}, 1]}, n \geq 1$. Then, clearly, $X_n \underset{n\to\infty}{\longrightarrow} 1$ pointwise, but $P(X_n = 0) = \lambda((0, \frac{1}{n})) = \frac{1}{n}$; i.e., $P(X_n = 0) > 0$ for all $n \geq 1$. #

7. It suffices to show that $\{X_n\}$ converges mutually a.s. That is, for every $\varepsilon > 0$ and every $\omega \in N^c$ with $P(N) = 0$, there exists $N = N(\varepsilon, \omega) > 0$ integer such that

$$|X_{n+\nu}(\omega) - X_n(\omega)| < \varepsilon,\ n \geq N,\ \text{uniformly in } \nu = 1, 2, \ldots$$

To this end, set $A_n = (|X_{n+1} - X_n| \geq \varepsilon), n \geq 1$. Then, by assumption $\sum_{n=1}^{\infty} P(A_n) < \infty$. Hence, by Exercise 3 here, $P(\bar{A}) = 0$, where $\bar{A} = \limsup_{n\to\infty} A_n = \bigcap_{n=1}^{\infty} \bigcup_{k=n}^{\infty} A_k$, or $P(\bar{A}^c) = 1$ where $\bar{A}^c = \bigcup_{n=1}^{\infty} \bigcap_{k=n}^{\infty} A_k^c$. Set $N^c = \bar{A}^c$, so that $P(N^c) = 1$. Now, for $\omega \in N^c$ or $\omega \in \bigcup_{n=1}^{\infty} \bigcap_{k=n}^{\infty} A_k^c$, we have that $\omega \in \bigcap_{k=n_1}^{\infty} A_k^c$ for some $n_1 = n_1(\omega)$. Thus, $\omega \in A_k^c, k \geq n_1$, so that $|X_{k+1}(\omega) - X_k(\omega)| < \varepsilon_k$ for all $k \geq n_1$, by the definition of A_k. The assumption $\sum_{n=1}^{\infty} \varepsilon_n < \infty$ implies that $\sum_{k=n}^{\infty} \varepsilon_n \underset{n\to\infty}{\longrightarrow} 0$, so that, for $\varepsilon > 0$ there exists $n_2 = n_2(\varepsilon) > 0$ integer such that $\sum_{k=n_2}^{\infty} \varepsilon_n < \varepsilon$. Set $N = N(\varepsilon, \omega) = \max\{n_1(\omega), n_2(\varepsilon)\}$. Then, for $n \geq N$, we have $|X_{n+1}(\omega) - X_n(\omega)| < \varepsilon_n$ for $\omega \in N^c$, and $\sum_{k=N}^{\infty} \varepsilon_n < \varepsilon$. Therefore, for $n \geq N$, and $\nu = 1, 2, \ldots,$

$$|X_{n+\nu}(\omega) - X_n(\omega)| \leq |X_{n+1}(\omega) - X_n(\omega)| + \ldots + |X_{n+\nu}(\omega) - X_{n+\nu-1}(\omega)| <$$

$\varepsilon_n + \ldots + \varepsilon_{n+\nu-1} \leq \varepsilon_n + \varepsilon_{n+1} + \ldots < \varepsilon$. That is, for every $\varepsilon > 0$ and each $\omega \in N^c$ (with $P(N^c) = 1$), there exists $N = N(\varepsilon, \omega) > 0$ integer such that, for $n \geq N$ and $\nu \geq 1$, $|X_{n+1}(\omega) - X_n(\omega)| < \varepsilon$. So, $\{X_n\}$ converges mutually a.s., and therefore $\{X_n\}$ converges a.s. to a r.v. #

8. To show that, for every $0 < \varepsilon(< 1)$ and $0 < \delta(< 1)$ there exists $N = N(\varepsilon, \delta) > 0$ integer such that $n \geq N$ implies $P[|g(X_n) - g(X)| \geq \varepsilon] < \delta$. Since for $0 < M \uparrow \infty$, $P(-M \leq X \leq M) \uparrow P(X \in \Re) = 1$, choose M sufficiently large (and >1) so that $P(|X| > M) < \frac{\delta}{2}$. Now g is uniformly continuous in the (closed) interval $[-2M, 2M]$, so that $|g(x') - g(x'')| < \varepsilon$ for all $x', x'' \in [-2M, 2M]$ with $|x' - x''| < \delta$. Next, $X_n \xrightarrow[n\to\infty]{P} X$ implies that there exists $N = N(\varepsilon, \delta) > 0$ integer such that $P(|X_n - X| \geq \delta) < \frac{\delta}{2}, n \geq N$. Below work with $n \geq N$ and set

$$A_1 = (|X| \leq M),\ A_2(n) = (|X_n - X| < \delta),\ A_3(n) = [|g(X_n) - g(X)| < \varepsilon].$$

On $A_1 \cap A_2(n)$, we have $-M \leq X \leq M$ and $X - \delta < X_n < X + \delta$, or $-2M \leq X \leq 2M$ and $-2M < X_n < 2M$ (since $\delta < 1$ and $M > 1$). So, on $A_1 \cap A_2(n)$, we have $X, X_n \in [-2M, 2M]$ and $|X_n - X| < \delta$. Hence $|g(X_n) - g(X)| < \varepsilon$ which implies that $A_1 \cap A_2(n) \subseteq A_3(n)$. Then $A_3^c(n) \subseteq A_1^c \cup A_2^c(n)$, so that $P[A_3^c(n)] \leq P(A_1^c) + P[A_2^c(n)] < \frac{\delta}{2} + \frac{\delta}{2} = \delta$. In other words, $P[|g(X_n) - g(X)| \geq \varepsilon] < \delta$ for $n \geq N(\varepsilon, \delta)$, which completes the proof.
Remark: An alternative approach is to use Exercise 5. To this end, set $Y_n = g(X_n), Y = g(Y)$ and show that $Y_n \xrightarrow[n\to\infty]{P} Y$. It suffices to show that for every subsequence $\{Y_n'\} \subseteq \{Y_n\}$ there is a further subsequence $\{Y_n''\} \subseteq \{Y_n'\}$ such that $Y_n'' \xrightarrow[n\to\infty]{a.s.} Y'$ and $P(Y' \neq Y) = 0$. The subsequence $\{Y_n'\}$ corresponds to the subsequence $\{X_n'\} \subseteq \{X_n\}$, and there exists $\{X_n''\} \subseteq \{X_n'\}$ such that $X_n'' \xrightarrow[n\to\infty]{a.s.} X'$ (with $P(X' \neq X) = 0$) by the fact that $X_n \xrightarrow[n\to\infty]{P} X$. Set $Y_n'' = g(X_n'')$. Then $\{Y_n''\} \subseteq \{Y_n'\}$ and $Y_n'' \xrightarrow[n\to\infty]{a.s.} Y' = g(X')$ by the fact that $X_n'' \xrightarrow[n\to\infty]{a.s.} X'$ and g is continuous. Also, $P(Y' \neq Y) = 0$ since $P(X' \neq X) = 0$. It follows that $Y_n \xrightarrow[n\to\infty]{P} Y$ or $g(X_n) \xrightarrow[n\to\infty]{P} g(X)$. #

9. Either apply in the plane the arguments used in Exercise 8 for the real line, or use an argument similar to the one employed in the Remark in the previous exercise. #

10. First, suppose that there exists $\{\varepsilon_n\}$ with $0 < \varepsilon_n \to 0$ for which

$$P\left[\bigcup_{k=n}^{\infty} (|X_k - X| \geq \varepsilon_k)\right] \underset{n\to\infty}{\longrightarrow} 0.$$

For every $\varepsilon > 0$, there exists an integer $k_\varepsilon > 0$ such that $k \geq k_\varepsilon$ implies $\varepsilon_k < \varepsilon$. Then, for $k \geq k_\varepsilon$, $(|X_k - X| \geq \varepsilon_k) \supseteq (|X_k - X| \geq \varepsilon)$, which for $n \geq k_\varepsilon$ implies

$$\bigcup_{k=n}^{\infty} (|X_k - X| \geq \varepsilon_k) \supseteq \bigcup_{k=n}^{\infty} (|X_k - X| \geq \varepsilon).$$

Hence, for $n \geq k_\varepsilon$,

$$P\left[\bigcup_{k=n}^{\infty}(|X_k - X| \geq \varepsilon_k)\right] \geq P\left[\bigcup_{k=n}^{\infty}(|X_k - X| \geq \varepsilon)\right].$$

Then

$$P\left[\bigcup_{k=n}^{\infty}(|X_k - X| \geq \varepsilon_k)\right] \underset{n\to\infty}{\longrightarrow} 0 \text{ implies } P\left[\bigcup_{k=n}^{\infty}(|X_k - X| \geq \varepsilon)\right] \underset{n\to\infty}{\longrightarrow} 0.$$

That is, for every $\varepsilon > 0$, $P\left[\cup_{k=n}^{\infty}(|X_k - X| \geq \varepsilon)\right] \underset{n\to\infty}{\longrightarrow} 0$ or $P\left[\cup_{\nu=0}^{\infty}(|X_{n+\nu} - X| \geq \varepsilon)\right] \underset{n\to\infty}{\longrightarrow} 0$, which is equivalent to $X_n \underset{n\to\infty}{\overset{a.s.}{\longrightarrow}} X$ by Theorem 4 in this chapter. Next, suppose that $X_n \underset{n\to\infty}{\overset{a.s.}{\longrightarrow}} X$. Then, by the theorem just cited,

$$P\left[\bigcup_{\nu=0}^{\infty}(|X_{n+\nu} - X| \geq \varepsilon)\right] \underset{n\to\infty}{\longrightarrow} 0 \text{ or } P\left[\bigcup_{k=n}^{\infty}(|X_k - X| \geq \varepsilon)\right] \underset{n\to\infty}{\longrightarrow} 0.$$

The last convergence implies the existence of an integer $n_\varepsilon > 0$ such that $n \geq n_\varepsilon$ implies

$$P\left[\bigcup_{k=n}^{\infty}(|X_k - X| \geq \varepsilon)\right] < \varepsilon.$$

Apply this by taking $\varepsilon = \frac{1}{m}$, $m \geq 1$. Then, for each such m, there exists an integer $n_m > 0$ such that $n \geq m_m$ implies

$$P\left[\bigcup_{k=n_m}^{\infty}\left(|X_k - X| \geq \frac{1}{m}\right)\right] < \frac{1}{m}.$$

Actually, let n_m be the smallest > 0 integer for which the above inequality is true, and without loss of generality, assume that $n_1 < n_2 < \ldots \to \infty$. For $k \geq n_1$, there exists n_m such that $n_k \leq k < n_{m+1}$. For this k and the subsequent ones which may lie in $[n_m, n_{m+1})$, set $\varepsilon_k = \frac{1}{m}$. Then $m \to \infty$ implies $\varepsilon_k \to 0$ and

$$P\left[\bigcup_{k=n_m}^{\infty}(|X_k - X| \geq \varepsilon_k)\right] = P\left[\bigcup_{k=n_m}^{\infty}\left(|X_k - X| \geq \frac{1}{m}\right)\right] < \frac{1}{m}.$$

For $n \in [n_m, n_{m+1})$, we, clearly, have

$$P\left[\bigcup_{k=n}^{\infty}(|X_k - X| \geq \varepsilon_k)\right] = P\left[\bigcup_{k=n}^{\infty}\left(|X_k - X| \geq \frac{1}{m}\right)\right] < \frac{1}{m} \underset{m\to\infty}{\longrightarrow} 0.$$

Since, clearly, $m \to \infty$ implies $n \to \infty$, we have $P\left[\cup_{k=n}^{\infty}(|X_k - X| \geq \varepsilon_k)\right] \underset{n\to\infty}{\longrightarrow} 0$, as was to be seen. #

11. Finiteness of μ implies (by Theorem 4 in this chapter) that $X_n \underset{n\to\infty}{\overset{a.s.}{\longrightarrow}} X$ if and only if $\mu\left(\cup_{\nu=0}^{\infty}|X_{n+\nu} - X| \geq \frac{1}{k}\right) \underset{n\to\infty}{\longrightarrow} 0$ for every $k \geq 1$. Let $\varepsilon > 0$ (to be kept fixed throughout) and, for each $k \geq 1$, consider $\varepsilon/2^k$. Then, for $\varepsilon/2^k$,there exists $N_k = N(\varepsilon, k) > 0$ integer such that

$$\mu\left[\bigcup_{\nu=0}^{\infty}\left(|X_{n+\nu} - X| \geq \frac{1}{k}\right)\right] < \frac{\varepsilon}{2^k}, \quad k \geq 1, \ n \geq N_k,$$

or

$$\mu\left[\bigcup_{n=N_k}^{\infty}\left(|X_n - X| \geq \frac{1}{k}\right)\right] < \frac{\varepsilon}{2^k}, \quad k \geq 1, \ n \geq N_k.$$

Setting $A_{\varepsilon,k} = \cup_{n=N_k}^{\infty}(|X_n - X| \geq \frac{1}{k})$ (where dependence of $A_{\varepsilon,k}$ on ε is through N_k), we have $\mu(A_{\varepsilon,k}) < \frac{\varepsilon}{2^k}, k \geq 1$. Let $A_\varepsilon = \cup_{k=1}^{\infty}A_{\varepsilon,k}$. Then $\mu(A_\varepsilon) = \mu\left(\cup_{k=1}^{\infty}A_{\varepsilon,k}\right) \leq \sum_{k=1}^{\infty}\mu\left(A_{\varepsilon,k}\right) \leq \varepsilon$. That is, $\mu(A_\varepsilon) < \varepsilon$. Consider $A_\varepsilon^c = \left(\cup_{k=1}^{\infty}A_{\varepsilon,k}\right)^c = \cap_{k=1}^{\infty}A_{\varepsilon,k}^c = \cap_{k=1}^{\infty}\cap_{n=N_k}^{\infty}(|X_n - X| < \frac{1}{k})$. Then, for $\omega \in A_\varepsilon^c$, we have: For every $k \geq 1$, there is $N_k(= N(k, \varepsilon)) > 0$ integer independent of ω, such that $|X_n(\omega) - X(\omega)| < \frac{1}{k}$ for all $n \geq N_k$. It follows that $X_n(\omega) \underset{n\to\infty}{\longrightarrow} X(\omega)$ and this convergence is uniform on A_ε^c. Thus, for every $\varepsilon > 0$, there exists A_ε with $\mu(A_\varepsilon) < \varepsilon$ such that $X_n(\omega) \underset{n\to\infty}{\longrightarrow} X(\omega)$ uniformly for $\omega \in A_\varepsilon^c$. It follows that $X_n \underset{n\to\infty}{\overset{a.u.}{\longrightarrow}} X$. #

12. Clearly, $A = \{\omega \in \Omega;\ \liminf_{n\to\infty}X_n(\omega) = \limsup_{n\to\infty}X_n(\omega)\}$, so that $A^c = \{\omega \in \Omega;\ \liminf_{n\to\infty}X_n(\omega) < \limsup_{n\to\infty}X_n(\omega)\}$. Thus, we have to show that

$$\{\omega \in \Omega;\ \liminf_{n\to\infty} X_n(\omega) < \limsup_{n\to\infty} X_n(\omega)\} =$$
$$\bigcup_{r,s}\{\omega \in \Omega;\ \liminf_{n\to\infty} X_n(\Omega) \leq r < s \leq \limsup_{n\to\infty} X_n(\omega)\}.$$

But this inequality is immediate since, if ω belongs in the left-hand side then $\liminf_{n\to\infty}X_n(\omega) < \limsup_{n\to\infty}X_n(\omega)$, and hence there exists rationals r and s with $r < s$, so that $\liminf_{n\to\infty}X_n(\omega) \leq r < s \leq \limsup_{n\to\infty}X_n(\omega)$, and thus ω belongs in the right-hand side. The other implications is immediate. #

13. **(i)** That $X_n \underset{n\to\infty}{\longrightarrow} X$ means that $X_n(\omega) \underset{n\to\infty}{\longrightarrow} X(\omega)$, $\omega \in \Omega$. So, for every $\omega \in \Omega$, the $\lim_{n\to\infty}X_n(\omega)$ exists, and this means that $\underline{\lim}_{n\to\infty}X_n(\omega) = \overline{\lim}_{n\to\infty}X_n(\omega)(= \lim_{n\to\infty}X_n(\omega)) = X(\omega)$. However, $X = \overline{\lim}_{n\to\infty}X_n = \inf_{n\geq 1}(\sup_{j\geq n}X_j) = \inf_{n\geq 1}Y_n$, where $Y_n = \sup_{j\geq n}X_j$, so that Y_n is defined in terms of $X_n, X_{n+1}, \ldots$ Next, $X = \inf_{n\geq 1}Y_n$, so that X is defined in terms of $Y_1, Y_2, \ldots$, and hence X is $\sigma(Y_1, Y_2, \ldots)$-measurable. However, $\sigma(Y_1, Y_2, \ldots) \subseteq \sigma(X_1, X_2, \ldots)$, since, for each $n \geq 1$, Y_n is a function of $X_n, X_{n+1}, \ldots$ Thus, X is $\sigma(X_1, X_2, \ldots)$-measurable.

(ii) The $\sigma(X_1, X_2, \ldots)$-measurability of X implies that $\sigma(X) \subseteq \sigma(X_1, X_2, \ldots)$, so that $\sigma(X_1, X_2, \ldots, X) = \sigma(\sigma(X_1, X_2, \ldots) \cup \sigma(X)) = \sigma(X_1, X_2, \ldots)$, as was to be seen. #

14. $\sum_{n=1}^{\infty} X_n$ converges if and only if $\sum_{n=m}^{\infty} X_n$ converges and this is true for every $m \geq 1$. On A (the set of convergence of $\sum_{n=1}^{\infty} X_n(\omega)$), $\sum_{n=m}^{\infty} X_n = \lim_{r\to\infty} \sum_{n=m}^{r} X_n = \lim_{r\to\infty} Y_{m,r}$, where $Y_{m,r} = \sum_{n=m}^{r} X_n$. However, $\{Y_{m,r}\}, r \geq 1$, converges if and only if it converges mutually, and the set of mutual convergence of $\{Y_{m,r}\}, r \geq 1$, is given by the expression

$$(Y_{m,r+\nu} - Y_{m,r} \underset{r\to\infty}{\longrightarrow} 0) = \bigcap_{k=1}^{\infty} \bigcup_{r=m}^{\infty} \bigcap_{\nu=1}^{\infty} \left(|X_{r+\nu} - X_r| < \frac{1}{k}\right).$$

This is so by Theorem 3. But the set on the right-hand side above is $\sigma(X_m, X_{m+1}, \ldots)$-measurable, being defined in terms of the r.v.s $X_m, X_{m+1}, \ldots$ Thus, the set A of convergence of $\sum_{n=1}^{\infty} X_n$ is $\sigma(X_m, X_{m+1}, \ldots)$-measurable for every $m \geq 1$. #

15. To show: that $\overline{\lim}_{n\to\infty} \frac{|X_n|}{n} \leq 1$ on a set A, say, with $P(A) = 1$. From the definition of the $\overline{\lim}_{n\to\infty}$ of a sequence of real numbers $\{x_n\}$, $\overline{\lim}_{n\to\infty} x_n \leq 1$ means that $x_n \leq 1$ for all but finitely many ns. So, what must be shown is that

$$P\left(\frac{|X_n|}{n} \leq 1 \text{ for all but finitely many } ns\right) = 1.$$

Setting $A_n = \left\{\omega \in \Omega;\ \frac{|X_n|}{n} > 1\right\}$, the last relation above becomes $P\left(\underline{\lim}_{n\to\infty} A_n^c\right) = 1$ (by Exercise 30(i) in Chapter 1). However,

$$P\left(\underline{\lim}_{n\to\infty} A_n^c\right) = P\left(\bigcup_{n=1}^{\infty} \bigcap_{k=n}^{\infty} A_k^c\right) = P\left[\left(\bigcap_{n=1}^{\infty} \bigcup_{k=n}^{\infty} A_k\right)^c\right],$$

so that

$$P\left(\underline{\lim}_{n\to\infty} A_n^c\right) = 1 \text{ if and only if } P\left[\left(\bigcap_{n=1}^{\infty} \bigcup_{k=n}^{\infty} A_k\right)^c\right] = 1,$$

if and only if

$$P\left(\bigcap_{n=1}^{\infty} \bigcup_{k=n}^{\infty} A_k\right) = 0, \text{ or } P\left(\overline{\lim}_{n\to\infty} A_n\right) = 0,$$

which follows by the assumption that $\sum_{n=1}^{\infty} P(A_n) < \infty$ (by Exercise 3 in this chapter). #

16. First, from the expression $||\mathbf{Z}_n - \mathbf{Z}|| = \left[\sum_{j=1}^{k} (Z_{nj} - Z_j)^2\right]^{\frac{1}{2}}$, we have that $Z_{nj} \underset{n\to\infty}{\overset{P}{\longrightarrow}} Z_j, j = 1, \ldots, k$, implies $||\mathbf{Z}_n - \mathbf{Z}|| \underset{n\to\infty}{\overset{P}{\longrightarrow}} 0$ (by Exercises 8 and 9). Next, for each $j = 1, \ldots, k$, $|Z_{nj} - Z_j| = [(Z_{nj} - Z_j)^2]^{\frac{1}{2}} \leq \left[\sum_{i=1}^{k} (Z_{ni} - Z_i)^2\right]^{\frac{1}{2}} = ||\mathbf{Z}_n - \mathbf{Z}|| \underset{n\to\infty}{\overset{P}{\longrightarrow}} 0$, so that $Z_{nj} \underset{n\to\infty}{\overset{P}{\longrightarrow}} Z_j, j = 1, \ldots, k$. #

17. It suffices to demonstrate it by means of a concrete example. To this effect, for $n = 1, 2, \ldots$, let the r.v.s X_n be defined on some probability space $(\Omega, \mathcal{A}, P)$ as follows: $X_{2n-1}(\omega) = p - \frac{1}{n}$, $X_{2n}(\omega) = p + \frac{1}{n}$ for some positive constant p and all $\omega \in \Omega$, Then, clearly, $X_n \underset{n\to\infty}{\longrightarrow} p$ pointwise and hence $X_n \overset{P}{\underset{n\to\infty}{\longrightarrow}} p$. Set $p = X$ and take $c = p$. Then, for all $\omega \in \Omega$,

$$\begin{aligned} \delta_c(X_{2n-1}(\omega)) &= 0 \text{ since } X_{2n-1}(\omega) = p - \frac{1}{n} < p\,(= c), \\ \delta_c(X_{2n}(\omega)) &= 1 \text{ since } X_{2n}(\omega) = p + \frac{1}{n} \geq p\,(= c), \\ \text{and } \delta_c(X(\omega)) &= 1 \text{ since } X(\omega) = p \geq p\,(= c). \end{aligned}$$

Therefore

$$\begin{aligned} |\delta_c(X_{2n-1}(\omega)) - \delta_c(X(\omega))| &= |0 - 1| = 1, \\ |\delta_c(X_{2n}(\omega)) - \delta_c(X(\omega))| &= |1 - 1| = 0. \end{aligned}$$

Thus, for $\varepsilon > 0$,

$$P[|\delta_c(X_n) - \delta_c(X)| > \varepsilon] = P[|\delta_c(X_{2n-1}) - \delta_c(X)| = 1] = 1,$$

so that $\delta_c(X_n) \overset{P}{\underset{n\to\infty}{\nrightarrow}} \delta_c(X)$. #

18. The convergence $X_n \overset{\mu}{\underset{n\to\infty}{\longrightarrow}} X$ implies the existence of a subsequence $\{X'_n\} \subseteq \{X_n\}$ such that $X'_n \overset{a.s.}{\underset{n\to\infty}{\longrightarrow}} X'$, a r.v. X', with $\mu(X' \neq X) = 0$ (by Theorem 5(ii)). Therefore $|X'| \leq M$ a.e. and this implies that $|X| \leq M$ a.e. #

19. Let $\omega_0 \in \Omega$, assume that $\{\omega_0\} \in \mathcal{A}$, and let $\mu(\{\omega_0\}) = 0$.

(i) For $n \geq 1$, let $X_n = 0$ on $\{\omega_0\}^c$, $X_n(\omega_0) = 1$, and let $X = 0$ on Ω. Then, clearly, $X_n \overset{a.e.}{\underset{n\to\infty}{\longrightarrow}} X$, and $X_n^{-1}(\mathcal{B}) = \{\varnothing, \{\omega_0\}, \{\omega_0\}^c, \Omega\}$, $X^{-1}(\mathcal{B}) = \{\varnothing, \Omega\}$. Since $\{\varnothing, \Omega\} \subset \{\varnothing, \{\omega_0\}, \{\omega_0\}^c, \Omega\}$, which is the same as the $\sigma(X_1, X_2, \ldots)$, it follows that X is $\sigma(X_1, X_2, \ldots)$-measurable.

(ii) Here, for $n \geq 1$, take $X_n = 0$ on Ω, and $X = 0$ on $\{\omega_0\}^c$, $X = 1$ on $\{\omega_0\}$. Then, again $X_n \overset{a.e.}{\underset{n\to\infty}{\longrightarrow}} X$, and $X_n^{-1}(\mathcal{B}) = \{\varnothing, \Omega\}$, $X^{-1}(\mathcal{B}) = \{\varnothing, \{\omega_0\}, \{\omega_0\}^c, \Omega\}$, which is the same as the $\sigma(X_1, X_2, \ldots)$. Since $X^{-1}(\mathcal{B}) \nsubseteq \sigma(X_1, X_2, \ldots)$, it follows that X is not $\sigma(X_1, X_2, \ldots)$-measurable.

(iii) Let N be the set over which $X_n \underset{n\to\infty}{\nrightarrow} X$, so that $\mu(N^c) = 0$. Next, set $X'_n = X_n$ on N^c and $X'_n = 0$ (for example) on N; also let $X' = X$ on N^c and $X' = 0$ on N. Then $X'_n \underset{n\to\infty}{\longrightarrow} X'$ pointwise and X' is $\sigma(X'_1, X'_2, \ldots)$-measurable. Furthermore, $\sum_{n=1}^{\infty} \mu(X'_n \neq X_n) = 0 = \mu(X' \neq X)$.

(iv) Clearly, $X_n \overset{a.e.}{\underset{n\to\infty}{\longrightarrow}} X$ because, although $\liminf_{n\to\infty} X_n(\omega_0) = 2 < 3 = \limsup_{n\to\infty} X_n(\omega_0)$, $\mu(\{\omega_0\}) = 0$. Setting, for example, $X'_n = X' = 0$ on

Ω, we have $X'_n \underset{n\to\infty}{\longrightarrow} X'$ pointwise (actually, $X'_n = X, n \geq 1$, on Ω), X' is $\sigma(X'_1, X'_2, \ldots)$-measurable, and $\sum_{n=1}^{\infty} \mu(X'_n \neq X_n) = 0 = \mu(X' \neq X)$. #

20. Let the events A and B be defined by:

$$A = \{\omega \in \Omega;\ X_m(\omega) - X_n(\omega) \underset{m,n\to\infty}{\longrightarrow} 0\},$$
$$B = \{\omega \in \Omega;\ X_{n+\nu}(\omega) - X_n(\omega) \underset{n\to\infty}{\longrightarrow} 0 \text{ uniformly in } \nu \geq 1\}.$$

Then $A = B$. Indeed, in the first place it is clear that $X_m(\omega) - X_n(\omega) \to 0$ as $m, n \to \infty$ is equivalent to $X_{n+i}(\omega) - X_{n+j}(\omega) \underset{n\to\infty}{\longrightarrow} 0$ uniformly in $0 \leq i < j$. So, for $\omega \in A$, $X_{n+i}(\omega) - X_{n+j}(\omega) \to 0$ uniformly in $0 \leq i < j$. By taking $i = 0$ and setting ν instead of j, we have $X_{n+\nu}(\omega) - X_n(\omega) \underset{n\to\infty}{\longrightarrow} 0$ uniformly in $\nu \geq 1$, so that $\omega \in B$ and hence $A \subseteq B$.

Next, let $\omega \in B$, so that $X_{n+\nu}(\omega) - X_n(\omega) \underset{n\to\infty}{\longrightarrow} 0$ uniformly, in $\nu \geq 1$, and observe that

$$|X_{n+i}(\omega) - X_{n+j}(\omega)| \leq |X_{n+i}(\omega) - X_n(\omega)| + |X_{n+j}(\omega) - X_n(\omega)|.$$

Then, by the fact that

$$X_{n+i}(\omega) - X_n(\omega) \underset{n\to\infty}{\longrightarrow} 0 \text{ uniformly in } i \geq 0$$

and

$$X_{n+j}(\omega) - X_n(\omega) \underset{n\to\infty}{\longrightarrow} 0 \text{ uniformly in } j \geq 0,$$

we get

$$X_{n+i}(\omega) - X_{n+j}(\omega) \underset{n\to\infty}{\longrightarrow} 0 \text{ uniformly in } 0 \leq i < j,$$

so that $\omega \in A$ and hence $B \subseteq A$. It follows that $A = B$. #

21. In the first place, for every $\varepsilon > 0$, it is clear that

$$\mu(|X_m - X_n| \geq \varepsilon) \longrightarrow 0 \text{ as } m, n \to \infty$$

is equivalent to

$$\mu(|X_{n+i} - X_{n+j}| \geq \varepsilon) \underset{n\to\infty}{\longrightarrow} 0 \text{ uniformly in } 0 \leq i < j.$$

Now, assume that $\mu(|X_{n+i} - X_{n+j}| \geq \varepsilon) \underset{n\to\infty}{\longrightarrow} 0$ uniformly in $0 \leq i < j$, take $i = 0$ and set ν instead of j to obtain $\mu(|X_{n+\mu} - X_n| \geq \varepsilon) \underset{n\to\infty}{\longrightarrow} 0$ uniformly in $\nu \geq 1$. Next, assume that $\mu(|X_{n+\nu} - X_n| \geq \varepsilon) \underset{n\to\infty}{\longrightarrow} 0$ uniformly in $\nu \geq 1$, and for $0 \leq i < j$, observe that

$$|X_{n+i} - X_{n+j}| \leq |X_{n+i} - X_n| + |X_{n+j} - X_n|,$$

so that

$$\left(|X_{n+i} - X_{n+j}| \geq \varepsilon\right) \subseteq \left(|X_{n+i} - X_n| \geq \frac{\varepsilon}{2}\right) \cup \left(|X_{n+j} - X_n| \geq \frac{\varepsilon}{2}\right).$$

However,

$$\mu\left(|X_{n+i} - X_n| \geq \frac{\varepsilon}{2}\right) \underset{n\to\infty}{\longrightarrow} 0 \text{ uniformly in } i \geq 0$$

and

$$\mu\left(|X_{n+j} - X_n| \geq \frac{\varepsilon}{2}\right) \underset{n\to\infty}{\longrightarrow} 0 \text{ uniformly in } j \geq 1,$$

and hence

$$\mu\left(|X_{n+i} - X_{n+j}| \geq \varepsilon\right) \underset{n\to\infty}{\longrightarrow} 0 \text{ uniformly in } 0 \leq i < j.$$

This completes the proof. #

Chapter 4

The Integral of a Random Variable and its Basic Properties

1. Let $(\Omega, \mathcal{A}, P) = ((0, 1], \mathcal{B}_{(0,1]}, \lambda)$ where λ is the Lebesgue measure, and define X_n and X by: $X_n = nI_{(0,\frac{1}{n}]}$, $n = 1, 2, \ldots$, $X = 0$. Then for any $\omega \in (0, 1]$, there exists $n_0 = n_0(\omega)$ such that $\omega > \frac{1}{n}$, $n \geq n_0$. Then $X_n(\omega) = 0$, $n \geq n_0$, so that, trivially, $X_n \to X$ pointwise. However, $EX_n = n \times \frac{1}{n} = 1$, $n \geq 1$, and $EX = 0$, so that $EX_n \underset{n\to\infty}{\nrightarrow} EX$. #

2. Let $A = (X > 0)$ and set $A_n = (X > \frac{1}{n})$, $n = 1, 2, \ldots$, so that $A_n \uparrow A$ and $\mu(A_n) \uparrow \mu(A)$. Then

$$\begin{aligned} 0 = \int X d\mu &= \int_{(X=0)} X d\mu + \int_{(X>0)} X d\mu \\ &= \int_{(X>0)} X d\mu = \int_A X d\mu = \int (XI_A) d\mu \\ &\geq \int (XI_{A_n}) d\mu \geq \frac{1}{n}\mu(A_n), \end{aligned}$$

so that $\mu(A_n) = 0$, $n \geq 1$, and hence $\mu(A) = 0$. #

3. Set $p_k = P(X = k), k \geq 1$. Then:

$$\begin{aligned}\sum_{k=1}^{\infty} P(X \geq k) &= \lim_{n\to\infty} \sum_{k=1}^{n} P(X \geq k)\\ &= \lim_{n\to\infty} \{p_1 + p_2 + \ldots + p_{n-1} + p_n\\ &\qquad + p_2 + \ldots + p_{n-1} + p_n\\ &\qquad + \ldots\ldots\\ &\qquad + p_{n-1} + p_n\\ &\qquad + p_n\}\\ &= \lim_{n\to\infty} (p_1 + 2p_2 + \ldots + np_n) = \lim_{n\to\infty} \sum_{k=1}^{n} kp_k\\ &= \sum_{k=1}^{\infty} kp_k = EX.\#\end{aligned}$$

4. **(i)** Clearly,

$$\begin{aligned}(XI_{A+B})^+ &= X^+ I_{A+B} = X^+(I_A + I_B) \text{ (by the fact that } A \cap B = \oslash)\\ &= X^+ I_A + X^+ I_B = (XI_A)^+ + (XI_B)^+.\end{aligned}$$

Similarly,

$$(XI_{A+B})^- = X^- I_{A+B} = X^- I_A + X^- I_B = (XI_A)^- + (XI_B)^-.$$

(ii)

$$\begin{aligned}(XI_{\sum_{i=1}^n A_i})^+ &= X^+ I_{\sum_{i=1}^n A_i} = X^+ \sum_{i=1}^{n} I_{A_i} \text{ (since } A_i \cap A_j = \oslash, i \neq j)\\ &= \sum_{i=1}^{n} X^+ I_{A_i} = \sum_{i=1}^{n} (XI_{A_i})^+, \text{ and similarly,}\\ (XI_{\sum_{i=1}^n A_i})^- &= X^- I_{\sum_{i=1}^n A_i} = X^- \sum_{i=1}^{n} I_{A_i} = \sum_{i=1}^{n} X^- I_{A_i}\\ &= \sum_{i=1}^{n} (XI_{A_i})^-.\#\end{aligned}$$

5. The existence of $\int_{\sum_{i=1}^n A_i} X = \int XI_{\sum_{i=1}^n A_i}$ implies that at least one of $\int (XI_{\sum_{i=1}^n A_i})^+$ or $\int (XI_{\sum_{i=1}^n A_i})^-$ is finite. Let $\int (XI_{\sum_{i=1}^n A_i})^+ < \infty$. Since $(XI_{\sum_{i=1}^n A_i})^+ = \sum_{i=1}^{n} (XI_{A_i})^+$ (by Exercise 4(ii)), it follows that

$\int (XI_{\sum_{i=1}^n A_i})^+ = \sum_{i=1}^n \int (XI_{A_i})^+$ by Theorem 4 (see also Remark 2), and similarly $\int (XI_{\sum_{i=1}^n A_i})^- = \sum_{i=1}^n \int (XI_{A_i})^-$. Therefore $\int (XI_{A_i})^+ < \infty$, $i = 1, \ldots, n$ and the integrals $\int XI_{A_i} = \int_{A_i} X_i$, $i = 1, \ldots, n$, exist. Furthermore,

$$\sum_{i=1}^{n} \int X^+ I_{A_i} - \sum_{i=1}^{n} \int X^- I_{A_i} = \int \sum_{i=1}^{n} (X^+ I_{A_i}) - \int \sum_{i=1}^{n} (X^- I_{A_i})$$

$$= \int (XI_{\sum_{i=1}^n A_i})^+ - \int (XI_{\sum_{i=1}^n A_i})^- = \int XI_{\sum_{i=1}^n A_i} = \int_{\sum_{i=1}^n A_i} X,$$

and

$$\sum_{i=1}^{n} \int X^+ I_{A_i} - \sum_{i=1}^{n} \int X^- I_{A_i} = \sum_{i=1}^{n} \left(\int X^+ I_{A_i} - \int X^- I_{A_i} \right)$$

$$= \sum_{i=1}^{n} \int XI_{A_i} = \sum_{i=1}^{n} \int_{A_i} X. \text{ That is,}$$

$\int_{\sum_{i=1}^n A_i} X = \sum_{i=1}^n \int_{A_i} X$. Similarly, if we assume that $\int (XI_{\sum_{i=1}^n A_i})^- < \infty$. #

6. Suppose that $\int X = \int Y = \infty$. From $\int X = \sum_{i=1}^m \alpha_i \mu(A_i) = \infty$, it follows that for those α_is with $\alpha_i < 0$, the corresponding A_is have $\mu(A_i) < \infty$. Then, since by convention $0 \times \infty = 0$, restrict attention to α_i s> 0. Hence there exists at least one i_0 such that $\alpha_{i_0} > 0$ and $\mu(A_{i_0}) = \infty$. Next, in summing the terms $(\alpha_i + \beta_j)\mu(A_i \cap B_j)$, restrict ourselves to $A_i \cap B_j \neq \oslash$. Now $\infty = \mu(A_{i_0}) = \sum_{j=1}^n \mu(A_{i_0} \cap B_j)$ implies that $\mu(A_{i_0} \cap B_{j_1}) = \infty$. for at least one j_1, which implies $\mu(B_{j_1}) = \infty$. Then, since $\int Y = \infty$, it follows that the corresponding $\beta_{j_1} \geq 0$. So, whenever $\mu(A_i \cap B_j) = \infty$, then $\alpha_i \geq 0$ and $\beta_j \geq 0$ for the respective α_is and β_js. It follows that $\sum_{i,j} (\alpha_i + \beta_i)\mu(A_i \cap B_j)$ is meaningful, and of course, $\sum_{i,j} (\alpha_i + \beta_j)\mu(A_i \cap B_j) = \sum_i \alpha_i \mu(A_i) + \sum_j \beta_j \mu(B_j)$. So, $\int (X + Y)$ exists and is equal to $\int X + \int Y$. Similarly, if $\int X = \int Y = -\infty$. #

7. Since $\int Xd(\mu_1 + \mu_2)$ exists, at least one of $\int X^+ d(\mu_1 + \mu_2)$ and $\int X^- d(\mu_1 + \mu_2)$ is finite. Let $\int X^+ d(\mu_1 + \mu_2) < \infty$, and let $0 \leq X_n$ simple $\underset{n\to\infty}{\uparrow} X^+$. Then $\int X_n d(\mu_1 + \mu_2) \underset{n\to\infty}{\uparrow} \int X^+ d(\mu_1 + \mu_2)$. Suppose $X_n = \sum_{i=1}^{r_n} \alpha_{ni} I_{A_{ni}}$. Then $\int X_n d(\mu_1 + \mu_2) = \sum_{i=1}^{r_n} \alpha_{ni} (\mu_1 + \mu_2)(A_{ni}) = \sum_{i=1}^{r_n} \alpha_{ni} \mu_1(A_{ni}) + \sum_{i=1}^{r_n} \alpha_{ni} \mu_2(A_{ni}) = \int X_n d\mu_1 + \int X_n d\mu_2 \underset{n\to\infty}{\uparrow} \int X^+ d\mu_1 + \int X^+ d\mu_2$. Thus, $\int X^+ d\mu_1 + \int X^+ d\mu_2 = \lim_{n\to\infty} \int X_n d(\mu_1 + \mu_2) = \int X^+ d(\mu_1 + \mu_2) < \infty$, so that $\int X^+ d\mu_1$ and $\int X^+ d\mu_2$ are $< \infty$. Then $\int X d\mu_1$ and $\int X d\mu_2$ exist, and from above,

$$\int X^+ d(\mu_1 + \mu_2) = \int X^+ d\mu_1 + \int X^+ d\mu_2.$$

Likewise,

$$\int X^- d(\mu_1 + \mu_2) = \int X^- d\mu_1 + \int X^- d\mu_2,$$

so that

$$\begin{aligned}\int Xd(\mu_1+\mu_2) &= \int X^+d(\mu_1+\mu_2) - \int X^-d(\mu_1+\mu_2)\\ &= (\int X^+d\mu_1 - \int X^-d\mu_1) + (\int X^+d\mu_2 - \int X^-d\mu_2)\\ &= \int Xd\mu_1 + \int Xd\mu_2. \#\end{aligned}$$

8. By Theorem 12, the integral $\int(X_1+X_2)$ exists and $\int(X_1+X_2) = \int X_1 + \int X_2$. Also, by Theorems 8(iv), 4 and 10,

$$|\int(X_1+X_2)| \le \int |X_1+X_2| \le \int |X_1| + \int |X_2| < \infty,$$

so that X_1+X_2 is integrable. Next, assume the theorem to be true for k and show it to be true for $k+1$. To this end, $\sum_{i=1}^{k+1} X_i = (\sum_{i=1}^{k} X_i) + X_{k+1}$, and by the induction hypothesis, $\sum_{i=1}^{k} X_i$ is integrable and $\int \sum_{i=1}^{k} X_i = \sum_{i=1}^{k} \int X_i$. Then by the first step, $(\sum_{i=1}^{k} X_i) + X_{k+1}$ is integrable and $\int[(\sum_{i=1}^{k} X_i) + X_{k+1}] = \int \sum_{i=1}^{k} X_i + \int X_{k+1} = \sum_{i=1}^{k} \int X_i + \int X_{k+1}$, or $\int \sum_{i=1}^{k+1} X_i = \sum_{i=1}^{k+1} \int X_i$. #

9. **(i)** Clearly, $0 \le \mathcal{E}\left(\frac{X-\mu_1}{\sigma_1} - \frac{Y-\mu_2}{\sigma_2}\right)^2 = 2 - \frac{2\,\mathrm{Cov}(X,Y)}{\sigma_1\sigma_2}$, or $\mathrm{Cov}(X,Y) \le \sigma_1\sigma_2$, with equality occurring if and only if $\mathcal{E}\left(\frac{X-\mu_1}{\sigma_1} - \frac{Y-\mu_2}{\sigma_2}\right)^2 = 0$, which implies that $P\left(\frac{X-\mu_1}{\sigma_1} - \frac{Y-\mu_2}{\sigma_2} = 0\right) = 1$ (by Exercise 2 here), or $P\left[Y = \mu_2 + \frac{\sigma_2}{\sigma_1}(X-\mu_1)\right] = 1$.

Likewise, $0 \le \mathcal{E}\left(\frac{X-\mu_1}{\sigma_1} + \frac{Y-\mu_2}{\sigma_2}\right)^2 = 2 + \frac{2\,\mathrm{Cov}(X,Y)}{\sigma_1\sigma_2}$, or $\mathrm{Cov}(X,Y) \ge -\sigma_1\sigma_2$, with equality occurring if and only if $\mathcal{E}\left(\frac{X-\mu_1}{\sigma_1} + \frac{Y-\mu_2}{\sigma_2}\right)^2 = 0$, which implies $P\left(\frac{X-\mu_1}{\sigma_1} + \frac{Y-\mu_2}{\sigma_2} = 0\right) = 1$, or $P\left[Y = \mu_2 - \frac{\sigma_2}{\sigma_1} \times (X-\mu_1)\right] = 1$.

(ii) The first expression in part (i) becomes $2-2\rho(X,Y) \ge 0$, or $\rho(X,Y) \le 1$, and $\rho(X,Y) = 1$ if and only if $P\left[Y = \mu_2 + \frac{\sigma_2}{\sigma_1}(X-\mu_1)\right] = 1$; whereas the second expression is $2 + 2\rho(X,Y) \ge 0$, or $\rho(X,Y) \ge -1$, and $\rho(X,Y) = -1$ if and only if $P\left[Y = \mu_2 - \frac{\sigma_2}{\sigma_1}(X-\mu_1)\right] = 1$. #

10. We have $X = -1 \times I_{\{-2,-1\}} + 1 \times I_{\{3,7\}}$, so that $\int_A Xd\mu = \int(XI_A)d\mu = \int[(-I_{\{-2,-1\}} + I_{\{3,7\}})I_{\{-2,3,7\}}]d\mu = \int(-I_{\{-2\}} + I_{\{3,7\}})d\mu = -\int(I_{\{-2\}}d\mu + \int I_{\{3,7\}})d\mu = -\mu(\{-2\}) + \mu(\{3,7\}) = -2+3+7 = 8$; i.e., $\int_A Xd\mu = 8$. #

11. The r.v. X can be written as follows:

$$X = \frac{1}{2}I_{(-5,2)} + \frac{1}{3}I_{\{2\}} + 1I_{(2,3]} + 0I_{(3,5)}.$$

Then

$$\int_A X d\mu = \int_{[-1,\,4]} X d\mu = \int (X I_{[-1,\,4]}) d\lambda$$
$$= \int \left[\frac{1}{2} I_{(-5,\,2)\cap[-1,\,4]} + \frac{1}{3} I_{\{2\}\cap[-1,\,4]} + I_{[-1,\,4]\cap(2,\,3]}\right] d\lambda$$
$$= \int \left[\frac{1}{2} I_{[-1,\,2)} + \frac{1}{3} I_{\{2\}} + I_{(2,\,3]}\right] d\lambda$$
$$= \frac{1}{2} \times 3 + \frac{1}{3} \times 0 + 1 \times 1 = \frac{3}{2} + 1 = \frac{5}{2}. \#$$

12. That $\mu(A) \geq 0$ and $\mu(\oslash) = 0$ are obvious. Next, let A_i be events with $A_i \cap A_j = \oslash, i \neq j$. Then $\mu(A_1 + A_2 + \ldots) =$ the number of ≥ 0 integers in $A_1 + A_2 + \ldots$, and it is clear that this is equal to: $\sum_{i=1}^{\infty}$(# of ≥ 0 integers in A_i) $= \sum_{i=1}^{\infty} \mu(A_i)$, so that μ is a measure. That μ is σ-finite is immediate, since $\Omega = \sum_{\omega=0}^{\infty}\{\omega\}$ and $\mu(\{\omega\}) = 1$. #

13. Indeed,

$$\mathcal{E}g(X) = \int_{(|X|\geq c)} g(X) dP + \int_{(|X|<c)} g(X) dP$$
$$\geq \int_{(|X|\geq c)} g(X) dP \text{ (since } g(x) > 0)$$
$$\geq g(c) \int_{(|X|\geq c)} dP \text{ (since } g \text{ is nondecreasing in } (0, \infty) \text{ and symmetric about 0)}$$
$$= g(c) P(|X| \geq c). \text{ Hence}$$

$P(|X| \geq c) \leq \mathcal{E}g(X)/g(c)$. #

14. Let $g(x) = \frac{|x|}{1+|x|}, x \in \Re$. Then $g(-x) = g(x)$ and $g(x)$ is nondecreasing in $(0, \infty)$ (because $g'(x) = \frac{1}{(1+x)^2} > 0$). Then Exercise 13 applies and gives: $P(|X| \geq c) \leq \frac{\mathcal{E}g(X)}{g(c)} = \frac{\mathcal{E}(\frac{|X|}{1+|X|})}{\frac{c}{1+c}} = \frac{1+c}{c}\mathcal{E}\left(\frac{|X|}{1+|X|}\right) = \left(1 + \frac{1}{c}\right)\mathcal{E}\left(\frac{|X|}{1+|X|}\right)$. #

15. Let A_{j_0} be an arbitrary but fixed member of the A_js. There is at least one of the B_is, call it B_{i_0}, intersecting A_{j_0}. Then $B_{i_0} = \alpha_{j_0}$ and, actually, the entire B_{i_0} lies within A_{j_0}. Thus, if k is the number of B_is intersecting A_{j_0}, then A_{j_0} is the sum of these B_is. The same holds true for each one of the remaining A_js. #

16. In the first place,

$$\mathcal{E}\left[X I_{(|X|\leq c)}\right] = \int_{\Omega} \left[X I_{(|X|\leq c)}\right] dP,$$

and

$$\mathcal{E}\left[Y I_{(|Y|\leq c)}\right] = \int_{\Omega} \left[Y I_{(|Y|\leq c)}\right] dP$$

exist, by the Corollary to Theorem 5, and are finite, by the fact that $|X|I_{(|X|\leq c)} \leq |X|$, $|Y|I_{(|Y|\leq c)} \leq |Y|$ and Theorem 10. Next, with P_X and P_Y standing for the probability distributions of X and Y, respectively, we have

$$\begin{aligned}
\mathcal{E}\left[XI_{(|X|\leq c)}\right] &= \int_\Omega \left[XI_{(|X|\leq c)}\right] dP \\
&= \int_\Re \left[xI_{(|X|\leq c)}\right] dP_X \qquad \text{(by Theorem 13)} \\
&= \int_\Re \left[xI_{(|X|\leq c)}\right] dP_Y \qquad \text{(since } X = Y \text{ in distribution)} \\
&= \int_\Re \left[yI_{(|Y|\leq c)}\right] dP_Y = \int_\Omega \left[YI_{(|Y|\leq c)}\right] dP \\
&= \mathcal{E}\left[YI_{(|Y|\leq c)}\right]. \quad \#
\end{aligned}$$

Chapter 5

Standard Convergence Theorems, the Fubini Theorem

1. Indeed,

$$nP(|X| \geq n) = \int [nI_{(|X|\geq n)}]dP \leq \int [|X|I_{(|X|\geq n)}]dP.$$

However,
$0 \leq |X|I_{(|X|\geq n)} \leq |X|$, independent of n and integrable, and $|X|I_{(|X|\geq n)} \underset{n\to\infty}{\longrightarrow} 0$. Then, by part (iii) of the Fatou-Lebesgue Theorem, it follows that $\int[|X|I_{(|X|\geq n)}]dP \underset{n\to\infty}{\longrightarrow} 0$, so that $nP(|X| \geq n) \underset{n\to\infty}{\longrightarrow} 0$.
For the converse, consider the following
Example: Let X take on the values $4, 5, \ldots,$ and let $x_n = \frac{\log n+1}{(n\log n)^2}$, $n \geq 4$, so that $\sum_{n=4}^{\infty} \frac{\log n+1}{(n\log n)^2} = \sum_{n=4}^{\infty} \frac{1}{n^2\log n} + \sum_{n=4}^{\infty} \frac{1}{(n\log n)^2} \leq 2\sum_{n=4}^{\infty} \frac{1}{n^2} < \infty$. Set $C = \sum_{n=4}^{\infty} x_n$ and take $p_n = P(X = n) = \frac{1}{C}\frac{\log n+1}{(n\log n)^2}$, $n \geq 4$. We shall show that $\mathcal{E}X = \infty$ and $nP(X \geq n) \underset{n\to\infty}{\longrightarrow} 0$. Indeed,

$$C\mathcal{E}X = \sum_{n=4}^{\infty} \frac{\log n+1}{n(\log n)^2} = \sum_{n=4}^{\infty} \frac{1}{n\log n} + \sum_{n=4}^{\infty} \frac{1}{n(\log n)^2},$$

and it suffices to show that one of the terms on the right-hand side above is ∞. To this end, set $g(x) = \frac{1}{x\log x}$, $x \geq 3.5$ and show that $\sum_{n=4}^{\infty} g(n) = \infty$. Clearly (see figure below), with $c = 3.5$,

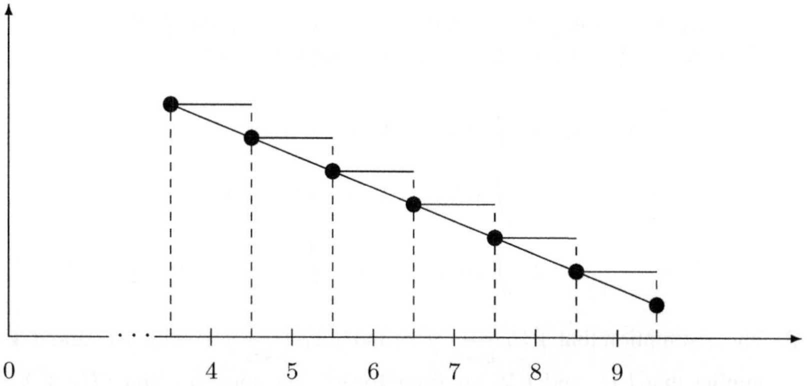

$$\sum_{n=4}^{\infty} g(n) \leq \int_c^{\infty} g(x)dx = \int_c^{\infty} \frac{dx}{x \log x} = \int_c^{\infty} \frac{d \log x}{\log x} = \log \log x|_c^{\infty} = \infty.$$

Thus, $\mathcal{E}X = \infty$.
Next, set $h(x) = \frac{\log x+1}{(xlogx)^2} = \frac{1}{x^2 \log x} + \frac{1}{(x \log x)^2}, x \geq c = 3.5$, and $\frac{1}{(x \log x)^2} \leq \frac{1}{x^2 \log x}$. Then, since $CnP(X \geq n) = n\sum_{k=n}^{\infty} \frac{1}{k^2 \log k} + n\sum_{k=n}^{\infty} \frac{1}{(k \log k)^2}$, it suffices to show that $n\sum_{k=n}^{\infty} \frac{1}{k^2 \log k} \underset{n\to\infty}{\longrightarrow} 0$. Clearly, $\sum_{k=n}^{\infty} \frac{1}{k^2 \log k} \leq \int_{n-\frac{1}{2}}^{\infty} \frac{dx}{x^2 \log x}$.

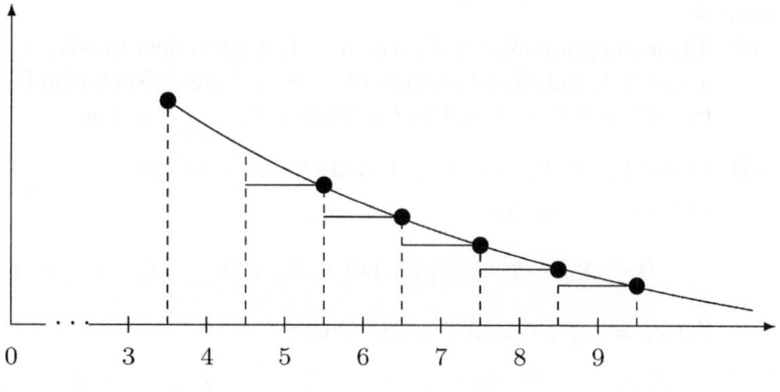

Set $\log x = y, x = e^y, dx = e^y dy, y : c = \log(n - \frac{1}{2}), \infty$. Then the integral becomes: $\int_c^{\infty} \frac{e^y dy}{e^{2y} \times y} = \int_c^{\infty} \frac{dy}{ye^y}$. However, $\int_c^{\infty} \frac{dy}{ye^y} = -\int_c^{\infty} \frac{de^{-y}}{y} = -\frac{1}{ye^y}|_c^{\infty} + \int_c^{\infty} \frac{dy^{-1}}{e^y} = \frac{1}{(n-\frac{1}{2})\log(n-\frac{1}{2})} - \int_c^{\infty} \frac{dy}{y^2 e^y}$. Hence $n\sum_{k=n}^{\infty} \frac{1}{k^2 \log k} \leq n\int_c^{\infty} \frac{dy}{ye^y} \leq \frac{n}{(n-\frac{1}{2})\log(n-\frac{1}{2})} \underset{n\to\infty}{\longrightarrow} 0$. This completes the proof. #

2. The integrability of X implies that both $\int X^+$ and $\int X^-$ are finite. Let $0 \leq X_n$ simple $\underset{n\to\infty}{\uparrow} X^+$ and $0 \leq Y_n$ simple $\underset{n\to\infty}{\uparrow} X^-$. Then $|X_n| \leq X^+$, independent of n and integrable, so that $\int |X_n - X^+| \underset{n\to\infty}{\longrightarrow} 0$, by part (iii) of the Fatou-Lebesgue Theorem, and likewise $\int |Y_n - X^-| \underset{n\to\infty}{\longrightarrow} 0$. Then, for every $\varepsilon > 0$, there exists

$n_0 = n(\varepsilon) > 0$ integer such that $\int |X_n - X^+| < \frac{\varepsilon}{2}$ and $\int |Y_n - X^-| < \frac{\varepsilon}{2}$. Set $X_\varepsilon = X_{n_0} - Y_{n_0}$. Then, clearly, X_ε is a simple r.v. and

$$\begin{aligned}\int |X - X_\varepsilon| &= \int |(X^+ - X^-) - (X_{n_0} - Y_{n_0})| \\ &= \int |(X_{n_0} - X^+) - (Y_{n_0} - X^-)| \\ &\leq \int |X_{n_0} - X^+| + \int |Y_{n_0} - X^-| < \tfrac{\varepsilon}{2} + \tfrac{\varepsilon}{2} = \varepsilon. \#\end{aligned}$$

3. The assumption that $\int U_n \underset{n\to\infty}{\longrightarrow} \int U$ and $\int V_n \underset{n\to\infty}{\longrightarrow} \int V$ with $\int U$ and $\int V$ finite implies that $\int U_n$ and $\int V_n$ are finite for $n \geq n_0$, some n_0. Since $U_n \leq X_n \leq V_n$ a.e., $n \geq 1$, it follows that $\int U_n \leq \int X_n \leq \int V_n, n \geq 1$, so that $\int X_n$ are finite for $n \geq n_0$. Next, $U_n \leq X_n$ a.e. implies $0 \leq X_n - U_n$ a.e. and therefore (by Theorem 2(i)), $(0 \leq) \int \liminf_{n\to\infty}(X_n - U_n) \leq \liminf_{n\to\infty} \int (X_n - U_n)$, or $\int (X - U) \leq \liminf_{n\to\infty} \int X_n - \int U$, or $\int X - \int U \leq \liminf_{n\to\infty} \int X_n - \int U$, or $\int X \leq \liminf_{n\to\infty} \int X_n$. Also, $X_n \leq V_n$ a.e. implies $X_n - V_n \leq 0$ a.e. and therefore (by Theorem 2(ii)) $\limsup_{n\to\infty} \int (X_n - V_n) \leq \int \limsup_{n\to\infty}(X_n - V_n)$, or $\limsup_{n\to\infty} \int X_n - \int V \leq \int (X - V)$, or $\limsup_{n\to\infty} \int X_n - \int V \leq \int X - \int V$, or $\limsup_{n\to\infty} \int X_n \leq \int X$. Thus $\int X \leq \liminf_{n\to\infty} \int X_n \leq \limsup_{n\to\infty} \int X_n \leq \int X$, so that $\lim_{n\to\infty} \int X_n$ exists and is equal to $\int X$ finite. #

4. **(i)** The assumption $|X_n| \leq U_n$ a.e., $n \geq 1$, is equivalent to $-U_n \leq X_n \leq U_n$ a.e., $n \geq 1$, and all assumptions of Exercise 3 are satisfied with U_n replaced by $-U_n$ and V_n replaced by U_n. Then $\int X_n \underset{n\to\infty}{\longrightarrow} \int X$ finite.

(ii) From $|X_n| \leq U_n$ a.e., $n \geq 1$, and $|X_n| \overset{a.e.}{\underset{n\to\infty}{\longrightarrow}} |X|$ and $U_n \overset{a.e.}{\underset{n\to\infty}{\longrightarrow}} U$, we get $|X| \leq U$ a.e., so that:

$$0 \leq |X_n - X| \leq |X_n| + |X| \leq U_n + |X| \leq U_n + U \text{ a.e., } n \geq 1.$$

Set $Z_n = U_n + U$ and $Z = 2U$. Then:

$$0 \leq |X_n - X| \leq Z_n \text{ a.e., } n \geq 1, \text{ and} \int Z_n \underset{n\to\infty}{\longrightarrow} \int Z \text{ finite.}$$

Since also $|X_n - X| \overset{a.e.}{\underset{n\to\infty}{\longrightarrow}} 0$, Exercise 3 applies and gives $\int |X_n - X| \underset{n\to\infty}{\longrightarrow} 0$.

(iii) Finally, for any $A \in \mathcal{A}$, $|\int_A X_n - \int_A X| = |\int_A (X_n - X)| \leq \int_A |X_n - X| \leq \int |X_n - X|$, independent of A and tending to 0 as $n \to \infty$ by part (ii). The proof is completed. #

5. **(i)** By the hint, it suffices to show that for each $\{m\} \subseteq \{n\}$ there exists $\{r\} \subseteq \{m\}$ such that $\int X_r \underset{r\to\infty}{\longrightarrow} \int X$. For $\{m\} \subseteq \{n\}$, look at $\{X_m\}$. Since $X_m \overset{\mu}{\underset{m\to\infty}{\longrightarrow}} X$, there exists $\{r\} \subseteq \{m\}$ such that $X_r \overset{a.e.}{\underset{r\to\infty}{\longrightarrow}} X$. Also, since $U_r \overset{\mu}{\underset{r\to\infty}{\longrightarrow}} U$, there exists $\{s\} \subseteq \{r\}$ such that $U_s \underset{s\to\infty}{\overset{a.e.}{\longrightarrow}} U$, and, of course, $X_s \underset{s\to\infty}{\overset{a.e.}{\longrightarrow}} X$. Thus: $|X_s| \leq U_s$ a.s., all s, $X_s \underset{s\to\infty}{\overset{a.e.}{\longrightarrow}} X$, $U_s \underset{s\to\infty}{\overset{a.e.}{\longrightarrow}} U$ and $\int U_s \underset{s\to\infty}{\longrightarrow} \int U$ finite. Then, by Exercise 4(i), $\int X_s \underset{s\to\infty}{\longrightarrow} \int X$ finite. That is, for each $\{m\} \subseteq \{n\}$ there exists $\{s\} \subseteq \{m\}$ such that $\int X_s \underset{s\to\infty}{\longrightarrow} \int X$ finite, so that $\int X_n \underset{n\to\infty}{\longrightarrow} \int X$ finite.

(ii) As in part (i), from $\{n\}$ pass to a subsequence $\{s\}$ for which $|X_s| \leq U_s$ a.e., all s, $|X_s| \underset{s\to\infty}{\overset{a.e.}{\longrightarrow}} |X|$, and $U_s \underset{s\to\infty}{\overset{a.e.}{\longrightarrow}} U$. Then work as in part (ii) of Exercise 4 in order to conclude that $\int |X_s - X| \underset{s\to\infty}{\longrightarrow} 0$. In turn, this implies, by the hint, that $\int |X_n - X| \underset{n\to\infty}{\longrightarrow} 0$.

(iii) Exactly as in Exercise 4(iii). #

6. Suppose that for any $\varepsilon > 0$, there exists $\delta = \delta(\varepsilon) > 0$ such that $\mu(A) < \delta$ implies $\phi(A) < \varepsilon$. Then, if $\mu(A) = 0$, it follows that, for every $\varepsilon > 0$ and any $\delta > 0$, we have $\mu(A) < \delta$ and $\phi(A) < \varepsilon$. Letting $\varepsilon \to 0$, we get $\phi(A) = 0$. So, $\mu(A) = 0$ implies $\phi(A) = 0$.

Next, assume that $\phi \ll \mu$, and suppose that the statement "for every $\varepsilon > 0$, there exists $\delta = \delta(\varepsilon) > 0$ such that $\mu(A) < \delta$ implies $\phi(A) < \varepsilon$" is violated. This would imply that, for some $\varepsilon > 0$, there does not exist a $\delta = \delta(\varepsilon) > 0$ such that $\mu(A) < \delta$ implies $\phi(A) < \varepsilon$. Equivalently, for this $\varepsilon > 0$ and every $\delta > 0$, there exists at least one A such that $\mu(A) < \delta$ implies $\phi(A) \geq \varepsilon$. Take $\delta_n = \frac{1}{2^n}$. Then there exists A_n with $\mu(A_n) < \frac{1}{2^n}$ and $\phi(A_n) \geq \varepsilon$. Set $A = \limsup_{n\to\infty} A_n = \cap_{n=1}^{\infty} \cup_{k=n}^{\infty} A_k$. Then:

$$\mu(A) = \mu\left(\bigcap_{n=1}^{\infty} \bigcup_{k=n}^{\infty} A_k\right) \leq \mu\left(\bigcup_{k=n}^{\infty} A_k\right) \leq \sum_{k=n}^{\infty} \mu(A_k) = \frac{1}{2^{n-1}} \underset{n\to\infty}{\longrightarrow} 0.$$

That is, $\mu(A) = 0$. However,

$$\phi(A) = \phi\left(\bigcap_{n=1}^{\infty} \bigcup_{k=n}^{\infty} A_k\right) = \phi\left(\lim_{n\to\infty} \bigcup_{k=n}^{\infty} A_k\right) = \lim_{n\to\infty} \phi\left(\bigcup_{k=n}^{\infty} A_k\right),$$

as follows from the σ-additivity of ϕ (see the proof of Theorem 2 in Chapter 2). Since $\phi\left(\cup_{k=n}^{\infty} A_k\right) \geq \phi(A_n)$, by taking the limits, we get:

$$\lim_{n\to\infty} \phi\left(\bigcup_{k=n}^{\infty} A_k\right) \geq \limsup_{n\to\infty} \phi(A_n) \geq \varepsilon.$$

Thus, $\mu(A) = 0$ whereas $\phi(A) \geq \varepsilon$, which is a contradiction to our assumption that $\phi \ll \mu$. #

7. **(i)** $\omega_2 \in E_{\omega_1}$ implies that $(\omega_1, \omega_2) \in E$, so that $(\omega_1, \omega_2) \in F$ and then $\omega_2 \in F_{\omega_1}$. Thus, $E_{\omega_1} \subseteq F_{\omega_1}$. Similarly, $E_{\omega_2} \subseteq F_{\omega_2}$.

(ii) If $E_{\omega_1} \cap F_{\omega_1} \neq \oslash$, there would exist $\omega_2 \in E_{\omega_1} \cap F_{\omega_1}$, so that $\omega_2 \in E_{\omega_1}$ and $\omega_2 \in F_{\omega_1}$ which imply that $(\omega_1, \omega_2) \in E$ and $(\omega_1, \omega_2) \in F$, a contradiction. Thus, $E_{\omega_1} \cap F_{\omega_1} = \oslash$, and similarly, $E_{\omega_2} \cap F_{\omega_2} = \oslash$.

(iii) $\omega_2 \in \left(\cup_{n=1}^{\infty} E_n\right)_{\omega_1}$ implies $(\omega_1, \omega_2) \in \cup_{n=1}^{\infty} E_n$, so that $(\omega_1, \omega_2) \in E_n$ for at least one n, n_0, say. Then $\omega_2 \in E_{n_0,\omega_1}$ and hence $\omega_2 \in \cup_{n=1}^{\infty} E_{n,\omega_1}$. Thus, $\left(\cup_{n=1}^{\infty} E_n\right)_{\omega_1} \subseteq \cup_{n=1}^{\infty} E_{n,\omega_1}$. Next, let $\omega_2 \in \cup_{n=1}^{\infty} E_{n,\omega_1}$. Then $\omega_2 \in E_{n,\omega_1}$ for at least one n, n_0, say, so that $(\omega_1, \omega_2) \in E_n$ and hence $(\omega_1, \omega_2) \in \cup_{n=1}^{\infty} E_n$. But then $\omega_2 \in \left(\cup_{n=1}^{\infty} E_n\right)_{\omega_1}$, so that $\cup_{n=1}^{\infty} E_{n,\omega_1} \subseteq \left(\cup_{n=1}^{\infty} E_n\right)_{\omega_1}$, and the result follows.

The special case follows from what it was just established and part (ii). Similarly, for the Ω_1-sections at ω_2.

(iv) If $\omega_2 \in \left(\cap_{n=1}^{\infty} E_n\right)_{\omega_1}$, then $(\omega_1, \omega_2) \in \cap_{n=1}^{\infty} E_n$, so that $(\omega_1, \omega_2) \in E_n, n \geq 1$, and then $\omega_2 \in E_{n,\omega_1}, n \geq 1$, or $\omega_2 \in \cap_{n=1}^{\infty} E_{n,\omega_1}$. Next, if $\omega_2 \in \cap_{n=1}^{\infty} E_{n,\omega_1}$, then $\omega_2 \in E_{n,\omega_1}, n \geq 1$, so that $(\omega_1, \omega_2) \in E_n, n \geq 1$, and $(\omega_1, \omega_2) \in \cap_{n=1}^{\infty} E_n$ or $\omega_2 \in (\cap_{n=1}^{\infty} E_n)_{\omega_1}$. The asserted equality follows, and similarly for the Ω_1-sections at ω_2.

(v) If $\omega_2 \in (E^c)_{\omega_1}$, then $(\omega_1, \omega_2) \in E^c$, so that $(\omega_1, \omega_2) \notin E$, and this implies that $\omega_2 \notin E_{\omega_1}$, so that $\omega_2 \in (E_{\omega_1})^c$. On the other hand, if $\omega_2 \in (E_{\omega_1})^c$, then $\omega_2 \notin E_{\omega_1}$, so that $(\omega_1, \omega_2) \notin E$ and hence $(\omega_1, \omega_2) \in E^c$, so that $\omega_2 \in (E^c)_{\omega_1}$. The asserted equality follows, and similarly for the Ω_1-sections at ω_2. #

8. **(i)** $\mu_0 \geq 0$ and $\mu_0(\emptyset) = 0$. It is also σ-additive since

$$\mu_0\left(\sum_{n=1}^{\infty} A_n\right) = \mu\left[\left(\sum_{n=1}^{\infty} A_n\right) \cap C\right] = \mu\left[\sum_{n=1}^{\infty} (A_n \cap C)\right]$$
$$= \sum_{n=1}^{\infty} \mu\left(A_n \cap C\right) = \sum_{n=1}^{\infty} \mu_0\left(A_n\right).$$

(ii) First, let $X = I_A$, $A \in \mathcal{A}$. Since $\int X d\mu$ exists, it follows that $\int_C X d\mu$ also exists (Corollary to Theorem 5 in Chapter 4), and

$$\int_C X d\mu = \int_C I_A d\mu = \mu(A \cap C) = \mu_0(A) = \int I_A d\mu_0 = \int X d\mu_0.$$

So, $\int X d\mu_0$ exists and $\int X d\mu_0 = \int_C X d\mu$. Next, let $X = \sum_{i=1}^{r} \alpha_i I_{A_i}$, where $\alpha_i \geq 0$ for $i = 1, \ldots, r$. Then

$$\int_C X d\mu = \int_C \left(\sum_{i=1}^{r} \alpha_i I_{A_i}\right) d\mu = \sum_{i=1}^{r} \alpha_i \int_C I_{A_i} d\mu$$
$$= \sum_{i=1}^{r} \alpha_i \int I_{A_i} d\mu_0 \text{ (by the part just proved)}$$
$$= \int \left(\sum_{i=1}^{r} \alpha_i I_{A_i}\right) d\mu_0 = \int X d\mu_0.$$

So, again, $\int X d\mu_0$ exists and equals $\int_C X d\mu$.
For $X \geq 0$, there exists $0 \leq X_n$ simple r.v.s $\underset{n\to\infty}{\uparrow} X$. Then $0 \leq X_n I_C$ simple r.v.s $\underset{n\to\infty}{\uparrow} X I_C$ and

$$\int_C X_n d\mu = \int (X_n I_C) d\mu \underset{n\to\infty}{\longrightarrow} \int (X I_C) d\mu, \quad \int X_n d\mu_0 \underset{n\to\infty}{\longrightarrow} \int X d\mu_0.$$

However, by the previous part, $\int_C X_n d\mu = \int X_n d\mu_0$ for all n, so that $\int_C X d\mu = \int X d\mu_0$. Once again, $\int X d\mu_0$ exists and equals $\int_C X d\mu$. Finally, for any X, consider $X = X^+ - X^-$, and suppose that $\int X^+ d\mu < \infty$. Then $\int_C X^+ d\mu$ is also finite, and $\int_C X^+ d\mu = \int X^+ d\mu_0$ by the previous part. Thus, $\int X d\mu_0$ exists. Since also $\int_C X^- d\mu = \int X^- d\mu_0$ (whether finite or not), it follows that

$$\int_C X d\mu = \int_C X^+ d\mu - \int_C X^- d\mu = \int X^+ d\mu_0 - \int X^- d\mu_0 = \int X d\mu_0.$$

Likewise, if we assume that $\int X^- d\mu < \infty$. #

9. Indeed, since $\int X d\mu$ exists, so does $\int_{A_i} X d\mu$ for all i. By Exercise 8, $\int_{A_i} X d\mu = \int X d\mu_i$. At this point, assume first that $X \geq 0$. Then

$$\begin{aligned}
\sum_{i=1}^{\infty} \int X d\mu_i = \sum_{i=1}^{\infty} \int_{A_i} X d\mu &= \sum_{i=1}^{n} \int (X I_{A_i}) d\mu \\
&= \int (\sum_{i=1}^{\infty} X I_{A_i}) d\mu \text{ (by Corollary 1 to Theorem 1)} \\
&= \int X d\mu \text{ (because } \{A_i, i \geq 1\} \text{ is a partition of } \Omega).
\end{aligned}$$

Next, for any X for which the integral $\int X d\mu$ exists, write $X = X^+ - X^-$, and suppose that $\int X^+ d\mu < \infty$. By the previous step,

$$\sum_{i=1}^{\infty} \int X^+ d\mu_i = \int X^+ d\mu \text{ and } \sum_{i=1}^{\infty} \int X^- d\mu_i = \int X^- d\mu.$$

The finiteness of $\int X^+ d\mu$ implies finiteness of $\int X^+ d\mu_i$ for each i. Thus, the integral $\int X^+ d\mu_i - \int X^- d\mu_i$ exist for all i, and

$$\begin{aligned}
\int X d\mu = \int X^+ d\mu - \int X^- d\mu &= \sum_{i=1}^{\infty} \int X^+ d\mu_i - \sum_{i=1}^{\infty} \int X^- d\mu_i \\
&= \sum_{i=1}^{\infty} \left(\int X^+ d\mu_i - \int X^- d\mu_i \right) \\
&= \sum_{i=1}^{\infty} \int (X^+ - X^-) d\mu_i
\end{aligned}$$

$$= \sum_{i=1}^{\infty} \int X d\mu_i, \text{ or}$$

$$\sum_{i=1}^{\infty} \int X d\mu_i = \int X d\mu.$$

Likewise, if we suppose that $\int X^- d\mu < \infty$. #

10. Let $X = \sum_{i=1}^{n} x_i I_{E_i}$ where $\{E_1, \ldots, E_n\}$ is a partition of $\Omega_1 \times \Omega_2$. Then $\int X d\lambda = \sum_{i=1}^{n} x_i \lambda(E_i)$ and its existence means that the x_is corresponding to E_is with $\lambda(E_i) = \infty$ (if such E_is exist) are either all positive or all negative. Next, for $\omega_2 \in \Omega_2$ and $\omega_1 \in \Omega_1$, consider the respective sections E_{i,ω_2} and E_{i,ω_1}, which (by Theorem 8) are measurable with respect to the appropriate σ-fields. Furthermore,

$$\Omega_1 = \sum_{i=1}^{n} E_{i,\omega_2}, \quad \Omega_2 = \sum_{i=1}^{n} E_{i,\omega_1};$$

this is so by Exercise 7(iii). Next, consider the (simple) r.v.s:

$$X(\cdot, \omega_2) = X_{\omega_2}(\cdot) = \sum_{i=1}^{n} x_i I_{E_{i,\omega_2}}, \quad X(\omega_1, \cdot) = X_{\omega_1}(\cdot) = \sum_{i=1}^{n} x_i I_{E_{i,\omega_1}}.$$

Then the integrals $\int X_{\omega_2}(\cdot) d\mu_1$ and $\int X_{\omega_1}(\cdot) d\mu_2$ exist, are $\mathcal{A}_2$-measurable and $\mathcal{A}_1$-measurable, respectively, and

$$\int X_{\omega_2}(\cdot) d\mu_1 = \sum_{i=1}^{n} x_i \mu_1(E_i, \omega_2), \quad \int X_{\omega_1}(\cdot) d\mu_2 = \sum_{i=1}^{n} x_i \mu_2(E_i, \omega_1).$$

The existence of $\int X_{\omega_2}(\cdot) d\mu_1$ follows from the fact that, if $\lambda(E_i) < \infty$, then $\mu_1(E_i, \omega_2) < \infty$ a.e. $[\mu_2]$, because otherwise $\int \mu_1(E_i, \omega_2) d\mu_2 = \lambda(E_i) = \infty$, a contradiction. So, $\mu_1(E_i, \omega_2)$ may be ∞ only if $\lambda(E_i) = \infty$. But then the respective x_is are either all positive or all negative, so that $\int X_{\omega_2}(\cdot) d\mu_1$ exists. The same for $\int X_{\omega_1}(\cdot) d\mu_2$. Their measurability follows from Theorem 10. Next,

$$\int [\int X_{\omega_2}(\cdot) d\mu_1] d\mu_2 = \sum_{i=1}^{n} x_i \int \mu_1(E_{i,\omega_2}) d\mu_2,$$

$$\int [\int X_{\omega_1}(\cdot) d\mu_2] d\mu_1 = \sum_{i=1}^{n} x_i \int \mu_2(E_{i,\omega_1}) d\mu_1,$$

and $\int \mu_1(E_{i,\omega_2}) d\mu_2 = \int \mu_2(E_{i,\omega_1}) d\mu_1 = \lambda(E_i)$, $i = 1, \ldots, n$, by Theorem 10 and the definition of λ. Thus,

$$\int [\int X_{\omega_2}(\cdot) d\mu_1] d\mu_2 = \sum_{i=1}^{n} x_i \int \mu_1(E_i, \omega_2) d\mu_2$$

$$= \sum_{i=1}^{n} x_i \int \mu_2(E_{i,\omega_1}) d\mu_1 \text{ (and the common value of the}$$

two expressions in the middle is $\sum_{i=1}^{n} x_i \lambda(E_i)$

$$= \int \int X(\omega_1, \omega_2) d\lambda)$$

$$= \int [\int X_{\omega_1}(\cdot) d\mu_2] d\mu_1. \#$$

11. Since $\int X d\lambda = \int X^+ d\lambda - \int X^- d\lambda$ and $\int X d\lambda$ exists, it follows that at least one of $\int X^+ d\lambda$ or $\int X^- d\lambda$ is finite. Let $\int X^+ d\lambda < \infty$. Since X^+ is integrable, it follows, by Theorem 12, that:

$$\int \int X^+ d\mu_1 d\mu_2 = \int \int X^+ d\mu_2 d\mu_1 = \int X^+ d\lambda,$$

and $\int \int X^+ d\mu_1 d\mu_2 < \infty, \int \int X^+ d\mu_2 d\mu_1 < \infty$. Also, since $X^- \geq 0$, it follows, by Theorem 12 again, that:

$$\int \int X^- d\mu_1 d\mu_2 = \int \int X^- d\mu_2 d\mu_1 = \int X^- d\lambda.$$

Combining these results, we get

$$\int \int X d\mu_1 d\mu_2 = \int \int X^+ d\mu_1 d\mu_2 - \int \int X^- d\mu_1 d\mu_2$$
$$= \int X^+ d\lambda - \int X^- d\lambda = \int X d\lambda,$$

and

$$\int \int X d\mu_2 d\mu_1 = \int \int X^+ d\mu_2 d\mu_1 - \int \int X^- d\mu_2 d\mu_1$$
$$= \int X^+ d\lambda - \int X^- d\lambda = \int X d\lambda;$$

i.e., the result holds. Similarly, if $\int X^- d\lambda < \infty$. #

12. We have $\frac{\sqrt{n}(\bar{X}_n - \mu)}{\sigma} \xrightarrow[n\to\infty]{d} Z$ and $\frac{\sigma}{\sqrt{n}} \xrightarrow[n\to\infty]{} 0$, so that, by Theorem 7 (ii), $\bar{X}_n - \mu \xrightarrow[n\to\infty]{d} 0$ or $\bar{X}_n \xrightarrow[n\to\infty]{d} \mu$, and hence $\bar{X}_n \xrightarrow[n\to\infty]{P} \mu$, by Theorem 6. #

13. The positive and negative parts of f are as follows:

$$f^+(x) = \begin{cases} 2n, & \frac{1}{2n+1} < x \leq \frac{1}{2n} \\ 0, & \text{otherwise} \end{cases} \qquad (n = 1, 2, \ldots)$$

$$f^-(x) = \begin{cases} 2n+1, & \frac{1}{2n+2} < x \leq \frac{1}{2n+1} \\ 0, & \text{otherwise.} \end{cases} \qquad (n = 0, 1, \ldots)$$

Then

$$\int_{(0,1]} f^+ d\lambda = \int_{\sum_{n=1}^{\infty}(\frac{1}{2n+1},\frac{1}{2n}]} (2n) d\lambda = \sum_{n=1}^{\infty} \int_{(\frac{1}{2n+1},\frac{1}{2n}]} (2n) d\lambda$$

(by Corollary 1(ii) to Theorem 1)

$$= \sum_{n=1}^{\infty} (2n) \times (\frac{1}{2n} - \frac{1}{2n+1}) = \sum_{n=1}^{\infty} \frac{2n}{(2n)(2n+1)}$$

$$= \sum_{n=1}^{\infty} \frac{1}{2n+1} = \infty.$$

Also,

$$\int_{(0,1]} f^- d\lambda = \int_{\sum_{n=0}^{\infty}(\frac{1}{2n+2},\frac{1}{2n+1}]} (2n+1) d\lambda = \sum_{n=0}^{\infty} \int_{(\frac{1}{2n+2},\frac{1}{2n+1}]} (2n+1) d\lambda$$

$$= \sum_{n=0}^{\infty} (2n+1) \times (\frac{1}{2n+1} - \frac{1}{2n+2})$$

$$= \sum_{n=0}^{\infty} \frac{2n+1}{2(n+1)(2n+1)} = \frac{1}{2} \sum_{n=0}^{\infty} \frac{1}{n+1} = \infty.$$

So, $\int_{(0,1]} f^+ d\lambda = \int_{(0,1]} f^- d\lambda = \infty$, and therefore the integral $\int_{(0,1]} f d\lambda$ does not exist.

Furthermore, since $|f| = f^+ + f^-$, it follows that $\int_{(0,1]} |f| d\lambda = \infty$. #

14. By Exercise 14 in Chapter 4,

$$P(|X_n - X| \geq \varepsilon) \leq (1 + \frac{1}{\varepsilon}) \mathcal{E}\left(\frac{|X_n - X|}{1 + |X_n - X|}\right).$$

Thus, if $\mathcal{E}\left(\frac{|X_n-X|}{1+|X_n-X|}\right) \underset{n\to\infty}{\longrightarrow} 0$, then $X_n \underset{n\to\infty}{\overset{P}{\longrightarrow}} X$. Next, let $X_n \underset{n\to\infty}{\overset{P}{\longrightarrow}} X$. Then $\frac{|X_n-X|}{1+|X_n-X|} \underset{n\to\infty}{\overset{P}{\longrightarrow}} 0$ (by Exercise 8 in Chapter 3) and $(0 \leq)\frac{|X_n-X|}{1+|X_n-X|} \leq 1$ for all n, integrable. Then by part (b) of Theorem 3, $\mathcal{E}\left(\frac{|X_n-X|}{1+|X_n-X|}\right) \underset{n\to\infty}{\longrightarrow} 0$, as was to be seen. #

Alternate proof (of one direction) without reference to Exercise 14 in Chapter 4:

That $\mathcal{E}\left(\frac{|X_n-X|}{1+|X_n-X|}\right) \underset{n\to\infty}{\longrightarrow} 0$ implies that $P(|X_n - X| \geq \varepsilon) \underset{n\to\infty}{\longrightarrow} 0$, may be seen as follows without employing Exercise 14 in Chapter 4. Namely, for $\varepsilon > 0$,

$$\mathcal{E}\left(\frac{|X_n - X|}{1 + |X_n - X|}\right) = \int_{(|X_n - X| \geq \varepsilon)} \frac{|X_n - X|}{1 + |X_n - X|} dP$$

$$+\int_{(|X_n-X|<\varepsilon)} \frac{|X_n - X|}{1+|X_n - X|}\, dP$$
$$\leq \int_{(|X_n-X|\geq\varepsilon)} dP + \frac{\varepsilon}{1+\varepsilon}$$

(since, for $x > 0$, the function $g(x) = \frac{x}{1+x} = 1 - \frac{1}{1+x}$ is increasing), and the right-hand side above

$$= P(|X_n - X| \geq \varepsilon) + \frac{\varepsilon}{1+\varepsilon}.$$

Letting $n \to \infty$, we get

$$0 = \lim \mathcal{E}\left(\frac{|X_n - X|}{1+|X_n - X|}\right) \leq \limsup P(|X_n - X| \geq \varepsilon) + \frac{\varepsilon}{1+\varepsilon}.$$

Letting $\varepsilon \to 0$, we obtain the desired result. #

15. All limits are taken as $n \to \infty$. Then:

$$\lim X_{2n+1} = 1 \text{ on } \left(0, \frac{1}{2}\right), \text{ and } \lim X_{2n+1} = 0 \text{ on } \left[\frac{1}{2}, 1\right).$$

Also,

$$\lim X_{2n+2} = 0 \text{ on } \left(0, \frac{1}{2}\right), \text{ and } \lim X_{2n+2} = 1 \text{ on } \left[\frac{1}{2}, 1\right).$$

It follows that

$$\liminf X_n = 0, \qquad \limsup X_n = 1,$$

and

$$\int \liminf X_n = 0, \qquad \int \limsup X_n = 1.$$

Next, $\int X_{2n+1} = 1 \times \frac{1}{2} = \frac{1}{2}$, $\int X_{2n+2} = 1 \times \frac{1}{2} = \frac{1}{2}$, so that $\int X_n = \frac{1}{2}$ for all n. It follows that:

$$\liminf \int X_n = \limsup \int X_n = \lim \int X_n = \frac{1}{2},$$

and

$$\int \liminf X_n = 0 < \frac{1}{2} = \liminf \int X_n,$$
$$\limsup \int X_n = \frac{1}{2} < 1 = \int \limsup X_n. \ \#$$

16. X is, indeed, a r.v. since $X^{-1}(\{1/2^\omega\}) = \{\omega\}$ is in $\mathcal{A}$, and

$$\int_\Omega X d\mu = \int_{\sum_{\omega=0}^\infty \{\omega\}} X d\mu = \sum_{\omega=0}^\infty \int_{\{\omega\}} X d\mu \text{ (by Corollary 1(ii) to Theorem 1)}$$
$$= \sum_{\omega=0}^\infty \frac{1}{2^\omega} \times 1 = \sum_{\omega=0}^\infty \frac{1}{2^\omega} = 1 + \frac{1}{2} + \frac{1}{2^2} + \cdots$$
$$= \frac{1}{1-\frac{1}{2}} = 2. \ \#$$

17. This exercise is a reformulation of Exercise 3 in this chapter in reference to the space $(\Re, \mathcal{B}^k, \lambda^k)$ rather than an abstract (σ-finite) measure space $(\Omega, \mathcal{A}, \mu)$. #

18.
(i) The convergences $X_n \xrightarrow[n\to\infty]{d} X$ and $Y_n - X_n \xrightarrow[n\to\infty]{P} c$ imply (by Theorem 7(i)) that $Y_n \xrightarrow[n\to\infty]{d} X + c$.

(ii) Again, $X_n \xrightarrow[n\to\infty]{d} X$ and $Y_n \xrightarrow[n\to\infty]{P} c$ imply (by Theorem 7(ii)) that $X_n Y_n \xrightarrow[n\to\infty]{d} cX$. #

19. Let $A_n \in \mathcal{A}$ such that $P(A_n) \xrightarrow[n\to\infty]{} 0$. Then, as $n \to \infty$, $P(I_{A_n} \geq \varepsilon) \leq \frac{1}{\varepsilon}\mathcal{E} I_{A_n} = \frac{1}{\varepsilon} P(A_n) \to 0$, and therefore $|X| I_{A_n} \xrightarrow{P} 0$. Next, $|X| I_{A_n} \leq |X|$ independent of n and integrable. Therefore, by the Dominated Convergence Theorem,

$$\int (|X|\, I_{A_n}) dP = \int_{A_n} |X| dP \xrightarrow[n\to\infty]{} 0.$$

So, if $P(A_n) \to 0$, then $\nu(A_n) \to 0$, as $n \to \infty$. Thus, for every $\varepsilon > 0$, there is $\delta = \delta(\varepsilon)(> 0)$ such that $P(A_n) < \delta$ implies $\nu(A_n) < \varepsilon$ for sufficiently large n. It follows that, if $P(A) < \delta$ then $\nu(A) < \varepsilon$, as was to be seen. #

20.
(i) Indeed,

$$\mathcal{E}X = \int_\Omega X dP = \int_0^\infty x dP_X = \int_0^\infty \left(\int_0^x dt\right) dP_X$$
$$= \int_0^\infty \left[\int_0^\infty I_{[0,x]}(t) dt\right] dP_X = \int_0^\infty \int_0^\infty I_{[0,x]}(t) dt dP_X$$
$$= \int_0^\infty \int_0^\infty I_{[0,x]}(t) dP_X dt \text{ (since the integrand is } \geq 0).$$

But $\int_0^\infty I_{[0,x]}(t) dP_X = \int_t^\infty dP_X = P(X \geq t)$, so that $\mathcal{E}X = \int_0^\infty P(X \geq t) dt$.

(ii) For the first case, $\mathcal{E}X = \int_0^\infty x\lambda e^{-\lambda x}dx = -\int_0^\infty x de^{-\lambda x} = -xe^{-\lambda x}\big|_0^\infty + \int_0^\infty e^{-\lambda x}dx = \int_0^\infty e^{-\lambda x}dx = -\frac{1}{\lambda}\int_0^\infty de^{-\lambda x} = -\frac{1}{\lambda}e^{-\lambda x}\Big|_0^\infty = \frac{1}{\lambda}$, whereas

$$P(X \geq t) = \int_t^\infty \lambda e^{-\lambda x}dx = -\int_t^\infty de^{-\lambda x} = -e^{-\lambda x}\Big|_0^\infty = e^{-\lambda t},$$

so that $\int_0^\infty P(X \geq t)dt = \int_0^\infty e^{-\lambda t}dt = \frac{1}{\lambda}\int_0^\infty e^{-\lambda x}d(\lambda x) = -\frac{1}{\lambda}e^{-\lambda x}\Big|_0^\infty = \frac{1}{\lambda}(= \mathcal{E}X)$. For the second case, $\mathcal{E}X = \frac{1}{2}$, whereas, for $0 \leq t \leq 1$,

$$P(X \geq t) = \int_t^1 dx = 1 - t,$$

so that

$$\int_0^\infty P(X \geq t)dt = \int_0^1 (1-t)dt = t\Big|_0^1 - \frac{t^2}{2}\Big|_0^1 = \frac{1}{2}(= \mathcal{E}X). \#$$

Chapter 6

Standard Moment and Probability Inequalities, Convergence in the *r*th Mean and its Applications

1. The inequality has been established for $n = 2$. Assume it to be true for $n = k$ and establish it for $n = k+1$. For $r_1, \ldots, r_{k+1} > 0$ with $\frac{1}{r_1} + \ldots + \frac{1}{r_{k+1}} = 1$, define r by: $\frac{1}{r} = \frac{1}{r_1} + \ldots + \frac{1}{r_k} (< 1)$, so that $\frac{1}{r} + \frac{1}{r_{k+1}} = 1$. Also, set $r_i' = \frac{r_i}{r}$, $i = 1, \ldots, k$, so that $r_i' > 0$, $i = 1, \ldots, k$, and $\frac{1}{r_1'} + \ldots + \frac{1}{r_k'} = r(\frac{1}{r_1} + \ldots + \frac{1}{r_k}) = r \times \frac{1}{r} = 1$. Then:

$$\begin{aligned}
\mathcal{E}|X_1 \ldots X_{k+1}| &= \mathcal{E}|(X_1 \ldots X_k)X_{k+1}| \leq \mathcal{E}^{\frac{1}{r}}|X_1 \ldots X_k|^r \mathcal{E}^{\frac{1}{r_{k+1}}}|X_{k+1}|^{r_{k+1}} \\
&= \mathcal{E}^{\frac{1}{r}}(|X_1|^r \ldots |X_k|^r)\mathcal{E}^{\frac{1}{r_{k+1}}}|X_{k+1}|^{r_{k+1}} \\
&\leq [\mathcal{E}^{\frac{1}{r_1'}}(|X_1|^r)^{r_1'} \ldots \mathcal{E}^{\frac{1}{r_k'}}(|X_k|^r)^{r_k'}]^{\frac{1}{r}}\mathcal{E}^{\frac{1}{r_{k+1}}}|X_{k+1}|^{r_{k+1}} \\
&\quad \text{(by the induction hypothesis)} \\
&= \mathcal{E}^{\frac{1}{r_1}}|X_1|^{r_1} \ldots \mathcal{E}^{\frac{1}{r_k}}|X_k|^{r_k}\mathcal{E}^{\frac{1}{r_{k+1}}}|X_{k+1}|^{r_{k+1}} \text{ (since } rr_i' = r_i, i = 1, \ldots, k).
\end{aligned}$$

Thus,

$$\mathcal{E}|X_1 \ldots X_{k+1}| \leq \mathcal{E}^{\frac{1}{r_1}}|X_1|^{r_1} \ldots \mathcal{E}^{\frac{1}{r_{k+1}}}|X_{k+1}|^{r_{k+1}}. \#$$

2. **(i)** For an arbitrary but fixed $x_0 \in I$, let $x_1, x_2 \in I$ be such that $x_1 < x_0 < x_2$ and set $\alpha = \frac{x_2 - x_0}{x_2 - x_1}$, $\beta = \frac{x_0 - x_1}{x_2 - x_1}$. Then $\alpha, \beta \geq 0$ and $\alpha + \beta = 1$. Also, $\alpha x_1 + \beta x_2 = x_0$. Then by convexity, $g(x_0) \leq \frac{x_2 - x_0}{x_2 - x_1}g(x_1) + \frac{x_0 - x_1}{x_2 - x_1}g(x_2)$.

By letting $x_2 \downarrow x_0$, we get: $g(x_0) \leq \underline{\lim}_{x_2 \downarrow x_0} g(x_2)$. Next, let $x_1 < x_2 < x_0$ and take $\alpha = \frac{x_0 - x_2}{x_0 - x_1}$, $\beta = \frac{x_2 - x_1}{x_0 - x_1}$, so that $\alpha, \beta \geq 0, \alpha + \beta = 1$, and $\alpha x_1 + \beta x_0 = x_2$. Therefore $g(x_2) \leq \frac{x_0 - x_2}{x_0 - x_1} g(x_1) + \frac{x_2 - x_1}{x_0 - x_1} g(x_0)$, and as $x_2 \uparrow x_0$, $\overline{\lim}_{x_2 \uparrow x_0} \leq g(x_0)$.
So, $\overline{\lim}_{x_2 \uparrow x_0} \leq g(x_0) \leq \underline{\lim}_{x_2 \downarrow x_0} g(x_0)$, so that $\lim_{x_2 \to x_0} g(x_2) = g(x_0)$ and g is continuous at x_0.

(ii) In $g(x_0) \leq \frac{x_2 - x_0}{x_2 - x_1} g(x_1) + \frac{x_0 - x_1}{x_2 - x_1} g(x_2)$ $(x_1 < x_2 < x_0)$, replace x_2 by x to get: $(x - x_1)g(x_0) \leq (x - x_0)g(x_1) + (x_0 - x_1)g(x)$, or $(x - x_1) \times g(x_0) - (x_0 - x_1)g(x_0) \leq (x - x_0)g(x_1) + (x_0 - x_1)g(x) - (x_0 - x_1) \times g(x_0)$, or $(x - x_0)g(x_0) \leq (x - x_0)g(x_1) + (x_0 - x_1)[g(x) - g(x_0)]$, or $(x_0 - x_1)[g(x) - g(x_0)] \geq (x - x_0)[g(x_0) - g(x_1)]$, or $g(x) - g(x_0) \geq \frac{g(x_0) - g(x_1)}{x_0 - x_1}(x - x_0)$. Take $x_1 = cx_0, c > 0$, so that $cx_0 \in I$. Then: $g(x) - g(x_0) = \lambda(x_0)(x - x_0)$, where $\lambda(x_0) = \frac{g(x_0) - g(x_1)}{x_0 - x_1}$. #

3. The inequality has been established for $n = 2$. Assume it to be true for $n = k$ and establish it for $n = k + 1$. Indeed,

$$\begin{aligned}
&\mathcal{E}^{\frac{1}{r}} |X_1 + \ldots + X_{k+1}|^r = \mathcal{E}^{\frac{1}{r}} |(X_1 + \ldots + X_k) + X_{k+1}|^r \\
&\leq \mathcal{E}^{\frac{1}{r}} |X_1 + \ldots + X_k|^r + \mathcal{E}^{\frac{1}{r}} |X_{k+1}|^r \\
&\leq (\mathcal{E}^{\frac{1}{r}} |X_1|^r + \ldots + \mathcal{E}^{\frac{1}{r}} |X_k|^r) + \mathcal{E}^{\frac{1}{r}} |X_{k+1}|^r \text{ (by the induction hypothesis)} \\
&\leq \mathcal{E}^{\frac{1}{r}} |X_1|^r + \ldots + \mathcal{E}^{\frac{1}{r}} |X_{k+1}|^r. \#
\end{aligned}$$

4. By the Markov inequality, $P(|X_n - X| \geq \varepsilon) \leq \varepsilon^{-r} \mathcal{E}|X_n - X|^r$, so that $\sum_{n=1}^{\infty} P(|X_n - X| \geq \varepsilon) \leq \varepsilon^{-r} \sum_{n=1}^{\infty} \mathcal{E}|X_n - X|^r < \infty$. Then Exercise 4(i) in Chapter 3 applies and gives the result. #

5. $X_n \xrightarrow[n\to\infty]{(r)} Y$ implies $X_n \xrightarrow[n\to\infty]{P} Y$. Also, $X_n \xrightarrow[n\to\infty]{a.s.} X$ implies $X_n \xrightarrow[n\to\infty]{P} X$. So, $X_n \xrightarrow[n\to\infty]{P} X$ and $X_n \xrightarrow[n\to\infty]{P} Y$, so that $P(X \neq Y) = 0$, by Theorem 1 in Chapter 3. #

6. Take $(\Omega, \mathcal{A}, P) = ((0, 1], \mathcal{B}_{(0,1]}, \lambda)$, where λ is the Lebesgue measure, and define X_n and X as follows:

$$X_n = \begin{cases} n \text{ on } (0, \frac{1}{n}] \\ 0 \text{ on } (\frac{1}{n}, 1] \end{cases}, n \geq 1; \quad X = 0 \text{ on } (0, 1].$$

Then, clearly, $X_n \xrightarrow[n\to\infty]{} 0$ pointwise and hence $X_n \xrightarrow[n\to\infty]{P} 0$. However, $\mathcal{E}|X_n - X|^r = \mathcal{E}X_n^r = n^r \times \frac{1}{n} = n^{r-1}$, so that $X_n \underset{n\to\infty}{\overset{(r)}{\nrightarrow}} 0$ for any $r \geq 1$. #

7. In reference to Exercise 2(ii) in Chapter 3, $Y_{nk} = 1$ on a set of probability $1/2^{n-1}$ and 0 otherwise. Set $X_n = Y_{mk}$ for some m and some $k = 1, \ldots, 2^{m-1}$. Then, as was seen in the exercise cited, $X_n \underset{n\to\infty}{\overset{a.s.}{\nrightarrow}} 0$ (indeed, it does not converge at any point), but $\mathcal{E}|X_n - X|^r = \mathcal{E}X_n^r = \mathcal{E}Y_{mk}^r = \frac{1}{2^{n-1}} \xrightarrow[n\to\infty]{} 0$ for any $r > 0$. #

8. Indeed, for $c > 0$, $\int_{(|X_n|^\alpha \geq c)} |X_n|^\alpha dP = \int_{(|X_n| \geq c^{\frac{1}{\alpha}})} |X_n|^\alpha dP = \int_{(|X_n| \geq c^{\frac{1}{\alpha}})} \times (|X_n|^\beta \times \frac{1}{|X_n|^{\beta-\alpha}}) dP \leq \frac{1}{c^{(\beta-\alpha)/\alpha}} \int |X_n|^\beta dP = \frac{1}{c^{(\beta-\alpha)/\alpha}} \mathcal{E}|X_n|^\beta \leq \frac{M}{c^{(\beta-\alpha)/\alpha}}$ independent of n and $\underset{n\to\infty}{\longrightarrow} 0$. #

9. For $M > 0$, we have:

$$\int_{(|X_n| \geq M)} |X_n| dP = \int_{(|X_n| \geq M)} X_n dP = \begin{cases} cn \times \frac{1}{n} = c \text{ for } M \leq cn \\ 0 \qquad\qquad \text{for } M > cn. \end{cases}$$

Then $\sup_{M>0} \int_{(|X_n| \geq M)} |X_n| dP = c$, so that $\limsup_{M\to\infty} \int_{(|X_n| \geq M)} |X_n| dP = c$, and this implies that $\int_{(|X_n| \geq M)} |X_n| d \underset{M\to\infty}{\not\rightarrow} 0$. Therefore the given r.v.s are not uniformly integrable. #

10. Here all limits are taken as $P(A) \to 0$, or we can pass to arbitrary sequences $\{A_n\}$ with $P(A_n) \underset{n\to\infty}{\longrightarrow} 0$. We elect the former. Now $P(I_A > \varepsilon) \leq \frac{EI_A}{\varepsilon} = \frac{P(A)}{\varepsilon} \to 0$, so that $I_A \overset{P}{\to} 0$ and hence $|X| I_A \overset{P}{\to} 0$. Also, $|X| I_A \leq |X|$ independent of A and integrable. Then, by Theorem 3(ii) in Chapter 5 (applied under assumption (b)), we have $\int |X| I_A \to 0$, or $\int_A |X| \to 0$. So, $\int_A |X| \to 0$ as $P(A) \to 0$. #

11. The assumptions $X_n \overset{P}{\to} X$ and $X_n \geq 0$ a.s. imply $X \geq 0$ a.s. (by passing to a subsequence $X_m \underset{m\to\infty}{\overset{a.s.}{\longrightarrow}} X$). Then the additional assumption that $\mathcal{E}X_n \underset{n\to\infty}{\longrightarrow} \mathcal{E}X$ finite implies that $X_n \underset{n\to\infty}{\overset{(1)}{\longrightarrow}} X$. This is so by Theorem 14. However, $0 \leq Y_n \leq X_n$ a.s. implies $0 \leq Y_n I_A \leq X_n I_A$ a.s. for any event A, and hence $\int_A Y_n dP \leq \int_A X_n dP$. But $X_n \underset{n\to\infty}{\overset{(1)}{\longrightarrow}} X$ implies that $\int X_n dP$ are uniformly continuous (by Theorem 13). Therefore so are $\int Y_n dP$. This fact along with the assumption $Y_n \underset{n\to\infty}{\overset{P}{\longrightarrow}} Y$ imply then that $Y_n \underset{n\to\infty}{\overset{(1)}{\longrightarrow}} Y$ (again by Theorem 13), so that $\mathcal{E}|Y_n - Y| \underset{n\to\infty}{\longrightarrow} 0$. #

12. Indeed, $\int |X|^r d\mu = \int_{(|X|^r \geq c)} |X|^r d\mu + \int_{(|X|^r < c)} |X|^r d\mu \geq \int_{(|X|^r \geq c)} |X|^r d\mu \geq c^r \mu(|X|^r \geq c)$, and hence $\mu(|X|^r \geq c) \leq c^{-r} \int |X|^r d\mu$. #

13. By the c_r-inequality,

$$\mathcal{E}|X-c|^r \leq c_r(\mathcal{E}|X|^r + |c|^r) \text{ and } \mathcal{E}|X|^r = \mathcal{E}|(X-c)+c|^r \leq c_r(\mathcal{E}|X-c|^r + |c|^r),$$

and the result follows. #

14. From the Tchebichev inequality, $P(|X-\mu| \geq c) \leq \frac{\sigma^2(X)}{c^2}$, we have $P(|X-\mu| \geq c) = 0$ for every $c > 0$, or $P(|X-\mu| < c) = 1$ for every $c > 0$. In particular, $P(|X-\mu| < \frac{1}{n}) = 1$ for every $n \geq 1$. Set $A_n = (|X-\mu| < \frac{1}{n})$, $n \geq 1$, and $A_0 = (|X-\mu| = 0) = (X = \mu)$. Then, clearly, $A_n \underset{n\to\infty}{\downarrow} A_0$, so that, by continuity of a probability measure, we have $P(A_n) \underset{n\to\infty}{\longrightarrow} P(A_0)$. Since $P(A_n) = 1$, $n \geq 1$, we obtain $P(A_0) = 1$. #

15. **(i)** If $|X_n| \underset{n\to\infty}{\overset{P}{\longrightarrow}} 0$, then $\frac{|X_n|}{1+|X_n|} \underset{n\to\infty}{\overset{P}{\longrightarrow}} 0$ since the function $g(x) = \frac{x}{1+x}, x \geq 0$, is continuous, so that Exercise 8 in Chapter 3 applies. Next, $\frac{|X_n|}{1+|X_n|} \underset{n\to\infty}{\overset{P}{\longrightarrow}} 0$ means that for every $(1 >)\varepsilon > 0$, $P\left(\frac{|X_n|}{1+|X_n|} > \varepsilon\right) \underset{n\to\infty}{\longrightarrow} 0$ or $P(|X_n| > \frac{\varepsilon}{1-\varepsilon}) \underset{n\to\infty}{\longrightarrow} 0$, so that $|X_n| \underset{n\to\infty}{\overset{P}{\longrightarrow}} 0$.

(ii) $|X_n| \underset{n\to\infty}{\overset{P}{\longrightarrow}} 0$ implies $\frac{|X_n|}{1+|X_n|} \underset{n\to\infty}{\overset{P}{\longrightarrow}} 0$, by part (i), whereas $\frac{|X_n|}{1+|X_n|} \leq 1$ independent of n and integrable. Then by the Dominated Convergence Theorem (Theorem 13, part (b), in Chapter 5), $\mathcal{E}\left(\frac{|X_n|}{1+|X_n|}\right) \underset{n\to\infty}{\longrightarrow} 0$. On the other hand, $\mathcal{E}\left(\frac{|X_n|}{1+|X_n|}\right) \underset{n\to\infty}{\longrightarrow} 0$ is equivalent to $\frac{|X_n|}{1+|X_n|} \underset{n\to\infty}{\overset{(1)}{\longrightarrow}} 0$, and this implies that $\frac{|X_n|}{1+|X_n|} \underset{n\to\infty}{\overset{P}{\longrightarrow}} 0$, by Theorem 8 in Chapter 6, and hence $|X_n| \underset{n\to\infty}{\overset{P}{\longrightarrow}} 0$, by part (i). #

16. **(i)** The function $g(x) = |x|^r, r > 1$, is convex (because the second derivative is positive). Consider a r.v. X taking on the values $x_1, \ldots, x_n$ with probability $\frac{1}{n}$ each. Then $\mathcal{E}X = \frac{1}{n}\sum_{j=1}^n x_j$, and by Jensen's inequality, $g(\mathcal{E}X) \leq \mathcal{E}g(X)$, or $|\frac{1}{n}\sum_{j=1}^n x_j|^r \leq \frac{1}{n}\sum_{j=1}^n |x_j|^r$.

(ii) In part (i), replace x_j by the r.v. X_j to get

$$\left|\frac{1}{n}\sum_{j=1}^n X_j\right|^r \leq \frac{1}{n}\sum_{j=1}^n |X_j|^r.$$

Taking expectations, we obtain the desired result; i.e.,

$$\mathcal{E}\left|\frac{1}{n}\sum_{j=1}^n X_j\right|^r \leq \frac{1}{n}\sum_{j=1}^n \mathcal{E}|X_j|^r. \#$$

17. The uniform integrability of $X_n, n \geq 1$, implies that $\mathcal{E}|X_n| \leq M(< \infty), n \geq 1$, and that $\int_A |X_n|dP, n \geq 1$, are uniformly continuous. This is so by Theorem 11. So, $\int_A |X_n|dP, n \geq 1$, are uniformly continuous and $X_n \underset{n\to\infty}{\overset{P}{\longrightarrow}} X$. Then $X_n \underset{n\to\infty}{\overset{(1)}{\longrightarrow}} X$, by Theorem 13(i). Therefore $|\int_A X_n dP - \int_A X dP| = |\int_A (X_n - X)dP| \leq \int_A |X_n - X|dP \leq \int |X_n - X|dP = \mathcal{E}|X_n - X|$ independent of A and $\underset{n\to\infty}{\longrightarrow} 0$. The desired result follows. #

18. Uniform integrability of $X_n, n \geq 1$, and $Y_n, n \geq 1$, implies that $\mathcal{E}|X_n| \leq M(< \infty), n \geq 1$, $\mathcal{E}|Y_n| \leq M, n \geq 1$, and $\int_A |X_n|dP \to 0$, $\int_A |Y_n|dP \to 0$, as $P(A) \to 0$, uniformly in $n \geq 1$. This is so by Theorem 11. Next, $\mathcal{E}|X_n + Y_n| \leq \mathcal{E}|X_n| + \mathcal{E}|Y_n| \leq 2M, n \geq 1$, and $\int_A |X_n + Y_n|dP \leq \int_A |X_n|dP + \int_A |Y_n|dP \to 0$, as $P(A) \to 0$, uniformly in $n \geq 1$. Then $X_n + Y_n, n \geq 1$, are uniformly integrable, by the same Theorem, Theorem 11. #

19. By Theorem 11, $|\bar{X}_n|, n \geq 1$, are uniformly integrable if and only if $\mathcal{E}|\bar{X}_n| \leq M(< \infty), n \geq 1$, and $\int_A |\bar{X}_n| dP \to 0$, as $P(A) \to 0$, uniformly in $n \geq 1$. However,

$$\mathcal{E}|\bar{X}_n| \leq \frac{1}{n}\sum_{j=1}^{n} \mathcal{E}|X_j| = \frac{1}{n} \times n\mathcal{E}|X_1| = \mathcal{E}|X_1| < \infty,\ n \geq 1,$$

and

$$\int_A |\bar{X}_n| dP \leq \frac{1}{n}\sum_{j=1}^{n} \int_A |X_j| dP \leq \frac{1}{n} \times n \int_A |X_1| dP = \int_A |X_1| dP$$

independent of n, and converging to 0 as $P(A) \to 0$. This is so by Exercise 6 in Chapter 5. #

20. We have $|X_n| \leq Y, n \geq 1$, and $X_n \underset{n\to\infty}{\overset{a.s.}{\longrightarrow}} X$. Thus, $|X| \leq Y$ a.s. or $|X|^r \leq Y^r$ a.s. and hence $\mathcal{E}|X|^r \leq \mathcal{E}Y^r < \infty$. Also, $X_n \underset{n\to\infty}{\overset{a.s.}{\longrightarrow}} X$ implies $X_n \underset{n\to\infty}{\overset{P}{\longrightarrow}} X$. Thus, we have: $|X_n| \leq Y, n \geq 1, \mathcal{E}Y^r < \infty$ and $X_n \underset{n\to\infty}{\overset{P}{\longrightarrow}} X$. Then Corollary 3 to Theorem 13 applies and gives $X_n \underset{n\to\infty}{\overset{(r)}{\longrightarrow}} X$. #

21. For $\varepsilon > 0$, apply the Tchebichev inequality to get:

$$P(|X_n - \mu_n| \geq \varepsilon) \leq \frac{\sigma_n^2}{\varepsilon^2} \underset{n\to\infty}{\longrightarrow} 0,$$

so that $X_n - \mu_n \underset{n\to\infty}{\overset{P}{\longrightarrow}} 0$. #

22. For $\omega \in (0, 1)$, we have that $X_n(\omega) = 0$ for $n > 1/\omega$, and the first assertion follows. Next, $\mathcal{E}|X_n|^r = (2^n)^r \times \frac{1}{n} = \frac{2^{nr}}{n} = \frac{\exp[(r\,log 2)n]}{n} \underset{n\to\infty}{\longrightarrow} \infty$, which completes the proof. #

23. **(i)** For $(1 >)\varepsilon > 0$, we have that $n^c > \varepsilon$, so that

$$P(|X_n| > \varepsilon) = P(X_n = n^c) = \frac{1}{n} \underset{n\to\infty}{\longrightarrow} 0.$$

Thus, $X_n \underset{n\to\infty}{\overset{P}{\longrightarrow}} 0$.

(ii) Here, $\mathcal{E}|X_n|^r = n^{cr-1} + n^{-(cr+1)}$, and therefore $\mathcal{E}|X_n|^r \underset{n\to\infty}{\longrightarrow} 0$ if $(0 <)cr < 1$, and $\mathcal{E}|X_n|^r \underset{n\to\infty}{\longrightarrow} \infty$ if $cr > 1$.

(iii) That $X_n \underset{n\to\infty}{\overset{(r)}{\longrightarrow}} 0$ for all $cr < 1$, follows by Theorem 14 on account of parts (i) and (ii). #

24. We have $\mathcal{E}|X_n - X|^2 \underset{n\to\infty}{\longrightarrow} 0$, which is equivalent to saying that $X_n \underset{n\to\infty}{\overset{(2)}{\longrightarrow}} X$, and this implies that $\mathcal{E}X_n^2 \underset{n\to\infty}{\longrightarrow} \mathcal{E}X^2 < \infty$ (by Theorem 7).

Next,

$$\begin{aligned}
\mathcal{E}|X_n^2 - X^2| &= \mathcal{E}|(X_n + X)(X_n - X)| \\
&\le \mathcal{E}^{\frac{1}{2}}|X_n + X|^2 \times \mathcal{E}^{\frac{1}{2}}|X_n - X|^2 \text{ (by the Cauchy-Schwarz inequality)} \\
&\le (\mathcal{E}^{\frac{1}{2}}X_n^2 + \mathcal{E}^{\frac{1}{2}}X^2)\mathcal{E}^{\frac{1}{2}}|X_n - X|^2 \text{ (by the Minkowski inequality)} \\
&\le 2M^{\frac{1}{2}}\mathcal{E}^{\frac{1}{2}}|X_n - X|^2 \underset{n\to\infty}{\longrightarrow} 0,
\end{aligned}$$

where M is a bound for $\mathcal{E}X^2$ and $\mathcal{E}X_n^2$ (for sufficient large n). #

25. In the first place,

$$\mathcal{E}|X_nY_n| \le \mathcal{E}^{\frac{1}{r}}|X_n|^r \times \mathcal{E}^{\frac{1}{s}}|Y_n|^s \le M < \infty,$$

for all sufficiently large n.
Next,

$$\begin{aligned}
\mathcal{E}|X_nY_n - XY| &= \mathcal{E}|X_n(Y_n - Y) + (X_n - X)Y| \\
&\le \mathcal{E}^{\frac{1}{r}}|X_n|^r\mathcal{E}^{\frac{1}{s}}|Y_n - Y|^s + \mathcal{E}^{\frac{1}{r}}|X_n - X|^r\mathcal{E}^{\frac{1}{s}}|Y|^s \\
&\le M(\mathcal{E}^{\frac{1}{r}}|X_n - X|^r + \mathcal{E}^{\frac{1}{s}}|Y_n - Y|^s) \\
&\qquad \text{(for all sufficiently large } n\text{)} \\
&\qquad \underset{n\to\infty}{\longrightarrow} 0, \text{ since } X_n \overset{(r)}{\underset{n\to\infty}{\longrightarrow}} X,\ Y_n \overset{(s)}{\underset{n\to\infty}{\longrightarrow}} Y. \#
\end{aligned}$$

26. Indeed, for any $r > 0$,

$$\mathcal{E}|X_n - X|^r = \mathcal{E}|X_n|^r = \mathcal{E}X_n^r = \frac{1}{n} \underset{n\to\infty}{\longrightarrow} 0.$$

By Theorem 4 in Chapter 3, $X_n \overset{a.s.}{\underset{n\to\infty}{\longrightarrow}} 0$ if and only if

$$\lambda\left[\bigcup_{\nu=1}^{\infty}\left(|X_{n+\nu}| \ge \frac{1}{k}\right)\right] \underset{n\to\infty}{\longrightarrow} 0 \text{ for } k = 1, 2, \ldots$$

Here

$$\bigcup_{\nu=1}^{\infty}\left(|X_{n+\nu}| \ge \frac{1}{k}\right) = \bigcup_{\nu=1}^{\infty}(X_{n+\nu} = 1) = \bigcup_{\nu=1}^{\infty}\left(0, \frac{1}{n+\nu}\right) = \left(0, \frac{1}{n+1}\right),$$

so that $\lambda\left[\bigcup_{\nu=1}^{\infty}\left(|X_{n+\nu}| \ge \frac{1}{k}\right)\right] = \lambda\left(\left(0, \frac{1}{n+1}\right)\right) = \frac{1}{n+1} \underset{n\to\infty}{\longrightarrow} 0$, and therefore $X_n \overset{a.s.}{\underset{n\to\infty}{\longrightarrow}} 0$. #

27. Indeed, $X_n \overset{(r)}{\underset{n\to\infty}{\longrightarrow}} X$ implies $X_n \overset{P}{\underset{n\to\infty}{\longrightarrow}} X$, and then (by Theorem 5(ii) in Chapter 3) there exists a subsequence as described. #

28. **(i)** For $\varepsilon > 0$, we have

$$\sum_{n=1}^{\infty} P(|X_n| > \varepsilon) = \sum_{n=1}^{\infty} P(X_n = e^n) = \sum_{n=1}^{\infty} \frac{1}{n^2} < \infty.$$

Therefore $X_n \underset{n\to\infty}{\overset{a.s.}{\longrightarrow}} 0$ (by Exercise 4(i) in Chapter 3), hence $X_n \underset{n\to\infty}{\overset{P}{\longrightarrow}} 0$.

(ii) Next, $\mathcal{E}|X_n|^r = e^{rn} \times \frac{1}{n^2}$ and this tends to ∞ as $n \to \infty$. So, $X_n \underset{n\to\infty}{\overset{P}{\longrightarrow}} 0$, and indeed, $X_n \underset{n\to\infty}{\overset{a.s.}{\longrightarrow}} 0$, but $\mathcal{E}|X_n|^r \underset{n\to\infty}{\not\to} 0$. So, $X_n \underset{n\to\infty}{\overset{(r)}{\not\to}} 0$, by Theorem 14. #

29. **(i)** For $\varepsilon > 0$, we have

$$\sum_{n=1}^{\infty} P(|X_n| > \varepsilon) = \sum_{n=1}^{\infty} P(|X_n| = 2^{cn}) = \sum_{n=1}^{\infty} \frac{1}{2^{n-1}}$$
$$= \frac{1}{1-\frac{1}{2}} = 2(< \infty).$$

Then $X_n \underset{n\to\infty}{\overset{a.s.}{\longrightarrow}} 0$ (by Exercise 4(i) in Chapter 3).

(ii) Next, $\mathcal{E}|X_n|^r = 2 \times 2^{cnr} \times \frac{1}{2^n} = \frac{1}{2^{(n-1)-cnr}}$ and this converges (to 0) if and only if $cr < 1 - \frac{1}{n}$ $(n \geq 2)$; i.e., $cr < 1$. Since $X_n \underset{n\to\infty}{\overset{a.s.}{\longrightarrow}} 0$ implies $X_n \underset{n\to\infty}{\overset{P}{\longrightarrow}} 0$, this result together with $\mathcal{E}|X_n|^r \underset{n\to\infty}{\longrightarrow} 0$, for $cr < 1$, imply that $X_n \underset{n\to\infty}{\overset{(r)}{\longrightarrow}} 0$. #

30. **(i)** From $|X| \geq cI(|X| \geq c)$, we get $\mathcal{E}|X| \geq cP(|X| \geq c)$, so that $P(|X| \geq c) \leq (\mathcal{E}|X|)/c$. The special case is immediate.

(ii) For $t > 0$, $P(X \geq c) = P(tX \geq tc) = P(e^{tX} \geq e^{tc}) \leq e^{-tc}\mathcal{E}e^{tX}$, by the Markov inequality.
For $t < 0$, $P(X \leq c) = P(tX \geq tc) = P(e^{tX} \geq e^{tc}) \leq e^{-tc}\mathcal{E}e^{tX}$ as above.

(iii) Clearly, $|X| = |X|I(|X| > 0) + |X|I(|X| = 0)$, so that

$$\mathcal{E}|X| = \mathcal{E}[|X|I(|X| > 0)] \leq (\mathcal{E}^{\frac{1}{2}}|X|^2)\mathcal{E}^{\frac{1}{2}}[I(|X| > 0)]^2$$
$$= (\mathcal{E}^{\frac{1}{2}}X^2)[P(|X| > 0)]^{\frac{1}{2}},$$

or $(\mathcal{E}|X|)^2 \leq (\mathcal{E}X^2)[P(|X| > 0)]$, and $P(|X| > 0) \geq \frac{(\mathcal{E}|X|)^2}{\mathcal{E}X^2}$. The special case is immediate. #

31. Indeed,

$$|X_n - \mu|^2 = |(X_n - \mu_n) + (\mu_n - \mu)|^2$$
$$\leq 2\left(|X_n - \mu_n|^2 + |\mu_n - \mu|^2\right) \quad \text{(by the } c_r\text{-inequality)},$$

so that

$$\mathcal{E}|X_n-\mu|^2 \le 2\left[\mathcal{E}(X_n-\mu_n)^2+(\mu_n-\mu)^2\right]$$
$$=2\sigma_n^2+2(\mu_n-\mu)^2 \underset{n\to\infty}{\longrightarrow} 0.$$

So, $\mathcal{E}(X_n-\mu)^2 \underset{n\to\infty}{\longrightarrow} 0$, and then by the Tchebichev inequality,

$$P(|X_n-\mu|\ge\varepsilon)\le\varepsilon^{-2}\mathcal{E}(X_n-\mu)^2 \underset{n\to\infty}{\longrightarrow} 0. \#$$

32. **(i)** For $a>0$,

$$\mathcal{E}[|X_n-Y_n|I_{(|X_n-Y_n|\ge a)}]\le\mathcal{E}[|X_n|I_{(|X_n-Y_n|\ge a)}]+\mathcal{E}[|Y_n|I_{(|X_n-Y_n|\ge a)}]$$
$$=\int_{(|X_n-Y_n|\ge a)}|X_n|\,dP+\int_{(|X_n-Y_n|\ge a)}|Y_n|\,dP.$$

However, for $c>0$,

$$\int_{(|X_n-Y_n|\ge a)}|X_n|\,dP=\int_{(|X_n-Y_n|\ge a)\cap|X_n|\ge c)}|X_n|\,dP$$
$$+\int_{(|X_n-Y_n|\ge a)\cap|X_n|<c)}|X_n|\,dP$$
$$\le\int_{(|X_n|\ge c)}|X_n|\,dP+cP(|X_n-Y_n|\ge a),$$

and $\int_{(|X_n|\ge c)}|X_n|\,dP\le\frac{\varepsilon}{2}$ for all n (by the uniform integrability of $|X_n|, n\ge 1$) and $P(|X_n-Y_n|\ge a)\le\frac{\varepsilon}{2c}, n\ge n_1=n_1(\varepsilon)$, so that

$$\int_{(|X_n-Y_n|\ge a)}|X_n|\,dP\le\varepsilon,\quad n\ge n_1.$$

Similarly,

$$\int_{(|X_n-Y_n|\ge a)}|Y_n|\,dP\le\varepsilon,\quad n\ge n_2=n_2(\varepsilon).$$

Thus, for $n\ge n_3=n_3(\varepsilon)=\max\{n_1,n_2\}$, we get that

$$\int_{(|X_n-Y_n|\ge a)}|X_n-Y_n|\,dP\le\varepsilon,\quad n\ge n_3.$$

Increasing a so as to make the n_3-1 integrals

$$\int_{(|X_n-Y_n|\ge a)}|X_n-Y_n|\,dP\le\varepsilon,$$

$n=1,\ldots,n_3-1$, we have that

$$\int_{(|X_n-Y_n|\ge a)}|X_n-Y_n|\,dP\le\varepsilon$$

for all n, as was to be proved.

(ii) Uniform integrability of $|X_n - Y_n|$, $n \geq 1$, implies their uniform continuity (by Theorem 11). Then uniform continuity and the assumption that $X_n - Y_n \underset{n\to\infty}{\overset{P}{\longrightarrow}} 0$ imply that $\mathcal{E}|X_n - Y_n| \underset{n\to\infty}{\longrightarrow} 0$ (by Theorem 13(ii)). #

33. **(i)** The slope at θ is

$$S(h) = h^{-1}[e^{\frac{1}{2}|x-\theta|-\frac{1}{2}|x-(\theta+h)|} - 1].$$

Let $x < \theta$. Then for small h (either $h > 0$, or $h < 0$), we have

$$S(h) = h^{-1}\left(e^{-h/2} - 1\right) \underset{h\to 0}{\longrightarrow} -\frac{1}{2}.$$

For $x > \theta$ and small h (either $h > 0$, or $h < 0$), we have

$$S(h) = h^{-1}(e^{h/2} - 1) \underset{h\to 0}{\longrightarrow} \frac{1}{2}.$$

The conclusion follows.

(ii) Since $P_\theta(X = \theta) = 0$, $\theta \in \Re$, it suffices to restrict attention to the cases that $X < \theta$ or $X > \theta$. From part (i), it follows that, for each $\theta \in \Re$, $S(h) \underset{h\to 0}{\longrightarrow} \dot{g}(X;\theta)$ with P_θ-probability 1, and hence $S(h) \underset{h\to 0}{\longrightarrow} \dot{g}(X;\theta)$ in P_θ-probability. In order to show that $S(h) \underset{h\to 0}{\overset{q.m.}{\longrightarrow}} \dot{g}(X;\theta)$, it suffices to show that $\mathcal{E}_\theta S^2(h) \underset{h\to 0}{\longrightarrow} \mathcal{E}_\theta\left[\dot{g}(X;\theta)\right]^2 < \infty$ (see Theorem 14 in this chapter). By the fact that $P_\theta(X > \theta) = P_\theta(X < \theta) = \frac{1}{2}$, it follows that $\mathcal{E}_\theta\left[\dot{g}(X;\theta)\right]^2 = \frac{1}{4}$.

Next,

$$\mathcal{E}_\theta\left\{h^{-1}\left[e^{\frac{1}{2}|X-\theta|-\frac{1}{2}|X-(\theta+h)|} - 1\right]\right\}^2 = \frac{2}{h^2}\left[1 - \mathcal{E}_\theta e^{\frac{1}{2}|X-\theta|-\frac{1}{2}|X-(\theta+h)|}\right]$$

(since

$$\mathcal{E}_\theta e^{|X-\theta|-|X-(\theta+h)|} = \mathcal{E}_\theta\frac{p(X;\theta+h)}{p(X;\theta)} = \int_\Re p(x;\theta+h)dx = 1),$$

and

$$I = \mathcal{E}_\theta e^{\frac{1}{2}|x-\theta|-\frac{1}{2}|x-(\theta+h)|} = \frac{1}{2}\int_{-\infty}^{\infty} e^{-\frac{1}{2}|x-\theta|-\frac{1}{2}|x-(\theta+h)|}\,dx.$$

For $h < 0$:

$$\begin{aligned}
I &= \frac{1}{2}\int_{-\infty}^{\theta+h} e^{-\frac{1}{2}|x-\theta|-\frac{1}{2}|x-(\theta+h)|}\,dx + \frac{1}{2}\int_{\theta+h}^{\theta} e^{-\frac{1}{2}|x-\theta|-\frac{1}{2}|x-(\theta+h)|}\,dx \\
&\quad + \frac{1}{2}\int_{\theta}^{\infty} e^{-\frac{1}{2}|x-\theta|-\frac{1}{2}|x-(\theta+h)|}\,dx \\
&= \frac{1}{2}(2-h)e^{h/2},
\end{aligned}$$

and for $h > 0$:

$$I = \frac{1}{2}\int_{-\infty}^{\theta} e^{-\frac{1}{2}|x-\theta|-\frac{1}{2}|x-(\theta+h)|}\,dx + \frac{1}{2}\int_{\theta}^{\theta+h} e^{-\frac{1}{2}|x-\theta|-\frac{1}{2}|x-(\theta+h)|}\,dx$$
$$+ \frac{1}{2}\int_{\theta+h}^{\infty} e^{-\frac{1}{2}|x-\theta|-\frac{1}{2}|x-(\theta+h)|}\,dx$$
$$= \frac{1}{2}(2+h)e^{-h/2}.$$

Therefore

$$\mathcal{E}_\theta\left\{h^{-1}\left[e^{\frac{1}{2}|X-\theta|-\frac{1}{2}|X-(\theta+h)|}-1\right]\right\}^2$$
$$= \begin{cases} \frac{2}{h^2}\left[1-\frac{1}{2}(2-h)e^{h/2}\right] & \text{for } h<0 \\ \frac{2}{h^2}\left[1-\frac{1}{2}(2+h)e^{-h/2}\right] & \text{for } h>0 \end{cases}$$
$$= \begin{cases} \frac{1}{2t^2}(1-e^t+te^t) & \text{for } t<0 \\ \frac{1}{2t^2}(1-e^{-t}-te^{-t}) & \text{for } t>0 \end{cases}$$

However, $\frac{1}{2t^2}(1-e^t+te^t) \to \frac{1}{4}$ as $t \to 0$, and likewise $\frac{1}{2t^2}(1-e^{-t}-te^{-t}) \to \frac{1}{4}$ as $t \to 0$. It follows that

$$\mathcal{E}_\theta\left\{h^{-1}\left[e^{\frac{1}{2}|X-\theta|-\frac{1}{2}|X-(\theta+h)|}-1\right]\right\}^2 \to \mathcal{E}_\theta\left[\dot{g}(X;\theta)\right]^2$$

as $h \to 0$, and the result follows. #

34. **(i)** This follows from Exercise 2 in Chapter 4.

(ii) $X \geq Y$ or $X - Y \geq 0$ implies $\mathcal{E}(X-Y) \geq 0$ and $\mathcal{E}(X-Y) = 0$ only if $P(X-Y=0) = 1$; equivalently, $\mathcal{E}X \geq \mathcal{E}Y$ and $\mathcal{E}X = \mathcal{E}Y$ only if $P(X=Y)=1$.

(iii) Indeed, $X > Y$ implies $X \geq Y$ and hence $\mathcal{E}X \geq \mathcal{E}Y$, and $\mathcal{E}X = \mathcal{E}Y$ only if $P(X=Y)=1$, which is a contradiction.

(iv) Strict convexity of g means that

$$g(Z) > g(\mathcal{E}Z) + \lambda(\mathcal{E}Z)(Z - \mathcal{E}Z),$$

and hence $\mathcal{E}g(Z) > g(\mathcal{E}Z)$ unless (by part (iii))

$$P\left[g(Z) = g(\mathcal{E}Z) + \lambda(\mathcal{E}Z)(Z-\mathcal{E}Z)\right]$$
$$= \left[g(\mathcal{E}Z) - \lambda(\mathcal{E}Z)(\mathcal{E}Z)\right] + \lambda(\mathcal{E}Z)Z$$
$$= a + bZ = 1.$$

However, this is not possible if $b \neq 0$, because, although the straight line $y = a + bz$ is convex, it is not strictly so. Since b need not be 0, the equality $P[g(Z) = a + bZ] = 1$ may only occur if $P(Z = \text{constant}) = 1$, as was to be seen. #

Chapter 7

The Hahn-Jordan Decomposition Theorem, the Lebesgue Decomposition Theorem, and the Radon-Nikodym Theorem

1. To show $A_{n_0} = A + \sum_{j=n_0}^{\infty}(A_j - A_{j+1})$. Indeed, if ω belongs to the right-hand side, then either $\omega \in A$ and hence $\omega \in A_{n_0}$, or $\omega \in (A_j - A_{j+1})$ for some $j \geq n_0$. Hence $\omega \in A_j$ (but $\omega \notin A_{j+1}$) and therefore $\omega \in A_{n_0}$ since $A_n \downarrow$. Next, let $\omega \in A_{n_0}$. Then, either $\omega \in A_j$ for all $j \geq n_0$, so that $\omega \in \cap_{j=n_0}^{\infty} A_j$, hence $\omega \in \cap_{j=1}^{\infty} A_j = A$, and thus, ω belongs to the right-hand side. Or there is at least one $j > n_0$ such that $\omega \notin A_j$. Let j_0 be the smallest $j > n_0$ for which this is true. That is, $\omega \in A_{j_0-1}$, but $\omega \notin A_{j_0}$ with $j_0 > n_0$ or $j_0 \geq n_0 + 1$ or $j_0 - 1 \geq n_0$. Then $\omega \in (A_{j_0-1} - A_{j_0})$, so that $\omega \in \sum_{j=n_0}^{\infty}(A_j - A_{j+1})$; i.e., ω belongs to the right-hand side.#
2. Define μ on $\mathcal{A}$ by: $\mu(A) = \sum_{n=1}^{\infty} \mu_n(A)/2^n\mu_n(\Omega)$. Then μ is finite, since $\mu(\Omega) = \sum_{n=1}^{\infty} 2^{-n} = 1$. It is a probability measure, since $\mu(\oslash) = 0$, $\mu(A) \geq 0$ for all $A \in \mathcal{A}$, and $\mu\left(\sum_{i=1}^{\infty} A_i\right) = \sum_{n=1}^{\infty} \frac{1}{2^n\mu_n(\Omega)}\mu_n\left(\sum_{i=1}^{\infty} A_i\right) = \sum_{n=1}^{\infty} \frac{1}{2^n\mu_n(\Omega)} \sum_{i=1}^{\infty} \mu_n(A_i) = \sum_{n=1}^{\infty}\sum_{i=1}^{\infty} \frac{\mu_n(A_i)}{2^n\mu_n(\Omega)} = \sum_{i=1}^{\infty}\sum_{n=1}^{\infty} \frac{\mu_n(A_i)}{2^n\mu_n(\Omega)}$ (because the terms are ≥ 0), and this is equal to $\sum_{i=1}^{\infty} \mu(A_i)$ (i.e., μ is σ-additive). Finally, $\mu(A) = 0$ is equivalent to $\sum_{n=1}^{\infty} \frac{\mu_n(A_i)}{2^n\mu_n(\Omega)} = 0$, which implies $\frac{\mu_n(A_i)}{2^n\mu_n(\Omega)} = 0$ for all n, or $\mu_n(A) = 0$ for all n. Thus, $\mu_n \ll \mu, n \geq 1$, and μ is a probability measure. #
3. **(i)** The first two properties are immediate, and the triangular inequality follows from the fact that $|P(A) - Q(A)| \leq |P(A) - R(A)| + |R(A) - Q(A)|$ for every $A \in \mathcal{A}$.

   (ii) For any $A \in \mathcal{A}$, set $B = (f - g > 0)$, $C = (g - f > 0)$. Then

$$\begin{aligned} |P(A) - Q(A)| &= \left|\int_A f\,d\mu - \int_A g\,d\mu\right| = \left|\int_A (f - g)\,d\mu\right| \\ &= \left|\int_{A\cap B} (f - g)\,d\mu + \int_{A\cap C^c} (f - g)\,d\mu\right| \\ &= \left|\int_{A\cap B} (f - g)\,d\mu - \int_{A\cap C} (g - f)\,d\mu\right|. \end{aligned}$$

   Since $f - g > 0$ on $A \cap B$, and $g - f > 0$ on $A \cap C$, it follows that the last quantity on the right-hand side above is maximized by maximizing $\int_{A\cap B}(f - g)d\mu$ and by minimizing $\int_{A\cap C}(g - f)d\mu$. This happens by taking $A = B = (f - g > 0)$, in which case $A \cap C = B \cap C = \oslash$. Thus, $\max\{|P(A) - Q(A)|;\ A \in \mathcal{A}\} = \int_B (f - g)d\mu$. Similarly, $\max\{|P(A) - Q(A)|;\ A \in \mathcal{A}\} = \int_C (g - f)d\mu$. Since

$$\int_\Omega |f - g|d\mu = \int_B (f - g)d\mu + \int_C (g - f)d\mu,$$

   the conclusion follows. #

4. Let P_n and Q_n be the probability measures corresponding to $U\left(-\frac{1}{n}, 1\right)$ and $U\left(0, 1+\frac{1}{n}\right)$, and let λ be the Lebesgue measure. Then

$$\frac{dP_n}{d\lambda} = f_n(x) = \frac{n+1}{n} I_{[-\frac{1}{n},1]}(x), \qquad \frac{dQ_n}{d\lambda} = g_n(x) = \frac{n+1}{n} I_{[0,1+\frac{1}{n}]}(x),$$

so that $|f_n(x) - g_n(x)| = \frac{n}{n+1}$ for $-\frac{1}{n} \leq x \leq 0$ or $1 \leq x \leq 1+\frac{1}{n}$, and 0 otherwise. It follows that $\int_{\Re} |f_n(x) - g_n(x)| d\lambda = \frac{n}{n+1} \times \frac{2}{n} = \frac{2}{n+1} \underset{n\to\infty}{\longrightarrow} 0$, and the result follows by Exercise 3. #

5. By Exercise 3(ii),

$$\|P - Q\| = 2\sup[|P(A) - Q(A)|; A \in \mathcal{A}] = \int_{\Omega} |f - g| d\mu.$$

Then, with $B = (f - g > 0)$, we have

$$\begin{aligned}
\int_{\Omega} |f - g| d\mu &= \int_{B} |f - g| d\mu + \int_{B^c} |f - g| d\mu \\
&= \int_{B} (f - g) d\mu + \int_{B^c} (g - f) d\mu \\
&= [P(B) - Q(B)] + [Q(B^c) - P(B^c)] = 2[P(B) - Q(B)],
\end{aligned}$$

so that

$$\|P - Q\| = 2[P(B) - Q(B)].$$

Set $C = (|Z| > \varepsilon)$. Then

$$\begin{aligned}
P(B) - Q(B) &= P(B \cap C) + P(B \cap C^c) - Q(B \cap C) - Q(B \cap C^c) \\
&\leq P(C) + P(B \cap C^c) - Q(B \cap C^c) \\
&= P(|Z| > \varepsilon) + P(B \cap C^c) - Q(B \cap C^c).
\end{aligned}$$

But

$$\begin{aligned}
Q(B \cap C^c) &= \int_{B \cap C^c} g \, d\mu \\
&= \int_{B \cap C^c} \frac{g}{f} f \, d\mu \text{ (since on } B,\ f > g \text{ and hence } f > 0) \\
&= \int_{B \cap C^c} \exp Z f \, d\mu \geq e^{-\varepsilon} P(B \cap C^c) \quad (\text{since } |Z| \leq \varepsilon \text{ on } C^c).
\end{aligned}$$

Combining results derived so far, we have

$$\begin{aligned}
\|P - Q\| &= 2[P(B) - Q(B)] \\
&\leq 2P(|Z| > \varepsilon) + 2P(B \cap C^c) - 2e^{-\varepsilon} P(B \cap C^c) \\
&= 2P(|Z| > \varepsilon) + 2P(B \cap C^c)(1 - e^{-\varepsilon}) \\
&\leq 2(1 - e^{-\varepsilon}) + 2P(|Z| > \varepsilon). \ \#
\end{aligned}$$

6. **(i)** Here

$$\rho = \int_{\Omega} (fg)^{1/2} d\mu = \int_{\Omega} (f^{1/2} g^{1/2}) d\mu$$
$$\leq \left(\int_{\Omega} f\, d\mu\right)^{1/2} \left(\int_{\Omega} g\, d\mu\right)^{1/2}$$
(by the Cauchy-Schwartz inequality)
$$= 1.$$

(ii) Clearly,

$$|f - g| = |f^{1/2} - g^{1/2}||f^{1/2} + g^{1/2}|$$
$$\geq |f^{1/2} - g^{1/2}|^2 = f + g - 2(fg)^{1/2}, \text{ so that}$$
$$\int_{\Omega} |f - g| d\mu \geq 2\left[1 - \int_{\Omega} (fg)^{1/2} d\mu\right] = 2(1 - \rho),$$

which is the left-hand side of the inequality. Next,

$$\int_{\Omega} |f - g| d\mu = \int_{\Omega} \left(|f^{1/2} + g^{1/2}||f^{1/2} - g^{1/2}|\right) d\mu$$
$$\leq \left(\int_{\Omega} |f^{1/2} + g^{1/2}|^2 d\mu\right)^{1/2} \left(\int_{\Omega} |f^{1/2} - g^{1/2}|^2 d\mu\right)^{1/2}$$
$$= \left[\int_{\Omega} (f + g + 2f^{1/2} g^{1/2}) d\mu\right]^{1/2}$$
$$\times \left[\int_{\Omega} (f + g - 2f^{1/2} g^{1/2}) d\mu\right]^{1/2}$$
$$= 2[(1 + \rho)(1 - \rho)]^{1/2} = 2(1 - \rho^2)^{1/2},$$

which is the right-hand side of the inequality.

(iii) This is immediate by the double inequality

$$2(1 - \rho_n) \leq d(P_n, Q_n) \leq 2(1 - \rho_n^2)^{1/2}. \#$$

7. The limits are taken as $\{n\}$ or subsequences thereof tend to ∞. Let Φ be the d.f. of the $N(0, 1)$ distribution with respective p.d.f. $(2\pi)^{-1/2} \exp(-x^2/2)$ (with respect to Lebesgue measure λ). Then, for $x_n \to 0$, we have:

$$\int_{\Re} |f_n(x + x_n)| d\Phi(x) = (2\pi)^{-1/2} \int_{\Re} |f_n(x + x_n)| e^{-x^2/2} d\lambda(x)$$
$$= \int_{\Re} (2\pi)^{-1/2} |f_n(y)| e^{-(y - x_n)^2/2} dy$$
(setting $x + x_n = y$).

Refer to Exercise 17 in Chapter 5, and set

$$h_n(y) = (2\pi)^{-1/2}|f_n(y)|e^{-(y-x_n)^2/2}, g_n(y) = 0,$$
$$G_n(y) = (2\pi)^{-1/2}Me^{-(y-x_n)^2/2}, \quad h(y) = g(y) = 0,$$
$$G(y) = (2\pi)^{-1/2}Me^{-y^2/2}.$$

Then the conditions of that exercise are satisfied, and therefore

$$\int_{\Re} h_n(y)dy \to 0 \text{ or } \int_{\Re} |f_n(x + x_n)|d\Phi(x) \to 0.$$

So, by the Markov inequality,

$$\Phi\text{-probability } (|f_n(x + x_n)| > \varepsilon) \leq \frac{1}{\varepsilon}\int_{\Re} |f_n(x + x_n)|d\Phi(x) \to 0,$$

so that $|f_n(x+x_n)| \to 0$ in Φ-measure. Hence there exists a subsequence $\{x_m\} \subseteq \{x_n\}$ such that $f_m(x+x_m) \to 0$ a.e. [Φ-measure]. However, it is clear that the Φ-measure is mutually absolutely continuous with respect to λ. Thus, $f_m(x+x_m) \to 0$ a.e. [λ], as was to be seen.
Remark: From the proof, it follows that the result also holds if $\Re$ is replaced by $\Re^k$, $k \geq 1$. #

8. Define g_n by $g_n = -f_n^-$, and observe that the g_n's satisfy the conditions of the f_n's in Exercise 7. Therefore, for any $x_n \underset{n\to\infty}{\longrightarrow} 0$, there exists $\{x_m\} \subseteq \{x_n\}$ such that $g_m(x + x_m) \underset{m\to\infty}{\longrightarrow} 0$. By their definition, $f_n(x + x_n) \geq g_n(x + x_n)$ for all n and $x \in \Re$. Therefore, as $n \to 0$ (or $m \to \infty$),

$$\limsup f_n(x + x_n) \geq \limsup g_n(x + x_n) \geq \lim g_m(x + x_m) = 0$$

a.e. [λ], as was to be seen. #

Chapter 8

Distribution Functions and their Basic Properties, Helly-Bray Type Results

1. In the first place, $0 \leq y \leq 1$. Next, for such ys and $t \in \Re$, we show that

$$F^{-1}(y) \leq t \text{ if and only if } y \leq F(t).$$

Indeed, let $F^{-1}(y) \leq t$. Then $F[F^{-1}(y)] \leq F(t)$. Also, from the definition of $F^{-1}(y)$, there exist $x_n \in \{x \in \Re;\ F(x) \geq y\}$ such that $x_n \downarrow F^{-1}(y)$, as $n \to \infty$. Hence, by the right continuity of F, $F(x_n) \downarrow F[F^{-1}(y)]$, as $n \to \infty$. However, $F(x_n) \geq y$ for all n, and therefore $F[F^{-1}(y)] \geq y$. Combining this result with the result $F[F^{-1}(y)] \leq F(t)$, we obtain $y \leq F(t)$. In the other way around, let

$y \leq F(t)$. Then, clearly, $t \in \{x \in \Re;\ F(x) \geq y\}$, and therefore $t \geq F^{-1}(y)$ by the definition of $F^{-1}(y)$. The justification of the assertion is completed. Now, the function $x = F^{-1}(y)$ is $\mathcal{B}$-measurable, because $\{y \in \Re;\ F^{-1}(y) \leq c\} = \{y \in \Re;\ y \leq F(c)\}$, by the assertion already established, and the last set is $(-\infty, F(c)]$, which is in $\mathcal{B}$ for all $c \in \Re$. Thus, if $Y \sim U(0, 1)$ and set $X = F^{-1}(Y)$, then X is a r.v. Finally, for $x \in \Re$:

$$F_X(x) = P(X \leq x) = P[F^{-1}(Y) \leq x] = P[Y \leq F(x)] = F(x),$$

as it was to be shown. #

2. **(i)** The following relations are self-explanatory. For every $\varepsilon > 0$,

$$\begin{aligned}(X_n \leq x) &= (X_n \leq x) \cap (|X_n - X| \geq \varepsilon) + (X_n \leq x) \cap (|X_n - X| < \varepsilon)\\ &\subseteq (|X_n - X| \geq \varepsilon) \cup (X_n \leq x) \cap (|X_n - X| < \varepsilon)\\ &= (|X_n - X| \geq \varepsilon) \cup (X_n \leq x) \cap (X - \varepsilon < X_n < X + \varepsilon)\\ &\subseteq (|X_n - X| \geq \varepsilon) \cup (X_n \leq x) \cap (X - \varepsilon < X_n)\\ &\subseteq (|X_n - X| \geq \varepsilon) \cup (X - \varepsilon \leq x)\\ &= (|X_n - X| \geq \varepsilon) \cup (X \leq x + \varepsilon).\end{aligned}$$

Hence

$$\begin{aligned}F_{X_n}(x) = P(X_n \leq x) &\leq P(|X_n - X| \geq \varepsilon) + P(X \leq x + \varepsilon)\\ &= P(|X_n - X| \geq \varepsilon) + F_X(x + \varepsilon).\end{aligned}$$

Let $n \to \infty$ and recall that $X_n \xrightarrow{P} X$ to obtain

$$\overline{\lim}\, F_{X_n}(x) \leq F_X(x + \varepsilon) \text{ for every } \varepsilon > 0.$$

Let $x \in C(F_X)$ and let $\varepsilon \to 0$ to obtain

$$\overline{\lim_{n\to\infty}}\, F_{X_n}(x) \leq F_X(x).$$

Next,

$$\begin{aligned}(X \leq x - \varepsilon) &= (X \leq x - \varepsilon) \cap (|X_n - X| \geq \varepsilon)\\ &\quad + (X \leq x - \varepsilon) \cap (|X_n - X| < \varepsilon)\\ &\subseteq (|X_n - X| \geq \varepsilon) \cup (X \leq x - \varepsilon)\\ &\quad \cap (X - \varepsilon < X_n < X + \varepsilon)\\ &\subseteq (|X_n - X| \geq \varepsilon) \cup (X \leq x - \varepsilon) \cap (X_n < X + \varepsilon)\\ &\subseteq (|X_n - X| \geq \varepsilon) \cup (X_n \leq x),\end{aligned}$$

so that

$$\begin{aligned}F_X(x - \varepsilon) = P(X \leq x - \varepsilon) &\leq P(|X_n - X| \geq \varepsilon) + P(X_n \leq x)\\ &= P(|X_n - X| \geq \varepsilon) + F_{X_n}(x).\end{aligned}$$

As $n \to \infty$,

$$F_X(x-\varepsilon) \le \underline{\lim} F_{X_n}(x) \text{ for every } \varepsilon > 0.$$

Taking $x \in C(F_X)$ and letting $\varepsilon \to 0$, we have $F_X(x) \le \underline{\lim}_{n\to\infty} F_{X_n}(x)$. Combining $\underline{\lim}$ and $\overline{\lim}$, we get $\lim F_{X_n}(x) = F_X(x)$.

(ii) Let $P(X=-1) = P(X=1) = \frac{1}{2}$ and let $X_n = -X$, $n \ge 1$, so that $X_n - X = -2X$ and $F_{X_n} = F_X$. However, for $0 < \varepsilon < \frac{1}{2}$, $P(|X_n - X| \ge \varepsilon) = P(|-2X| \ge \varepsilon) = P(|X| \ge \frac{\varepsilon}{2}) = 1$, so that $X_n \overset{P}{\nrightarrow} X$. #

3. We have $Var \times F_n = Var \times F = 1, n \ge 1$, and $F_n \underset{n\to\infty}{\Rightarrow} F$. Then, by Theorem 3(iii), $F_n(\pm\infty) \underset{n\to\infty}{\longrightarrow} F(\pm\infty)$. But $F_n(\infty) = 1$ and $F_n(-\infty) = 0, n \ge 1$. Thus, $F(\infty) = 1$ and $F(-\infty) = 0$, so that F is the d.f. of a r.v. #

4. Observe that the proof of Theorem 5 depends on the following facts: First, that $F(x)$ is bounded (not necessarily by 1), so that, e.g., the sequence $\{F_n(x_1)\}$ has a convergent subsequence $\{F_{n1}(x_1)\}$. The same argument is repeated throughout the proof. The second fact is Proposition 1, whose justification, however, uses only property #2 in Definition 1, in order to argue that F has only jumps as its discontinuities, if any. The third fact is Proposition 2, whose justification also holds, if F is defined thus: $F : \Re \to [0, B]$. #

5. Regarding the proof of Theorem 6, the only point where the boundedness $F(x) \le 1$ and $F_n(x) \le 1$ enters the picture is the following:

$$\left| \int_{(\alpha,\beta]} (g - g_m) dF \right| \le \sup_{x\in(\alpha,\beta]} |g_m(x) - g(x)| \, [F(\beta) - F(\alpha)],$$

which would be $\le B\sup_{x\in(\alpha,\beta]}|g_m(x) - g(x)|$ rather than $\le \sup_{x\in(\alpha,\beta]}|g_m(x) - g(x)|$, should F be bounded by B. The same for $|\int_{(\alpha,\beta]}(g - g_m)dF_n|$.

The proof of Theorem 7 hinges upon the following points: Finiteness of the integrals $\int_\Re g dF_n$ and $\int_\Re g dF$, which is ensured by the boundedness of F_n and F. Relation (8.4), which holds on account of Theorem 6; relation (8.5), which holds on account of the boundedness of F; and relation (8.6), where now the bound is $B \times \sup_{x\in(\alpha,\beta]} g(x)$, which is $< \varepsilon/3$ for sufficiently small α and sufficiently large β. Finally, the proof of Theorem 8 depends on the following facts: That the integrals $\int_\Re g dF_n$ and $\int_\Re g dF$ are finite, which is ensured by the boundedness of F_n and F. That the boundedness

$$\left| \int_\Re g dF_n - \int_\Re g dF \right| \le M\{Var.F_n - [F_n(\beta) - F_n(\alpha)]\}$$

holds regardless of the bound of F_n and F. That (8.7) holds, because all quantities figuring there are finite, due to the boundedness of F_n and F. Also, relations (8.8) and (8.9) hold by the finiteness of the integrals involved, and finally, relation (8.10) holds on account of Theorem 6. #

6. In all that follows, $n \to \infty$ unless otherwise explicitly stated. Now $|X_n| \le Y, n \ge 1$, implies $|X_n|^r \le Y^r, n \ge 1$, and hence $\mathcal{E}|X_n|^r \le \mathcal{E}Y^r < \infty, n \ge 1$.

Also, $P(A) \to 0$ implies $\int_A Y^r dP \to 0$ (Exercise 6, Chapter 5), and hence $\int_A |X_n|^r dP \to 0$ uniformly in $n \geq 1$. Then Theorem 11, Chapter 6, gives

$$\int_{(|X_n|>c)} |X_n|^r dP \left(= \int_{(|X_n|>c)} |X_n|^r dF(x)\right) \leq \varepsilon, \quad n \geq 1. \tag{8.1}$$

$$\begin{aligned} \mathcal{E}|X_n|^r &= \int |X_n|^r dP = \int_{(|X_n|>c)} |X_n|^r dP + \int_{(|X_n|\leq c)} |X_n|^r dP \\ &= \int_{(|x|>c)} |x|^r dF_n(x) + \int_{(|x|\leq c)} |x|^r dF_n(x). \end{aligned} \tag{8.2}$$

By (8.2) and (8.1) here,

$$0 \leq \mathcal{E}|X_n|^r - \int_{(|x|\leq c)} |x|^r dF_n(x) \left(= \int_{(|x|>c)} |x|^r dF_n)x\right) \leq \varepsilon, \quad n \geq 1. \tag{8.3}$$

With $-c, c \in C(F)$, and by Theorem 6 here,

$$\int_{(|x|\leq c)} |x|^r dF_n(x) \underset{n\to\infty}{\longrightarrow} \int_{(|x|\leq c)} |x|^r dF(x). \tag{8.4}$$

Taking the limits in (8.3) as $n \to \infty$, and using (8.4), we get

$$\begin{aligned} 0 &\leq \underline{\lim}\, \mathcal{E}|X_n|^r - \int_{(|x|\leq c)} |x|^r dF(x) \\ &= \underline{\lim}\, \mathcal{E}|X_n|^r - \int \left[|x|^r I_{(|x|\leq c)}\right] dF(x) \\ &\leq \overline{\lim}\, \mathcal{E}|X_n|^r - \int \left[|x|^r I_{(|x|\leq c)}\right] dF(x) \leq \varepsilon. \end{aligned} \tag{8.5}$$

Also,

$$\begin{aligned} \int \left[|x|^r I_{(|x|\leq c)}\right] dF_n(x) &= \mathcal{E}\left[|X_n|^r I_{(|X_n|\leq c)}\right] \\ &\leq \mathcal{E}|X_n|^r \leq \mathcal{E}Y^r < \infty, \quad n \geq 1. \end{aligned} \tag{8.6}$$

By (8.6) and (8.4), and with $-c, c$ in $C(F)$,

$$\int \left[|x|^r I_{(|x|\leq c)}\right] dF(x) \leq \mathcal{E}Y^r. \tag{8.7}$$

Since $0 \leq |x|^r I_{(|x|\leq c)}$, it follows (by the Fatou-Lebesgue Theorem, Theorem 2 in Chapter 5) that, as $c \uparrow \infty$,

$$\int \underline{\lim} \left[|x|^r I_{(|x|\leq c)}\right] dF(x) \leq \underline{\lim} \int \left[|x|^r I_{(|x|\leq c)}\right] dF(x).$$

However,

$$\int \underline{\lim}\left[|x|^r I_{(|x|\le c)}\right] dF(x) = \int \lim\left[|x|^r I_{(|x|\le c)}\right] dF(x)$$
$$= \int |x|^r dF(x),$$

and

$$\underline{\lim} \int \left[|x|^r I_{(|x|\le c)}\right] dF(x) = \lim \int \left[|x|^r I_{(|x|\le c)}\right] dF(x),$$

since the integrals $\int \left[|x|^r I_{(|x|\le c)}\right] dF(x) \uparrow$ (as $c \uparrow \infty$). It follows that

$$\int |x|^r dF(x) \le \lim \int \left[|x|^r I_{(|x|\le c)}\right] dF(x). \tag{8.8}$$

From (8.8) and (8.7), it follows that

$$\int |x|^r dF(x) < \infty. \tag{8.9}$$

From $|x|^r I_{(|x|\le c)} \le |x|^r$ independent of c and integrable (by (8.9)), and the fact that $|x|^r I_{(|x|\le c)} \to |x|^r$ (as $c \uparrow \infty$ with $-c, c$ in $C(F)$), we have (by the Dominated Convergence Theorem)

$$\int \left[|x|^r I_{(|x|\le c)}\right] dF(x) \to \int |x|^r dF(x). \tag{8.10}$$

Finally, taking the limits in (8.5), as $c \uparrow \infty$ (with $-c, c$ in $C(F)$), and using (8.10), we have

$$0 \le \underline{\lim}\mathcal{E}|X_n|^r - \int |x|^r dF(x) \le \overline{\lim}\mathcal{E}|X_n|^r - \int |x|^r dF(x) \le \varepsilon.$$

Letting $\varepsilon \to 0$, we get that

$$\underline{\lim}\mathcal{E}|X_n|^r = \overline{\lim}\mathcal{E}|X_n|^r = \lim \mathcal{E}|X_n|^r = \int |x|^r dF(x)$$
$$= \mathcal{E}|X|^r. \quad \#$$

7. **(i)** Let $x_n \downarrow 0$ as $n \to \infty$, and define the events $A_n = (X \le x - x_n)$, $A = (X < x)$. Then, clearly, as $n \to \infty$, $A_n \uparrow$ and $\lim_{n\to\infty} A_n = \cup_{n=1}^{\infty} A_n = A$. Then, by the continuity of a probability measure,

$$\lim_{n\to\infty} P(A_n) = P(\lim_{n\to\infty} A_n) = P(A),$$

or

$$\lim_{n\to\infty} P(X \le x - x_n) = P(X < x).$$

However, $\lim_{n\to\infty} P(X \le x - x_n) = F(x-)$. Thus, $P(X < x) = F(x-)$.

(ii) By part (i),

$$P(X = x) = P(X \le x) - P(X < x) = F(x) - F(x-).$$

So, if F is continuous at x, then $F(x) = F(x-)$, and hence $P(X = 0)$. If $P(X = x) = 0$, then $F(x) = F(x-)$, so that F is continuous at x. #

8. F has got to be nondecreasing, continuous from the right, and $F(-\infty) = 0$, $F(\infty) = 1$. For $(0 <)x < y$, we must have $F(x) \le F(y)$ or $\alpha + \beta e^{-x^2/2} \le \alpha + \beta e^{-y^2/2}$ or $\beta e^{-x^2/2} \le \beta e^{-y^2/2}$. Now $x < y$ is equivalent to $e^{-y^2/2} < e^{-x^2/2}$, and therefore the previous inequality holds if $\beta < 0$.
Next, for $x < 0$ and $x > 0$, $F(x)$ is continuous, so it remains to examine the case $x = 0$. Since $F(0) = 0$, we must have $F(x) \to 0$ as $x \downarrow 0$, or $\lim_{x\downarrow 0}(\alpha + \beta e^{-x^2/2}) = \alpha + \beta \lim_{x\downarrow 0} e^{-x^2/2} = \alpha + \beta = 0$, so that $\beta = -\alpha$.
Finally, $1 = F(\infty) = \lim_{x\to\infty} F(x) = \lim_{x\to\infty}(\alpha + \beta e^{-x^2/2}) = \alpha$. Summarizing the above conclusions, we have then $\alpha = 1, \beta = -1$, so that $F(x) = 1 - e^{-x^2/2}, x > 0$. #

9. For $\varepsilon > 0$, select $a < b$, so that

$$F(a) < \frac{\varepsilon}{3}, \quad F(\infty) - F(b) < \frac{\varepsilon}{3}. \tag{1}$$

Next, F being continuous, is uniformly continuous in $[a, b]$. Then we can select a partition $a = x_0 < x_1 < \ldots < x_{k-1} < x_k = b$, so that

$$F(x_j) - F(x_{j-1}) < \frac{\varepsilon}{3}, \quad j = 1, \ldots, k. \tag{2}$$

Since $F_n(x) \underset{n\to\infty}{\longrightarrow} F(x)$ for all x, we have

$$-\frac{\varepsilon}{3} < F_n(x_j) - F(x_j) < \frac{\varepsilon}{3} \text{ for } n \ge N' \text{ independent of } j. \tag{3}$$

Finally, since $F_n \underset{n\to\infty}{\overset{c}{\longrightarrow}} F$, Theorem 3(iii) implies that $F_n(\pm\infty) \underset{n\to\infty}{\longrightarrow} F(\pm\infty)$, so that

$$-\frac{\varepsilon}{3} < F_n(\infty) - F(\infty) < \frac{\varepsilon}{3}, \quad n \ge N''. \tag{4}$$

Next, we proceed as follows. First, let $x \in [a, b]$, so that $x \in [x_{j-1}, x_j]$ for exactly one j. Then, for $n \ge N = \max\{N', N''\}$,

$$\begin{aligned}
F_n(x) - F(x) &\le F_n(x_j) - F(x_{j-1}) && (\text{since } x_{j-1} \le x \le x_j)\\
&\le F(x_j) + \frac{\varepsilon}{3} - F(x_{j-1}) && (\text{by (3)})\\
&= [F(x_j) - F(x_{j-1})] + \frac{\varepsilon}{3}\\
&< \frac{\varepsilon}{3} + \frac{\varepsilon}{3} = \frac{2\varepsilon}{3} < \varepsilon && (\text{by (2)}).
\end{aligned}$$

Also,

$$\begin{aligned} F_n(x) - F(x) &\geq F_n(x_{j-1}) - F(x_j) \qquad \text{(since } x_{j-1} \leq x \leq x_j) \\ &\geq F(x_{j-1}) - \frac{\varepsilon}{3} - F(x_j) \qquad \text{(by (3))} \\ &= -[F(x_j) - F(x_{j-1})] - \frac{\varepsilon}{3} \\ &> -\frac{\varepsilon}{3} - \frac{\varepsilon}{3} = -\frac{2\varepsilon}{3} > -\varepsilon \qquad \text{(by (2))}. \end{aligned}$$

Therefore, for $x \in [a, b]$,

$$|F_n(x) - F(x)| < \varepsilon, \ \ n \geq N. \tag{5}$$

Next, let $x < a$, then, for $n \geq N, 0 \leq F(x) \leq F(a) < \frac{\varepsilon}{3}$; i.e., $F(x) \leq \frac{\varepsilon}{3}$. Also,

$$\begin{aligned} 0 \leq F_n(x) \leq F_n(a) &< F(a) + \frac{\varepsilon}{3} \text{ (by (3))} \\ &< \frac{\varepsilon}{3} + \frac{\varepsilon}{3} = \frac{2\varepsilon}{3} < \varepsilon \text{ (by (2)); i.e.,} \end{aligned}$$

$F_n(x) < \frac{2\varepsilon}{3}$. Therefore $F_n(x) - F(x) < \frac{2\varepsilon}{3} - 0 = \frac{2\varepsilon}{3} < \varepsilon$, and $F_n(x) - F(x) > 0 - \frac{\varepsilon}{3} = -\frac{\varepsilon}{3} > -\varepsilon$, so that

$$|F_n(x) - F(x)| < \frac{\varepsilon}{3}, \ \ n \geq N \ (x < a). \tag{6}$$

Finally, let $x > b$. Then, for $n \geq N$,

$$\begin{aligned} F(\infty) - \frac{\varepsilon}{3} &< F(b) \qquad \text{(by (1))} \\ &\leq F(x) \leq F(\infty), \end{aligned}$$

and, by (1), (3), and (4), $F(\infty) - \frac{2\varepsilon}{3} < F(b) - \frac{\varepsilon}{3} < F_n(b) \leq F_n(x) \leq F_n(\infty) \leq F(\infty) + \frac{\varepsilon}{3}$; i.e., $F(\infty) - \frac{\varepsilon}{3} \leq F(x) \leq F(\infty)$ and $F(\infty) - \frac{2\varepsilon}{3} \leq F_n(x) \leq F(x) + \frac{\varepsilon}{3}$. Therefore

$$F_n(x) - F(x) \leq F(\infty) + \frac{\varepsilon}{3} - F(\infty) + \frac{\varepsilon}{3} = \frac{2\varepsilon}{3} < \varepsilon, \tag{7}$$

and

$$F_n(x) - F(x) \geq F(\infty) - \frac{2\varepsilon}{3} - F(\infty) = -\frac{2\varepsilon}{3} > -\varepsilon. \tag{8}$$

Relations (7) and (8) yield,

$$|F_n(x) - F(x)| < \varepsilon, \ \ n \geq N \ (x > b). \tag{9}$$

The result follows from relations (5), (6) and (9). #

10. **(i)** In the first place, $P(Y = C) = P(X > C) = 1 - F_X(C)$. Next,

$$\begin{aligned} F_Y(y) &= P(Y \leq y) = P(Y \leq y|X \leq C)P(X \leq C) \\ &\quad + P(Y \leq y|X > C)P(X > C) \\ &= P(X \leq y|X \leq C)P(X \leq C) + P(C \leq y|X > C)P(X > C) \\ &= P(X \leq y, X \leq C) + P(C \leq y, X > C). \end{aligned}$$

Then, for $y < C$, $F_Y(y) = P(X \leq y) + 0 = F_X(y)$, whereas, for $y \geq C$, $F_Y(y) = P(X \leq C) + P(X > C) = 1$. So, $F_Y(y) = F_X(y)$ for $y < C$, and $F_Y(y) = 1$ for $y \geq C$.

(ii) Define F_1 and F_2 as follows:

$$F_1(x) = \begin{cases} F_X(x), & x \leq C \\ F_X(C), & x > C \end{cases}, \qquad F_2(x) = \begin{cases} 0, & x \leq C \\ 1 - F_X(C), & x > C \end{cases}.$$

Then, clearly, F_1 is a continuous d.f., F_2 is a step function, and $F_1 + F_2 = F_Y$. #

11. For $n \geq 1$, let X_n be r.v.s such that $P(X_n = -n) = P(X_n = n) = \frac{1}{2}$, so that the respective d.f.s are given by:

$$F_n(x) = \begin{cases} 0, & x < -n \\ 1/2, & -n \leq x < n \\ 1, & x \geq n \end{cases}, \qquad n \geq 1.$$

Then, clearly, $F_n(x) \underset{n\to\infty}{\longrightarrow} \frac{1}{2}$ for every $x \in \Re$. Thus, if $F(x) = \frac{1}{2}$, $x \in \Re$, then F is a d.f., $F_n \underset{n\to\infty}{\Longrightarrow} F$, but F is not the d.f. of a r.v. #

12. To show that $\int_{(|X_n| \geq a)} |X_n| dP \to 0$ uniformly in n as $(0 <)a \to \infty$, or $\int_{(|x| \geq a)} |x| dF_n \to 0$ uniformly in n as $a \to \infty$. For $0 < c < a$, we have

$$\begin{aligned} \int_{(|x| \geq a)} |x| dF_n &\leq \int_{(-\infty,-c]} |x| dF_n + \int_{(c,\infty)} |x| dF_n \\ &= \int_{-\infty}^{\infty} |x| dF_n - \int_{(-c,c]} |x| dF_n. \end{aligned}$$

Thus, it suffices to show that the right-hand side above $\to 0$ uniformly in n as $c \to \infty$. We have

$$\begin{aligned} &(0 \leq) \int_{-\infty}^{\infty} |x| dF_n - \int_{(-c,c]} |x| dF_n \\ &\leq \left(\int_{-\infty}^{\infty} |x| dF_n - \int_{-\infty}^{\infty} |x| dF \right) - \left(\int_{(-c,c]} |x| dF_n - \int_{(-c,c]} |x| dF \right) \\ &\quad + \int_{(|x| \geq c)} |x| dF \\ &\leq |\mathcal{E}|X_n| - \mathcal{E}|X|| + \int_{(|x| \geq c)} |x| dF + \left| \int_{(-c,c]} |x| dF_n - \int_{(-c,c]} |x| dF \right|. \end{aligned}$$

However, for $\varepsilon > 0$,

$$|\mathcal{E}|X_n| - \mathcal{E}|X|| < \frac{\varepsilon}{3}, \quad n \geq n_1 = n_1(\varepsilon), \text{ (since } \mathcal{E}|X_n| \underset{n\to\infty}{\longrightarrow} \mathcal{E}|X|\text{)},$$

$$\int_{(|x|\geq c)} |x| dF < \frac{\varepsilon}{3} \text{ for sufficiently large } c \text{ (since } \mathcal{E}|X| < \infty\text{)},$$

$$\text{and } \left| \int_{(-c,c]} |x| dF_n - \int_{(-c,c]} |x| dF \right| < \frac{\varepsilon}{3}$$

for c as above and also such that c and $-c$ are continuity points of F, $n \geq n_2 = n_2(\varepsilon)$ (by Theorem 6 here). Therefore, if $n \geq n_3 = \max\{n_1, n_2\}$, and c and $-c$ are as above, the right-hand side above becomes $< \varepsilon$. Hence $\int_{(|x|\geq a)} |x| dF_n < \varepsilon$, $n \geq n_3$. Increasing a to make the $n_3 - 1$ integrals $\int_{(|x|\geq a)} |x| dF_n, n = 1, \ldots, n_3 - 1 < \varepsilon$, the result follows. #

13. In the sequel, all limits are taken as $n \to \infty$. Using assumptions (i)-(iv), we have the following chain of relations. Let $c, d \in \Re$ with $c < d$ and always continuity points of F. Then:

$$\begin{aligned}
&\left| \int_{\Re} f_n dF_n - \int_{\Re} f dF \right| \\
&= \left| \left[\int_{(-\infty,c]} f_n dF_n + \int_{(c,d]} f_n dF_n + \int_{(d,\infty)} f_n dF_n \right] \right. \\
&\quad \left. - \left[\int_{(-\infty,c]} f dF + \int_{(c,d]} f dF + \int_{(d,\infty)} f dF \right] \right| \\
&= \left| \left[\int_{(-\infty,c]} f_n dF_n + \int_{(d,\infty)} f_n dF_n \right] - \left[\int_{(-\infty,c]} f dF + \int_{(d,\infty)} f dF \right] \right. \\
&\quad \left. + \left[\int_{(c,d]} f_n dF_n - \int_{(c,d]} f dF \right] \right| \\
&\leq \int_{(-\infty,c]} |f_n| dF_n + \int_{(d,\infty)} |f_n| dF_n + \int_{(-\infty,c]} |f| dF + \int_{(d,\infty)} |f| dF \\
&\quad + \left| \int_{(c,d]} f_n dF_n - \int_{(c,d]} f dF \right| \\
&= \int_{(-\infty,c]} |f_n| dF_n + \int_{(d,\infty)} |f_n| dF_n + \int_{(-\infty,c]} |f| dF + \int_{(d,\infty)} |f| dF \\
&\quad + \left| \left[\int_{(c,d]} f_n dF_n - \int_{(c,d]} f dF_n \right] + \left[\int_{(c,d]} f dF_n - \int_{(c,d]} f dF \right] \right| \\
&\leq \int_{(-\infty,c]} g dF_n + \int_{(d,\infty)} g dF_n + \int_{(-\infty,c]} |f| dF + \int_{(d,\infty)} |f| dF \\
&\quad + \int_{(c,d]} |f_n - f| dF_n + \left| \int_{(c,d]} f dF_n - \int_{(c,d]} f dF \right| \text{ (by (ii))}
\end{aligned}$$

$$= \left[\int_{\Re} g\,dF_n - \int_{(c,d]} g\,dF_n\right] + \int_{(-\infty,c]} |f|\,dF + \int_{(d,\infty)} |f|\,dF$$
$$+ \int_{(c,d]} |f_n - f|\,dF_n + \left|\int_{(c,d]} f\,dF_n - \int_{(c,d]} f\,dF\right|.$$

From (ii), (iii), and $\int_{\Re} g\,dF < \infty$ (by (iv)), it follows that $\int_{(-\infty,c]} |f|\,dF < \infty$ and $\int_{(d,\infty)} |f|\,dF < \infty$. Therefore, for $\varepsilon > 0$, c sufficiently small, d sufficiently large and always points of $C(F)$, we have

$$\int_{(-\infty,c]} |f|\,dF < \frac{\varepsilon}{6}, \quad \int_{(d,\infty)} |f|\,dF < \frac{\varepsilon}{6}.$$

Also, $\int_{\Re} g\,dF_n \to \int_{\Re} g\,dF$ (by (iv)), and $\int_{(c,d]} g\,dF_n \to \int_{(c,d]} g\,dF$ (by boundedness of g on $[c,d]$, its continuity on $(c,d]$, assumption (i), and Theorem 6), so that

$$\int_{\Re} g\,dF_n - \int_{(c,d]} g\,dF_n \to \int_{\Re} g\,dF - \int_{(c,d]} g\,dF.$$

It follows that, for all sufficiently large n,

$$\int_{\Re} g\,dF_n - \int_{(c,d]} g\,dF < \frac{2\varepsilon}{6}.$$

Also, for all sufficiently large n,

$$\int_{(c,d]} |f_n - f|\,dF_n < \frac{\varepsilon}{6} \text{ by ((iii))},$$

and

$$\left|\int_{(c,d]} f\,dF_n - \int_{(c,d]} f\,dF\right| < \frac{\varepsilon}{6} \text{ (by Theorem 6)}.$$

Combining all results obtained above, we have that for all sufficiently large n,

$$\left|\int_{\Re} f_n\,dF_n - \int_{\Re} f\,dF\right| < \varepsilon. \#$$

Chapter 9

Conditional Expectation and Conditional Probability, and Related Properties and Results

1. Indeed,

$$\int_{\sum_{i=1}^{\infty} A_i} X\, dP = \int \left(X I_{\sum_{i=1}^{\infty} A_i}\right) dP = \int \left(X \sum_{i=1}^{\infty} I_{A_i}\right) dP,$$

because $I_{\sum_{i=1}^{\infty} A_i} = \sum_{i=1}^{\infty} I_{A_i}$ as is easily seen,

$$= \int \left(\sum_{i=1}^{\infty} X I_{A_i}\right) dP,$$

because $X \sum_{i=1}^{\infty} I_{A_i} = \sum_{i=1}^{\infty} X I_{A_i}$ as is easily argued,

$$= \int \left(\lim_{n\to\infty} \sum_{i=1}^{n} X I_{A_i}\right) dP$$

$$= \lim_{n\to\infty} \int \left(\sum_{i=1}^{n} X I_{A_i}\right) dP\,,$$

because, clearly, $\left|\sum_{i=1}^{n} X I_{A_i}\right| \leq |X|$

independent of n and integrable, so that the Dominated Convergence Theorem applies,

$$= \lim_{n\to\infty} \sum_{i=1}^{n} \int (X I_{A_i})\, dP\,,$$

by Exercise 8 in Chapter 4,

$$= \sum_{i=1}^{\infty} \int (X I_{A_i})\, dP$$

$$= \sum_{i=1}^{\infty} \int_{A_i} X\, dP,$$

as was to be seen. #

2. By following the standard four steps, we have:

(a) Let $X = I_C, C \in \mathcal{A}$. Then $\int X dP(\cdot|B) = \int I_C dP(\cdot|B) = P(C|B) = \frac{P(C\cap B)}{P(B)} = \frac{1}{P(B)} \int_B I_C dP = \frac{1}{P(B)} \int_B X dP$.

(b) Let $X = \sum_{i=1}^{n} \alpha_i I_{C_i}$. Then $\int X dP(\cdot|B) = \int (\sum_i \alpha_i I_{C_i}) dP(\cdot|B) = \sum_i \alpha_i \times \int I_{C_i} dP(\cdot|B) = \sum_i \alpha_i \frac{1}{P(B)} \int_B I_{C_i} dP$ (by (a)) $= \frac{1}{P(B)} \int_B (\sum_i \alpha_i I_{C_i}) dP = \frac{1}{P(B)} \int_B X dP$.

(c) For $X \geq 0$, there exist $0 \leq X_n$ simple $\underset{n\to\infty}{\uparrow} X$ and $0 \leq X_n I_B \underset{n\to\infty}{\uparrow} X I_B$. Hence $\int X_n dP(\cdot|B) = \frac{1}{P(B)} \int_B X_n dP$ (by (b)) $= \frac{1}{P(B)} \int (X_n I_B) dP \underset{n\to\infty}{\longrightarrow} \frac{1}{P(B)} \int (X I_B) dP$ (by the Lebesgue Monotone Convergence Theorem) $= \frac{1}{P(B)} \times \int_B X dP$; i.e., $\int X_n dP(\cdot|B) \underset{n\to\infty}{\longrightarrow} \frac{1}{P(B)} \int_B X dP$. Also, $\int X_n dP(\cdot|B) \underset{n\to\infty}{\longrightarrow} \int X dP(\cdot|B)$, so that $\int X dP(\cdot|B) = \frac{1}{P(B)} \int_B X dP$.

(d) For any (integrable) X,

$$\begin{aligned}\int X dP(\cdot|B) &= \int (X^+ - X^-) dP(\cdot|B)\\ &= \int X^+ dP(\cdot|B) - \int X^- dP(\cdot|B)\\ &= \frac{1}{P(B)} \int_B X^+ dP - \frac{1}{P(B)} \int_B X^- dP \text{ (by (c))}\\ &= \frac{1}{P(B)} \int_B (X^+ - X^-) dP = \frac{1}{P(B)} \int_B X dP. \#\end{aligned}$$

3. Set $B_1 = [0, \frac{1}{4}]$, $B_2 = (\frac{1}{4}, \frac{2}{3}]$, $B_3 = (\frac{2}{3}, 1]$. Then $\{B_1, B_2, B_3\}$ form a partition of Ω. Therefore, by the "Special Case" right after Definition 1, we have:

$$\mathcal{E}(X|\mathcal{F}) = (\mathcal{E}_{B_1} X) I_{B_1} + (\mathcal{E}_{B_2} X) I_{B_2} + (\mathcal{E}_{B_3} X) I_{B_3},$$

where

$$\mathcal{E}_{B_1} X = \frac{1}{P(B_1)} \int_{B_1} X dP = 4 \int_0^{\frac{1}{4}} x^2 dx = \frac{1}{48},$$

$$\mathcal{E}_{B_2} X = \frac{1}{P(B_2)} \int_{B_2} X dP = \frac{12}{5} \int_{\frac{1}{4}}^{\frac{2}{3}} x^2 dx = \frac{97}{432},$$

$$\mathcal{E}_{B_3} X = \frac{1}{P(B_3)} \int_{B_3} X dP = 3 \int_{\frac{2}{3}}^{1} x^2 dx = \frac{19}{27}.$$

Thus, $\mathcal{E}(X|\mathcal{F}) = \frac{1}{48} I_{[0,\frac{1}{4}]} + \frac{97}{432} I_{(\frac{1}{4},\frac{2}{3}]} + \frac{19}{27} I_{(\frac{2}{3},1]}$. #

4. $\int_B Y dP \leq \int_B X dP$ or equivalently, $\int_B (Y - X) dP \leq 0$ or $\int_B Z dP \leq 0$, where $Z = Y - X$. Thus, to show $\int_B Z dP \leq 0$ for every B implies $Z \leq 0$ a.s. Let $C = (Z \leq 0)$, $D = (Z > 0)$ $(= C^c)$. Then, it suffices to show that $P(D) = 0$. However, by taking $B = D$, $\int_D Z dP \leq 0$ and $\int_D Z dP = \int (Z I_D) dP \geq 0$ since $Z > 0$ on D, so that $\int_D Z dP = 0$. Then, we shall show that, if for a r.v. Z with $D = (Z > 0)$, it holds $\int_D Z dP = 0$, then $P(D) = 0$. To this end, we follow the familiar four steps.

(a) Let $Z = I_A$. Then necessarily $D = A$. Hence $0 = \int_A ZdP = \int_A I_A dP = P(A) = P(D)$; i.e., $P(D) = 0$.

(b) Let $Z = \sum_{i=1}^{n} \alpha_i I_{A_i}$ and let $\alpha_{i_1}, \ldots, \alpha_{i_k}$ be > 0 and $\alpha_{j_1}, \ldots, \alpha_{j_l} \leq 0$ $(k + l = n)$. Then $D = (Z > 0) = A_{i_1} + \ldots + A_{i_k}$ and

$$
\begin{aligned}
0 &= \int_D ZdP = \int (ZI_D)dP = \int (ZI_{\sum_{r=1}^{k} A_{i_r}})dP \\
&= \int \left(\sum_{r=1}^{k} ZI_{A_{i_r}}\right) dP = \sum_{r=1}^{k} \int (ZI_{A_{i_r}})dP = \sum_{r=1}^{k} \int_{A_{i_r}} ZdP \\
&= \sum_{r=1}^{k} \int_{A_{i_r}} \alpha_{i_r} dP = \sum_{r=1}^{k} \alpha_{i_r} P(A_{i_r}).
\end{aligned}
$$

Hence $\alpha_{i_r} P(A_{i_r}) = 0$ for all r, and since $\alpha_{i_r} > 0$ for all r, it follows that $P(A_{i_r}) = 0$ for all r; consequently, $P(D) = 0$.

(c) For $Z \geq 0$, $0 \leq X_n$ simple $\underset{n\to\infty}{\uparrow} X$, where we take $X_n = \sum_{j=1}^{n2^n} \frac{j-1}{2^n} I_{A_{nj}} + nI_{A_n}$, $A_{nj} = (\frac{j-1}{2^n} < Z \leq \frac{j}{2^n})$, $A_n = (Z > n)$. For $\omega \in D$, we have $Z(\omega) > 0$, so that $X_n(\omega) > 0$ for all sufficiently large n, $n \geq n_0 = n_0(\omega)$, say. Working with such ns, we have $0 < X_n I_D$ simple $\underset{n\to\infty}{\uparrow} ZI_D$ and $\int (X_n I_D)dP \underset{n\to\infty}{\uparrow} \int (ZI_D)dP = \int_D ZdP = 0$, so that $\int (X_n I_D)dP = 0$ (for all $n \geq n_0$). Since $X_n = \sum_{j=1}^{k_n} \alpha_j I_{A_j}$ with $\alpha_j > 0$, we have $\int (X_n I_D)dP = \sum_j \alpha_j P(A_j \cap D) = 0$. Then, as in (b), $P(D) = 0$.

(d) For any Z, $\int_D ZdP = \int_D Z^+ dP - \int_D Z^- dP$, where $Z^+ = Z$ if $Z > 0$ and $Z^+ = 0$ if $Z \leq 0$. Then $\int_D ZdP = \int_D Z^+ dP = 0$, and hence $P(D) = 0$ by (c). #

5. **(i)** $0 = \int_A XdP = \int_{A\cap C} XdP + \int_{A\cap C^c} XdP = \int_{A\cap C} XdP = \int_D XdP$, where $D = A \cap C$ and $C = (X > 0)$. So, $X > 0$ on D and $\int_D XdP = 0$ must imply that $P(D) = 0$. This would mean that $X = 0$ a.s.

   (a) Let $X = I_B$. Then it must be that $B = C$, so that $B \supseteq D$. Then $0 = \int_D XdP = \int_D I_B dP = P(B \cap D) = P(D) = 0$.

   (b) Let $X = \sum_{i=1}^{n} \alpha_i I_{A_i}$ with $\alpha_{i_1}, \ldots, \alpha_{i_k} > 0$ (and $\alpha_{j_1}, \ldots, \alpha_{j_l} = 0$, $k + l = n$). Then $0 = \int_D XdP = \sum_{r=1}^{k} \alpha_{i_r} P(A_{i_r} \cap D)$, so that $\alpha_{i_r} P(A_{i_r} \cap D) = 0$ for all r, and hence $P(A_{i_r} \cap D) = 0$ for all r. Since $A_{i_1}, \ldots, A_{i_k}$ form a partition of D, we have $P(D) = 0$.

   (c) For $X \geq 0$, $0 \leq X_n$ simple $\underset{n\to\infty}{\uparrow} X$, and as in Exercise 4(c), take $X_n > 0$, $n \geq n_0$. Then conclude, as in that part, that $P(D) = 0$.

   (ii) $X \geq 0$ implies $\mathcal{E}^{\mathcal{B}} X \geq 0$ a.s. Also, for $B \in \mathcal{B}$, $\int_B XdP = \int_B \mathcal{E}^{\mathcal{B}} XdP_{\mathcal{B}}$. In particular, since $A \in \mathcal{B}$, $\int_{B\cap A} XdP = \int_{B\cap A} \mathcal{E}^{\mathcal{B}} XdP_{\mathcal{B}}$. But $0 \leq \int_{B\cap A} XdP \leq \int_A XdP = 0$ (since $X \geq 0$). Thus, $\int_{B\cap A} XdP = 0$ and hence $\int_{B\cap A} \mathcal{E}^{\mathcal{B}} XdP_{\mathcal{B}} = 0$. However, $0 = \int_{B\cap A} \mathcal{E}^{\mathcal{B}} XdP_{\mathcal{B}} = \int_B (I_A \mathcal{E}^{\mathcal{B}} X)dP_{\mathcal{B}} =$

$\int_B 0dP_{\mathcal{B}}$. So, $\int_B (I_A \mathcal{E}^{\mathcal{B}} X) dP_{\mathcal{B}} = \int_B 0dP_{\mathcal{B}}$ for all $B \in \mathcal{B}$, and both $I_A \mathcal{E}^{\mathcal{B}} X$ and 0 are $\mathcal{B}$-measurable. Hence $I_A \mathcal{E}^{\mathcal{B}} X = 0$ a.s. Since $I_A = 1$ on A, in order for $I_A \mathcal{E}^{\mathcal{B}} X = 0$ a.s. either $P(A) = 0$ (which is not) or $\mathcal{E}^{\mathcal{B}} X = 0$ a.s. on A. #

6. The function $g^*(x) = x^r$ $(x \geq 0, r \geq 1)$ is convex (since $\frac{d^2}{dx^2} g^*(x) = r(r-1) \times x^{r-2} \times \geq 0$) and $|x|$ is symmetric about 0. Thus, $g(x) = |x|^r$, $x \in \Re$, is convex. Then, by Jensen inequality, $g(\mathcal{E}^{\mathcal{B}} X) \leq \mathcal{E}^{\mathcal{B}} g(X)$ a.s. or $|\mathcal{E}^{\mathcal{B}} X|^r \leq \mathcal{E}^{\mathcal{B}} |X|^r$ a.s. #

7. From $\mathcal{E} X \in \Re$, it follows that $\mathcal{E} X^+$ and $\mathcal{E} X^- < \infty$, and also $\int_B X^+$ and $\int_B X^- < \infty$ for every $B \in \mathcal{B}$. Since $\int_B X^+ dP = \int_B \mathcal{E}^{\mathcal{B}} X^+ dP_{\mathcal{B}}$ and $\int_B X^- dP = \int_B \mathcal{E}^{\mathcal{B}} X^- dP_{\mathcal{B}}$, it follows that $\int_B \mathcal{E}^{\mathcal{B}} X^+ dP_{\mathcal{B}}$ and $\int_B \mathcal{E}^{\mathcal{B}} X^- dP_{\mathcal{B}} < \infty$ for every $B \in \mathcal{B}$. Let $C = \{\mathcal{E}^{\mathcal{B}} X^+ = \infty\}$. Then $C = \cup_{n=1}^{\infty} \{\mathcal{E}^{\mathcal{B}} X^+ \geq n\}$, so that $C \in \mathcal{B}$. Hence $\infty > \int_C X^+ dP = \int_C \mathcal{E}^{\mathcal{B}} X^+ dP_{\mathcal{B}}$, and hence $P(C) = 0$. Likewise, if $D = \{\mathcal{E}^{\mathcal{B}} X^- = \infty\}$, then $D \in \mathcal{B}$ and $P(D) = 0$. It follows that $\mathcal{E}^{\mathcal{B}} X$ is finite except on the set $C + D$ with $P(C + D) = P(C) + P(D) = 0$; i.e., $\mathcal{E}^{\mathcal{B}} X$ is a.s. finite. #

8. The inequality is true for $n = 2$ (by definition of convexity). Assume (*) to be true for k and establish it for $k + 1$. Without loss of generality, assume $\alpha_{k+1} < 1$. Then:

$$
\begin{aligned}
& g(\alpha_1 x_1 + \ldots + \alpha_k x_k + \alpha_{k+1} x_{k+1}) = \\
& g\left((1-\alpha_{k+1})\left(\frac{\alpha_1}{1-\alpha_{k+1}} x_1 + \ldots + \frac{\alpha_k}{1-\alpha_{k+1}} x_k\right) + \alpha_{k+1} x_{k+1}\right) \leq \\
& (1-\alpha_{k+1}) g\left(\frac{\alpha_1}{1-\alpha_{k+1}} x_1 + \ldots + \frac{\alpha_k}{1-\alpha_{k+1}} x_k\right) + \alpha_{k+1} g(x_{k+1}) \\
& \quad (\text{since } 1-\alpha_{k+1},\ \alpha_{k+1} \geq 0 \text{ and } (1-\alpha_{k+1}) + \alpha_{k+1} = 1) \\
& \leq (1-\alpha_{k+1})\left[\frac{\alpha_1}{1-\alpha_{k+1}} g(x_1) + \ldots + \frac{\alpha_k}{1-\alpha_{k+1}} g(x_k)\right] + \alpha_{k+1} g(x_{k+1}) \\
& \quad \left(\text{by the induction hypothesis, since } \frac{\alpha_i}{1-\alpha_{k+1}} \geq 0,\ i = 1, \ldots, k\right. \\
& \quad \left.\text{and their sum is 1}\right) \\
& = \alpha_1 g(x_1) + \ldots + \alpha_k g(x_k) + \alpha_{k+1} g(x_{k+1}).\ \#
\end{aligned}
$$

9. As already shown in the proof of Lemma 2, $\int_{B'} g(y) dP'_Y = \int_B g(Y) dP_Y$ for $g = I_{A'}$, $A' \in \mathcal{B}'_Y$.

Next, let $g = \sum_{i=1}^n \alpha_i I_{A'_i}$, where $\{A'_1, \ldots, A'_n\}$ is a (measurable) partition of Ω', and let $A_i = Y^{-1}(A'_i)$, $i = 1, \ldots, n$, so that $g(Y) = \sum_{i=1}^n \alpha_i I_{A_i}$, and let $B' \in \mathcal{B}'_Y$, $B = Y^{-1}(B')$. Then by the linearity of the integral and the previous step,

$$
\int_{B'} g(y) dP'_Y = \sum_{i=1}^n \alpha_i \int_{B'} I_{A'_i} P'_Y = \sum_{i=1}^n \alpha_i \int_B I_{A_i} P_Y = \int_B g(Y) dP_Y.
$$

Now, let g be nonnegative. Then there exist $0 \leq g_n$ simple $\underset{n\to\infty}{\uparrow} g$; i.e., $g_n = \sum_{i=1}^{r_n} \alpha_{ni} I_{A'_{ni}}$ ($\alpha_{ni} \geq 0, i = 1, \ldots, r_n$), which implies that $0 \leq g_n(Y) = \sum_{i=1}^{r_n} \alpha_{ni} \times I_{A_{ni}}$ simple $\underset{n\to\infty}{\uparrow} g(Y)$, where $A_{ni} = Y^{-1}(A'_{ni}), i = 1, \ldots, r_n$. Then, for every $B' \in \mathcal{B}'_Y$ and $B = Y^{-1}(B)$, we have

$$\int_{B'} g_n(y)dP'_Y \underset{n\to\infty}{\longrightarrow} \int_{B'} g(y)dP'_Y \text{ and } \int_B g_n(Y)dP_Y \underset{n\to\infty}{\longrightarrow} \int_B g(Y)dP_Y,$$

whereas by the previous step, $\int_{B'} g_n(y)dP'_Y = \int_B g_n(Y)dP_Y$, $n \geq 1$. It follows that $\int_{B'} g(y)dP'_Y = \int_B g(Y)dP_Y$. Finally, for any g, write $g = g^+ - g^-$, which implies $g(Y) = g^+(Y) - g^-(Y)$. Now, if $\int_B g(Y)dP_Y$ exists, it follows that either $\int_B g^+(Y)dP_Y < \infty$ or $\int_B g^-(Y)dP_Y < \infty$ or both. Since $\int_B g^+(y)dP_Y = \int_{B'} g^+(Y)dP'_Y$ and $\int_B g^-(y)dP_Y = \int_{B'} g^-(Y)dP'_Y$, by the previous step, it follows that either $\int_{B'} g^+(y)dP'_Y < \infty$ or $\int_{B'} g^-(Y)dP'_Y < \infty$ or both. Thus, $\int_{B'} g(y)dP'_Y$ exists and

$$\begin{aligned}\int_{B'} g(y)dP'_Y &= \int_{B'} g^+(y)dP'_Y - \int_{B'} g^-(y)dP'_Y \\ &= \int_B g^+(Y)dP_Y - \int_B g^-(Y)dP_Y = \int_B g(Y)dP_Y.\end{aligned}$$

Likewise, the existence of $\int_{B'} g(y)dP'_Y$ implies the existence of $\int_B g(Y)dP_Y$ and their equality. #

10. Let $x_i, i \geq 1, y_j, j \geq 1$ be the values of X and Y, respectively, and without loss of generality assume that $P(X = x_i) > 0, i \geq 1, P(Y = y_j) > 0, j \geq 1$. Let D_Y be the discrete σ-field generated by $\{y_j, j \geq 1\}$. Since, clearly, $\{y_j\} \in \mathcal{B}'_Y$ (because $Y^{-1}(\{y_j\}) \in \mathcal{A}$), it follows that $D_Y \subset \mathcal{B}'_Y$. Next, set $f_Y(y_j) = P(Y = y_j) = p_j$. Then $f_Y^{-1}(\{p_j\})$ is a member of D_Y containing y_j and hence $f_Y^{-1}(\{p_j\}) \in \mathcal{B}'_Y$. It follows that $f_Y(\cdot)$ is $\mathcal{B}'_Y$-measurable.
Next, let $\mathcal{B}'_{X,Y}$ and $D_{X,Y}$ be defined in a way similar to that of $\mathcal{B}'_Y$ and D_Y, and set $f_{X,Y}(x_i, y_j) = P(X = x_i, Y = y_j) = p_{ij}$. Then, as above, $f_{X,Y}(\cdot)$ is $D_{X,Y}$-measurable (and hence $\mathcal{B}'_{X,Y}$-measurable). For any fixed x_i, x_{i_0}, say, $f_{X,Y}(x_{i_0}, \cdot)$ is the $\{y_j, j \geq 1\}$-section of $f_{X,Y}(\cdot, \cdot)$ at x_{i_0}, and hence it is D_Y-measurable. Thus, $f_{X,Y}(x_{i_0}, \cdot)$ and $f_Y(\cdot)$ are both $\mathcal{B}'_Y$-measurable, and then so is their ratio $P(X = i_0 | Y = \cdot)$, as was to be seen. #

11. We can always take $\mathcal{B} = \mathcal{A}$ and consider a sequence $\{X_n\}$ such that $X_n \overset{P}{\underset{n\to\infty}{\longrightarrow}} X$, some r.v., but $X_n \overset{a.s.}{\underset{n\to\infty}{\nrightarrow}} X$. Then $\mathcal{E}^{\mathcal{B}} X_n = X_n$ a.s. and $\mathcal{E}^{\mathcal{B}} X = X$ a.s. with $\mathcal{E}^{\mathcal{B}} X_n \overset{P}{\underset{n\to\infty}{\longrightarrow}} \mathcal{E}^{\mathcal{B}} X$, but $\mathcal{E}^{\mathcal{B}} X_n \overset{a.s.}{\underset{n\to\infty}{\nrightarrow}} \mathcal{E}^{\mathcal{B}} X$.
As another possibility, let $X_n \overset{P}{\underset{n\to\infty}{\longrightarrow}} c$ but $X_n \overset{a.s.}{\underset{n\to\infty}{\nrightarrow}} c$, and take $\mathcal{B} = \sigma(X_1, X_2, \ldots)$. Then X_n is $\mathcal{B}$-measurable, $n \geq 1$ (and so is c), and $\mathcal{E}^{\mathcal{B}} X_n = X_n$ a.s., $\mathcal{E}^{\mathcal{B}} c = c$ a.s. Therefore $\mathcal{E}^{\mathcal{B}} X_n \overset{P}{\underset{n\to\infty}{\longrightarrow}} \mathcal{E}^{\mathcal{B}} c$, but $\mathcal{E}^{\mathcal{B}} X_n \overset{a.s.}{\underset{n\to\infty}{\nrightarrow}} \mathcal{E}^{\mathcal{B}} c$. #

12. **(i)** The first step is immediate, and the second goes like this:

$$\left(\frac{y-\mu_2}{\sigma_2}\right)-\rho\left(\frac{x-\mu_1}{\sigma_1}\right)=\frac{y-\mu_2}{\sigma_2}-\frac{1}{\sigma_2}\times\rho\sigma_2\frac{x-\mu_1}{\sigma_1}$$
$$=\frac{1}{\sigma_2}\left[y-\left(\mu_2+\rho\sigma_2\frac{x-\mu_1}{\sigma_1}\right)\right]=\frac{1}{\sigma_2}(y-b)=\frac{y-b}{\sigma_2}.$$

(ii) Observe that the second factor on the right-hand side is the p.d.f. of a Normal distribution with mean b and variance $\sigma_2^2(1-\rho^2)$. Thus, integration with respect to y, leaves us with the first factor which is the p.d.f. of $N(\mu_1,\sigma_1^2)$. #

13. The expression $p_{X,Y}(x,y)/p_X(x)$, where p_X is the p.d.f. of X, leaves us with the second expression on the right-hand side of $p_{X,Y}(x,y)$ in Exercise 12 (ii), which is the p.d.f. of a Normal distribution as described. #

14. **(i)** We have:

$$\begin{aligned}\mathcal{E}(XY)&=\mathcal{E}[\mathcal{E}(XY|X)]=\mathcal{E}[X\mathcal{E}(Y|X)]\text{ (by Theorem 4)}\\&=\mathcal{E}(Xb)\text{ (by Exercise 13)}\\&=\mathcal{E}\{X[\mu_2+\frac{\rho\sigma_2}{\sigma_1}(X-\mu_1)]\}\text{ (by the expression of } b)\\&=\mu_2\mu_1+\frac{\rho\sigma_2}{\sigma_1}(\mathcal{E}X^2-\mu_1^2)\\&=\mu_1\mu_2+\frac{\rho\sigma_2}{\sigma_1}\times\sigma_1^2=\mu_1\mu_2+\rho\sigma_1\sigma_2.\end{aligned}$$

(ii) $\text{Cov}(X,Y)=\mathcal{E}(XY)-(\mathcal{E}X)(\mathcal{E}Y)=(\mu_1\mu_2+\rho\sigma_1\sigma_2)-\mu_1\mu_2=\rho\sigma_1\sigma_2.$

(iii) $\rho(X,Y)=\frac{\text{Cov}(X,Y)}{\sigma_1\sigma_2}=\frac{\rho\sigma_1\sigma_2}{\sigma_1\sigma_2}=\rho.$ #

15. From the transformations $u=(x_1-\mu_1)/\sigma_1$, $v=(y-\mu_2)/\sigma_2$, we get $x=\mu_1+\sigma_1 u$, $y=\mu_2+\sigma_2 v$, so that the Jacobian of the transformation is $J=\sigma_1\sigma_2$. Since $p_{X,Y}(x,y)=\frac{1}{2\pi\sigma_1\sigma_2\sqrt{1-\rho^2}}e^{-q/2}$, where

$$q=\frac{1}{1-\rho^2}\left[\left(\frac{x-\mu_1}{\sigma_1}\right)^2-2\rho\left(\frac{x-\mu_1}{\sigma_1}\right)\left(\frac{y-\mu_2}{\sigma_2}\right)+\left(\frac{y-\mu_2}{\sigma_2}\right)^2\right],$$

we get:

$$\begin{aligned}p_{U,V}(u,v)&=\frac{1}{2\pi\sigma_1\sigma_2\sqrt{1-\rho^2}}\exp\left[-\frac{1}{2(1-\rho^2)}(u^2-2\rho uv+v^2)\right]\times\sigma_1\sigma_2\\&=\frac{1}{2\pi\sqrt{1-\rho^2}}\exp\left[-\frac{1}{2(1-\rho^2)}(u^2-2\rho uv+v^2)\right].\end{aligned}$$

Thus, the r.v.s U and V have the Bivariate Normal distribution with parameters 0,0,1,1, and ρ. #

16. **(i)** We have

$$\begin{aligned}\mathrm{Var}(X|Y) &= \mathcal{E}\{[X-\mathcal{E}(X|Y)]^2|Y\} \text{ (by definition)}\\ &= \mathcal{E}\{X^2+[\mathcal{E}(X|Y)]^2-2X\mathcal{E}(X|Y)|Y\}\\ &= \mathcal{E}(X^2|Y)+[\mathcal{E}(X|Y)]^2-2[\mathcal{E}(X|Y)]^2 \text{ a.s.}\\ &= \mathcal{E}(X^2|Y)-[\mathcal{E}(X|Y)]^2.\end{aligned}$$

(ii) From part (i), we have

$$\begin{aligned}\mathcal{E}[\mathrm{Var}(X|Y)] &= \mathcal{E}\{\mathcal{E}(X^2|Y)-[\mathcal{E}(X|Y)]^2\}\\ &= \mathcal{E}X^2-\mathcal{E}[\mathcal{E}(X|Y)]^2.\end{aligned}$$

On the other hand,

$$\begin{aligned}\mathrm{Var}[\mathcal{E}(X|Y)] &= \mathcal{E}[\mathcal{E}(X|Y)]^2-\{\mathcal{E}[\mathcal{E}(X|Y)]\}^2\\ &= \mathcal{E}[\mathcal{E}(X|Y)]^2-(\mathcal{E}X)^2.\end{aligned}$$

Adding up the above two expressions, we get $\mathcal{E}X^2-(\mathcal{E}X)^2=\mathrm{Var}(X)=\mathcal{E}[\mathrm{Var}(X|Y)]+\mathrm{Var}[\mathcal{E}(X|Y)]$. #

17. **(i)** Set $B_j=(N=j)$, $j=1,2,\ldots$ Then it is clear that $X=X_1+\ldots+X_j$ on the event B_j, $j\geq 1$. It is also clear that, for any $B\in\mathcal{B}$,

$$\begin{aligned}(X\in B) &= \sum_{j=1}^{\infty}\{B\cap B_j\cap[(X_1+\ldots+X_j)\in B]\}\\ &= B\cap\sum_{j=1}^{\infty}\{B_j\cap[(X_1+\ldots+X_j)\in B]\},\end{aligned}$$

whereas the set on the right-hand side above is an event. Then so is $(X\in B)$.

(ii) On the basis of the Special Case right after Definition 1, we have

$$\mathcal{E}(X|N)=\sum_{j=1}^{\infty}(\mathcal{E}_{B_j}X)I_{B_j},\ \mathcal{E}_{B_j}X=\frac{1}{P(B_j)}\int_{B_j}XdP.$$

However,

$$\begin{aligned}\mathcal{E}_{B_j}X &= \frac{1}{P(B_j)}\mathcal{E}(XI_{B_j})=\frac{1}{P(B_j)}\mathcal{E}[(X_1+\ldots+X_j)I_{B_j}]\\ &= \frac{1}{P(B_j)}[\mathcal{E}(X_1I_{B_j})+\ldots+\mathcal{E}(X_jI_{B_j})]\\ &= \frac{1}{P(B_j)}[(\mathcal{E}X_1)P(B_j)+\ldots+(\mathcal{E}X_j)P(B_j)]\\ &\quad \text{(by independence of } N \text{ and the } X_is)\\ &= j\mu.\end{aligned}$$

That is, $\mathcal{E}_{B_j}X = j\mu$, and therefore

$$\mathcal{E}(X|N) = \sum_{j=1}^{\infty}(j\mu)I_{B_j} = \mu\sum_{j=1}^{\infty} jI_{B_j} = \mu N,$$

since, clearly, $N = \sum_{j=1}^{\infty} jI_{B_j}$. Therefore

$$\mathcal{E}X = \mathcal{E}[\mathcal{E}(X|N)] = \mathcal{E}(\mu N) = \mu(\mathcal{E}N).$$

(iii) Next, working as in part (ii),

$$\mathcal{E}(X^2|N) = \sum_{j=1}^{\infty}(\mathcal{E}_{B_j}X^2)I_{B_j}, \ \mathcal{E}_{B_j}X^2 = \frac{1}{P(B_j)}\int_{B_j} X^2 dP,$$

and

$$\begin{aligned}\mathcal{E}_{B_j}X^2 &= \frac{1}{P(B_j)}\mathcal{E}[(X_1 + \ldots + X_j)^2 I_{B_j}] \\ &= \frac{1}{P(B_j)}\mathcal{E}\left[\left(\sum_{i=1}^{j} X_i^2 + \ 2\sum_{1\le k<l\le j} X_k X_l\right) I_{B_j}\right] \\ &= \frac{1}{P(B_j)}P(B_j)[j\mathcal{E}X_1^2 + j(j-1)(\mathcal{E}X_1)^2] \\ &= j(\sigma^2 + \mu^2) + j(j-1)\mu^2 = j\sigma^2 + j^2\mu^2,\end{aligned}$$

so that,

$$\mathcal{E}(X^2|N) = \sum_{j=1}^{\infty}(j\sigma^2 + j^2\mu^2)I_{B_j} = \sigma^2 N + \mu^2 N^2.$$

It follows that

$$\text{Var}(X|N) = \mathcal{E}(X^2|N) - [\mathcal{E}(X|N)]^2 = \sigma^2 N.$$

(iv) Since $\mathcal{E}(X|N) = \mu N$ and $\text{Var}(X|N) = \sigma^2 N$, then part (ii) of Exercise 16 yields:

$$\begin{aligned}\text{Var}(X) &= \mathcal{E}[\text{Var}(X|N)] + \text{Var}[\mathcal{E}(X|N)] = \mathcal{E}(\sigma^2 N) + \text{Var}(\mu N) \\ &= \sigma^2(\mathcal{E}N) + \mu^2\text{Var}(N). \ \#\end{aligned}$$

18. For every $B \in \mathcal{B}$, it follows that $\int_B X dP_{\mathcal{B}}$ is either 0 or $\mathcal{E}X$, according to whether $P(B) = 0$ or $P(B) = 1$; or $\int_B X dP_{\mathcal{B}} = \int_B (\mathcal{E}X) dP$. However, $\int_B (\mathcal{E}X) dP = \int_B (\mathcal{E}X) dP_{\mathcal{B}}$, since $\mathcal{E}X$ is $\mathcal{B}$-measurable (by being a constant). Therefore X and $\mathcal{E}X$ are both $\mathcal{B}$-measurable and $\int_B X dP_{\mathcal{B}} = \int_B (\mathcal{E}X) dP_{\mathcal{B}}$ for every $B \in \mathcal{B}$. Hence $X = \mathcal{E}X$ a.s. #

19. Consider the r.v.s $X_1, \ldots, X_n$ taking on the respective values $x_{i1}, \ldots, x_{in}$, where $x_{ij} \geq 0$, $j = 1, \ldots, n$, all i, and $x_{i1} + \cdots + x_{in} = t$ (which imply, of course, that $0 \leq x_{ij} \leq t$, $j = 1, \ldots, n$, all i). Now, look at the r.v.s $X_1, \ldots, X_n$ as n distinct cells, and also consider t indistinguishable balls. The the number of n-tuples $(x_{i1}, \ldots, x_{in})$ with $0 \leq x_{ij} \leq t$, $j = 1, \ldots, n$, all i, and $x_{i1} + \cdots + x_{in} = t$ is the same as the number of ways that the t (indistinguishable) balls are distributed into the n (distinct) cells, so that the jth cell contains x_{ij} balls (that is, $X_j = x_{ij}$, $j = 1, \ldots, n$). However, this number is equal to $\binom{n+t-1}{t}$ (see Theorem 9(iii) of Chapter 2 in Roussas (1997)). #

Chapter 10

Independence

1. If the events $A_1, \ldots, A_n$ are independent, we wish to show that the events $A'_1, \ldots, A'_n$ are also independent, where A'_i is either A_i or A^c_i, $i = 1, \ldots, n$. For the events $A'_1, \ldots, A'_n$ to be independent, we have to show that:
For any $2 \leq k \leq n$ and any $i_1, \ldots, i_k$ with $1 \leq i_1 < i_2 < \ldots < i_k \leq n$, we have:

$$P(A'_{i_1} \cap \ldots A'_{i_k}) = P(A'_{i_1}) \ldots P(A'_{i_k}). \tag{1}$$

Without loss of generality and for easier writing, instead of (1), it suffices to show that:

$$P(A'_1 \cap \ldots \cap A'_k) = P(A'_1) \ldots P(A'_k). \tag{2}$$

We establish (2) by showing that:

(i) The factorization in (2) holds when only one of the A_is appears as a complement, and without loss of generality and for easier writing, assume it to be A^c_k.

(ii) Assume the factorization in (2) to hold true when only m of the A_is appear in their complements, and without loss of generality and for easier writing, assume them to be the last $k - m$; i.e., $A^c_{k-m+1}, \ldots, A^c_k$.

(iii) We prove the factorization in (2) to be true when only $m + 1$ of the A_is appear in their complements, and without loss of generality and for easier writing, we assume them to be the last $k - (m + 1) = k - m - 1$; i.e., $A^c_{k-m}, \ldots, A^c_k$.

Proof of (i): We have:

$$\begin{aligned} P(A_1 \cap \ldots \cap A_{k-1} \cap A^c_k) &= P[(A_1 \cap \ldots \cap A_{k-1}) - (A_1 \cap \ldots \cap A_k)] \\ &\quad (\text{since } A \cap B^c = A - (A \cap B)) \\ &= P(A_1 \cap \ldots \cap A_{k-1}) - P(A_1 \cap \ldots \cap A_k) \end{aligned}$$

$$
\begin{aligned}
&= P(A_1)\ldots P(A_{k-1}) - P(A_1)\ldots P(A_k)\\
&\quad \text{(by assumption)}\\
&= P(A_1)\ldots P(A_{k-1})[1 - P(A_k)]\\
&= P(A_1)\ldots P(A_{k-1})P(A_k^c), \text{ which shows (i).}
\end{aligned}
$$

By the (induction) hypothesis made in (ii), we have:

$$
\begin{aligned}
&P(A_1 \cap \ldots \cap A_{k-m} \cap A_{k-m+1}^c \cap \ldots \cap A_k^c)\\
&\quad = P(A_1)\ldots P(A_{k-m})P(A_{k-m+1}^c)\ldots P(A_k^c). \qquad (3)
\end{aligned}
$$

Proof of (iii): We have:

$$
\begin{aligned}
&P(A_1 \cap \ldots \cap A_{k-m-1} \cap A_{k-m}^c \cap \ldots \cap A_k^c)\\
&= P[(A_1 \cap \ldots \cap A_{k-m-1} \cap A_{k-m+1}^c \cap \ldots \cap A_k^c)\\
&\quad -(A_1 \cap \ldots \cap A_{k-m-1} \cap A_{k-m} \cap A_{k-m+1}^c \cap \ldots \cap A_k^c)]\\
&\quad \text{(since } A \cap B^c = A - (A \cap B) \text{ with } A = A_1 \cap \ldots \cap A_{k-m-1}\\
&\quad \cap A_{k-m+1}^c \cap \ldots \cap A_k^c \text{ and } B = A_{k-m})\\
&= P(A_1 \cap \ldots \cap A_{k-m-1} \cap A_{k-m+1}^c \cap \ldots \cap A_k^c)\\
&\quad -P(A_1 \cap \ldots \cap A_{k-m-1} \cap A_{k-m} \cap A_{k-m+1}^c \cap \ldots \cap A_k^c)\\
&= P(A_1)\ldots P(A_{k-m-1})P(A_{k-m+1}^c)\ldots P(A_k^c)\\
&\quad -P(A_1)\ldots P(A_{k-m-1})P(A_{k-m})P(A_{k-m+1}^c)\ldots P(A_k^c)\\
&\quad \text{(by the (induction) hypothesis in (ii))}\\
&= P(A_1)\ldots P(A_{k-m-1})[1 - P(A_{k-m})]P(A_{k-m+1}^c)\ldots P(A_k^c)\\
&= P(A_1)\ldots P(A_{k-m-1})P(A_{k-m}^c)P(A_{k-m+1}^c)\ldots P(A_k^c).
\end{aligned}
$$

The proof of (iii), and hence of the asserted independence, is completed. #

2. Suppose A and B are of the form $A = (A_{11} \cup A_{21}) \cup \cdots \cup (A_{1m} \cup A_{2m})$, $B = (B_{11} \cup B_{21}) \cup \cdots \cup (B_{1n} \cup B_{2n})$ with $A_{1i}, B_{1i} \in \mathcal{F}_1, i = 1, \ldots, m$, $A_{2j}, B_{2j} \in \mathcal{F}_2, j = 1, \ldots, n$. Then $A \cup B$ is of the same form and hence belongs in $\mathcal{F}$. The same happens if, e.g., $A = (A_{11} \cap A_{21}) \cup \cdots \cup (A_{1m} \cap A_{2m})$, $B = (B_{11} \cap B_{21}) \cup \cdots \cup (B_{1n} \cap B_{2n})$. If, e.g., $A = (A_{11} \cup A_{21}) \cup \cdots \cup (A_{1m} \cup A_{2m})$, $B = (B_{11} \cap B_{21}) \cup \cdots \cup (B_{1n} \cap B_{2n})$, then again $A \cup B \in \mathcal{F}$, since $A \cup B$ is of the form described in the definition of $\mathcal{F}$. Thus $\mathcal{F}$ is closed under unions.
Next, let $A = (A_{11} \cup A_{21}) \cup \cdots \cup (A_{1m} \cup A_{2m})$, so that $A^c = (A_{11}^c \cap A_{21}^c) \cap \cdots \cap (A_{1m}^c \cap A_{2m}^c) = (A_{11}^c \cap \cdots \cap A_{1m}^c) \cap (A_{21}^c \cap \cdots \cap A_{2m}^c)$ with $A_{11}^c \cap \cdots \cap A_{1m}^c \in \mathcal{F}_1$ and $A_{21}^c \cap \cdots \cap A_{2m}^c \in \mathcal{F}_2$. Therefore $A^c \in \mathcal{F}$. Now, let $A = (A_{11} \cap A_{21}) \cup (A_{12} \cap A_{22})$, so that

$$
\begin{aligned}
A^c &= (A_{11}^c \cup A_{21}^c) \cap (A_{12}^c \cup A_{22}^c)\\
&= (A_{11}^c \cap A_{12}^c) \cup (A_{12}^c \cap A_{21}^c) \cup (A_{11}^c \cap A_{22}^c) \cup (A_{21}^c \cap A_{22}^c)
\end{aligned}
$$

and this is of the form of sets lying in $\mathcal{F}$. Thus $A^c \in \mathcal{F}$. By induction, it holds that $A^c \in \mathcal{F}$ if $A = A_m = (A_{11} \cap A_{21}) \cup \cdots \cup (A_{1m} \cap A_{2m})$. Indeed, for

$A_{m+1} = (A_{11} \cap A_{21}) \cup \cdots \cup (A_{1m} \cap A_{2m}) \cup (A_{1,m+1} \cap A_{2,m+1})$, we have

$$
\begin{aligned}
A^c_{m+1} &= [(A_{11} \cap A_{21}) \cup \cdots \cup (A_{1m} \cup A_{2m})]^c \cap (A^c_{1,m+1} \cup A^c_{2,m+1}) \\
&= \left[\bigcup_i (A_i \cap B_i)\right] \cap (A^c_{i,m+1} \cup A^c_{2,m+1}) \quad \text{(with } A_i \in \mathcal{F}_1,\ B_i \in \mathcal{F}_2, \text{ by the induction hypothesis)} \\
&= \bigcup_i \left\{\left[(A_i \cap A^c_{i,m+1}) \cap B_i\right] \cup \left[A_i \cap (B_i \cap A^c_{2,m+1})\right]\right\}
\end{aligned}
$$

(with $A_i \cap A^c_{i,m+1}$, $A_i \in \mathcal{F}_1$, and B_i, $B_i \cap A^c_{2,m+1} \in \mathcal{F}_2$), so that $A^c_{m+1} \in \mathcal{F}$. Finally, suppose that $C = A \cup B$, where, e.g., $A = (A_{11} \cup A_{21}) \cup \cdots \cup (A_{1m} \cup A_{2m})$, $B = (B_{11} \cap B_{21}) \cup \cdots \cup (B_{1n} \cap B_{2n})$. Then

$$
\begin{aligned}
C^c &= A^c \cap B^c = (D \cap E) \cap \left[\bigcup_i (A_i \cap B_i)\right] \quad \text{(with } D,\ A_i \in \mathcal{F}_1 \text{ and } E,\ B_i \in \mathcal{F}_2, \text{ by previous steps)} \\
&= \bigcup_i \left[(D \cap A_i) \cap (E \cap B_i)\right] \quad \text{(with } D \cap A_i \in \mathcal{F}_1,\quad E \cap B_i \in \mathcal{F}_2),
\end{aligned}
$$

so that $C^c \in \mathcal{F}$. It follows that $\mathcal{F}$ is a field. #

3. In the first place, $e^x \geq 1 + x$, $x \in \Re$. (Indeed, with $g(x) = e^x - x - 1$, we have $g'(x) = e^x - 1 = 0$, for $x = 0$, and $g''(x) = e^x > 0$. Thus, $\min_{x \in \Re} g(x) = g(0) = 0$ and hence $e^x \geq x + 1$.) In the inequality $e^x \geq x + 1$, replace x by $-x$ to get $e^{-x} \geq -x + 1$ and apply it for $x = p_i$, $i = m, \ldots, n$ to obtain $e^{-p_i} \geq 1 - p_i$, so that $\prod_{i=m}^n e^{-p_i} \geq \prod_{i=m}^n (1 - p_i)$, or $e^{-\sum_{i=m}^n p_i} \geq \prod_{i=m}^n (1 - p_i)$, or $1 - \exp(-\sum_{i=m}^n p_i) \leq 1 - \prod_{i=m}^n (1 - p_i)$, which is the left-hand side inequality.

As for the right-hand inequality, we have that, for $n = m$, it becomes: $1 - (1 - p_m) = p_m$, so that the inequality holds. Next, assume the inequality to be true for $n = m + k$ and establish it for $n = m + k + 1$. To this end,

$$
\begin{aligned}
1 - \prod_{i=m}^{m+k+1} (1 - p_i) &= 1 - \prod_{i=m}^{m+k} (1 - p_i) \times (1 - p_{m+k+1}) \\
&= 1 - \prod_{i=m}^{m+k} (1 - p_i) + p_{m+k+1} \times \prod_{i=m}^{m+k} (1 - p_i) \\
&= 1 - \prod_{i=m}^{m+k} (1 - p_i) - p_{m+k+1} \left[1 - \prod_{i=m}^{m+k} (1 - p_i)\right] \\
&\quad + p_{m+k+1} \\
&= \left[1 - \prod_{i=m}^{m+k} (1 - p_i)\right] (1 - p_{m+k+1}) + p_{m+k+1} \\
&\leq \left[1 - \prod_{i=m}^{m+k} (1 - p_i)\right] + p_{m+k+1} \quad (\text{since } 1 - p_{m+k+1} \leq 1)
\end{aligned}
$$

$$\leq \left(\sum_{i=m}^{m+k} p_i\right) + p_{m+k+1} \text{ (by the induction hypothesis)}$$

$$= \sum_{i=m}^{m+k+1} p_i, \text{ which is the right-hand side inequality. \#}$$

4. **(i)** Set $\bar{A} = \limsup_{m\to\infty} A_m = \cap_{m=1}^{\infty} \cup_{i=m}^{\infty} A_i$. Then, as m and $n \to \infty$,

$$P((\bar{A})^c) = P\left(\bigcup_{m=1}^{\infty}\bigcap_{i=m}^{\infty} A_i^c\right) = P\left(\lim_m \bigcap_{i=m}^{\infty} A_i^c\right) = \lim_m P\left(\bigcap_{i=m}^{\infty} A_i^c\right)$$

$$= \lim_m P\left(\lim_n \bigcap_{i=m}^{n} A_i^c\right) = \lim_m \lim_n P\left(\bigcap_{i=m}^{n} A_i^c\right)$$

$$= \lim_m \lim_n \prod_{i=m}^{n} P(A_i^c) = \lim_m \lim_n \prod_{i=m}^{n}(1-p_i) \text{ where } p_i = P(A_i).$$

That is, $P((\bar{A})^c) = \lim_m \lim_n \prod_{i=m}^{n} P(A_i^c)$, $p_i = P(A_i) = 1 - P(A_i^c)$. Apply the left-hand side inequality in Exercise 3 with p_i as defined here to get:

$$1 - \exp\left(-\sum_{i=m}^{n} p_i\right) \leq 1 - \prod_{i=m}^{n} P(A_i^c).$$

First, suppose that $\sum_{i=1}^{\infty} p_i = \infty$ (which implies that $\sum_{i=m}^{\infty} p_i = \infty$ for every m), and take the limits in the inequality above, first as $n \to \infty$ and then as $m \to \infty$, to get: $1 - 0 \leq 1 - P((\bar{A})^c)$ on the basis of what has been established above. Thus, $P(\bar{A}) \geq 1$, so that $P(\bar{A}) = 1$.

Next, let $P(\bar{A}) = 1$. Then $\sum_{i=1}^{\infty} p_i = \infty$ because otherwise, $P(\bar{A}) = 0$ by Exercise 3 in Chapter 3. Thus, $P(\limsup_{n\to\infty} A_n) = 1$ if and only if $\sum_{n=1}^{\infty} P(A_n) = \infty$.

(ii) Now, with $\bar{A}$ as in part (i), suppose that $P(\bar{A}) = 0$. Then $\sum_{n=1}^{\infty} P(A_n) < \infty$ because otherwise, $P(\bar{A}) = 1$ by part (i). On the other hand, $\sum_{n=1}^{\infty} P(A_n) < \infty$ implies $P(\bar{A}) = 0$ by Exercise 3 in Chapter 3. Thus, $P(\limsup_{n\to\infty} A_n) = 0$ if and only if $\sum_{n=1}^{\infty} P(A_n) < \infty$. #

5. Here $\bar{A} = \limsup_{n\to\infty} A_n = A$ since $\lim_{n\to\infty} A_n$ exists. Then, by Exercise 4, $P(A) = 0$ if and only if $\sum_{n=1}^{\infty} P(A_n) < \infty$, and $P(A) = 1$ if and only if $\sum_{n=1}^{\infty} P(A_n) = \infty$. Thus, $P(A)$ is either 0 or 1. #

6. $P(|\bar{X}_n - p| \geq \varepsilon) \leq \frac{pq}{\varepsilon^2} \times \frac{1}{n} \underset{n\to\infty}{\longrightarrow} 0$, and $\sum_{k=1}^{\infty} P(|\bar{X}_{k^2} - p| \geq \varepsilon) \leq \frac{pq}{\varepsilon^2} \sum_{k=1}^{\infty} \frac{1}{k^2} < \infty$, so that by Exercise 4(i) and Theorem 4 in Chapter 3, $\bar{X}_{k^2} \underset{k\to\infty}{\overset{a.s.}{\longrightarrow}} p$. #

7. For an arbitrary, but fixed $\varepsilon > 0$, set $A_n = (|X_n| \geq \varepsilon)$. Independence of X_ns implies independence of A_ns, $n \geq 1$. Then, by Exercise 4(ii), $P(\limsup_{n\to\infty} A_n) = 0$ if and only if $\sum_{n=1}^{\infty} P(A_n) < \infty$. But

$$\limsup_{n\to\infty} A_n = \bigcap_{n=1}^{\infty}\bigcup_{\nu=n}^{\infty} A_\nu = \bigcap_{n=1}^{\infty}\bigcup_{\nu=n}^{\infty}(|X_\nu| \geq \varepsilon) = \bigcap_{n=1}^{\infty}\bigcup_{\nu=1}^{\infty}(|X_{n+\nu}| \geq \varepsilon),$$

and $X_n \underset{n\to\infty}{\overset{a.s.}{\longrightarrow}} 0$ if and only if $P\left(\cap_{n=1}^{\infty} \cup_{\nu=1}^{\infty} (|X_{n+\nu}| \geq \varepsilon)\right) = 0$ for every $\varepsilon > 0$; this is so by Theorem 4 in Chapter 3. Thus, $P(\limsup_{n\to\infty} A_n) = P\left(\cap_{n=1}^{\infty} \cup_{\nu=1}^{\infty} (|X_{n+\nu}| \geq \varepsilon)\right) = 0$ if and only if $\sum_{n=1}^{\infty} P(A_n) = \sum_{n=1}^{\infty} P(|X_n| \geq \varepsilon) < \infty$, or $X_n \underset{n\to\infty}{\overset{a.s.}{\longrightarrow}} 0$ if and only if $\sum_{n=1}^{\infty} P(|X_n| \geq \varepsilon) < \infty$ for every $\varepsilon > 0$. #

8. For a concrete example, for $n \geq 1$, let X_n be independent r.v.s $\sim B(1, 1/n)$; i.e.,

$$X_n = \begin{cases} 1, & 1/n \\ 0, & 1 - 1/n \end{cases}, n \geq 1.$$

Then, clearly, $X_n \underset{n\to\infty}{\overset{P}{\longrightarrow}} 0$. However, $X_n \underset{n\to\infty}{\overset{a.s.}{\nrightarrow}} 0$. Indeed, for $\varepsilon > 0$, let $A_n = (|X_n| \geq \varepsilon) = (X_n = 1)$. Then $P(A_n) = P(X_n = 1) = \frac{1}{n}$, and hence $\sum_{n=1}^{\infty} P(A_n) = \infty$. Since the events A_n, $n \geq 1$, are independent, Exercise 4(i) implies that $P(\overline{\lim}_{n\to\infty} A_n) = 1$. However,

$$P\left(\overline{\lim_{n\to\infty}} A_n\right) = P\left(\bigcap_{n=1}^{\infty} \bigcup_{k=n}^{\infty} A_k\right) = P\left(\bigcap_{n=1}^{\infty} \bigcup_{k=0}^{\infty} X_{n+k} \geq \varepsilon\right)$$

$$= P\left(\bigcap_{n=1}^{\infty} \bigcup_{k=0}^{\infty} |X_{n+k} - 0| \geq \varepsilon\right).$$

Since this probability is $\neq 0$, it follows that $X_n \underset{n\to\infty}{\overset{a.s.}{\nrightarrow}} 0$. Next, let $\mathcal{B} = \sigma(X_1, X_2, \ldots)$. Then $\mathcal{E}^{\mathcal{B}} X_n = X_n$ a.s., $\mathcal{E}^{\mathcal{B}} X = \mathcal{E}^{\mathcal{B}} 0 = 0$ a.s. Thus, $\mathcal{E}^{\mathcal{B}} X_n \underset{n\to\infty}{\overset{a.s.}{\nrightarrow}} \mathcal{E}^{\mathcal{B}} X$. #

9. If the r.v.s are independent, then the desired factorization follows from relation (10.4) with $B_j = (a_j, b_j]$, a_j, b_j in $\Re$ and $a_j < b_j$, $j = 1, \ldots, k$.

Next, suppose that

$$P(a_j < X_j \leq b_j, j = 1, \ldots, k) = \prod_{j=1}^{k} P(a_j < X_j \leq b_j),$$

and we wish to show that $X_1, \ldots, X_k$ are independent. By Theorem 1, it suffices to show that the factorization in relation (10.5) holds. To this end, we show that

$$P(X_1 \leq x_1, a_j < X_j \leq b_j, j = 2, \ldots, k)$$
$$= P(X_1 \leq x_1) \prod_{j=2}^{k} P(a_j < X_j \leq b_j).$$

Indeed, with $(x_1 >) y_m \downarrow -\infty$ as $m \to \infty$, we have:

$$P(X_1 \leq x_1, a_j < X_j \leq b_j, j = 2, \ldots, k)$$
$$= P(X_1 \in \bigcup_m (y_m, x_1], a_j < X_j \leq b_j, j = 2, \ldots, k)$$
$$= P(\bigcup_m (X_1 \in (y_m, x_1]), a_j < X_j \leq b_j, j = 2, \ldots, k)$$
$$= P(\bigcup_m (X_1 \in (y_m, x_1], a_j < X_j \leq b_j, j = 2, \ldots, k))$$

$$
\begin{aligned}
&= P(\lim_m (y_m < X_1 \le x_1, a_j < X_j \le b_j, j = 2, \ldots, k)) \\
&= \lim_m P(y_m < X_1 \le x_1, a_j < X_j \le b_j, j = 2, \ldots, k) \\
&= \lim_m P(y_m < X_1 \le x_1) P(a_2 < X_2 \le b_2) \ldots P(a_k < X_k \le b_k) \\
&= P(X_1 \le x_1) P(a_2 < X_2 \le b_2) \ldots P(a_k < X_k \le b_k).
\end{aligned}
$$

This process can be repeated, clearly, with each one of the remaining $k-2$ r.v.s, upon the completion of which we obtain the factorization in relation (10.5). #

10. The existence of $\int X dP$ implies the existence of $\int_A X dP$ for every $A \in \mathcal{A}$ (by the Corollary to Theorem 5 in Chapter 4), and $\int_A X dP = \int (X I_A) dP = \mathcal{E}(X I_A)$. Independence of X and Y means independence of $\mathcal{A}_X$ and $\mathcal{A}_Y$, of the σ-fields induced by X and Y. Since $A \in \mathcal{A}_Y$, it follows that $\mathcal{A}_X$ and $\{\oslash, A, A^c, \Omega\}$ are also independent, and hence X and I_A are independent. Then, by Lemma 1, $\mathcal{E}(X I_A) = (\mathcal{E}X)(\mathcal{E}I_A) = (\mathcal{E}X)P(A)$. #

11. Consider two r.v.s X and Y, and let $\mathcal{A}_X$ and $\mathcal{A}_Y$ be the σ-fields induced by them. Then X and Y are said to be independent if $\mathcal{A}_X$ and $\mathcal{A}_Y$ are independent; i.e., $P(A_1 \cap A_2) = P(A_1)P(A_2)$ for all $A_1 \in \mathcal{A}_X$ and all $A_2 \in \mathcal{A}_Y$. According to this, X and X are independent if $P(A_1 \cap A_2) = P(A_1)P(A_2)$ for every A_1, A_2 in $\mathcal{A}_X$. This is also true for $A_1 = A_2 = A$, say. That is, $P(A \cap A) = P(A)P(A)$ or $P(A) = [P(A)]^2$, which occurs only if $P(A) = 0$ or $P(A) = 1$. So, for every $B \in \mathcal{B}$, $P(X \in B)$ is either 0 or 1; or $\mathcal{A}_X$ is equivalent to $\{\oslash, \Omega\}$. Next, for sufficiently small a, $P(X < a) = 0$, because otherwise $P(X = -\infty) = 1$, by letting $a \to -\infty$. Likewise, $P(X > b) = 0$ for sufficiently large b. Thus, $P(a \le X \le b) = 1$. Set $I_0 = [a, b]$, and let I_1 be that half of I_0 for which $P(X \in I_1) = 1$. Next, let I_2 be that half of I_1 for which $P(X \in I_2) = 1$, etc. It is clear that, as $n \to \infty$, $I_n \downarrow \{c\}$, some $c \in (a, b)$, and then $P(X \in I_n) \to P(X = c)$. Since $P(X \in I_n) = 1$ for all n, it follows that $P(X = c) = 1$. #

12. **(i)** By Lemma 1 in Chapter 9,

$$
\begin{aligned}
\operatorname{Cov}(X, Y) &= \mathcal{E}[(X - \mathcal{E}X)(Y - \mathcal{E}Y)] = \mathcal{E}(XY) - (\mathcal{E}X)(\mathcal{E}Y) \\
&= (\mathcal{E}X)(\mathcal{E}Y) - (\mathcal{E}X)(\mathcal{E}Y) = 0,
\end{aligned}
$$

so that $\rho(X, Y) = 0$ also.

(ii) By Exercise 14(iii) in Chapter 9, $\rho(X, Y) = \rho$, so that if $\rho = 0$, we have by the expression of $p_{X,Y}$ (see Exercise 12 in Chapter 9):

$$
\begin{aligned}
p_{X,Y}(x, y) &= \frac{1}{2\pi\sigma_1\sigma_2} \exp\left[-\left(\frac{x-\mu_1}{\sigma_1}\right)^2 - \left(\frac{y-\mu_2}{\sigma_2}\right)^2\right] \\
&= \frac{1}{\sqrt{2\pi}\sigma_1} \exp\left[-\left(\frac{x-\mu_1}{\sigma_1}\right)^2\right] \\
&\quad \times \frac{1}{\sqrt{2\pi}\sigma_2} \exp\left[-\left(\frac{y-\mu_2}{\sigma_2}\right)^2\right] \\
&= p_X(x) p_Y(y), \text{ so that } X \text{ and } Y \text{ are independent. \#}
\end{aligned}
$$

13. Indeed,

$$\begin{aligned} e^{it(X_1+X_2)} &= e^{itX_1} \times e^{itX_2} \\ &= [\cos(tX_1) + i\sin(tX_1)][\cos(tX_2) + i\sin(tX_2)] \\ &= [\cos(tX_1)\cos(tX_2) - \sin(tX_1)\sin(tX_2)] \\ &\quad + i[\cos(tX_1)\sin(tX_2) + \sin(tX_1)\cos(tX_2)], \end{aligned}$$

and

$$\begin{aligned} \mathcal{E}e^{it(X_1+X_2)} &= [\mathcal{E}\cos(tX_1)][\mathcal{E}\cos(tX_2)] - [\mathcal{E}\sin(tX_1)][\mathcal{E}\sin(tX_2)] \\ &\quad + i[\mathcal{E}\cos(tX_1)][\mathcal{E}\sin(tX_2)] + [\mathcal{E}\sin(tX_1)][\mathcal{E}\cos(tX_2)] \\ &= [\mathcal{E}\cos(tX_1) + i\mathcal{E}\sin(tX_1)][\mathcal{E}\cos(tX_2) + i\mathcal{E}\sin(tX_2)] \\ &= (\mathcal{E}e^{itX_1})(\mathcal{E}e^{itX_2}), \text{ as was to be seen.} \end{aligned}$$

Next,

$$\begin{aligned} \mathcal{E}e^{it(X_1+\ldots+X_{k+1})} &= \mathcal{E}[e^{it(X_1+\ldots+X_k)}e^{itX_{k+1}}] \\ &= \mathcal{E}e^{it(X_1+\ldots+X_k)} \times e^{itX_{k+1}} \text{ (by the previous step)} \\ &= \mathcal{E}e^{itX_1} \times \ldots \times \mathcal{E}e^{itX_k} \times \mathcal{E}e^{itX_{k+1}} \\ &\qquad \text{(by the induction hypothesis)} \\ &= \mathcal{E}e^{itX_1} \times \ldots \times \mathcal{E}e^{itX_{k+1}}. \ \# \end{aligned}$$

14. (i) Here $p_{X,Y} = \frac{1}{2\pi\sigma^2}\exp\left(-\frac{x^2+y^2}{2\sigma^2}\right)$, and the transformations $u = x + y$, $v = x - y$ produce the Jacobian $-\frac{1}{2}$. Thus,

$$\begin{aligned} p_{U,V}(u,v) &= \frac{1}{2\pi \times 2\sigma^2}\exp\left(-\frac{u^2+v^2}{4\sigma^2}\right) \\ &= \frac{1}{\sqrt{2\pi}(\sigma\sqrt{2})}\exp\left[-\frac{u^2}{2(\sigma\sqrt{2})^2}\right] \\ &\quad \times \frac{1}{\sqrt{2\pi}(\sigma\sqrt{2})}\exp\left[-\frac{v^2}{2(\sigma\sqrt{2})^2}\right] \end{aligned}$$

and the desired result follows.

(ii) The r.v.s $X^* = X - \mu_1$ and $Y^* = Y - \mu_2$ are independent, distributed as $N(0, \sigma^2)$, so that, by part (i), $U = X^* + Y^*$ and $V = X^* - Y^*$ are independent, distributed as $N(0, 2\sigma^2)$. Equivalently, $X + Y - (\mu_1 + \mu_2)$ and $X - Y - (\mu_1 - \mu_2)$ are independent, distributed as $N(0, 2\sigma^2)$, or $X + Y$ and $X - Y$ are independent, distributed as $N(\mu_1 + \mu_2, 2\sigma^2)$ and $N(\mu_1 - \mu_2, 2\sigma^2)$, respectively. #

15. (i) It is clear that X takes on the values $0, 1, \ldots, n$, and that in order to show that X is a r.v., it suffices to show that $X^{-1}(\{k\}) \in \mathcal{A}$, $k = 0, 1, \ldots, n$. We have,

$$(X = k) = \cup(A_{i_1} \cap \ldots \cap A_{i_k} \cap A^c_{j_1} \cap \ldots \cap A^c_{j_l}),$$

where the union is taken over all $i_1, \ldots, i_k$ with $1 \leq i_1 < \ldots < i_k \leq n$ and all $j_1 < \ldots < j_l$ distinct from all $i_1, \ldots, i_k$ and $k + l = n$. However, this union is in $\mathcal{A}$, and hence $(X = k) \in \mathcal{A}$.

(ii) In part (i), the members of the union are $\binom{n}{k}$, and each such member consists of n independent events; this is so by the assumed independence of $A_1, \ldots, A_n$ and Exercise 1. Since

$$P(A_{i_1} \cap \ldots \cap A_{i_k} \cap A^c_{j_1} \cap \ldots \cap A^c_{j_l})$$
$$= P(A_{i_1}) \ldots P(A_{i_k}) P(A^c_{j_1}) \ldots P(A^c_{j_l})$$
$$= p^k(1-p)^{n-k}, \text{ we have } P(X = k) = \binom{n}{k} p^k (1-p)^{n-k}, k = 0, 1,$$
$$\ldots, n, \text{ so that } X \sim B(n, p).\#$$

16. Let $\mathcal{A}_n = X_n^{-1}(\mathcal{B}) = \{\emptyset, A_n, A_n^c, \Omega\}$. Then $X_1, X_2, \ldots$ are independent if $\mathcal{A}_1, \mathcal{A}_2, \ldots$ are independent. Next, $\mathcal{A}_1, \mathcal{A}_2, \ldots$ are independent if any finite member of them is a collection of independent σ-fields. Without loss of generality, consider the collection $\mathcal{A}_1, \ldots, \mathcal{A}_n$. Then $A_1, \ldots, A_n$ are independent if and only if $\mathcal{A}_1, \ldots, \mathcal{A}_n$ are independent. Indeed, if $A_1, \ldots, A_n$ are independent, then $A'_1, \ldots, A'_n$ are also independent, where A'_j is either A_j or A^c_j, $j = 1, \ldots, n$. This is so by Exercise 1. But this implies that $\mathcal{A}_1, \ldots, \mathcal{A}_n$ are independent, since for every choice $B_j \in \mathcal{A}_j$, we have $P(B_1 \cap \ldots \cap B_n) = P(B_1) \ldots P(B_n)$. On the other hand, if $\mathcal{A}_1, \ldots, \mathcal{A}_n$ are independent, then $A_1, \ldots, A_n$ are independent, since $P(A_{j_1} \cap \ldots \cap A_{j_k}) = P(A_{j_1}) \ldots P(A_{j_k})$ for all $2 \leq k \leq n$ and $1 \leq j_1 < \ldots < j_k \leq n$, and $A_{j_r} \in \mathcal{A}_{j_r}$, $r = 1, \ldots, k$. This is so, because $P(B_1 \cap \ldots \cap B_n) = P(B_1) \ldots P(B_n)$, $B_i \in \mathcal{A}_i$, $i = 1, \ldots, n$, and we may choose $B_{j_i} = A_{j_i}$, $i = 1, \ldots, k$, and $B_j = \Omega$ for $j \neq j_1, \ldots, j_k$. This completes the proof. #

17. We have $X = \#$ of Hs, and let $Y = \#$ of Ts. Then $X = Y + r$ and, of course, $X + Y = n$, so that $Y = n - X$. Then:

$$P(X = Y + r) = P(X = n - X + r)$$
$$= P(2X = n + r) = P\left(X = \frac{n+r}{2}\right)$$
$$= \begin{cases} 0 & \text{if } n + r \text{ is odd} \\ \binom{n}{\frac{n+r}{2}} p^{\frac{n+r}{2}} (1-p)^{n - \frac{n+r}{2}} & \text{if } n + r \text{ is even. \#} \end{cases}$$

18. Consider the quantities D_n, D_n^+, D_n^- and y_i, $i = 1, \ldots, n$ mentioned in the hint, and let $y_0 = -\infty$, $y_{n+1} = \infty$. Then observe that

$$D_n^+ = \sup_{x \in \Re}[F_n(x, \cdot) - F(x)] = \max_{0 \leq i \leq n} \sup_{y_i \leq x < y_{i+1}} [F_n(x, \cdot) - F(x)]$$
$$= \max_{0 \leq i \leq n} \sup_{y_i \leq x < y_{i+1}} \left[\frac{i}{n} - F(x)\right]$$

$$= \max_{0\le i\le n}\left[\frac{i}{n} - \inf_{y_i\le x<y_{i+1}} F(x)\right]$$

$$= \max_{0\le i\le n}\left[\frac{i}{n} - F(y_i)\right] \text{ (since } F \text{ is right-continuous)}$$

$$= \max\{\max_{0\le i\le n}\left[\frac{i}{n} - F(y_i)\right], 0\} \left(\text{since for } i=0,\ \frac{i}{n} - F(y_0) = 0\right).$$

That is,

$$D_n^+ = \max\left\{\max_{0\le i\le n}\left[\frac{i}{n} - F(y_i)\right], 0\right\}.$$

Likewise,

$$D_n^- = \sup_{x\in\Re}[F(x) - F_n(x,\cdot)] = \max_{1\le i\le n+1}\ \sup_{y_{i-1}\le x<y_i}[F(x) - F_n(x,\cdot)]$$

$$= \max_{1\le i\le n+1}\ \sup_{y_{i-1}\le x<y_i}\left[F(x) - \frac{i-1}{n}\right]$$

$$= \max_{1\le i\le n+1}\left[\sup_{y_{i-1}\le x<y_i} F(x) - \frac{i-1}{n}\right]$$

$$= \max_{1\le i\le n+1}\left[F(y_i - 0) - \frac{i-1}{n}\right]$$

$$= \max\left\{\max_{1\le i\le n}\left[F(y_i - 0) - \frac{i-1}{n}\right], 0\right\}$$

$$\left(\text{since for } i = n+1,\ F(y_{n+1} - 0) - \frac{i-1}{n} = 0\right).$$

That is,

$$D_n^- = \max\left\{\max_{1\le i\le n}\left[F(y_i - 0) - \frac{i-1}{n}\right], 0\right\}.$$

The quantities D_n^+ and D_n^- are r.v.s because they are expressed in terms of finitely many r.v.s Furthermore, it is clear that $D_n = \max\{D_n^+, D_n^-\}$. Therefore D_n is a r.v. #

19. For $0 < p < 1$, let $x_p = \inf\{x \in \Re;\ F(x) \ge p\}$. Then $x \ge x_p$ implies $F(x) \ge p$, whereas $x < x_p$ implies $F(x) < p$ so that $F(x_p - 0) \le p$. With this in mind, proceed as follows. Let

$$x_{ki} = \inf\left\{x \in \Re;\ F(x) \ge \frac{i}{k}\right\},\quad i = 0, 1, \ldots, k,$$

so that $-\infty \le x_{k0} < x_{k1}$ and $x_{k,k-1} < x_{kk} \le \infty$. From the observation at the beginning, we have that $x \in [\frac{i}{k}, \frac{i+1}{k})$ and $i = 0, 1, \ldots, k-1$ imply

$$\frac{i}{k} \le F(x_{ki}) \le F(x) \le F(x_{k,i+1} - 0) \le \frac{i+1}{k},$$

so that

$$F(x_{k,i+1}-0)-F(x_{ki})\leq\frac{1}{k}. \quad (1)$$

Therefore, for $x \in [x_{ki}, x_{k,i+1})$, $i = 1, \ldots, k-2$,

$$\begin{aligned}F_n(x)-F(x) &\leq F_n(x_{k,i+1}-0)-F(x_{ki})\\ &= [F_n(x_{k,i+1}-0)-F(x_{k,i+1}-0)]\\ &+ [F(x_{k,i+1}-0)-F(x_{ki})]\\ &\leq [F_n(x_{k,i+1}-0)-F(x_{k,i+1}-0)]+\frac{1}{k}\ \text{(by (1))}. \quad (2)\end{aligned}$$

Likewise,

$$\begin{aligned}F_n(x)-F(x) &\geq F_n(x_{ki})-F(x_{k,i+1}-0)\\ &= [F_n(x_{ki})-F(x_{ki})]\\ &- [F(x_{k,i+1}-0)-F(x_{ki})]\\ &\geq [F_n(x_{ki})-F(x_{ki})]-\frac{1}{k}\ \text{(by (1))}. \quad (3)\end{aligned}$$

From relations (2) and (3), and for $x \in [x_{ki}, x_{k,i+1})$, $i = 1, \ldots, k-2$, it follows that

$$\begin{aligned}|F_n(x)-F(x)| \leq \max\{&|F_n(x_{k,i+1}-0)-F(x_{k,i+1}-0)|,\\ &|F_n(x_{ki})-F(x_{ki})|\}+\frac{1}{k},\end{aligned}$$

so that for $x \in [x_{k1}, x_{k,k-1})$,

$$\begin{aligned}|F_n(x)-F(x)| \leq \max\{&|F_n(x_{k,i+1}-0)-F(x_{k,i+1}-0)|,\\ &|F_n(x_{ki})-F(x_{ki})|;\ i=1,\ldots,k-2\}+\frac{1}{k}. \quad (4)\end{aligned}$$

In particular, for $i = 0$, relation (1) gives $F(x_{k1}) \leq \frac{1}{k}$, and therefore (3) and (4) become, for $x < x_{k1}$,

$$F_n(x)-F(x)\leq[F_n(x_{k1}-0)-F(x_{k1}-0)]+\frac{1}{k},$$

$$F_n(x)-F(x)\geq[F_n(x_{k0})-0]-\frac{1}{k},$$

so that (4) still holds for $i = 0$ and $x < x_{k1}$. Next, relation (1) gives, for $i = k-1$, $F(x_{kk}-0)-F(x_{k,k-1}) \leq \frac{1}{k}$, and therefore (3) and (4) become, for $x \geq x_{k,k-1}$,

$$F_n(x)-F(x)\leq[F_n(x_{k,k}-0)-F(x_{k,k}-0)]+\frac{1}{k},$$

$$F_n(x)-F(x)\geq[F_n(x_{k,k-1})-F(x_{k,k-1})]-\frac{1}{k},$$

so that (4) still holds for $i = k - 1$ and $x \geq x_{k,k+1}$. In other words, relation (4) holds for $i = 0, 1, \ldots, k-1$ and $x \in \Re$.
Now, from the definition of $F_n(x) = F_n(x, \omega)$ in Exercise 4, we have that

$$F_n(x) = \frac{1}{n}\sum_{i=1}^{n} Y_i, \text{ where } Y_i = Y_i(x, \omega) = \begin{cases} 1, & X_i(\omega) \leq x \\ 0, & X_i(\omega) > x, \end{cases}$$

so that $\sum_{i=1}^{n} Y_i \sim B(n, F(x))$ and by the SLLN, $F_n(x) \xrightarrow[n\to\infty]{a.s.} F(x)$. Likewise, if $Z_i = Z_i(x, \omega)$ are defined by

$$Z_i = \begin{cases} 1, & X_i(\omega) < x \\ 0, & X_i(\omega) \geq x, \end{cases}$$

then $\sum_{i=1}^{n} Z_i \sim B(n, F(x-0))$ and $F_n(x-0) = \frac{1}{n}\sum_{i=1}^{n} Z_i$. Therefore, again by the SLLN, $F_n(x-0) \xrightarrow[n\to\infty]{a.s.} F(x-0)$. Next, in relation (4) take the $\sup_{x\in\Re}$ when the left-hand side produces the r.v. $\sup_{x\in\Re}|F_n(x) - F(x)|$ (by Exercise (4)), whereas the right-hand side is

$$\max\{|F_n(x_{k,i+1}-0) - F(x_{k,i+1}-0)|, |F_n(x_{ki}) - F(x_{ki})|;\ i = 0, 1, \ldots, k-1\}. \tag{5}$$

Each term within the square bracket in (5) tends a.s. to 0, and then so does their max. In other words: $\sup_{x\in\Re}|F_n(x) - F(x)| \leq \frac{1}{k}$ a.s. Letting $k \to 0$, we get then the desired result. #

Chapter 11

Topics from the Theory of Characteristic Functions

1. Set $z = re^{i\theta}$ $(r \geq 0, 0 \leq \theta < 2\pi)$ for the representation of $g(x) + ih(x)$ in polar coordinates, and $\int_\Re [g(x) + ih(x)]d\mu = \int_0^{2\pi}\int_0^{\infty} z dr d\theta$ (or just $\int z$) $= \rho e^{i\alpha}$. Then: $|z| = r$, $|\int z| = \rho$. Therefore (real)$|\int z| = \rho = \rho e^{i\alpha} \times e^{-i\alpha} = (\int z)e^{-i\alpha} = \int(e^{-i\alpha}z) = \int(e^{-i\alpha} \times re^{i\theta}) = \int re^{i(\theta-\alpha)} = \int r\cos(\theta - \alpha)$ (since real), and this is $\leq \int r = \int |z|$; i.e., $|\int z| \leq \int |z|$.
Alternatively, without loss of generality, suppose that $\int_\Re [g(x) + ih(x)]d\mu \neq 0$ (since the inequality is true if it is $= 0$), and set z for the complex number $z = |\int_\Re [g(x) + ih(x)]d\mu| / \int_\Re [g(x) + ih(x)]d\mu$. Then

$$\begin{aligned} \text{real} = \left|\int_\Re [g(x) + ih(x)]d\mu\right| &= \text{Re}\left\{z\int_\Re [g(x) + ih(x)]d\mu\right\} \\ &= \text{Re}\int_\Re z[g(x) + ih(x)]d\mu = \int_\Re \text{Re}\{z[g(x) + ih(x)]\}d\mu \end{aligned}$$

$$\leq \left| \int_{\Re} \text{Re}\{z[g(x) + ih(x)]\} d\mu \right| \leq \int_{\Re} |\text{Re}\{z[g(x) + ih(x)]\} d\mu|$$

$$\leq \int_{\Re} |z[g(x) + ih(x)]| \, d\mu = \int_{\Re} |z||g(x) + ih(x)| \, d\mu$$

$$= \int_{\Re} |g(x) + ih(x)| \, d\mu \text{ (since } |z| = 1).$$

Thus, $|\int_{\Re}[g(x) + ih(x)]d\mu| \leq \int_{\Re} |g(x) + ih(x)|d\mu$.

For the special case, $Z = re^{i\theta}$ where r and θ are r.v.s with $r \geq 0$ and $0 \leq \theta < 2\pi$. Also, let $\mathcal{E}Z = \rho e^{i\alpha}$, so that $|Z| = r$ and $|\mathcal{E}Z| = \rho$. Then real$= |\mathcal{E}Z| = \rho = \rho e^{i\alpha} \times e^{-i\alpha} = (\mathcal{E}Z)e^{-i\alpha} = \mathcal{E}(e^{-i\alpha}Z)$

$$= \mathcal{E}(e^{-i\alpha} \times re^{i\theta}) = \mathcal{E}\left[re^{i(\theta-\alpha)}\right]$$

$$= \mathcal{E}[r \cos(\theta - \alpha)] \text{ (since real)}$$

$$\leq |\mathcal{E}[r \cos(\theta - \alpha)]| \leq \mathcal{E}|r \cos(\theta - \alpha)| = \mathcal{E}r = \mathcal{E}|Z|;$$

i.e., $|\mathcal{E}Z| \leq \mathcal{E}|Z|$.

Another approach for the special case is the following. Set $Z = X + iY$. Then $\mathcal{E}Z = \mathcal{E}X + i\mathcal{E}Y$, $|\mathcal{E}Z| = \sqrt{(\mathcal{E}X)^2 + (\mathcal{E}Y)^2}$, $|Z| = \sqrt{X^2 + Y^2}$. Consider the function $g(x, y) = \sqrt{x^2 + y^2}$. Then $g(x, y)$ is convex with respect to one of its variables, the other being kept fixed (this is so because the second order derivatives are ≥ 0). Then look at $g(\cdot, \mathcal{E}Y)$ and apply Jensen's inequality to obtain:

$$g(\mathcal{E}X, \mathcal{E}Y) \leq \mathcal{E}g(X, \mathcal{E}Y) = \int_{\Re} g(x, \mathcal{E}Y)dP_X, \tag{1}$$

where P_X is the probability distribution of the r.v. X.

For each x, look at $g(x, \cdot)$ and apply Jensen's inequality again to obtain:

$$g(x, \mathcal{E}Y) \leq \mathcal{E}g(x, Y) = \mathcal{E}[g(X, Y)|X = x].$$

Hence

$$\int_{\Re} g(x, \mathcal{E}Y)dP_X \leq \int_{\Re} \mathcal{E}[g(X, Y)|X = x]dP_X$$
$$= \mathcal{E}\{\mathcal{E}[g(X, Y)|X]\} = \mathcal{E}g(X, Y)$$
$$= \mathcal{E}\sqrt{X^2 + Y^2} = \mathcal{E}|Z|. \tag{2}$$

Since $g(\mathcal{E}X, \mathcal{E}Y) = \sqrt{(\mathcal{E}X)^2 + (\mathcal{E}Y)^2} = |\mathcal{E}Z|$ and $\mathcal{E}|Z| = \mathcal{E}\sqrt{X^2 + Y^2} = \mathcal{E}g(X, Y)$, relations (1) and (2) yield the result. #

2. In the expansion $e^{ix} = 1 + ix - \frac{x^2}{2}e^{ix^*}$, $|x^*|$ bounded, replace x by $-ta$ and $-tb$ successively to obtain: $e^{-ita} = 1 - ita - \frac{(ta)^2}{2}e^{it_1^*}$, $e^{-itb} = 1 - itb - \frac{(tb)^2}{2}e^{it_2^*}$, $|t_1^*|$ and $|t_2^*|$ bounded. Hence

$$e^{-ita} - e^{-itb} = it(b - a) + \frac{t^2}{2}(b^2e^{it_2^*} - a^2e^{it_1^*}),$$

and therefore

$$(e^{-ita} - e^{-itb})/(it) = (b - a) + t(b^2 e^{it_2^*} - a^2 e^{it_1^*})/2i.$$

Letting $t \to 0$, we have that the limit is $b-a$. Since also $\lim_{t\to 0} f(t) = f(0) = 1$, it follows that $\lim_{t\to 0} g(t) = b - a$. #

3. The continuity of g at 0 implies that, for every $\varepsilon > 0$, there exists $\delta = \delta(\varepsilon) > 0$ such that $|g(v) - g(0)| < \varepsilon$ for $|v - 0| = |v| < \delta$. Hence, for $0 < v < t < \delta$ (assuming that $t > 0$; similarly for $t < 0$),

$$\left|\frac{1}{t}\int_0^t [g(v) - g(0)]dv\right| \le \frac{1}{t}\int_0^t |g(v) - g(0)|dv \le \frac{1}{t}\int_0^t \varepsilon dv = \frac{\varepsilon t}{t} = \varepsilon,$$

or

$$\left|\frac{1}{t}\int_0^t [g(v) - g(0)]dv\right| = \left|\frac{1}{t}\int_0^t g(v)dv - g(0)\right| < \varepsilon.$$

Since ε is arbitrary, it follows that $\frac{1}{t}\int_0^t g(v)dv \underset{t\to\infty}{\longrightarrow} g(0)$.

Alternatively, apply L'Hôpital's Rule, which states that, if f and h are two functions such that $f(x_0) = h(x_0) = 0$, f and h are differentiable in an open interval I containing x_0, and $h'(x) \ne 0$ for $x \in I$, then $\lim_{x\to x_0}\frac{f(x)}{h(x)} = \lim_{x\to x_0}\frac{f'(x)}{h'(x)}$, provided this limit exists (or is ∞ or $-\infty$). Here take $f(t) = \int_0^t g(v)dv$, $h(t) = t$, and $x_0 = 0$ to obtain:

$$\lim_{t\to 0}\frac{1}{t}\int_0^t g(v)dv = \lim_{t\to 0}\frac{\int_0^t g(v)dv}{t} = \lim_{t\to 0}\frac{g(t)}{1} = \lim_{t\to 0} g(t) = g(0). \#$$

4. $|e^{it} - 1| = |\cos t + i\sin t - 1| = |(\cos t - 1) + i\sin t| = \sqrt{(\cos t - 1)^2 + \sin^2 t} = \sqrt{2}\sqrt{1-\cos t} = \sqrt{2}\sqrt{2\frac{\sin^2 t}{2}} = 2\left|\frac{\sin t}{2}\right| \le 2\left|\frac{t}{2}\right|$, since $|\sin t| \le t$; i.e., $|e^{it} - 1| \le |t|$. That $|\sin t| \le t$ is seen as follows:

Recall that: $\sin t = \sum_{n=0}^{\infty}(-1)^n\frac{t^{2n+1}}{(2n+1)!}$, $\cos t = \sum_{n=0}^{\infty}(-1)^n\frac{t^{2n}}{(2n)!}$. We have:

$$|\sin t| = \left|\sum_{n=0}^{\infty}(-1)^n\frac{t^{2n+1}}{(2n+1)!}\right| = \left|\frac{t}{2n+1}\sum_{n=0}^{\infty}(-1)^n\frac{t^{2n}}{(2n)!}\right| = \left|\frac{t}{2n+1}\cos t\right| \le |t| \text{ (since } 2n+1 \ge 1). \#$$

5. We know that $\log(1+z) = z(1+\theta z)$, $|\theta| = |\theta(z)| \le 1$, $|z| \le \frac{1}{2}$. Therefore,

$$\log\left(1+\frac{c_n}{n}\right)^n = n\log\left(1+\frac{c_n}{n}\right) = n\times\frac{c_n}{n}\left(1+\theta\frac{c_n}{n}\right) = c_n\left(1+\theta\frac{c_n}{n}\right)\underset{n\to\infty}{\longrightarrow} c,$$

and hence

$$\left(1+\frac{c_n}{n}\right)^n = e^{n\log(1+\frac{c_n}{n})} = e^{c_n(1+\theta\frac{c_n}{n})}\underset{n\to\infty}{\longrightarrow} e^c. \#$$

6. **(i)** $\int_{-\infty}^{\infty} xp(x)dx = \frac{1}{\pi}\int_{-\infty}^{\infty}\frac{xdx}{1+x^2} = \frac{1}{2\pi}\int_{-\infty}^{\infty}\frac{d(1+x^2)}{1+x^2} = \frac{1}{2\pi}\log(1+x^2)|_{-\infty}^{\infty} = \infty - \infty$, so that $\mathcal{E}X$ does not exist.

(ii) $f_X(t) = \int_{-\infty}^{\infty} e^{itx}p(x)dx = \frac{1}{\pi}\int_{-\infty}^{\infty}\frac{e^{itx}}{1+x^2}dx = \frac{1}{\pi}\int_{-\infty}^{\infty}\frac{\cos(tx)}{1+x^2}dx + i\frac{1}{\pi}\int_{-\infty}^{\infty}\frac{\sin(tx)}{1+x^2}dx = \frac{1}{\pi}\int_{-\infty}^{\infty}\frac{\cos(tx)}{1+x^2}dx$ (since $\frac{\sin(tx)}{1+x^2}$ is an odd function), and this is $= \frac{2}{\pi}\int_0^{\infty}\frac{\cos(tx)}{1+x^2}dx$ (since $\frac{\cos(tx)}{1+x^2}$ is an even function), and this is $= \frac{2}{\pi}\times\frac{\pi}{2}e^{-|t|} = e^{-|t|}$ (by the reference given).

(iii) $f_{\frac{S_n}{n}}(t) = \prod_{j=1}^{n} f_{X_j}(\frac{t}{n}) = [f_{X_1}(\frac{t}{n})]^n = (e^{-|\frac{t}{n}|})^n = e^{-|t|}$.

(iv) $\frac{S_n}{n} \xrightarrow[n\to\infty]{P} 0$ is equivalent to $\frac{S_n}{n} \xrightarrow[n\to\infty]{d} 0$ which does not happen, since the ch.f. of $\frac{S_n}{n}$ is $e^{-|t|}$ and the ch.f. of 0 is $e^{it\times 0} = e^0 = 1$. #

7. We have $\frac{[(2n)!]^{1/2n}}{[(2n-1)!]^{1/(2n-1)}} = \frac{[(2n-1)!]^{1/2n}(2n)^{1/2n}}{[(2n-1)!]^{1/(2n-1)}} = (2n)^{1/2n}\times[(2n-1)!]^{-1/2n(2n-1)} = (2n)^{1/2n}\times\left[\frac{(2n)!}{2n}\right]^{-1/2n(2n-1)}$. Now $\log(2n)^{1/2n} = \frac{1}{2n}\log(2n) \xrightarrow[n\to\infty]{} 0$, so that $(2n)^{1/2n} \to e^0 = 1$. Next, Stirling's formula states that: $\frac{n!}{\sqrt{2\pi}\times n^{n+\frac{1}{2}}\times e^{-n}} \xrightarrow[n\to\infty]{} 1$.

Therefore $\left[\frac{(2n)!}{2n}\right]^{-1/2n(2n-1)} = \left[\frac{(2n)!/\sqrt{2\pi}(2n)^{2n+\frac{1}{2}}e^{-2n}}{2n/\sqrt{2\pi}(2n)^{2n+\frac{1}{2}}e^{-2n}}\right]^{-1/2n(2n-1)}$ and $[(2n)!/\sqrt{2\pi}(2n)^{2n+\frac{1}{2}}e^{-2n}]^{-1/2n(2n-1)} \xrightarrow[n\to\infty]{} 1$, since the bracket $\xrightarrow[n\to\infty]{} 1$ and $1/2n(2n-1) \xrightarrow[n\to\infty]{} 0$. Furthermore, $\left[\frac{1}{2n/\sqrt{2\pi}(2n)^{2n+\frac{1}{2}}e^{-2n}}\right]^{-1/2n(2n-1)} = \left[\frac{(2n)e^{2n}}{\sqrt{2\pi}(2n)^{2n+\frac{1}{2}}}\right]^{1/2n(2n-1)} = \frac{e^{1/(2n-1)}}{\left(\sqrt{2\pi}\right)^{1/2n(2n-1)}}\times\left[\frac{2n}{(2n)^{2n+\frac{1}{2}}}\right]^{1/2n(2n-1)}$ and the first term $\xrightarrow[n\to\infty]{} 1$. Finally, setting A_n for the second term, we have: $\log A_n = \frac{-2n+\frac{1}{2}}{2n(2n-1)}\times\log(2n) = \frac{-1+\frac{1}{4n}}{1-\frac{1}{2n}}\times\frac{\log(2n)}{2n}$ and the first term $\xrightarrow[n\to\infty]{} -1$ whereas the second term $\xrightarrow[n\to\infty]{} 0$. Thus, $\log A_n \xrightarrow[n\to\infty]{} 0$, so that $A_n \xrightarrow[n\to\infty]{} 1$. #

8. The series converges (for $-\rho < t < \rho$) if $\rho^{-1} \overset{def}{=} \overline{\lim}_{n\to\infty}\left[\frac{|m^{(n)}|}{n!}\right]^{1/n} < \infty$. We have: $\left|\frac{|m^{(n)}|}{n!}\right|^{1/n} = \left|\frac{\mathcal{E}X^n}{n!}\right|^{1/n} \le \left(\frac{\mathcal{E}|X|^n}{n!}\right)^{1/n} = \left[\frac{\mu^{(n)}}{n!}\right]^{1/n} = \left[\frac{\mu^{(n)}/\sqrt{2\pi}n^{n+\frac{1}{2}}e^{-n}}{n!/\sqrt{2\pi}n^{n+\frac{1}{2}}e^{-n}}\right]^{1/n} = \left[\frac{\mu^{(n)}}{\sqrt{2\pi}n^{n+\frac{1}{2}}e^{-n}}\right]^{1/n}\times\left(\frac{1}{n!/\sqrt{2\pi}n^{n+\frac{1}{2}}e^{-n}}\right)^{1/n}$ and the second term $\xrightarrow[n\to\infty]{} 1$ by Stirling's formula and the fact that $1/n \xrightarrow[n\to\infty]{} 0$. The first term is equal to: $\frac{[\mu^{(n)}]^{1/n}}{n}\times\frac{1}{(\sqrt{2\pi})^{1/n}}\times\frac{1}{e^{-1}}\times\frac{1}{n^{1/2n}}$ and: $\frac{1}{(\sqrt{2\pi})^{1/n}} \xrightarrow[n\to\infty]{} 1$, $\log(n^{1/2n}) = \frac{1}{2n}\log n \xrightarrow[n\to\infty]{} 0$, so that $n^{1/2n} \xrightarrow[n\to\infty]{} 1$. Therefore, if $\overline{\lim}_{n\to\infty}\frac{[\mu^{(n)}]^{1/n}}{n} < \infty$, then $\overline{\lim}_{n\to\infty}\left|\frac{|m^{(n)}|}{n!}\right|^{1/n} < \infty$. #

9. Set $Y_n = \frac{S_n - n\mu}{\sigma\sqrt{n}}$ and show that $\{Y_n\}$ does not converge mutually in probability by showing that $\{Y_{2n} - Y_n\}$ does not converge in probability to 0, as $n \to \infty$. We have $Y_{2n} - Y_n = \frac{S_{2n}-2n\mu}{\sigma\sqrt{2n}} - \frac{S_n - n\mu}{\sigma\sqrt{n}} = \frac{1}{\sigma\sqrt{2n}}[S_{2n} - 2n\mu - \sqrt{2}(S_n - n\mu)]$.
But

$$S_{2n} - 2n\mu - \sqrt{2}(S_n - n\mu) = \sum_{j=1}^{2n} X_j - 2n\mu - \sqrt{2}\left(\sum_{j=1}^{n} X_j - n\mu\right)$$

$$= \left(\sum_{j=1}^{n} X_j - n\mu\right) + \left(\sum_{j=n+1}^{2n} X_j - n\mu\right) - \sqrt{2}\left(\sum_{j=1}^{n} X_j - n\mu\right)$$

$$= (1-\sqrt{2})\left(\sum_{j=1}^{n} X_j - n\mu\right) + \left(\sum_{j=n+1}^{2n} X_j - n\mu\right),$$

and hence

$$Y_{2n} - Y_n = \frac{1-\sqrt{2}}{\sqrt{2}} \times \frac{\sum_{j=1}^{n} X_j - n\mu}{\sigma\sqrt{n}} + \frac{1}{\sqrt{2}} \times \frac{\sum_{j=n+1}^{2n} X_j - n\mu}{\sigma\sqrt{n}}$$

$$= \frac{1-\sqrt{2}}{\sqrt{2}} U_n + \frac{1}{\sqrt{2}} V_n \text{ and } U_n, V_n \text{ are independent.}$$

Then:

$$f_{\frac{1-\sqrt{2}}{\sqrt{2}}U_n}(t) \underset{n\to\infty}{\longrightarrow} e^{-\left(\frac{1-\sqrt{2}}{\sqrt{2}}\right)^2 t^2/2}, \quad f_{\frac{1}{\sqrt{2}}V_n}(t) \underset{n\to\infty}{\longrightarrow} e^{-\left(\frac{1}{\sqrt{2}}\right)^2 t^2/2},$$

so that

$$f_{Y_{2n}-Y_n}(t) = f_{\frac{1-\sqrt{2}}{\sqrt{2}}U_n + \frac{1}{\sqrt{2}}V_n}(t) \underset{n\to\infty}{\longrightarrow} e^{-(2-\sqrt{2})t^2/2}.$$

Thus, the asymptotic distribution of $Y_{2n} - Y_n$ is $N(0, 2-\sqrt{2})$ or $\frac{Y_{2n}-Y_n}{\sqrt{2-\sqrt{2}}}$ is asymptotically distributed as $N(0,1)$ and then $P\left(\left|\frac{Y_{2n}-Y_n}{\sqrt{2-\sqrt{2}}}\right| > \varepsilon\right) \underset{n\to\infty}{\longrightarrow} 2 - 2\Phi(\varepsilon) \neq 0$. #

10. First, $\mathcal{E}X_1^+ (= \mathcal{E}X_1^-) = c\sum_{n=3}^{\infty} \frac{1}{n\log n}$, and $\sum_{n=3}^{\infty} \frac{1}{n \log n} \geq \int_{2.5}^{\infty} \frac{dx}{(x+0.5)\log(x+0.5)} = \int_3^{\infty} \frac{dt}{t \log t} = \int_3^{\infty} \frac{d\log t}{\log t} = \log\log t|_3^{\infty} = \infty$ (see also the figure below), so that the EX_1 does not exist.

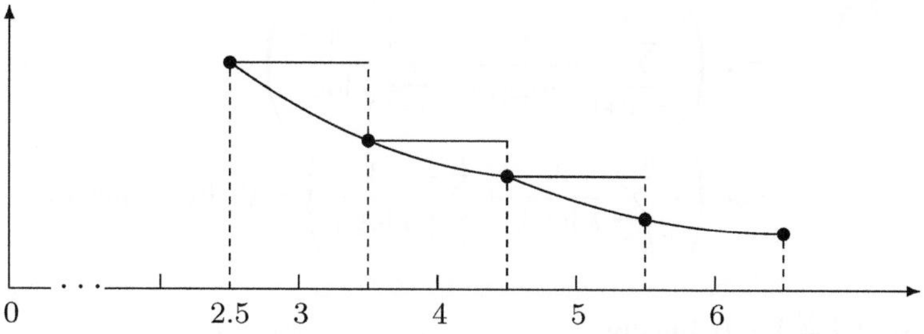

Next, $P\left(\frac{S_n}{n} \neq \frac{S_n^*}{n}\right) = P(S_n \neq S_n^*) \leq P(X_j \neq X_{nj}$ for some $j = 1, \ldots, n) \leq \sum_{j=1}^{n} P(X_j \neq X_{nj}) = \sum_{j=1}^{n} P(|X_j| \geq n) = nP(|X_1| \geq n) = nc\sum_{j=n}^{\infty} \frac{1}{j^2 \log j} \leq \frac{nc}{\log n}\sum_{j=n}^{\infty}\frac{1}{j^2}$, and $\sum_{j=n}^{\infty}\frac{1}{j^2} \leq \frac{1}{n-1}$, because $\sum_{j=n}^{\infty}\frac{1}{j^2} \leq \int_{n-0.5}^{\infty} \frac{dx}{(x-0.5)^2} = \int_{n-1}^{\infty}\frac{dt}{t^2} = -\frac{1}{t}\Big|_{n-1}^{\infty} = \frac{1}{n-1}$ (see also the figure below), so that $\frac{nc}{\log n}\sum_{j=n}^{\infty}\frac{1}{j^2} \leq \frac{nc}{(n-1)\log n} \underset{n\to\infty}{\longrightarrow} 0$, and hence $P\left(\frac{S_n}{n} \neq \frac{S_n^*}{n}\right) \underset{n\to\infty}{\longrightarrow} 0$, or $\frac{S_n}{n} - \frac{S_n^*}{n} \underset{n\to\infty}{\overset{P}{\longrightarrow}} 0$.

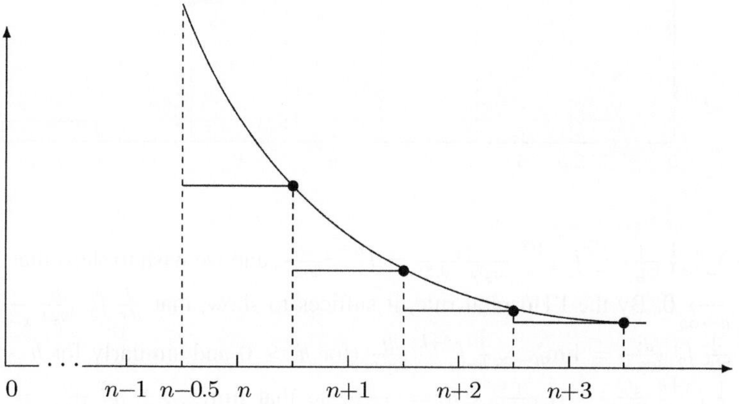

On the other hand,

$$\begin{aligned}
\mathcal{E}\left(\frac{S_n^*}{n}\right) &= \frac{1}{n} \times n\int_{(|X_1|<n)} X_1 dP = \int_{(|X_1|<n)} X_1 dP = \int_{(-n+1\leq X_1\leq n-1)} X_1 dP \\
&= \int_{(-n+1\leq X_1\leq -3)} X_1 dP + \int_{(3\leq X_1\leq n-1)} X_1 dP \\
&= c\left(\sum_{j=-n+1}^{-3} \frac{j}{j^2 \log|j|} + \sum_{j=3}^{n-1} \frac{j}{j^2 \log j}\right)
\end{aligned}$$

$$= c\left(\sum_{j=-n+1}^{-3}\frac{1}{j\log|j|}+\sum_{j=3}^{n-1}\frac{1}{j\log j}\right)$$

$$= c\left(-\sum_{k=3}^{n-1}\frac{1}{k\log k}+\sum_{j=3}^{n-1}\frac{1}{j\log j}\right) = 0, \text{ by setting } j=-k.$$

So, $\mathcal{E}\left(\frac{S_n^*}{n}\right)=0$. Finally,
$\mathrm{Var}\left(\frac{S_n^*}{n}\right) = \mathcal{E}\left(\frac{S_n^*}{n}\right)^2 = \frac{1}{n^2}\times n\mathcal{E}X_{n1}^2 = \frac{1}{n}\int_{(|X_1|<n)}X_1^2dP = \frac{2c}{n}\sum_{j=3}^{n-1}\frac{1}{\log j}$
as above. However (see also figure below),

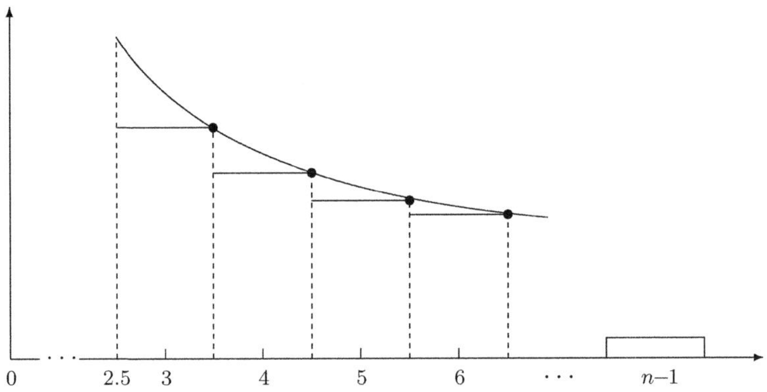

$\sum_{j=3}^{n-1}\frac{1}{\log j} \le \int_{2.5}^{n-0.5}\frac{dx}{\log(x-0.5)} = \int_2^{n-1}\frac{dt}{\log t}$, and we wish to show that $\mathrm{Var}\left(\frac{S_n^*}{n}\right) \underset{n\to\infty}{\longrightarrow} 0$. By the l'Hôspital rule, it suffices to show, that $\frac{d}{dx}\int_2^x\frac{dt}{\log t}\underset{x\to\infty}{\longrightarrow}0$. But $\frac{d}{dx}\int_2^x\frac{dt}{\log t} = \lim_{h\to\infty}\frac{1}{h}\int_x^{x+h}\frac{dt}{\log t}$ (for $h>0$ and similarly for $h<0$), and $\frac{1}{h}\int_x^{x+h}\frac{dt}{\log t} \le \frac{1}{h}\frac{1}{\log x}\times h = \frac{1}{\log x}$, so that $\lim_{x\to\infty}\frac{d}{dx}\int_2^x\frac{dt}{\log t} = \lim_{x\to\infty}\lim_{h\to\infty}\frac{1}{h}\int_x^{x+h}\frac{dt}{\log t} \le \lim_{x\to\infty}\frac{1}{\log x} = 0$. Therefore $\mathrm{Var}\left(\frac{S_n^*}{n}\right)\underset{n\to\infty}{\longrightarrow}0$, and hence $\frac{S_n^*}{n}\underset{n\to\infty}{\overset{P}{\longrightarrow}}0$. This convergence together with $\frac{S_n}{n}-\frac{S_n^*}{n}\underset{n\to\infty}{\overset{P}{\longrightarrow}}0$ complete the proof. #

In all of Exercises 11-15 below, i is treated as a real number subject, of course, to the requirement that $i^2=-1$.

11. Since $p_X(x)=\binom{n}{x}p^xq^{n-x}$, $x=0,1,\ldots,n$, we have:

$$f_X(t)=\mathcal{E}e^{itX}=\sum_{x=0}^{n}e^{itx}\binom{n}{x}p^xq^{n-x}=\sum_{x=0}^{n}\binom{n}{x}(pe^{it})^xq^{n-x}$$

$$=(pe^{it}+q)^n. \#$$

12. Here $p_X(x) = e^{-\lambda}\frac{\lambda^x}{x!}$, $x = 0, 1, \ldots$, so that

$$f_X(t) = \mathcal{E}e^{itX} = \sum_{x=0}^{\infty} e^{itx} e^{-\lambda}\frac{\lambda^x}{x!} = e^{-\lambda}\sum_{x=0}^{\infty}(\lambda e^{it})^x$$
$$= e^{-\lambda}e^{\lambda e^{it}} = e^{\lambda e^{it}-\lambda}. \#$$

13.

(i) Since $p_Z(z) = \frac{1}{\sqrt{2\pi}}e^{-z^2/2}$, we have:

$$f_Z(t) = \mathcal{E}e^{itZ} = \int_{-\infty}^{\infty} e^{itz}\frac{1}{\sqrt{2\pi}}e^{-z^2/2}dz$$
$$= \int_{-\infty}^{\infty}\frac{1}{\sqrt{2\pi}}e^{-\frac{z^2-2itz}{2}}dz = \int_{-\infty}^{\infty}\frac{1}{\sqrt{2\pi}}e^{-\frac{(z-it)^2}{2}} \times e^{-\frac{t^2}{2}}dz$$
$$= e^{-\frac{t^2}{2}}\int_{-\infty}^{\infty}\frac{1}{\sqrt{2\pi}}e^{-\frac{(z-it)^2}{2}}dz = e^{-\frac{t^2}{2}} \times 1 = e^{-\frac{t^2}{2}},$$

since the integrand can be viewed as the p.d.f. of the Normal distribution with mean it and variance 1.

(ii) Here

$$e^{-\frac{t^2}{2}} = f_Z(t) = f_{\frac{X-\mu}{\sigma}}(t) = \mathcal{E}e^{it(X-\mu)/\sigma} = e^{-i\mu t/\sigma}\mathcal{E}e^{i(t/\sigma)X}$$
$$= e^{-i\mu t/\sigma}f_X\left(\frac{t}{\sigma}\right), \text{ so that } f_X\left(\frac{t}{\sigma}\right) = e^{\frac{it\mu}{\sigma}-\frac{t^2}{2}}.$$

Replacing $\frac{t}{\sigma}$ by t, we get: $f_X(t) = e^{i\mu t-\frac{\sigma^2t^2}{2}}$. #

14. Here $f_X(t) = \mathcal{E}e^{itX} = \int_0^{\infty} e^{itx}\frac{1}{\Gamma(\alpha)\beta^{\alpha}}x^{\alpha-1}e^{-x/\beta}dx = \frac{1}{\Gamma(\alpha)\beta^{\alpha}}\int_0^{\infty}x^{\alpha-1}$ $\times e^{-(1-i\beta t)x/\beta}dx = \frac{1}{\Gamma(\alpha)\beta^{\alpha}}\int_0^{\infty}\frac{\beta^{\alpha-1}y^{\alpha-1}e^{-y}}{(1-i\beta t)^{\alpha-1}}\frac{\beta}{1-i\beta t}dy$ (by setting $\frac{(1-i\beta t)x}{\beta} = y$)$=$ $\frac{1}{(1-i\beta t)^{\alpha}} \times \frac{1}{\Gamma(\alpha)}\int_0^{\infty}y^{\alpha-1}e^{-y}dy = \frac{1}{(1-i\beta t)^{\alpha}} \times \frac{\Gamma(\alpha)}{\Gamma(\alpha)} = \frac{1}{(1-i\beta t)^{\alpha}}$. The special cases follow immediately. #

15.

(i) For $\mu_1 = \mu_2 = 0$ and $\sigma_1 = \sigma_2 = 1$, we have:

$$p_{X,Y}(x,y) = \frac{1}{2\pi\sqrt{1-\rho^2}}\exp\left[-\frac{1}{2(1-\rho^2)}(x^2-2\rho xy+y^2)\right]$$
$$= \frac{1}{\sqrt{2\pi}}\exp\left(-\frac{x^2}{2}\right) \times \frac{1}{\sqrt{2\pi}\sqrt{1-\rho^2}}\exp\left[-\frac{(y-\rho x)^2}{2(1-\rho^2)}\right]$$

(by Exercise 12(ii) in Chapter 9),

so that

$$f_{X,Y}(t_1,t_2)=\mathcal{E}e^{it_1X+it_2Y}=\int_{-\infty}^{\infty}\int_{-\infty}^{\infty}e^{it_1x+it_2y}p_{X,Y}(x,y)dxdy$$
$$=\left[\int_{-\infty}^{\infty}e^{it_1x}\frac{1}{\sqrt{2\pi}}e^{-\frac{x^2}{2}}\right.$$
$$\left.\times\int_{-\infty}^{\infty}e^{it_2y}\frac{1}{\sqrt{2\pi}\sqrt{1-\rho^2}}e^{-\frac{(y-\rho x)^2}{2(1-\rho^2)}}dy\right]dx.$$

However, the second integral on the right-hand side above is the ch.f. of a r.v. distributed as $N(\rho x, 1-\rho^2)$, and therefore, by Exercise 13(ii), is equal to: $e^{i(\rho x)t_2-\frac{(1-\rho^2)t_2^2}{2}}$. Therefore

$$f_{X,Y}(t_1,t_2)=\int_{-\infty}^{\infty}e^{it_1x}\frac{1}{\sqrt{2\pi}}e^{-\frac{x^2}{2}}e^{i(\rho x)t_2-\frac{(1-\rho^2)t_2^2}{2}}dx$$
$$=e^{-\frac{(1-\rho^2)t_2^2}{2}}\int_{-\infty}^{\infty}\frac{1}{\sqrt{2\pi}}e^{it_1x-\frac{x^2}{2}+i\rho t_2x}dx$$
$$=e^{-\frac{(1-\rho^2)t_2^2}{2}}\int_{-\infty}^{\infty}\frac{1}{\sqrt{2\pi}}e^{-\frac{x^2-2i(t_1+\rho t_2)x}{2}}dx.$$

However,

$$x^2-2i(t_1+\rho t_2)x=x^2+[i(t_1+\rho t_2)]^2-2i(t_1+\rho t_2)x+(t_1+\rho t_2)^2$$
$$=[x-i(t_1+\rho t_2)]^2+(t_1+\rho t_2)^2,$$

so that

$$f_{X,Y}(t_1,t_2)=e^{-\frac{(1-\rho^2)t_2^2}{2}-\frac{(t_1+\rho t_2)^2}{2}}\int_{-\infty}^{\infty}\frac{1}{\sqrt{2\pi}}e^{-[x-i(t_1+\rho t_2)]^2/2}dx$$
$$=\exp\left\{-\frac{1}{2}[(1-\rho^2)t_2^2+(t_1+\rho t_2)]^2]\right\}$$
$$=\exp\left[-\frac{1}{2}(t_1^2+2\rho t_1t_2+t_2^2)\right],$$

as was to be seen.

(ii) Clearly, $\mathcal{E}U=\mathcal{E}V=0$, $\text{Var}(U)=\text{Var}(V)=1$, and $\rho(U,V)=\text{Cov}(U,V)=\mathcal{E}(UV)=\frac{1}{\sigma_1\sigma_2}\mathcal{E}[(X-\mu_1)(Y-\mu_2)]=\frac{\text{Cov}(X,Y)}{\sigma_1\sigma_2}=\rho(X,Y)=\rho$. Furthermore, $U\sim N(0,1)$ and $V\sim N(0,1)$. Next,

$X = \sigma_1 U + \mu_1, Y = \sigma_2 V + \mu_2$, so that

$$\begin{aligned} f_{X,Y}(t_1, t_2) &= \mathcal{E} e^{t_1 X + t_2 Y} = e^{it_1\mu_1 + it_2\mu_2} \mathcal{E} e^{i(t_1\sigma_1)U + i(t_2\sigma_2)V} \\ &= e^{it_1\mu_1 + it_2\mu_2} f_{U,V}(t_1\sigma_1, t_2\sigma_2) \\ &= e^{it_1\mu_1 + it_2\mu_2 - \frac{1}{2}(t_1^2\sigma_1^2 + 2\rho\sigma_1\sigma_2 t_1 t_2 + t_2^2\sigma_2^2)} \\ &= exp\left[i\mu_1 t_1 + i\mu_2 t_2 - \frac{1}{2}(\sigma_1^2 t_1^2 + 2\rho\sigma_1\sigma_2 t_1 t_2 + \sigma_2^2 t_2^2)\right]. \end{aligned}$$

This is so because the joint distribution of U and V is the Bivariate Normal with parameters 0,0,1,1 and ρ (by Exercise 15 in Chapter 9), so that part (i) applies. #

16. **(i)** That $\mathcal{E}U = \mu_1 + \mu_2, \mathcal{E}V = \mu_1 - \mu_2$, $\text{Var}(U) = \sigma_1^2 + \sigma_2^2 + 2\rho\sigma_1\sigma_2$, and $\text{Var}(V) = \sigma_1^2 + \sigma_2^2 - 2\rho\sigma_1\sigma_2$ are immediate. Also,

$$\begin{aligned} \text{Cov}(U, V) &= \mathcal{E}(UV) - (\mathcal{E}U)(\mathcal{E}V) \\ &= \mathcal{E}[(X+Y)(X-Y)] - (\mu_1 + \mu_2)(\mu_1 - \mu_2) \\ &= \mathcal{E}X^2 - \mathcal{E}Y^2 - \mu_1^2 + \mu_2^2 \\ &= (\mathcal{E}X^2 - \mu_1^2) - (\mathcal{E}Y^2 - \mu_2^2) = \sigma_1^2 - \sigma_2^2. \end{aligned}$$

(ii) We have

$$\begin{aligned} &f_{X,Y}(t_1 + t_2, t_1 - t_2) \\ &\quad = \exp\{i\mu_1(t_1 + t_2) + i\mu_2(t_1 - t_2) \\ &\quad\quad - \frac{1}{2}[\sigma_1^2(t_1 + t_2)^2 + 2\rho\sigma_1\sigma_2(t_1 + t_2)(t_1 - t_2) + \sigma_2^2(t_1 - t_2)^2]\Big\}. \end{aligned}$$

Doing the algebra in the exponent, we get:

$$\begin{aligned} &i(\mu_1 + \mu_2)t_1 + i(\mu_1 - \mu_2)t_2 - \frac{1}{2}[(\sigma_1^2 + \sigma_2^2 + 2\rho\sigma_1\sigma_2)t_1^2 \\ &\quad +2(\sigma_1^2 - \sigma_2^2)t_1 t_2 + (\sigma_1^2 - 2\rho\sigma_1\sigma_2 + \sigma_2^2)t_2^2], \end{aligned}$$

and the result follows.

(iii) Immediate from part (ii) and Exercise 15.

(iv) Follows from part (iii), since $\rho(U, V) = 0$ if and only if $\sigma_1 = \sigma_2$. #

17. **(i)** By Exercise 13 in Chapter 10,

$$\begin{aligned} f_X(t) &= \mathcal{E}e^{itX} = \mathcal{E}e^{it(X_1 + \ldots + X_k)} = \mathcal{E}e^{itX_1} \times \ldots \times \mathcal{E}e^{itX_k} \\ &= (pe^{it} + q)^{n_1} \times \ldots \times (pe^{it} + q)^{n_k} = (pe^{it} + q)^n, \end{aligned}$$

which is the ch.f. of the $B(n, p)$ distribution.

(ii) As above,

$$f_X(t) = \mathcal{E}e^{itX_1} \times \ldots \times \mathcal{E}e^{itX_k}$$
$$= e^{\lambda_1 e^{it} - \lambda_1} \times \ldots \times e^{\lambda_k e^{it} - \lambda_k} = e^{\lambda e^{it} - \lambda},$$

which is the ch.f. of the $P(\lambda)$ distribution.

(iii) Again, as above,

$$f_X(t) = \mathcal{E}e^{itX_1} \times \ldots \times \mathcal{E}e^{itX_k}$$
$$= e^{i\mu_1 t - \frac{\sigma_1^2 t^2}{2}} \times \ldots \times e^{i\mu_k t - \frac{\sigma_k^2 t^2}{2}}$$
$$= e^{i\mu t - \frac{\sigma^2 t^2}{2}},$$

which is the ch.f. of the $N(\mu, \sigma^2)$ distribution. Furthermore, independence of $X_1, \ldots, X_k$ implies that of $c_1 X_1, \ldots, c_k X_k$, and so

$$f_{c_1 X_1 + \ldots + c_k X_k}(t) = \mathcal{E}e^{it(c_1 X_1 + \ldots + c_k X_k)}$$
$$= \mathcal{E}e^{it(c_1 X_1)} \times \ldots \times \mathcal{E}e^{it(c_k X_k)}$$
$$= \mathcal{E}e^{i(c_1 t)X_1} \times \ldots \times \mathcal{E}e^{i(c_k t)X_k}$$
$$= f_{X_1}(c_1 t) \times \ldots \times f_{X_k}(c_k t)$$
$$= e^{i\mu_1 c_1 t - \frac{\sigma_1^2 c_1^2 t^2}{2}} \times \ldots \times e^{i\mu_k c_k t - \frac{\sigma_k^2 c_k^2 t^2}{2}}$$
$$= e^{i(c_1\mu_1 + \ldots + c_k\mu_k)t - \frac{(c_1^2\sigma_1^2 + \ldots + c_k^2\sigma_k^2)t^2}{2}},$$

which is the ch.f. of the $N(c_1\mu_1 + \ldots + c_k\mu_k, c_1^2\sigma_1^2 + \ldots + c_k^2\sigma_k^2)$ distribution.

(iv) Here

$$f_X(t) = \mathcal{E}e^{itX_1} \times \ldots \times \mathcal{E}e^{itX_k}$$
$$= \frac{1}{(1 - i\beta t)^{\alpha_1}} \times \ldots \times \frac{1}{(1 - i\beta t)^{\alpha_k}}$$
$$= \frac{1}{(1 - i\beta t)^{\alpha}},$$

which is the ch.f. of the Gamma distribution with parameters α and β. For the special cases, we have: $\alpha_1 = \cdots = \alpha_k = 1$ (so that $\alpha = k$), $\beta = 1/\lambda$, so that $f_X(t) = \frac{1}{(1 - \frac{it}{\lambda})^k}$, which is the ch.f. of the Gamma distribution with parameters $\alpha = k, \beta = 1/\lambda$. For $\alpha_j = r_j/2, \beta = 2$, we have: $f_X(t) = \frac{1}{(1-2it)^{r/2}}$, which is the ch.f. of the χ_r^2 distribution. #

18. Suppose the r.v.s $X_1, \ldots, X_k$ are independent. Referring to the derivations in Exercise 13 in Chapter 10, it is clear that the expression $it(X_1 + X_2)$ can be replaced by $i(t_1 X_1 + t_2 X_2)$. Then we obtain

$$\mathcal{E}e^{i(t_1 X_1 + t_2 X_2)} = (\mathcal{E}e^{it_1 X_1})(\mathcal{E}e^{it_2 X_2}),$$

and by induction, (as in the exercise just cited),

$$\mathcal{E}e^{i(t_1X_1+\ldots+t_kX_k)} = (\mathcal{E}e^{it_1X_1}) \times \ldots \times (\mathcal{E}e^{it_kX_k}),$$

or

$$f_{X_1,\ldots,X_k}(t_1,\ldots,t_k) = f_{X_1}(t_1) \times \ldots \times f_{X_k}(t_k)$$

for all $t_1, \ldots, t_k$ in $\Re$.

Next, assume that this factorization, and we wish to show that the r.v.s $X_1, \ldots, X_k$ are independent. By the assumed factorization, the expression in Theorem 2′ becomes as follows for continuity points $\mathbf{a} = (a_1, \ldots, a_k)$ and $\mathbf{b} = (b_1, \ldots, b_k)$ of the joint d.f. $F = F_{X_1,\ldots,X_k}$ of the r.v.s $X_1, \ldots, X_k$,

$$\begin{aligned}
&P(a_j < X_j \leq b_j,\ j = 1, \ldots, k)\\
&= \lim_{T_1\to\infty}\left[\frac{1}{2\pi}\int_{-T_1}^{T_1}\frac{e^{-it_1a_1} - e^{-it_1b_1}}{it_1}f_{X_1}(t_1)dt_1\right] \times \ldots \times\\
&\quad \lim_{T_k\to\infty}\left[\frac{1}{2\pi}\int_{-T_k}^{T_k}\frac{e^{-it_ka_k} - e^{-it_kb_k}}{it_k}f_{X_k}(t_k)dt_k\right]\\
&= [F_{X_1}(b_1) - F_{X_1}(a_1)] \times \ldots \times [F_{X_k}(b_k) - F_{X_k}(a_k)]\\
&\quad \text{(by Theorem 2 as it becomes for continuity points}\\
&\quad a_j \text{ and } b_j \text{ of } F_{X_j},\ j = 1, \ldots, k)\\
&= P(a_1 < X_1 \leq b_1) \times \ldots \times P(a_k < X_k \leq b_k).
\end{aligned}$$

That is,

$$P(a_j < X_j \leq b_j,\ j = 1, \ldots, k) = \prod_{j=1}^{k} P(a_j < X_j \leq b_j)$$

for continuity point. Since any point can be approached by a sequence of continuity points, it follows that the last relation above holds for any points $(a_1, \ldots, a_k)$ and $(b_1, \ldots, b_k)$ with $a_j < b_j$, $j = 1, \ldots, k$. Then Exercise 9 in Chapter 10 applies and yields the desired independence of the r.v.s $X_1, \ldots, X_k$. #

19. Indeed, if F is the d.f. corresponding to f, then

$$\begin{aligned}
\sum_{k=1}^{n}\sum_{l=1}^{n} f(t_k - t_l)z_k\bar{z}_l &= \sum_{k=1}^{n}\sum_{l=1}^{n}\left[\int_{\Re} e^{i(t_k-t_l)x}dF(x)\right]z_k\bar{z}_l\\
&= \int_{\Re}\left[\sum_{k=1}^{n}\sum_{l=1}^{n} e^{i(t_k-t_l)x}z_k\bar{z}_l\right]dF(x)\\
&= \int_{\Re}\left(\sum_{k=1}^{n}\sum_{l=1}^{n} z_ke^{it_kx} \times \bar{z}_le^{-it_lx}\right)dF(x)
\end{aligned}$$

$$= \int_{\Re}\left[\left(\sum_{k=1}^{n} z_k e^{it_k x}\right)\left(\sum_{l=1}^{n} \bar{z}_l e^{-it_l x}\right)\right] dF(x)$$

$$= \int_{\Re}\left[\left(\sum_{k=1}^{n} z_k e^{it_k x}\right)\left(\overline{\sum_{l=1}^{n} z_l e^{it_l x}}\right)\right] dF(x)$$

$$= \int_{\Re}\left|\sum_{k=1}^{n} z_k e^{it_k x}\right|^2 dF(x) \geq 0,$$

where we recall that $|z| = |x + iy| = (x^2 + y^2)^{\frac{1}{2}}$. #

20. The ch.f. of X_n is : $f_{X_n}(t) = e^{ne^{it} - n}$. Then

$$f_{Y_n}(t) = f_{\frac{X_n - n}{\sqrt{n}}}(t) = e^{-it\sqrt{n}} f_{X_n}\left(\frac{t}{\sqrt{n}}\right)$$
$$= e^{-it\sqrt{n}} e^{n\left(e^{\frac{it}{\sqrt{n}}} - 1\right)}$$
$$= e^{-it\sqrt{n} + n\left(e^{\frac{it}{\sqrt{n}}} - 1\right)}.$$

Use the expansion: $e^{iz} = 1 + iz - \frac{z^2}{2} - \frac{iz^3}{6} e^{iz^*}$, for some complex number z^*, to obtain: $e^{\frac{it}{\sqrt{n}}} = 1 + \frac{it}{\sqrt{n}} - \frac{t^2}{2n} - \frac{it^3}{6n\sqrt{n}} e^{it_n}$ ($|e^{it_n}| = 1$), and hence $-it\sqrt{n} + n(e^{\frac{it}{\sqrt{n}}} - 1) = -\frac{t^2}{2} - \frac{it^3}{6\sqrt{n}} e^{it_n}$. It follows that the right-hand side above tends to $-\frac{t^2}{2}$ as $n \to \infty$, so that $f_{Y_n}(t) \underset{n \to \infty}{\longrightarrow} e^{-\frac{t^2}{2}}$, which is the ch.f. of $Z \sim N(0, 1)$. This completes the proof. #

21. **(i)** Indeed,

$$G(x) = \int_{-\infty}^{\infty} F_1(x - y) dF_2(y) = \int_{-\infty}^{\infty}\left[\int_{-\infty}^{x-y} p_1(u) du\right] dF_2(y)$$
$$= \int_{-\infty}^{\infty}\left[\int_{-\infty}^{x} p_1(v - y) dv\right] dF_2(y) \text{ (by setting } y + u = v)$$
$$= \int_{-\infty}^{\infty}\int_{-\infty}^{x} p_1(v - y) dv dF_2(y)$$
$$= \int_{-\infty}^{x}\int_{-\infty}^{\infty} p_1(v - y) dF_2(y) dv \text{ (by the Fubini Theorem).}$$

Next, observe that the function $g(v) = \int_{-\infty}^{\infty} p_1(v - y) dF_2(y)$ is nonnegative and

$$\int_{-\infty}^{\infty} g(v) dv = \int_{-\infty}^{\infty}\int_{-\infty}^{\infty} p_1(v - y) dF_2(y) dv$$

$$= \int_{-\infty}^{\infty}\int_{-\infty}^{\infty} p_1(v-y)dv dF_2(y) \text{ (by the Fubini Theorem)}$$
$$= \int_{-\infty}^{\infty}\left[\int_{-\infty}^{\infty} p_1(v-y)dv\right] dF_2(y)$$
$$= \int_{-\infty}^{\infty}\left[\int_{-\infty}^{\infty} p_1(w)dw\right] dF_2(y) \text{ (by setting } v-y=w)$$
$$= \int_{-\infty}^{\infty} dF_2(y) = 1.$$

Since also $G(x) = \int_{-\infty}^{x} g(v)dv$, it follows that g is the density of G; i.e., $g = p$. Furthermore, $p(u) = \int_{-\infty}^{\infty} p_1(v-y)dF_2(y)$.

(ii) This part is immediate, because

$$p(v) = \int_{-\infty}^{\infty} p_1(v-y)dF_2(y)$$
$$= \int_{-\infty}^{\infty} p_1(v-y)p_2(y)dy$$

(by the Corollary to the Fubini Theorem). #

22. We have $f_{X-Y}(t) = f_X(t)f_{-Y}(t) = f_X(t)f_Y(-t) = f_X(t)\overline{f_Y(t)} = f(t)\overline{f(t)} = |f(t)|^2$. #

23. Recall that a r.v. is symmetric about 0 if and only if its ch.f. is real. Next, $f_{X_1+\ldots+X_n}(t) = f_{X_1}(t)\ldots f_{X_n}(t)$ (by independence of the X_j s) and the right-hand side is real since each of $f_{X_j}(t)$, $j = 1,\ldots,n$, is real. Thus, $f_{X_1+\ldots+X_n}(t)$ is real and hence $X_1 + \ldots + X_n$ is symmetric about 0. #

24. All limits are taken as $\{n\}$ or subsequences thereof tend to ∞. Continuous convergence implies that $g_n(x_n) \to g(x)$ whenever $x_n \to x$. Indeed, if g were not continuous at x, the implication would be that there is a sequence $x_n \to x$ and some $\varepsilon > 0$ such that $|g(x_n) - g(x)| > \varepsilon$. Again, continuous convergence implies that $g_n(x) \to g(x)$ by taking $\{x_n\} = \{x\}$. Applying this for x_1, we have that there exists m_1 such that $|g_{m_1}(x_1) - g(x_1)| < \frac{\varepsilon}{2}$. Also, for x_2 there exists $m_2 > m_1$ such that $|g_{m_2}(x_2) - g(x_2)| < \frac{\varepsilon}{2}$. Continuing like this, we have that there exists $\{m_n\} \subseteq \{n\}$ such that $|g_{m_n}(x_n) - g(x_n)| < \frac{\varepsilon}{2}$. Setting $x_n = y_{m_n}$, this inequality becomes $|g_{m_n}(y_{m_n}) - g(y_{m_n})| < \frac{\varepsilon}{2}$. Also, $|g(y_{m_n}) - g(x)| > \varepsilon$. Therefore, we have:

$$\varepsilon < |g(y_{m_n}) - g(x)| = |[g(y_{m_n}) - g_{m_n}(y_{m_n})] + [g_{m_n}(y_{m_n}) - g(x)]|$$
$$\leq |g_{m_n}(y_{m_n}) - g(y_{m_n})| + |g_{m_n}(y_{m_n}) - g(x)|.$$

However, $|g_{m_n}(y_{m_n}) - g(x)| < \frac{\varepsilon}{2}$. Therefore $|g_{m_n}(y_{m_n}) - g(y_{m_n})|$ must be $> \frac{\varepsilon}{2}$, and this contradicts continuous convergence. Thus, g is continuous at every $x \in E$. #

25. (i) Let $x, x_n \in E$ with $x_n \to x$ as $n \to \infty$ here and in the remaining of the discussion. Then

$$|g_n(x_n) - g(x)| = |[g_n(x_n) - g(x_n)] + [g(x_n) - g(x)]|$$
$$\leq |g_n(x_n) - g(x_n)| + |g(x_n) - g(x)| \to 0,$$

since $g_n(x_n) - g(x_n) \to 0$ by uniformity, and $g(x_n) \to g(x)$ by continuity.

(ii) Continuous convergence implies continuity of g (by Exercise 24). If the convergence were not uniform, there would exist $\{m\} \subseteq \{n\}$ and $x_m \in E$ such that $|g_m(x_m) - g(x_m)| > \varepsilon$ for some $\varepsilon > 0$. By the compactness of E, there exists $\{x_r\} \subseteq \{x_m\}$ with $x_r \to x \in E$, and $|g_r(x_r) - g(x_r)| > \varepsilon$. However,

$$\varepsilon < |g_r(x_r) - g(x_r)| \leq |g_r(x_r) - g(x)| + |g(x_r) - g(x)|,$$

and $|g(x_r) - g(x)| < \varepsilon/2$ by continuity of g. Therefore $|g_r(x_r) - g(x)| > \varepsilon/2$, which contradicts the assumed continuous convergence. #

26. Expand $g(X_n, Y_n)$ around (d_1, d_2) according to the Taylor formula by using terms involving first order derivatives to get

$$g(X_n, Y_n) = g(d_1, d_2) + (X_n - d_1 \quad Y_n - d_2)\left(\dot{g}_x(\mathbf{Z}_n) \quad \dot{g}_y(\mathbf{Z}_n)\right)',$$
$$= g(d_1, d_2) + \left[(X_n - d_1)\dot{g}_x(\mathbf{Z}_n) + (Y_n - d_2)\dot{g}_y(\mathbf{Z}_n)\right],$$

where $\mathbf{Z}_n$ is a 2-dimensional random vector lying between (d_1, d_2) and (X_n, Y_n) on the line segment determined by them. Then

$$\begin{aligned} &c_n[g(X_n, Y_n) - g(d_1, d_2)] \\ &\quad = c_n(X_n - d_1)[\dot{g}_x(\mathbf{Z}_n) - \dot{g}_x(d_1, d_2)] \\ &\qquad + c_n(Y_n - d_2)[\dot{g}_y(\mathbf{Z}_n) - \dot{g}_y(d_1, d_2)] \\ &\qquad + \dot{g}_x(d_1, d_2)c_n(X_n - d_1) + \dot{g}_y(d_1, d_2)c_n(Y_n - d_2). \end{aligned}$$

However, as $n \to \infty$, $c_n(X_n - d_1) \xrightarrow{d} X$, $\dot{g}_x(\mathbf{Z}_n) - \dot{g}_x(d_1, d_2) \to 0$, and $c_n(Y_n - d_2) \xrightarrow{d} 0$, $\dot{g}_y(\mathbf{Z}_n) - \dot{g}_y(d_1, d_2) \to 0$ (the latter by the assumed continuity of the partial derivatives at (d_1, d_2)), so that

$$c_n(X_n - d_1)[\dot{g}_x(\mathbf{Z}_n) - \dot{g}_x(d_1, d_2)] \xrightarrow{P} 0,$$

and

$$c_n(Y_n - d_2)[\dot{g}_y(\mathbf{Z}_n) - \dot{g}_y(d_1, d_2)] \xrightarrow{P} 0.$$

Next, the fact that

$$c_n(X_n - d_1, \quad Y_n - d_2) \xrightarrow{d} (X, Y)$$

implies that any linear combination of the two components converges in distribution to the corresponding linear combination of (X, Y) (by the Cramér-Wold device, Theorem 4 in this chapter). Accordingly,

$$\dot{g}_x(d_1, d_2)c_n(X_n - d_1) + \dot{g}_y(d_1, d_2)c_n(Y_n - d_2) \xrightarrow{d} \dot{g}_x(d_1, d_2)X + \dot{g}_y(d_1, d_2)Y,$$

and hence

$$c_n[g(X_n, Y_n) - g(d_1, d_2)] \xrightarrow{d} \dot{g}_x(d_1, d_2)X + \dot{g}_y(d_1, d_2)Y,$$

as was to be seen.
Remark: From the process of the proof, it is quite clear that the result holds if the 2-dimensional random vector (X_n, Y_n) is replaced by a k-dimensional random vector. #

Chapter 12

The Central Limit Problem: The Centered Case

1. $X_j \sim U(-j, j)$, $j \geq 1$, implies $\mathcal{E}X_j = 0$, $\sigma_j^2 = \sigma^2(X_j) = \frac{j^2}{3}$, so that $s_n^2 = \frac{n(n+1)(2n+1)}{18}$. Also, $p_j(x) = \frac{1}{2j}$ for $-j \leq x \leq j$, and 0 otherwise.
Next, for n sufficiently large, $n < \varepsilon s_n$ or $n^2 < \varepsilon^2 s_n^2$, or $n^2 < \varepsilon^2 \frac{n(n+1)(2n+1)}{18}$, or $\frac{\varepsilon^2}{18}(n+1)(2+\frac{1}{n}) > 1$, which is correct. Thus, for $j \leq n$ sufficiently large, $\int_{(|x| \geq \varepsilon s_n)} x^2 p_j(x)dx = 0$, because $p_j(x) = 0$ outside $[-j, j] \subset (-\varepsilon s_n, \varepsilon s_n)$. Then, for sufficiently large n,

$$g_n(\varepsilon) = \frac{1}{s_n^2} \sum_{j=1}^{n} \int_{(|x| \geq \varepsilon s_n)} x^2 p_j(x)dx = 0,$$

so that the Lindeberg condition is satisfied. #
2. Here $\mathcal{E}X_j = 0$, $\sigma_j^2 = 2j^{2\alpha} \times \frac{1}{j^\beta} = 2j^{2\alpha-\beta}$ and $s_n^2 = \sum_{j=2}^{n} 2j^{2\alpha-\beta}$. Then, for every $\varepsilon > 0$, $g_n(\varepsilon) = 0$ if the set of js with $j = 2, \ldots, n$ and $|\pm j^\alpha| \geq \varepsilon s_n$ is $\oslash$; or if for all js with $j = 2, \ldots, n$, it happens that $|\pm j^\alpha| < \varepsilon s_n$ or $j^\alpha < \varepsilon s_n$, which is implied by $n^\alpha < \varepsilon s_n$, or $\frac{n^\alpha}{s_n} < \varepsilon$, and this is equivalent to saying that $\frac{n^\alpha}{s_n} \underset{n\to\infty}{\longrightarrow} 0$, or $\frac{s_n}{n^\alpha} \underset{n\to\infty}{\longrightarrow} \infty$, or $\frac{s_n^2}{n^{2\alpha}} \underset{n\to\infty}{\longrightarrow} \infty$.

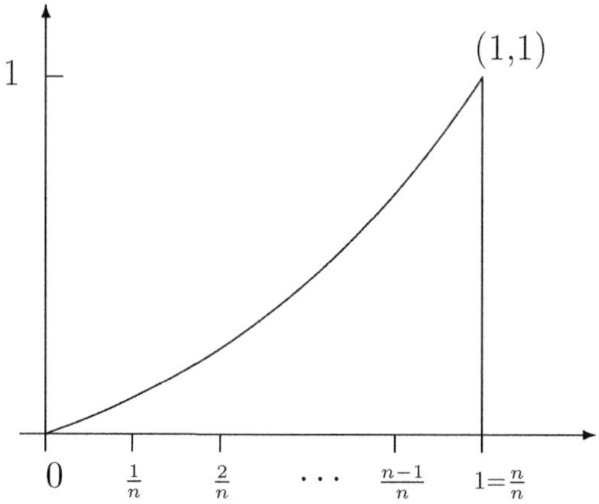

However, $\frac{s_n^2}{n^{2\alpha}} = \sum_{j=2}^{n} 2(\frac{j}{n})^{2\alpha} \times \frac{1}{j^{\beta}} > \frac{2}{n^{\beta}} \sum_{j=2}^{n}(\frac{j}{n})^{2\alpha}$, and: $\sum_{j=2}^{n}(\frac{j}{n})^{2\alpha} = n \times \sum_{j=2}^{n}(\frac{j}{n})^{2\alpha} \times \frac{1}{n} = nT_n$, where $T_n = \sum_{j=2}^{n}(\frac{j}{n})^{2\alpha} \times \frac{1}{n} = \sum_{j=1}^{n}\left(\frac{j}{n}\right)^{2\alpha} \times \frac{1}{n} - \left(\frac{1}{n}\right)^{2\alpha} \times \frac{1}{n}$ and $\sum_{j=1}^{n}\left(\frac{j}{n}\right)^{2\alpha} \times \frac{1}{n}$ is a partial sum converging to the integral $\int_0^1 x^{2\alpha}dx = \frac{1}{2\alpha+1}$. That is, $T_n \underset{n\to\infty}{\longrightarrow} \frac{1}{2\alpha+1}$, so that $T_n = \frac{1}{2\alpha+1} + o(1)$ ($o(1) \to 0$ as $n \to \infty$). Therefore $\frac{s_n^2}{n^{2\alpha}} > \frac{2n}{n^{\beta}} \times T_n = 2n^{1-\beta}T_n = 2n^{1-\beta}(\frac{1}{2\alpha+1} + o(1)) \underset{n\to\infty}{\longrightarrow} \infty$, provided $1 - \beta > 0$ or $\beta < 1$. #

3. Here $\mathcal{E}X_j = 0$, $\sigma_j^2 = \sigma^2(X_j) = \mathcal{E}X_j^2 = \frac{2j^{2\alpha}}{6j^{2(\alpha-1)}} = \frac{1}{3}j^2$, so that $s_n^2 = \frac{n(n+1)(2n+1)}{18}$. Consider js with $j = 1, \ldots, n$ and $|\pm j|^{\alpha} < \varepsilon s_n$ or $j^{2\alpha} < \varepsilon^2 s_n^2$. Then $g_n(\varepsilon) = s_n^{-2}\sum_{j=1}^{n}\int_{(|x|\geq \varepsilon s_n)} x^2 dF_j(x) = 0$ (since there are no values of X_j, other than 0, in the region of integration). Next, the inequality $j^{2\alpha} < \varepsilon^2 s_n^2$ is implied by $n^{2\alpha} < \varepsilon^2 s_n^2$, or $n^{2\alpha} < \varepsilon^2 n(n+1)(2n+1)/18$, or $\frac{18}{\varepsilon^2} < \frac{n(n+1)(2n+1)}{n^{2\alpha}} = \frac{(1+\frac{1}{n})(2+\frac{1}{n})}{n^{2\alpha-3}}$ which $\underset{n\to\infty}{\longrightarrow} \infty$, provided $\alpha < \frac{3}{2}$.
So, for $j = 1, \ldots, n$ and $\alpha < \frac{3}{2}$, it follows that $|\pm j|^{\alpha} < \varepsilon s_n$ and $g_n(\varepsilon) = 0$. Next, let

$$|\pm j|^{\alpha} \geq \varepsilon s_n \text{ or } j^{\alpha} \geq \varepsilon s_n \text{ or } j^{\alpha} \geq \varepsilon s_n, \tag{1}$$

and set $k = k_n = [(\varepsilon s_n)^{1/\alpha}]$ (the integer part of $(\varepsilon s_n)^{1/\alpha}$), so that $k \to \infty$ (since $s_n \underset{n\to\infty}{\longrightarrow} \infty$) and $k/(\varepsilon s_n)^{1/\alpha} \underset{n\to\infty}{\longrightarrow} 1$, or

$$\frac{k^{2\alpha}}{\varepsilon^2 s_n^2} \underset{n\to\infty}{\longrightarrow} 1. \tag{2}$$

For j with $j \geq k$,

$$
\begin{aligned}
g_n(\varepsilon) &\geq \frac{1}{s_n^2}\sum_{j=k}^{n}\frac{j^2}{3} = \frac{1}{s_n^2}\left(\frac{1}{3}\sum_{j=1}^{n} j^2 - \frac{1}{3}\sum_{j=1}^{k-1} j^2\right) \\
&= \frac{1}{s_n^2}\left[s_n^2 - \frac{k(k-1)(2k-1)}{18}\right] = 1 - \frac{\varepsilon^2}{18} \times \frac{k(k-1)(2k-1)}{\varepsilon^2 s_n^2} \\
&= 1 - \frac{\varepsilon^2}{18} \times \left(1 - \frac{1}{k}\right)\left(2 - \frac{1}{k}\right) \times \frac{k^3}{\varepsilon^2 s_n^2} \\
&= 1 - \frac{\varepsilon^2}{18} \times \left(1 - \frac{1}{k}\right)\left(2 - \frac{1}{k}\right) \times \frac{k^{2\alpha}}{\varepsilon^2 s_n^2} \times k^{3-2\alpha}. \qquad (3)
\end{aligned}
$$

Then, by means of (2), we have that, as $n \to \infty$, the expression on the right-hand side of (3) tends to:

$$
\begin{array}{ll}
1 - \dfrac{\varepsilon^2}{18} \times 1 \times 2 \times 1 \times 1 & \text{if } \alpha = 3/2 \\
1 & \text{if } \alpha > 3/2.
\end{array}
$$

In summary then, $g_n(\varepsilon) \underset{n\to\infty}{\longrightarrow} 0$ (indeed, $g_n(\varepsilon) = 0$) if $\alpha < 3/2$ (because of (1)), and $g_n(\varepsilon) \underset{n\to\infty}{\nrightarrow} 0$ for $\alpha \geq 3/2$. The conclusion follows. #

4. We have $\mathcal{E}X_j = 0$, $\sigma_j^2 = \sigma^2(X_j) = \mathcal{E}X_j^2 = j^4 \times \frac{1}{12j^2} \times 2 + j^2 \times \frac{1}{6} \times 2 = \frac{j^2}{3}$, so that $s_n^2 = \sum_{j=1}^{n} \frac{j^2}{3} = \frac{n(n+1)(2n+1)}{18}$. Next,

$$
\begin{aligned}
g_n(\varepsilon) &= \frac{1}{s_n^2}\sum_{j=1}^{n}\int_{(|x|\geq\varepsilon s_n)} x^2 dF_j(x) = \frac{1}{s_n^2}\sum_{j=1}^{n} 2 \times \frac{j^4}{12j^2} I_{(|j^2|\geq\varepsilon s_n)} + \\
&\quad \frac{1}{s_n^2}\sum_{j=1}^{n} 2 \times \frac{j^2}{12} I_{(|j|\geq\varepsilon s_n)} = \frac{1}{6s_n^2}\sum_{j=1}^{n} j^2 I_{(j^2\geq\varepsilon s_n)} + \frac{1}{6s_n^2}\sum_{j=1}^{n} j^2 I_{(j\geq\varepsilon s_n)}.
\end{aligned}
$$

However, $\varepsilon s_n = \frac{\varepsilon}{3\sqrt{2}}[n(n+1)(2n+1)]^{1/2} \sim n^{3/2}$ (is of the order of $n^{3/2}$), so that $n < \varepsilon s_n$ (for large n) and $I_{(j\geq\varepsilon s_n)} = 0$, $j = 1, \ldots, n$. Thus, $\frac{1}{6s_n^2}\sum_{j=1}^{n} j^2 I_{(j\geq\varepsilon s_n)} = o(1)$. Next, set $k_n = k = [(\varepsilon s_n)^{1/2}]$, so that k is of the order of $n^{3/4}$ ($k \sim n^{3/4}$), and

$$
\begin{aligned}
\frac{1}{6s_n^2}\sum_{j=1}^{n} j^2 I_{(j^2\geq\varepsilon s_n)} &= \frac{1}{6s_n^2}\sum_{j=1}^{n} j^2 I_{(j\geq k)} = \frac{1}{6s_n^2}\left(\sum_{j=1}^{n} j^2 - \sum_{j=1}^{k-1} j^2\right) \\
&= \frac{1}{6s_n^2}\left[3s_n^2 - \frac{k(k-1)(2k-1)}{6}\right] \\
&= \frac{1}{2} - \frac{1}{36} \times \frac{k(k-1)(2k-1)}{s_n^2}.
\end{aligned}
$$

At this point, observe that $k(k-1)(2k-1)$ is of the order of $n^{9/4}$ and s_n^2 is of the order of $n^3 = n^{12/4}$. Therefore $\frac{k(k-1)(2k-1)}{s_n^2} = o(1)$ and $g_n(\varepsilon) \underset{n\to\infty}{\longrightarrow} \frac{1}{2}$, not to 0. #

5. From $M_n = o(s_n)$, it follows that, for $\varepsilon > 0$, there exists $n_0 = n(\varepsilon)$ such that $M_n < \varepsilon s_n, n > n_0$. For $n > n_0$, $\frac{S_n - \mathcal{E}S_n}{s_n} = \frac{1}{s_n}\sum_{j=1}^{n_0}(X_j - \mathcal{E}X_j) + \frac{1}{s_n}\sum_{j=n_0+1}^{n}(X_j - \mathcal{E}X_j)$ with $\frac{1}{s_n}\sum_{j=1}^{n_0}(X_j - \mathcal{E}X_j) \underset{n\to\infty}{\longrightarrow} 0$ since $s_n \underset{n\to\infty}{\longrightarrow} \infty$. Next, with $\tau_n^2 = s_n^2 - s_{n_0}^2$,

$$\frac{1}{s_n}\sum_{j=n_0+1}^{n}(X_j - \mathcal{E}X_j) = \frac{\tau_n}{s_n}\sum_{j=n_0+1}^{n}\frac{X_j - \mathcal{E}X_j}{\tau_n}.$$

Thus, if we set $Y_{nj} = \frac{X_j - \mathcal{E}X_j}{\tau_n}$, $j = n_0+1, \ldots, n$, then $\mathcal{E}Y_{nj} = 0$, $\sum_{j=n_0+1}^{n}\sigma^2(Y_{nj}) = \frac{s_n^2 - s_{n_0}^2}{\tau_n^2} = \frac{\tau_n^2}{\tau_n^2} = 1$. Also, for $j > n_0$, $|Y_{nj}| \le \frac{2M_j}{\tau_n} \le \frac{2\varepsilon s_j}{\tau_n} \le \frac{2\varepsilon s_n}{\tau_n}$ (since $s_n \underset{n\to\infty}{\uparrow}$). That is, $|Y_{nj}| \le \frac{2\varepsilon s_n}{\tau_n}$, $j = n_0+1, \ldots, n$. Then $\sum_{j=n_0+1}^{n}\mathcal{E}|Y_{nj}|^3 = \sum_{j=n_0+1}^{n}\mathcal{E}|Y_{nj}^2 \times Y_{nj}| \le \frac{2\varepsilon s_n}{\tau_n}\sum_{j=n_0+1}^{n}\sigma^2(Y_{nj}) = \frac{2\varepsilon s_n}{\tau_n} \times 1 = 2\varepsilon\sqrt{\frac{s_n^2}{s_n^2 - s_{n_0}^2}} = 2\varepsilon\sqrt{\frac{1}{1 - \frac{s_{n_0}^2}{s_n^2}}}$ and $\frac{s_{n_0}^2}{s_n^2} \underset{n\to\infty}{\longrightarrow} 0$ (since $s_n^2 \underset{n\to\infty}{\longrightarrow} \infty$). So, for every $\varepsilon > 0$, there exists $n_0 = n(\varepsilon)$ such that $\sum_{j=n_0+1}^{n}\mathcal{E}|Y_{nj}|^3 <$ a multiple of ε, and this implies that $\sum_{j=n_0+1}^{n}\mathcal{E}|Y_{nj}|^3 \underset{n\to\infty}{\longrightarrow} 0$. Then (by Theorem 3 in this chapter) $\sum_{j=n_0+1}^{n}\frac{X_j - \mathcal{E}X_j}{s_n} \underset{n\to\infty}{\Rightarrow} N(0,1)$ (since $\frac{\tau_n}{s_n} \underset{n\to\infty}{\longrightarrow} 1$) and hence $\frac{S_n - \mathcal{E}S_n}{s_n} \underset{n\to\infty}{\Rightarrow} N(0,1)$. #

6. We assume that $S_n \overset{d}{\to} N(0,1)$ and we shall show that $\sum_{j=1}^{n}X_{nj}^2 \underset{n\to\infty}{\overset{P}{\longrightarrow}} 1$. Since, under the assumptions made, $S_n \underset{n\to\infty}{\overset{d}{\longrightarrow}} N(0,1)$ is equivalent to $\sum_{j=1}^{n}\int_{(|x|\ge\varepsilon)}x^2 \times dF_{nj}(x) \underset{n\to\infty}{\longrightarrow} 0$ for every $\varepsilon > 0$, where F_{nj} is the d.f. of X_{nj}, it suffices to assume this latter convergence. Our assumption implies:

$$\sum_{j=1}^{n}\sigma^2(|X_{nj}|I_{(|X_{nj}|\ge 1)}) \underset{n\to\infty}{\longrightarrow} 0 \text{ and } \sum_{j=1}^{n}P(|X_{nj}| \ge \varepsilon) \underset{n\to\infty}{\longrightarrow} 0. \qquad (1)$$

The first of these convergences is immediate since (for $\varepsilon = 1$):

$$\begin{aligned}\sum_{j=1}^{n}\sigma^2(|X_{nj}|I_{(|X_{nj}|\ge 1)}) &= \sum_{j=1}^{n}\mathcal{E}\left[X_{nj}^2 I_{(|X_{nj}|\ge 1)}\right] - \sum_{j=1}^{n}\{\mathcal{E}\left[|X_{nj}|I_{(|X_{nj}|\ge 1)}\right]\}^2 \\ &\le \sum_{j=1}^{n}\mathcal{E}\left[X_{nj}^2 I_{(|X_{nj}|\ge 1)}\right] \\ &= \sum_{j=1}^{n}\int_{(|x|\ge 1)}x^2 dF_{nj}(x) \underset{n\to\infty}{\longrightarrow} 0,\end{aligned}$$

and the second convergence follows thus:

$$\sum_{j=1}^{n} P(|X_{nj}| \geq \varepsilon) = \sum_{j=1}^{n} \int_{(|x| \geq \varepsilon)} dF_{nj}(x) = \frac{1}{\varepsilon^2} \sum_{j=1}^{n} \int_{(|x| \geq \varepsilon)} \varepsilon^2 dF_{nj}(x)$$
$$\leq \frac{1}{\varepsilon^2} \sum_{j=1}^{n} \int_{(|x| \geq \varepsilon)} x^2 dF_{nj}(x) \underset{n \to \infty}{\longrightarrow} 0.$$

At this point, set:

$$Y_{nj} = X_{nj}^2, \quad Z_{nj} = Y_{nj} I_{(Y_{nj} < 1)}, \quad T_n = \sum_{j=1}^{n} Y_{nj}, \text{ and } V_n = \sum_{j=1}^{n} Z_{nj}. \tag{2}$$

Then we show that:

$$\sum_{j=1}^{n} \sigma^2(Z_{nj}) \underset{n \to \infty}{\longrightarrow} 0. \tag{3}$$

Indeed, for $0 < \varepsilon < 1$,

$$0 \leq \sum_{j=1}^{n} \sigma^2(Z_{nj}) = \sum_{j=1}^{n} \mathcal{E} Z_{nj}^2 - \sum_{j=1}^{n} (\mathcal{E} Z_{nj})^2 \leq \sum_{j=1}^{n} \mathcal{E} Z_{nj}^2$$
$$= \sum_{j=1}^{n} \mathcal{E}[Y_{nj}^2 I_{(Y_{nj} < 1)}] = \sum_{j=1}^{n} \mathcal{E}[Y_{nj}^2 I_{(Y_{nj} < \varepsilon)}] + \sum_{j=1}^{n} \mathcal{E}[Y_{nj}^2 I_{(\varepsilon \leq Y_{nj} < 1)}]$$
$$\leq \varepsilon \sum_{j=1}^{n} \mathcal{E}[Y_{nj} I_{(Y_{nj} < \varepsilon)}] + \sum_{j=1}^{n} \mathcal{E} I_{(\varepsilon \leq Y_{nj} < 1)}$$
$$\leq \varepsilon \sum_{j=1}^{n} \mathcal{E}[Y_{nj} I_{(Y_{nj} < \varepsilon)}] + \sum_{j=1}^{n} P(Y_{nj} \geq \varepsilon)$$
$$= \varepsilon \left[1 - \sum_{j=1}^{n} \int_{(|x| \geq \varepsilon)} x^2 dF_{nj}(x) \right] + \sum_{j=1}^{n} P(|X_{nj}| \geq \sqrt{\varepsilon}); \text{ i.e.,}$$

$$0 \leq \sum_{j=1}^{n} \sigma^2(Z_{nj}) \leq \varepsilon \left[1 - \sum_{j=1}^{n} \int_{(|x| \geq \varepsilon)} x^2 dF_{nj}(x) \right] + \sum_{j=1}^{n} P(|X_{nj}| \geq \sqrt{\varepsilon}). \tag{4}$$

Taking the limits in (4) (as $n \to \infty$) and utilizing our assumption and (1), we obtain (3).

Next,

$$P(|V_n - \mathcal{E}V_n| \geq \varepsilon) = P\left[\left|\sum_{j=1}^{n}(Z_{nj} - \mathcal{E}Z_{nj})\right| \geq \varepsilon\right]$$

$$\leq \frac{1}{\varepsilon^2}\sigma^2\left(\sum_{j=1}^{n}(Z_{nj} - \mathcal{E}Z_{nj})\right)$$

$$= \frac{1}{\varepsilon^2}\sum_{j=1}^{n}\sigma^2(Z_{nj}) \underset{n\to\infty}{\longrightarrow} 0 \text{ by (3). That is,}$$

$$V_n - \mathcal{E}V_n \underset{n\to\infty}{\overset{P}{\longrightarrow}} 0. \tag{5}$$

However,

$$\mathcal{E}V_n = \sum_{j=1}^{n}\mathcal{E}Z_{nj} = \sum_{j=1}^{n}\mathcal{E}\left[Y_{nj}I_{(Y_{nj}<1)}\right] = \sum_{j=1}^{n}\mathcal{E}\left[X_{nj}^2 I_{(|X_{nj}|<1)}\right]$$

$$= 1 - \sum_{j=1}^{n}\mathcal{E}\left[X_{nj}^2 I_{(|X_{nj}|\geq 1)}\right]$$

$$= 1 - \sum_{j=1}^{n}\int_{(|x|\geq 1)} x^2 dF_{nj}(x) \underset{n\to\infty}{\longrightarrow} 1 - 0 = 1.$$

This result along with (3) yield then

$$V_n \underset{n\to\infty}{\overset{P}{\longrightarrow}} 1. \tag{6}$$

Finally,

$$P(V_n \neq T_n) = P\left(\sum_{j=1}^{n} Z_{nj} \neq \sum_{j=1}^{n} Y_{nj}\right) \leq P\left[\bigcup_{j=1}^{n}(Z_{nj} \neq Y_{nj})\right]$$

$$\leq \sum_{j=1}^{n} P(Z_{nj} \neq Y_{nj}) = \sum_{j=1}^{n} P\left[Y_{nj}I_{(Y_{nj}<1)} \neq Y_{nj}\right]$$

$$= \sum_{j=1}^{n} P(Y_{nj} \geq 1) = \sum_{j=1}^{n} P(X_{nj}^2 \geq 1) = \sum_{j=1}^{n} P(|X_{nj}| \geq 1)$$

$$\underset{n\to\infty}{\longrightarrow} 0, \text{ by (1).}$$

This implies that, for every $\varepsilon > 0$,

$$P(|V_n - T_n| \geq \varepsilon) \leq P(V_n \neq T_n) \underset{n\to\infty}{\longrightarrow} 0,$$

so that

$$V_n - T_n \underset{n\to\infty}{\overset{P}{\longrightarrow}} 0. \tag{7}$$

Relations (6) and (7) complete the proof. #

7. **(i)** By definition, $\triangle g(u) = g(u+h) - g(u-h)$, call it $g_1(u)$. Then

$$\begin{aligned}\triangle^{(2)} g(u) &= \triangle g_1(u) = g_1(u+h) - g_1(u-h)\\ &= g(u+2h) - 2g(u) + g(u-2h)\\ &= \sum_{r=0}^{2}(-1)^r \binom{2}{r} g(u+(2-2r)h),\end{aligned}$$

so that the formula is true for $n = 2$. Next, assume it to be true for m; i.e., we assume:

$$\triangle^{(m)} g(u) = \sum_{r=0}^{m}(-1)^r \binom{m}{r} g(u+(m-2r)h),$$

and we shall establish it for $m+1$; i.e., we shall show that:

$$\triangle^{(m+1)} g(u) = \sum_{r=0}^{m+1}(-1)^r \binom{m+1}{r} g(u+((m+1)-2r)h).$$

We have:

$$\begin{aligned}&\triangle^{(m+1)} g(u) = \triangle\triangle^{(m)} g(u)\\ &= \sum_{r=0}^{m}(-1)^r \binom{m}{r} g(u+(m-2r)h)|_{u+h}\\ &\quad - \sum_{r=0}^{m}(-1)^r \binom{m}{r} g(u+(m-2r)h)|_{u-h}\\ &= \sum_{r=0}^{m}(-1)^r \binom{m}{r} g(u+((m+1)-2r)h)\\ &\quad - \sum_{r=0}^{m}(-1)^r \binom{m}{r} g(u+((m-1)-2r)h)\\ &= g(u+(m+1)h) + \sum_{r=1}^{m}(-1)^r \binom{m}{r} g(u+(m+1-2r)h)\end{aligned}$$

$$-\sum_{r=0}^{m-1}(-1)^r\binom{m}{r}g(u+(m-1-2r)h)-(-1)^m g(u+(-m-1)h)$$

$$=g(u+(m+1)h)+\sum_{k=0}^{m-1}(-1)^{k+1}\binom{m}{k+1}g(u+(m+1-2k-2)h)$$

$$-\sum_{r=0}^{m-1}(-1)^r\binom{m}{r}g(u+(m-1-2r)h)+(-1)^{m+1}g(u+(-m-1)h)$$

$$=g(u+(m+1)h)-\sum_{r=0}^{m-1}(-1)^r\binom{m}{r+1}g(u+(m-1-2r)h)$$

$$-\sum_{r=0}^{m-1}(-1)^r\binom{m}{r}g(u+(m-1-2r)h)+(-1)^{m+1}g(u-(m+1)h)$$

$$=g(u+(m+1)h)-\sum_{r=0}^{m-1}(-1)^r\left[\binom{m}{r+1}+\binom{m}{r}\right]$$

$$\times g(u+(m-1-2r)h)+(-1)^{m+1}g(u-(m+1)h)$$

$$=g(u+(m+1)h)-\sum_{r=0}^{m-1}(-1)^r\binom{m+1}{r+1}g(u+(m-1-2r)h)$$

$$+(-1)^{m+1}g(u-(m+1)h)\ \ (\text{since}\ \binom{m}{r+1}$$

$$+\binom{m}{r}=\binom{m+1}{r+1})$$

$$=g(u+(m+1)h)-\sum_{r=0}^{m-1}(-1)^r\binom{m+1}{r+1}g(u+(m+1-2(r+1))h)$$

$$+(-1)^{m+1}g(u-(m+1)h)$$

$$=g(u+(m+1)h)-\sum_{k=1}^{m}(-1)^{k-1}\binom{m+1}{k}g(u+(m+1-2k)h)$$

$$+(-1)^{m+1}g(u-(m+1)h)$$

$$=g(u+(m+1)h)+\sum_{k=1}^{m}(-1)^{k}\binom{m+1}{k}g(u+(m+1-2k)h)$$

$$+(-1)^{m+1}g(u-(m+1)h)$$

$$=\sum_{k=0}^{m+1}(-1)^{k}\binom{m+1}{k}g(u+(m+1-2k)h),\ \text{as was to be seen.}$$

(ii) By part (i),

$$\Delta^{(2n)} f(0) = \sum_{r=0}^{2n} (-1)^r \binom{2n}{r} f(2(n-r)h), \quad h \in \Re. \tag{1}$$

Expand $f(u+t)$ around u up to terms of order $2n$ to get:

$$f(u+t) = \sum_{k=0}^{2n} \frac{t^k}{k!} f^{(k)}(u) + o(t^{2n}). \tag{2}$$

In (2), take $u = 0$ and replace t by $2(n-r)h$ to obtain:

$$f(2(n-r)h) = \sum_{k=0}^{2n} \frac{(2(n-r)h)^k}{k!} f^{(k)}(0) + o((2(n-r)h)^{2n}). \tag{3}$$

By means of (3), relation (1) becomes:

$$\begin{aligned}
&\Delta^{(2n)} f(0) \\
&= \sum_{r=0}^{2n} (-1)^r \binom{2n}{r} \left[\sum_{k=0}^{2n} \frac{(2(n-r)h)^k}{k!} f^{(k)}(0) + o((2(n-r)h)^{2n}) \right] \\
&= \sum_{r=0}^{2n} \sum_{k=0}^{2n} (-1)^r \binom{2n}{r} \frac{(2(n-r)h)^k}{k!} f^{(k)}(0) \\
&\quad + \sum_{r=0}^{2n} (-1)^r \binom{2n}{r} o((2(n-r)h)^{2n}).
\end{aligned} \tag{4}$$

The first term on the right-hand side in (4) is equal to:

$$\begin{aligned}
&\sum_{k=0}^{2n} \frac{h^k}{k!} f^{(k)}(0) \sum_{r=0}^{2n} (-1)^r \binom{2n}{r} (2n-2r)^k \\
&= \sum_{k=0}^{2n-1} \frac{h^k}{k!} f^{(k)}(0) \sum_{r=0}^{2n} (-1)^r \binom{2n}{r} (2n-2r)^k \\
&\quad + \frac{h^{2n}}{(2n)!} f^{(2n)}(0) \sum_{r=0}^{2n} (-1)^r \binom{2n}{r} (2n-2r)^{2n} \\
&= 0 + \frac{h^{2n}}{(2n)!} f^{(2n)}(0) \times 2^{2n} (2n)! \text{ (according to the hint)} \\
&= (2h)^{2n} f^{(2n)}(0).
\end{aligned} \tag{5}$$

Next, for the second term on the right-hand side of (4), we have

$$\begin{aligned}
&\left|\sum_{r=0}^{2n}(-1)^r\binom{2n}{r}o((2(n-r)h)^{2n})\right| \\
&\quad=\left|\sum_{r=0}^{2n}(-1)^r\binom{2n}{r}(2(n-r)h)^{2n}o(1)\right| \\
&\quad\leq (2h)^{2n}\times o(1)\times\sum_{r=0}^{2n}\binom{2n}{r}(n-r)^{2n} \\
&\quad\leq (2h)^{2n}\times o(1)\times n^{2n}\sum_{r=0}^{2n}\binom{2n}{r} \\
&\quad= (2h)^{2n}\times o(1)\times n^{2n}\times 2^{2n}=(2h)^{2n}\times o(1)\times(2n)^{2n} \\
&\quad= o((2h)^{2n})\ \text{(since } n \text{ is constant).} \qquad (6)
\end{aligned}$$

By way of (5) and (6), relation (4) becomes: $\triangle^{(2n)}f(0) = (2h)^{2n} f^{(2n)}(0) + o((2h)^{2n})$, and hence $\frac{\triangle^{(2n)}f(0)}{(2h)^{2n}} = f^{(2n)}(0) + \frac{o((2h)^{2n})}{(2h)^{2n}} = f^{(2n)}(0) + \frac{o(h^{2n})}{h^{2n}}$. #

8. To show that, for $m = 1, 2, \ldots,$

$$\sum_{r=0}^{m}(-1)^r\binom{m}{r}(m-2r)^k=\begin{cases}0 & \text{for } k<m\\ 2^m\times m! & \text{for } k=m.\end{cases} \qquad (1)$$

Relation (1) is easily checked to be true for $m = 1, 2, 3$. Assume (1) to be true for some value of m and establish it for $m + 1$; i.e., we shall show that

$$\sum_{r=0}^{m+1}(-1)^r\binom{m+1}{r}(m+1-2r)^k=\begin{cases}0 & \text{for } k<m+1\\ 2^{m+1}\times(m+1)! & \text{for } k=m+1.\end{cases} \qquad (2)$$

The proof of (2) is split into two parts. First, it is shown that

$$I=\sum_{r=0}^{m+1}(-1)^r\binom{m+1}{r}(m+1-2r)^k=0,\ k\leq m, \qquad (3)$$

and then that

$$\sum_{r=0}^{m+1}(-1)^r\binom{m+1}{r}(m+1-2r)^{m+1}=2^{m+1}\times(m+1)! \qquad (4)$$

In establishing (3), we use the familiar identity

$$\binom{m+1}{r}=\binom{m}{r}+\binom{m}{r-1},$$

so that

$$\begin{aligned}
I &= \sum_{r=0}^{m+1}(-1)^r\binom{m}{r}(m+1-2r)^k + \sum_{r=0}^{m+1}(-1)^r\binom{m}{r-1}(m+1-2r)^k\\
&= \sum_{r=0}^{m}(-1)^r\binom{m}{r}(m+1-2r)^k + \sum_{r=1}^{m+1}(-1)^r\binom{m}{r-1}(m+1-2r)^k\\
&= \sum_{r=0}^{m}(-1)^r\binom{m}{r}(m+1-2r)^k + \sum_{l=0}^{m}(-1)^{l+1}\binom{m}{l}(m-1-2l)^k\\
&= \sum_{r=0}^{m}(-1)^r\binom{m}{r}(m+1-2r)^k - \sum_{r=0}^{m}(-1)^r\binom{m}{r}(m-1-2r)^k \stackrel{def}{=} I_1 - I_2.
\end{aligned}$$

Now, $(m+1-2r)^k = [(m-2r)+1]^k = \sum_{l=0}^{k}\binom{k}{l}(m-2r)^l$, so that

$$\begin{aligned}
I_1 &= \sum_{r=0}^{m}(-1)^r\binom{m}{r}\sum_{l=0}^{k}\binom{k}{l}(m-2r)^l = \sum_{l=0}^{k}\binom{k}{l}\sum_{r=0}^{m}(-1)^r\binom{m}{r}(m-2r)^l\\
&= \sum_{l=0}^{k-1}\binom{k}{l}\sum_{r=0}^{m}(-1)^r\binom{m}{r}(m-2r)^l + \sum_{r=0}^{m}(-1)^r\binom{m}{r}(m-2r)^k. \qquad (5)
\end{aligned}$$

For $k < m$, both terms in (5) are equal to 0, by (1), whereas for $k = m$, the first term in (5) is equal to 0, and the second term equals $2^m \times m!$, again by (1). Thus,

$$I_1 = \begin{cases} 0 & \text{for } k < m\\ 2^m \times m! & \text{for } k = m. \end{cases} \qquad (6)$$

Next,

$$(m-1-2r)^k = [(m-2r)-1]^k = \sum_{l=0}^{k}\binom{k}{l}(m-2r)^l(-1)^{k-l},$$

so that

$$\begin{aligned}
I_2 &= \sum_{r=0}^{m}(-1)^r\binom{m}{r}\sum_{l=0}^{k}\binom{k}{l}(m-2r)^l(-1)^{k-l}\\
&= \sum_{l=0}^{k}\binom{k}{l}(-1)^{k-l}\sum_{r=0}^{m}(-1)^r\binom{m}{r}(m-2r)^l\\
&= \sum_{l=0}^{k-1}\binom{k}{l}(-1)^{k-l}\sum_{r=0}^{m}(-1)^r\binom{m}{r}(m-2r)^l\\
&\quad + \sum_{r=0}^{m}(-1)^r\binom{m}{r}(m-2r)^k. \qquad (7)
\end{aligned}$$

As in (5), $\sum_{r=0}^{m}(-1)^r\binom{m}{r}(m-2r)^l = 0$ for $l = 0, 1, \ldots, k-1$, so that the first term in (7) is 0, and the second term is $2^m \times m!$ Therefore

$$I_2 = \begin{cases} 0 & \text{for } k < m \\ 2^m \times m! & \text{for } k = m, \end{cases} \tag{8}$$

so that, by (7) and (8), $I = I_1 - I_2 = 0, k \leq m$, which is (3). Next, to establish (4). To this end.

$$\begin{aligned} I^* &= \sum_{r=0}^{m+1}(-1)^r\binom{m+1}{r}(m+1-2r)^{m+1} \\ &= \sum_{r=0}^{m+1}(-1)^r\binom{m+1}{r}(m+1-2r)^m[(m+1)-2r] \\ &= (m+1)\sum_{r=0}^{m+1}(-1)^r\binom{m+1}{r}(m+1-2r)^m \\ &\quad -2\sum_{r=0}^{m+1}(-1)^r r\binom{m+1}{r}(m+1-2r)^m \\ &\overset{def}{=} (m+1)I_3 - 2I_4. \end{aligned} \tag{9}$$

However,

$$\begin{aligned} I_4 &= \sum_{r=0}^{m+1}(-1)^r r\binom{m+1}{r}(m+1-2r)^m \\ &= (m+1)\sum_{r=0}^{m+1}(-1)^r\binom{m}{r-1}(m+1-2r)^m \\ &\quad \left(\text{since } r\binom{m+1}{r} = (m+1)\binom{m}{r-1}\right) \\ &= (m+1)\sum_{r=1}^{m+1}(-1)^r\binom{m}{r-1}(m+1-2r)^m \\ &= (m+1)\sum_{l=0}^{m}(-1)^{l+1}\binom{m}{l}(m-1-2l)^m \\ &= -(m+1)\sum_{r=0}^{m}(-1)^r\binom{m}{r}(m-1-2r)^m \\ &= -(m+1)\sum_{r=0}^{m}(-1)^r\binom{m}{r}[(m-2r)-1]^m \end{aligned}$$

$$
\begin{aligned}
&= -(m+1)\sum_{r=0}^{m}(-1)^r\binom{m}{r}\sum_{l=0}^{m}\binom{m}{l}(m-2r)^l(-1)^{m-l}\\
&= -(m+1)\sum_{l=0}^{m}\binom{m}{l}(-1)^{m-l}\sum_{r=0}^{m}(-1)^r\binom{m}{r}(m-2r)^l\\
&= -(m+1)\sum_{l=0}^{m-1}\binom{m}{l}(-1)^{m-l}\sum_{r=0}^{m}(-1)^r\binom{m}{r}(m-2r)^l\\
&\quad -(m+1)\sum_{r=0}^{m}(-1)^r\binom{m}{r}(m-2r)^m\\
&= -(m+1)(2^m\times m!), \text{ and therefore}\\
&\quad -2I_4 = 2(m+1)(2^m\times m!) = 2^{m+1}\times(m+1)! \qquad (10)
\end{aligned}
$$

Next,

$$
\begin{aligned}
I_3 &= \sum_{r=0}^{m+1}(-1)^r\binom{m+1}{r}(m+1-2r)^m\\
&= \sum_{r=0}^{m+1}(-1)^r\binom{m}{r}(m+1-2r)^m + \sum_{r=0}^{m+1}(-1)^r\binom{m}{r-1}(m+1-2r)^m\\
&= \sum_{r=0}^{m}(-1)^r\binom{m}{r}(m+1-2r)^m + \sum_{r=1}^{m+1}(-1)^r\binom{m}{r-1}(m+1-2r)^m\\
&\stackrel{def}{=} I_5 + I_6.
\end{aligned}
$$

But

$$
\begin{aligned}
I_5 &= \sum_{r=0}^{m}(-1)^r\binom{m}{r}[(m-2r)+1]^m = \sum_{r=0}^{m}(-1)^r\binom{m}{r}\sum_{l=0}^{m}\binom{m}{l}(m-2r)^l\\
&= \sum_{l=0}^{m}\binom{m}{l}\sum_{r=0}^{m}(-1)^r\binom{m}{r}(m-2r)^l\\
&= \sum_{l=0}^{m-1}\binom{m}{l}\sum_{r=0}^{m}(-1)^r\binom{m}{r}(m-2r)^l + \sum_{r=0}^{m}(-1)^r\binom{m}{r}(m-2r)^m\\
&= 2^m\times m!,
\end{aligned}
$$

and

$$
I_6 = \sum_{r=1}^{m+1}(-1)^r\binom{m}{r-1}(m+1-2r)^m = \sum_{l=0}^{m}(-1)^{l+1}\binom{m}{l}(m-1-2l)^m
$$

$$= -\sum_{r=0}^{m}(-1)^r\binom{m}{r}(m-1-2r)^m$$
$$= -\sum_{r=0}^{m}(-1)^r\binom{m}{r}[(m-2r)-1]^m$$
$$= -\sum_{r=0}^{m}(-1)^r\binom{m}{r}\sum_{l=0}^{m}\binom{m}{l}(m-2r)^l(-1)^{m-l}$$
$$= -\sum_{l=0}^{m}\binom{m}{l}(-1)^{m-l}\sum_{r=0}^{m}(-1)^r\binom{m}{r}(m-2r)^l$$
$$= -\sum_{l=0}^{m-1}\binom{m}{l}(-1)^{m-l}\sum_{r=0}^{m}(-1)^r\binom{m}{r}(m-2r)^l$$
$$-\sum_{r=0}^{m}(-1)^r\binom{m}{r}(m-2r)^m$$
$$= -2^m \times m!,$$

so that $I_3 = I_5 + I_6 = 0$. From this result and (10), relation (9) yields $I^* = 2^{m+1} \times (m+1)!$ which is (4). The proof is completed. #

9. We have

$$P\left(\max_{1\le j\le k_n}|X_{nj}|\ge\varepsilon\right) = 1 - P\left(\max_{1\le j\le k_n}|X_{nj}|<\varepsilon\right)$$
$$= 1 - P[\bigcap_{j=1}^{k_n}(|X_{nj}|<\varepsilon)]$$
$$= 1 - \prod_{j=1}^{k_n}P(|X_{nj}|<\varepsilon) \text{ (by rowwise independence)}$$
$$= 1 - \prod_{j=1}^{k_n}[1 - P(|X_{nj}|\ge\varepsilon)]$$
$$= 1 - \prod_{j=1}^{k_n}[1 - \int_{(|x|\ge\varepsilon)} dF_{n_j}(x)].$$

It follows that

$$P\left(\max_{1\le j\le k_n}|X_{nj}|\ge\varepsilon\right)\underset{n\to\infty}{\longrightarrow}0 \qquad (1)$$

if and only if

$$\prod_{j=1}^{k_n}[1 - \int_{(|x|\ge\varepsilon)} dF_{n_j}(x)]\underset{n\to\infty}{\longrightarrow}1. \qquad (2)$$

The left-hand side inequality in Exercise 3 of Chapter 10 is equivalent to

$$\prod_{j=1}^{n}(1-x_j) \leq \exp\left(-\sum_{j=1}^{n} x_j\right),$$

whereas the right-hand side inequality in the same exercise is equivalent to

$$1-\sum_{j=1}^{n} x_j \leq \prod_{j=1}^{n}(1-x_j).$$

Combining these two inequalities, we obtain

$$1-\sum_{j=1}^{n} x_j \leq \prod_{j=1}^{n}(1-x_j) \leq \exp\left(-\sum_{j=1}^{n} x_j\right), \quad 0 \leq x_j \leq 1.$$

Taking $x_j = \int_{(|x|\geq\varepsilon)} dF_{n_j}(x)$, $j = 1, \ldots, k_n$, we get

$$\begin{aligned} 1-\sum_{j=1}^{k_n}\int_{(|x|\geq\varepsilon)} dF_{n_j}(x) &\leq \prod_{j=1}^{k_n}\left[1-\int_{(|x|\geq\varepsilon)} dF_{n_j}(x)\right] \\ &\leq \exp\left[-\sum_{j=1}^{k_n}\int_{(|x|\geq\varepsilon)} dF_{n_j}(x)\right] \quad (\leq 1). \end{aligned} \tag{3}$$

Then from (2) and (3), we have that (5) is equivalent to (4) below.

$$\sum_{j=1}^{k_n}\int_{(|x|\geq\varepsilon)} dF_{n_j}(x) \underset{n\to\infty}{\longrightarrow} 0. \tag{4}$$

Therefore (1) is equivalent to (4). Next,

$$\begin{aligned} (0 \leq) \max_{1\leq j\leq k_n} P(|X_{nj}| \geq \varepsilon) &= \max_{1\leq j\leq k_n}\int_{(|x|\geq\varepsilon)} dF_{n_j}(x) \\ &\leq \sum_{j=1}^{k_n}\int_{(|x|\geq\varepsilon)} dF_{n_j}(x) \underset{n\to\infty}{\longrightarrow} 0 \text{ (by (4))}. \end{aligned}$$

This completes the proof. #

10. Indeed, for a complex number z, we have $z = |z|e^{i\theta}$ with $-\pi < \theta \leq \pi$, so that $z^2 = |z|^2 e^{2i\theta}$ and $|z^2| = |z|^2\left|e^{2i\theta}\right| = |z|^2 \times 1 = |z|^2$. #

11. The first difference of f at u is $\Delta^{(1)} f(u) = f(u+h) - f(u-h)$, and its second difference is:

$$\begin{aligned} \Delta^{(2)} f(u) &= (f(u+h) - f(u-h))(h) - (f(u+h) - f(u-h))(-h) \\ &= f(u+2h) - 2f(u) + f(u-2h), \end{aligned}$$

which, for $u = 0$, becomes:

$$\begin{aligned}\Delta^{(2)} f(0) &= f(2h) - 2f(0) + f(-2h)\\ &= \int_{\Re} e^{i(2h)x} dF(x) - 2\int_{\Re} e^{i(0)x} dF(x) + \int_{\Re} e^{i(-2h)x} dF(x)\\ &= \int_{\Re} \left(e^{ihx} - e^{-ihx}\right)^2 dF(x)\\ &= \int_{\Re} [2i \sin(hx)]^2 dF(x) = -4\int_{\Re} [\sin(hx)]^2 dF(x).\end{aligned}$$

Hence $\dfrac{\Delta^{(2)} f(0)}{(2h)^2} = -\displaystyle\int_{\Re}\left[\frac{\sin(hx)}{hx}\right]^2 x^2 dF(x)$, and $f^{(2)}(0) = \displaystyle\lim_{h\to 0}\frac{\Delta^{(2)} f(0)}{(2h)^2} = -\lim_{h\to 0}\int_{\Re}\left[\frac{\sin(hx)}{hx}\right]^2 x^2 dF(x)$. However, $0 \leq \left[\frac{\sin(hx)}{hx}\right]^2$, and therefore, by the Fatou-Lebesgue Theorem, as $h \to 0$,

$$\begin{aligned}\lim \int_{\Re}\left[\frac{\sin(hx)}{hx}\right]^2 x^2 dF(x) &= \liminf \int_{\Re}\left[\frac{\sin(hx)}{hx}\right]^2 x^2 dF(x)\\ &\geq \int_{\Re} \liminf \left[\frac{\sin(hx)}{hx}\right]^2 x^2 dF(x)\\ &= \int_{\Re} x^2 dF(x),\end{aligned}$$

since $\lim \frac{\sin(hx)}{hx} = 1$. Thus, $\int_{\Re} x^2 dF(x) = \mathcal{E}X^2 < \infty$. #

Chapter 13

The Central Limit Problem: The Noncentered Case

1. We have $\sum_{j=1}^{k_n}\int_{(|x|\geq \varepsilon s_n)} x^2 d\bar{F}_{nj}(x) \underset{n\to\infty}{\longrightarrow} 0$ if and only if $\frac{1}{s_n^2}\sum_{j=1}^{k_n}\int_{(|x|\geq \varepsilon s_n)} x^2 d\bar{F}_{nj}(x) \underset{n\to\infty}{\longrightarrow} 0$ (since $s_n^2 \underset{n\to\infty}{\longrightarrow} \sigma^2$), and by Theorem 4 in Chapter 12, this happens if and only if

$$\frac{\bar{S}_n}{s_n} \underset{n\to\infty}{\Rightarrow} N(0,1) \text{ (and } \max_{1\leq j\leq k_n} \frac{\sigma_{nj}^2}{s_n^2} \underset{n\to\infty}{\longrightarrow} 0)$$

(since $\mathcal{E}\bar{X}_{nj} = 0, \sigma^2(\bar{X}_{nj}) = \sigma_{nj}^2$). But $\frac{\bar{S}_n}{s_n} \underset{n\to\infty}{\Rightarrow} N(0,1)$ implies $\bar{S}_n \underset{n\to\infty}{\Rightarrow} N(0,\sigma^2)$, and then $\bar{S}_n + \alpha_n \underset{n\to\infty}{\Rightarrow} N(\mu,\sigma^2)$, where $\alpha_n = \sum_{j=1}^{k_n}\alpha_{nj} \underset{n\to\infty}{\longrightarrow} \mu$. Also, $\bar{S}_n + \alpha_n = \sum_{j=1}^{k_n} X_{nj} = S_n$. Thus, for every $\varepsilon > 0$, $\sum_{j=1}^{k_n}\int_{(|x|\geq \varepsilon s_n)} x^2 d\bar{F}_{nj}(x) \underset{n\to\infty}{\longrightarrow} 0$ implies $S_n \underset{n\to\infty}{\Rightarrow} N(\mu,\sigma^2)$. #

2. See Exercise 9 in Chapter 8. #

Chapter 14

Topics from Sequences of Independent Variables

1. By Exercise 6(i) in Chapter 11, the $\mathcal{E}X$ does not exist. Hence, by Theorem 9 in this chapter, $\{\frac{S_n}{n}\}$ is unbounded with probability 1. This implies that, as $n \to \infty$, $\{\frac{S_n}{n}\}$ cannot converge in probability to 0. #
2. By Lemma 1, $\sum_{n=1}^{\infty} P(|X| \geq n) \leq \mathcal{E}|X| \leq 1 + \sum_{n=1}^{\infty} P(|X| \geq n)$. For some fixed $c > 0$, consider the r.v. $Y = X/c$. Then

$$\sum_{n=1}^{\infty} P(|Y| \geq n) \leq \mathcal{E}|Y| \leq 1 + \sum_{n=1}^{\infty} P(|Y| \geq n),$$

or

$$\sum_{n=1}^{\infty} P(|X| \geq nc) \leq \frac{1}{c}\mathcal{E}|X| \leq 1 + \sum_{n=1}^{\infty} P(|X| \geq nc).$$

Thus, $\mathcal{E}|X| < \infty$ if and only if $\sum_{n=1}^{\infty} P(|X| \geq nc) < \infty$. #
3. Since X is $\mathcal{A}_n = \sigma(X_n, X_{n+1}, \ldots)$-measurable, it is measurable with respect to the tail σ-field, $\mathcal{T}$, induced by the X_ns. However, by Theorem 10, $\mathcal{T}$ is equivalent to $\{\oslash, \Omega\}$. Hence X is $\{\oslash, \Omega\}$-measurable and therefore $P(X = c) = 1$ for some (finite) constant c. (See also the discussion in Exercise 11 in Chapter 10.) #
4. By setting $Y_n = |X_n|$, we will show that $\sum_{n=1}^{\infty} Y_n$ converges a.s. This is done by means of Theorem 11 (Three Series Criterion). To this end, for some fixed $c > 0$, define Y_n^c by: $Y_n^c = Y_n I_{(Y_n < c)}$. Then: $\sum_{n=1}^{\infty} P(Y_n \geq c) \leq \frac{1}{c}\sum_{n=1}^{\infty} \mathcal{E}Y_n < \infty$; next, $Y_n^c \leq Y_n$, so that $\sum_{n=1}^{\infty} \mathcal{E}Y_n^c \leq \sum_{n=1}^{\infty} \mathcal{E}Y_n < \infty$; finally, $\sigma^2(Y_n^c) = \mathcal{E}(Y_n^c)^2 - (\mathcal{E}Y_n^c)^2 \leq \mathcal{E}(Y_n^c)^2 \leq \mathcal{E}(cY_n^c) \leq c\mathcal{E}Y_n$, so that $\sum_{n=1}^{\infty} \sigma^2(Y_n^c) \leq c\sum_{n=1}^{\infty} \mathcal{E}Y_n < \infty$. The conclusion follows. #
5. Given any events $A_n, n \geq 1$, recall that $\limsup_{n\to\infty} A_n = \cap_{n=1}^{\infty} \cup_{\nu=n}^{\infty} A_\nu$. Here $A_n = \left(|X_n| \geq \frac{1}{k}\right)$ (for each arbitrary but fixed $k = 1, 2, \ldots$), and observe that

$$\begin{aligned}\bigcap_{n=1}^{\infty} \bigcup_{\nu=0}^{\infty} \left(|X_{n+\nu}| \geq \frac{1}{k}\right) &= (A_1 \cup A_2 \cup A_3 \cup \cdots) \cap \\ &\quad (A_2 \cup A_3 \cup \cdots\cdots\cdots) \cap \\ &\quad \cdots\cdots\cdots\cdots\cdots\cdots \\ &\quad (A_n \cup A_{n+1} \cup \cdots\cdots) \cap \\ &\quad \cdots\cdots\cdots\cdots\cdots\cdots \\ &= \bigcap_{n=1}^{\infty} \bigcup_{\nu=n}^{\infty} A_\nu = \limsup_{n\to\infty} A_n.\end{aligned}$$

The result follows. #
6. If the events $A_n, n \geq 1$, are independent and $P\left(\limsup_{n\to\infty} A_n\right) = 0$, then $\sum_{n=1}^{\infty} P(A_n) < \infty$. This is so, because, if $\sum_{n=1}^{\infty} P(A_n) = \infty$, then $P\left(\limsup_{n\to\infty} A_n\right) = 1$, by Theorem 4(ii), a contradiction. #

7. **(a)** Observe that, for any $\omega \in (0, 1)$, $X_n(\omega) = 0$ for all sufficiently large n ($n \geq n(\omega)$, say). Then $X_n \underset{n\to\infty}{\longrightarrow} 0$ pointwise, and hence $X_n \underset{n\to\infty}{\longrightarrow} 0$ a.s. Then, by Exercise 5, $P\left(\limsup_{n\to\infty} A_n\right) = 0$, where $A_n = \left(|X_n| \geq \frac{1}{k}\right)$ for each arbitrary but fixed $k = 1, 2, \ldots$. However, $P\left(|X_n| \geq \frac{1}{k}\right) = P\left(X_n \geq \frac{1}{k}\right) = P(X_n = 1) = 1/n$, so that $\sum_{n=1}^{\infty} P(A_n) = \sum_{n=1}^{\infty} 1/n = \infty$.

(b) As in (a), $X_n \underset{n\to\infty}{\longrightarrow} 0$ a.s., and $P\left(\limsup_{n\to\infty} A_n\right) = 0$. Also, $P\left(|X_n| \geq \frac{1}{k}\right) = P\left(X_n \geq \frac{1}{k}\right) = 1/n^2$, so that $\sum_{n=1}^{\infty} P(A_n) = \sum_{n=1}^{\infty} 1/n^2 < \infty$.

Remark: It is to be observed that, both in (a) and (b), the r.v.s (and hence the events) are dependent, since they are ordered: $X_n > X_{n+1}, n \geq 1$. #

Chapter 15

Topics from Ergodic Theory

1. For measurability of T, it suffices to show that $T^{-1}((u, v]) \in \mathcal{B}_{(0,1]}$ for $0 < u < v \leq 1$. This is so, by Exercise 6 in Chapter 1. That this is true is seen as follows: Let $0 < u < v \leq \frac{1}{2}$. Then $u < x - \frac{1}{2} \leq v$ or $u + \frac{1}{2} < x \leq v + \frac{1}{2}$, so that $T^{-1}((u, v]) = (u + \frac{1}{2}, v + \frac{1}{2}]$. Next, let $\frac{1}{2} < u < v \leq 1$. Then $u < x + \frac{1}{2} \leq v$ or $u - \frac{1}{2} < x \leq v - \frac{1}{2}$, so that $T^{-1}((u, v]) = (u - \frac{1}{2}, v - \frac{1}{2}]$. Finally, let $0 < u < v \leq 1$. Then $(u, v] = (u, \frac{1}{2}] + (\frac{1}{2}, v]$ (assuming of course, that $v > \frac{1}{2}$), and $u < x - \frac{1}{2} \leq \frac{1}{2}$ or $u + \frac{1}{2} < x \leq 1$, and $\frac{1}{2} < x + \frac{1}{2} \leq v$ or $0 < x \leq v - \frac{1}{2}$, so that $T^{-1}((u, v]) = (u + \frac{1}{2}, 1] + (0, v - \frac{1}{2}] = (0, v - \frac{1}{2}] + (u + \frac{1}{2}, 1]$. In all three cases, $T^{-1}((u, v]) \in \mathcal{B}$, so that T is measurable.

Next, to show that T is measure-preserving. Let $\mathcal{C}$ be the field as described in Exercise 7(i) of Chapter 1. Then by Exercise 8 in the same chapter, $\sigma(\mathcal{C}_{(0,1]}) = \mathcal{B}_{(0,1]}$. Thus, by Proposition 5 in this chapter, it suffices to show the measure-preserving property of T for members of $\mathcal{C}_{(0,1]}$, or just for $(0 <)u < v(\leq 1)$. That $\lambda((u, v]) = v - u = \lambda(T^{-1}(u, v])$ is immediate, because in all three possible cases considered above, we have: $(v + \frac{1}{2}) - (u + \frac{1}{2}) = v - u$, $(v - \frac{1}{2}) - (u - \frac{1}{2}) = v - u$, and $[(v - \frac{1}{2}) - 0] + [1 - (u + \frac{1}{2})] = v - \frac{1}{2} + 1 - u - \frac{1}{2} = v - u$. #

2. For every $0 < x \leq 1$, clearly, $T^{-1}((c^2x, cx]) = (cx, x]$, and if a probability measure P is measure-preserving, then $P((c^2x, cx]) = P((cx, x])$. Also, for $n \geq 2$, $T^{-1}((c^n x, c^{n-1}x]) = (c^{n-1}x, c^{n-2}x]$ and $P((c^n x, c^{n-1}x]) = P((c^{n-1}x, c^{n-2}x]) = \ldots = P((cx, x])$. But $(c^n x, c^{n-1}x] \downarrow (0, 0] = \emptyset$, as $n \to \infty$, so that $P((c^n x, c^{n-1}x]) \to 0$, and hence $P((cx, x]) = 0$. However, $\{x\} \subset (cx, x]$, and hence $P(\{x\}) = 0$. Thus, if P is measure-preserving, then $P(\{x\}) = 0$ for every x with $0 < x \leq 1$. #

3. By Definition 8, T is ergodic if and only if every invariant set A has λ-probability either 0 or 1. Consider, e.g., the set $A = A_1 + A_2$, where $A_1 = (\frac{2}{6}, \frac{3}{6}]$ and

$A_2 = (\frac{5}{6}, \frac{6}{6}]$. Then $T^{-1}(A_1) = A_2$ and $T^{-1}(A_2) = A_1$, so that $T^{-1}(A) = A$, and hence A is invariant. However, $\lambda(A) = \frac{2}{6} = \frac{1}{3} \neq$ from 0 or 1. Hence T is not ergodic. #

4. At the start of the proof of the inverse of Proposition 16, it was seen that $X(T\omega) = X(\omega)$, for all $\omega \in \Omega$, when $X = I_A$, $A \in \mathcal{J}$. Next, let $X = \sum_{j=1}^{r} \alpha_j I_{A_j}$, where $\{A_j;\ j = 1, \ldots, r\}$ is a $\mathcal{J}$-measurable partition of Ω. Then $X(T\omega) = \sum_{j=1}^{r} \alpha_j I_{A_j}(T\omega) = \sum_{j=1}^{r} \alpha_j I_{A_j}(\omega) = X(\omega)$, since $I_{A_j}(T\omega) = I_{A_j}(\omega)$, $j = 1, \ldots, r$, by the previous step. Now, let $0 \leq X_n$ simple r.v.s $\underset{n\to\infty}{\uparrow} X$. Then $X_n = \sum_{j=1}^{r_n} \alpha_{nj} I_{A_n j}$, where, for each n, $\{A_{nj},\ j = 1, \ldots, r_n\}$ is a $\mathcal{J}$-measurable partition of Ω, and $0 \leq X_n(T\omega) \underset{n\to\infty}{\uparrow} X(T\omega)$. However, $X_n(T\omega) = X_n(\omega)$, $n \geq 1$, by the previous step. Then $X(T\omega) = X(\omega)$. Finally, for any $\mathcal{J}$-measurable r.v. X, we have $X(T\omega) = X^+(T\omega) - X^-(T\omega) = X^+(\omega) - X^-(\omega)$, by the previous step, and this is equal to $X(\omega)$. #

5. At the start of the proof of Lemma 2, it was seen that $\mathcal{E}X(\omega) = \mathcal{E}X(T\omega)$ when $X = I_A$, $A \in \mathcal{A}$. Next, let $X = \sum_{j=1}^{r} \alpha_j I_{A_j}$, where $\{A_j;\ j = 1, \ldots, r\}$ is a (measurable) partition of Ω. Then $\mathcal{E}I_{A_j}(T\omega) = \mathcal{E}I_{A_j}(\omega)$, $j = 1, \ldots, r$, by the previous step, and $\sum_{j=1}^{r} \alpha_j \mathcal{E}I_{A_j}(\omega) = \sum_{j=1}^{r} \alpha_j \mathcal{E}I_{A_j}(T\omega)$, or $\mathcal{E}X(\omega) = \mathcal{E}X(T\omega)$.
Now, let $0 \leq X_n$ simple r.v.s $\underset{n\to\infty}{\uparrow} X$ a nonnegative r.v. That is, as $n \to \infty$, $X_n(\omega) \to X(\omega)$ and $X_n(T\omega) \to X(T\omega)$. However, $\mathcal{E}X_n(\omega) = X_n(T\omega)$, $n \geq 1$, by the previous step, and hence $\mathcal{E}X(\omega) = \mathcal{E}X(T\omega)$. Finally, for any r.v. X for which the $\mathcal{E}X$ exists, we have $\mathcal{E}X(\omega) = \mathcal{E}X^+(\omega) - \mathcal{E}X^-(\omega) = \mathcal{E}X^+(T\omega) - \mathcal{E}X^-(T\omega)$, by the previous step, and this is equal to $\mathcal{E}X(T\omega)$. #

6. Let $\mathcal{J}$ be the class of sets $A \in \mathcal{A}$ for which $T^{-1}A = A$. Then $T^{-1}\Omega = \Omega$, so that $\mathcal{J} \neq \oslash$. Next, for $A \in \mathcal{J}$, we have $T^{-1}A^c = (T^{-1}A)^c = A^c$, so that $A^c \in \mathcal{J}$. Finally, for $A_n \in \mathcal{J}$, $n = 1, 2, \ldots$, we have $T^{-1}(\cup_{n=1}^{\infty} A_n) = \cup_{n=1}^{\infty} T^{-1}A_n = \cup_{n=1}^{\infty} A_n$, so that $\cup_{n=1}^{\infty} A_n \in \mathcal{J}$. It follows that $\mathcal{J}$, is, indeed, a σ-field. #

7.

(i) Z_2 is a function of Z_1; i.e., if $\omega, \omega' \in \Omega$, with $\omega \neq \omega'$ and $Z_1(\omega) = Z_1(\omega')$, then $Z_2(\omega) = Z_2(\omega')$.
Set $Z_2(\omega) = \omega_2$, $Z_2(\omega') = \omega_2'$ and suppose that $\omega_2 \neq \omega_2'$. By assumption, $\{\omega_2\}, \{\omega_2'\} \in \mathcal{A}_2'$, and let $C = Z_2^{-1}(\{\omega_2\})$, $C' = Z_2^{-1}(\{\omega_2'\})$. Then $C, C' \in \mathcal{A}_2$ and $C \cap C' = \oslash$ since $\omega_2 \neq \omega_2'$. By assumption, $\mathcal{A}_2 \subseteq \mathcal{A}_1$, so that $C, C' \in \mathcal{A}_1$. Then there exists $B, B' \in \mathcal{A}_1'$ such that $C = Z_1^{-1}(B)$ and $C' = Z_1^{-1}(B')$, and $B \cap B' = \oslash$ since $C \cap C' = \oslash$. Next, $\omega \in C$, $\omega' \in C'$, $Z_1(\omega) = Z_1(\omega')$, and $Z_1(\omega) \in B$, $Z_1(\omega') \in B'$. Then $B \cap B' \neq \oslash$, a contradiction. Thus, $Z_2(\omega) = Z_2(\omega')$.

(ii) On $Z_1(\Omega)$, define Z by: $Z(\omega_1) = Z_2(\omega; Z_1(\omega) = \omega_1)$, so that $Z_2(\omega) = Z(Z_1(\omega))$ or $Z_2 = Z(Z_1)$. (That is, for $\omega_1 \in Z_1(\Omega)$, consider $B = \{\omega \in \Omega;\ Z_1(\omega) = \omega_1\}$. Since for any $\omega, \omega' \in B$ with $\omega \neq \omega'$ we have $Z_1(\omega) = Z_1(\omega')(= \omega_1)$, it follows, by part (i), that $Z_2(\omega) = Z_2(\omega')$, so that Z, as defined above, is well defined.) To show that it is unique.

Let $Z' : Z_1(\Omega) \to \Omega_2$ be another mapping such that $Z_2 = Z'(Z_1)$. To show $Z(\omega_1) = Z'(\omega_1)$ for every $\omega_1 \in Z_1(\Omega)$. For $\omega_1 \in Z_1(\Omega)$, we have $Z_2(\omega) = Z(Z_1(\omega)) = Z(\omega_1)$ for some $\omega \in \Omega$; also, $Z_2(\omega) = Z'(Z_1(\omega)) = Z'(\omega_1)$. Hence $Z(\omega_1) = Z'(\omega_1)$, as was to be seen.

(iii) To show that Z of part (ii) is $\mathcal{A}_1' \cap Z_1(\Omega)$-measurable. Let $D \in \mathcal{A}_2'$, and set $Z_2^{-1}(D) \overset{def}{=} A$ which is in $\mathcal{A}_2$. Also, $Z_2^{-1}(D) = [Z(Z_1)]^{-1}(D) = Z_1^{-1}[Z^{-1}(D)] = Z_1^{-1}(B)$, where we have set $B = Z^{-1}(D)$. So, $A = Z_1^{-1}(B)$ and $B \subseteq Z_1(\Omega)$ (because $Z : Z_1(\Omega) \to \Omega_2$). Since $A \in \mathcal{A}_2$, it follows (by assumption) that $A \in \mathcal{A}_1$. Then there exists $C \in \mathcal{A}_1'$ such that $A = Z_1^{-1}(C)$. So, $A = Z_1^{-1}(C)$ with $C \in \mathcal{A}_1'$ and $A = Z_1^{-1}(B)$ with $B \subseteq Z_1(\Omega)$. Therefore $B = C \cap Z_1(\Omega)$. That is, Z is $\mathcal{A}_1' \cap Z_1(\Omega)$-measurable. #

8. **(i)** In the first place, $\mathbf{X}$ is stationary, by Proposition 3. Next, let A be invariant (relative to $\mathbf{X}$) (see Definition 11). Then $A = (X_n, X_{n+1}, \ldots)^{-1}B$ for some $B \in \mathcal{B}^\infty$ and all $n \geq 1$. It follows that $A \in \sigma(X_n, X_{n+1}, \ldots)$ for all $n \geq 1$, and hence $A \in \mathcal{T}$, where $\mathcal{T}$ is the tail σ-field defined on $\mathbf{X}$ (see Definition 1 and discussion just prior to it in Chapter 14). Thus, $\mathcal{J} \subseteq \mathcal{T}$. For independent r.v.s (whether identically distributed or not), Theorem 10 in Chapter 4 gives that $P(A)$ is either 0 or 1 for all $A \in \mathcal{T}$. It follows that $P(A)$ is either 0 or 1 for all $A \in \mathcal{J}$, and hence $\mathbf{X}$ is ergodic.

(ii) It follows immediately from part (i), and Corollary 2 to Theorem 3. #

Index

G

H

I

N

O

P

Q

R

U

V

W

Printed and bound by CPI Group (UK) Ltd, Croydon, CR0 4YY
08/05/2025
01864902-0002